AF588040

Ordinary Differential Equations: Concepts, Methods, and Models

Leigh C. Becker

Ordinary Differential Equations: Concepts, Methods, and Models

A Course after Calculus 2

Leigh C. Becker
Germantown, TN, USA

ISBN 978-3-032-15149-0 ISBN 978-3-032-15150-6 (eBook)
https://doi.org/10.1007/978-3-032-15150-6

© The Editor(s) (if applicable) and The Author(s), under exclusive license to Springer Nature Switzerland AG 2026

This work is subject to copyright. All rights are solely and exclusively licensed by the Publisher, whether the whole or part of the material is concerned, specifically the rights of translation, reprinting, reuse of illustrations, recitation, broadcasting, reproduction on microfilms or in any other physical way, and transmission or information storage and retrieval, electronic adaptation, computer software, or by similar or dissimilar methodology now known or hereafter developed.
The use of general descriptive names, registered names, trademarks, service marks, etc. in this publication does not imply, even in the absence of a specific statement, that such names are exempt from the relevant protective laws and regulations and therefore free for general use.
The publisher, the authors and the editors are safe to assume that the advice and information in this book are believed to be true and accurate at the date of publication. Neither the publisher nor the authors or the editors give a warranty, expressed or implied, with respect to the material contained herein or for any errors or omissions that may have been made. The publisher remains neutral with regard to jurisdictional claims in published maps and institutional affiliations.

This Springer imprint is published by the registered company Springer Nature Switzerland AG
The registered company address is: Gewerbestrasse 11, 6330 Cham, Switzerland

If disposing of this product, please recycle the paper.

To Lynne Marie,
my wonderful wife and best friend
And our children:
Lily, Lexi, Liesl, and Leighton

Denkt daran, daß die wunderbaren Dinge, die ihr in euren Schulen kennenlernt, das Werk vieler Generationen sind, das in allen Ländern der Erde in begeistertem Streben und mit großer Mühe geschaffen worden ist. All dies wird als euer Erbe in eure Hände gelegt, damit ihr es empfanget, ehret, weiterbildet und treulich euren Kinder einst übermittelt. So sind wir Sterbliche in dem unsterblich, was wir an bleibenden Werken gemeinsam schaffen.[1]

Bear in mind that the wonderful things you learn in your school are the work of many generations, produced by enthusiastic effort and infinite labor in every country of the world. All this is put into your hands as your inheritance in order that you may receive it, honor it, add to it, and one day faithfully hand it on to your children. Thus do we mortals achieve immortality in the permanent things which we create in common.[2]

Albert Einstein

[1] The passage is from the section "Lehrer und Schüler" in the book *Mein Weltbild* [34] by Albert Einstein.

[2] This particular translation of the preceding German passage is found in "About Education: Teachers and Pupils" in *Ideas and Opinions* [33], which is a translation of *Mein Weltbild*.

Preface

This textbook is an introduction to the subject of ordinary differential equations. An *ordinary differential equation* (ODE for short) is an equation containing one independent variable; one or more dependent variables, each representing an unknown function of the independent variable; and ordinary derivatives of one or more of the dependent variables. As the title indicates, this book is intended for the student at the sophomore level who has successfully completed the first two semesters of a typical college or university calculus sequence and who intends to enroll in engineering or mathematics courses or any one of the physical sciences courses involving ODEs—but it will also serve the student who will be taking courses in other disciplines that occasionally employ ODEs, such as actuarial science, econometrics, computational science, and operational research. Some thoughts about teaching an ODEs course as well as important (and some novel) features of this book are addressed in the following paragraphs.

Review of Basic Integration Formulas and Integration Techniques

Since students tend to be rusty at calculus after a summer recess, reviews of calculus concepts inextricably tied to ODEs are interspersed throughout the book at the appropriate places. For example, in Chap. 1 students are shown or reminded, as the case may be, how to solve the simplest ODEs, namely, equations of the form $dy/dx = f(x)$. Since the right-hand side of such an equation depends only on the independent variable x, each of its antiderivatives (with respect to x) is a solution. In other words, for a given function $f(x)$, all solutions of $dy/dx = f(x)$ can be obtained by simply finding the indefinite integral of $f(x)$. The point to be made here is that since students already know this (or should) from integral calculus, the first chapter is the ideal place to engage them right away in solving ODEs by presenting a diverse array of examples of this type of equation for the twofold purpose of illustrating how to carry out the requisite integrations in order to obtain the solutions and at the same time reviewing the basic integration formulas and the techniques of integration, notably substitution, integration by parts, completion of the square, and partial fractions. Table 1.4 lists the basic integral formulas that regularly turn up in the examples and problem sets in this book. Students who have not already memorized these integral formulas from calculus courses should do so immediately, certainly before Chap. 2. In doing so, it is more likely that a student will see how to begin solving an ODE and carry out whatever integrations are needed without having to lose time by consulting an integral table or resorting to a calculator or computer algebra system.

Solutions of Differential Equations

It is imperative that newcomers to this subject understand that a solution (explicit) of a differential equation is a function that satisfies the equation on some interval. Right away in Chap. 1, even before students are tasked with solving ODEs, some of the problems simply present an ODE along with a function that allegedly is a solution of it. However, this function may or may not be a solution. The point of such a problem is to make students aware that it is usually easy to ascertain whether an "alleged solution" really is a solution. The hope is that students will come to the realization that it is not just about churning out a solution but also about checking whether it really is a solution: someone's "solution" may not be a solution at all but merely the unintended consequence of some error; hence, the need for verification. It should be the goal of instructors to encourage students to get into the habit of checking what they claim to be solutions, especially solutions of the differential equations that they turn in for homework grades. And to point out that if there is no time during an exam to check solutions, to at least ask oneself the question "Do my solutions appear correct and make sense?" It is also crucial to explain the difference between explicit and implicit solutions, which is addressed at various places in this book commencing with separable differential equations in Chap. 2.

Methods and Concepts

The topics in an introductory differential equations textbook should address the needs of students who sooner or later will encounter ODEs in courses in their curricula for which prior knowledge of differential equations course is essential. In this book, we introduce the types of ODEs that students ought to be conversant with in order to succeed in these courses. The presentation of a method for solving a particular type of ODE, be it an analytical method or a numerical algorithm, usually begins with a couple of examples of equations of that type. The examples are contrived in the sense that the equations are relatively simple, their solutions are somehow already known or are easily obtainable by inspection, and they can be experimented with in order to ascertain what sequence of steps of computations and integrations are needed in order to arrive at their solutions. These steps are analyzed for mathematical correctness and to determine their limitations, if any. If it is determined that a given sequence of steps always leads to solutions, then these steps are elevated to the status of constituting a method for solving this type of ODE. Of course, all of the methods that we come up with in this way are already well-known. Generally, but not always, the steps of a method are listed inside a box with a name for the method preceding the box, such as the *Integrating Factor Method* in Chap. 5 and the *Method for Solving Homogeneous Cauchy-Euler Equations* in Chap. 10.

The discussion of every method concludes with examples of ODEs that are more complicated than the introductory ones. The solutions of these equations are worked out in detail in order to illustrate how to effectively use the method. Despite the emphasis on being familiar with and knowing how to use some of the most well-known and established methods for solving the kinds of ODEs that students will encounter in their undergraduate courses, this book is by no means merely a collection of "recipes." Conceptual understanding of the mathematics undergirding every method, situation, and example is the primary objective of this book.

Models and Applications

Simple models and applications of differential equations are introduced right away in Chap. 1—not put off until later chapters. A number of well-known mathematical models of phenomena involving matter and energy are presented at appropriate places in this book, such as the *Tsiolkovsky rocket equation*, an ODE that is the result of using Newton's second law of motion to model the velocity of a rocket in deep space, and the *radioactive decay law*, which is an ODE that is based on an experimentally verifiable hypothesis about the probability of the decay of a single nucleus during a short time interval. The raison d'être for the inclusion of mathematical models in this book is to motivate readers and to incentivize the study of ODEs—and to answer the usual classroom question (whether or not it is actually asked): "What good is all this stuff?" Another reason for studying classic mathematical models is due to their import and commonality to everyone involved in the applied sciences and engineering. This is in keeping with Albert Einstein's musings and advice to students about the wonderful things learned in school: "All this is put into your hands as your inheritance in order that you may receive it, honor it, add to it, and one day faithfully hand it on to your children."[3]

Different scenarios involving people who employ ODE models (or could) to make decisions are described in this book, for example: the quality control engineer at an orange juice processing plant who needs to know at any given moment the weight percentage of soluble solids in a tank of orange juice as this juice is being pumped out of the tank while at the same time orange juice from other sources is being pumped into the tank; the field biologist whose responsibility includes estimating the population of a colony of white-tailed prairie dogs that are being continually preyed on by coyotes, hawks, and snakes; and the finance major enrolled in an ODEs course who has already studied and used the "amortization of loans formula" and is familiar with its derivation using the present value of an annuity formula but who is curious as to whether this formula could also be derived by means of an ODEs model[4] without having to know anything about annuities.

[3] See page vi.

[4] The answer is yes; see the last section in Chap. 7. The derivation begins with modeling the rate of change of the balance of a home loan by means of a linear ODE, which has a term representing the accrual of interest and a finite sequence of terms involving so-called *Dirac delta functions* modeling the periodic payments needed to amortize (pay off) the home mortgage.

These and other models and applications in this book were selected to be comprehensible and accessible to students at the sophomore level who have completed the equivalent of two semesters of college calculus and who are currently enrolled in an introductory college physics course or who have already completed it or taken an equivalent advanced high school physics course. So no apologies are made for including models that require a basic understanding of elementary physics. Since students may be taking their first physics course concurrently with their first ODEs course, all the physics that is really needed to read this book is a working knowledge of Newton's second law of motion—and even that is introduced and discussed in the last section of Chap. 1. To be sure, mathematical models at this stage of learning should not be too involved or overly complicated because few topics, if any, exist that are a part of every student's academic experience who is enrolled in their first differential equations course. So the applications and models in this book were chosen with this in mind, not to burden students with having to spend a lot of time on material that is unfamiliar to them, for instance, knowing about the electrical components of an RLC circuit and the laws associated with the names of Ohm, Faraday, and Kirchhoff in order to understand the derivation of the ODEs model for the flow of current through the circuit. Considering the time that it would take to discuss all of this, it seems this model is perhaps best left to the second or third semester of an introductory physics course or to an introductory circuit analysis course.

Abrupt Changes
In introductory differential equation textbooks, step functions and the Dirac delta function generally make their first appearances at some point after a study of the Laplace transform and its properties. But a different approach is taken in this book. Chapter 7 ("Modeling Abrupt Changes") introduces step functions; rectangular, triangular, and sawtooth pulses and waves; staircase functions and the like; impulsive forces; and the properties and applications of the Dirac delta function. By the time students are introduced to Laplace transform methods later on in Chap. 11, they will be at ease with these functions and not distracted by them as they focus on the Laplace transform and how it is well-suited for solving a linear differential equation with constant coefficients whose forcing term has discontinuities. Even if a syllabus excludes Laplace transform methods, this chapter or parts of it can still be covered to show how to solve linear equations with discontinuous forcing terms using the methods discussed prior to Chap. 7. In keeping with the subject matter in this chapter, examples of linear ODE models involving the Dirac delta function are included, such as models of the population of *E. coli* bacteria in a flask when batches of the bacteria are removed from the flask at regularly occurring intervals and the balance of an interest-bearing bank account when equal periodic withdrawals from the account are made.

Laplace Transforms
The Laplace transform is an integral part of engineering curricula as exemplified by the courses for which it is a prerequisite, such as electrical and mechanical network analysis, design of feedback control systems, analog and digital communication

systems, applied partial differential equations, system theory, etc. System theory may also be found in the curricula of some other disciplines, such as economics, political science, biology, and psychology. Chapter 11 is an introduction to the Laplace transform with detailed explanations of its properties and how to use it to solve linear differential equations with constant coefficients as well as Volterra integral and integro-differential equations of convolution type. Generally speaking, Laplace transform methods cannot be used to solve linear differential equations with variable coefficients; however, it is shown that it is possible to do so when all of the coefficients are polynomials in the independent variable.

The syllabus of an ordinary differential equations course should include Chap. 11 if its enrollees are students in departments with courses that rely on Laplace transform methods. The goal of Sect. 11.1 of this chapter is not only to define the Laplace transform and then leave it at that but also to present some ideas as to why it is defined in the way that it is. It is highly recommended that students commit to memory the basic Laplace transforms which are given in Table 11.1 because if they are firmly implanted in a student's mind and the student is well-versed in the fundamental properties of the Laplace transform in Sect. 11.6, then that student will be much more adept at finding both Laplace and inverse Laplace transform of functions.

This chapter contains all the topics on Laplace transforms that typically are included in introductory differential equations textbooks—and then some. For instance, besides the standard theorem in these textbooks (see Theorem 11.6.7) that is used to help obtain the Laplace transform of the solution of an initial value problem when the differential equation has a periodic forcing term, this chapter also includes a complementary result not to be found in other textbooks, namely Theorem 11.7.6, that can be used to find the inverse Laplace transform of the transformed solution, such as delineated in Example 11.7.22. Furthermore, this theorem facilitates not only solving initial value problems with periodic forcing terms but also some initial value problems with nonperiodic forcing functions that are expressed as an infinite series. Finally, it is worth pointing out here that Corollary 11.7.7 justifies the interchange of the inverse Laplace transform operator $\mathcal{L}^{-1}$ and the summation operator Σ for the types of series that are typically found in engineering textbooks.

Exact Differential Equations and Special Integrating Factors (Previous Exposure to Partial Derivatives Is Not Needed to understand This Material)

Students enrolled in physical science and engineering courses should know how to solve exact differential equations, and time permitting, inexact differential equations, namely, those equations that are not exact but which (by means of an integrating factor) can be changed into ones that are. After all, exact equations have a way of cropping up now and then in physical science and engineering applications. For example, in thermodynamics, the equation of state of a real gas can be derived from an exact equation when certain rates of change are known. In electrostatics, finding surfaces in space that have the same electrical potential involve exact equations. Solving exact equations and finding integrating factors require a basic understanding of the meaning of a partial derivative; however, students with only two semesters of calculus under their belt have probably never even seen a partial derivative much less know how to use one. Since this book is intended for students who have not yet taken a multivariable calculus course, whatever they need to know about partial derivatives in order to understand how to solve these types of equations is gently introduced in Chap. 1 and then covered more extensively in Chap. 8, where exact differential equations and integrating factors are dealt with.

Planar Autonomous Homogeneous Linear Systems and the Eigenvalue-Eigenvector Method (Familiarity with Matrices and Linear Algebra Is Not a Prerequisite)
Planar (two-dimensional) autonomous linear systems of first-order differential equations with constant coefficients are covered in Chaps. 12 and 13. Proofs of theorems involving linear algebra are limited to two-dimensional linear systems because most students who enroll in ODEs courses immediately after completing two semesters of calculus have little or no knowledge of elementary linear algebra, which is essential to an understanding of the underlying mathematics of n-dimensional linear systems; moreover, whatever knowledge of matrices a student may have picked up from high school or in an introductory engineering course is minimal. For this reason, this book has been structured so that it can be read without any previous exposure to matrices and prior knowledge of linear algebra: any terminology, results, and theorems from linear algebra that are needed to study planar linear systems have been directly incorporated into the chapters rather than relegated to an appendix. Moreover, relevant theorems are proven for 2×2 matrices, not just stated, and most of these proofs are adaptable to $n \times n$ matrices. Chapter 12 begins by showing how to convert planar linear systems into vector differential equations and is accompanied by concurrent introductions to vector and matrix notation, matrix-vector products, matrix algebra, nonsingular and invertible matrices, and so on. These preliminaries set the stage for introducing the concept of an eigenvalue and its corresponding eigenvectors, which eventually culminates in the eigenvalue-eigenvector method for solving planar linear systems.

Jordan Canonical Matrices
The topics covered in Chap. 12 engender new questions about linear systems, which pave the way for the topics covered in Chap. 13. For example, consider the vector differential equation $\mathbf{x}' = A\mathbf{x}$. Since e^{at} is a solution of the scalar differential equation $x' = ax$, is e^{At} a solution of the vector differential equation? But hold on: A is a matrix; so what would e^{At} even mean? And if it can be defined in some way that is natural and compatible with the definition of its scalar counterpart e^{at}, can it be differentiated and is its derivative Ae^{At}? But since A is a matrix, is that the same as $e^{At}A$? The answers to questions like these begin with the definitions of the fundamental and principal matrix solutions of $\mathbf{x}' = A\mathbf{x}$ in Sect. 13.1 of Chap. 13, the rules for differentiating matrix functions, and the properties of principal matrix solutions. That prepares the way for defining the matrix exponential function (that is, e^{At}) in Section 2 and investigating its properties. In Section 3, it is shown that every 2×2 coefficient matrix A can be transformed into a matrix belonging to

one of four basic (canonical) forms. These are known as the *Jordan canonical forms of the matrix A*. Some of the other topics covered in Chap. 13 are: similarity transformations; finding solutions of $\mathbf{x}' = A\mathbf{x}$ using the Jordan canonical form of A, which is elucidated in Theorem 13.3.4; the power series expansion of e^{At}; and a detailed discussion of the phase portraits of Jordan canonical linear systems. The chapter concludes with a derivation of the *variation of parameters formula* and detailed examples of using it to solve nonhomogeneous linear systems.

By and large, Jordan canonical forms of matrices and similarity transformations are not covered in introductory ODE textbooks; but when they are, the short shrift given them is barely noticeable. That is understandable considering the depth of knowledge of linear algebra that is needed to understand the how and why of obtaining a Jordan canonical form of an $n \times n$ matrix. However, because this book deals only with 2×2 matrices, it makes sense to include this topic. A benefit of this is that those students who will eventually take a linear algebra course or a more advanced differential equations course will be more at ease with Jordan canonical forms, similarity transformations, and related topics having already been exposed to them in Chap. 13.

Sample Syllabus

The following syllabus is intended for students in engineering, physical science, and mathematics. It includes some applications and mathematical modeling.

Chapters	Sections
Chapter 1. The Wonderful World of Differential Equations	• Ordinary and Partial Derivatives • Differential Equations • Integration and Solutions • Simple Differential Equations Models
Chapter 2. Separable and Related Equations	• Separable Equations • Separation of Variables • Initial Value Problems • Homogeneous Equations (if time permits)
Chapter 3. Direction Fields, Trajectories, and Solution Curves	• Direction Fields and Trajectories • Solution Curves • Vertical Motion of a Body (if time permits)
Chapter 4. Introduction to Numerical Methods	• First-Order Initial Value Problems • Tangent Line Approximations • Euler's Method • Improved Euler's Method • Classical Runge-Kutta Method

(continued)

Chapters	Sections
Chapter 5. First-Order Linear and Related Equations	• Standard Form • Solutions • Functions Defined by Integrals • Bernoulli's Equation • Riccati's Equation (if time permits)
Chapter 6. Modeling with First-Order Equations	• Mathematical Models • Orange Juice Problem • One-Compartment Systems
Chapter 7. Modeling Abrupt Changes	• One-Armed Bandits • Step Functions • The Dirac Delta Function • Abrupt Population Changes and Amortized Loans (if time permits)
Chapter 8. Exact and Related Equations	• Initial Observations • Exact Equations • Total Derivatives and Differentials • Second-Order Partial Derivatives • Solutions of Exact Equations • Criterion for Exactness • First Integrals and Integral Curves (if time permits) • Special Integrating Factors
Chapter 10. Second-Order Linear Equations	• Horizontal Mass-Spring System • Second-Order Homogeneous Linear Equations • Characteristic Roots and Solutions • Complex Exponential Functions • Cauchy-Euler Equations • Homogeneous Linear Equations of Higher Order • Nonhomogeneous Linear Equations • Method of Undetermined Coefficients • Method of Variation of Parameters
Chapter 11. Laplace Transforms	• Definition of the Laplace Transform • Laplace Transforms of Basic Functions • Linearity of the Laplace Transform • First-Order Linear Equations • Existence of the Laplace Transform • Properties of the Laplace Transform • Inverse Laplace Transforms (Omit periodic functions and infinite series unless time permits) • Second-Order Linear Equations

There are plenty of other sections and chapters from which to choose other topics. As a result, this syllabus could be modified to meet the particular needs and interests of a department or those of a class and its instructor. For example, Chap. 10 (Second-Order Linear Equations) could be omitted and replaced with Chap. 12 (Autonomous Homogeneous Linear Systems), after which it could be supplemented with a lecture or handout explaining how a second-order homogeneous linear differential equation with constant coefficients can be solved by first converting it to an equivalent "companion" homogeneous linear system and then solving this companion system. Then the last three sections of Chap. 10 could be covered to show how second-order nonhomogeneous linear differential equation with constant coefficients can be solved using the methods of undetermined coefficients and variation of parameters.

To the Student

A key ingredient in the recipe for success in a mathematics course, or for that matter in any technical course with mathematical underpinnings, is homework. Homework is not only about working assigned problems but also about studying pertinent sections in the textbook. Homework includes preparatory work so that you can begin working the problems with some sense of the way things should go: this involves going over the examples in the textbook, not only reading them but also working through them—with pencil in hand and paper and calculator at hand. It entails going over the examples worked in class by the instructor. At times, it means reviewing forgotten material and boning up on the basics. Homework includes checking answers to see if they make sense, correcting them when they do not. Homework is also about effective communication: making sure you say what you mean—and mean what you say. Even a certain little bear by the name of Winnie-the-Pooh knows that what you think you thought and said may be one thing, but quite another thing to someone else:

> ... when you are a Bear of Very Little Brain, and you Think of Things, you find sometimes that a Thing which seemed very Thingish inside you is quite different when it gets out into the open and has other people looking at it.
>
> —A. A. Milne, *The House at Pooh Corner* (1928)

Doing homework is an absolutely, positively necessary step toward understanding concepts, becoming proficient in working problems, learning the tricks of the mathematical trade, preparing for exams, and feeling a sense of achievement. By now you have heard the admonition "Do your homework!" so many times that it has become clichéd and tiresome to hear. So it goes in one ear and out the other. Homework may seem to you an unnecessary bother and a small matter; but that does not lessen the consequences of consistently failing to do the homework.

The legendary circumstances surrounding the last battle of the "Wars of the Roses" and the demise of an English king illustrate how seemingly trifling matters have a way of affecting more important matters and how easily things can get out of hand. In 1483, Richard, Duke of Gloucester, crowned himself King Richard III.[5] However, the rightful heir to the English throne was his nephew, the twelve-year-old Edward V; but he and his younger brother were imprisoned and murdered in the Tower of London, most likely on Richard's orders. The nobility revolted and eventually King Richard III was defeated and slain at the Battle of Bosworth Field in 1485 by the forces of Henry Tudor, who afterwards was crowned King Henry VII. The defeat of Richard III is portrayed in William Shakespeare's *The Tragedy of King Richard III*. According to the play, the horse that Richard III was riding was slain on the battlefield forcing him to fight on foot. While desperately looking for another horse, he shouts those immortal words attributed to him by Shakespeare: "A horse! A horse! My kingdom for a horse!" But it is the version told in the short story "The Horseshoe Nails" by James Baldwin (1841–1925), an American educator and prolific author of children's books, which we recount here.[6] In hurriedly readying the mount of Richard III for the upcoming battle, a blacksmith made and fitted four horseshoes to the horse's hooves. After he nailed on two shoes, he found that he had only six nails left for the other two hooves but needed ten more. But since the battle was about to begin, he was only able to use three nails to fasten each of the other two shoes. This want of nails sets into motion a sequence of events during the course of the battle. Richard III noticed from the front lines that some of his troops had fallen back. Seeing that the rest of his army was losing courage, he rode back to rally them. But then his horse lost two shoes causing it to stumble and throw the king to the ground. The frightened horse galloped away. Before he could find another horse, Henry's troops closed in. Richard III's troops, upon seeing this, turned and ran. Richard III was slain. The battle was over. One can almost imagine the otherworldly moans and groans of Richard III's ghost as it haunts Bosworth Field, tormented by the sequence of events described in the elegy:

> For the want of a nail, the shoe was lost;
> For the want of a shoe, the horse was lost;
> For the want of a horse, the battle was lost;
> For the failure of battle, the kingdom was lost;
> And all for the want of a horseshoe nail.

Granted, comparing a blacksmith who improperly nails on horseshoes to a student who does little or no homework is a bit of a stretch. (But it does make for a good story!) Even though failure to do one's homework will not lead to the loss of a kingdom, there could still be some repercussions.

---

[5] Will Durant, *The Story of Civilization*, Vol. 6, p. 108, Simon and Schuster, New York, 1957.

[6] "The Horseshoe Nails" is one of the stories in *Fifty Famous People: A Book of Short Stories* by James Baldwin, which was published in 1912 by the American Book Company.

A Student's Lament

For the want of doing homework, my understanding was bad;
For the want of understanding, my exam grades were bad;
For the want of good exam grades, my course grades were bad;
For the want of good course grades, my GPA was bad;
For the want of a good GPA, my job was bad;
For the failure I felt, I was sad;
And all for the want of doing homework.

Germantown, TN, USA Leigh C. Becker

Contents

List of Figures

List of Tables

Chapter 1
The Wonderful World of Differential Equations

The essential fact is simply that all the pictures which science now draws of nature, and which alone seem capable of according with observational fact, are mathematical pictures.
...

Sir James Jeans[1]

Abstract This chapter begins with a collection of classic examples of ordinary differential equations (ODEs) from different scientific and technical fields, such as physics and chemistry, engineering, demography, and chaos theory. A short description of these equations is provided to point out their importance in mathematically modeling physical phenomena and real-world situations. Hopefully, these examples will help students realize that ODEs are relevant and important in their respective fields of study.

This textbook, as its title suggests, is written for students who have completed Calculus I and II (single-variable calculus). However, after a summer recess, students tend to become rusty at calculus. Acknowledging that reality, basic calculus concepts inextricably tied to ODEs, such as the definitions of the *average rate of change* and the *derivative of a function*, are reviewed in this chapter.

Techniques of integration are also reviewed in numerous examples, where it is explained in detail how to go about finding solutions of the simplest ODEs, namely, first-order equations of the form

$$\frac{dy}{dx} = f(x). \tag{1}$$

[1] See [47, ch. 5]. Sir James Jeans (1877–1946) was a British mathematical physicist, Cambridge University lecturer, Princeton University professor of applied mathematics, and author of a number of popular works of science, of which *The Mysterious Universe* [47] was one of his most famous. His treatise, *Problems of Cosmogony and Stellar Dynamics* (1917), on the behavior of fluids in space contributed to a greater understanding of the origin and evolution of the universe.

© The Author(s), under exclusive license to Springer Nature Switzerland AG 2026

L. C. Becker, *Ordinary Differential Equations: Concepts, Methods, and Models*,
https://doi.org/10.1007/978-3-032-15150-6_1

Since the right-hand side only involves the independent variable x, solutions of (1) for a given function $f(x)$ are obtained by finding its indefinite integral. But students are already familiar with this from their Calculus I and II courses. So this is the ideal time to engage and immerse them right away in ODEs by assigning them the problems at the end of this chapter, most of which involve ODEs of the form (1). By solving these problems, students will also be reviewing the basic integration formulas and techniques of integration: substitution, integration by parts, completion of the square, and partial fractions. This will even help those students who have already taken Calculus III (multivariable calculus) since they tend to forget the integration techniques that are part of a Calculus II course.

This textbook does not presuppose any familiarity with partial derivatives; however, Chap. 8 covers exact and related differential equations. In preparation for that chapter, first-order partial derivatives are introduced via an example about the temperature variations on the surface of an unevenly heated copper plate. In addition, there are examples and problems concerned with finding solutions of simple partial differential equations of the form

$$\frac{\partial u}{\partial x} = f(x, y) \quad \text{and} \quad \frac{\partial u}{\partial y} = f(x, y). \tag{2}$$

Simple models and applications of differential equations are introduced in this chapter—not put off until later chapters. For instance, examples and problems of differential equations arising from proportionality statements are presented. An example is the observation that under certain conditions, the rate at which ethane (C_2H_6) decomposes is proportional to its concentration. This observation expressed as an ODE is

$$\frac{d\,[C_2H_6]}{dt} = -k\,[C_2H_6]\,, \tag{3}$$

where the concentration of ethane is denoted by enclosing its molecular formula in brackets and $k > 0$ is the constant of proportionality. Some other examples involving proportionality assumptions are Newton's law of cooling and Poiseuille's law for the volumetric flow rate of a fluid though a cylindrical pipe.

Since Newton's laws of motion are used to introduce a number of different topics in later chapters, a detailed introduction to Newton's second law of motion is presented so that students who have not had a physics course will have no trouble following these topics. Examples in this chapter include modeling the vertical motion of a rock released from the top of a cliff and a derivation of Tsiolkovsky's rocket equation.

The raison d'être for these models and applications is to motivate students and to incentivize the study of ODEs—and to answer the typical classroom question (whether or not it is actually asked): "What good is all this stuff?"

1.1 Ordinary and Partial Derivatives

Just what are differential equations? Following the wisdom of the old Chinese proverb that "one picture is worth more than a thousand words," we defer answering this until we have shown a picture of sorts: Table 1.1 is a collage of some differential equations. These are well-known equations from diverse scientific and technical disciplines. A sense of their importance is realized from the efficacy with which they provide mathematical descriptions of certain physical phenomena and real-world situations, enabling those who use them to predict how observable events will unfold. These equations come from the disciplines of demography, ecology, chemical kinetics, architecture, physics, mechanical engineering, quantum mechanics, electrical engineering, civil engineering, meteorology, epidemiology, and from a relatively new field of study called *chaos theory*. The same differential equation may be important to several disciplines, although for different reasons. For example, demographers, ecologists, and mathematical biologists would immediately recognize the simple equation

$$\frac{dp}{dt} = rp$$

(the first entry in Table 1.1) as the *Malthusian law of population growth*, which is used to predict the populations of certain kinds of organisms reproducing under ideal conditions. But physicists, chemists, and nuclear engineers would point out that this equation is a mathematical portrayal, or *model*, of radioactive decay. Many economists and mathematically-minded investors would also recognize this differential equation but in the context of a formula that is used to calculate future balances of investments earning interest at rates compounded continuously.

Another example is the *van der Pol equation*

$$\frac{d^2x}{dt^2} - \varepsilon(1 - x^2)\frac{dx}{dt} + x = 0,$$

which resulted from modeling oscillations of currents in nonlinear electrical circuits of the first commercial radios. For many years, it was the subject of research by electrical engineers and mathematicians alike.

Even though the equations in Table 1.1 are drawn from a number of diverse fields, they do have some common features. The foremost feature shared by all of them is that they have at least one derivative, which is precisely what makes them differential equations in the first place. We note this here with the following definition of what is meant by a differential equation.

Table 1.1 Differential equation models

Differential equation	Description
$\frac{dp}{dt} = rp$	The *Malthusian law of population growth* is used to model the populations of certain kinds of organisms living in ideal environments for limited lengths of time. It gives the rate at which a population p changes with respect to the time t. The value of the constant r depends on the type of organism.
$\frac{dx}{dt} = k(A - x)^2$	This *second-order reaction rate* law gives the rate at which a single chemical species combines to produce a new species, such as methyl radicals combining in a gas to form ethane molecules. See Atkins [3, p. 134].
$\frac{d^2y}{dx^2} = \frac{C}{L}\sqrt{\left(\frac{AC}{L}\right)^2 + \left(\frac{dy}{dx}\right)^2}$	The graph of the solution of this equation describes the shape of the *Gateway Arch* in St. Louis, Missouri (www.nps.gov/jeff). The variable y gives the height of the arch at a distance x from one end of its base. The constants A, C, and L relate the lengths of the base, top, and centroid. The graph is a so-called *inverted catenary*, which is the curve that has the shape of a chain suspended from two points at the same level.
$m\frac{d^2x}{dt^2} + b\frac{dx}{dt} + kx = F(t)$	This differential equation models the motion of a damped mass-spring system subjected to a time-dependent force $F(t)$.
$y'' - xy = 0$	This is *Airy's differential equation*. It is named after the British astronomer and mathematician G. B. Airy, who found a solution of this equation, called the Airy integral, which he used to explain rainbow phenomena.
$\frac{\hbar^2}{2m} \cdot \frac{d^2\psi}{dx^2} + (E - \frac{1}{2}kx^2)\psi = 0$	This equation from quantum mechanics is the time-independent *Schrödinger's equation* for the one-dimensional simple harmonic oscillator. The constant $\hbar$ is defined in terms of Planck's constant h by $\hbar = h/2\pi$.
$x'' - \varepsilon(1 - x^2)x' + x = 0$	*Van der Pol's equation*, another eponymous equation, models the current x at time t in an electrical circuit with nonlinear resistance.
$EI\frac{d^4y}{dx^4} = w(x)$	The solution of this equation gives the vertical displacement y of a point located a distance x from the fixed end of a beam of uniform cross section, where E and I are constants and $w(x)$ represents the load at x.
$\frac{dx}{dt} = \sigma(y - x)$ $\frac{dy}{dt} = rx - y - xz$ $\frac{dz}{dt} = xy - bz$	This system of three differential equations is known as the *Lorenz system*. It is an overly simplified version of a complicated system of twelve equations used to model convection in the atmosphere. This particular system models the chaotic rotational motion of a wheel with leaking compartments of water symmetrically positioned around its rim. See Appendix C.

(continued)

Table 1.1 (continued)

Differential equation	Description
$S' = \Lambda - \mu S - \beta S \dfrac{I}{N}$ $E' = \beta S \dfrac{I}{N} - (\mu + \epsilon)E$ $I' = \epsilon E - (\gamma + \mu + \alpha)I$ $R' = \gamma I - \mu R$	This system of four equations, the *SEIR model*, is used to model the spread of an infectious disease. The variables $S(t)$, $E(t)$, $I(t)$, and $R(t)$ denote the number of individuals who are (i) susceptible, (ii) exposed, (iii) infected or infectious, and (iv) recovered, respectively, at time t. Many studies of the spread of the disease COVID-19, such as reference [4], are based on this model.

Differential Equation

Definition 1.1.1 A ***differential equation*** is an equation that involves one or more derivatives of some unknown function or functions.

To complicate matters, there are various types of differential equations, such as ordinary differential equations, partial differential equations, and integro-differential equations. The equations in Table 1.1 are examples of *ordinary differential equations* since they only involve ordinary derivatives. Ordinary derivatives are the kinds of derivatives that are covered in an introductory (single-variable) calculus course.

Partial differential equations involve so-called partial derivatives. How a partial derivative differs from an ordinary derivative will be discussed later on, after a review of ordinary derivatives in the next section. To give us an inkling of what partial differential equations look like, here is a classic example:

$$\frac{\partial u}{\partial t} = k \frac{\partial^2 u}{\partial x^2}.$$

This particular partial differential equation is used to model the conduction of heat through an extremely thin metal bar, where $u(x, t)$ is the temperature at the point x in the bar at time t.

Integro-differential equations involve not only derivatives of unknown functions but also their integrals. In Chap. 11 we will solve integro-differential equations of the form

$$x'(t) = f(t) + \int_0^t k(t - u)x(u)\, du.$$

This book is devoted to a study of ordinary differential equations. Even so, there will be brief forays from time to time into topics involving very simple partial differential equations, integral equations, and integro-differential equations.

Before we formally define what is meant by an ordinary differential equation, let us point out some other features that the equations in Table 1.1 have in common. First observe that each equation in Table 1.1 contains a single ***independent variable*** and one or more ***dependent variables***. It is a relatively simple matter to tell these two types of variables apart from the derivative itself, since differentiation always takes place with respect to the independent variable. Obviously then, the other variable, the one being differentiated, is the dependent variable.

Example 1.1.1 The first entry in Table 1.1 is the Malthusian law of population growth:

$$\frac{dp}{dt} = rp.$$

Translated into words, this equation says that the rate at which the current population p of an organism changes with respect to the time t is equal to the product of a constant r and the current population p. The time t is the independent variable and the population p is the dependent variable. ♦

Example 1.1.2 The differential equation

$$EI\frac{d^4y}{dx^4} = w(x)$$

in Table 1.1 models the vertical displacement of a beam. Since y is differentiated with respect to x, the independent variable is x and the dependent one is y. ♦

The term *dependent variable* is apt because it describes the type of variable it is, namely, that it depends in some functional way on the independent variable as prescribed by the differential equation, although this dependence may not always be possible. For example,

$$y^2 + \left(\frac{dy}{dx}\right)^2 = -1$$

is a differential equation; even so, no real-valued function[2] can fulfill the prescription that the sum of its square and the square of its derivative is equal to a negative number.

[2] A function is ***real-valued*** when every evaluation of it results in a real number. Even though the function $i \sin x$, where $i^2 = -1$, satisfies the differential equation, it is a complex-valued solution, not a real-valued solution.

Space on a page can be saved by replacing ***Leibniz's notation*** for the derivative, which employs the Latin "*d*" to indicate differentiation, such as

$$\frac{dp}{dt}, \quad \frac{d^2x}{dt^2}, \quad \frac{d^4y}{dx^4},$$

with a more concise notation that uses the prime symbol ($'$). With ***prime notation*** (also called ***Lagrange's notation***), the derivatives

$$\frac{dy}{dx} \quad \text{and} \quad \frac{dp}{dt}$$

are written as

$$y' \quad \text{and} \quad p',$$

respectively. A shortcoming of this notation is that the name of the independent variable is not given here. Take, for instance, the derivative z'. Without more information or context, z' by itself does not tell us the name of the independent variable—whether it is x, t, or something else. Now consider the simple differential equation

$$y' = 2x.$$

Since y is the dependent variable, the other variable in this equation, namely x, is the independent variable. Of course, we know from studying calculus that the function $y = x^2 + k$, for any constant k, satisfies this differential equation. Contrast that with the equation

$$y' = 2y.$$

Since the name of the independent variable is not provided, we are free to choose one. For instance, if the independent variable represents time, then t is an appropriate choice. A function satisfying this differential equation is $y = ke^{2t}$, where k denotes any constant.

Another concise way to indicate differentiation is to use ***overdot notation***, where a dot ($\dot{}$) is placed over the dependent variable. This is also known as ***Newton's notation*** and usually reserved for derivatives taken with respect to time. For example, suppose a lab experiment is conducted with the goal of being able to predict the population p of E. coli bacteria in a flask of nutrient broth at any given time. If the letter t is chosen to represent time, then $\dot{p}$ means dp/dt. In overdot notation, the Malthusian law of population growth (see Table 1.1 and Example 1.1.1)

$$\frac{dp}{dt} = rp$$

becomes

$$\dot{p} = rp.$$

It will become clear in later chapters that we have to be aware of the *orders* of the derivatives appearing in differential equations. The derivatives

$$\frac{dy}{dx}, \quad \dot{p}, \quad z'$$

are examples of ***first-order derivatives***, whereas the derivatives

$$\frac{d^2x}{dt^2}, \quad y'', \quad \ddot{p}$$

are ***second-order derivatives***. A differential equation may have derivatives of even higher order. For example, it could contain ***third-order derivatives***, such as

$$y''', \quad \frac{d^3x}{dt^3},$$

or ***fourth-order derivatives***, such as

$$\frac{d^4x}{dt^4},$$

or even higher order derivatives. It is easy to lose count of the number of primes or overdots when the order is more than three. In such cases, it is customary to use either Leibniz's notation or to use superscripts enclosed in parentheses to denote such derivatives: for instance, d^4y/dx^4 or $y^{(4)}$ is preferred over y''''. The nth derivative of y with respect to x is written as d^ny/dx^n or as $y^{(n)}$.

All of the derivatives up to this point have been *ordinary derivatives*. When we are asked to take the (ordinary) derivative of a function, the term *ordinary* indicates that we are dealing with a function of one variable. In other words, an ***ordinary derivative*** is a derivative of a function of a single variable with respect to that variable. The word *ordinary* qualifies the word *derivative*, distinguishing between the derivatives of single-variable functions (such as $f(x) = 5x^2 + \sin x$) from the derivatives of multivariable functions (such as $f(x, y) = 5y^2 + \sin x$). Single variable calculus (Calculus 1 and 2) deals only with functions of one variable. Multivariable calculus (Calculus 3) deals with functions of two or more variables;

their derivatives are called *partial derivatives*. The definition of a partial derivative will be introduced later; but for now, let us review the definition and meaning of the *ordinary derivative of a function*.

1.1.1 Ordinary Derivatives

Let us review the definition of an *ordinary derivative* by way of an example. Suppose a very thin copper wire of length 25 cm is completely stretched out in a straight line. Let the line serve as the x-axis and the left end of the wire designate the location of the origin. Suppose the wire is heated unevenly in such a way that each of its points eventually settles at a constant temperature but that the temperature generally varies from point to point. Even though in reality the wire is a three-dimensional object, its very thinness suggests that variations in temperature along the y- and z-directions are negligible. Thus, the wire may be viewed as a one-dimensional mathematical object: the line segment extending from $x = 0$ cm to $x = 25$ cm. Table 1.2 lists the results of temperature measurements at 5-cm intervals along the wire, accurate to the ten-thousandth decimal place. The ***average rate of change*** of the temperature T of the wire as x changes from x_1 to x_2 is given by the ***difference quotient*** $\Delta T/\Delta x$, where $\Delta T = T(x_2) - T(x_2)$ and $\Delta x = x_2 - x_1$. For example, if x changes from $x_1 = 10$ cm to $x_2 = 15$ cm, then the average rate of change of the temperature is

$$\frac{\Delta T}{\Delta x} = \frac{T(15) - T(10)}{15 - 10} = \frac{98.8720 - 99.4980}{5} = -0.1252.$$

Thus, the temperature of the wire decreases at an average rate of approximately 0.1 °C per cm as x increases from 10 to 15 cm. As another example, let us calculate the average rate of change of the temperature when x decreases from 10 to 5 cm. In that case, the difference quotient is

$$\frac{\Delta T}{\Delta x} = \frac{T(5) - T(10)}{5 - 10} = \frac{99.8740 - 99.4980}{-5} = -0.0752.$$

The negative sign is the result of the temperature increasing as x is decreasing; so one way to interpret this is that T increases on average about 0.08 °C per cm as x decreases from 10 to 5 cm. Or, equivalently, as x increases from 5 to 10 cm, the temperature decreases roughly 0.08 °C per cm.

Table 1.2 Temperatures at points of an unevenly heated wire

x (cm)	0	5	10	15	20	25
T (°C)	100.0000	99.8740	99.4980	98.8720	97.9960	96.8700

If no other temperature data are available other than what is provided in Table 1.2, one of the two previously calculated values might be used as a rough estimate of the instantaneous rate of change[3] of the temperature at *at the point* $x = 10$ cm—or better yet, the average of the two calculated values. However, a change of 5 cm is a big jump for estimating the rate of change at a specific point. Even better estimates could be obtained with smaller jumps. For instance, suppose we also knew the temperature of the wire at $x = 11$ cm. Then an estimate of the rate of change of the temperature at $x = 10$ cm is given by the difference quotient

$$\frac{\Delta T}{\Delta x} = \frac{T(11) - T(10)}{11 - 10}.$$

Suppose the temperature of the wire at $x = 11$ cm is 99.3928 °C. Then,

$$\frac{\Delta T}{\Delta x} = \frac{99.3928 - 99.498}{1} = -0.1052.$$

This provides us with a new and better estimate of the rate at which the temperature decreases at $x = 10$ cm, namely, about 0.1 °C per cm. This is an improvement over previous estimates since Δx is now smaller by a factor of 5.

Of course, better estimates could be found with more temperature data. Then we could choose even smaller values for Δx. Ideally, if we knew the temperature at each of the points of the wire, then we could compute the (*instantaneous*) *rate of change* of the temperature at $x = 10$ cm by taking the limit of the difference quotient $\Delta T/\Delta x$ as Δx approaches 0, where $\Delta x = x - 10$ and ΔT denotes the accompanying change in the temperature from $x = 10$ cm to $x = 10 + \Delta x$ cm. In mathematical notation, this is expressed by

$$\lim_{\Delta x \to 0} \frac{\Delta T}{\Delta x} = \lim_{\Delta x \to 0} \frac{T(10 + \Delta x) - T(10)}{\Delta x}.$$

This limit is called the *derivative* of the temperature at $x = 10$ cm and is denoted by the symbol $T'(10)$. For example, suppose the temperature at every point of the wire is given by the function

$$T(x) = 100 - 0.0002x - 0.005x^2.$$

Thus, the temperature change ΔT from $x = 10$ cm to $x = 10 + \Delta x$ cm is

$$\begin{aligned}\Delta T &= T(10 + \Delta x) - T(10)\\ &= 100 - 0.0002(10 + \Delta x) - 0.005(10 + \Delta x)^2 - 99.4980\\ &= -0.1002\Delta x - 0.005(\Delta x)^2.\end{aligned}$$

[3] The word "instantaneous" is frequently omitted. The phrase "rate of change" without the qualifier "average" means "instantaneous rate of change."

It follows that the difference quotient is

$$\frac{\Delta T}{\Delta x} = -0.1002 - 0.005\Delta x.$$

Since this quotient approaches $-0.1002\,^\circ\mathrm{C/cm}$ as $\Delta x \to 0$, the derivative of $T(x)$ at $x = 10$ cm is

$$T'(10) = -0.1002\,^\circ\mathrm{C/cm}.$$

In other words, the temperature of the wire decreases at a rate of 0.1002 °C per cm at $x = 10$ cm. Notice, however, that since we now have a formula for $T(x)$ instead of just a table, the computation of $T'(10)$ can be carried out more quickly using the derivative rules and formulas of calculus. For example,

$$T'(x) = \frac{d}{dx}\left(100 - 0.0002x - 0.005x^2\right) = -0.0002 - 0.01x.$$

And so we have, as before,

$$T'(10) = -0.0002 - 0.01(10) = -0.1002\,^\circ\mathrm{C/cm}.$$

Frequently, in engineering, the mathematical sciences, and other related disciplines, computations involving rates of change have to be dealt with. The previous example provided a way of reviewing the definition and the meaning of the derivative. Here is a summary. Computing the rate of change of a function f with respect to x starts with the ***difference quotient***

$$\frac{f(x + \Delta x) - f(x)}{\Delta x}.$$

This quotient gives the ***average rate of change*** of f over the interval $[x, x + \Delta x]$. The ***ordinary derivative*** of f at x gives the ***instantaneous rate of change*** of f at this point.[4] It is defined as the limit of the difference quotient of f as Δx approaches 0, provided that this limit exists. This is expressed succinctly by writing

$$\lim_{\Delta x \to 0} \frac{f(x + \Delta x) - f(x)}{\Delta x}. \tag{1.1.1}$$

The notation for this limit is $f'(x)$ or, if y is the name of the dependent variable, by dy/dx. The result of computing (1.1.1), if the limit exists, is an expression in x. It is in fact a function of x—just as f is—however, its domain may differ from that of f. The domain of f' consists of all x-values for which the limit (1.1.1)

[4] We usually omit the word "ordinary" and simply say the ***derivative*** of f at x.

exists. For example, the domain of the real-valued function $f(x) = \sqrt{x}$ consists of all nonnegative numbers ($x \geq 0$). For these values of x, aside from $x = 0$, the difference quotient (1.1.1) converges to $1/(2\sqrt{x})$. We conclude that

$$f'(x) = \frac{1}{2\sqrt{x}}$$

and that its domain is $(0, \infty)$—that is, the set of all positive numbers.

When x is assigned a value, say $x = a$, where a denotes a specific number, then $f'(a)$ denotes a number that is the instantaneous rate of change of f at $x = a$. For instance, the instantaneous rate of change of the function $f(x) = \sqrt{x}$ at $x = 9$ is

$$f'(9) = \frac{1}{2\sqrt{9}} = \frac{1}{6}.$$

Generally speaking, the quickest way to obtain derivatives of functions is to use the differentiation rules and formulas of calculus instead of the definition of the derivative given by (1.1.1). A list of those rules, such as the Quotient Rule and Chain Rule, and a short table of basic derivatives can be found in Appendix A.

1.1.2 Ordinary Differential Equations

Now that we have finished reviewing the definition of the ordinary derivative, we can state what is meant by an ordinary differential equation.

Ordinary Differential Equation

Definition 1.1.2 An ***ordinary differential equation*** is an equation containing one independent variable; one or more dependent variables, each representing an unknown function of the independent variable; and ordinary derivatives of one or more of the dependent variables.

The equations in Table 1.1 fit the description in Definition 1.1.2; accordingly, all of them are ordinary differential equations. Notice, however, that the mathematical models of the situations described in the last two entries of the table require more than one differential equation. For example, the model in the next-to-last entry consists of a system of three differential equations.

$$\begin{aligned} \dot{x} &= \sigma(y - x) \\ \dot{y} &= rx - y - xz \\ \dot{z} &= xy - bz. \end{aligned} \tag{1.1.2}$$

When a steady, uniformly distributed shower of water falls on a wheel that has leaky compartments symmetrically positioned around its rim, it can be shown that these equations model the rotational motion of the wheel.[5] Observe the equations in (1.1.2) are inextricably linked with each other through the dependent variables x, y, z. Together they constitute the ***Lorenz system*** of ordinary differential equations. This system is iconic in being a catalyst for initiating a branch of mathematics and science called *chaos theory*. The equation $\dot{y} = rx - y - xz$ is still regarded as an ordinary differential equation, even though it contains all three dependent variables. The reason for this is that $\dot{y}$ is an ordinary derivative and all three dependent variables depend only on the single independent variable t. This is implied from the form of the three equations in the system. Likewise, the other two equations are also ordinary differential equations. The Lorenz system, in sum, consists of three ordinary differential equations.

One characteristic of a differential equation is the order of the highest derivative appearing in the equation. For example, if a differential equation contains a derivative of second order (a second derivative) but none of higher order, then we say it is of order 2 or that it is a second-order differential equation. Airy's differential equation listed in Table 1.1 is such an equation. The Lorenz system consists of three first-order equations. Other examples are:

(a) $\dfrac{dx}{dt} = k(a - x^2)$ is an equation of order 1 (or, first-order equation);
(b) $2xyy' + (yy')^2 = y^2$ is a first-order equation;
(c) $EI\dfrac{d^4y}{dx^4} = w(x)$ is an equation of order 4 (or, fourth-order equation);
(d) $4xy^2\big(y^{(4)}\big)^3 - 3x^4y^5\big(y''\big)^6 = \cos^9(x^{10})$ is a fourth-order equation.

Order of an Ordinary Differential Equation

Definition 1.1.3 The ***order*** of an ordinary differential equation is said to be n if the order of the highest derivative appearing in the equation is n.

Besides an independent variable and a dependent variable (or variables), most of the differential equations listed in Table 1.1 contain quantities known as *parameters*. A ***parameter*** does not change in value with changes in the value of the independent variable; however, its value may change when the situation or experiment that it models is modified. For example, consider the simple differential equation

$$\frac{dp}{dt} = rp,$$

[5] For more information and a derivation of these equations, see Appendix C.

which is known as the *Malthusian law* when it is used to predict the population p of a certain types of species at time t. The quantity r is a constant for a given species; its value does not change with time. Yet its value will most likely change if it is used to model a different species. Another example is the damped mass-spring system

$$m\frac{d^2x}{dt^2} + b\frac{dx}{dt} + kx = F(t),$$

which has three parameters: m, b, and k. The parameter m denotes the mass of a body to which a spring is attached. The parameter k measures the stiffness of the spring, and b measures the retardation in the motion of the body due to damping forces, such as friction. None of these parameters depends on time, yet their values would change if the body and spring were replaced by some other body and spring.

1.1.3 Solutions

Generally speaking, when one is dealing with a differential equation, the goal is to find out as much as possible about its solutions. However, the meaning of the word "solution" in the context of differential equations is sometimes misunderstood. The reason for this goes back to its meaning in algebra, trigonometry, and calculus—in these worlds, solutions are numbers. But in the world of differential equations, solutions are not numbers! Let us illustrate this by comparing a solution of an algebraic equation to that of a differential equation. The solution of

$$2x - 3 = x + 7$$

is "10", a number. Why is it a solution? The answer, of course, is that "10" satisfies this equation. Substitution of "10" for the unknown "x" results in the left- and right-hand sides of the equation being equal: the left-hand side (LHS) becomes

$$\text{LHS} = 2x - 3 = 2(10) - 3 = 17,$$

which equals the right-hand side (RHS)

$$\text{RHS} = x + 7 = 10 + 7 = 17.$$

By contrast, the solutions of differential equations are functions, not numbers. A simple example is provided by the differential equation

$$\frac{dy}{dx} = 2x.$$

A solution is $y = x^2$. In fact, every function of the form $y = x^2 + C$, where C is a constant, is a solution. The reason for this is that these functions satisfy the equation.

When we substitute "$x^2 + C$" for the unknown y and differentiate, we obtain "$2x$", which is precisely the right-hand side of the equation:

$$\text{LHS} = \frac{d}{dx}y = \frac{d}{dx}(x^2 + C) = 2x \quad \text{equals} \quad \text{RHS} = 2x.$$

No other functions have derivatives equal to $2x$, aside from those of the form "$x^2 + C$". Consequently, these are the only solutions of the differential equation. Let us take a look at some more examples.

Example 1.1.3 Suppose someone alleges that $y = \ln x$ is a solution of

$$xy'' = -y'. \tag{1.1.3}$$

Determine if it really is a solution.

Solution We have not yet learned any methods for solving differential equations. Even so, we can still answer this question. All that we have to do is to substitute the first and second derivatives of $\ln x$ into the equation to see if a true statement results. With these substitutions, the left-hand side of the equation is

$$\text{LHS} = xy'' = x\frac{d^2}{dx^2}(\ln x) = x\frac{d}{dx}\left(\frac{1}{x}\right) = x\left(-\frac{1}{x^2}\right) = -\frac{1}{x}$$

and the right-hand side is

$$\text{RHS} = -\frac{d}{dx}(\ln x) = -\frac{1}{x}.$$

Since both the left- and right-hand sides turn out to be equal to $-x^{-1}$, the function $y = \ln x$ is a solution. There is also another matter to consider. Since a solution is a function, we need to be aware of the domain of its definition and where it solves the differential equation. As for this example, we know from calculus that the domain of $\ln x$ is the interval $(0, \infty)$. Since $\ln x$ also satisfies (1.1.3) on this interval, we say that the maximum interval for which $y = \ln x$ is a solution of (1.1.3) is $(0, \infty)$. ♦

Example 1.1.4 The function $y = x^2$ is also alleged to be a solution of (1.1.3). Is it?

Solution Substituting, we find that

$$\text{LHS} = x\frac{d^2}{dx^2}(x^2) = x\frac{d}{dx}(2x) = 2x,$$

whereas

$$\text{RHS} = -\frac{d}{dx}(x^2) = -2x.$$

Since the LHS $\neq$ RHS, we conclude that $y = x^2$ is not a solution of the differential equation. ♦

Example 1.1.5 Is $y = 1$ a solution of (1.1.3)?

Solution This might be interpreted as: Is "1" a solution? But that would be incorrect. The question really asks: Does the constant function $y(x) = 1$ satisfy the equation? Now this may seem like quibbling over semantics but there is a point to be made:

> *Solutions of algebraic equations are numbers. Solutions of differential equations are functions!*

Now the answer: Since both the first and second derivatives of this function are equal to 0 at all values of x, it satisfies the equation for all $-\infty < x < \infty$. Thus $y(x) \equiv 1$ is indeed a solution.[6] ♦

Before presenting any more examples, let us summarize what the previous examples have taught us about what is meant by a *solution* of a differential equation.

Solution of an Ordinary Differential Equation

Definition 1.1.4 A ***solution*** of an ordinary differential equation with one dependent variable is a differentiable function of the independent variable that satisfies the equation on some interval. In other words, if we substitute the function for the dependent variable, we obtain a result that is valid on the interval.

The interval on which a function is a solution of an ordinary differential equation is called the ***interval of existence of the solution*** or the ***domain of the solution***. Depending on the differential equation, this interval could be a finite interval, such as an open interval (a, b) or a closed interval $[a, b]$, or an infinite interval, such as (a, ∞) or the set of all real numbers (∞, ∞).

Example 1.1.6 Is $y = \sin x$ a solution of the equation $y^2 + (y')^2 = 1$?

Solution If we substitute $\sin x$ for y, we obtain $y^2 + (y')^2 = \sin^2 x + \cos^2 x = 1$ because of the Pythagorean identity of trigonometry. Since this is true for all values of x, this function is a solution on the interval $(-\infty, \infty)$. ♦

An algebraic equation may not have any real-valued solutions, such as $x^2 = -1$. The same may be true of a differential equation. Consider, for instance,

$$y^2 + (y')^2 = -1.$$

[6] The symbol "$\equiv$" stands for the phrase "is identically equal to." For instance, $y(x) \equiv 1$ is a more succinct way to say that $y(x) = 1$ for all $-\infty < x < \infty$.

Since the left-hand side of this equation is nonnegative no matter what real-valued, differentiable function is substituted for y, there is no real-valued function that solves this equation.

Example 1.1.7 The differential equation

$$(x-1)\frac{dy}{dx} = \frac{y^2-3}{y} \tag{1.1.4}$$

clearly has the constant solutions $y = \sqrt{3}$ and $y = -\sqrt{3}$. Since each of these satisfies the equation for $-\infty < x < \infty$, we express this by saying that the interval of existence of both solutions is $(-\infty, \infty)$. Equation (1.1.4) also has nonconstant solutions. One of them is

$$y = \sqrt{3-(x-1)^2}. \tag{1.1.5}$$

Verify this and determine the interval of existence of this solution.

Solution If y is given by the function (1.1.5), the left-hand side of the differential equation is

$$\text{LHS} = (x-1)\frac{d}{dx}\sqrt{3-(x-1)^2} = \frac{-(x-1)^2}{\sqrt{3-(x-1)^2}}.$$

And the right-hand side is

$$\text{RHS} = \frac{y^2-3}{y} = \frac{3-(x-1)^2-3}{\sqrt{3-(x-1)^2}} = \frac{-(x-1)^2}{\sqrt{3-(x-1)^2}}.$$

Thus, for the given function, we see that LHS = RHS wherever $3-(x-1)^2 > 0$. Solving this inequality, we obtain $1-\sqrt{3} < x < 1+\sqrt{3}$. We conclude that (1.1.5) is a solution of (1.1.4) on the interval $(1-\sqrt{3}, 1+\sqrt{3})$. ♦

1.1.4 *Partial Derivatives*

We will introduce *partial derivatives* much in the same way as we did with our review of *ordinary derivatives* by considering temperature variations in an unevenly heated copper object. This time, however, instead of a filament-like copper wire, let us imagine heating a very thin, rectangular copper plate with a length of 25 cm and a width of 5 cm. Due to the thinness of the plate, we will ignore its thickness in the ensuing discussion; in effect, we are using a two-dimensional rectangle to model the real, three-dimensional copper plate. Let us orient the plate so that two of its adjoining edges are along the x- and y-axes as depicted in Fig. 1.1. As in the

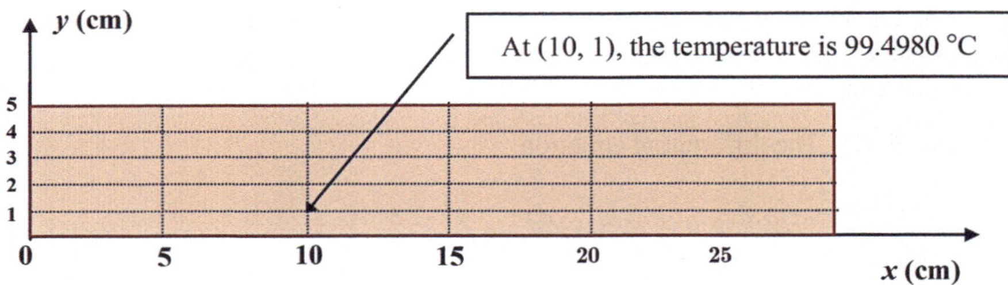

Fig. 1.1 Unevenly heated copper plate

Table 1.3 Temperatures (°C) at points of an unevenly heated plate

		x (cm)					
		0	5	10	15	20	25
y (cm)	0	99.0000	98.8750	**98.5000**	97.8750	97.0000	95.8750
	1	**100.0000**	**99.8740**	**99.4980**	**98.8720**	**97.9960**	**96.8700**
	2	101.0000	100.8710	**100.4920**	99.8630	98.9840	97.8550
	3	102.0000	101.8660	**101.4820**	100.8480	99.9640	98.8300
	4	103.0000	102.8590	**102.4680**	101.8270	100.9360	99.7950
	5	104.0000	103.8500	**103.4500**	102.8000	101.9000	100.7500

example with the wire, let us suppose the plate is heated unevenly in such a way that the temperature at each of its points remains constant but that it generally changes from point to point. Consider Table 1.3, which shows the temperatures of the plate at a number of points (x, y) to the ten-thousandth decimal place. Let us use the symbol $T(x, y)$ to denote the temperature at (x, y). In other words, T is the temperature function. Since it depends on both x and y, it is a function of two variables, not just one. Let us select a point on the plate, say $(10, 1)$, for the purpose of examining the temperatures near it. The temperatures in the row and column containing the temperature at $(10, 1)$, namely $99.4980\,^\circ$C, are in boldface. The common feature shared by the temperatures in the boldfaced row is that all of them are at points with the same value of y, to wit: $y = 1$. So if we restrict our attention to the temperatures in this row, then T may be viewed as a function of the single variable x. Let $r(x)$ denote the temperature at a point x belonging to this row. The derivative

$$r'(10) = \lim_{\Delta x \to 0} \frac{r(10 + \Delta x) - r(10)}{\Delta x} \tag{1.1.6}$$

gives the instantaneous rate of change of the temperature T at the point $(10, 1)$ when x is allowed to change but y is held fixed at the value 1. The symbol $T_x(10, 1)$ is one of several that are used to denote this derivative. Since $r(x) = T(x, 1)$, the limit of the difference quotient (1.1.6) expressed in terms of T is

$$T_x(10, 1) = \lim_{\Delta x \to 0} \frac{T(10 + \Delta x, 1) - T(10, 1)}{\Delta x}. \tag{1.1.7}$$

This is called the *partial derivative of T with respect to x* at the point (10, 1). Since we do not have function that gives the temperature at every point of the plate, the best that we can do is to estimate $T_x(10, 1)$ with an average rate of change. For example, if we choose $\Delta x = 5$, then from Table 1.3, we have

$$\frac{T(15, 1) - T(10, 1)}{5} = \frac{98.8720 - 99.4980}{5} = -0.1252.$$

So $T_x(10, 1)$ is about -0.1 °C per cm.

Since (1.1.7) gives the instantaneous rate of change of T at the point (10, 1) when the value of y is held fixed at 1, it is natural to ask how one would go about finding the instantaneous rate of change of the temperature T at the point (10, 1) when y is allowed to change but x is held fixed at the value 10. This time we use the temperatures in the boldfaced column. Let $c(y)$ denote these temperatures. To estimate the rate of change of T at the point (10, 1), we could approximate it with the average rate of change of $c(y)$ over the interval [1, 2] (or over [0, 1]; better yet, compute both and use their average as an estimate). Over [1, 2], the average rate of change of $c(y)$ is

$$\frac{\Delta c}{\Delta y} = \frac{c(1 + \Delta y) - c(1)}{\Delta y} = \frac{c(2) - c(1)}{2 - 1} = \frac{100.4920 - 99.4980}{1} = 0.9940.$$

Since $c(y) = T(10, y)$ and $\Delta y = 1$, this translates to

$$\begin{aligned} \frac{\Delta T}{\Delta y} &= \frac{T(10, 1 + \Delta y) - T(10, 1)}{\Delta y} \\ &= \frac{T(10, 2) - T(10, 1)}{2 - 1} = \frac{100.4920 - 99.4980}{1} = 0.9940. \end{aligned} \tag{1.1.8}$$

So the temperature increases about 0.99 °C when y increases from 1 to 2 cm and x is held fixed at 10 cm. The instantaneous rate of change at the point (10, 1) with x held fixed at the value 10 cm is defined to be the limit

$$\lim_{\Delta y \to 0} \frac{\Delta T}{\Delta y} = \lim_{\Delta y \to 0} \frac{T(10, 1 + \Delta y) - T(10, 1)}{\Delta y}. \tag{1.1.9}$$

However, there is a fly in the ointment: If we do not know the temperatures at all of the points in the plate that have their x-coordinates equal to 10, then the limit (1.1.9) can only be estimated with a difference quotient, such as the one in (1.1.8).

The limit given by (1.1.9) is called the *partial derivative of T with respect to y* at the point (10, 1) and is denoted by the symbol $T_y(10, 1)$. For example, suppose the function

$$T(x, y) = 99 + y - 0.0002xy^2 - 0.005x^2 \tag{1.1.10}$$

gives the temperature at every point of the copper plate. The partial derivative $T_y(10, 1)$ is found by taking the limit of the difference quotient in (1.1.9) as follows:

$$\lim_{\Delta y \to 0} \frac{T(10, 1+\Delta y) - T(10, 1)}{\Delta y} = \lim_{\Delta y \to 0} \frac{\Delta y - 0.004\Delta y - 0.002(\Delta y)^2}{\Delta y}$$

$$= \lim_{\Delta y \to 0} (0.996 - 0.002\Delta y) = 0.996.$$

Therefore, $T_y(10, 1) = 0.996\,^\circ\text{C/cm}$.

The point of the heated copper plate example was to introduce the definition of a *partial derivative*. The rate of change of a function of two variables with respect to one of its variables, as the other one is held constant, is known as a partial derivative of the function. Since there are two variables, there are two first-order partial derivatives.[7] A partial derivative of a function of two variables is obtained by taking the limit of a difference quotient for that function, just as an ordinary derivative of a function of one variable is obtained by taking the limit of its difference quotient. We just have to state precisely what is meant by the differences quotients for functions of two variables and how to take their limits. Referring to the computations involving the temperature function (1.1.10), the next definition should come as no surprise.

First-Order Partial Derivatives

Definition 1.1.5 Let F be a function of x and y. The ***partial derivative of*** F ***with respect to*** x, denoted by F_x, is defined by

$$F_x(x, y) = \lim_{\Delta x \to 0} \frac{F(x+\Delta x, y) - F(x, y)}{\Delta x}, \tag{1.1.11}$$

for values of x and y for which the limit exists. Likewise, the ***partial derivative of*** F ***with respect to*** y, denoted by F_y, is defined by

$$F_y(x, y) = \lim_{\Delta y \to 0} \frac{F(x, y+\Delta y) - F(x, y)}{\Delta y}, \tag{1.1.12}$$

for values of x and y for which the limit exists.

Example 1.1.8 Find the partial derivative of $F(x, y) = 5x^2y$ with respect to x.

[7] There are also higher-order partial derivatives, just as there are higher-order ordinary derivatives. In this chapter, we only consider first-order partial derivatives. However, in a future chapter, we will need to talk about second-order partial derivatives but that can wait for now.

Solution Referring to (1.1.11), we have

$$\begin{aligned} F_x(x, y) &= \lim_{\Delta x \to 0} \frac{5(x + \Delta x)^2 y - 5x^2 y}{\Delta x} \\ &= \lim_{\Delta x \to 0} \frac{5x^2 y + 10x\Delta x y + 5(\Delta x)^2 y - 5x^2 y}{\Delta x} \\ &= \lim_{\Delta x \to 0} (10xy + 5\Delta x y) = 10xy. \end{aligned}$$ ♦

As we already know, Leibniz's notation uses the Latin d to represent derivatives, as in dy/dx. Similarly, the symbol ∂ (***curly d***) is used to denote partial derivatives, as in $\partial F/\partial x$ and $\partial F/\partial y$. In other words, besides F_x and F_y, we can write $\partial F/\partial x$ and $\partial F/\partial y$, respectively.

When we examine Definition 1.1.5 and look at the result of the previous example, we see that taking the partial derivative of F with respect to x amounts to nothing more than taking the (ordinary) derivative of F with respect to x, if at the same time we view y as held fixed at some arbitrary constant value. In short,

$$F_x(x, y) = \frac{d}{dx} F(x, y = \text{constant}).$$

Similarly, taking the partial derivative of F with respect to y can be viewed as

$$F_y(x, y) = \frac{d}{dy} F(x = \text{constant}, y).$$

Looking at it this way, it is a relatively simple matter to take the partial derivatives of two-variable functions, such as $F(x, y) = 5x^2 y$. The partial derivative of this function with respect to x is

$$F_x(x, y) = \frac{\partial}{\partial x}\left(5x^2 y\right) = \frac{d}{dx} F(x, y = \text{constant}) = \frac{d}{dx}\left.(5x^2 y)\right|_{y=\text{constant}} = 10xy,$$

as we had already determined in Example 1.1.8. Likewise, the partial derivative of F with respect to y is

$$\begin{aligned} F_y(x, y) &= \frac{\partial}{\partial y}\left(5x^2 y\right) = \frac{d}{dy} F(x = \text{constant}, y) \\ &= \frac{d}{dy}\left.(5x^2 y)\right|_{x=\text{constant}} = 5x^2. \end{aligned} \tag{1.1.13}$$

The notation

$$\frac{d}{dx} F(x, y = \text{constant}) \quad \text{and} \quad \frac{d}{dy} F(x = \text{constant}, y)$$

is not used in practice but is merely a pedagogic aid for newcomers to this topic. For example, in our mind's eye we can visualize the steps in (1.1.13) but only write down

$$F_y(x, y) = \frac{\partial}{\partial y}\left(5x^2y\right) = 5x^2.$$

Example 1.1.9 Evaluate the first-order partial derivatives of the temperature function

$$T(x, y) = 99 + y - 0.0002xy^2 - 0.005x^2$$

when $x = 10$ cm and $y = 1$ cm, where T is measured in degrees Celsius.

Solution First we find the partial derivative of T with respect to x as follows:

$$T_x(x, y) = \frac{\partial}{\partial x}T(x, y) = \frac{\partial}{\partial x}(99+y-0.0002xy^2-0.005x^2) = -0.0002y^2-0.01x.$$

Next we evaluate the result at the point (10, 1):

$$\begin{aligned} T_x(10, 1) &= (-0.0002y^2 - 0.01x)\big|_{(10,1)} \\ &= -0.0002(1)^2 - 0.01(10) = -0.1002. \end{aligned}$$

Thus, $T_x(10, 1) = -0.1002\,°\text{C/cm}$. The partial derivative of T with respect to y is

$$T_y(x, y) = \frac{\partial}{\partial y}T(x, y) = \frac{\partial}{\partial y}(99 + y - 0.0002xy^2 - 0.005x^2) = 1 - 0.0004xy.$$

Consequently,

$$T_y(10, 1) = (1 - 0.0004xy)\big|_{(10,1)} = 1 - 0.0004(10)(1) = 0.996.$$

So $T_y(10, 1) = 0.996\,°\text{C/cm}$. Recall this was obtained earlier by taking the limit of a difference quotient. Note the ease with which we can find partial derivatives by simply using the rules of differentiation that we already know from single variable calculus. ♦

1.1.5 Partial Differential Equations

Up to now we have explained what ordinary differential equations are and shown a number of examples. They are simply equations containing ordinary derivatives. Likewise, equations containing partial derivatives (but not ordinary derivatives) are called *partial differential equations*. In this book we are not concerned with these kinds of equations per se; nevertheless, we will still need to know a little about them. As it turns out, some methods for solving certain kinds of ordinary differential equations involve partial derivatives. We will see this in Chap. 8. Let us begin by defining what is meant by a partial differential equation.

Partial Differential Equation

Definition 1.1.6 A ***partial differential equation*** is an equation containing more than one independent variable, one or more dependent variables, and partial derivatives of one or more of these dependent variables.

A well-known example of a partial differential equation from classical physics is

$$\frac{\partial u}{\partial t} = k\frac{\partial^2 u}{\partial x^2}.$$

It models the conduction of heat through an extremely thin metal bar, where $u(x, t)$ is the temperature at the point x in the bar at time t. This equation is known as the *one-dimensional heat equation*.[8] There are two independent variables: the spatial variable x and the temporal variable t. The value of u depends on the values of both of them; so it is the dependent variable. The parameter k is called the *diffusivity* of the bar. What makes this a partial differential equation is that it is an equation containing partial derivatives. Whereas $\partial u/\partial t$ is a first-order partial derivative, $\partial^2 u/\partial x^2$ is a second-order partial derivative, the analog of a second-order ordinary derivative. Therefore, this is an example of a second-order partial differential equation. We will say a little more about higher-order partial derivatives in Chap. 8.

The only types of partial differential equations that we will encounter in this chapter and later on are first-order equations of the form

$$\frac{\partial u}{\partial x} = f(x, y) \quad \text{and} \quad \frac{\partial u}{\partial y} = g(x, y). \tag{1.1.14}$$

An example of an equation of the first form is

$$\frac{\partial u}{\partial x} = 2xy - \sin x.$$

[8] For more information, see Churchill [20].

An example of the second form is

$$\frac{\partial u}{\partial y} = 5y^4 + \frac{2xy}{x^2 + y^2} + 10.$$

In Sect. 1.2.3, we will discuss how to find solutions of partial differential equations of the two forms shown in (1.1.14). For the sake of brevity, it is common to use the abbreviations "ODE" for "ordinary differential equation" and "PDE" for "partial differential equation."

1.2 Differential Equations

Section 1.1 points out that differential equations arise when mathematics is employed to model certain real-world situations and natural phenomena. Equations judged to be good models imitate reality closely, provide insight and understanding, and predict well. Finding just the right equation (or equations) could be quite complicated—not only because of the mathematics but also because other disciplines are involved as well, such as physics and engineering. In spite of this, we will ease our way into a study of differential equations by starting with some of the more elementary equations that arise from modeling relatively simple situations. In the rest of this chapter, we will present examples of elementary differential equations that result when

(a) the rate of change of a quantity is known or can easily be determined;
(b) the rate of change of a quantity is conjectured to be proportional to itself or some related quantity;
(c) the motion of a body is modeled by Newton's second law of motion.

1.2.1 Rates of Change

Mathematical modeling frequently involves rates of change of quantities. If the rate of change of a quantity can somehow be determined or is already known, then this can be expressed with an ordinary differential equation. Consider the following examples.

Bathtub
Suppose it has been determined that water flows out of a spout into a bathtub at the rate of 3 gallons per minute. Translated into the succinct language of mathematics, this verbal statement becomes the differential equation

$$\frac{dN}{dt} = 3,$$

where $N(t)$ denotes the number of gallons of water that has flowed into the bathtub after t minutes.

Temperature Along a Heated Wire

In the previous section, there is the example of an unevenly heated thin copper wire whose temperature, denoted by T, changes at a rate of $-0.0002 - 0.01x$ degrees Celsius per centimeter, where x is the distance in centimeters from the left end of the wire. The equivalent mathematical statement is

$$\frac{dT}{dx} = -0.0002 - 0.01x.$$

Marginal Cost

An economic decision may be based in part on the increment in cost that will be incurred if one more unit of a product is manufactured. Some economists call this cost increment the *marginal cost*. It can usually be approximated by the derivative that gives the instantaneous rate of change of the total cost function with respect to the number of manufactured units, say x, of the product. For this reason, many economists prefer defining the ***marginal cost*** (abbreviated *MC*) to be this derivative. Let us use this latter definition. As an example, suppose it is stated that the marginal cost to manufacture x widgets[9] is given by the function $MC = 10x + 5000$. Letting $C(x)$ denote the total cost to produce x widgets, this statement can be expressed with the differential equation

$$\frac{dC}{dx} = 10x + 5000.$$

Speed of a Falling Object

An object is dropped from the top story of the Leaning Tower of Pisa, the famous freestanding, eight-story bell tower of the cathedral of Pisa, Italy. Let s denote the distance, in feet, the object has fallen t seconds after being dropped. The ***speed*** of the dropped object is the rate at which its distance increases with time, that is, the time derivative $\dot{s}$. According to the laws of elementary mechanics, the speed after t seconds is approximately $32t$ feet per second, provided the counteracting resistance of the air pushing upward on the object is negligible. Therefore, the speed of the

[9] A ***widget*** is a substitute for the name of some device or gadget, usually used when its real name is not known or temporally forgotten, in other words, a thingamajig. Here we use it to mean a fictitious, manufactured product. In this way, we can easily fabricate marginal cost functions to illustrate the mathematics and economics, without also having to consider whether or not they represent reality.

falling object is modeled by the differential equation $\dot{s} = 32t$ or

$$\frac{ds}{dt} = 32t,$$

which is valid up to the time it hits the ground.

Arc Length
Finding lengths of curves is one of the many applications of definite integrals. Let us look at this from the perspective of differential equations. But first, let us review what is meant by a *smooth* function. A function $y = F(x)$, defined for $\alpha \le x \le \beta$, where α and β are real numbers, is said to be ***smooth*** if $F'(x)$ exists and is continuous at every point of $[\alpha, \beta]$. For each value of $x \in [\alpha, \beta]$, there is a corresponding point $(x, y) = (x, F(x))$ on the graph of the function. We can envision a point particle starting out at the ***initial point*** $(\alpha, F(\alpha))$ in the xy-plane and tracing out a curve as x increases from α to β until it reaches the ***terminal point*** $(\beta, F(\beta))$. If $F'(x)$ is continuous, the tangent line to the curve exists at every point and turns continuously as x varies in value. As a result, it is a ***smooth curve***: it has no gaps, corners, or cusps.

Now consider a pair of equations of the form

$$x = f(t), \quad y = g(t) \tag{1.2.1}$$

for $\alpha \le t \le \beta$. Unlike above, the variable x is no longer an independent variable; rather it, as does y, depends on the independent variable t. Each value of t corresponds to a point $(x, y) = (f(t), g(t))$ in the xy-plane. As before, let us envision a point particle starting out at the ***initial point*** $(f(\alpha), g(\alpha))$ and tracing out a curve as t increases from α to β until it reaches the ***terminal point*** $(f(\beta), g(\beta))$. Equations of the form (1.2.1) are called ***parametric equations*** and the independent variable t is called the ***parameter***. The curve defined by these pair of equations is said to be ***smooth*** if the derivatives of f and g exist, are continuous, and never simultaneously zero for all values of t in the interval $[\alpha, \beta]$. Smoothness guarantees that

1. there is a unique tangent line at every point of the curve, and
2. $\theta(t)$, the angle of inclination of the tangent line at the point $(f(t), g(t))$, is defined for all values of $t \in [\alpha, \beta]$ and is a continuous function.

Consequently, there are no gaps in the curve nor does it have any corners or cusps. Moreover, no portion of the curve is retraced as t increases. In other words, a point particle starting out at the *initial point* $(f(\alpha), g(\alpha))$ will never stop and then move in the reverse direction along the curve before reaching the *terminal point* $(f(\beta), g(\beta))$. With that said, we are now ready to consider the length (***arc length***) of a smooth curve and to determine the rate at which the arc length changes with respect to t.

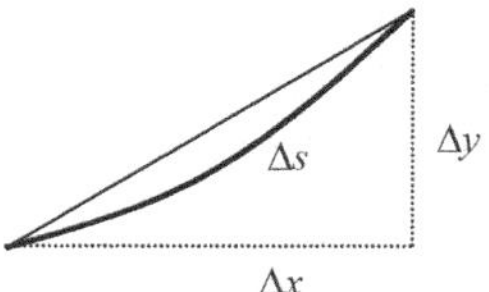

Fig. 1.2 Approximating the length of Δs

Let $s(t)$ denote the length of a curve from the initial point $(f(\alpha), g(\alpha))$ to a point $(f(t), g(t))$. For $t_1 \geq \alpha$, let Δs denote the length of the portion of the curve that corresponds to the values of $t \in [t_1, t_1 + \Delta t]$. The length of this portion when $\Delta t \approx 0$ is approximately the length of the hypotenuse of the right triangle depicted in Fig. 1.2. By the Pythagorean theorem,

$$\Delta s \approx \sqrt{(\Delta x)^2 + (\Delta y)^2}.$$

Since the difference quotient

$$\frac{\Delta x}{\Delta t} = \frac{f(t_1 + \Delta t) - f(t_1)}{\Delta t}$$

approximates the derivative $f'(t_1)$ for values of Δt near zero, $\Delta x \approx f'(t_1)\Delta t$. Likewise, $\Delta y \approx g'(t_1)\Delta t$. Thus,

$$\Delta s \approx \sqrt{[f'(t_1)\Delta t]^2 + [g'(t_1)\Delta t]^2} = \Delta t\sqrt{[f'(t_1)]^2 + [g'(t_1)]^2}$$

or

$$\frac{\Delta s}{\Delta t} \approx \sqrt{[f'(t_1)]^2 + [g(t_1)]^2}.$$

As $\Delta t \to 0$, this approximation gets better and better. We conclude that the rate of change of the length s of the curve (1.2.1) at $t = t_1$ is given by the value of

$$\frac{ds}{dt} = \sqrt{[f'(t)]^2 + [g'(t)]^2} = \sqrt{\left(\frac{dx}{dt}\right)^2 + \left(\frac{dy}{dt}\right)^2} \tag{1.2.2}$$

at $t = t_1$. In applications where t represents time and $(x, y) = (f(t), g(t))$ the position of a moving body at time t, the derivative $\dot{s}$ is the rate of change of the distance of the body from its initial position, that is, the ***speed*** of the body.

Finally, note that with $x = t$ and $y = F(t)$, Eq. (1.2.2) becomes

$$\frac{ds}{dx} = \sqrt{1 + [F'(x)]^2},$$

which gives the rate of change of s with respect to x for the function $y = F(x)$. The well-known *arc length formula* can be quickly obtained from this equation by integrating both of its sides with respect to x over an interval $[\alpha, \beta]$ (see Problem 81).

1.2.2 The Simplest Differential Equations

Each of the differential equations in the previous examples resulted from knowing the rate at which some quantity changes. Note that the form of those equations is a derivative that is equal to either a constant or to an expression involving the independent variable but not the dependent variable. This is succinctly conveyed by the notation

$$\frac{dy}{dx} = f(x). \tag{1.2.3}$$

As we shall see below and in Sect. 1.3, differential equations looking like (1.2.3) are the simplest to solve. In fact, we already know how to find their solutions from studying integrals in calculus.

Whatever is said about finding solutions of (1.2.3) does not generally carry over to a differential equation of the form

$$\frac{dy}{dx} = f(y) \tag{1.2.4}$$

since the function in this equation involves the dependent variable, not the independent variable, nor does it carry over to

$$\frac{dy}{dx} = f(x, y). \tag{1.2.5}$$

An ordinary differential equation having the form of (1.2.3) is the simplest type of differential equation to solve because all that it takes is an integration with respect to the independent variable. For a given function $f(x)$, a ***solution*** of (1.2.3) is a function whose derivative with respect to x is the function $f(x)$. From calculus we know there are infinitely many of these functions and that they all differ from one another by only a constant—and that there are no other functions besides these. This infinite collection of functions is known as the ***indefinite integral*** of the function f and is denoted by the symbol

$$\int f(x)\,dx. \tag{1.2.6}$$

In other words, all of the solutions of (1.2.3) are the antiderivatives of the function f. For example, consider the simple differential equation

$$\frac{dy}{dx} = 2x. \tag{1.2.7}$$

In the notation of (1.2.3), $f(x) = 2x$. Since

$$\frac{d}{dx}x^2 = 2x,$$

the solutions of (1.2.7) are the antiderivatives of $2x$, namely, $x^2 + C$. In other words, we can find all of the solutions of (1.2.7) by merely carrying out the integration indicated by (1.2.6):

$$y = \int f(x)\,dx = \int 2x\,dx = x^2 + C.$$

At times we may have to delve into the recesses of our minds to come up with an antiderivative, or we may have to consult a table of integrals. For example, the solutions of the differential equation

$$\frac{dy}{dx} = \frac{1}{1+x^2}$$

are the functions

$$y = \int \frac{dx}{1+x^2} = \tan^{-1} x + C$$

since

$$\frac{d}{dx}(\tan^{-1} x) = \frac{1}{1+x^2}.$$

Besides tables of integrals, there are technological tools that can be used to obtain integration formulas, such as graphing calculators with symbolic algebra capabilities, the web-based computational knowledge engine *Wolfram|Alpha*, and computer algebra systems.[10] Nonetheless, the formulas of the indefinite integrals listed in Table 1.4 are **basic integration formulas** (or **standard integral forms**) that should be memorized! These particular integrals pop up so often that committing them to

[10] A ***computer algebra system*** (***CAS***) is an interactive computer program that can carry out mathematical computations involving symbolic expressions, such as yielding the result x^2 when the appropriate command to compute the integral of $2x$ is entered. Some well-known computer algebra systems are *Maple*, *Mathematica*, and *MATLAB*. Also, some handheld calculator models have a built-in CAS that can perform symbolic manipulations, such as the *TI-Nspire* calculators.

Table 1.4 Basic integration formulas

$f(u)$	$\int f(u)\,du$
0	C
k	$ku + C$
$u^n \quad (n \neq -1)$	$\dfrac{u^{n+1}}{n+1} + C$
$u^{-1} = \dfrac{1}{u}$	$\ln\|u\| + C = \begin{cases} \ln u + C, & \text{if } u > 0 \\ \ln(-u) + C, & \text{if } u < 0 \end{cases}$
e^u	$e^u + C$
$b^u \quad (b > 0)$	$\dfrac{b^u}{\ln b}$
$\cos u$	$\sin u + C$
$\sin u$	$-\cos u + C$
$\sec^2 u$	$\tan u + C$
$\csc^2 u$	$-\cot u + C$
$\sec u \tan u$	$\sec u + C$
$\csc u \cot u$	$-\csc u + C$
$\sec u$	$\ln\|\sec u + \tan u\| + C$
$\csc u$	$-\ln\|\csc u + \cot u\| + C$
$\cot u$	$\int (\cos u/\sin u)\,du = \ln\|\sin u\| + C$
$\tan u$	$\int (\sin u/\cos u)\,du = -\ln\|\cos u\| + C$
$\dfrac{1}{1+u^2}$	$\tan^{-1} u + C$
$\dfrac{1}{\sqrt{1-u^2}}$	$\sin^{-1} u + C$
$\dfrac{1}{u\sqrt{u^2-1}}$	$\sec^{-1}\|u\| + C$

memory will actually save time and effort in the long run—just think of the time it takes to locate a table and then look up an integral or to turn on a computer and enter the appropriate commands. Moreover, these integration formulas are so common and well-known that it may prove somewhat embarrassing not to have them memorized. Would not the mathematical competency of someone who had to consult a table or use a computer or calculator to evaluate $\int x^2\,dx$ be called into question?

Table 1.4 lists the basic integration formulas that will regularly turn up in the text and problem sets in this book. The letters b, C, k, and n in this table denote constants. By memorizing the formulas in this table, the reader will be able to work nearly all of the problems in this book without having to resort to tables, calculators, or computer algebra systems. For a function $f(u)$ listed in the table, the symbol $\int f(u)\,du$ denotes its indefinite integral. Recall from calculus that the indefinite integral is obtained by first finding an antiderivative of $f(u)$ and then adding a constant of integration, say C, to it. The result is an expression that represents all of the antiderivatives of $f(u)$. Equivalently, this represents all solutions of the differential equation

$$\frac{dy}{du} = f(u).$$

Unfortunately, the integration formula in the last row of the table may be slightly different in another book, table of integrals, or from what is obtained using a CAS or some other technological tool. That is because there is no universally accepted definition of the inverse secant function. This is the integration formula if the ***inverse secant function*** is defined by

$$y = \sec^{-1} u \ (|u| \geq 1) \quad \text{if and only if} \quad \sec y = u \text{ and } y \in \left[0, \tfrac{\pi}{2}\right) \cup \left(\tfrac{\pi}{2}, \pi\right].$$

However, some other book, table, or CAS may choose another range. For instance, if $y \in \left[0, \frac{\pi}{2}\right) \cup \left[\pi, \frac{3\pi}{2}\right)$ is chosen, then the corresponding integration formula is

$$\int \frac{1}{u\sqrt{u^2 - 1}}\, du = \sec^{-1} u + C$$

rather than

$$\int \frac{1}{u\sqrt{u^2 - 1}}\, du = \sec^{-1} |u| + C.$$

See also the *Caveat* in Appendix A.2 following Table A.1.

1.2.3 Simple Partial Differential Equations

A class of differential equations, known as *exact equations*, is important to physics, mathematics, engineering, and other related fields of study. It is too early to explain what exact equations are except to say that an entire chapter[11] is devoted to them and some related equations—and even though they are ordinary differential equations, the method for solving them depends on knowing how to solve simple partial differential equations of the form

$$\frac{\partial u}{\partial x} = f(x, y) \tag{1.2.8}$$

and

$$\frac{\partial u}{\partial y} = f(x, y). \tag{1.2.9}$$

[11] See Chap. 8.

Fortunately, just as with the simplest ordinary differential equations, integration is all that is required to find their solutions.

A solution of (1.2.8) is a function whose partial derivative with respect to x is equal to $f(x, y)$. Solving this equation means finding all such functions. Recall that partial differentiation with respect to x amounts to nothing more than ordinary differentiation with respect to x with y held fixed. So solutions are found by reversing the operation of partial differentiation, that is, by integration with respect to x with y held fixed. This is indicated symbolically in the following way:

$$u = \int f(x, y)\, dx.$$

As an example, let us find solutions of the partial differential equation

$$\frac{\partial u}{\partial x} = 2xy - \sin x. \tag{1.2.10}$$

Integrating with respect to x, we obtain the solutions

$$u = \int (2xy - \sin x)\, dx = x^2 y + \cos x + C.$$

Or so we might think! But let us reconsider. Are these really all of the solutions of (1.2.10)? When we think about it, the so-called constant of integration "C" is really more than just a constant here—any function that depends only on y should be included as well. Of course, the reason for this is that the partial derivative of a function of y alone (i.e., of y but not x) with respect to x is equal to 0. If the letter g is chosen to represent such a function, we write

$$\frac{\partial}{\partial x} g(y) = 0.$$

Therefore,

$$u = x^2 y + \cos x + g(y). \tag{1.2.11}$$

Let us check this by substituting $x^2 y + \cos x + g(y)$ for the dependent variable u in (1.2.10) to verify that it truly satisfies this equation. Since

$$\frac{\partial u}{\partial x} = \frac{\partial}{\partial x}\left[x^2 y + \cos x + g(y)\right] = 2xy - \sin x,$$

we conclude that the complete set of solutions of (1.2.10) includes all functions of the form (1.2.11).

For an example of a partial differential equation of the form (1.2.9), we can simply replace the partial derivative $\partial u/\partial x$ in (1.2.10) with $\partial u/\partial y$.

Example 1.2.1 Find solutions of the partial differential equation

$$\frac{\partial u}{\partial y} = 2xy - \sin x. \tag{1.2.12}$$

Solution This time, to undo the partial differentiation in (1.2.12), let us integrate with respect to y. By doing so we get the solutions

$$u = \int (2xy - \sin x)\, dy = xy^2 - y \sin x + h(x),$$

where h denotes a function of x alone. ♦

For another example, consider the heated copper plate depicted in Fig. 1.1. This time, however, instead of being given a function that models the temperature at each of the points of the plate and asked to find the rates at which the temperature changes along the x and y directions, suppose we are given one of these rates and asked to find the temperature function.

Example 1.2.2 Suppose it is known that the rate at which the temperature $T(x, y)$ of the copper plate in Fig. 1.1 changes with respect to y is

$$\frac{\partial T}{\partial y} = 1 - 0.0004xy \text{ °C/cm}.$$

What can be said about the temperature function $T(x, y)$ itself?

Solution Since we are given the partial derivative of T with respect to y, let us integrate the right-hand side of the equation with respect to this variable. Thus,

$$T(x, y) = \int (1 - 0.0004xy)\, dy = y - 0.0002xy^2 + h(x). \tag{1.2.13}$$

Unfortunately, this is all that we can say about $T(x, y)$ since no other information has been provided that would allow us to determine $h(x)$.

To illustrate how we would go about finding $h(x)$ had we been given additional information, suppose that we also know that temperatures along the vertical line $y = 1$ are given by the function

$$\varphi(x) = 100 - 0.0002x - 0.005x^2.$$

If (1.2.13) is to model the temperature at all points of the plate, then it follows that $T(x, 1) = \varphi(x)$; accordingly,

$$1 - 0.0002x + h(x) = 100 - 0.0002x - 0.005x^2.$$

Solving for h, we have

$$h(x) = 99 - 0.005x^2.$$

Substituting this into (1.2.13), we find that the temperature in degrees Celsius at each point (x, y) in the plate is[12]

$$T(x, y) = 99 + y - 0.0002xy^2 - 0.005x^2. \qquad \blacklozenge$$

The next example is a bit different from the previous examples in that we are asked to find the solutions of a system of two partial differential equations, provided there are any. It is a foretaste of the method that is used to solve exact ordinary differential equations (cf. Sect. 8.5 in Chap. 8).

Example 1.2.3 Determine if there are any functions that solve both of the following partial differential equations

$$\frac{\partial u}{\partial x} = x \quad \text{and} \quad \frac{\partial u}{\partial y} = y. \tag{1.2.14}$$

Solution Let us begin by supposing that both of these equations have some solutions in common. Integration of the first equation with respect to x yields

$$u(x, y) = \frac{1}{2}x^2 + g(y).$$

The partial derivative of this with respect to y is

$$\frac{\partial}{\partial y}u(x, y) = \frac{\partial}{\partial y}\left(\frac{1}{2}x^2 + g(y)\right) = g'(y).$$

This agrees with the second equation in (1.2.14) if

$$g'(y) = y.$$

Thus, $g(y) = \frac{1}{2}y^2 + C$, where C denotes any constant. As a result, we have

$$u(x, y) = \frac{1}{2}x^2 + g(y) = \frac{1}{2}x^2 + \frac{1}{2}y^2 + C. \tag{1.2.15}$$

Although it appears that we are done, let us make sure that these functions satisfy both of the equations in (1.2.14). Taking the partial derivative of (1.2.15) with

[12] Compare this to Example 1.1.9.

respect to x, we have

$$\frac{\partial}{\partial x}u(x, y) = \frac{\partial}{\partial x}\left(\frac{1}{2}x^2 + \frac{1}{2}y^2 + C\right) = x.$$

Likewise,

$$\frac{\partial}{\partial y}u(x, y) = \frac{\partial}{\partial y}\left(\frac{1}{2}x^2 + \frac{1}{2}y^2 + C\right) = y.$$

This confirms that (1.2.15), irrespective of the value of the constant C, is a solution of both equations. ♦

Before leaving this topic, we give an example of a system of partial differential equations that has no solutions.

Example 1.2.4 Determine if there are any functions that solve both

$$\frac{\partial u}{\partial x} = x \quad \text{and} \quad \frac{\partial u}{\partial y} = x + 2y. \tag{1.2.16}$$

Solution Since the first equation is the same as in the previous example, we have

$$u(x, y) = \frac{1}{2}x^2 + g(y) \quad \text{and} \quad u_y(x, y) = g'(y).$$

This agrees with the second equation in (1.2.16) provided

$$g'(y) = x + 2y.$$

Integrating this with respect to y, we get

$$g(y) = \int (x + 2y)\, dy = xy + y^2 + C,$$

where C is the constant of integration. So if there is a function that is a solution of both of the equations, it would have to be of the form

$$u(x, y) = \frac{1}{2}x^2 + xy + y^2 + C. \tag{1.2.17}$$

Let us check whether such a function is a solution. Taking the partial derivative of (1.2.17) with respect to x and then y, we obtain

$$\frac{\partial}{\partial x}u(x, y) = \frac{\partial}{\partial x}\left(\frac{1}{2}x^2 + xy + y^2 + C\right) = x + y$$

and

$$\frac{\partial}{\partial y}u(x, y) = \frac{\partial}{\partial y}\left(\frac{1}{2}x^2 + xy + y^2 + C\right) = x + 2y,$$

respectively. From this we see that any function looking like (1.2.17) is a solution of the second equation in (1.2.16) but not of the first equation. In fact, it follows from the equation in the second sentence that the system has no solution since the equation is contradictory: g' cannot be a function of y alone and at the same time depend on both x and y. ♦

1.3 Integration and Solutions

Familiarity with the most commonly used integration formulas, thorough memorization of the most basic of these, and a facility for applying the techniques of integration are indispensable when it comes to finding solutions of ordinary differential equations. Table 1.4 is a list of some of the most basic formulas of integration. These are the ones that occur most often in this and other introductory differential equations textbooks. Next to the functions in the left-hand column are their antiderivatives in the right-hand column. Since these integration formulas are the ones that are most likely to be needed to solve the problems in this book, they should be committed to memory. However, the integrals that arise in most of these problems probably will not look exactly like any of those listed in the table—in that case, we will have to draw upon the integration techniques learned in calculus, such as *substitution*, *integration by parts*, and *partial fractions*. These are some of the most important techniques for transforming an integral into one of the basic integrals listed in Table 1.4. In this section, we review some of these integration techniques by presenting examples of simple differential equations of the form $y' = f(x)$ that will require the use of one or more of these techniques to solve them. As you work your way through the examples, it would be a good idea to refresh your memory of integration and hone your skills by opening up your favorite calculus book to the chapter dealing with the techniques of integration and to review your past calculus class notes and worked-out problem sets.

1.3.1 Integration by Substitution

Let us take a look at some examples of differential equations whose solutions can be found with the help of a *substitution*, i.e., by making an appropriate *change of variable*.

Example 1.3.1 Find all solutions of the differential equation

$$\frac{dy}{dx} = \frac{5}{x^2 + 9}.$$

Solution Solutions of this differential equation can be obtained by simply integrating its right-hand side (see (1.2.6) and the discussion preceding it). That is, the solutions are

$$y = \int \frac{5}{x^2+9}\,dx = 5\int \frac{1}{9+x^2}\,dx.$$

Since the integrand most nearly resembles the integrand $1/(1+u^2)$ in Table 1.4, let us rewrite the denominator of the integrand so that it leads off with the digit 1. Factoring out 9, substituting u for $x/3$, and replacing the differential dx with $3\,du$, we have

$$\int \frac{1}{9+x^2}\,dx = \int \frac{1}{9\left[1+\left(\frac{x}{3}\right)^2\right]}\,dx = \frac{1}{9}\int \frac{1}{1+u^2}\,3\,du.$$

Therefore, the solutions are

$$y = \frac{5}{9}\int \frac{1}{1+u^2}\,3\,du = \frac{5}{3}\int \frac{du}{1+u^2} = \frac{5}{3}\tan^{-1}u + C = \frac{5}{3}\tan^{-1}\left(\frac{x}{3}\right) + C. \quad \blacklozenge$$

Example 1.3.2 Find the solutions of $\dfrac{dy}{dx} = \dfrac{5x}{x^2+9}$.

Solution The solutions of this equation are

$$y = \int \frac{5x}{x^2+9}\,dx.$$

The difference between the integrand here and the one in the previous example is the "x" in the numerator. What we should observe is that, aside from a numerical factor, the numerator is the derivative of the denominator. A standard integral results if we change variables from x to u by letting $u = x^2+9$. Then, $du = 2x\,dx$. And so

$$y = \int \frac{5x}{x^2+9}\,dx = \frac{5}{2}\int \frac{2x\,dx}{x^2+9} = \frac{5}{2}\int \frac{du}{u} = \frac{5}{2}\ln|u| + C.$$

Since $|u| = |x^2+9| = x^2+9$, the solutions are

$$y = \frac{5}{2}\ln\left(x^2+9\right) + C. \quad \blacklozenge$$

Example 1.3.3 Find the solutions of $\dfrac{dy}{dx} = \dfrac{x}{x^2+2x+1}$.

Solution Once again, as the right-hand side of the differential equation is a function of the independent variable x alone, the solutions of the equation are

$$y = \int \frac{x}{x^2 + 2x + 1}\,dx.$$

Now we have to figure out how to carry out the integration. As a rule of thumb, we should always try substitution first. Clearly, the only choice is $u = x^2 + 2x + 5$. Since the differential of u is

$$du = (2x + 2)\,dx = 2(x + 1)\,dx,$$

this will not work since the numerator is not a constant multiple of $x + 1$. At this point, we may be at a loss of what to do next. But then we notice that the denominator is a perfect square trinomial. Replacing it with $(x + 1)^2$, we have

$$y = \int \frac{x}{(x + 1)^2}\,dx.$$

This suggest the substitution $u = x + 1$. Then, as $du = dx$ and $x = u - 1$, we get

$$y = \int \frac{u - 1}{u^2}\,du = \int \left(u^{-1} - u^{-2}\right)\,du = \ln|u| + \frac{1}{u} + C.$$

Therefore, the solutions are

$$y = \ln|x + 1| + \frac{1}{x + 1} + C.$$ ♦

Example 1.3.4 Find the solutions of $\dfrac{dy}{dx} = \dfrac{x}{x^2 + 2x + 5}$.

Solution This is very much like the differential equation in Example 1.3.3. However, the quadratic expression in the denominator cannot be replaced by a perfect square. But let us see what happens if we complete the square of the first two terms:

$$x^2 + 2x + 5 = (x^2 + 2x + 1) - 1 + 5 = (x + 1)^2 + 4.$$

Thus,

$$y = \int \frac{x}{x^2 + 2x + 5}\,dx = \int \frac{x}{(x + 1)^2 + 4}\,dx.$$

With the substitution $u = x + 1$, we have

$$y = \int \frac{u - 1}{u^2 + 4}\,du = \int \frac{u}{u^2 + 4}\,du - \int \frac{1}{u^2 + 4}\,du$$

$$= \frac{1}{2}\int \frac{2u}{u^2+4}\,du - \frac{1}{4}\int \frac{1}{1+\left(\frac{u}{2}\right)^2}\,du$$

$$= \frac{1}{2}\ln\left(u^2+4\right) - \frac{1}{2}\tan^{-1}\left(\frac{u}{2}\right) + K.$$

Consequently, the solutions are

$$y = \frac{1}{2}\ln(x^2+2x+5) - \frac{1}{2}\tan^{-1}\left(\frac{x+1}{2}\right) + K. \qquad \blacklozenge$$

As a final example of using the method of substitution to carry out an integration, let us find the solutions of a simple partial differential equation.

Example 1.3.5 Find the solutions of $\dfrac{\partial u}{\partial x} = y\sec(3x)$.

Solution The equation itself indicates that the independent variables are x and y and that u is dependent on both of them. Obviously, the way in which to solve for u is to undo the partial differentiation by integrating u_x with respect to x:

$$u = \int \frac{\partial u}{\partial x}\,dx = \int y\sec(3x)\,dx = y\int \sec(3x)\cdot\frac{1}{3}\cdot 3\,dx = \frac{y}{3}\int \sec v\,dv,$$

where $v = 3x$. From Table 1.4, we have

$$u = \frac{y}{3}\ln|\sec v + \tan v| + g(y).$$

Therefore, the solutions of the partial differential equation are

$$u = \frac{y}{3}\ln|\sec(3x) + \tan(3x)| + g(y). \qquad \blacklozenge$$

1.3.2 Integration by Parts

Let us move on to a couple of examples of differential equations of the form $y' = f(x)$, where *integration by parts* is needed to solve the equations.

Example 1.3.6 Find the solutions of $\dfrac{dy}{dx} = x\cos 2x$.

Solution As in the previous examples, the right-hand side of the equation depends on x alone. So its solutions are

$$y = \int x\cos 2x\,dx.$$

Integrating the product of a power function and a sinusoidal function (sine or cosine function) calls for the method of integration by parts. Recall the ***integration by parts formula***:

$$\int u\,dv = uv - \int v\,du. \tag{1.3.1}$$

To apply this formula to $\int x\cos 2x\,dx$, let $u = x$ and $dv = \cos 2x\,dx$. Then

$$du = dx \left(\text{as } \frac{du}{dx} = 1\right) \quad \text{and} \quad v = \frac{1}{2}\sin 2x \left(\text{as} \int \cos 2x\,dx = \frac{1}{2}\sin 2x + K\right).$$

As a result, we have

$$y = uv - \int v\,du = \frac{1}{2}x\sin 2x - \frac{1}{2}\int \sin 2x\,dx = \frac{x}{2}\sin 2x - \frac{1}{2}\left(-\frac{1}{2}\cos 2x\right) + C.$$

Therefore, the solutions are

$$y = \frac{x}{2}\sin 2x + \frac{1}{4}\cos 2x + C. \qquad \blacklozenge$$

It is important to learn when and how to use the integration by parts formula (1.3.1). If the original integral, represented by $\int u\,dv$, is difficult or impossible to integrate, the purpose of (1.3.1) is to present an alternative that is easier to deal with, namely, the integral $\int v\,du$. The art of integrating by parts is in the selection of u and dv so that $\int v\,du$ is easier to integrate than is $\int u\,dv$. Achieving this is a matter of common sense, trial and error, and experience. However, if you have difficulty in choosing u and dv the acronym *LIATE*[13] is a useful device. *LIATE* is a mnemonic for remembering five types of functions in the following order:

<u>L</u>ogarithmic, <u>I</u>nverse trig, <u>A</u>lgebraic, <u>T</u>rig, <u>E</u>xponential.

It is precisely this order that makes *LIATE* work. When the integrand is the product of any two of these types of functions and substitution does not work, the integration by parts formula should be tried. Let the type of function appearing first in the order given by *LIATE* (from left to right) be the "u", while the other one together with the differential in the integral, be the "dv". Then apply the integration by parts formula. Hopefully, $\int v\,du$ will be easier to integrate than $\int u\,dv$. Take, for instance, the integral $\int x\cos 2x\,dx$ from the previous example. Its integrand is the product of the two functions x and $\cos 2x$, where x is an algebraic function and $\cos 2x$ is a trigonometric function. So let $u = x$ since the type *Algebraic* comes before the type *Trig* in *LIATE*. Then the rest of the integrand is dv, that is, $dv = \cos 2x\,dx$. We illustrate the use of *LIATE* in the next example.

---

[13] See Kasube [48, pp. 210–211].

Example 1.3.7 Find all solutions of $\dfrac{dy}{dx} = \sin^{-1}(2x)$.

Solution The solutions of this differential equation are

$$y(x) = \int \sin^{-1}(2x)\,dx.$$

The substitution $u = 2x$ does not lead anywhere unless the formula for

$$\int \sin^{-1} u\,du$$

is already at hand.[14] It also looks like integration by parts will not work here by virtue of the absence of a product of two functions belonging to *LIATE*. But there really is one: the trivial product $1\cdot\sin^{-1}(2x)$. Since the constant function "1" belongs to the *Algebraic* type and is preceded by the *Inverse Trig* type in *LIATE*, let $u = \sin^{-1}(2x)$. This leaves 1 and the differential dx; so $dv = 1 \cdot dx = dx$.Hence,

$$du = u'(x)\,dx = \left(\frac{d}{dx}\sin^{-1}(2x)\right)dx = \frac{2}{\sqrt{1-(2x)^2}}\,dx; \qquad v = x.$$

It follows from the integrations by parts formula that the solutions are

$$\begin{aligned}
y(x) &= \int \underbrace{\sin^{-1}(2x)}_{u}\ \underbrace{dx}_{dv=1\cdot dx} \\
&= \underbrace{\sin^{-1}(2x)}_{u}\cdot\underbrace{x}_{v} - \int \underbrace{x}_{v}\ \underbrace{\frac{2}{\sqrt{1-(2x)^2}}\,dx}_{du} \\
&= x\sin^{-1}(2x) - \left(\frac{2}{-8}\right)\int \frac{-8x\,dx}{\sqrt{1-4x^2}} = x\sin^{-1}(2x) + \frac{1}{2}\sqrt{1-4x^2} + C.
\end{aligned}$$

♦

1.3.3 Partial Fractions

Suppose we are confronted with an integral whose integrand is a *rational function*;[15] however neither substitution nor completion of the square (as in Examples 1.3.3

[14] This integral can be found in a table of integrals or computed with a CAS or with a handheld calculator that can perform symbolic computations. However, the point of this section is to review integration and to hone the skills already acquired in a calculus course to solve relatively simple integrals without having to bother with tables or technology.

[15] Recall that a ***rational function*** is the quotient of two polynomials. Constants, such as 6 and π, are also considered polynomials.

and 1.3.4) are of any help in carrying out the integration. Fortunately, all is not lost if the denominator of the rational function can be factored; for then we can rewrite the rational function as a sum of simpler rational functions, called *partial fractions*, and then attempt to integrate these. Consider the following two examples.

Example 1.3.8 Find the solutions of $\dfrac{dy}{dx} = \dfrac{6}{x^2-9}$.

Solution As is the case with all the equations in this section, this one has the form $y' = f(x)$. So its solutions are obtained by integrating its right-hand side. Although the function $f(u) = 1/(u^2 - b^2)$ could have been included in Table 1.4, it was not since there was no need to. That is because the denominator $u^2 - b^2$ (or $x^2 - 9$ in the example) can be factored, which provides the clue for turning to *partial fractions*.

A quadratic polynomial is said to be ***reducible over the reals*** or ***factorable*** when it can be expressed as a product of two linear polynomials with real coefficients. Otherwise, it is said to be ***irreducible over the reals***. For the sake of brevity, we will omit the phrase "over the reals." Thus, $x^2 - 9$ is reducible since it reduces to the product of the linear polynomials $x - 3$ and $x + 3$. That is, it is equal to the product $(x-3)(x+3)$. On the other hand, x^2+9 is irreducible,[16] since it cannot be expressed it as a product of two linear polynomials with real coefficients. An important result of algebra states that it is always possible to express a nonconstant polynomial with real coefficients as a product of linear and irreducible quadratic polynomials with real coefficients. So, as a rule of thumb, when faced with integrating a rational function, first factor its denominator into a product of linear and irreducible quadratic polynomials—that is, unless the numerator is a constant multiple of the derivative of the denominator; in which case the method of substitution will work, as demonstrated in Example 1.3.2.

Now let us apply the method of partial fractions to evaluate the integral

$$\int \frac{6}{x^2-9}\,dx.$$

As pointed out above, its integrand is a rational function with the reducible denominator $x^2 - 9$. After factoring, we expand the integrand by writing it as a sum of two simpler rational functions with linear polynomials as their denominators:

$$\frac{6}{(x-3)(x+3)} = \frac{A}{x-3} + \frac{B}{x+3}. \tag{1.3.2}$$

The fractions

$$\frac{A}{x-3} \quad \text{and} \quad \frac{B}{x+3}$$

[16] That is, it is irreducible over the reals (meaning the set of real numbers). However, it is reducible over the set of complex numbers since $x^2 + 9 = (x - 3i)(x + 3i)$, where i is defined by $i^2 = -1$.

are called ***partial fractions*** because their denominators contain part of the original denominator $x^2 - 9$ but not all of it. The expansion of the integrand in (1.3.2) is known as its ***partial fraction expansion***. The coefficients A and B are called ***undetermined coefficients*** until values are found making (1.3.2) an identity. To determine the values of A and B, clear out the denominators in (1.3.2) by multiplying both sides by the denominator of the integrand with the result

$$A(x+3) + B(x-3) = 6.$$

Now we can quickly find the value of A by setting $x = 3$ since this eliminates the term involving B:

$$x = 3 \quad \Rightarrow \quad 6A + 0B = 6 \quad \Rightarrow \quad A = 1.$$

Likewise, setting $x = -3$ eliminates the term involving A and yields $B = -1$. Thus,

$$\frac{6}{x^2-9} = \frac{1}{x-3} - \frac{1}{x+3}.$$

Now integration is a piece of cake:

$$\begin{aligned} y &= \int \frac{6}{x^2-9}\,dx = \int \frac{dx}{x-3} - \int \frac{dx}{x+3} \\ &= \ln|x-3| - \ln|x+3| + C = \ln\left|\frac{x-3}{x+3}\right| + C. \end{aligned}$$ ♦

In the next example, factorization of the denominator of the rational function results in the product of a linear factor and an irreducible quadratic factor.

Example 1.3.9 Find the solutions of $\dfrac{dy}{dx} = \dfrac{18}{x^3+9x}$.

Solution A change of variable is of no help in integrating the right-hand side of the equation due to the absence of the factor

$$\frac{d}{dx}(x^3 + 9x) = 3x^2 + 9$$

in the numerator. Let us see how the method of partial fractions fares here. First we have to factor the denominator completely. Factoring out an x, we have

$$x^3 + 9x = x(x^2 + 9).$$

But this is as far as we can go since the quadratic factor is irreducible. Since the denominator consists of two factors, the partial fraction expansion of the integrand consists of two partial fractions. A terse rule for writing the form of each fraction in a partial fraction expansion is to remember:

Constants go over linear factors while linear factors go over irreducible quadratic factors.

As a result, the form of the expansion is

$$\frac{18}{x^3+9x} = \frac{A}{x} + \frac{Bx+C}{x^2+9}.$$

We clear out the denominators of the expansion by multiplying both of its sides by $x(x^2+9)$, obtaining

$$A(x^2+9) + (Bx+C)x = 18.$$

Grouping like terms together gives

$$(A+B)x^2 + Cx + 9A = 18.$$

This equation is satisfied for all values of x if A, B, and C have values satisfying the equations:

$$A+B=0; \quad C=0; \quad 9A=18.$$

Consequently, $A=2$, $B=-2$, and $C=0$. And so

$$y = \int \frac{18}{x^3+9x}\,dx = \int \frac{2}{x}\,dx - \int \frac{2x}{x^2+9}\,dx = 2\ln|x| - \ln|x^2+9| + C.$$

Therefore, the solutions are $y = \ln\left(\frac{x^2}{x^2+9}\right) + C$. ♦

1.4 Simple Differential Equation Models

In this section we present several relatively simple mathematical models of situations involving differential equations, where each is the result of conjecturing that the rate of change of a quantity is proportional to itself or some related quantity. We will also discuss Newton's second law of motion, which is an ordinary differential equation whose solutions model the motion of a body subjected to external forces.

1.4.1 Proportions

The statement "u ***is (directly) proportional to*** v" verbally expresses that

$$u = kv$$

for some constant k. The use of the word "directly" is optional; it is frequently omitted. The constant k is called the ***constant of proportionality***. The notation $u \propto v$ is another way to express that u is proportional to v.

Example 1.4.1 The assertion that "a city's average daily garbage collection G is proportional to its population p" can be expressed succinctly by writing $G \propto p$, which means that

$$G = kp$$

for some constant k. If there is any truth to this statement, then the value of k would have to be determined experimentally by determining the population of the city and keeping track of its daily amount of garbage. ♦

Example 1.4.2 A more important example is a postulate by the famous physicist Max Planck, which aided in the development of a field of physics called quantum mechanics. In 1900, in an attempt to reconcile the theory of black body radiation with experimental results, Planck postulated that energy is not radiated continuously but rather in discrete amounts called ***quanta*** and that a quantum of energy E is proportional to the frequency ν of the radiation.[17] That is,

$$E = h\nu$$

where h denotes the proportionality constant. By adjusting the value of h, he was able to reconcile his theory with experimental results. In fact, h is now called ***Planck's constant***. Its current accepted value is 6.63×10^{-34} joule·second. ♦

Example 1.4.3 Consider the wry comment of a good-natured hostess, as she watched a guest's toast land on her newly laid carpet, that the likelihood of toast landing jelly-side down is directly proportional to the cost of the carpet. With L denoting the likelihood, or probability, of this happening and C the cost, her comment translates to the mathematical statement

$$L = \gamma C,$$

where γ denotes the constant of proportionality.[18] ♦

Example 1.4.4 At one point in C. S. Forester's novel entitled *Admiral Hornblower in the West Indies*, Rear Admiral Lord Hornblower's ship is pursuing another ship called the *Estrella*. During the pursuit the admiral says, "There's all the difference in the world between six knots, which she's (the *Estrella*) making now close-hauled, and twelve knots, which she'll make when she puts up her helm." He goes on to say, "Mr. Spendlove here will tell you that the water resistance is a function of the square

[17] The symbol ν is the lower case Greek letter *nu*.

[18] The symbol γ is the lower case Greek letter *gamma*.

of the speed. Isn't that so, Mr. Spendlove?"[19] If by his statement Hornblower means that water resistance "is proportional" to the square of the speed, then he is saying

$$F_r = k\left(\frac{ds}{dt}\right)^2$$

where F_r denotes the force acting on the *Estrella* opposing its motion, s is the distance along its path of motion, measured say from the start of the pursuit, and k is the constant of proportionality. ♦

If two variables u and v are not directly proportional but are related by the equation

$$u = \frac{k}{v}$$

for some constant k, then we say "***u is inversely proportional to** v*" or "***u varies inversely with** v*."

Example 1.4.5 In the kinetic theory of gases, there is the empirical result known as ***Boyle's Law***. It states:

> *The volume V of a certain amount of gas confined to a container and held at a constant temperature is inversely proportional to the pressure P on the gas.*

Mathematically, this is expressed as

$$V = \frac{k}{P} \quad \text{or} \quad PV = k,$$

where $k > 0$ is the constant of proportionality. ♦

Example 1.4.6 A Wall Street rule of thumb is that airline stock prices increase whenever petroleum stock prices go down and vice versa. Hence,

$$AP = C,$$

where A and P are the prices of the airline and petroleum stocks, respectively, and C is the constant of proportionality. ♦

Example 1.4.7 Every now and then we may encounter the use of the term "inversely proportional" in everyday life or other settings. Take, for instance, the owner of a brand-new car who has angst about it getting dinged because life experiences seem to suggest that the likelihood of something getting damaged is inversely proportional to its age. Or consider a statement from Fr. David Knight, a spiritual author, who in one of his books (see [49, p. 80]) writes "The fundamental

[19] See Forester [35, p. 132].

rule is that our ability to recognize the voice of God is in inverse proportion to our attachment to the things of this world."

Since this book is all about differential equations, the proportional relationships that we consider from now on will involve derivatives. The following examples are taken from physics and chemistry.

1.4.2 Newton's Law of Cooling

It is patently clear to anyone that an ice-cold can of soda left on a patio will eventually warm up to the outside temperature and a cup of hot chocolate set on a kitchen table will cool down to room temperature. Experimental evidence indicates that for moderate temperature differences between a body and its surroundings, the rate of change of the temperature T of the body with respect to t (time) is proportional to the difference in the temperatures of the body and its surroundings. Using mathematical notation, this evidence is expressed by writing

$$\frac{dT}{dt} \propto T - T_a,$$

where T_a denotes the ***ambient temperature***, or the temperature of the surroundings. As a result, we end up with the differential equation

$$\frac{dT}{dt} = c(T - T_a), \tag{1.4.1}$$

where c is the constant of proportionality. This model of temperature change is known as ***Newton's law of cooling***.

Everyday experiences, such as the two mentioned earlier, indicate that if the temperature $T(t)$ of a body exceeds the ambient temperature T_a, then $T'(t) < 0$ and $T(t)$ decreases; whereas if $T(t) < T_a$, then $T'(t) > 0$ and $T(t)$ increases. Notice that both of these cases imply that the constant c in (1.4.1) is negative. Since it is customary in the sciences to keep physical constants and parameters positive, let us replace c with $-k$, where $k > 0$. Accordingly, (1.4.1) becomes

$$\frac{dT}{dt} = -k(T - T_a) \quad (k > 0). \tag{1.4.2}$$

Note that Newton's law of cooling is not the result of any attempt to explain the physical processes taking place, such as heat transfer between a body and its surroundings by conduction, convection, and radiation. So in a sense all of the unexplained physical processes have been subsumed into the constant of proportionality k. This is asking a lot of a constant and would explain the experimental evidence that Newton's law of cooling is not always valid, such as in the case of extreme temperature differences.

1.4.3 Rates of Chemical Reactions

The gas ethane is a hydrocarbon: a compound consisting of only carbon and hydrogen. An ethane molecule is composed of two carbon atoms and six hydrogen atoms; so its molecular formula is C_2H_6. Ethane decomposes when it is strongly heated in the absence of air. Under certain experimental conditions, it has been observed that the rate at which this decomposition takes place over time is proportional to the concentration of ethane.[20] In chemistry, it is customary to denote the concentration of a compound by enclosing its molecular formula with brackets: thus, $[C_2H_6]$ denotes the concentration of ethane. In this notation, the observation that ethane decomposes at a rate proportional to its concentration can be expressed as:

$$\text{Rate of decomposition of ethane} = k[C_2H_6],$$

where the constant of proportionality k is called a ***reaction rate constant***. It is a parameter, positive in value, which must be determined experimentally. Since ethane decomposes, the time rate of change (the rate of change with respect to time) in its concentration is a negative quantity. Thus, as $k > 0$,

$$\frac{d\,[C_2H_6]}{dt} = -k\,[C_2H_6]\,. \tag{1.4.3}$$

Since the mathematical manipulation of bracketed quantities is cumbersome when dealing with differential equations involving them, it is convenient to use lower case letters instead. For example, by letting $x(t) = [C_2H_6]\,(t)$ denote the concentration of ethane at time t, Eq. (1.4.3) simplifies to

$$\frac{dx}{dt} = -kx.$$

As for a second example, consider one of several chemical reactions that have been conjectured by some scientists to take place in the earth's stratosphere to explain the alleged depletion of the ozone layer in the polar regions.[21] In this reaction, an oxygen atom (O) collides with an ozone molecule (O_3) to produce two oxygen molecules (O_2):

$$O + O_3 \rightarrow 2O_2.$$

It is believed that the rate at which the concentration of oxygen molecules increases is proportional to the product of the concentrations of oxygen atoms and ozone

[20] See Atkins [3, p. 131].

[21] See Atkins [3, p. 140].

molecules. Translating this statement in the language of mathematics, we obtain the differential equation

$$\frac{d[O_2]}{dt} = K[O][O_3],$$

where K is the constant of proportionality. Since the derivative is positive, K must be positive—of course, its value can only be determined experimentally.

1.4.4 Classical Mechanics

Important applications of differential equations are found in the branch of physics called ***classical mechanics***, which is the study of the motion of bodies and particles. A ***body*** possesses both mass[22] and extent, whereas a ***particle*** is the idealized notion of a geometric point possessing mass. Classical mechanics is based on three famous laws of motion formulated by Sir Isaac Newton (1643–1727); so it is also called ***Newtonian mechanics***. A proper study of the laws of motion rightfully belongs to a physics or engineering course on classical mechanics. Nevertheless, the second of these laws, known as *Newton's second law of motion*, is a virtual treasure-trove of differential equations problems that are instructive and motivational, yet not too difficult. For this reason, we use it in this book as our primary source for problems and examples and to demonstrate the indispensability of differential equations for modeling physical phenomenon. We will discuss Newton's second law of motion in the remainder of this section, but only briefly. It is the fundamental law of classical mechanics. It may be only a slight exaggeration to say that any classical mechanics course is essentially a study of how to apply the second law to a whole host of different kinds of situations and problems.

1.4.4.1 Newton's Second Law of Motion

A body accelerates when external forces act on it provided they do not cancel each other out. We can add these forces, but we have to take into account that they may act in different directions. Force is a ***vector***. That is, it is a directed quantity,[23] which means that it has a direction as well as magnitude. Some other examples of vectors are *displacements* and *velocities*. The (vector) addition of two or more vectors results in a single vector, called the ***vector sum***, that is equivalent to all the other vectors acting concurrently. The vector sum of all the external forces acting on

[22] The ***mass*** of a body is a measure of the amount of matter making up the body. In order to understand what this really means, consult an introductory college level physics textbook.

[23] Directed quantities also have to obey certain rules of combination before they can legitimately be considered vectors.

a body is called the ***resultant force*** or ***resultant***. It is the single force that can replace the original set of external forces and still cause the body to move in precisely the same way.

Newton's second law of motion is an equation relating the mass of a body, the resultant force acting on the body, and its acceleration. It is derived from the following three experimental observations:

1. A body accelerates in the direction of the resultant force.
2. The magnitude of the acceleration of a body of constant mass is proportional to the magnitude of the resultant force.
3. For a constant resultant force, the magnitude of the acceleration is inversely proportional to the mass of the body.

Letting m denote the mass of the body, $\mathbf{a}$ its acceleration, and $\mathbf{F}$ the resultant of all the external forces acting on the body, we can encapsulate the above observations with the proportionality statement:

$$\mathbf{a} \propto \frac{\mathbf{F}}{m}.$$

Thus, we end up with the vector equation

$$m\mathbf{a} = \lambda \mathbf{F}, \tag{1.4.4}$$

where λ is a constant of proportionality. The letters for the acceleration and force are in boldface to indicate that they are vector quantities. However, the letter m is not in boldface since mass is a ***scalar***, a quantity that has magnitude but no direction associated with it.

If units of measurement (such as time, mass, and length) are chosen so that the value of λ is 1, then (1.4.4) simplifies to

$$\mathbf{F} = m\mathbf{a}. \tag{1.4.5}$$

In words:

> *The resultant force acting on a body is equal to the product of the mass of the body and its acceleration.*

This is ***Newton's second law of motion*** for the accelerated motion of a body subjected to external forces when its mass remains constant.

Since ***acceleration*** is the instantaneous rate at which velocity changes with time, Newton's second law can be written as

$$\mathbf{F} = m\frac{d\mathbf{v}}{dt}, \tag{1.4.6}$$

where $\mathbf{v}$ is the ***velocity***, that is, the instantaneous rate at which the position of a body changes with time t. This statement of Newton's second law is only valid if the mass of a body is constant. Note that this is a first-order differential equation.

The ***momentum*** of a body is defined as the vector quantity $m\mathbf{v}$. If the mass of a body changes, then we have to use the general form of ***Newton's second law of motion***:

> *The resultant force acting on a body is equal to the rate of change of the momentum of the body.*

This is expressed mathematically by the vector equation

$$\mathbf{F} = \frac{d}{dt}(m\mathbf{v}) \qquad \text{or} \qquad \mathbf{F} = \frac{d\mathbf{p}}{dt}, \tag{1.4.7}$$

where $\mathbf{p} := m\mathbf{v}$. Observe that (1.4.7) reduces to (1.4.6) when m is constant.

1.4.4.2 Rectilinear Motion

For those situations where all of the forces acting on a body point only in one direction or the opposing direction, such as those pointing vertically (upward or downward) or those pointing horizontally (to the right or to the left), a positive or negative sign is affixed to the magnitude of the force to indicate its direction. For forces acting horizontally, we use the convention that positive forces point to the right and negative forces point to the left. Likewise, for forces acting vertically, we regard upward as positive and downward as negative. The resultant force is obtained by simply adding all of the forces acting on the body. The resulting motion of the body is ***rectilinear***, that is, motion along a straight line. Since there is no need for vector notation here, we dispense with the boldface notation and simply write (1.4.5) as

$$F = ma \tag{1.4.8}$$

and (1.4.7) as

$$F = \frac{d}{dt}(mv) \qquad \text{or} \qquad F = \frac{dp}{dt}. \tag{1.4.9}$$

1.4.4.3 Units of Measurement

Let us say a few words about units of measurement for the quantities m, a, and F. Unfortunately there is no single set of units that everyone agrees on. There is of course the metric system and then there are systems involving English units, such as the *foot* and the *pound*. Generally speaking, however, all of the systems of units have one thing in common: units are chosen so that the constant of proportionality λ in (1.4.4) is equal to 1. Such a set of units is known as ***absolute*** or ***consistent units***. As a result, (1.4.4) simplifies to (1.4.5) and for rectilinear motion to (1.4.8). One set of consistent units uses *second* (for time) and the metric units *kilogram* (for mass) and *meter* (for length). Consequently, velocity and acceleration inherit the units *meters per second* and *meters per second squared* (i.e., *meters per second per second*), respectively. Setting $m = 1$ kg (kilogram) and $a = 1\ \mathrm{m/s^2}$ (meter per second squared) in (1.4.8), we obtain the corresponding force

$$F = (1\,\mathrm{kg}) \cdot \left(1\,\frac{\mathrm{m}}{\mathrm{s}^2}\right) = 1\,\frac{\mathrm{kg}\cdot\mathrm{m}}{\mathrm{s}^2}.$$

The combination of units "kg · m/s^2" is called a ***newton*** (abbr. N). So if a 1-N force acts on a body of mass 1 kg, it will cause the body to accelerate at a rate of 1 meter per second squared. For more details, consult a textbook on classical mechanics, such as Osgood [61, pp. 51–52] or Becker [12, p. 25].

1.4.4.4 Gravity

We will use Newton's second law shortly (see Example 1.4.8) to model the vertical motion of a body of constant mass near the earth's surface. All bodies in the universe are subject to the gravitational attractive force of the earth. In fact, according to Newton's law of universal gravitation, every body in the universe exerts a gravitational force on every other body in the universe. This includes everybody too! The force that the earth exerts on a body is called the ***force due to gravity***. Experimentally it is found that when various bodies—those whose motion is not appreciably affected by the resistive force of air[24] and other factors[25]—are released from the same point above the earth's surface, each of them falls with the same acceleration. This constant acceleration is called the ***acceleration due to gravity***.

Actually the acceleration due to gravity varies from place to place. However, close to the earth's surface it is nearly constant. Its precise value, as well as approximate values, is denoted by the letter g. At sea level and mid-latitudes, measurements show that g is approximately 32.2 ft/s^2 in the system of units known

[24] We will consider resistive forces such as air resistance in Chap. 3. Perhaps you have seen the rather striking demonstration of a feather falling as rapidly as a steel ball bearing inside a long, empty glass cylinder from which most of the air has been pumped out with a vacuum pump. However, when both bodies are removed from the confines of the glass cylinder, it is quite a different story. The air resistance on the feather retards its motion considerably, on the other hand, its effect on the ball bearing is barely noticeable.

[25] For example, the earth's rotation also contributes slightly to the acceleration of a falling body.

as the *U. S. Customary System.*[26] In another set of units called the *SI units*, short for the French *"Système International d'Unités"* and commonly referred to as the metric system, g is approximately 9.81 m/s^2. In the SI system, g is defined precisely as 9.80665 m/s^2 and in the U.S. Customary System as 32.174049 ft/s^2. These precise values are referred to as the ***standard acceleration of gravity*** (or ***standard gravity***).

The U.S. Customary System can be confusing at times. For instance, consider the word *pound*. It can designate a unit of mass or a unit of force. To distinguish between these two different uses, the terms *pound-mass* and *pound-force* are often used. The pound-mass is legally defined in terms of the kilogram: the ***pound-mass*** is precisely 0.45359237 kg. Contrast this with the definition of pound-force. The ***pound-force*** is the force that gives a body of mass 0.45359237 kg an acceleration equal to the standard acceleration of gravity, namely, 32.174049 ft/s^2.

When "pound" is used to express weight, it is the "pound-force" that is meant. The ***weight*** of a body is defined as the gravitational force exerted on it by the earth at a particular location on earth. If in (1.4.8) we set a equal to the standard acceleration of gravity g, then we will have the weight of a body of mass m at some imagined spot on earth where the acceleration due to gravity is precisely 32.174049 ft/s^2. In other words, the formula for the weight W of a body of mass m at this spot is

$$W = -mg. \tag{1.4.10}$$

We use the negative sign to indicate that the force W is directed downward—toward the center of the earth.

Since weight is a force, an appropriate unit of force must be used. In the SI system it is clear-cut: the unit of force is the newton. However, in the U.S. Customary System, it is more complicated. If we assign F a value of 1 lb$_F$ (pound-force) and m a value of 1 lb$_m$ (pound-mass), then we have from (1.4.8) that

$$a = \frac{F}{m} = \frac{1 \text{ lb}_F}{1 \text{ lb}_m} = 1 \ \frac{\text{lb}_F}{\text{lb}_m}.$$

However, referring back to the definition of pound-force, we should get 32.174049 ft/s^2 instead of the above result. The reason for the disparity is that the pound-force and pound-mass are not consistent units. That is, in order to retain both units, we would have to use

$$F = \lambda ma$$

rather than $F = ma$ and adjust λ accordingly. To avoid this, let us retain the pound-force but replace the pound-mass with a new unit of mass based on (1.4.8) where $\lambda = 1$. This new unit of mass is called the ***slug*** and is defined as the mass of a body

[26] The U.S. Customary System is the American version of the British Imperial System. Both are commonly referred to as the English or British system of units.

whose acceleration is 1 ft/s^2 when the force acting on the body is 1 pound. Since we are dispensing with the pound-mass, we will from now revert back to using "pound" instead of "pound-force." Thus, a body with a mass of $m = 1$ slug has a weight of about

$$W = -mg \approx -(1\,\text{slug}) \times (32\,\text{ft/sec}^2) = -32\,\text{pounds}.$$

Weight, being a force, is a vector quantity. However, in everyday usage only the magnitude of the weight is stated, such as in stating that a bag of sugar weighs 5 pounds.

Example 1.4.8 Model the vertical motion of a body of constant mass m, such as a baseball thrown straight upward or a rock released from the top of a tall cliff. Assume that the force of gravity is the only significant force acting on the body.

Solution In reality there are other forces besides gravity, such as air resistance, acting on vertically moving bodies. However, for the sake of simplicity, let us assume their influence on the motion of the body in this example is negligible. Let us set up a vertical frame of reference with the positive y-axis pointing straight upward, where the point from which the body begins its upward or downward motion is chosen to be the origin ($y = 0$).[27] Let $y(t)$ denote the vertical position of the body at time t relative to the origin. Then, as velocity (by definition) is the instantaneous rate of change of the position of the body with respect to time, the vertical velocity v of the body is given by the derivative

$$v = \frac{dy}{dt}.$$

At first glance, v may appear to be a scalar quantity. However, it is actually a vector quantity because the sign of dy/dt indicates the direction in which the body is moving. If, at some instant, $dy/dt > 0$, then the First Derivative Test of calculus tells us that y is increasing at that very instant. So the body is moving upward. In other words, $v > 0$ informs us that the direction of the velocity is upward. On the other hand, $v < 0$ means that the velocity is directed downward; so the body is moving downward and y is decreasing. Finally, recall the difference between velocity and speed: ***speed*** is the absolute value of velocity.

Now let us apply Newton's second law of motion for constant mass to find the *equation of motion* for a vertically moving body. Since we are assuming that gravity is the only significant force that is acting on the body, the resultant force is given by (1.4.10). Consequently, it follows from (1.4.8) that the equation of motion is

$$m\frac{dv}{dt} = -mg,$$

[27] Or, depending on one's preference, it may make sense to choose some other point that is located directly above or below this point as the origin.

which simplifies to

$$\frac{dv}{dt} = -g. \tag{1.4.11}$$

Integrating (1.4.11) with respect to t, we obtain

$$v(t) = -\int g\,dt = -gt + C.$$

Setting $t = 0$, we have

$$v(0) = -g \cdot 0 + C = C.$$

In other words, the constant of integration C represents the velocity $v(0)$, that is, the velocity of the body at the instant that it is released or thrown upward or downward. This particular velocity is usually denoted by the symbol v_0 and is called the ***initial velocity***. Therefore,

$$v(t) = v_0 - gt. \tag{1.4.12}$$

Replacing v with dy/dt, we obtain the differential equation

$$\frac{dy}{dt} = v_0 - gt.$$

Integrating this equation with respect to t, we obtain

$$y(t) = \int (v_0 - gt)\,dt = v_0 t - \frac{1}{2}gt^2 + K.$$

According to this formula, the position of the body at $t = 0$ is $y(0) = K$. Thus,

$$K = y_0,$$

where $y_0 = y(0)$ denotes the initial position of the body relative to the origin. We conclude that the position of the body y at time t (before it hits the ground) is

$$y = y_0 + v_0 t - \frac{1}{2}gt^2, \tag{1.4.13}$$

where y_0 and v_0 are its initial position and initial velocity, respectively. ♦

1.4.4.5 Constant Acceleration

If a body is moving horizontally with constant acceleration a, then its velocity v at time t is given by the formula

$$v = v_0 + at. \tag{1.4.14}$$

The derivation of this is essentially the same as the derivation of (1.4.12). Likewise, a slight alteration in the derivation of (1.4.13) yields the formula

$$x = x_0 + v_0 t + \frac{1}{2}at^2 \tag{1.4.15}$$

for the position x of the body at time t. For some types of problems, it is better to regard v as a function of x rather than of t. We leave it to the reader to derive the formula (see Problem 77)

$$v^2 = v_0^2 + 2a(x - x_0). \tag{1.4.16}$$

Another useful formula is

$$x = x_0 + \bar{v}t, \tag{1.4.17}$$

where $\bar{v}$ is defined by

$$\bar{v} = \frac{v_0 + v}{2}. \tag{1.4.18}$$

Note that $\bar{v}$ is the average velocity of the body between the times 0 and t.

The term *uniform* is used to describe the motion of a body whose velocity or acceleration does not change. To be precise, by the ***uniform velocity*** of a body is meant that the magnitude and direction of the velocity of the body remain constant throughout its motion. Since there is no change in velocity, there is no acceleration. Likewise, ***uniform acceleration*** means that the magnitude and direction of acceleration remain constant.

Example 1.4.9 A bowling ball with a mass of 6 kg is rolling in a straight line on a bowling lane at a constant speed of 3 meters per second. Imagine this bowing lane to be perfect: no forces act on the ball to impede it motion. However, at some point, a constant force suddenly acts on the bowling ball causing it to decelerate uniformly and stop in 10 m. Find this force.

Solution Since the deceleration (negative acceleration) is uniform, we can use formula (1.4.16) to calculate it. Setting $v_0 = 3$ m/s, $v = 0$ m/s, and $\Delta x = 10$ m, we have

$$a = \frac{v^2 - v_0^2}{2\Delta x} = \frac{(0\text{ m/s})^2 - (3\text{ m/s})^2}{2(10\text{ m})} = -0.45\ \frac{\text{m}}{\text{s}^2}.$$

By Newton's second law, the force is

$$F = ma = (6\,\text{kg})\left(-0.45\,\frac{\text{m}}{\text{s}^2}\right) = -2.7\,\frac{\text{kg}\cdot\text{m}}{\text{s}^2} = -2.7\,\text{N}.$$

We conclude that if at some point during the rolling of the bowling ball a constant force of 2.7 N opposes its forward motion, the ball would cease rolling in 10 m. ♦

1.4.4.6 Variable Mass

The masses of the moving bodies in Examples 1.4.8 and 1.4.9 do not change. But of course there are situations where the masses of bodies do change with time as they move. Such is the case with rockets. Let us model the motion of a conventional rocket, such as a Russian Soyuz vehicle or one of the NASA space shuttles, which were retired in 2011.

Rocket Propulsion A conventional rocket in deep space is moving along a straight line with respect to some inertial reference frame.[28] Suppose it is accelerating because its engines are burning a propellant consisting of a fuel and oxidizer. This produces combustion products, primarily superheated gases, that are ejected from nozzles at the rear of the rocket. It is the ejection of these combustion products that provides the thrust that accelerates the rocket. Assume that the velocity of the ejected combustion products measured relative to the nozzles is constant. Also, assume that all external forces acting on the rocket, such as gravitational forces, either cancel each other out or are negligible. Determine the equation of motion for this rocket.

Derivation Since the mass of the rocket is changing, we will us use the momentum form of Newton's second law of motion, namely (1.4.9), to derive the equation of motion of the rocket. By assumption, virtually no external force acts on the rocket. Accordingly, it follows from (1.4.9) that

$$\frac{dp}{dt} = 0.$$

This implies that the momentum p does not change. So the momentum at an arbitrary time t is equal to the momentum at some later time $t + \Delta t$. That is,

$$p(t) = p(t + \Delta t). \tag{1.4.19}$$

But we have to be careful to interpret and use this equation correctly. The left-hand side of (1.4.19) denotes the momentum of the rocket and the unburned

[28] An ***inertial reference frame*** is a reference frame in which Newton's first law holds. This means that if the resultant force acting on a body is zero, then the body has no acceleration in such a reference frame. Consult any university physics textbook for more information.

propellant that it contains at time t. However, the right-hand side not only denotes the momentum of the rocket and its unburned propellant at time $t + \Delta t$ but also the momentum of the combustion products resulting from the burned propellant that have been ejected from the rocket's nozzles during the time interval Δt. In other words,

> *The momentum of the rocket at time t is equal to the momentum of the rocket at some later time $t + \Delta t$ plus the momentum of the combustion products ejected during the time interval Δt.*

At an arbitrary time t, let v denote the velocity of the rocket with respect to the inertial reference frame mentioned earlier. Let m denote the ***total rocket mass*** (i.e., the mass of the rocket together with the unburned propellant). Hence, the momentum of the rocket at time t is

$$p(t) = mv. \tag{1.4.20}$$

Later on, at time $t + \Delta t$, the combustion products ejected from the rocket resulting from the burning of some of the propellant during the time interval Δt reduces the mass of the propellant remaining inside the rocket. As a result, the total rocket mass is now $m + \Delta m$, where Δm is a negative quantity. Since the velocity of the rocket increases by an amount Δv, the momentum of the rocket at time $t + \Delta t$ is

$$(m + \Delta m)(v + \Delta v). \tag{1.4.21}$$

But we must not forget to include the momentum of the ejected combustion products, which from now on we will call the *exhaust*. And we have to be careful with handling the signs of the quantities. The mass of the exhaust ejected during the time interval Δt is not Δm but rather $-\Delta m$ as Δm is negative. Let v_e denote the ***exhaust velocity*** (assumed constant), that is, the velocity of the exhaust measured with respect to the nozzles. Note that $v_e < 0$ since the motion of the exhaust is opposite to that of the rocket. The velocity of the exhaust with respect to an inertial reference frame is

$$\text{velocity of the exhaust} = (v + \Delta v) + v_e.$$

And so the momentum of the exhaust is

$$-\Delta m(v + \Delta v + v_e). \tag{1.4.22}$$

Substitution of the quantities (1.4.20), (1.4.21), and (1.4.22) in Eq. (1.4.19) yields

$$mv = (m + \Delta m)(v + \Delta v) - \Delta m(v + \Delta v + v_e). \tag{1.4.23}$$

This simplifies to

$$m\Delta v - v_e \Delta m = 0,$$

which upon division by Δt becomes

$$m\frac{\Delta v}{\Delta t} - v_e\frac{\Delta m}{\Delta t} = 0.$$

As $\Delta t \to 0$, both difference quotients approach their respective derivatives. That is to say, $\Delta v/\Delta t \to \dot{v}$ and $\Delta m/\Delta t \to \dot{m}$. Therefore,

$$m\frac{dv}{dt} - v_e\frac{dm}{dt} = 0 \tag{1.4.24}$$

is the equation of motion of the rocket. ♦

We will now use (1.4.24) to derive two fundamental rocket equations. With the first rocket equation, we will be able to calculate the acceleration of a rocket at some given instant if we know (i) the rate at which propellant is burned, (ii) the exhaust velocity, and (iii) the total rocket mass at that instant. With the second one, we can calculate its velocity.

Rocket Equation 1 Recall that m denotes the total rocket mass (the combined mass of the rocket and the unburned propellant). Let r be defined by

$$r = -\frac{dm}{dt}.$$

This gives the rate of *fuel consumption*, that is, the rate at which propellant is burned. It is positive because $\dot{m} < 0$. Let a denote the acceleration of the rocket. Expressed in terms of a and r, Eq. (1.4.24) is

$$ma = -rv_e. \tag{1.4.25}$$

Rocket Equation 2 Since we now have a formula for the rocket's acceleration, let us also derive a formula for its velocity. It follows from (1.4.24) that

$$\frac{dv}{dt} = \frac{v_e}{m}\frac{dm}{dt}. \tag{1.4.26}$$

Now let us change our perspective by regarding v as depending directly on m instead of on t. Then, applying the Chain Rule of differential calculus, we have

$$\frac{dv}{dt} = \frac{dv}{dm}\frac{dm}{dt}. \tag{1.4.27}$$

With this, (1.4.26) becomes

$$\frac{dv}{dm}\frac{dm}{dt} = \frac{v_e}{m}\frac{dm}{dt},$$

which simplifies to

$$\frac{dv}{dm} = \frac{v_e}{m}. \tag{1.4.28}$$

Since the right-hand side depends only on the independent variable m, we integrate with respect to m in order to solve for v, thereby obtaining

$$v = \int \frac{v_e}{m}\,dm = v_e \ln m + C. \tag{1.4.29}$$

Let v_0 and m_0 denote the velocity and total rocket mass, respectively, at some earlier time. Then we have that

$$C = v_0 - v_e \ln m_0.$$

Substituting this into (1.4.29), we obtain

$$v = v_e \ln m + v_0 - v_e \ln m_0 = v_0 - v_e \ln\left(\frac{m_0}{m}\right)$$

or

$$\Delta v = -v_e \ln\left(\frac{m_0}{m}\right). \tag{1.4.30}$$

where $\Delta v = v - v_0$ (change in the rocket's velocity). Keep in mind two things here: $m_0/m > 1$ since the burning of propellant decreases the total rocket mass and, as noted earlier, $v_e < 0$. Equation (1.4.30) is called the ***classical rocket equation*** or the ***Tsiolkovsky rocket equation***.[29] We conclude our discussion of rockets by applying the classical rocket equation to the following hypothetical situation.

Example 1.4.10 The mass of a certain single-stage rocket without propellant is M. Before launch, it is filled with a propellant that has a mass of $9M$.[30] Consequently, the rocket's total mass at launch is $10M$. Suppose at some time t_1 after launch the total rocket mass is $5M$ and that from this point in time to a later time t_2, the exhaust

[29] Konstantin Tsiolkovsky (1857–1935) was a Russian physicist and aviation engineer who published more than 400 works on rocketry, space travel, and related subjects. He published equation (1.4.30) in a Russian aviation magazine in 1903.

[30] The mass of the propellant at launch typically constitutes 85–91% of the total rocket mass. For more information, see https://www.nasa.gov and related websites.

velocity remains a constant $v_e = -10{,}000$ mph. If from t_1 to t_2 the rocket burns half of the propellant that it had at t_1, calculate the change in the rocket's velocity.

Solution During the time interval t_1 to t_2, the total rocket mass has decreased from $m_0 = 5M$ to $m = M + 2M = 3M$. According to the Tsiolkovsky rocket equation, the change in the rocket's velocity is

$$\Delta v = -v_e \ln\left(\frac{m_0}{m}\right) = 10{,}000 \ln\left(\frac{5M}{3M}\right) = 10{,}000 \ln\left(\frac{5}{3}\right) \approx 5108 \text{ mph.} \quad \blacklozenge$$

Problems

Heigh-ho, Heigh-ho,
It's off to work we go.
(Whistle)
Heigh-ho, Heigh-ho, Heigh-ho,
Heigh-ho, Heigh-ho,
It's off to work we go ...

Heigh-Ho (song in the 1937 Walt Disney movie Snow White and the Seven Dwarfs), lyrics by Larry Morey and music by Frank Churchill

Parts of an Ordinary Differential Equation
In Problems 1 through 8, an ODE is given. Find the independent variable, the dependent variable, and the parameter (or parameters if there is more than one). Also, state the order of the equation.

1. $\dot{N} = rN\left(1 - \frac{N}{k}\right)$
2. $\frac{d^2p}{ds^2} + \lambda\left(p^3 - 1\right)\frac{dp}{ds} = -10\sqrt{s^4 - 2}$
3. $\frac{d^4x}{dt^4} - a\frac{dx}{dt} + \frac{b}{5}x^5 = -10\sin^7(\gamma t)$
4. $\frac{d^2y}{dx^2} = \frac{C}{L}\sqrt{\left(\frac{AC}{L}\right)^2 + \left(\frac{dy}{dx}\right)^2}$
5. $EIy^{(4)} + py'' + ky = c(1 - x)$
6. $\ddot{x} - \varepsilon(1 - x^2)\dot{x} + x = 0$
7. $\frac{\hbar^2}{2m} \cdot \frac{d^2\psi}{dx^2} + \left(E - \frac{1}{2}kx^2\right)\psi = 0$
8. $6y''' + \eta(y'')^4 - 2\rho\omega xyy' = \cos(xy^5)$

First-Order Partial Derivatives
In Problems 9–12, find $\partial f/\partial x$ and $\partial f/\partial y$.

9. $f(x, y) = 15x^2 - 3x^4y^3 + \frac{2}{3}y^5$
10. $f(x, y) = \frac{x^2y}{x + 4y}$
11. $f(x, y) = x + 5e^{2x}\sin(xy)$
12. $f(x, y) = 2xy^2 - e^{4y}\ln x$

Alleged Solutions of ODEs
In Problems 13 through 18, an ordinary differential equation is given along with a function alleged to be its solution. Determine whether the alleged solution is truly a solution by means of direct substitution of the function and its derivative(s) into the equation.

13. $\frac{dy}{dx} = xy, \quad y = 4e^{x^2/2}$
14. $x^2y' = y^2, \quad y = \frac{x}{1 + Cx}$
15. $\frac{dy}{dx} = \frac{x}{y}, \quad y = -\sqrt{4 - x^2}$
16. $\frac{dy}{dx} = \frac{x^2 + xy + y^2}{x^2}$, $y = x\tan(\ln x)$, where $x > 0$
17. $2y'' - 7y' + 3y = 0, \quad y = e^{2x}$
18. $\frac{d^2y}{dx^2} + 16y = e^{3x}, \quad y = \frac{1}{25}e^{3x}$

Solutions of the Simplest ODEs
In Problems 19 through 44, find the solutions of each differential equation. (All integrations can be carried out using the formulas in Table 1.4 and the integration techniques that were reviewed in Sect. 1.3.)

19. $\frac{dy}{dx} = \cos(3x)$
20. $\frac{dy}{dx} = \sin\left(\frac{x}{5}\right)$
21. $\frac{dy}{dx} = \frac{1}{x^2 + 5}$
22. $\frac{dy}{dx} = \frac{\ln x^2}{x}$
23. $\frac{dy}{dx} = \tan(5x)$
24. $\frac{dy}{dx} = \sec^2(7x)$
25. $\frac{dy}{dx} = xe^{2x}$
26. $\frac{dy}{dx} = x\sin(3x)$
27. $\frac{dy}{dx} = x^2\sin(2x)$
28. $\frac{dy}{dx} = e^{2x}\sin\left(\frac{x}{3}\right)$
29. $\frac{dy}{dx} = \frac{3}{x^2 + 4x}$
30. $\frac{dy}{dx} = \frac{x}{x^2 - 5x + 6}$
31. $\frac{dy}{dx} = \frac{4x - 10}{x^2 - 5x + 6}$
32. $\frac{dy}{dx} = \frac{1}{x^3 + 9x}$

33. $\dfrac{dy}{dx} = \dfrac{3}{x^2 + 4x + 5}$

34. $\dfrac{dy}{dx} = \sec(3x)\tan(3x)$

35. $\dfrac{dy}{dx} = \dfrac{\sin(2x)}{\cos^3(2x)}$

36. $\dfrac{dy}{dx} = 5x^2\sqrt{1 + 4x^3}$

37. $\dfrac{dy}{dx} = \dfrac{5}{\sqrt{4 - x^2}}$

38. $\dfrac{dy}{dx} = \dfrac{x^2 - 5}{x^3 - 3x^2 + 4x - 12}$

39. $\dfrac{dy}{dx} = \dfrac{1}{2x\sqrt{x^2 - 9}}$

40. $\dfrac{dy}{dx} = \sqrt{x}\,\ln x$

41. $\dfrac{dy}{dx} = \frac{1}{2}\sin(\sqrt{x})$

42. $\dfrac{dy}{dx} = 3\,e^{\sqrt{x}}$

43. $\dfrac{dy}{dx} = \dfrac{\cos x}{\sqrt{1 - \sin^2 x}}$ $(-\pi/2 \le x \le \pi/2)$

44. $\dfrac{dy}{dx} = \dfrac{\cos x}{\sqrt{1 - \sin^2 x}}$ $(\pi/2 \le x \le 3\pi/2)$

Solutions of Simple PDEs

In Problems 45 through 50, find the solutions of each partial differential equation.

45. $\dfrac{\partial u}{\partial x} = 6x^2y + \dfrac{2}{y}$

46. $\dfrac{\partial u}{\partial y} = 6x^2y + \dfrac{2}{y}$

47. $\dfrac{\partial u}{\partial y} = 5 + \dfrac{x}{4 + y^2}$

48. $\dfrac{\partial u}{\partial x} = 2y - 5y^3\csc^2 x$

49. $\dfrac{\partial u}{\partial x} = 2x\cos(\sqrt{y})$

50. $\dfrac{\partial u}{\partial y} = 2x\cos(\sqrt{y})$

ODEs from Rates of Change

In Problems 51–56, translate the given statement into an ordinary differential equation. Use appropriate mathematical notation. Identify every letter (variable or constant) used in the equation.

51. Water is leaking out of a city swimming pool at the rate of 25 gallons per hour.

52. Let $f(v)$ be the fuel efficiency in mpg (miles per gallon) when a car is traveling at a speed of v mph (miles per hour). When the speed is 70 mph, the fuel efficiency of the car is decreasing by 0.30 mpg per mph.

53. A sewage treatment tank contains 10,000 gallons of polluted water. The tank removes 5% of the pollutants in the water per minute.

54. The index of refraction of a substance (such as water, flint glass, acetone, etc.) is the ratio of the speed of light in a vacuum to the speed of light in the substance. Use the variables t and s, where s denotes the distance that the light has traveled through the substance at time t.

55. A patient recovering from surgery is fed glucose intravenously at the rate of b milligrams per minute, where b is a constant.

56. For a given drop in pressure along a cylindrical pipe, the volumetric rate of flow of natural gas through the pipe is 4.5 times the fourth power of the radius of the pipe.

ODEs from Proportions

In Problems 57–70, translate the given statement into an ordinary differential equation using appropriate mathematical notation. Write the equation in a form so that the constant of proportionality is positive. Identify every letter (variable or constant) appearing in the equation.

57. The number of squirrels in a forest preserve increases at a rate that is proportional to their number.

58. A basic electrical circuit that is usually considered in elementary physics courses consists of a switch that can be opened or closed, a battery, a resistor, and a capacitor connected in series. When the switch is closed, an electrical current begins flowing through the circuit; however, it immediately begins to decrease to zero. The rate of change of the current at a given moment is proportional to its value at that moment.

59. The rate at which the volume of a melting snowball changes with time is proportional to its surface area. Given that the volume of a snowball of radius r is $4\pi r^3/3$ and its surface area is $4\pi r^2$, find the differential equation that expresses the rate at which the radius changes with time.
60. A rancher is tracking a wolf headed straight toward a snow-covered mountain. The speed with which he is able to pursue the wolf is inversely proportional to the depth of the snow.
61. There are 1000 people in Mayberry. Whenever a rumor is started by the town's gossip, the time rate of change of the number of people who have heard the rumor is proportional to the number of people who have not yet heard the rumor. (Write the differential equation describing this situation in terms of two variables.)
62. An epidemic of rubella (German measles) breaks out in a remote, mountainous region in Argentina. Assume that P people live in this region, that this number does not change during the course of the epidemic, and that the time rate of change of the number of people infected with rubella is proportional to the product of the number who are infected and the number who are not. (Write the differential equation for the time rate of change of the number of infected people in terms of two variables.)
63. As a spherical raindrop evaporates, its volume changes at a rate proportional to its surface area. (Write the equation so that it involves only the variables volume and time.)
64. The tank of a certain toilet has a constant cross-sectional area. After it is flushed and all of the water has drained from the tank, water flows from the filler tube into the tank at the rate of 3 liters per minute. However, due to a defective valve seat, waters leaks from the tank at a rate that is proportional to the square root of the depth of water in the tank. (Use this information to write a differential equation for the volume of water in the tank as it is filling. Write the equation so that the only variables that explicitly appear in it are "volume" and "time". Define every letter, whether it is a variable or a constant, that you use to come up with this equation.)
65. A tank with a constant cross-sectional area is filled with water; however, the water leaks through a small hole in its bottom.
    (a) According to ***Torricelli's law***[31] of fluid flow, the speed of the water exiting from the hole is proportional to the square root of the depth of the water in the tank.
    (b) An alternate form of Torricelli's law states that the rate with which the depth of the water in the tank decreases is proportional to the square root of the depth.
    (c) Before Torricelli formulated his law, it was thought that the depth of the water in a leaking tank would decrease at a rate proportional to the depth in the tank.[32]
66. In Aristotelian physics, objects of different weights fall at different speeds. It was believed that an object falls at a speed proportional to its weight.
67. According to Newton's law of universal gravitation, the acceleration of a body caused by the earth's gravitational pull is directed toward the center of the earth and its magnitude is inversely proportional to the square of the distance between the body and the center of the earth. Assuming all other external forces are negligible, use this information to

---

[31] Besides formulating this law of fluid mechanics, Evangelista Torricelli (1608–1647), an Italian mathematician and physicist, is also remembered for inventing the mercury barometer in 1643. In Florence, he served briefly as Galileo Galilei's assistant and secretary. He inherited Galileo's appointment, after the latter's death in 1642, as philosopher and chief mathematician to the court of Grand Duke Ferdinando II of Tuscany.

[32] See Driver [29, p. 454].

find a first-order differential equation relating the variables, distance and velocity, of a freely falling body.

68. A body is falling to earth. Apart from the force due to gravity, assume that the only other non-negligible force acting on the body is the drag force. The force due to gravity points downward. In the U. S. Customary System, it has a magnitude of $mg = 32m$, where m is the mass of the body. The drag force results from the force exerted by the air on the body opposing its motion—and so points upward. Assume that its magnitude is proportional to the velocity of the body. The product of the mass of the body and the rate of change in its velocity is equal to the sum of the drag force and the force due to gravity.

69. The rate with which water (H_2O) forms in the reaction

$$OH + H_2 \to H_2O + H,$$

is proportional to the product of the concentrations of hydrogen molecules (H_2) and hydroxide ions (OH).

70. One reaction among several reactions that may take place in the decomposition of ozone (O_3) in the stratosphere is

$$O_3 \to O_2 + O.$$

The rate at which the concentration of ozone is decreasing at any instant is proportional to its concentration at that very instant.

Newton's Second Law of Motion

In Problems 71–77, translate the given statement into a differential equation using appropriate mathematical notation. Identify every letter (variable or constant) appearing in the equation.

71. A body with mass m is moving along the positive x-axis due to a force that attracts it to the origin with a magnitude proportional to its distance from the origin. The frictional force opposing the motion of the body is proportional to the body's weight. Use Newton's second law of motion to find the second-order differential equation governing the motion of the body along the x-axis.

72. Experiments confirm that a good model for the drag force on a parachutist is a force that is proportional to the square of the velocity of the parachutist. The weight of the parachutist is mg, where g is the acceleration due to gravity and m is the total mass of the parachutist, including the parachute and other equipment. Other than the weight and drag force, ignore all other forces acting on the parachutist. Find the rate of change of the velocity of the parachutist using Newton's second law of motion.

73. Imagine a tunnel bored into the earth connecting Madrid, Spain to its antipodal point on the opposite side of the earth, which is Weber, New Zealand. Suppose a bowling ball is dropped in the opening of the tunnel at Madrid. If, for the sake of simplicity, it is assumed that the earth is a homogeneous sphere, then it can be shown with physics, trigonometry, and calculus that the gravitational force exerted on the bowling ball is directly proportional to its distance from the center of the earth.

    (a) Translate the above information about the gravitational force exerted on the bowling ball into a mathematical formula. Express it in terms of the bowling ball's position from the center of the earth.

    (b) If all other forces exerted on the ball are ignored, determine the differential equation that models the motion of the bowling ball using Newton's second law.

74. A baseball is thrown vertically upward. The thrower's hand is y_0 feet from the ground when the ball is released. The baseball attains a height of h feet t_1 seconds after its release. After reaching a maximum height, the baseball then falls back to this point at some time $t_2 > t_1$.

    (a) Show that the velocity of the baseball at the moment of release is the product of g and the average of t_1 and t_2.

(b) Show that $h = y_0 + \frac{1}{2}gt_1t_2$.

75. The Greek philosopher Aristotle (384–322 B.C.) taught that an object falls at a speed proportional to its weight. This was the prevailing view of university professors, even as late as the seventeenth century. To demonstrate the falsity of this notion, Galileo,[33] as legend has it, dragged cannonballs and musket balls of different weights up the spiral staircase of the Leaning Tower of Pisa and dropped them from the top story, a height of approximately 180 feet.

    (a) Use Newton's second law of motion to estimate the time it takes for a 10-pound cannonball to fall to the ground. Assume air resistance has a negligible effect on the ball's motion.

    (b) According to Aristotelian physics, how long would it take a one-pound musket ball to reach the ground? Base your answer on part (a) and compare it with the time predicted by Newtonian physics.

76. A two-euro coin is accidentally dropped from the Eiffel Tower's highest observation deck, which is 276 m (906 ft) above the ground. If air resistance has a negligible effect on the coin's motion, show that its velocity v after having fallen a distance of y meters is

$$v = \sqrt{2gy}.$$

Hint. Apply the Chain Rule of differential calculus (similar to how (1.4.26) was handled).

77. Derive the four kinematic formulas (1.4.14) to (1.4.17) for rectilinear motion of bodies undergoing constant acceleration a.

Lengths of Curves

78. Use the differential equation (1.2.2) to find a general formula for the length of a plane curve defined by the parametric equations (1.2.1).

79. The parametric equations for a circle of radius r are

$$x = r\cos t,\ \ y = r\sin t.$$

Derive the circumference of a circle of radius r using these equations. *Hint.* Use the result of Problem 78.

80. The parametric equations for an ellipse are

$$x = a\cos t,\ \ y = b\sin t.$$

For $a > b > 0$, show that the perimeter s of an ellipse is

$$s = 4a\int_0^{\pi/2}\sqrt{1 - k^2\cos^2\theta}\,d\theta,$$

where $k^2 = 1 - b^2/a^2$. *Hint.* Use the result of Problem 78. Use the trig identity

$$\cos\theta = \sin(\pi/2 - \theta)$$

and a change of variable in the integral to show that the perimeter can also be calculated using the formula

$$s = 4aE(k),$$

where $E(k)$ denotes the ***complete elliptic integral of the second kind***, which is defined by

$$E(k) = \int_0^{\pi/2}\sqrt{1 - k^2\sin^2\theta}\,d\theta.$$

The values of this integral can be found in a table of elliptic integrals, such as in [1], or it can be computed with a *CAS*.

81. Use the result of Problem 78 to find the general formula for the length of a curve defined by the function $y = f(x)$ for $\alpha \le x \le \beta$. *Hint.* Set $x = t$ so that $y = f(t)$.

[33] A narrative of the life of Galileo Galilei (1564–1642), Italian astronomer and physicist extraordinaire, is wonderfully told in the book *Galileo's Daughter* by Dava Sobel [66]. Letters written to him by his eldest daughter, Suor Maria Celeste, a cloistered nun of the Order of the Poor Clares, are woven masterfully into the story of the life of this incredible genius.

Stopping Distances

82. A BMW coupe is barreling along Bundesautobahn 95 (Federal Expressway 95) toward Munich at 200 kilometers per hour. Seeing a tractor-trailer truck overturned on the autobahn and realizing there is no way of driving around it, the driver slams on the brakes. Suppose it takes the driver 1.5 seconds to perceive the danger ahead and to apply the brakes. Also, suppose the speed of the coupe decreases a constant 8 meters per second every second. If the truck is 350 m away the instant the driver perceives the danger, will the coupe collide with the truck?

83. A 150-car freight train is approaching Carbondale, Illinois at a constant speed of 50 feet per second (approximately 34 mph). As the train nears a railroad crossing, the locomotive engineer sees a car stalled on the tracks. It takes him 4 seconds to react before he applies the brakes. It then takes the train another 1.25 minutes to come to a full stop. Assume that the deceleration of the train is constant while the brakes are being applied.

    (a) What is the deceleration of the train?

    (b) How far does the train travel from the moment the engineer sees the car until the time the train comes to a full stop? Search the Internet for actual stopping distances of trains and compare them to your answer.

Poiseuille's Law

84. The volumetric rate of flow of a fluid, such as water or natural gas, through a pipe of circular cross section is determined by a number of variables: the fluid's viscosity, the pipe's radius and length, and the difference in pressure between the ends of the pipe. In parts (a) through (d), convert the given proportional relationship into a differential equation.

    (a) The volumetric rate of flow is proportional to the difference in pressure (p) between the ends of the pipe when all other variables are held constant.

    (b) The volumetric rate of flow is inversely proportional to the length (l) of the pipe when all other variables are held constant.

    (c) The volumetric rate of flow is inversely proportional to the fluid's viscosity (η) when all other variables are held constant.

    (d) The volumetric rate of flow is proportional to the pipe's radius (r) to the fourth power when all other variables are held constant.

    (e) Combine the proportional relationships given in parts (a) through (d) into one differential equation. The resulting equation is known as ***Poiseuille's law***.

    (f) Explain why the air ducts for ventilating buildings generally have a large radius.

    (g) If arteriosclerosis reduces the effective radius of a person's artery by 10%, by what percentage does the volumetric rate of flow of the blood through the artery decrease?

Computer Algebra System Problems

Use a Computer Algebra System, such as Maple, Mathematica, *or* MATLAB, *to work the following problems.*

85. Find the solutions of some of the differential equations in Problems 19 through 44 with the aid of a CAS. Compare these solutions with the solutions that you obtained using only pencil and paper. For some problems, the CAS may appear to give different answers; when that happens, use algebra or trigonometry to show that the answers are really equivalent.

86. The rate of change of a quantity Q with respect to the time t is equal to

$$2\pi\sqrt{t}e^{-t} - 5t^4 \ln\sqrt{3+t^4}.$$

Use a CAS to determine Q up to a constant.

87. Evaluate $\int_0^1 \ln x \ln(1-x)\,dx$.

Chapter 2
Separable and Related Equations

To be, or not to be? That is the question—Whether 'tis nobler in the mind to suffer the slings and arrows of outrageous fortune, or to take arms against a sea of troubles, and by opposing end them?

Hamlet's soliloquy in *The Tragedy of Hamlet* (Act 3) by William Shakespeare

To separate, or not to separate? Alas, that is the question.

Student's musings on a differential equation

Abstract This chapter is about *separable equations*, namely, first-order differential equations $y' = f(x, y)$ that can be rewritten in the form

$$\frac{dy}{dx} = g(x)h(y). \tag{1}$$

The *method of separation of variables* is introduced in this chapter by playing around and tinkering with simple examples of (1) and "discovering" this method, after which it is completely justified using an argument involving the indefinite integrals of $g(x)$ and $1/h(y)$ and the chain rule.

The distinction between explicit and implicit solutions of (1) is discussed and illustrated with many examples, some of which are accompanied by graphs of several solutions. Since there is always the possibility of making mistakes in algebra or integration, readers are encouraged to always check the functions they allege to be solutions by verifying that these functions truly satisfy the original form of Eq. (1). It is also shown how to check implicit solutions using the method of implicit differentiation. Intervals of existence of solutions are also examined.

Even though most initial value problems involving equations of the form (1) can be solved using indefinite integration, a couple of examples suggest that it may be more convenient or even necessary to use definite integration since there are initial

© The Author(s), under exclusive license to Springer Nature Switzerland AG 2026
L. C. Becker, *Ordinary Differential Equations: Concepts, Methods, and Models*,
https://doi.org/10.1007/978-3-032-15150-6_2

value problems, such as

$$\frac{dy}{dx} = \frac{e^{-x^2}}{2y}, \quad y(1) = -2, \tag{2}$$

whose solution must be expressed in terms of a definite integral. Such is the case with (2) since the function $\exp(-x^2)$ does not have an elementary antiderivative.

Applications involving separable equations considered in this chapter include using Newton's law of cooling to predict temperatures, such as a cup of hot coffee left on a table, and a derivation of the *Cobb-Douglas production function*, which is a formula used by economists to model the total production of goods in an economic system.

The concluding section is about *homogeneous first-order equations*, which are equations that can be written in the form

$$\frac{dy}{dx} = F\left(\frac{y}{x}\right), \tag{3}$$

where F denotes a function of the ratio y/x. An equation that looks like (3) can be solved by replacing y/x with a new dependent variable, say z, which transforms it into a separable equation. A convenient test to ascertain whether a differential equation $y' = f(x, y)$ is homogeneous without having to expend time and effort rewriting it as (3) is also discussed.

2.1 Separable Equations

In some of the problems at the end of Chap. 1, an ordinary differential equation is presented together with a function that is alleged to be its solution. The problem is to ascertain if the "alleged solution" is really a solution—in other words, to determine whether or not the given function actually satisfies the differential equation on some interval. However, this is not the way it is in the real world. Whether you are majoring in engineering, mathematics, mathematical economics, or in one of the physical sciences, you will have to deal with differential equations now and then; however, generally speaking, these equations will not be accompanied by their solutions or even hints as to what their solutions might be. Rather it is you who will somehow have to come up with their solutions. And so it is incumbent upon you to learn how to solve the different types of differential equations that you are most likely to encounter in your chosen field of study.

The type of equations that we will introduce and learn how to solve in this chapter are known as *separable differential equations*. In fact, the differential equations in Sect. 1.2.2 that we referred to as the "simplest type of differential equations" are separable. So what exactly is a separable differential equation?

Separable Equation

Definition 2.1.1 A first-order differential equation

$$\frac{dy}{dx} = f(x, y) \tag{2.1.1}$$

is said to be ***separable*** if

$$f(x, y) = g(x) \cdot h(y), \tag{2.1.2}$$

where, as the notation indicates, g depends only on x and h depends only on y.

In other words, Eq. (2.1.1) is separable if f, a function of two variables, can be separated by algebraic means into a product of two single-variable functions. For example, the right-hand side of

$$\frac{dy}{dx} = \frac{4\cos(2x)}{y+1} \tag{2.1.3}$$

in the notation of Definition 2.1.1 is

$$f(x, y) = \frac{4\cos(2x)}{y+1}.$$

Equation (2.1.3) is separable since f can be rewritten as the product of the function

$$g(x) = 4\cos(2x),$$

which depends only on x, and the function

$$h(y) = \frac{1}{y+1},$$

which depends only on y. By the way, it does not matter to which of the functions the constant 4 is attached—making it a part of h instead of g accomplishes the same thing as far as separability is concerned.

Sometimes we may not recognize a differential equation to be separable until we do something with the function $f(x, y)$: it may involve factoring it, using trigonometric identities, the laws of exponents, and so on. For example, the equation

$$\frac{dy}{dx} = \frac{e^x \ln x^2}{xe^x + xye^{x+y}} \tag{2.1.4}$$

does not appear to be separable at first. However, if we replace e^{x+y} with $e^x \cdot e^y$ and factor, then the right-hand side simplifies to

$$\frac{e^x \ln x^2}{xe^x + xye^{x+y}} = \frac{\ln x^2}{x(1 + ye^y)} = \frac{\ln x^2}{x} \cdot \frac{1}{1 + ye^y}.$$

So the equation is separable since its right-hand side can be rewritten as the product of two functions, one of which depends only on x while the other depends only on y.

At this point in the discussion, do not jump to the false conclusion that all first-order equations are separable. In fact, most are not! Obviously,

$$\frac{dy}{dx} = y \sin x$$

is separable as is

$$\frac{dy}{dx} = x \sin y.$$

On the other hand, the equation

$$\frac{dy}{dx} = \sin(xy)$$

is ***inseparable*** (not separable) because its right-hand side cannot be split into x- and y-parts; that is, $\sin(xy)$ cannot be expressed as the product of two functions with one depending only on x and the other only on y. Another example of an inseparable equation is

$$\frac{dy}{dx} = x - y.$$

Just as Hamlet poses his famous question in his soliloquy in Act 3 of *The Tragedy of Hamlet* in an attempt to find a solution to his predicament, so too must we ask ourselves a question when confronted with a first-order equation (albeit one with less dire consequences) and that is whether the equation is separable or not—for only then can we use the methods described in this chapter.

Now that we are able to discern separable equations from other kinds of differential questions, the question to be answered is this: How do we go about finding solutions of separable equations? Let us begin with an equation whose solutions are already well-known to us from calculus. This will help us to come up with a method for solving separable differential equations.

Example 2.1.1 Find solutions of the differential equation

$$\frac{dy}{dx} = y. \tag{2.1.5}$$

Solution Is this equation separable? At first this question might be confusing due to the absence of the variable x on the right-hand side of the equation since (2.1.1) and (2.1.2) in the definition could be misunderstood as meaning that both independent variables have to be present. Yes, it is true that the symbol $f(x, y)$ denotes a function whose domain is a subset of the xy-plane and whose value usually depends on both x and y; but it could also denote a function that depends on only one of the variables. Or it could be a constant function, such as $f(x, y) = 10$. In other words, the function (2.1.2) includes the cases where $g(x)$ or $h(y)$ (or both) is a constant function. So we can regard the right-hand side of (2.1.5) as the product of the constant function $g(x) \equiv 1$ and the function $h(y) = y$. With this viewpoint, (2.1.5) is a separable equation—now to solve it. But there is not much to do. We already know from studying the natural exponential function in calculus that $y = e^x$ is a solution. In fact, any function of the form $y = Ce^x$, where C is a constant, is a solution. By that we mean that if we imagine assigning C every possible real numerical value, we will generate all real-valued solutions of (2.1.5). We express this by stating that $y = Ce^x$ is the ***general solution*** of (2.1.5). ♦

Obviously, we would like to come up with a method that will handle any separable differential equation. Let us see if we can come up with something by playing around with (2.1.5). Perhaps expressing it in another form might somehow lead to its general solution, which we know to be $y = Ce^x$. We could try dividing both sides of the equation by y and see where that takes us. By doing this, we get

$$\frac{1}{y} \cdot \frac{dy}{dx} = 1. \tag{2.1.6}$$

Assume $y \neq 0$ so that this step is legitimate. By the chain rule of differentiation, we can replace the left-hand side by the derivative of the (natural) logarithmic function; so (2.1.6) becomes

$$\frac{d}{dx} \ln|y| = 1.$$

Note that the use of the absolute value bars makes this equation equivalent to (2.1.5) on intervals where $y < 0$ as well as where $y > 0$. Why? Recall that on intervals where $y < 0$ that

$$\frac{d}{dx} \ln|y| = \frac{d}{dx} \ln(-y) = \frac{1}{-y} \cdot \frac{d}{dx}(-y) = \frac{1}{y} \cdot \frac{dy}{dx}.$$

Since the derivative of x is also 1, it follows that $\ln|y|$ and x differ at most by a constant K. Hence,

$$\ln|y| = x + K. \tag{2.1.7}$$

Since the exponential function is the inverse of the logarithmic function, we can solve for $|y|$ by taking the exponential of both sides of (2.1.7) as follows:

$$e^{\ln|y|} = e^{x+K} = e^x \cdot e^K.$$

Note that we have also used one of the laws of exponents. So now we have

$$|y| = e^K \cdot e^x.$$

This implies that

$$y = \pm e^K \cdot e^x.$$

Observe that any positive number can be obtained from e^K by assigning K an appropriate value. Likewise, any negative number can be obtained from its additive inverse, namely, $-e^K$. For the sake of brevity, let us replace $\pm e^K$ with a single letter, such as C. That is, let

$$C := \pm e^K.$$

So C represents all positive and negative real numbers—just not zero since e^K is never zero, regardless of the value of K. Up to this point, we have found that the family of exponential functions

$$y = Ce^x$$

are solutions of (2.1.5), where C is any nonzero real number.

The preceding work precludes C from having the value 0. But let us reconsider. At the outset of this example, we assumed that $y \neq 0$ so that we could legitimately divide both sides of (2.1.5) by y. But just as we may lose solutions of algebraic equations when dividing by quantities involving an unknown variable, solutions of differential equations may also be lost if division is involved. For example, the equation

$$z(z-1) = 2(z-1)$$

has two solutions: $z = 1$ and $z = 2$. But, if we foolishly divide both sides of this equation by $z-1$, we end up with $z = 2$, thereby losing the solution $z = 1$. Likewise, the same can happen with differential equations if we are not careful. It is obvious that the constant function $y \equiv 0$ (that is, $y = 0$ for all x) is a solution of (2.1.5). However, when we divided (2.1.5) by y, we lost this solution. The upshot of this is that we may remove the restriction that C is nonzero. Thus, let C represent all values. In summary, all solutions of (2.1.5) are of the form

$$y = Ce^x, \tag{2.1.8}$$

where C denotes any real number. Or, we may express this by saying that this is a general solution of (2.1.5).

Let us revisit the method that we used to find the general solution of (2.1.5). Note that we could also obtain (2.1.7) by integrating both sides of (2.1.6) with respect to x:

$$\int \left(\frac{1}{y} \cdot \frac{dy}{dx} \right) dx = \int 1\, dx.$$

Since $\frac{d}{dx} \ln |y| = \frac{1}{y} \cdot \frac{dy}{dx}$, the left-hand side is

$$\int \left(\frac{1}{y} \cdot \frac{dy}{dx} \right) dx = \ln |y| + K_1.$$

And as $\int dx = x + K_2$, we have

$$\ln |y| + K_1 = x + K_2 \quad \text{or} \quad \ln |y| = x + (K_2 - K_1). \tag{2.1.9}$$

Since K_1 and K_2 are arbitrary constants, we can replace $K_2 - K_1$ by another arbitrary constant K obtaining

$$\ln |y| = x + K,$$

which is (2.1.7) again. As before, we can solve for y by taking the exponential of both sides of this equation. ♦

The previous example demonstrates that it is tedious and an unnecessary chore to write down all of the constants of integration since they end up being combined and replaced by a single constant anyway. So in the interest of efficiency, we will invoke the following convention whenever indefinite integrals are involved in solving differential equations.

Indefinite Integral Convention (When Solving ODEs)
The indefinite integral $\int g(x)\, dx$ denotes any (single) antiderivative of $g(x)$. In other words, it is any function $G(x)$ with the derivative

$$\frac{d}{dx} G(x) = g(x).$$

Let us clarify this with an example. If $g(x) = 2x$, then with this convention

$$\int 2x\, dx = x^2 \quad \text{because} \quad \frac{d}{dx} x^2 = 2x.$$

On the other hand,

$$\int 2x\,dx = x^2 + 10, \quad \int 2x\,dx = x^2 - 5\tfrac{2}{17}, \quad \int 2x\,dx = x^2 + 0.15\pi,$$

and so on are also correct. At first this convention appears to contravene the definition of the indefinite integral that is instilled in calculus students, namely, that the symbol

$$\int g(x)\,dx$$

represents the set of all antiderivatives of $g(x)$, which is what is meant by writing

$$\int 2x\,dx = x^2 + C.$$

However, if the above convention is adopted to solve differential equations, nothing really changes as we just saw in the previous example: we will always achieve the same result without having to deal with several constants of integration that we would end up combining anyway. As a matter of fact, computer algebra systems, such as *Maple* and those found on fancy calculators, employ this convention.

This convention will expedite the process of obtaining solutions. Let us explain by taking a look at Example 2.1.1 one more time. Recall that the general solution $y = Ce^x$ was obtained by exponentiating both sides of Eq. (2.1.7), to wit:

$$\ln|y| = x + K.$$

So once we have this equation, differential equation (2.1.5) can be considered solved. Let us summarize the steps that are needed to obtain (2.1.7). First, separate the y's (y and dy) in the differential equation from the x's (here just dx) by algebraically manipulating the equation so that the y's are on the left-hand side of the equation and dx is on the right-hand side, thereby ending up with

$$\frac{1}{y}\,dy = dx.$$

Second, integrate each side with respect to its variable, which is indicated by writing

$$\int \frac{1}{y}\,dy = \int dx + K. \tag{2.1.10}$$

Since we are abiding by the indefinite integral convention that we just established, we have to include the constant K, which represents combining the two constants of integration. Finally, replace each indefinite integral by its simplest antiderivative. The result is (2.1.7). We explore this procedure further in the next example.

Example 2.1.2 Solve

$$\frac{dy}{dx} = \frac{4\cos(2x)}{y+1}. \tag{2.1.11}$$

Solution First, separate the variables:

$$(y+1)\,dy = 4\cos(2x)\,dx.$$

Next, indicate that integrations are to take place by writing

$$\int (y+1)\,dy = \int 4\cos(2x)\,dx + K.$$

Then carry them out:

$$\frac{y^2}{2} + y = 2\sin(2x) + K. \tag{2.1.12}$$

Now note that we can express y explicitly in terms of x by multiplying both sides of the equation by 2 and completing the square as follows:

$$y^2 + 2y + 1 = 4\sin(2x) + 2K + 1 \quad \text{or} \quad (y+1)^2 = 4\sin(2x) + C,$$

where we have replaced $2K + 1$ by C. Solving for y, we obtain the solutions

$$y = -1 \pm \sqrt{4\sin(2x) + C}. \tag{2.1.13}$$

That both of these functions are truly solutions can be verified by substituting them and their derivatives back into (2.1.11). ♦

We managed to find solutions of the differential equations in the previous examples by separating the x's from the y's. But does this always work? Does it always yield the solutions of separable equations? In the next section, we will justify the technique of separating variables and delineate the steps of a procedure for solving separable equations. This procedure is known as the ***method of separation of variables***.

2.2 Separation of Variables

The purpose of this section is to formulate a general method that we can employ to find all solutions of a separable differential equation, namely, an equation that can be written in the form

$$\frac{dy}{dx} = g(x)\,h(y). \tag{2.2.1}$$

Once we have come up with a method, we will show how to use it by taking a look at several examples.

The steps that we used to solve Eqs. (2.1.5) and (2.1.11) will guide us in finding a general method. But before we start, observe that (2.2.1) may have constant solutions. If there is a real number c with $h(c) = 0$, then it follows that $y = c$ is a solution of (2.2.1) on the interval $(-\infty, \infty)$. For example, the equation

$$\frac{dy}{dx} = xy(y-1)$$

has the constant solutions $y \equiv 0$ and $y \equiv 1$. That is, the constant functions $y(x) = 0$ and $y(x) = 1$ satisfy this equation for all $x \in (-\infty, \infty)$

Now that we realize (2.2.1) may have constant solutions, let us focus on finding nonconstant solutions of this equation. To that end, suppose $y = y(x)$ is a nonconstant solution of (2.2.1) on some interval I, where $h(y(x)) \neq 0$ for all $x \in I$. So, for this interval, we can rewrite the equation as

$$\frac{1}{h(y)} \cdot \frac{dy}{dx} = g(x). \tag{2.2.2}$$

Define $r(y)$ by[1]

$$r(y) := \frac{1}{h(y)}.$$

Then (2.2.2) becomes

$$r(y) \cdot \frac{dy}{dx} = g(x). \tag{2.2.3}$$

Let $G(x)$ be an antiderivative of $g(x)$; that is, G is any function whose derivative is g. This is expressed by writing $G'(x) = g(x)$. Alternatively, because of the indefinite integral convention introduced in Sect. 2.1, we can also write

$$G(x) = \int g(x)\,dx.$$

Similarly, let

$$R(y) = \int r(y)\,dy.$$

[1] The symbol " :=" means that what is on its left side "is defined by" what is on its right side. Thus, $r(y) := 1/h(y)$ means that $r(y)$ is defined by the expression $1/h(y)$. The computer algebra system *Maple* also uses := in this sense.

That is, $R(y)$ denotes any function such that $R'(y) = r(y)$. Thus, we can rewrite (2.2.3) as

$$\frac{d}{dy}R(y) \cdot \frac{dy}{dx} = \frac{d}{dx}G(x). \tag{2.2.4}$$

At this point, the dependent variable y is unknown; but the goal is to replace it with a function $y(x)$ that will satisfy the differential equation. Keep in mind that y denotes a function of x, even though we may not always write $y(x)$. The differential "dy" on the left-hand side of (2.2.4), appearing in both derivatives of the product, in the denominator of the first derivative and in the numerator of the second one, is a clarion call to use the chain rule. So let us replace the left-hand side of (2.2.4) by the derivative of the composite function $R[y(x)]$ with respect to x. As a result, (2.2.4) becomes

$$\frac{d}{dx}R[y(x)] = \frac{d}{dx}G(x).$$

Since $R[y(x)]$ and $G(x)$ have the same derivative, they differ at most by a constant K. In other words, $R[y(x)] = G(x) + K$. The constant K may be equal to zero, but most likely not; it depends on the initial condition. We have now arrived at the point where we have eliminated all derivatives; in other words, any solution of the differential equation must satisfy the equation

$$R(y) = G(x) + K, \tag{2.2.5}$$

Expressed in terms of integrals, (2.2.5) is

$$\int r(y)\,dy = \int g(x)\,dx + K. \tag{2.2.6}$$

Unless $r(y) \equiv 1$, Eq. (2.2.6) will not be of the form $y = \varphi(x)$. So, generally speaking, it is not a function. Rather, it is a ***relation*** between the variables x and y, that is, an equation that relates x and y (e.g., $x^2 + y^2 = 4$), which is different from saying that y is a function of x (e.g., $y = -\sqrt{4 - x^2}$). We have already seen examples of this with (2.1.7) and (2.1.12). As a rule of thumb, we will solve for the dependent variable y whenever it is possible and reasonable to do so—as we were able to do with both (2.1.7) and (2.1.12) in obtaining the respective solutions (2.1.8) and (2.1.13) of the differential equations (2.1.5) and (2.1.11). These solutions are examples of explicit solutions. An ***explicit solution*** of an ordinary differential equation in x is a function $y = \varphi(x)$ that satisfies the equation on some interval.

It turns out that for many differential equations, the mathematics involved in trying to solve (2.2.6) for y may be too difficult—or simply not possible. The result of separating the variables in Eq. (2.1.4) and integrating will attest to this as we shall see later on in Example 2.2.4 (see (2.2.19)). For such equations, (2.2.6) is as far as we can go in obtaining an expression relating the independent variable x

and the dependent variable y. If, for a given value of K, Eq. (2.2.6) defines one or more explicit solutions on some interval I, then it is said to be an ***implicit solution*** of (2.2.1) on I. Finally, let us just remark that if finding an explicit solution of a differential equation is not feasible or possible, then one can resort to numerical methods.[2]

A summary of the steps that we used to obtain (2.2.6) yields the following method for solving separable equations.

Method of Separation of Variables

To find all solutions of $\dfrac{dy}{dx} = g(x)h(y)$, execute the following steps:

1. Find all constant solutions (if any) either by inspection of the differential equation or by solving the equation $h(y) = 0$.
2. Separate the variables of the differential equation. Multiply each side of the equation by dx and $r(y) := 1/h(y)$ to express it in the form

$$r(y)\,dy = g(x)\,dx.$$

3. Carry out the integrations

$$\int r(y)\,dy = \int g(x)\,dx + K$$

to obtain a relation defining the set of implicit solutions.
4. Whenever feasible, solve the relation for y in order to express the solutions solely in terms of the independent variable x.
5. If possible, incorporate the constant solutions (if any) into the set of nonconstant solutions.

We will illustrate this method with several examples. Example 2.2.3 will elucidate the final step of the method where a couple of constant solutions are incorporated into the set of nonconstant solutions of the differential equation in that example.

Example 2.2.1 Solve the differential equation

$$x^3 y' - 2(y-1)^2 = 0 \tag{2.2.7}$$

and determine the intervals of existence of the solutions.

[2] See Chap. 4 for an introduction to numerical methods.

Solution By inspection, this equation has the constant solution $y \equiv 1$ on $(-\infty, \infty)$. That is to say, a mere glance at (2.2.7) reveals that the constant function $y(x) = 1$ is a solution for all values of x.

Now let us find nonconstant solutions of (2.2.7). Separating variables, we have

$$\frac{dy}{(y-1)^2} = \frac{2}{x^3}\,dx.$$

Then carrying out the integrations

$$\int (y-1)^{-2}\,dy = \int 2x^{-3}\,dx + K,$$

we obtain

$$-\frac{1}{y-1} = -\frac{1}{x^2} + K$$

or

$$\frac{1}{y-1} = \frac{1}{x^2} - K = \frac{1-Kx^2}{x^2}.$$

Solving for y, we find that

$$y = 1 + \frac{x^2}{1-Kx^2}. \tag{2.2.8}$$

For a given value of the constant K, the function (2.2.8) is a solution of (2.2.7) on any interval that excludes points where $1 - Kx^2 = 0$, unless we have erred somewhere in the preceding computations. In order to be one hundred percent sure, let us verify that (2.2.8) is a solution. At points where $1 - Kx^2 \neq 0$,

$$y' = \frac{d}{dx}\left[1 + \frac{x^2}{1-Kx^2}\right] = \frac{(1-Kx^2)\dfrac{d}{dx}x^2 - x^2\dfrac{d}{dx}(1-Kx^2)}{(1-Kx^2)^2}$$

$$= \frac{2x}{(1-Kx^2)^2}. \tag{2.2.9}$$

Substituting (2.2.8) and (2.2.9) in the left-hand side of (2.2.7), we have

$$x^3y' - 2(y-1)^2 = x^3\left[\frac{2x}{(1-Kx^2)^2}\right] - 2\left(\frac{x^2}{1-Kx^2}\right)^2 = 0.$$

This confirms that the function (2.2.8) is a solution on any interval that excludes points where $1 - Kx^2 = 0$.

Let us elaborate on what we are saying about intervals by examining how the value of K affects the interval of existence of (2.2.8). If $K \le 0$, then $1 - Kx^2 \ge 1$ for all values of x; and so it is never zero. Consequently, (2.2.8) is a solution of (2.2.7) on the interval $(-\infty, \infty)$—or, of course, on any subinterval of $(-\infty, \infty)$. In particular, if we take $K = 0$, then we obtain the solution

$$y = 1 + x^2.$$

Taking $K = -1$, we have the solution

$$y = 1 + \frac{x^2}{1 + x^2}.$$

Both of these functions are solutions on $(-\infty, \infty)$. The graph of the latter one is the continuous curve in Fig. 2.1 that approaches the horizontal line $y = 2$ as $x \to \pm\infty$.

If $K > 0$, then (2.2.8) is continuous except where $1 - Kx^2 = 0$, which occurs at the points $x = \pm 1/\sqrt{K}$. Consequently, we can say that (2.2.8) is a solution on any interval that excludes these two points. For example, if $K = 1$, then (2.2.8) becomes

$$y = 1 + \frac{x^2}{1 - x^2} = \frac{1}{1 - x^2}. \tag{2.2.10}$$

This particular function is a solution of the differential equation (2.2.7) on any interval that excludes the discontinuities at $x = -1$ and $x = 1$. Its graph on the interval $(-\infty, -1)$ is shown in Fig. 2.1; it is the continuous curve in Quadrant III. The continuous curve in Quadrant IV is the graph of (2.2.10) on $(1, \infty)$. And the

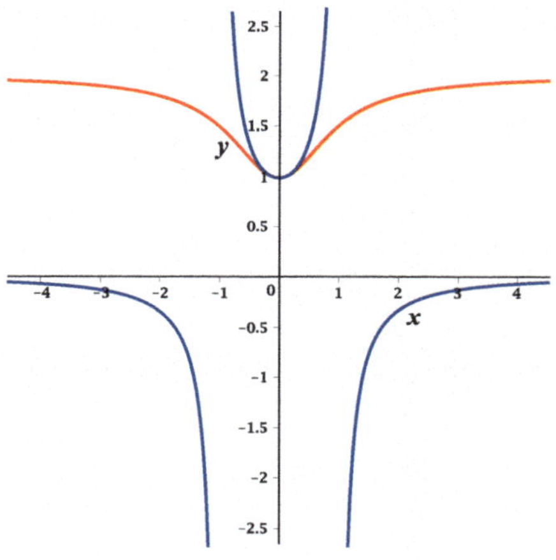

Fig. 2.1 Graphs of some solutions of $x^3y' - 2(y-1)^2 = 0$

continuous curve situated between the vertical lines $x = -1$ and $x = 1$ is the graph of the solution (2.2.10) on the interval $(-1, 1)$.

Finally, note that no value can be assigned to the constant K that will yield $y = 1$. So the constant solution $y \equiv 1$ cannot be incorporated into (2.2.8). In sum, all solutions of (2.2.7) are functions of the form (2.2.8) and the constant solution $y \equiv 1$. ♦

Example 2.2.2 Find the solutions of the equation

$$xyy' = 2\ln x \tag{2.2.11}$$

when $x > 0$.

Solution Clearly this equation has no constant solutions. In other words, there is no function of the form $y = c$ satisfying it, regardless of the value of the constant c. So let us move on to find nonconstant solutions. Separating variables, we have

$$y\,dy = 2\,\frac{\ln x}{x}\,dx.$$

Then integrate both sides, which we indicate by writing

$$\int y\,dy = \int 2\,\frac{\ln x}{x}\,dx + K. \tag{2.2.12}$$

The integral on the right-hand side is easily evaluated by letting $u = \ln x$. As the differential of this is $du = x^{-1}\,dx$, we have

$$\frac{y^2}{2} = \int 2u\,du + K \quad \Rightarrow \quad y^2 = 2u^2 + 2K \quad \Rightarrow \quad y^2 = 2(\ln x)^2 + 2K.$$

Since K represents any real number, so does $2K$. It is customary to replace the product of a number and an arbitrary constant with a single letter, such as $C := 2K$. As a result, we now have

$$y^2 = 2(\ln x)^2 + C.$$

Thus,

$$y = \pm\sqrt{2(\ln x)^2 + C} \tag{2.2.13}$$

are solutions. In other words, for every value of C, the functions

$$y = \sqrt{2(\ln x)^2 + C} \quad \text{and} \quad y = -\sqrt{2(\ln x)^2 + C} \tag{2.2.14}$$

satisfy the differential equation (2.2.11). It is left as an exercise to verify this. ♦

Example 2.2.3 Solve the differential equation

$$(xy - y)y' = y^2 - 3. \tag{2.2.15}$$

Solution By inspection, $y = \sqrt{3}$ and $y = -\sqrt{3}$ are constant solutions of (2.2.15) on the interval $(-\infty, \infty)$. With the coefficient of y' factored, we see that

$$y(x-1)\frac{dy}{dx} = y^2 - 3$$

is a separable equation. Proceeding with separating variables and carrying out the integrations

$$\int \frac{y}{y^2-3}\,dy = \int \frac{1}{x-1}\,dx + C_1,$$

we have

$$\frac{1}{2}\ln|y^2-3| = \ln|x-1| + C_1.$$

After multiplying both sides by 2 and replacing $2C_1$ with C_2, we then have

$$\ln|y^2-3| = \ln(x-1)^2 + C_2.$$

With solving for y the objective, let us take the exponential of each side obtaining

$$|y^2-3| = e^{C_2}\cdot e^{\ln(x-1)^2} = e^{C_2}(x-1)^2.$$

From this it follows that

$$y^2 - 3 = \pm e^{C_2}(x-1)^2.$$

For reasons of economy, replace $\pm\, e^{C_2}$ with the arbitrary constant C. Since $\pm\, e^{C_2}$ is never equal to zero irrespective of the value of C_2, the constant C can assume any real value except for $C = 0$. So now we have

$$y^2 = 3 + C(x-1)^2.$$

where $C \neq 0$. Solving for y, we obtain the solutions

$$y = \pm\sqrt{3 + C(x-1)^2} \qquad (C \neq 0). \tag{2.2.16}$$

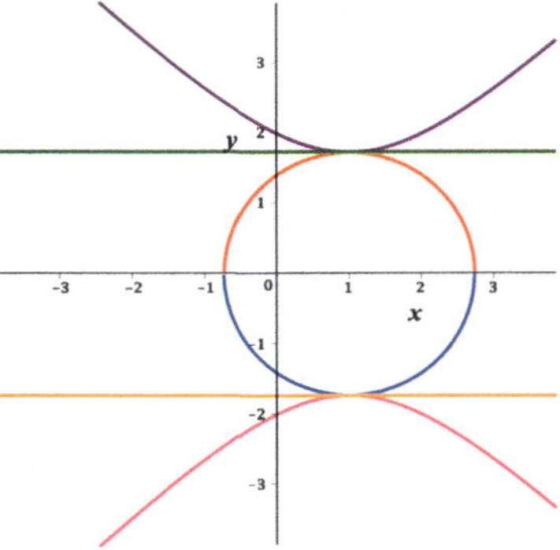

Fig. 2.2 Graphs of six solutions of $(xy - y)y' = y^2 - 3$

Note that by lifting the restriction $C \neq 0$, the constant solutions $y = \pm\sqrt{3}$ are also included in (2.2.16). Consequently, the family of functions

$$y = \sqrt{3 + C(x-1)^2} \quad \text{and} \quad y = -\sqrt{3 + C(x-1)^2}, \tag{2.2.17}$$

where C is allowed to have any value, represent all solutions of the differential equation (2.2.15). The graphs of six of these solutions corresponding to $C = -1$, $C = 0$, and $C = 1$ are shown in Fig. 2.2. The graphs of the two solutions corresponding to $C = -1$ are semicircles since the graph of

$$(x-1)^2 + y^2 = 3$$

is a circle of radius $\sqrt{3}$ whose center is at $(1, 0)$. The graphs of the two solutions corresponding to $C = 0$ are the horizontal lines $y = \pm\sqrt{3}$. The graphs of the two solutions corresponding to $C = 1$ are the two branches of the hyperbola

$$y^2 - (x-1)^2 = 3$$

shown in Fig. 2.2. ♦

Example 2.2.4 Solve

$$\frac{dy}{dx} = \frac{e^x \ln x^2}{xe^x + xye^{x+y}}. \tag{2.2.18}$$

Solution As we explained earlier in Sect. 2.1, this equation is separable since it can be rewritten as

$$\frac{dy}{dx} = \frac{\ln x^2}{x} \cdot \frac{1}{1 + ye^y}$$

if $x \neq 0$. Separating variables and replacing x^2 with $|x|^2$ and then integrating, we have

$$\int (1 + ye^y)\, dy = \int \frac{\ln |x|^2}{x}\, dx + C.$$

To carry out the integration on the right-hand side of the equation, use the change of variable $z = \ln |x|$. Then

$$\int \frac{\ln |x|^2}{x}\, dx = \int \frac{2\ln |x|}{x}\, dx = \int 2z\, dz = z^2 = (\ln |x|)^2.$$

As for $\int ye^y\, dy$, integrate by parts with $u = y$ and $dv = e^y\, dy$:

$$\int ye^y\, dy = uv - \int v\, du = ye^y - \int e^y\, dy = ye^y - e^y.$$

The result of carrying out these integrations is

$$y + ye^y - e^y - (\ln |x|)^2 = C. \tag{2.2.19}$$

Before continuing, let us make sure that we have not made any mistakes in algebra or integration in deriving (2.2.19). Rather than retracing our steps and possibly not catching a mistake (if there is one), let us differentiate each side of (2.2.19) implicitly with respect to x. Bear in mind we are presupposing that (2.2.19) defines y as a function of x (more on this below). Remembering to use the chain rule, we have

$$\frac{dy}{dx} + \underbrace{y \cdot e^y \frac{dy}{dx} + e^y \frac{dy}{dx}}_{\frac{d}{dx}(ye^y)} - e^y \frac{dy}{dx} - 2(\ln |x|) \cdot \frac{1}{x} = \frac{d}{dx} C$$

or

$$y'(1 + ye^y) - 2\frac{\ln |x|}{x} = 0.$$

Solving for y', we get

$$y' = \frac{2\ln |x|}{x\,(1 + ye^y)} = \frac{e^x \cdot \ln x^2}{e^x \cdot x\,(1 + ye^y)} = \frac{e^x \ln x^2}{xe^x + xye^{x+y}},$$

which is Eq. (2.2.18). Therefore, (2.2.19) is correct.

Now let us continue from where we left off. We would like to solve Eq. (2.2.19) for y. But alas, that is not possible. This is as far as we can go. So now we have to ask if (2.2.19) for a given value of C defines at least one differentiable function that would satisfy the differential equation (2.2.18) on some interval. To be more specific, suppose we would like to know if there is such a function whose graph contains the point $(1, 0)$. Setting $x = 1$ and $y = 0$ in (2.2.19), we find that $C = -1$. As a result of this and as the question concerns the existence of an explicit solution on the positive x-axis (recall $x \neq 0$), Eq. (2.2.19) becomes

$$y + ye^y - e^y - (\ln x)^2 + 1 = 0. \tag{2.2.20}$$

As it turns out, the left-hand side of (2.2.20) satisfies the hypotheses of a classical theorem in mathematical analysis called the ***implicit function theorem***.[3] For this situation, it guarantees that for some $\alpha > 0$ a unique continuous function $\varphi(x)$ exists that satisfies Eq. (2.2.20) on the interval $(1-\alpha, 1+\alpha)$ and such that $\varphi(1) = 0$. (Note $\alpha < 1$ as (2.2.20) is not defined for $x \leq 0$.) Furthermore, $\varphi(x)$ also satisfies the differential equation (2.2.18) on this interval. As a result, we can say that (2.2.20) is an implicit solution of (2.2.18). From now on, we will not use the implicit function theorem since it is an advanced calculus topic. Wherever implicit differentiation is called for, the reader is to assume that this way of computing the derivative can be justified. ♦

Before moving on to the next section, let us say something about checking one's work. If implicit differentiation had not yielded (2.2.18), it would mean that we had erred somewhere. In that case, we would have to backtrack and check the correctness of each step—or start again from scratch. Checking a solution and considering whether or not it is reasonable is also part of the solution process. Anyone can make a mistake. What is unacceptable, and sometimes tragic,[4] is to allow a claim or statement to go unchallenged, especially if there is a way of checking it, such as in the previous example where all it took was differentiation and some algebra. Yes, this requires more effort and additional time that seems to keep you from moving on to other things. In the same vein, consider the following quotation from C. S. Lewis:[5]

> We all want progress. But progress means getting nearer to the place where you want to be. And if you have taken a wrong turning, then to go forward does not get you any nearer. If you are on the wrong road, progress means doing an about-turn and walking back to the right road; and in that case the man who turns back soonest is the most progressive man.

---

[3] A version of this theorem for a function of two variables along with a proof is given in an advanced calculus textbook by Friedman [36, p. 236]. The theorem is also stated in a differential equations textbook by Driver [30, p. 458] but with no proof.

[4] In 1981, engineering and architectural errors caused the death of 114 people and injury to more than 200 others when two walkways in the atrium of a Hyatt Regency Hotel in Kansas City, Missouri collapsed.

[5] This quotation is taken from the chapter "We Have Cause to Be Uneasy" in *Mere Christianity*, published by Macmillan, New York, 1952. C. S. Lewis (1898–1963) was a professor of Medieval and Renaissance literature at Cambridge University. Some of his other well-known works are *The Screwtape Letters*, *Space Trilogy*, and *The Lion, the Witch, and the Wardrobe*.

We have all seen this when doing arithmetic. When I have started a sum the wrong way, the sooner I admit this and go back and start over again, the faster I shall get on. There is nothing progressive about being pigheaded and refusing to admit a mistake.

2.3 Initial Value Problems

The objective in this section is to explain how to go about finding the solution of a separable differential equation so that it satisfies a given initial condition. An ***initial condition*** stipulates the value a solution is to have at a specified value of the independent variable. For example, if x denotes the independent variable and $y(x)$ a solution, then an initial condition is indicated by writing $y(x_0) = y_0$, where x_0 is a real number belonging to the domain of the solution and y_0 is any real number. This notation means that y_0 is the value of the solution $y(x)$ at $x = x_0$. A differential equation together with an initial condition is known as an ***initial value problem***, which is abbreviated ***IVP***. To illustrate this, let us take a look at Eq. (2.2.11) again but this time accompanied by an initial condition.

Example 2.3.1 Solve the initial value problem

$$xyy' = 2\ln x, \quad y(1) = -2\,. \tag{2.3.1}$$

Solution In Example 2.2.2, we determined that

$$y = \pm\sqrt{2(\ln x)^2 + C}.$$

So all that remains is to determine the value for C so that the initial condition is satisfied. Setting $x = 1$ and $y = -2$, we get

$$-2 = \pm\sqrt{2(\ln 1)^2 + C} = \pm\sqrt{C}.$$

Hence, $C = 4$ and

$$y = \pm\sqrt{2(\ln x)^2 + 4}\,.$$

So which is it: $y = -\sqrt{2(\ln x)^2 + 4}$ or $y = \sqrt{2(\ln x)^2 + 4}$? Obviously the latter does not satisfy the initial condition. Therefore,

$$y = -\sqrt{2(\ln x)^2 + 4}$$

is the solution of (2.3.1). ♦

2.3.1 Solutions

Example 2.3.1 illustrates how to solve an initial value problem involving a separable differential equation using indefinite integrals. Definite integration is an alternative and is better suited for certain problems as we will see in Example 2.3.4. Before showing any examples, let us explain the procedure. Consider the initial value problem

$$\frac{dy}{dx} = g(x)h(y), \quad y(x_0) = y_0 \tag{2.3.2}$$

where x_0 and y_0 denote given numbers. The goal is to find a differentiable function $y(x)$ that satisfies the differential equation on some interval and the initial condition.

First separate variables. With $r(y) := 1/h(y)$, we have

$$r(y)\,dy = g(x)\,dx.$$

If $y(x)$ is a solution of the differentiation equation, then

$$r(y(x))y'(x)\,dx = g(x)\,dx.$$

Now integrate both sides of this equation from $x = x_0$ to $x = x_1$. That is to say,

$$\int_{x_0}^{x_1} r(y(x))y'(x)\,dx = \int_{x_0}^{x_1} g(x)\,dx. \tag{2.3.3}$$

As in Sect. 2.2, let $R(y)$ denote any antiderivative of $r(y)$. By the Chain Rule,

$$\frac{d}{dx}R(y(x)) = R'(y(x))y'(x) = r(y(x))y'(x).$$

Consequently,

$$\int_{x_0}^{x_1} r(y(x))y'(x)\,dx = R(y(x))\Big|_{x_0}^{x_1} = R(y(x_1)) - R(y(x_0)).$$

Replacing the left-hand side of (2.3.3) with this, it becomes

$$R(y(x_1)) - R(y(x_0)) = \int_{x_0}^{x_1} g(x)\,dx.$$

Also, as $R'(y) = r(y)$,

$$\int_{y(x_0)}^{y(x_1)} r(y)\,dy = R(y)\Big|_{y(x_0)}^{y(x_1)} = R(y(x_1)) - R(y(x_0)).$$

Consequently,

$$\int_{y(x_0)}^{y(x_1)} r(y)\,dy = \int_{x_0}^{x_1} g(x)\,dx.$$

Now replace $y(x_0)$ with y_0. Also, replace x_1 with x and $y(x_1)$ with $y(x)$ or simply y. This necessitates using letters other than x and y for the variables of integration in order to avoid a conflict in names. For example, we could change both to u, in which case we have

$$\int_{y_0}^{y} r(u)\,du = \int_{x_0}^{x} g(u)\,du.$$

We conclude that if $y = y(x)$ is a solution of the initial value problem (2.3.2), then it can be found by separating variables and executing the following steps.

Separation of Variables (IVPs)

To find the solution of the separable equation $\dfrac{dy}{dx} = g(x)h(y)$ satisfying the initial condition $y(x_0) = y_0$, proceed as follows:

1. First determine if $y(x) \equiv y_0$ is the solution. If it is not a solution, then go on to the next step.
2. Multiply each side of the differential equation by dx and $r(y) := 1/h(y)$ to separate variables:

$$r(y)\,dy = g(x)\,dx.$$

3. Integrate as follows:

$$\int_{y_0}^{y} r(u)\,du = \int_{x_0}^{x} g(u)\,du.$$

4. If it is feasible, solve for the dependent variable y.

Example 2.3.2 We illustrate this method by revisiting the initial value problem (2.3.1).

Solution The first step is to determine if $y \equiv -2$ is the solution. Obviously it is not, so let us go on to the next step. Separating the variables of $xyy' = 2\ln x$, we get

$$y\,dy = \frac{2\ln x}{x}\,dx$$

as before. The third step is to integrate each side of the separated equation:

$$\int_{-2}^{y} u\,du = \int_{1}^{x} \frac{2\ln u}{u}\,du.$$

(The lower limits of integration come from the initial condition $y(1) = -2$.) To evaluate the integral on the right-hand side, let $s = \ln u$. When $u = 1$, $s = 0$. And when $u = x$, $s = \ln x$. Thus,

$$\left.\frac{u^2}{2}\right|_{-2}^{y} = \int_{0}^{\ln x} 2s\,ds \quad \Rightarrow \quad \frac{y^2}{2} - 2 = (\ln x)^2.$$

The last step is to solve for the dependent variable y:

$$y = \pm\sqrt{2(\ln x)^2 + 4}.$$

Since $y = -2$ when $x = 1$, the solution of the IVP is

$$y = -\sqrt{2(\ln x)^2 + 4},$$

Note that this is what we had obtained earlier in Example 2.3.1. ♦

An important differential equation that turns up in a number of applications is

$$\frac{dy}{dx} = ry, \tag{2.3.4}$$

where r is a parameter. It was pointed out in Chap. 1 that this equation is used to model the population of certain organisms reproducing under ideal circumstances—in this context, it is known as the *Malthusian law of population growth*. If we wish to be more descriptive with our choice of variables, we could write (2.3.4) as $\dot{p} = rp$. In another context, this equation models the decay of radioactive elements and isotopes. Both of these models and how they are derived will be discussed later on in Chap. 6. For now, we are merely interested in demonstrating how to use the definite integration version of the method of separation of variables to solve (2.3.4) together with the initial condition $y(x_0) = y_0$.

Example 2.3.3 Solve the initial value problem

$$\frac{dy}{dx} = ry, \quad y(x_0) = y_0, \tag{2.3.5}$$

where r, x_0, and y_0 denote given constants.

Solution Separating the variables, we have

$$\frac{dy}{y} = r\,dx.$$

Then carrying out the integrations

$$\int_{y_0}^{y} \frac{du}{u} = \int_{x_0}^{x} r\, du,$$

we get

$$\ln\left|\frac{y}{y_0}\right| = r(x - x_0).$$

Exponentiation yields

$$\left|\frac{y}{y_0}\right| = e^{r(x-x_0)}.$$

Thus,

$$\frac{y}{y_0} = \pm e^{r(x-x_0)}.$$

Note that $y = y_0 e^{r(x-x_0)}$ satisfies the initial condition, whereas $y = -y_0 e^{r(x-x_0)}$ does not. Consequently, the solution of the initial value problem is

$$y = y_0 e^{r(x-x_0)}. \tag{2.3.6}$$

From this we see that solutions of (2.3.5) grow exponentially when $r > 0$ and decay exponentially when $r < 0$, unless of course $C = 0$. Consequently, the differential equation $y' = ry$ is often referred to as the ***exponential growth equation*** when $r > 0$ and the ***exponential decay equation*** when $r < 0$. ♦

Example 2.3.4 Solve the initial value problem

$$\frac{dy}{dx} = \frac{e^{-x^2}}{2y}, \quad y(1) = -2.$$

Solution After separating variables and using the indefinite integration version of the method of separation of variables, we obtain

$$\int 2y\, dy = \int e^{-x^2}\, dx + C,$$

only to realize that we are now at an impasse due to the integral on the right-hand side. This is due to the integrand, $\exp(-x^2)$, which does not have an **elementary antiderivative**. That is, it does not have an antiderivative that is one of the familiar functions (or constant multiples thereof) of calculus or one that results from combining such functions a finite number of times using the operations of addition, subtraction, multiplication, division, raising to powers, extraction of roots, and the composition of functions. More details about elementary antiderivatives can be found in Chap. 5.

Even though the use of indefinite integrals fails here, we still have recourse to the definite integration approach. Integrating and incorporating the initial condition that $y = -2$ when $x = 1$, we have

$$\int_{-2}^{y} 2u\,du = \int_{1}^{x} e^{-u^2}\,du.$$

Thus,

$$y^2 - 4 = \int_{1}^{x} e^{-u^2}\,du.$$

Taking into account the initial condition, we conclude the solution is

$$y = -\sqrt{4 + \int_{1}^{x} e^{-u^2}\,du}.$$ ♦

2.3.2 *Applications*

2.3.2.1 Temperature of Coffee

Example 2.3.5 The temperature inside a coffee shop is kept at 72 °F. The shop serves hot beverages at temperatures ranging anywhere from 160 to 185 °F. Suppose a cup of freshly poured coffee has a temperature of 182 °F when it is set on a table, where it is left undisturbed. After 5 minutes, the temperature of the coffee is measured and found to be 159 °F. Use Newton's law of cooling to predict the temperature of the coffee after 10 minutes.

Solution Let $T(t)$ represent the temperature of the cup of coffee, in degrees Fahrenheit, t minutes after it has been set on the table. It is given that $T(0) = 182$ and $T(5) = 159$. Since we are to use Newton's law of cooling, the differential equation modeling the temperature (cf. Sect. 1.4.2) is

$$\frac{dT}{dt} = -k(T - 72)$$

where $k > 0$ is the constant of proportionality. Clearly this is a separable equation. Although the indefinite integration version of the method of separation of variables can be used to solve this initial value problem, we will once again illustrate how to obtain the solution with definite integrals. Separating the variables and integrating,

we have

$$\int_{182}^{T} \frac{ds}{s-72} = \int_0^t -k\,ds. \tag{2.3.7}$$

where the lower limit of integration 182 is the temperature of the coffee corresponding to the time $t = 0$ and the upper limit of integration T denotes the temperature corresponding to a later time t. With the change of variable $u = s - 72$ and the limits of integration adjusted accordingly, (2.3.7) becomes

$$\int_{110}^{T-72} \frac{du}{u} = -kt. \tag{2.3.8}$$

Thus,

$$\ln|T-72| - \ln 110 = -kt.$$

Since $T(t) > 72$ for all $t > 0$, we can dispense with the absolute bars. And so

$$\ln\left(\frac{T-72}{110}\right) = -kt. \tag{2.3.9}$$

The value of k can be found from (2.3.9) by setting $t = 5$ and $T = 159$:

$$k = -\frac{1}{5}\ln\left(\frac{87}{110}\right) = 0.0469144\ldots. \tag{2.3.10}$$

Exponentiating both sides of (2.3.9), then solving for T and using the approximation $k \approx 0.0469$, we predict that the temperature of the coffee after t minutes will be

$$T(t) = 72 + 110e^{-0.0469t}. \tag{2.3.11}$$

According to this model, the coffee's temperature after 10 minutes will be

$$T(10) = 72 + 110\,e^{-0.0469(10)} = 140.8\,°\text{F}.$$

Alternatively, from (2.3.10), we have

$$e^{-k} = e^{\frac{\ln(87/110)}{5}} = \left(\frac{87}{110}\right)^{1/5}.$$

Thus,

$$T(t) = 72 + 110e^{-kt} = 72 + 110(e^{-k})^t = 72 + 110\left[\left(\frac{87}{110}\right)^{1/5}\right]^t.$$

or

$$T(t) = 72 + 110\left(\frac{87}{110}\right)^{t/5}. \tag{2.3.12}$$

Once again we find that

$$T(10) = 72 + 110\left(\frac{87}{110}\right)^{2} = 140.8\,^{\circ}\text{F}.$$

♦

2.3.2.2 A Production Model from Economics

Example 2.3.6 A formula used by economists to model the total production of goods in an economic system is called the *Cobb-Douglas production function*. It also serves as the foundation for obtaining even better and more realist economic models. The basic idea is that production depends only on two variables: the amounts of labor and capital investment. *Total production*, which we denote by P, is the monetary value of all goods that are produced in the economic system during the course of a given year. For this particular year, let L represent the amount of *labor*, namely, the total number of man-hours worked. Let K denote the amount of *capital investment*, which is the total monetary value of all accumulated goods (materials, machinery, buildings, etc.) that are used for the production of new goods. We express that a functional relationship exists (at this point unknown) between the variables by writing $P = f(L, K)$. The goal is to come up with reasonable assumptions that will enable us to determine a function f that can serve as a good model of production.

Economists call the partial derivative $\partial P/\partial L$ the **marginal productivity of labor**. This is the rate at which production changes with respect to the amount of labor when capital investment does not change. The partial derivative $\partial P/\partial K$ is called the **marginal productivity of capital**. Likewise, it is the rate at which production changes with respect to the amount of capital investment when the amount of labor is kept at some fixed level.

As stated above, the total production P in a year is viewed as some function of the total amounts of labor L and capital investment K. In order to find such a function, call it f, let us assume the following:

(i) Marginal productivity of labor is proportional to the amount of production per unit labor. That is,

$$\frac{\partial P}{\partial L} = \alpha \frac{P}{L}, \tag{2.3.13}$$

where $\alpha > 0$ is the constant of proportionality.

(ii) Marginal productivity of capital is proportional to the amount of production per unit capital. That is,

$$\frac{\partial P}{\partial K} = \beta \frac{P}{K}. \tag{2.3.14}$$

where $\beta > 0$ is the constant of proportionality.

(iii) If the total amounts of labor and capital investment both increase by a factor of λ, then total production will also increase by a factor of λ.

Let us find the functional relationship $P = f(L, K)$ that is implied by the foregoing assumptions.

Solution For a fixed value of K, (2.3.13) becomes the ordinary differential equation

$$\frac{dP}{dL} = \alpha \frac{P}{L}. \tag{2.3.15}$$

Separating variables and integrating, we obtain

$$\int \frac{dP}{P} = \alpha \int \frac{dL}{L} \quad \Rightarrow \quad \ln P = \alpha \ln L + C_1(K).$$

Note that C_1, the so-called constant of integration, depends on the value of K. Exponentiation of both sides yields

$$P = e^{C_1(K)} \cdot e^{\alpha \ln L}$$

which simplifies to

$$P = g(K)L^{\alpha}, \tag{2.3.16}$$

where g denotes some function of K. From this we have,

$$\frac{\partial P}{\partial K} = g'(K)L^{\alpha}. \tag{2.3.17}$$

Then it follows from (2.3.14), (2.3.16), and (2.3.17) that

$$g'(K)L^{\alpha} = \frac{\beta}{K} g(K)L^{\alpha}.$$

Simplifying and separating variables, we obtain

$$\frac{g'(K)}{g(K)} = \frac{\beta}{K}$$

or

$$\frac{d}{dK} \ln g(K) = \frac{\beta}{K}.$$

An integration with respect to K yields

$$\ln g(K) = \beta \ln K + C_2.$$

Exponentiating and replacing e^{C_2} with b, we get

$$g(K) = bK^\beta.$$

It follows from this and (2.3.16) that

$$P(L, K) = bL^\alpha K^\beta. \tag{2.3.18}$$

According to this, if the amount of labor and capital investment increases by a factor of λ, then the total production is

$$P(\lambda L, \lambda K) = b(\lambda L)^\alpha (\lambda K)^\beta = b\lambda^{\alpha+\beta} L^\alpha K^\beta = \lambda^{\alpha+\beta} P(L, K).$$

On the other hand, according to (iii),

$$P(\lambda L, \lambda K) = \lambda P(L, K).$$

Comparing both, we conclude that $\alpha + \beta = 1$. (This implies $\alpha < 1$ since α and β are assumed to be positive.) As a result, (2.3.18) becomes

$$P(L, K) = bL^\alpha K^{1-\alpha}, \tag{2.3.19}$$

where b and α are positive constants and $\alpha < 1$. This function is known as the **Cobb-Douglas production function**.

The Cobb-Douglas production function appeared in a paper entitled "A Theory of Production" co-authored by Charles W. Cobb[6] and Paul H. Douglas,[7] which was published in 1928 in a prestigious economics journal called the *American Economic Review*. The paper can be found at the website: assets.aeaweb.org/asset-server/journals/aer/top20/18.1.139-165.pdf. ♦

---

[6] Cobb was a mathematician at Amherst College from 1908 to 1948.

[7] Douglas was an economist at the University of Chicago and later served as a U.S. Senator from Illinois from 1949 to 1967.

2.4 Homogeneous Equations

Homogeneous Equation

Definition 2.4.1 A first-order differential equation

$$\frac{dy}{dx} = f(x, y) \tag{2.4.1}$$

is said to be ***homogeneous*** if the right-hand side can be expressed as a function of the ratio y/x alone, in which case the equation becomes

$$\frac{dy}{dx} = F\left(\frac{y}{x}\right) \tag{2.4.2}$$

where F, as indicated, denotes a function of y/x.

Example 2.4.1 The differential equation

$$\frac{dy}{dx} = \frac{x^2 + y^2}{xy}$$

is homogeneous since its right-hand side can be expressed as a function of the ratio y/x as follows:

$$\frac{x^2 + y^2}{xy} = \frac{x}{y} + \frac{y}{x} = \left(\frac{y}{x}\right)^{-1} + \frac{y}{x}.$$

Now the differential equation looks exactly like (2.4.2), where

$$F\left(\frac{y}{x}\right) = \left(\frac{y}{x}\right)^{-1} + \frac{y}{x}.$$ ♦

Remark In the context of first-order differential equations, the word "homogeneous" also has another meaning. The term "homogeneous differential equation" may refer to the type of equation described above in Definition 2.4.1 or it may refer to an equation of the form

$$a_1(x)\frac{dy}{dx} + a_0(x)y = 0, \tag{2.4.3}$$

which is discussed in Chap. 5. To distinguish between these two types of equations, we will refer to (2.4.3) as a *first-order homogeneous linear equation*—with "linear" being the operative word.

What is important to remember about the homogeneous equation (2.4.2) is that it can be transformed into a separable equation by letting

$$z := \frac{y}{x}. \tag{2.4.4}$$

Then as

$$y = xz \quad \text{and} \quad y' = xz' + z, \tag{2.4.5}$$

(2.4.2) becomes

$$xz' + z = F(z)$$

or

$$\frac{dz}{dx} = \frac{F(z) - z}{x},$$

which clearly is a separable equation.

Example 2.4.2 Solve the differential equation

$$\frac{dy}{dx} = \frac{x^2 + y^2}{xy}$$

Solution Note that this is the differential equation in Example 2.4.1. Employing the substitutions (2.4.4) and (2.4.5), this equation becomes

$$xz' + z = \frac{1}{z} + z,$$

which simplifies to

$$xz' = \frac{1}{z}.$$

Separating variables and integrating, we have

$$\int z\,dz = \int \frac{dx}{x} + K,$$

from which we obtain

$$\frac{z^2}{2} = \ln|x| + K.$$

In terms of the original dependent variable y, this is

$$\left(\frac{y}{x}\right)^2 = \ln x^2 + 2K.$$

Multiplying both sides by x^2 and letting $C := 2K$, we get

$$y^2 = x^2 \ln x^2 + Cx^2.$$

Consequently,

$$y = \pm\sqrt{x^2 \ln x^2 + Cx^2}$$

are solutions on any open interval that excludes $x = 0$, such as $(-\infty, 0)$ and $(0, \infty)$. ♦

Sometimes it may not be quite so obvious, as it was in the previous example, that a differential equation is homogeneous. In this regard, the following theorem comes in handy.

Criterion for Homogeneity

Theorem 2.4.2 *Suppose for every $x \neq 0$, $y \neq 0$, and $\lambda > 0$ that*

$$f(\lambda x, \lambda y) = f(x, y).$$

Then the differential equation

$$\frac{dy}{dx} = f(x, y)$$

is homogeneous.

Proof For $x > 0$ and $y \neq 0$, let $z = y/x$. Then

$$f(x, y) = f(x, xz) = f(1, z) = f(1, y/x)$$

since x plays the role of λ. Thus, for this case, f can be expressed as a function of the ratio y/x. Now suppose $x < 0$ and $y \neq 0$. Again let $z = y/x$. Then, as

$x = -|x|$,

$$f(x, y) = f(x, xz) = f(-|x|, -|x|z) = f(-1, -z) = f(-1, -y/x),$$

where this time $|x|$ plays the role of λ. Therefore, f can be expressed in terms of the ratio y/x for all $x \neq 0$ and $y \neq 0$. ■

Example 2.4.3 Use Theorem 2.4.2 to establish that

$$y\frac{dy}{dx} = -x + \sqrt{x^2 + y^2} \tag{2.4.6}$$

is a homogeneous differential equation.

Solution First, rewrite the equation as $dy/dx = f(x, y)$, where

$$f(x, y) = \frac{-x + \sqrt{x^2 + y^2}}{y}.$$

For $\lambda > 0$, $x \neq 0$, and $y \neq 0$,

$$f(\lambda x, \lambda y) = \frac{-\lambda x + \sqrt{(\lambda x)^2 + (\lambda y)^2}}{\lambda y} = \frac{-\lambda x + \lambda\sqrt{x^2 + y^2}}{\lambda y} = f(x, y).$$

Therefore, the equation is homogeneous. ♦

Example 2.4.4 Determine the solutions of (2.4.6) and their intervals of existence.

Solution By inspection of (2.4.6), the constant function $y = 0$ satisfies this equation if

$$-x + \sqrt{x^2} = -x + |x| = 0.$$

Since this is true for $x \geq 0$, we conclude that $y \equiv 0$ is a solution on the interval $[0, \infty)$.

Next, we seek nonconstant solutions of (2.4.6). It has already been established in Example 2.4.3 that this equation is homogeneous. As explained earlier, the trick for solving this type of equation is to transform it to a separable equation, which can be done by replacing y with a new variable z, where z is defined by the equation $y = xz$. Expressed in terms of x and z, (2.4.6) is

$$xz(xz' + z) = -x + \sqrt{x^2 + (xz)^2}.$$

Let us look for solutions of this equation on intervals where $x > 0$ and $z \neq 0$ (which implies $y \neq 0$). Dividing both sides of the equation by xz, we have

$$x\frac{dz}{dx} + z = -\frac{1}{z} + \frac{\sqrt{x^2(1+z^2)}}{xz} = -\frac{1}{z} + \frac{|x|}{x} \cdot \frac{\sqrt{1+z^2}}{z}. \tag{2.4.7}$$

Since we are currently restricting our search for solutions to intervals where $x > 0$, $|x|/x = 1$. For these intervals, (2.4.7) simplifies to

$$x\frac{dz}{dx} + z = -\frac{1}{z} + \frac{\sqrt{1+z^2}}{z}.$$

With the goal of separating the variables x and z, let us rewrite this as

$$x\frac{dz}{dx} = -z + \frac{-1+\sqrt{1+z^2}}{z} = \frac{-(1+z^2)+\sqrt{1+z^2}}{z}. \tag{2.4.8}$$

Separating the variables and integrating, we obtain

$$\int \frac{z\,dz}{\sqrt{1+z^2}-(1+z^2)} = \int \frac{dx}{x} + C_1$$

or

$$\int \frac{z\,dz}{\sqrt{1+z^2}\left(1-\sqrt{1+z^2}\right)} = \ln x + C_1. \tag{2.4.9}$$

To evaluate the integral, let $u = \sqrt{1+z^2}$. Then

$$\int \frac{z\,dz}{\sqrt{1+z^2}\left(1-\sqrt{1+z^2}\right)} = \int \frac{du}{1-u} = -\ln|1-u|.$$

Also noting that $u > 1$, (2.4.9) becomes

$$-\ln(u-1) = \ln x + C_1$$

or

$$\ln\left[x(u-1)\right] = -C_1.$$

Exponentiating both sides, we have

$$e^{\ln\left[x(u-1)\right]} = e^{-C_1},$$

which simplifies to

$$x(u - 1) = C$$

where $C := e^{-C_1} > 0$. In terms of the original dependent variable y, this is

$$x\left(\sqrt{1 + \left(\frac{y}{x}\right)^2} - 1\right) = C,$$

which simplifies to

$$\sqrt{x^2 + y^2} = C + x.$$

Thus,

$$y^2 = (C + x)^2 - x^2 = C^2 + 2Cx \qquad (2.4.10)$$

where $C > 0$.

Recall that we obtained (2.4.10) by looking for solutions on intervals where $x > 0$ and $y \neq 0$. Now let us look for solutions on those intervals where $x < 0$ and $y \neq 0$, should any exist. Referring back to (2.4.7), we see that the equation simplifies to

$$x\frac{dz}{dx} = \frac{-(1 + z^2) - \sqrt{1 + z^2}}{z}$$

because $|x|/x = -1$ when $x < 0$. Solutions of this equation are also given by (2.4.10). The details showing this are left as an exercise.

Up to this point, we have shown that any solution of the differential equation (2.4.6) on an interval that excludes the point $x = 0$ and those points where $y = 0$ satisfies the equation

$$y^2 = C^2 + 2Cx \qquad (2.4.11)$$

for some $C > 0$. For a given positive value of C, the graph of (2.4.11) is a parabola that opens up to the right with its vertex at the point $(-C/2, 0)$. See Fig. 2.3.

Solving (2.4.11) for y, we obtain

$$y = \pm\sqrt{C^2 + 2Cx} \qquad (C > 0). \qquad (2.4.12)$$

It is left to the reader to verify that both of these functions for a given value of $C > 0$ are solutions of (2.4.6) on the interval $[-C/2, \infty)$.

Recall that $C > 0$. However, we can amend this to also include $C = 0$ since for this particular value (2.4.12) simplifies to $y = 0$ on the interval $[0, \infty)$, which is the trivial solution that we had observed at the very outset of this example.

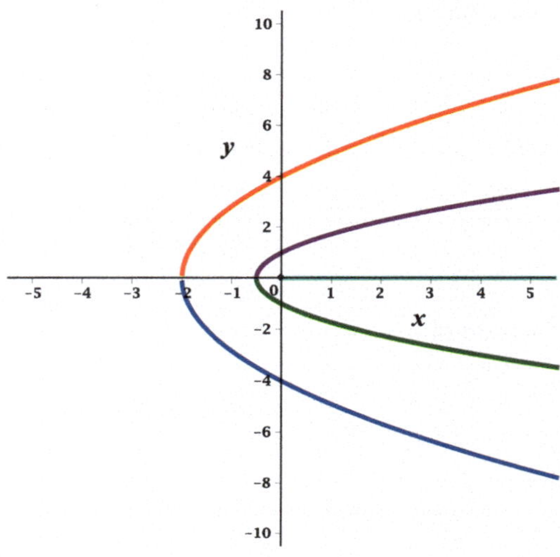

Fig. 2.3 Graphs of five solutions of $yy' = -x + \sqrt{x^2 + y^2}$

In summary, for a given value of $C \geq 0$, the function

$$y = \sqrt{C^2 + 2Cx} \tag{2.4.13}$$

is an explicit solution of (2.4.6) on the interval $[-C/2, \infty)$. Graphs of (2.4.13) corresponding to the values $C = 1$ and $C = 4$ are the upper halves of the parabolas in Fig. 2.3. Furthermore,

$$y = -\sqrt{C^2 + 2Cx} \tag{2.4.14}$$

for a given value of $C \geq 0$, is also a solution on $[-C/2, \infty)$. The graphs of (2.4.14) corresponding to the values $C = 1$ and $C = 4$ are the lower halves of the parabolas. The graph of the trivial solution $y = 0$ on the interval $[0, \infty)$ is also shown. ♦

Problems

You know my methods. Apply them, and it will be instructive to compare results.

(Mr. Sherlock Holmes to Dr. Watson)

Sir Arthur Conan Doyle,[8] *The Sign of the Four* (1890) [28, ch. 6]

Separable Equations

In Problems 1 through 4, determine whether or not the differential equation is separable.

1. $\dfrac{dy}{dx} = \sqrt{x+y}$
2. $\dfrac{dy}{dx} = \sqrt{xy}$
3. $\dfrac{dy}{dx} = \cos(x+y)$
4. $\dfrac{dy}{dx} = \dfrac{e^{x-y}}{6+y^2}$

Alleged Solutions of ODEs

In Problems 5 through 10, an ordinary differential equation is given along with a function (relation) alleged to be its explicit (implicit) solution. Determine whether the alleged solution is truly a solution by means of direct substitution of the function and its derivative into the equation, or if that is not possible, use implicit differentiation as in Example 2.2.4.

5. $\dfrac{dy}{dx} + 5y = 4e^{3x}, \quad y = 0.5e^{3x} - 2e^{-5x}$
6. $y' + 2x^2 = 2x^{-1}y, \quad y = x^2 - 4x^3$
7. $\dfrac{dy}{dx} = \dfrac{x}{y}, \quad x^2 + y^2 = 4$
8. $\dfrac{dy}{dx} = \dfrac{2x}{1-\cos y}, \quad y - \sin y = x^2 + 1$
9. $\dfrac{dy}{dx} = \dfrac{5x^4 - y - 3y^2 + 1}{x}, \quad xy + y^3 = x^5 + x$
10. $\dfrac{dy}{dx} = \dfrac{1-\sin y}{1 + x\cos y}, \quad y + x\sin y = x$

[8] Arthur Conan Doyle first gave life to the great detective Sherlock Holmes in the book *A Study in Scarlet* in 1887. *The Sign of the Four* was the second Sherlock Holmes story. It was first published in *Lippincott's*, a monthly magazine, in February, 1890.

Separation of Variables

In Problems 11 through 46, solve each of the following equations. Whenever possible, find explicit solutions. Check each explicit solution by showing that it satisfies the equation. Use the method of implicit differentiation to check implicit solutions.

11. $\frac{dy}{dx} = \frac{3x^2}{2y}$
12. $\frac{dy}{dx} = y \sin x$
13. $\frac{1}{x}\frac{dy}{dx} = y \sin x$
14. $x^3 y' = 2y^2$
15. $\frac{dx}{dy} = \frac{2ye^x}{1-x}$
16. $\frac{dy}{dx} = \frac{2xy - 2y}{xy + 2x}$
17. $y' = 3y^2\sqrt{x}$
18. $xy' = \frac{x+1}{x+2}$
19. $x\frac{dy}{dx} = 5 - x^2 \sin\frac{x}{2}$
20. $\frac{dx}{dt} = 2\sqrt{x+5}\tan t$
21. $\cos x \frac{dy}{dx} = y \sin x$
22. $\frac{ds}{dt} = 2s - s\cos t$
23. $6xy^2\frac{dy}{dx} = 5y - 3xy\sec^2 x$
24. $y' = t \sin y$
25. $\frac{2}{y}\frac{dy}{dx} = 1 + \frac{1}{y^2}$
26. $\frac{dy}{dx} = \frac{y+1}{\sqrt{1-x^2}}$
27. $y' = \frac{e^{2y}}{x^2+4}$
28. $\frac{dy}{dx} = \frac{xe^{-2y}}{4+x^2}$
29. $\frac{dy}{dx} = y - y^2$
30. $\frac{dy}{dx} = -\frac{x}{y}$
31. $\frac{dx}{dt} = 1 - t + x^2 - tx^2$
32. $tp' = 2 + t^2 e^{-t}$
33. $\frac{dp}{dt} = p - 2p^2$
34. $\sqrt{4-x^2}\,\frac{dy}{dx} = 4$
35. $\frac{dy}{dx} = \frac{1+y^2}{\cos^2 x}$
36. $\frac{dx}{dt} = \frac{6tx - 3x}{\sqrt{4-t^2}}$
37. $4yy'\sqrt{9-x^2} = 3\csc y$
38. $\frac{dy}{dx} = \frac{9y}{x^2+4}$
39. $\frac{dy}{dx} = \frac{xy+x}{x^2+4}$
40. $xy'\sqrt{9-x^2} = 6x^2$
41. $(x+2)\frac{dy}{dx} = \frac{3xy^2}{x-1}$
42. $2\frac{dx}{dt} = \frac{3}{1+x+t^2+xt^2}$
43. $\frac{dy}{dx} = 2xy^2 + 8x$
44. $\frac{dx}{dt} = \frac{x^2\cos t}{\sin^2 t}$
45. $\tan x\,\frac{dy}{dx} = \sqrt{1+2y}$
46. $\cos y\,\frac{dy}{dx} = \frac{\sin y}{\sqrt{1-x^2}}$

Initial Value Problems

In Problems 47 through 55, solve the initial value problem.

47. $\frac{dx}{dt} = \frac{t}{\sqrt{x}},\ x(2) = 9$
48. $y' = 2x\csc y,\ y(-1) = \pi$
49. $y' = 2xy - 10x,\ y(0) = 2$
50. $\frac{dx}{dt} = 6e^{2t-x},\ x(0) = 0$
51. $\frac{dy}{dx} = -\frac{\sin x}{2y\cos^2 x},\ y(\pi) = -4$
52. $\frac{dy}{dx} = -\frac{x}{y},\ y(-3) = 4$
53. $y' = 2xy^2 + y^2\sin x,\ y(\pi) = -1$
54. $\frac{dy}{dx} = 98\frac{x^2 - 3x + 12}{y(x^2 - 4x + 11)^2}$, where $y = 0$ when $x = 2$
55. $\frac{dy}{dx} = -\frac{5e^{\sin x}\cos x}{2y},\ y(\pi) = -5$

Sewage Treatment

56. Suppose at some point in time a sewage treatment tank contains 50,000 gallons of a water-sediment mixture. If there are 1250 pounds of sediment mixed in the water and the tank removes 6% of the sediment per minute, how much sediment will remain in the tank after one hour?

Traffic Bottleneck

57. A car is traveling eastward along I-40, an east-west Interstate highway, toward Nashville, Tennessee. Road construction near an off-ramp to Bucksnort, a small, unincorporated community, has caused a bottleneck that slows down the traffic. The speed of the car from a point 5 miles west of the bottleneck to 5 miles east of the bottleneck is

$$s^2 + 10,$$

where s is the distance in miles from the bottleneck and t is the time in hours. How long does it take the car to travel this total distance of 10 miles?

Melting Snowball

58. A snowball has a radius of 3 inches. Assume the rate with which the volume of the snowball melts is proportional to its surface area. If, after 1 hour, the radius of the snowball is 2.9 inches, predict what its radius will be after one day.

Quartic Polynomial

59. Find the quartic polynomial $P(x)$ that is divisible by its derivative $P'(x)$ so that $P(1) = 5$ and $P(2) = 0$.

Snow Removal

60. Once upon a time in the wee hours of a wintry morning in Lake Wobegon, Minnesota, it began to snow heavily and steadily. At 6:00 A.M., a snowplow started clearing a country road. By 8:00 A.M., it had cleared two miles. By 10:00 A.M., it had only cleared an additional mile. It snowed unabatedly throughout the entire morning. Assuming that the snowplow clears the road at a rate that is inversely proportional to the depth of the snow, when had it begun to snow?

Falling Raindrop

61. A raindrop is falling through the atmosphere. Apart from the force of gravity, assume that all other forces acting on the raindrop are negligible. Suppose the atmosphere is saturated with water vapor and that as a result of condensation, the mass of the raindrop is increasing at a rate proportional to its surface area. Assume it is always spherical in shape and its density ρ is constant. Let r_0 denote its radius and v_0 its velocity at time $t = 0$.

(a) Show that the radius r of the raindrop increases linearly with time.

(b) Express the raindrop's velocity v in terms of its radius r. Start with the general form of Newton's second law on motion (see (1.4.7) in Chap. 1):

$$\frac{d}{dt}(mv) = -mg.$$

Note 1. Newton's second law of motion expressed in the form

$$F = ma$$

is not applicable because the mass of the raindrop is not constant.

Note 2. Care must be taken in using (1.4.7) because the momentum of the elements of mass that are gained (or lost, as in a rocket propelled by gases) must also be considered. However, that is not the case here if we assume that the water being added does not impart momentum to the raindrop.

(c) For $r_0 = 0$ and $v_0 = 0$, find the formula for the velocity v of the raindrop in terms of the time t. What is the acceleration of the raindrop?

Interest-Bearing Account

62. The CFO (chief financial officer) of a small company deposits \$100,000 in an interest-bearing account. The account is structured so that the amount of money in the account at any given point in time increases at a rate proportional to the amount at that moment and that after one year from the day of deposit the amount will accrue to a total of \$105,000. Find a formula for the amount of money t years from the day of deposit by setting up the relevant IVP and then solving it.

Rowboat

63. An oarsman rowing in still water keeps a boat moving at a uniform velocity of 2 miles per hour in an eastward direction. Suppose the resistive force of the water on the boat is proportional to the velocity v of the boat. Use Newton's second law of motion to find the initial value problem modeling the velocity of the boat after the oarsman ships the oars (that is, places the oar in a resting position in the boat). Solve the initial value problem for v expressing it in terms of the time t and the combined masses m of the boat, its contents, and the oarsman.

Newton's Law of Cooling

64. Predict how long it will take for the temperature of the coffee to cool down to 100 °F in Example 2.3.5.

Cobb-Douglas Model

65. Suppose the Cobb-Douglas production function

$$P = bL^{\alpha}K^{1-\alpha}$$

is used to model the economy of Genovia, a tiny fictitious European kingdom that is the setting of *The Princess Diaries* (a movie that was released in 2001). Five years ago, the total production of goods (P) in Genovia had a monetary value of 101 billion Genovian francs, while the monetary values of labor (L) and capital investment (K) were each 100 billion francs. The current values of the Genovian economy are $P = 145$ billion francs, $L = 150$ billion francs, and $K = 126$ billion francs. Use this data to estimate the values of α and b. Then use this model to predict the value of P when $L = 160$ billion francs, and $K = 130$ billion francs.

66. Show that the Cobb-Douglas production function satisfies the equation

$$L\frac{\partial P}{\partial L} + K\frac{\partial P}{\partial K} = P.$$

Homogeneous Equations

In Problems 67 through 70, solve each of the following differential equations.

67. $\dfrac{dy}{dx} = \dfrac{2xy - y^2}{x^2}$

68. $2xy\dfrac{dy}{dx} = 3x^2 - 4y^2$

69. $\dfrac{dy}{dx} = \dfrac{y + x\sec(y/x)}{x}$

70. $xy\,dx - (x^2 + y^2)\,dy = 0$

Tethered Dog in Pursuit of Deer

71. A round grain bin on a family farm has a radius of a feet. One end of a tie-out dog cable of length $2\pi a$ is hooked to the base of the bin at point B, which has coordinates $(a, 0)$. The other end is attached to the collar of a Cavalier King Charles Spaniel by the name of Scholar. While lapping up water from a bowl located at point A with coordinates $(a, 2\pi a)$, Scholar spots deer emerging from woods that are southwest of the silo. Seeing that the deer are headed in a northeasterly direction, Scholar runs toward them while continually pulling on the cable so that it is always tangent to the circular base of the bin. If Scholar continues the pursuit, he will eventually run out of cable and end up at point B.

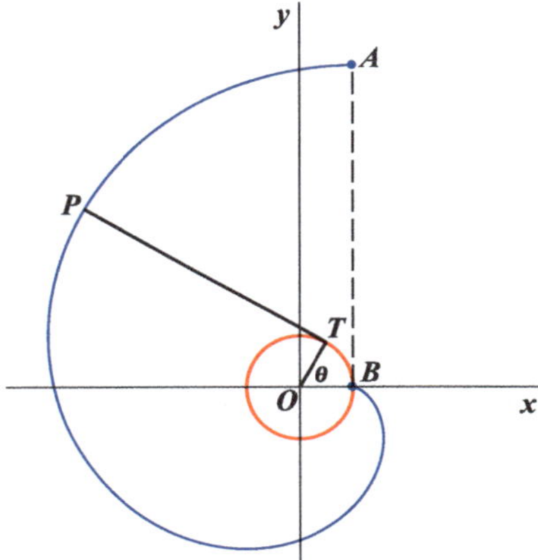

(a) Let P represent any point on the path taken by Scholar, that is, the curve shown in the figure passing through the points A, P, and B. Since the cable remains tangent to the circular base of the bin, the line segment PT is perpendicular to the radius segment OT. Find the parametric equations of the path by expressing the coordinates x and y of P in terms of a and θ, where a is the radius of the bin (length of OT) and the parameter θ denotes the positive angle between the positive x-axis and OT.

(b) Modify the parametric equations found in part (a) for the path by changing the parameter from θ to $t := 2\pi - \theta$. Note that this yields the same path but reverses the direction of Scholar from point B back to A.

(c) For what value of t does the path cross the negative x-axis?

(d) Determine the length of Scholar's path through A, P, and B by using the parametric equations in either part (a) or part (b) and the differential equation for arc length (cf. (1.2.2)):

$$\frac{ds}{dt} = \sqrt{\left(\frac{dx}{dt}\right)^2 + \left(\frac{dy}{dt}\right)^2}$$

(e) Show that the reflection of Scholar's path through A, P, and B in the x-axis is the graph of the parametric equations

$$x = a(\cos t + t \sin t), \quad y = a(\sin t - t \cos t)$$

for $0 \leq t \leq 2\pi$. *Remark.* For $t \geq 0$, this graph is known as the ***involute*** of a circle of radius a.

(f) While Scholar is tethered to the grain bin at point B, compute the total area of the region over which Scholar can roam.

Computer Algebra System Problems

72. Solve Problem 55 by hand and then by using a CAS. Depending on the CAS, the two solutions may appear to differ. If that turns out to be the case, reconcile them.
73. Use a CAS to reproduce the path of Scholar through the points A, P, and B that is shown in Problem 71. Also, draw the boundary of the region in part (f) of the problem.

Chapter 3
Direction Fields, Trajectories, Solution Curves

I find the great thing in this world is not so much where we stand, as in what direction we are moving: To reach the port of heaven, we must sail sometimes with the wind and sometimes against it—but we must sail, and not drift, nor lie at anchor.

Oliver Wendell Holmes,[1] *The Autocrat of the Breakfast Table* (1858), ch. 4

Abstract This chapter introduces the concepts of *direction fields*, *trajectories*, and *solution curves* with illustrative examples, such as the motion of a yacht adrift on a lake and the motion of a vertically moving object with a velocity-dependent drag force acting on it.

The chapter begins with the scenario of a yacht on a lake drifting with the currents, where the motion of the yacht is governed by a simple ordinary differential equation. A *direction field* for the differential equation is constructed by covering a map of the lake with a rectangular array of regularly spaced points and then assigning a direction arrow to each of them to indicate the direction in which the yacht would move were it located at one of these points. Examples, such as the one described here, are used to demonstrate that the direction field for a differential equation is a useful tool for gaining some insight into the qualitative behavior of the solutions of the equation, even if closed-form solutions of the equation cannot be found.

Besides direction fields, even more useful information about *solution curves* (graphs of solutions) can be obtained by using the curve-sketching techniques of calculus. One of the examples presented in this chapter is to determine all of the important features of the graph of the solution of the innocuous-looking initial value

---

[1] Oliver Wendell Holmes (1809–1894) was an author and professor of anatomy and physiology at Harvard. He is also the father of Oliver Wendell Holmes, Jr., jurist and associate justice of the U.S. Supreme Court from 1902 to 1932. *The Autocrat of the Breakfast Table* is a collection of humorous essays originally published in the *Atlantic Monthly*.

© The Author(s), under exclusive license to Springer Nature Switzerland AG 2026

L. C. Becker, *Ordinary Differential Equations: Concepts, Methods, and Models*,
https://doi.org/10.1007/978-3-032-15150-6_3

problem

$$y' = x^2 + y^2 - 1, \quad y(0) = 0. \tag{1}$$

One might be inclined to think that (1) has a relatively simple solution. But as it turns out, its solution cannot be expressed in terms of the familiar functions of calculus. Even so, all of the important features of its graph can be gleaned from the sign changes of y' and y'', such as intervals of increase and decrease, concavity, end behavior, and whether there are any local extrema and points of inflection.

The difference between *solution curves* and *trajectories* is explained and illustrated with their graphs by imagining that the position (x, y) at time t of a small piece of driftwood floating in a body of water can be modeled by the system of differential equations

$$\frac{dx}{dt} = y, \quad \frac{dy}{dt} = -\omega^2 x, \tag{2}$$

where ω denotes some positive constant. Graphs of some of the trajectories (ellipses) of (2) in the phase plane (xy-plane) are compared to their counterparts (elliptical helices) in txy-space.

3.1 Direction Fields and Trajectories

Envision an idyllic, deep-blue lake nestled in a mountainous region of the San Juan National Forest in Colorado. Imagine being there one summer day sunbathing on the deck of a small yacht as it drifts with the currents in the lake. At some point, bored and hoping to find something intellectually stimulating to read, you step into the cabin. Espying a book on differential equations, you pick it up. While thumbing through it, you come across a section about *direction fields* (which just happens to be one of the topics in this chapter) and begin reading. After finishing and mulling over this new term, you wonder if it were applicable to the motion of a yacht.

In order to introduce the concept of direction fields, we will analyze the motion of a yacht in several different scenarios. Since a complete description of its position and motion requires a reference frame, let us visualize a Cartesian coordinate system on the surface of the lake with its origin located at some fixed point—say a stationary buoy—and its positive x- and y-axes pointing eastward and northward, respectively. Then, at any given time, the position of the yacht relative to the buoy can be given.

3.1.1 Yacht Adrift

In our first scenario, let us imagine that the yacht's engine is turned off and the yacht is drifting toward the buoy. Let $t = 0$ denote the moment it grazes the buoy. In other words, the yacht's position is $(x, y) = (0, 0)$ at time $t = 0$. Suppose every second thereafter the yacht drifts 10 feet eastward and 10 feet northward due to the currents in the lake. So the yacht's speed in the eastward direction is $\dot{x} = 10$ ft/sec and its speed northward is $\dot{y} = 10$ ft/sec. A way of looking at this is to say that the motion of the yacht is governed by the pair, or system, of differential equations

$$\frac{dx}{dt} = 10, \quad \frac{dy}{dt} = 10. \tag{3.1.1}$$

Clearly the yacht's position at time t (in feet), is given by the following parametric equations:

$$x(t) = 10t, \quad y(t) = 10t, \tag{3.1.2}$$

where t (time) is the parameter. We can say that this pair of functions is a solution of the system because they satisfy both of the equations constituting the system as well as the initial condition: $(x(0), y(0)) = (0, 0)$.

Between the times of $t = 0$ and $t = T$, the yacht traces out a curve on the lake's surface consisting of the points

$$(x(t), y(t)) = (10t, 10t)$$

for $0 \leq t \leq T$. This curve is known as the *trajectory* of the yacht. The terms *path* and *orbit* are also used. Generally speaking, the ***trajectory*** (***path***) of a body in motion is the curve that it makes as it moves, such as the drifting yacht or a slapped hockey puck in two-dimensional space or a punted football in flight in three-dimensional space. In two-dimensional space, the trajectory of a body (the yacht for instance) consists of a set of points, such as

$$\{(x(t), y(t)) : t_1 \leq t \leq t_2\}, \tag{3.1.3}$$

where $(x(t), y(t))$ is the location of the body at time t. In other words, it is the plot of the points $(x(t), y(t)$ in the xy-plane over the interval $[t_1, t_2]$. In the context of moving bodies and their trajectories, the xy-plane is called the ***phase plane***.

The direction in which the yacht is drifting at any given instant $t = \tau$ can be ascertained from the value of the derivative dy/dx at τ since it is equal to the slope of the line tangent to the path of the yacht at the point $(x(\tau), y(\tau)) = (10\tau, 10\tau)$. We can compute dy/dx by first expressing the y-coordinate of the position of the yacht directly in terms of its x-coordinate. Eliminating the parameter t in (3.1.2), we have

$$y = x.$$

Therefore, at any point in the phase plane

$$\frac{dy}{dx} = 1. \tag{3.1.4}$$

This corroborates the obvious: the yacht always drifts northeastward.

Let us alter the scenario somewhat by moving the starting position of the yacht: from the buoy to a point $(0, y_0)$ either directly north or south of the buoy. Accordingly, the yacht's trajectory also changes: from (3.1.2) to

$$x(t) = 10t, \quad y(t) = 10t + y_0. \tag{3.1.5}$$

Again, it is easy to eliminate t from (3.1.5). Solving each equation for $10t$ and equating the two results yields

$$y = x + y_0.$$

Despite the alteration, the derivative dy/dx is still given by (3.1.4). So, regardless of where the yacht starts on the y-axis, it will still drift in the northeast direction but along a line parallel to $y = x$.

It is worth noting that the derivative dy/dx can also be computed without having to eliminate t by invoking the chain rule:

$$\frac{dy}{dt} = \frac{dy}{dx} \cdot \frac{dx}{dt}.$$

Since $dx/dt \neq 0$, it follows from this and (3.1.5) that

$$\frac{dy}{dx} = \frac{dy/dt}{dx/dt} = \frac{10}{10} = 1.$$

Resorting to the chain rule to compute the derivative is the way to proceed if it is impossible, or too onerous a task, to eliminate the parameter.

A map of the lake together with arrows distributed at various spots on the map indicating the direction in which the yacht would move were it located at one of these spots is an example of a ***direction field***, such as the one shown in Fig. 3.1. This direction field also includes the coordinate system described earlier. It shows that part of the lake east and north of the stationary buoy, which is located at the origin. It was constructed by covering the map in Fig. 3.1 with a rectangular array of regularly spaced points and assigning a direction arrow to each of them. Since the yacht always drifts in the northeast direction, irrespective of its location, all of the arrows point northeastward. They were all given the same length since the goal for now is to only show the yacht's direction of motion and nothing more. Since the motion of the yacht is governed by the system of differential equations (3.1.1), we can say that Fig. 3.1 shows a *direction field for this system* or, for that matter, a *direction field for the differential equation* (3.1.4) due to the relationship between (3.1.1) and (3.1.4). Anyone looking at this direction field can readily see that a yacht starting at the origin would drift along the line $y = x$ in the northeast direction—or, along a line parallel to $y = x$ if it starts elsewhere on the y-axis. Two paths, one with the yacht starting at the origin (location of the buoy) and the other with it starting 200 feet north of the buoy, are shown in Fig. 3.2.

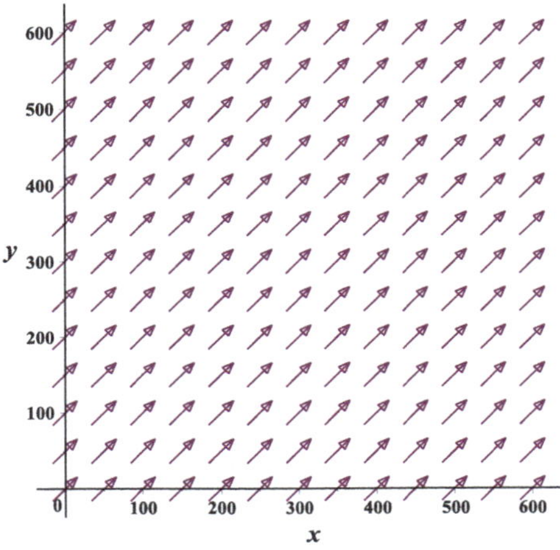

Fig. 3.1 A direction field for system (3.1.1)

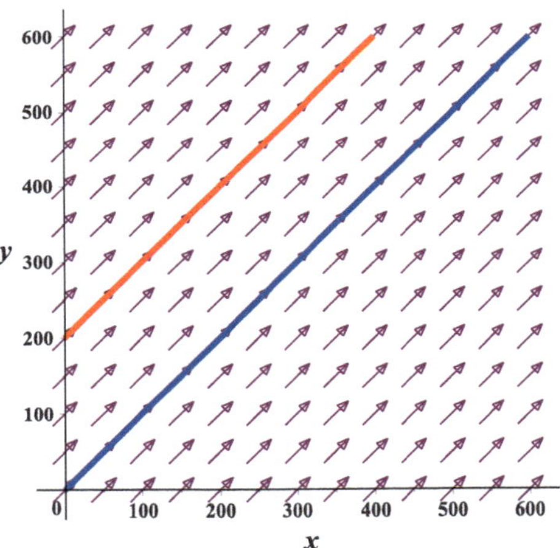

Fig. 3.2 Two trajectories of the drifting yacht

3.1.2 Yacht Accelerating

Now consider another scenario. Imagine that the yacht's engine is idling and its propeller motionless. So, as in the first scenario, the yacht is drifting 10 feet northward and 10 feet eastward every second. This time, however, suppose that at the very moment the yacht grazes the buoy someone engages forward gear and steers the yacht in the eastward direction while throttling the engine so that the yacht accelerates eastward at a constant rate of 2 feet per second per second. With $v_x = \dot{x}$ denoting the yacht's eastward speed relative to the stationary buoy, the acceleration in the eastward direction is

$$\frac{d}{dt}v_x = 2.$$

Integrating this with respect to t, we find that

$$v_x = \int 2\,dt = 2t + C.$$

Since the yacht's eastward speed at $t = 0$ is 10 ft/sec, the value of C is 10. And so

$$\dot{x} = 2t + 10.$$

If the constant acceleration of 2 ft/sec^2 is maintained for a full 20 seconds, the eastward speed will increase from 10 ft/sec to 50 ft/sec, which is about 34 mph or 29.6 knots.[2]

The motion of the yacht is now governed by the system of differential equations

$$\frac{dx}{dt} = 2t + 10, \quad \frac{dy}{dt} = 10, \tag{3.1.6}$$

rather than by (3.1.1). Integrating both of these equations with respect to t and choosing the constants of integration so that $(x(0), y(0)) = (0, 0)$, we find that the yacht's position at time t is given by the parametric equations

$$x(t) = t^2 + 10t, \quad y(t) = 10t. \tag{3.1.7}$$

Bear in mind that this is the trajectory of the yacht if it starts accelerating at the buoy. However, if it starts north or south of the buoy, say at the point $(0, y_0)$, then we must replace (3.1.7) with

$$x(t) = t^2 + 10t, \quad y(t) = 10t + y_0. \tag{3.1.8}$$

By the chain rule,

$$\frac{dy}{dx} = \frac{\dot{y}}{\dot{x}} = \frac{10}{2t + 10} = \frac{5}{t + 5}.$$

To express dy/dx in terms of x, solve $t^2 + 10t = x$ for t by completing the square:

$$(t + 5)^2 = x + 25 \quad \Rightarrow \quad t = -5 + \sqrt{x + 25}.$$

[2] A ***knot*** is a unit of speed equal to 1 nautical mile per hour. A ***nautical mile*** is a unit of distance that historically was defined to be the length of an arc along a line of longitude subtending an angle of one-sixtieth of a degree. According to the International Union of Geodesy and Geophysics, the earth's mean radius is about 3959 (statute) miles. Using this value, the nautical mile is

$$s = r\theta \approx 3959 \cdot (1/60) \cdot (\pi/180) \approx 1.15 \text{ miles}.$$

The current definition is that a nautical mile is exactly 1852 m, or approximately 1.1508 statute miles.

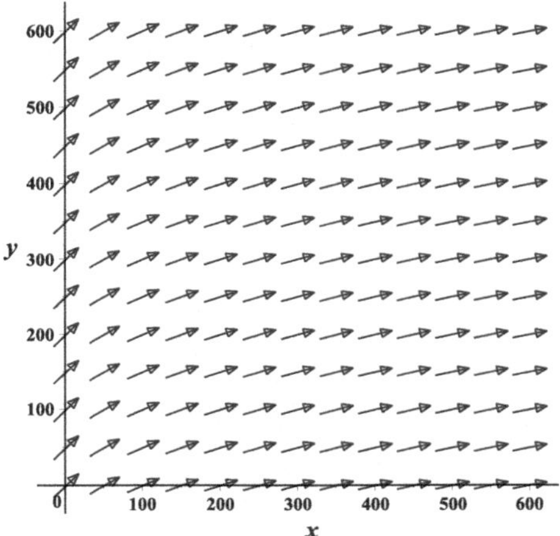

Fig. 3.3 A direction field for $\frac{dy}{dx} = \frac{5}{\sqrt{x+25}}$

Thus,

$$\frac{dy}{dx} = \frac{5}{(-5 + \sqrt{x+25}) + 5} = \frac{5}{\sqrt{x+25}}. \tag{3.1.9}$$

A direction field for (3.1.9), just east and north of the buoy, is shown in Fig. 3.3. It is only valid when the yacht begins its accelerated motion at some point on the y-axis (see Problem 15).The slopes of the direction arrows in the figure can be easily calculated with (3.1.9). For instance, when $x = 0$, then $dy/dx = 1$. So all of the direction arrows on the positive y-axis are pointing precisely in the northeast direction. When $x = 600$, then

$$\frac{dy}{dx} = \left.\frac{5}{\sqrt{x+25}}\right|_{x=600} = \frac{5}{\sqrt{625}} = \frac{5}{25} = \frac{1}{5}.$$

Note that since

$$\frac{dy}{dx} \to 0 \quad \text{as} \quad x \to \infty,$$

the yacht moves less northward and more eastward as x increases.

3.1.3 Trajectories

The curve-sketching procedure for graphing functions that is part of the syllabus of every standard calculus course is an invaluable tool for sketching trajectories of moving bodies. We will use this tool to determine the various trajectories a yacht describes as it begins its acceleration from various starting locations on the y-axis. The lake's currents, the yacht being steered eastward, and the driving force of its rotating propeller are the factors contributing to increasing both coordinates of the yacht's position as it threads its way, so to speak, along the direction field. From the standpoint of calculus, $x(t)$ increases in value with time because its derivative

$$\dot{x} = 2t + 10$$

is positive for $t \geq 0$.[3] Moreover, as $x(0) = 0$, it follows that $x(t)$ is positive for $t > 0$. Consequently, the right-hand side of (3.1.9) is positive. And so the derivative, dy/dx, is positive. This explains why all of the arrows shown in Fig. 3.3 have positive slope. It also confirms that the solutions of (3.1.9) are increasing functions; so the y-coordinate along a trajectory increases with increasing x. We can determine the concavity of a trajectory using the second derivative

$$\frac{d^2y}{dx^2} = \frac{d}{dx}\left[5(x+25)^{-1/2}\right] = -\frac{5}{2\left(\sqrt{x+25}\right)^3}.$$

Since d^2y/dx^2 is always negative for $x \geq 0$, every trajectory is concave downward. This information coupled with the direction field makes it easy to sketch trajectories of the yacht. For the trajectory starting at the origin (buoy), we draw an increasing, concave downward curve tangent to every arrow with which it comes into contact. The result is the lower trajectory superimposed on the direction field shown in Fig. 3.4. From starting points elsewhere on the y-axis, we still draw increasing, concave downward curves tangent to all of the direction arrows. The second trajectory starting 200 feet north of the buoy is an example.

The equations of these trajectories are easily obtained by integrating (3.1.9):

$$y = \int \frac{5}{\sqrt{x+25}}\,dx + C = 10\sqrt{x+25} + C.$$

If the yacht starts accelerating at $(0, y_0)$ on the y-axis, then $y(0) = y_0$, which makes $C = y_0 - 50$. Accordingly,

$$y = 10\sqrt{x+25} - 50 + y_0. \tag{3.1.10}$$

Thus, the trajectories corresponding to $y_0 = 0$ and $y_0 = 200$ shown in Fig. 3.4 are

$$y = 10\sqrt{x+25} - 50 \quad \text{and} \quad y = 10\sqrt{x+25} + 150,$$

respectively.

---

[3] Bear in mind that this is true while the yacht is accelerating.

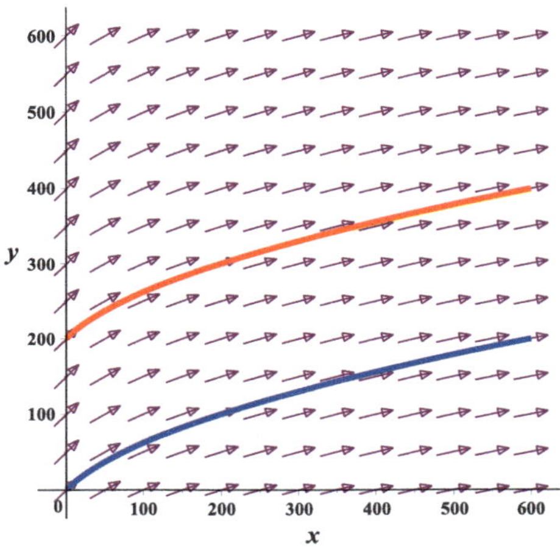

Fig. 3.4 Trajectories of the yacht starting from the points (0, 0) and (0, 200)

3.2 Solution Curves

In the previous section, we looked at several scenarios involving a yacht moving on a lake as a way to introduce and explain the concepts of direction fields and trajectories. In one of the scenarios, we considered a yacht whose motion is governed by the system of differential equations (see (3.1.6))

$$\dot{x} = 2t + 10, \quad \dot{y} = 10 \tag{3.2.1}$$

and determined that

$$(x(t), y(t)) = (t^2 + 10t, 10t) \tag{3.2.2}$$

is the solution satisfying the initial conditions: $x(0) = 0$, $y(0) = 0$. In order to have a visual representation of this solution that includes the temporal variable t as well as the spatial variables x and y, let us move out of the two-dimensional plane into a three-dimensional space by adding a time axis (t-axis) and positioning it perpendicular to the xy-plane at the point (0, 0) as depicted in Fig. 3.5. So for a given value of t, an ordered triple $(t, x(t), y(t))$ of numbers, such as $(2, x(2), y(2)) = (2, 24, 20)$, can be depicted by a point in this three-dimensional space, which we call txy-space. Consequently, the locus of the set of points

$$(t, x(t), y(t)) = (t, t^2 + 10t, 10t)$$

describes a curve in txy-space. This curve, of course, is the graph of the solution (3.2.2). A portion of this graph between the points (0, 0, 0) and (2, 24, 20) is shown in Fig. 3.5. The graph of a solution is also called a ***solution curve***.

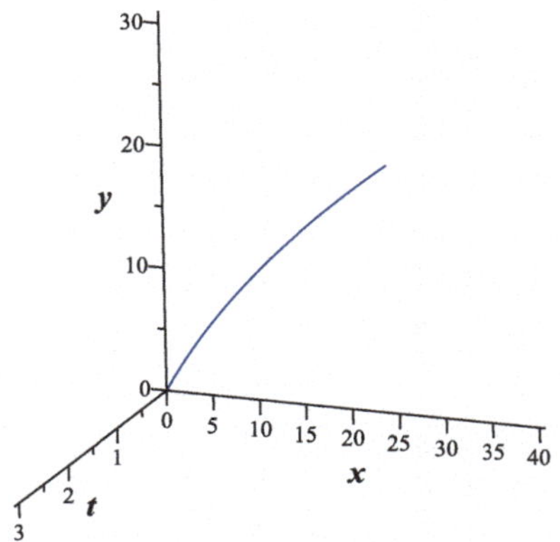

Fig. 3.5 A solution curve of (3.1.6)

In our discussions involving system (3.2.1), we introduced the terms: solution curve and trajectory. So what is the difference between them? The answer is simply this. For a system of two ordinary differential equations, where t designates the independent variable and x and y the dependent variables, a ***solution curve*** is just another name for the graph of a solution of this system in txy-space. For example, Fig. 3.5 depicts the solution curve of system (3.2.1) though the origin $(0, 0, 0)$. A ***trajectory*** is the result of projecting a solution curve onto the xy-plane. An example of this is the curve through the origin $(0, 0)$ in Fig. 3.4: it is the trajectory corresponding to the solution curve in Fig. 3.5.

The discussion up to this point might give someone the impression that direction fields are meant only for systems of the form

$$\dot{x} = f(t), \quad \dot{y} = g(t),$$

where the functions f and g exclude the dependent variables x and y, such as the system (3.1.7). But that is not the case. Generally speaking, physical processes are modeled with systems of differential equations that do not depend explicitly on time since it is assumed that the physical laws governing these processes are the same at any moment in time.[4] These types of systems are called autonomous systems. An ***autonomous system*** with two dependent variables, say x and y, has the form

$$\frac{dx}{dt} = f(x, y), \quad \frac{dy}{dt} = g(x, y). \tag{3.2.3}$$

As the notation indicates, the dependent variables x and y may appear explicitly in the functions f and g but not the independent variable t. That is, the expression for each of the functions may involve x and y (or just one of these variables or neither)

[4] See Driver [30, p. 401].

but t is not present. An example of such a system is

$$\frac{dx}{dt} = y, \quad \frac{dy}{dt} = -\omega^2 x. \tag{3.2.4}$$

Despite its simplicity, this system of two differential equations is used to model the oscillatory motion of a body attached to one end of a spring and whose other end is fastened to a stationary support as depicted in Fig. 10.1 in Chap. 10. In addition, the motion takes place in a hypothetical setting in which there are no damping forces, such as friction, impeding the body's motion. Under these circumstances, a spring with an attached mass (body) is known as an ***undamped mass-spring system*** (see Sect. 10.1 for details). In this context, the variable x represents the position of the body, y its velocity, and ω is a parameter whose value depends on the body's mass and the stiffness of the spring. Another form of this model more familiar to physicists and engineers is the second-order linear differential equation

$$\frac{d^2x}{dt^2} + \omega^2 x = 0. \tag{3.2.5}$$

It can be obtained from (3.2.4) by differentiating the first equation there with respect to t and then setting the right-hand side of the result equal to the second equation.

So as to get a feel for the shapes of the trajectories for the system of equations (3.2.4), let us start off with the direction field. First we cover the xy-plane with a rectangular array of regularly spaced points. Then, as with the examples involving a moving yacht, we will place direction arrows of the same length at these points. In order to determine their directions, let us think of the xy-plane as lying on the surface of a sea, say the Sargasso Sea,[5] with its origin located near the middle of the sea and its axes oriented so that the positive x-axis points due east[6] and the positive y-axis due north. Even though we usually think of (3.2.4) as modeling the position and velocity (speed and direction) of the body in an undamped mass-spring system, let us imagine that one day, for a few hours at least, it also models the motion of a small piece of driftwood as it is being carried by the currents in the vicinity of the origin.

As we shall see, the precise direction in which the piece of driftwood is moving when it is at a point (x, y) in the sea can be determined from the signs of the derivatives $\dot{x}$ and $\dot{y}$ and the slope of the tangent line to the driftwood's trajectory at this point. It follows from (3.2.4) and the chain rule that this slope is

$$\frac{dy}{dx} = \frac{\dot{y}}{\dot{x}} = -\frac{\omega^2 x}{y}. \tag{3.2.6}$$

---

[5] The Sargasso Sea, located within the North Atlantic Ocean, is the only sea with no shorelines. How is that possible? For an answer, see [31].

[6] *Due east* is a precise direction. Here it emphasizes that the x-axis points directly east, not slightly north or south of east but directly east.

However, no direction arrow can be placed at the origin since both $\dot{x}$ and $\dot{y}$ are equal to 0 at this point. So if the piece of driftwood happens to be at the origin at $t = 0$ (the moment this model takes effect), it will remain there (as long as the currents do not change). That is to say, $(x(t), y(t)) = (0, 0)$ for $t \geq 0$. Note that this is the solution of (3.2.4) satisfying the initial condition $(x(0), y(0)) = (0, 0)$. A constant-valued solution like this one is known as an ***equilibrium solution***. Its trajectory consists of the single point $(0, 0)$, which is called an ***equilibrium point*** (other terms are *stationary point*, *critical point*, and *rest position*, to name a few).

Suppose at some moment the driftwood is located at a point (x_1, y_1) in the northeast quadrant of the xy-plane. Then from (3.2.4), we have

$$\dot{x} = y_1, \ \ \dot{y} = -\omega^2 x_1.$$

Since both x_1 and y_1 are positive, $\dot{x} > 0$ and $\dot{y} < 0$, which means the driftwood is moving eastward and southward at this moment—so, more or less in a southeasterly direction. The precise direction can be determined from the slope of the trajectory at (x_1, y_1), which from (3.2.6) is

$$\frac{dy}{dx} = -\frac{\omega^2 x_1}{y_1}.$$

For example, suppose $y_1 = \omega^2 x_1$. Then

$$\frac{dy}{dx} = -1,$$

which implies that all of the arrows at the points $(x_1, \omega^2 x_1)$, where $x_1 > 0$, point due southeast (that is, directly towards the southeast direction).

Since the driftwood is always moving southward when it is in the northeast quadrant, it will eventually cross the x-axis. At that instant, it moves directly southward since $\dot{x} = 0$. However, as soon as it moves into the southeast quadrant, it turns westward since $\dot{x}$ becomes negative. After that, the driftwood moves westward as it continues its southward journey since at any point (x_2, y_2) in this quadrant, $\dot{x} = y_2 < 0$ and $\dot{y} = -\omega^2 x_2 < 0$. Consequently, the arrows in this quadrant point toward the southwest direction. But as soon as the driftwood lands on y-axis, it ceases its southward movement since $\dot{y} = 0$ at all points on the y-axis.

Continuing with this analysis, we find that all arrows in the southwest quadrant point toward the northwest direction but less to the west and more to the north near the negative x-axis. Those on the negative x-axis point directly north. Finally we come to northwest quadrant. The arrows in this quadrant point toward the northeast direction since $\dot{x} = y > 0$ and $\dot{y} = -\omega^2 x > 0$. As $x \to 0$, $\dot{y} \to 0$. So at the very instant the driftwood crosses the y-axis, it is moving due east.

At this juncture we can use this analysis to get a rough idea of what the direction field for (3.2.4) looks like by proceeding as follows: First, assign ω a small value, such as $\omega = 2$. Next, select four points, one in each quadrant that are reflections

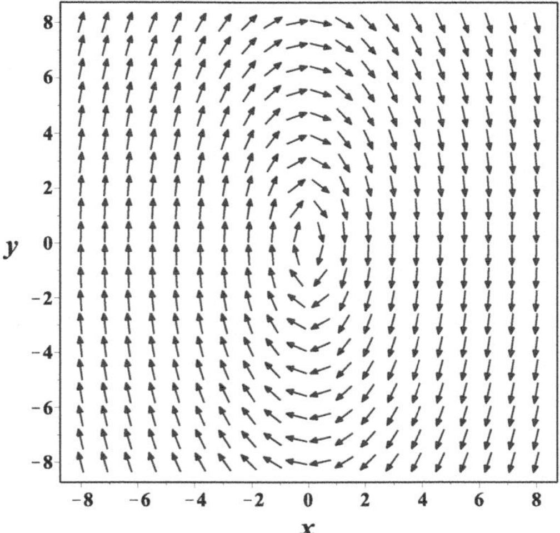

Fig. 3.6 Direction field for the undamped mass-spring system

of each other across the axes, such as $(2, \pm 2)$ and $(-2, \pm 2)$, and four points on the positive and negative axes that are equidistant from the origin, such as $(\pm 3, 0)$ and $(0, \pm 3)$. Use (3.2.6) to calculate the slopes of the trajectories at these points and then place arrows at each of them using the calculated slopes, pointing them in the directions obtained in the analysis. The resulting sketch indicates that the direction field looks like the one in Fig. 3.6, which was obtained using the computer algebra system *Maple*.

The direction field suggests that all of the trajectories of (3.2.4), other than the equilibrium point $(0, 0)$, are closed curves encircling the origin. The curves are in fact closed since no energy is lost in an undamped mass-spring system. Consequently, we know from physics that the mass in such a system returns periodically to its original state. Also, the trajectories appear to be oval-shaped. Actually, they are concentric ellipses centered at the origin, which we show next.

In Chap. 12, we will investigate how to obtain solutions of the autonomous system (3.2.3) when f and g are functions of the form

$$f(x, y) = ax + by, \quad g(x, y) = cx + dy,$$

where a, b, c, d are real numbers. Likewise, our investigations in Chap. 10 will enable us to obtain solutions of the second-order linear differential equation

$$a\frac{d^2x}{dt^2} + b\frac{dx}{dt} + cx = 0,$$

where, as before, a, b, c denote real numbers. The results of these investigations will end up being called theorems, which we can then employ to solve an initial value problem, such as

$$\ddot{x} + 4x = 0; \quad x(0) = 2, \ \dot{x}(0) = 4\sqrt{3}, \tag{3.2.7}$$

where we have set $\omega = 2$ in (3.2.5). This is equivalent to the initial value problem

$$\dot{x} = y, \ \dot{y} = -4x; \quad x(0) = 2, \ y(0) = 4\sqrt{3}. \tag{3.2.8}$$

Since it would not make sense at this point to use these theorems, let it suffice for now to state that the solution of (3.2.7) is (cf. Example 10.3.7)

$$x(t) = 4\cos(2t - \pi/3) \tag{3.2.9}$$

and the solution of (3.2.8) is

$$x(t) = 4\cos(2t - \pi/3), \quad y(t) = -8\sin(2t - \pi/3). \tag{3.2.10}$$

(It is left as an exercise for the reader to verify these are truly solutions, to wit: $x(t)$ satisfies the initial value problem (3.2.7) while the pair of functions $(x(t), y(t))$ satisfies the initial value problem (3.2.8).) An important observation to make is that

$$\frac{x^2(t)}{16} + \frac{y^2(t)}{64} = 1.$$

As a result of this observation, we see that the trajectory, namely, the graph of

$$(x(t), y(t)) = (4\cos(2t - \pi/3), -8\sin(2t - \pi/3))$$

in the xy-plane, is an ellipse centered at the origin with its vertices at $(0, \pm 8)$. The point $(2, 4\sqrt{3})$, of course, is on this ellipse. What happens if the initial conditions are changed to $x(0) = a$ and $y(0) = b$, where (a, b) are a different pair of real numbers? It is left to the reader to show that the trajectory corresponding to these initial conditions is an ellipse with center $(0, 0)$ and vertices $(0, \pm\sqrt{4a^2 + b^2})$ and which contains the point (a, b). Since the values of a and b are arbitrary, we conclude that all of the trajectories of the system

$$\dot{x} = y, \ \dot{y} = -4x$$

are concentric ellipses centered at the origin. Nine of these elliptical trajectories are depicted in Fig. 3.7. Their graphs are superimposed on the direction field plotted in Fig. 3.6. The elliptical helix shown on the right is the graph of the solution

$$(t, x(t), y(t)) = (t, 4\cos(2t - \pi/3), -8\sin(2t - \pi/3))$$

for $t \geq 0$. It is the solution curve of (3.2.8) corresponding to the outermost trajectory (the one with the vertices $(0, \pm 8)$).

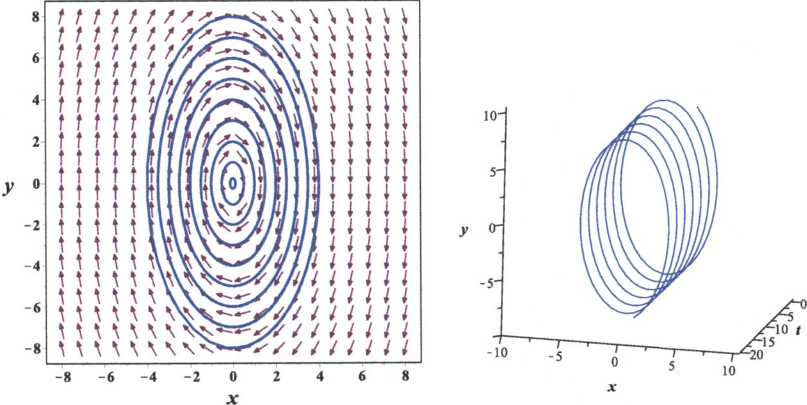

Fig. 3.7 Undamped mass-spring system: trajectories and a solution curve

For relatively simple differential equations, direction fields can be constructed by hand using pencil and paper, such as for the equations in Sects. 3.1 and 3.1.3 governing the motion of a drifting and accelerating yacht. For more complicated equations, such as the pair of equations modeling the motion of the piece of driftwood in Sect. 3.2, the tedium of calculating the slopes of direction arrows and positioning them correctly at all of the points on a grid is best left to and computers and graphing calculators. Another commonly used term for direction field is ***slope field***. Computer algebra systems, such as *Maple*, give the user the option of using short line segments in lieu of arrows. Graphing calculators as well as many textbooks use line segments. We prefer the terms *direction field* when direction arrows are used and *slope field* when line segments are used. Whether arrows or line segments are used, a direction field provides a means for visualizing the solutions of a differential equation without actually having to come up with them. This is an effective tool for depicting and analyzing the behavior of solutions, especially for a differential equation whose solutions are known to exist but which cannot be expressed in terms of the standard functions of calculus or for which there is no known formula for the general solution of this equation.

3.2.1 First-Order Equations

The remainder of this section is dedicated to reviewing and using the curve-sketching techniques taught in introductory calculus courses to sketch solutions of the first-order differential equation

$$\frac{dy}{dx} = f(x, y), \tag{3.2.11}$$

where the right-hand side represents a given function; and, as the notation indicates, this function generally involves both the independent variable x and the dependent variable y. Recall that the graph of a solution of a differential equation is also known as a ***solution curve***. That is to say, if a function $\varphi(x)$ is a solution of (3.2.11) on an interval J, then the locus of the set of points

$$\{(x, \varphi(x)) : x \in J\} \tag{3.2.12}$$

is a solution curve. The beauty of sketching a solution of (3.2.11) is that we are able to do so without actually knowing what the solution is, provided we are given its value at one point and f is a relatively simple function (i.e., no more complicated than the functions in a standard calculus course). We only need to know the derivative of the solution, which we already have! All of what we have just said is predicated on the solutions of the equation existing and being unique. Consider the following examples.

Example 3.2.1 Determine if there are any constant solutions of $y' = 5 - y$, and if so, graph them. Use the curve-sketching techniques of calculus to sketch the solution curves passing through the origin and the points $(0, -10)$ and $(0, 10)$.

Solution Observe that the right-hand side of the differential equation is equal to 0 if and only if $y = 5$. Consequently, $y(x) \equiv 5$ is the only constant solution. The right-hand side of the equation also tells us that $y' < 0$ when $y > 5$. This implies that the solution curves lying above the line $y = 5$ fall (from left to right). Those lying below this line rise since $y' > 0$ when $y < 5$.

Now let us use the second derivative to obtain information about the concavity of the solution curves. Since

$$y'' = \frac{d}{dx}(5 - y) = -y' = y - 5,$$

$y'' > 0$ when $y > 5$ and $y'' < 0$ when $y < 5$. So the solution curves lying above the line $y = 5$ are concave up while those lying below this line are concave down. With this information, we can easily sketch enough of them by hand to get a sense of what they all look like. In Fig. 3.8 the graphs of four solution curves passing through the points $(0, 10)$, $(0, 5)$, $(0, 0)$, and $(0, -10)$ are shown. For the sake of appearance and expediency, most graphs in this book, including those in Fig. 3.8 were drawn with *Maple*. However, the point of this example is that neither *Maple* nor any other CAS is needed to sketch such simple solution curves and that curve-sketching techniques promote an understanding of why solutions behave as they do.

With the method of separation of variables, it is easy to show that all solutions of $y' = 5 - y$ are of the form

$$y = 5 + Ce^{-x},$$

where C denotes a constant. Since different values of C always give different values of y for each value of x, it follows that no two solution curves can ever touch or

Fig. 3.8 Four solution curves of $y' = 5 - y$

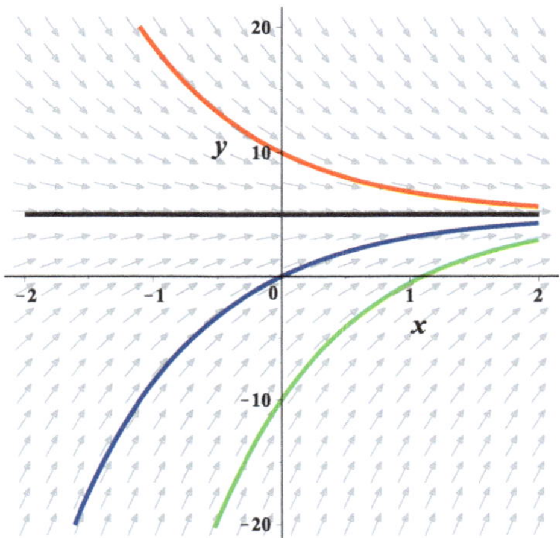

cross each other, even though Fig. 3.8 may suggest otherwise. Setting $x = x_0$ and $y = y_0$ and solving for C, we find that the solution satisfying the initial condition $y(x_0) = y_0$ is

$$y = 5 + (y_0 - 5)e^{-(x-x_0)}.$$

Therefore, the solution curves in Fig. 3.8 passing through the points (0, 10), (0, 5), (0, 0), and (0, −10) are graphs of the solutions

$$y = 5 + 5e^{-x}, \quad y = 5, \quad y = 5 - 5e^{-x}, \quad y = 5 - 15e^{-x},$$

respectively. ♦

Example 3.2.2 Sketch several solution curves of $y' = x^2(y - 5)$.

Solution First observe that $y' = 0$ if and only if $y = 5$. Thus $y(x) \equiv 5$ is the only constant solution. Solution curves lying above the line $y = 5$ rise (from left to right) as $x^2(y - 5) > 0$ and so $y' > 0$. But those lying below the line fall since $y' < 0$.

In order to determine the concavity of these curves, consider the second derivative:

$$\begin{aligned}\frac{d^2y}{dx^2} &= \frac{d}{dx}\big[x^2(y-5)\big] = x^2\frac{d}{dx}(y-5) + (y-5)\frac{d}{dx}x^2\\ &= x^2\frac{dy}{dx} + 2x(y-5) = x^2\big[x^2(y-5)\big] + 2x(y-5)\\ &= x^4(y-5) + 2x(y-5) = (x^4+2x)(y-5).\end{aligned}$$

Thus, $y'' = x(x^3 + 2)(y - 5)$. It is clear from this factored form that y'' may change sign at $x = 0$ or $x = \sqrt[3]{-2}$. So let us construct a sign chart for solution curves lying above the line $y = 5$ and then one for solution curves lying below this line. For $y > 5$, we have

	$x < \sqrt[3]{-2}$	$\sqrt[3]{-2} < x < 0$	$x > 0$
Sign of $y'' = x(x^3 + 2)(y - 5)$	Positive	Negative	Positive
Concavity	Concave up	Concave down	Concave up

The sign chart reveals changes in concavity at $x = \sqrt[3]{-2} \approx -1.26$ and $x = 0$. As a result, solution curves above the line $y = 5$ have inflection points at these values of x. For $y < 5$, we have

	$x < \sqrt[3]{-2}$	$\sqrt[3]{-2} < x < 0$	$x > 0$
Sign of $y'' = x(x^3 + 2)(y - 5)$	Negative	Positive	Negative
Concavity	Concave down	Concave up	Concave down

So we see that solution curves below the line $y = 5$ also have inflection points at $x = \sqrt[3]{-2}$ and $x = 0$. Five solution curves are shown in Fig. 3.9. ♦

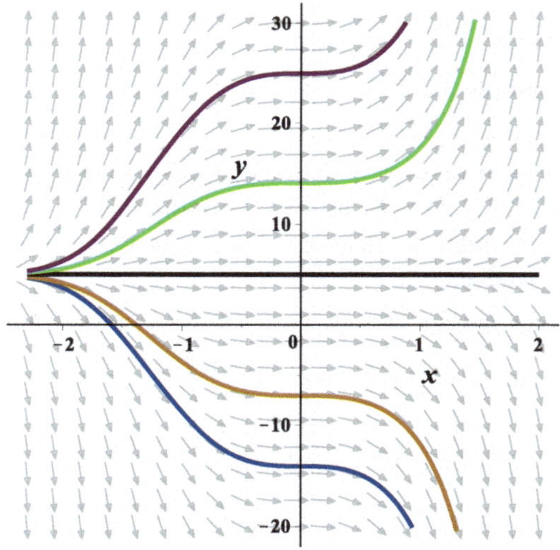

Fig. 3.9 Five solution curves of $y' = x^2(y - 5)$

Example 3.2.3 Given the information that the initial value problem

$$\frac{dy}{dx} = x^2 + y^2 - 1, \quad y(0) = 0, \tag{3.2.13}$$

has a solution, and only one solution, use calculus to determine important features of this solution, such as intervals of increase, decrease, and concavity and whether there are any local extrema and points of inflection.

Solution Let $y = f(x)$ designate the unique solution of (3.2.13). For now, let us focus on the behavior of this solution for $x \geq 0$. When the solution curve (graph of this solution) is inside the unit circle $x^2 + y^2 = 1$, it falls as x increases because y' is negative. This and $f(0) = 0$ implies that the solution curve will cross the unit circle at some point $(a, -\sqrt{1-a^2})$, where $0 < a < 1$. As soon as it crosses the unit circle, the solution curve begins to rise and continues to rise since y' is positive outside this circle. Since y' changes its sign from negative to positive at $x = a$, the solution $f(x)$ has a local minimum at this point.

Now consider the concavity of the solution curve for $x > 0$. Inside the unit circle,

$$y'' = 2x + 2yy' = 2[x + y(x^2 + y^2 - 1)] > 0$$

since $x > 0$, $y < 0$, and $x^2 + y^2 - 1 < 0$. Thus the solution curve is concave up on the interval $(0, 1)$. Moreover, it remains concave up outside the unit circle because y increases as x increases; so $y' = x^2 + y^2 - 1$ increases, which implies $y'' > 0$.

So far, we have determined that the graph of the solution $y = f(x)$ is concave up for $x > 0$; and that at some point $x = a$, where $0 < a < 1$, the solution decreases for $0 < x < a$ but then increases for $x > a$. The solution curve has a local minimum at the point $(a, f(a))$, which is located somewhere on the unit circle in Quadrant IV.

We could determine the behavior of the solution of (3.2.13) for $x < 0$ as above; however, the direction field for the differential equation (try imagining the solution curve in Fig. 3.10 not being there) suggests that the solution curve through the point $(0, 0)$ may be symmetric about this point. If that is truly the case, then we should be able to show that $f(-x) = -f(x)$. To that end, consider the function $g(x) := -f(u)$, where $u = -x$. By the chain rule, its derivative is

$$\begin{aligned}\frac{d}{dx}g(x) &= -\frac{d}{dx}f(u) = -\frac{d}{du}f(u)\cdot\frac{du}{dx} = \frac{d}{du}f(u) = u^2 + [f(u)]^2 - 1\\ &= (-x)^2 + [f(-x)]^2 - 1 = x^2 + [-g(x)]^2 - 1 = x^2 + g^2(x) - 1.\end{aligned}$$

In other words, the function $g(x)$ satisfies the differential equation in (3.2.3). Not only that, it also satisfies the initial condition since $g(0) = -f(0) = 0$. This and the additional information that the solution of (3.2.13) is unique implies that $g(x) \equiv f(x)$, where $g(x) = -f(-x)$. And so $-f(-x) = f(x)$, which confirms what we suspected earlier from viewing the direction field.

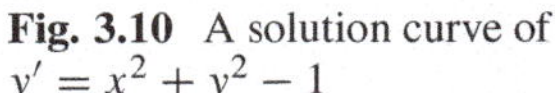

Fig. 3.10 A solution curve of $y' = x^2 + y^2 - 1$

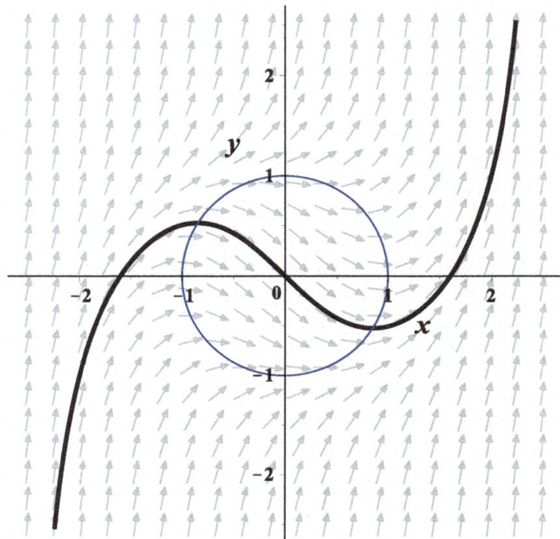

To summarize, we have shown that for some $a \in (0, 1)$, the solution curve through the origin rises for $x < -a$, falls on the interval $(-a, a)$, and rises for $x > a$. It has a point of inflection at the origin: the solution curve is concave down for $x < 0$ and concave up for $x > 0$. The point $(-a, -f(a))$, which is located on the unit circle in Quadrant II, is a local maximum and the point $(a, f(a))$, which is located on the unit circle in Quadrant IV, is a local minimum. See Fig. 3.10. ♦

Remark Judging from the direction field of $y' = x^2 + y^2 - 1$ and its solution curve through the origin, one might be inclined to think that this nondescript differential equation has relatively simple solutions. After all, the solution curve looks like the graph of a cubic polynomial. But it is not. One indicator of this is the rapid increase of the slopes of the tangent lines to the solution curve in Quadrant I. Recall that outside the unit circle, y increases as x increases, which causes $x^2 + y^2 - 1 = y'$ to increase very rapidly. In fact, so much so that the solution curve approaches infinity as $x \to b^-$, for a value of $b \in (2, 3)$, which can be seen by graphing the solution curve over a larger interval, such as $(-3, 3)$. Moreover, the solution curve approaches negative infinity as $x \to -b^+$. It turns out that solutions of this differential equation cannot be expressed in terms of the familiar functions of algebra, trigonometry, and elementary calculus. This equation is a member of the family of equations of the form

$$y' = a(x)y^2 + b(x)y + c(x),$$

which are known as *Riccati equations*. Generally speaking, solutions of these equations are quite complicated. Some of the more tractable of these are considered in Sect. 5.7 in Chap. 5.

3.2.2 Existence and Uniqueness of Solutions

When we sketched solution curves of the differential equations in the previous examples, we proceeded from a couple of tacit assumptions. One is that solutions exist. Another is that their graphs never touch or cross. Are these assumptions correct? A brief, albeit vague, answer is that *solutions of "nice" first-order equations exist and are unique*; so two solution curves can ever touch or cross each other. But what exactly are "nice" equations? One answer is contained in the next theorem, which is not proven here since a proof would divert attention from the current topic. However, a proof can be found in Chap. 9.

Existence and Uniqueness of Solutions

Theorem 3.2.1 *Let $f(x, y)$ be a function that is continuous on a rectangular region*

$$\mathcal{R} = \{ (x, y) : |x - x_0| \leq a, |y - y_0| \leq b \}, \tag{3.2.14}$$

where $a, b, x_0, y_0 \in \mathbb{R}$ and $a, b > 0$. Suppose there is a constant $L > 0$ such that

$$|f(x, y) - f(x, z)| \leq L|y - z| \tag{3.2.15}$$

for all (x, y) and (x, z) in $\mathcal{R}$. Then, for some $h \in (0, a]$, there exists a unique solution $y(x)$ of the initial value problem

$$\frac{dy}{dx} = f(x, y), \quad y(x_0) = y_0 \tag{3.2.16}$$

on the interval $[x_0 - h, x_0 + h]$.

Remarks The symbol $\mathbb{R}$ designates the set of all real numbers. A function $f(x, y)$ satisfying condition (3.2.15) is said to satisfy a ***Lipschitz condition*** with respect to y on the region $\mathcal{R}$ with ***Lipschitz constant*** L.

Example 3.2.4 In view of Theorem 3.2.1, let us take another look at the equation

$$y' = 5 - y \tag{3.2.17}$$

that was considered in Example 3.2.1. Use Theorem 3.2.1 to prove that a solution curve passes through each point of the xy-plane and that it is unique.

Solution Let (x_0, y_0) be an arbitrary point in the xy-plane. Suppose we ask if there is a solution satisfying the initial condition $y(x_0) = y_0$. In the notation of Theorem 3.2.1, $f(x, y) = 5 - y$. For a given pair of positive numbers a and b, let $\mathcal{R}$ designate the rectangular region

$$\mathcal{R} = \{ (x, y) : |x - x_0| \leq a, |y - y_0| \leq b \}.$$

Clearly, f is continuous everywhere in the xy-plane. Since

$$|f(x, y) - f(x, z)| = |(5 - y) - (5 - z)| = |y - z|,$$

for all $(x, y), (x, z) \in \mathcal{R}$, the function f satisfies a Lipschitz condition on $\mathcal{R}$ with Lipschitz constant $L = 1$. Hence, Theorem 3.2.1 implies (3.2.17) has a unique solution $y(x)$ on some interval $[x_0 - h, x_0 + h]$ satisfying the initial condition $y(x_0) = y_0$. Unfortunately the theorem, as stated here, does not say how to determine the value of h from the dimensions of the rectangular region $\mathcal{R}$—only that its value lies somewhere between 0 and a. So this rules out finding the interval of maximum length over which this solution exists by increasing the dimensions of $\mathcal{R}$. That information then would have to be obtained by other means. Nevertheless, the theorem is valuable in that it guarantees the existence of a unique solution of (3.2.17) satisfying the initial condition $y(x_0) = y_0$, irrespective of the values of x_0 and y_0. This means a solution curve passes through every point of the xy-plane and that no two solution curves can ever touch or cross each other. So even though it may appear that the solution curves in Fig. 3.8 might eventually run into each other, they will not.

Although Theorem 3.2.1 provides no information about the maximum lengths of the intervals over which solutions are defined, we already have the answer from Example 3.2.1. There it was pointed out that all solutions of (3.2.17) are of the form:

$$y = 5 + (y_0 - 5)e^{-(x - x_0)}.$$

It is left as an exercise to verify that this satisfies (3.2.17) for all $x \in (-\infty, \infty)$. ♦

Example 3.2.5 Prove that the solution of the initial value problem

$$y' = x^2(y - 5), \quad y(x_0) = y_0 \tag{3.2.18}$$

exists and is unique, regardless of the values of x_0 and y_0.

Solution Let $f(x, y) = x^2(y - 5)$. Let (x_0, y_0) be a point in the xy-plane. For any $a, b > 0$, let $\mathcal{R}$ be the rectangular region given by (3.2.14). Let M denote the maximum value of x^2 on the interval $[x_0 - a, x_0 + a]$. For any (x, y) and (x, z) in $\mathcal{R}$, we have

$$|f(x, y) - f(x, z)| = |x^2(y - 5) - x^2(z - 5)| = x^2|y - z| \leq M|y - z|.$$

Thus f satisfies a Lipschitz condition on $\mathcal{R}$. This and the continuity of f fulfills the hypotheses of Theorem 3.2.1. Therefore, (3.2.18) has a unique solution. ♦

3.3 Vertical Motion of a Body

In this section we begin with a derivation of a differential equation model of the vertical motion of a body as it moves either upward or downward in the earth's atmosphere near sea level assuming that the only force acting on the body is due to gravity. After that, we will consider more realistic models involving drag forces that retard the motion of bodies moving in the air.

3.3.1 Absence of a Drag Force

Consider a body that is moving either upward or downward in the atmosphere near sea level. Let y denote its height in feet above the ground, v its velocity, and t the time in seconds. From the definition of velocity, $v = dy/dt$. If $v > 0$, then $dy/dt > 0$. Thus from calculus we know that y is increasing and the body is moving upward. If $v < 0$, then the body is moving downward. For instance, $v = -100$ ft/sec means the body is moving downward at a speed of 100 ft/sec.

Suppose the force exerted by the air on the moving body retarding its motion is negligible or virtually absent, as would be the case of a bowling ball dropped from the top of a building or a feather falling inside a glass tube from which the air had been pumped out. Then the only significant force affecting the vertical motion of the body is due to gravity. Consequently (see Example 1.4.8),

$$\frac{dv}{dt} = -g. \tag{3.3.1}$$

Since $g \approx 32$ ft/sec near sea level, a direction field for (3.3.1) consists of the set of arrows shown in Fig. 3.11, where the slope of each arrow is -32. The direction field clearly depicts straight-line solution (or velocity) curves. A formula for these lines can be obtained by simply integrating (3.3.1) with the result

$$v = -gt + v_0 = -32t + v_0, \tag{3.3.2}$$

where $v_0 = v(0)$ denotes the initial velocity. Two of these lines are drawn on the direction field. The upper line is the graph of (3.3.2) when a body is initially propelled directly upward with a velocity of 128 ft/sec, such as a baseball thrown straight up by a pitcher; consequently, (3.3.2) becomes $v = -32t + 128$. After 4 seconds, we can see from the line that the body's velocity will be 0 ft/sec, which

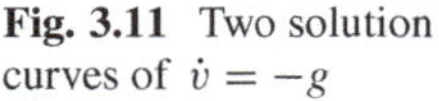
Fig. 3.11 Two solution curves of $\dot{v} = -g$

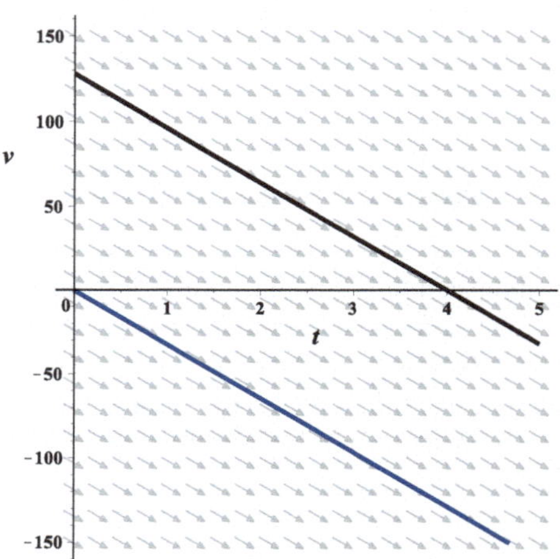

corresponds to the maximum height attained by the body before it begins falling back to the ground.

The lower line in the figure is the graph of (3.3.2) when its initial velocity is 0 ft/sec. This would be the case of a body being simply dropped rather than propelled upward. Thus, its initial velocity is $v_0 = 0$. Its velocity at time t is $v = -32t$. This line shows that the body's velocity will be -128 ft/sec after 4 seconds (unless it has already hit the ground). In other words, the body will be falling at a speed of 128 ft/sec when $t = 4$ seconds.

Let us conclude by also deriving a formula for the height of the body at any given time. Since the velocity v is the rate at which the height changes with time, equation $v = -32t$ can be written as the differential equation

$$\frac{dy}{dt} = -32t. \tag{3.3.3}$$

Since $\dot{y}(t) < 0$ and $\ddot{y}(t) = -32 < 0$ for $t > 0$, the solution curves of (3.3.3) are decreasing and concave down.[7] Integration of (3.3.3) yields

$$y = -16t^2 + y_0,$$

where $y_0 = y(0)$ denotes the height from which the body is dropped. Thus, a solution curve is the right half of a parabola opening downward with its vertex located at $(0, y_0)$. Figure 3.12 shows the direction field for (3.3.3) and two solution (or height) curves that give the heights of bodies dropped from initial heights of $y_0 = 100$ and $y_0 = 200$ feet.

[7] The overdot notation for derivatives, first introduced in Chap. 1, is commonly used in physics and engineering to indicate differentiation with respect to time.

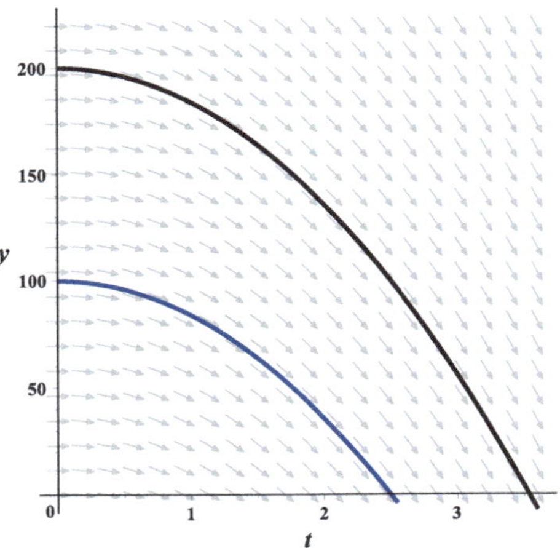

Fig. 3.12 Two solution curves of $\dot{y} = -32t$

3.3.2 *Velocity-Dependent Drag Force*

We just looked at a mathematical model for the vertical motion of bodies when the counteracting effect of air resistance was either absent, as in the case of a feather falling inside a long glass tube from which the air had been evacuated, or simply ignored, as in the case of a baseball thrown straight up in the air. Now we will consider a more realistic model for the baseball and other bodies whose motions are noticeably impeded, or retarded, by the air. Moreover, we will consider a model for bodies moving in a more viscous medium than air, such as a steel ball bearing dropped into a tall cylinder containing a heavy grade of motor oil. The retardation in the motion of a body is caused by an upward-pointing ***drag force*** which the medium exerts on the body. It has been determined experimentally that the drag force on a relatively small body moving at a speed less than roughly 80 feet per second is approximately proportional to the velocity of the body.[8] In other words, the quantity $-kv$, where k is a positive constant of proportionality, is a suitable model for drag forces at low speeds. The constant k is called the ***drag coefficient***; clearly, its value for a given body varies from medium to medium.

Let us see how the inclusion of a drag force changes the equation of motion (3.3.1). First recall from Sect. 1.4.4 that Newton's second law of motion for a body of constant mass m is

$$\mathbf{F} = m\frac{d\mathbf{v}}{dt},$$

where $\mathbf{F}$ is the resultant force, or vector sum of all the forces, acting on the body. In this section we will drop the vector notation, represented here by boldface letters,

[8] See Marion [57, p. 65] or Long and Weiss [53, p. 128].

since the force due to gravity is directed downward while the drag force is directed downward (resp. upward) if the body is moving upward (resp. downward). As we have been doing all along, we view upward as the positive direction and downward as the negative direction. Accordingly, adding the drag force $-kv$ to the gravitational force $-mg$ gives the resultant force

$$F = -mg - kv. \tag{3.3.4}$$

Consequently, the equation of motion of the body is

$$m\frac{dv}{dt} = -mg - kv. \tag{3.3.5}$$

Note that $-kv$ always gives the correct direction. If the motion of a body is downward, $v < 0$; and so $-kv > 0$, which indicates that the drag force points upward. The directions are reversed for upward motion.

We can obtain valuable qualitative information about the vertical motion of a body, without actually having to solve (3.3.5), by analyzing the signs of the first and second derivatives of v. A good way to begin is to divide both sides of (3.3.5) by m and then write it in the form

$$\frac{dv}{dt} = -\frac{k}{m}\left(v + \frac{mg}{k}\right). \tag{3.3.6}$$

Clearly the variables in (3.3.6) can be separated; consequently, this equation can be solved by using the method of separation of variables (see Chap. 2). There is also another way to solve the equation because it is also a member of the family of *first-order linear equations*, which we will be introduced to later on in Chap. 5. There we will learn that every first-order linear equation subject to a given initial condition has a solution. Moreover, this solution is unique in that it is the only solution which satisfies both the equation and the initial condition. Having this information and by analyzing the sign of $\dot{v}$, we can deduce the following:

- If $v = -mg/k$, then $\dot{v} = 0$. Clearly, $v(t) = -mg/k$ is a constant solution of (3.3.6). It follows from uniqueness of solutions that if the initial velocity of the body is $-mg/k$, then its velocity remains a constant $-mg/k$ for as long as the body is in motion.
- If $v > -mg/k$, then $\dot{v} < 0$; that is, the body is decelerating. Consequently, v is decreasing. If $v > 0$, this means that its speed in the upward direction is decreasing. On the other hand if $v < 0$, its speed in the downward direction is increasing.
- If $v < -mg/k$, then $\dot{v} > 0$ and so the body is accelerating. Hence its velocity v increasing. Since v is negative, the body is moving downward. An increasing v means that v is becoming less negative. In other words, the body is slowing down. Put another way, its speed is decreasing as it moves downward.

To determine the concavity of the velocity curves, we first differentiate (3.3.6) to obtain

$$\frac{d^2v}{dt^2} = -\frac{k}{m} \cdot \frac{dv}{dt}.$$

From this we see that $\ddot{v} > 0$ when $\dot{v} < 0$. Recall that $\dot{v} < 0$ when $v > -mg/k$. Also note that if the initial velocity of the body in greater than $-mg/k$, then its velocity will remain greater than $-mg/k$ for as long as the body is in motion by uniqueness of solutions. We conclude:

- The graph of v is concave upward if the velocity of the body is greater than $-mg/k$.

Since $\ddot{v} < 0$ when $\dot{v} > 0$, we also conclude:

- The graph of v is concave downward if the velocity of the body is less than $-mg/k$.

Note that with the curve-sketching tools of calculus we were able to obtain the above information without needing numerical values for the parameters m, k, and g. Now we have enough information to sketch velocity curves by hand. See Fig. 3.13. Actually the velocity curves in the figure were obtained by setting $g = 32\,\text{ft/sec}^2$ and $k/m = 8\,\text{sec}^{-1}$ and using *Maple* to draw the curves. With these values,

$$\frac{mg}{k} = \frac{1}{8\,\text{sec}^{-1}} \cdot \frac{32\,\text{ft}}{\text{sec}^2} = 4\,\text{ft/sec}$$

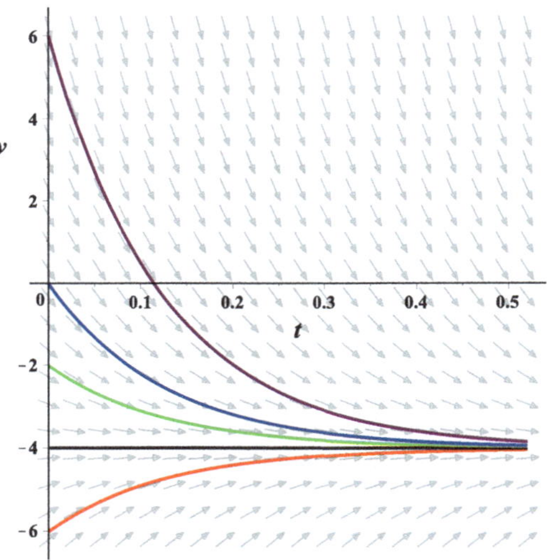

Fig. 3.13 Five velocity curves of $\dot{v} = -32 - 8v$

and (3.3.6) is

$$\frac{dv}{dt} = -8(v+4) = -32 - 8v.$$

Five solution curves corresponding to initial velocities of 6, 0, −2, −4, and −6 feet per second are shown superimposed on a direction field for this differential equation in Fig. 3.13. We can see that this model predicts that the velocity of the body will approach a limiting velocity of −4 ft/sec, regardless of its initial velocity. Moreover, if the body's initial velocity is $v_0 = -4$ ft/sec, its velocity will remain at this value.

It is worth noting here that the solution of equation (3.3.6) satisfying the initial condition $v(0) = v_0$ can be derived since the equation is separable. It is left as an exercise (see Problem 16) to show that the velocity v of the falling body at time t is

$$v(t) = -\frac{mg}{k} + \left(v_0 + \frac{mg}{k}\right) e^{-\frac{kt}{m}}. \tag{3.3.7}$$

This formula is well known and can be found in classical mechanics and engineering dynamics textbooks. It is evident from the formula that the velocity of a body will always approach the velocity $-mg/k$, regardless of the initial velocity v_0. This corroborates what we observed in our hypothetical example. This limiting velocity is known as the ***terminal velocity*** of the body.

Since a body that has attained its terminal velocity is no longer accelerating, the net force acting on it must be zero. So the quickest way to find the terminal velocity of a body is to solve the equation

$$F = 0$$

for the velocity v. Applying this to (3.3.4), we obtain $v = -mg/k$ without the need for the velocity formula (3.3.7).

3.3.3 Velocity- and Time-Dependent Drag Force

Let us imagine a situation where the drag force acting on a body falling through a medium depends not only on the body's velocity but on the time as well. For example, suppose a very tall transparent cylinder containing a warm viscous substance, such as a heavy grade of motor oil, is placed in something very cold, such as dry ice. See Fig. 3.14.

Suppose we were to drop a steel ball bearing into the cylinder in order to observe it falling through the oil. If the wall of the cylinder is a good conductor, heat energy will flow from the warm viscous oil to the colder environment outside the cylinder. As a result, the oil will cool. As it does, the viscosity of the oil increases (think about starting a car in midwinter). Finally, suppose it were possible to carry out this experiment with a cylinder tall enough so that the oil's viscosity increases noticeably

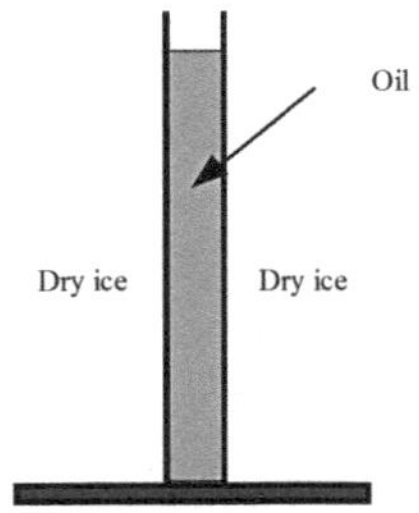

Fig. 3.14 Tall cylinder filled with motor oil

as the ball bearing is falling. It may not be worth the effort or even feasible to actually carry out this experiment—even so, we can try to imagine it in our mind's eye and suggest a mathematical model for the motion of the ball bearing. This kind of mental experiment is known as a ***thought experiment*** or ***Gedankenexperiment***.[9]

Now that the drag coefficient k is no longer constant, let us modify (3.3.5) by replacing the constant k with a time-dependent drag coefficient $\kappa(t)$, thereby obtaining

$$\frac{dv}{dt} = -g - \frac{\kappa(t)}{m}v. \tag{3.3.8}$$

So as to have a concrete example of (3.3.8) and a direction field for it along with some velocity curves, let us set $g = 32$ ft/sec^2 and suppose that $\kappa(t)/m = 8 + t$.[10] Then (3.3.8) becomes

$$\frac{dv}{dt} = -32 - (8 + t)v. \tag{3.3.9}$$

Clearly, this is not a separable equation, but it is a *first-order linear equation*. However, at this point in the book, we do not know yet how to solve this type of equation. If we did, then we could derive a formula for v to determine the velocity of the ball bearing at any time t, that is, until the ball bearing hits the bottom of the cylinder (see Problems 39 and 67 in Chap. 5). Nevertheless, we can still see how its solutions will behave over the course of time by looking at the direction field. The graphs of five velocity curves with the same initial velocities as in the previous subsection (so as to compare both models: see Figs. 3.13 and 3.15) are drawn on

[9] Albert Einstein developed his theory of relativity with the help of thought experiments that modeled the motion of bodies moving near the speed of light. Being German-born, he used the German word for these kinds of experiments, namely, *Gedankenexperimente*. Not only Einstein but other theoretical scientists as well, such as quantum physicists, rely on them. ***Gedanke*** is a German word for *thought* or *idea*.

[10] This is not an actual experimental model of the drag coefficient—keep in mind our purpose is to teach the mathematics of differential equations, even if that means dreaming up some stuff now and then.

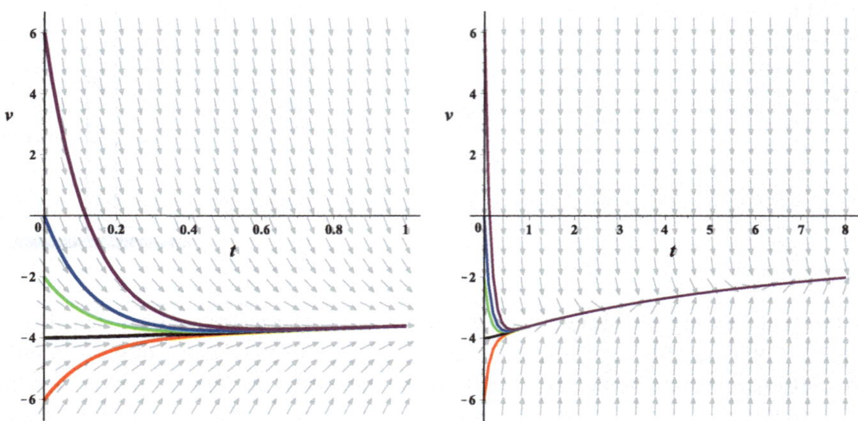

Fig. 3.15 Five velocity curves of $\dot{v} = -32 - (8 + t)v$

the direction field for (3.3.9) in Fig. 3.15. The velocity curves resemble those in Fig. 3.13; but, instead of approaching a terminal velocity of -4 ft/sec, the curves continue to rise and quickly merge so that they become virtually indistinguishable from one another. In fact, it can be shown that they will approach a limiting value of 0 ft/sec. Graphs of the velocity curves on the left in the figure are shown for $0 \leq t \leq 1$ while these same velocity curves are shown on the right for $0 \leq t \leq 8$.

Problems

> ...wisdom will come into your heart, and knowledge will be pleasant to your soul; discretion will watch over you; understanding will guard you ...
>
> Proverbs 2: 10–11
> New Oxford Annotated Bible (RSV)

Direction Fields

1. A direction field for $y' = (1 - x)y - x$ is shown in Fig. 3.16.
   (a) Calculate the slope of the direction arrow at the point (0, 1).
   (b) Use a CAS to reproduce the direction field and print it out. Sketch the solution curve passing through the point (0, 1).

2. A direction field for $\dfrac{dy}{dx} = \dfrac{x^2}{2} - \dfrac{y^2}{5}$ is shown in Fig. 3.17.
   (a) Calculate the slope of the direction arrow at the point $(-2, 4)$.
   (b) Use a CAS to reproduce the direction field and print it out. Sketch the solution curve passing through the point $(-2, 4)$.

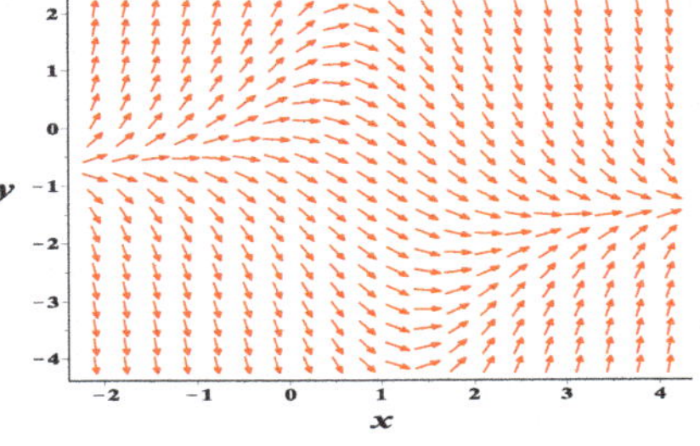

Fig. 3.16 Direction field for Problem 1

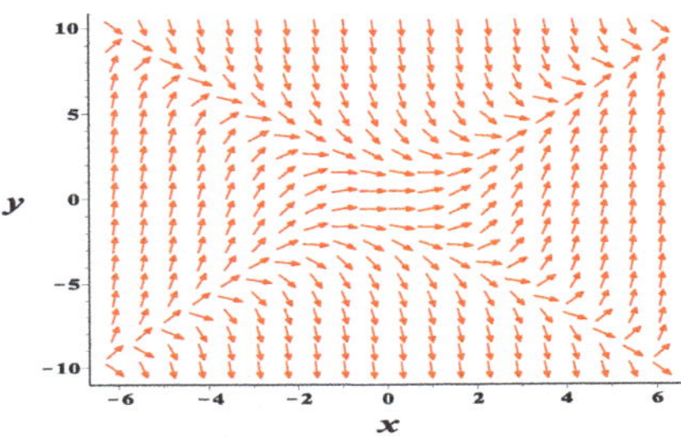

Fig. 3.17 Direction field for Problem 2

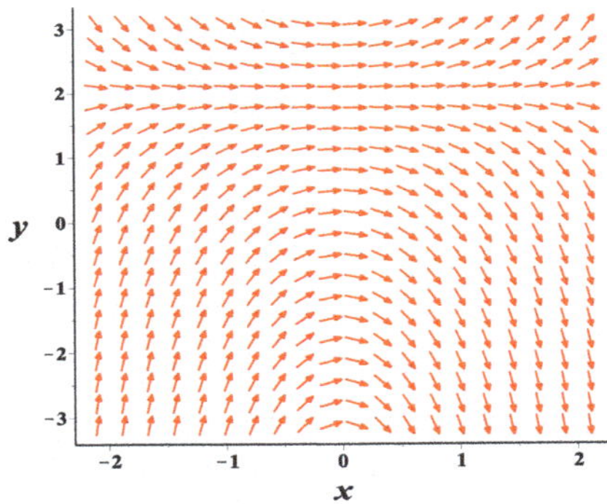

Fig. 3.18 Direction field for Problem 3

3. A direction field for $y' = xy - 2x$ is shown in Fig. 3.18.

   (a) Use a CAS to reproduce the direction field and print it out. Sketch the solution curve passing through the origin.

   (b) What is the slope of the direction arrow at the point $(2, -1)$?

   (c) Is there a constant solution? If so, find it and include its graph in the printout of the direction field.

4. A direction field for $y' = x^2 + y^2$ is shown in Fig. 3.19.

   (a) Sketch by hand the solution curve passing through $(0, 0)$ using calculus.

   (b) Calculate the slopes of the direction arrows at the points $(0, 0)$ and $(1, -1)$.

   (c) Use a CAS to reproduce the direction field and print it out. Sketch the solution curves passing through the points $(0, 0)$ and $(1, -1)$.

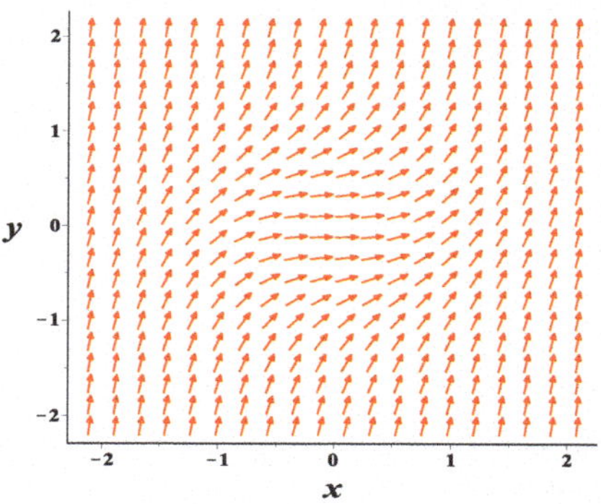

Fig. 3.19 Direction field for Problem 4

Solution Curves

In Problems 5 through 14, the phrase "sketch the solution curve of $y' = f(x, y)$" means to use the curve-sketching techniques of calculus to sketch it by hand. This entails using the first derivative $y' = dy/dx$ to determine where the solution is increasing and decreasing and to find any extrema and using the second derivative y'' to determine the concavity of the solution curve and to locate points of inflection if there are any. Furthermore, use any other important information that can be gleaned from the differential equation, such as the asymptotic behavior of solutions. Wherever $f(x_0, y_0)$ is defined in the following problems, it may be assumed that there is a unique solution curve passing through the point (x_0, y_0).

5. The following parts pertain to the differential equation $y' = y^2 + 4y$.
   (a) Find the slope of the direction arrow at the point $(x, y) = (0, -10)$.
   (b) Are there any constant solutions? If so, graph and label them.
   (c) Is the first derivative y' positive or negative when $y < -4$?
   (d) Is the second derivative y'' positive or negative when $y < -4$?
   (e) Sketch the solution curve crossing the y-axis at -10 on the set of axes used in part (b).
6. (a) Graph all constant solutions of $y' = 1 - y^2$.
   (b) Sketch the graph of the solution passing through the origin.
7. Follow the directions in Problem 6 for $y' = (y - 3)^2$.
8. (a) Graph any constant solutions of $y' = y^2 - y - 2$.
   (b) Sketch the solution curve that passes through the point (0, 1).
9. (a) For the differential equation $y' = x(4 - y)$, compute y''. Simplify the result and express it only in terms of x and y.
   (b) Sketch (on the same set of axes) some solutions of $y' = x(4 - y)$. Include any constant solutions. Clearly indicate where solutions are increasing and decreasing and where their graphs are concave up and concave down.
10. Follow the directions in Problem 9 for $y' = xy + 2x$.
11. Follow the directions in Problem 9 for $y' = e^{-x}(5 - y)$.
12. The following parts pertain to $y' = xy$.
    (a) Compute y''. Simplify and express it in terms of x and y.
    (b) Do not actually solve this differential equation; instead, use the first and second derivative tests of calculus to sketch (on the same set of axes) the solution curves that

pass through the points (0, 5), (0, −5), and (0, 0). Clearly state where the solutions are increasing and decreasing and where their graphs are concave up and concave down.

(c) Find the equation of the solution curve that passes through the point (0, 5). That is, solve the initial value problem: $y' = xy$, $y(0) = 5$.

13. Follow the directions in Problem 12 for $y' = x^2 y$.

14. Use a CAS to draw a direction field for $\dfrac{dy}{dx} = \dfrac{2xy - 5x}{x - y}$.

(a) If there are any constant solutions, include them in the direction field.

(b) Use the CAS to graph the solution curve through the origin.

(c) Use the CAS to graph the solution curve through the point (1, 4).

Motion of a Falling Body and Related Problems

15. If the yacht in Sect. 3.1 begins its acceleration x_0 feet directly west of the stationary buoy instead of directly north or south of it, how would this alter equations (3.1.8) and (3.1.9)?

16. Use the method of separation of variables to solve the differential equation (3.3.6) to verify that (3.3.7) is its solution.

17. The terminal speed of a certain skydiver, before he opens his parachute, is approximately 170 miles per hour. His total jumping weight, including equipment, is 260 pounds. Assume that the drag force exerted on him at any given instant is proportional to his velocity at that instant.[11] Find the value of the constant of proportionality k, the drag coefficient, for the 260-pound jumper with units included. Imagine that a year later, after going on a strenuous weight reduction program, his total jumping weight is down to 200 pounds. Use the calculated value for k to estimate his new terminal speed in miles per hour.

18. The differential equation $m\dot{v} = -mg - kv$ models the vertical motion of a body through a medium that exerts a drag force on the body that is approximately proportional to its velocity. For bodies moving faster than 80 ft/sec but slower than the speed of sound (1100 ft/sec), the drag force is approximately proportional to the square of the velocity of the body.[12]

(a) Using Newton's second law of motion, find the differential equation that models the vertical motion of a body when the drag force is proportional to the square of the body's velocity. Assume that the body begins its motion moving downward.

(b) Modify the differential equation in (a) to model the vertical motion of a body when it is moving upward.

(c) The ***signum function***[13] (also known as the ***sign function***) is defined by

$$\operatorname{sgn}(x) = \begin{cases} +1, & \text{if } x > 0 \\ 0, & \text{if } x = 0 \\ -1, & \text{if } x < 0. \end{cases}$$

Observe that $\operatorname{sgn}(x) = x/|x|$ when $x \neq 0$ and that it gives the sign of x. For example,

$$\operatorname{sgn}(0.5) = +1, \quad \operatorname{sgn}(1 - \sqrt{2}) = -1, \quad \text{and} \quad \operatorname{sgn}(\pi - 3) = +1.$$

---

[11] This assumption is actually inaccurate. Experiments show that the drag force on a skydiver is more or less proportional to the square of the skydiver's velocity. See Long and Weiss [53, p. 128].

[12] See Marion [57, p. 65] or Long and Weiss [53, p. 128].

[13] *Signum* is the Latin word for *sign*.

Use this function to combine the two differential equations found in (a) and (b) into one differential equation. This differential equation is the ***quadratic drag model*** of free fall.

(d) Find a constant solution of the quadratic drag model. Graph this constant solution. Include sketches of some of the other solutions using the curve-sketching techniques of calculus. What can you say about the long-term behavior of the solutions?

19. In a speed skating race, Liliya Ekaterina (Lily Catherine) skates down the straight portion of the home stretch, crosses the finish line with speed v_0, and then coasts. While Liliya is coasting to a stop, assume the resistive force F opposing her forward motion when her speed is v is

$$F = -bv^2,$$

where b is a positive constant. Let m denote her mass (including costume, accessories, and skates). Show that a model of Liliya's speed t units after crossing the finish line is

$$v(t) = \frac{mv_0}{m + bv_0 t}.$$

20. A high-altitude balloon carrying instrumentation is released from a weather station. Its combined mass is 5 kilograms. It ascends to an altitude of approximately 25 km. Eventually the balloon will burst and the assemblage of balloon and instrumentation will fall back to earth. When this happens, assume that the drag force opposing the downward motion of the assemblage will be proportional to its speed. Also, assume that the assemblage will reach half of its terminal speed in 10 seconds. How far will it have traveled during the first 10 seconds?

21. A skier, starting from rest, coasts down a steep slope of a downhill course that is located on the Hahnenkamm (German: rooster's comb) mountain in Austria.[14] Suppose the resistive force acting on the skier is a velocity-dependent drag force F_d that can be approximated by the formula

$$F_d = -20v$$

when the skier's speed is v meters per second, but due to favorable ski conditions the frictional force acting on the skis is negligible. If the mass of the skier is 80 kg and the grade of this particular slope of the course is a constant 65%, what is the fastest speed (kph or mph) that the skier can reach by merely coasting down the slope?

Computer Algebra System Problems

22. Plot a direction field of the quadratic drag model and the graphs of some of the velocity solutions to support your pencil-and-paper work in part (d) of Problem 18. First assign some convenient values to the parameters.

23. Plot a direction field for the yacht beginning its acceleration 200 feet directly west of the buoy instead of directly north or south of it. See Problem 15.

[14] The Hahnenkamm is part of the ski resort of Kitzbühel, Austria.

Chapter 4
Introduction to Numerical Methods

Why do I have to take differential equations?

Engineering major

It is agonizing to have to interrupt one's work to learn a new discipline, but scientists do it when we have to. At various times I have managed to take time off from what I was doing to learn all sort of things I needed to know ....

Steven Weinberg[1]

Abstract This chapter is an introduction to numerical methods featuring *Euler's method* and the *improved Euler's method* (often called *Heun's method*) and the *classical Runge-Kutta method* (*RK4 method*). The author, who taught at a university with a large number of engineering and science majors and was himself an erstwhile physics major, contends that part of the curriculum of an introductory differential equations class should include some basic numerical methods for approximating solutions of ODEs—simply because they are commonly used in undergraduate engineering and physics courses and are essential to modeling a wide spectrum of engineering systems. Furthermore, questions related to these methods appear now and then on the FE (Fundamentals of Engineering) exam, which is one of the first steps in the process of becoming a licensed professional engineer.

The narrative in Chap. 3 involving an accelerating yacht whose position in a lake is governed by a differential equation is used to introduce Euler's method. A sequence of connected line segments whose slopes are obtained from the differential equation is used to construct an *Euler polygonal curve* that approximates the actual path of the yacht. This example eventually culminates with *Euler's recursive formula* and the numerical algorithm known as Euler's method. This is followed by

[1] See [76, ch.VII]. Steven Weinberg (born 1933), an American theoretical physicist and the author of several popular books on physics, was one of the 1979 recipients of the Nobel prize in physics for his contributions to electroweak theory, a unified theory of the weak and electromagnetic forces. This theory predicted the existence of the W and Z particles, which were discovered in 1983 and 1984, respectively.

© The Author(s), under exclusive license to Springer Nature Switzerland AG 2026

L. C. Becker, *Ordinary Differential Equations: Concepts, Methods, and Models*,
https://doi.org/10.1007/978-3-032-15150-6_4

examples, where the first several numerical calculations are shown in detail along with comparisons of the relative errors of the approximations as the step size is reduced.

Due to time constraints and the fact that this textbook is written for readers who have only completed the first two calculus courses, the usual derivations of the Runge-Kutta methods of order two, such as the improved Euler's method, and the higher-order Runge-Kutta methods, such as the RK4 method (both of which involve Taylor polynomials in two variables and the attendant error analysis) is best left to a numerical analysis course. Instead, a less formal introduction to the improved Euler's method involving the trapezoidal rule of integration is used to improve upon Euler's method. Similarly, the RK4 method is introduced using Simpson's rule in order to enhance the accuracy of the improved Euler's method.

4.1 First-Order Initial Value Problems

Nearly all of the differential equations we have dealt with up to this point have been first-order equations of the form $y' = f(x, y)$, whose solutions can be obtained by employing the standard techniques of integration that are part of the syllabi of second semester calculus courses and which, by the way, were also reviewed in Chap. 1. In this chapter, we consider initial value problems of the form

$$\frac{dy}{dx} = f(x, y), \quad y(x_0) = y_0, \tag{4.1.1}$$

where x_0 and y_0 are given values, and the function f is "nice" in the sense that a unique solution of (4.1.1) is known to exist. (For example, see the existence and uniqueness theorem in Sect. 3.2.) However, let us add a new twist that has not come up before: suppose we simply do not know how to get an explicit or implicit formula for the solution of some initial value problem, or even if we do, suppose it is far too complicated to be of any practical use—or it simply does not exist. If that is the case, then we still have another option: a numerical approximation of the solution.

4.2 Tangent Line Approximations

We begin with an example of a yacht moving on a body of water (cf. Sect. 3.1) whose position is given by the solution $y(x)$ of an initial value problem of the form (4.1.1). Our immediate goal is to introduce a simple geometric procedure that we can use to numerically approximate values of this solution at a finite number of equally spaced points on the x-axis, which we label $x_0, x_1, x_2, \ldots, x_N$. We will approximate the values of the solution $y(x)$ at these points beginning with x_1 since the value of the solution at x_0 is already known from the initial condition, namely that $y(x_0) = y_0$. As we will see shortly, the procedure consists of repetitive computations whereby approximations of the solution at the points x_n are calculated using a so-called recursive formula, which is a formula that is used over and over again until an approximation of the solution at the last point x_N is obtained. The end result will be numerical approximations to the actual values of $y(x_n)$ for $n = 1, 2, \ldots, N$. A procedure for numerically approximating the solution of an initial value problem at a finite number of discrete points, such as the one we are about to consider, by means of a repetitive succession of computations using a recursive formula (or formulas) is known as a **numerical algorithm**.

Imagine a yacht departing from some fixed location (which we regard as the origin of a rectangular coordinate system) whose subsequent locations are determined by the differential equation

$$\frac{dy}{dx} = \frac{5}{\sqrt{x+25}}. \tag{4.2.1}$$

This equation also governed the motion of the accelerating yacht in Chap. 3. It gives the slope of the direction arrow at (x, y), thereby indicating the instantaneous direction in which the yacht is moving at that point. Recall that the solution of (4.2.1) satisfying the initial condition $y(0) = 0$ is (see (3.1.10))

$$y = 10\sqrt{x+25} - 50. \tag{4.2.2}$$

A graph of this solution is shown in Fig. 3.4 (lower curve).

Up to now, we have solved differential equations using algebraic operations and integration techniques, which resulted in a function (or an expression implicitly defining one) that satisfies the equation on some interval. This kind of solution is called an ***analytical solution***. For instance, (4.2.2) is an analytical solution of (4.2.1). However, if the right-hand side of (4.2.1) were changed to make the equation nonseparable, then we would be at a loss attempting to find an analytical solution since we have only been introduced to one method so far for solving differential equations, to wit: the method of separation of variables. Worse yet, if the right-hand side of (4.2.1) were altered, it could result in an equation whose solution still exists in the sense that it determines a unique y-value for each x-value—but which cannot be expressed by a formula. In such situations, if quantitative information about the solutions of a differential equation are needed, then the only option is to abandon *analytical methods* and turn to *numerical methods*.

The simplest numerical method for (4.1.1) is called ***Euler's method***.[2] It is easily explained with geometry; however, it is the least accurate numerical method and thereby rarely ever used in practice. Nevertheless, it is the basis for more sophisticated and accurate numerical methods. As we shall see shortly, it can be used to approximate the actual graph of a solution of a differential equation with a

[2] The method is named in honor of the Swiss mathematician Leonhard Euler (1707–1783). His surname rhymes with "oiler."

sequence of connected line segments, each of which is parallel to one of the arrows in the direction field. This collection of connected line segments is called an ***Euler polygonal curve***.

We introduce Euler's method by describing a graphical and numerical scheme for handling the differential equation (4.2.1), pretending for now that we do not know how to solve it analytically even though we really do as attested by (4.2.2), namely, the solution satisfying the initial condition $y(0) = 0$. Suppose we wish to approximate (roughly) the y-coordinates (in feet) of the yacht's positions corresponding to the following x-coordinates (in feet): 0, 100, 200, 300, … . The slope of the direction arrow at the point $(0, 0)$ is the value of the derivative of the solution of (4.2.1) at $x = 0$:

$$\left.\frac{dy}{dx}\right|_{x=0} = \left.\frac{5}{\sqrt{x+25}}\right|_{x=0} = \frac{5}{\sqrt{0+25}} = 1.$$

Let us approximate the path of the yacht from $x = 0$ to $x = 100$ with a straight line segment with slope 1, even though we understand that this is a crude approximation of the path, not the actual path. Now let us use the endpoint of this line segment to approximate the y-coordinate of the yacht's position at $x_1 = 100$. Note that this line segment is tangent to the actual path at the starting point $(0, 0)$. For this problem (and all future ones), we will denote the starting (initial) point by (x_0, y_0). In this example, $x_0 = 0$ and $y_0 = 0$. Since the slope m of a line is the ratio of its rise to its run, we have

$$\frac{\text{rise}}{\text{run}} = \frac{y - y_0}{x - x_0} = m,$$

where (x, y) denotes any point on the line segment distinct from (x_0, y_0). Multiplying both sides of the equation by $x - x_0$, we obtain the *point-slope form* of the equation of the line segment, namely,

$$y - y_0 = m(x - x_0).$$

Since $(x_0, y_0) = (0, 0)$ and $m = 1$, we have

$$y - 0 = 1 \cdot (x - 0) \quad \Rightarrow \quad y = x.$$

So the y-coordinate of the yacht's position is roughly 100 feet when its x-coordinate is 100 feet. Let y_1 denote this value. Then we can repeat the foregoing steps, using that $y_1 = 100$, to approximate the y-coordinate of the yacht's position at $x_2 = 200$ feet. Denote this approximation by y_2. The slope m_1 of the direction arrow at the point $(x_1, y_1) = (100, 100)$ is

$$m_1 = \left.\frac{dy}{dx}\right|_{x=100} = \left.\frac{5}{\sqrt{x+25}}\right|_{x=100} = \frac{5}{5\sqrt{5}} = \frac{\sqrt{5}}{5}.$$

This is the same as the slope of the tangent line to the graph of the solution passing through (100, 100). Hence, by the point-slope formula, the equation of the tangent line with this slope is

$$y - y_1 = m_1(x - x_1) \quad \Rightarrow \quad y - 100 = \frac{\sqrt{5}}{5}(x - 100).$$

Letting $x = 200$, we find that

$$y_2 = 100 + \frac{\sqrt{5}}{5}(200 - 100) = 100 + 20\sqrt{5} \approx 145 \text{ ft.}$$

Now repeat this process using x_2 and y_2 to approximate the y-coordinate of the yacht's position at $x = x_3$ feet, that is, at $x = 300$ feet. The slope of the direction arrow at the point $(x_2, y_2) = (200, 100 + 20\sqrt{5})$ is

$$m_2 = \left.\frac{dy}{dx}\right|_{x=200} = \left.\frac{5}{\sqrt{x+25}}\right|_{x=200} = \frac{5}{15} = \frac{1}{3}.$$

So the equation of the tangent line to the solution curve through this point is

$$y - y_2 = m_2(x - x_2) \quad \Rightarrow \quad y - (100 + 20\sqrt{5}) = \frac{1}{3}(x - 200).$$

Thus, the y-coordinate of the yacht's position at $x_3 = 300$ is roughly

$$y_3 = 100 + 20\sqrt{5} + \frac{1}{3}(300 - 200) = \frac{400}{3} + 20\sqrt{5} \approx 178 \text{ ft.}$$

Continuing this process up to $x_7 = 700$ feet, we obtain the approximations y_n of the values of $y(x_n)$ for $n = 0, 1, 2, \ldots, 7$. See Table 4.1.

Table 4.1 Tangent line approximations

n	x_n	y_n
0	0	0
1	100	100
2	200	$100 + 20\sqrt{5} \approx 145$
3	300	$\frac{400}{3} + 20\sqrt{5} \approx 178$
4	400	$\frac{400}{3} + 20\sqrt{5} + \frac{100}{13}\sqrt{13} \approx 206$
5	500	$\frac{400}{3} + 20\sqrt{5} + \frac{100}{13}\sqrt{13} + \frac{100}{17}\sqrt{17} \approx 230$
6	600	$\frac{400}{3} + 20\sqrt{5} + \frac{100}{13}\sqrt{13} + \frac{100}{17}\sqrt{17} + \frac{100}{21}\sqrt{21} \approx 252$
7	700	$\frac{460}{3} + 20\sqrt{5} + \frac{100}{13}\sqrt{13} + \frac{100}{17}\sqrt{17} + \frac{100}{21}\sqrt{21} \approx 272$

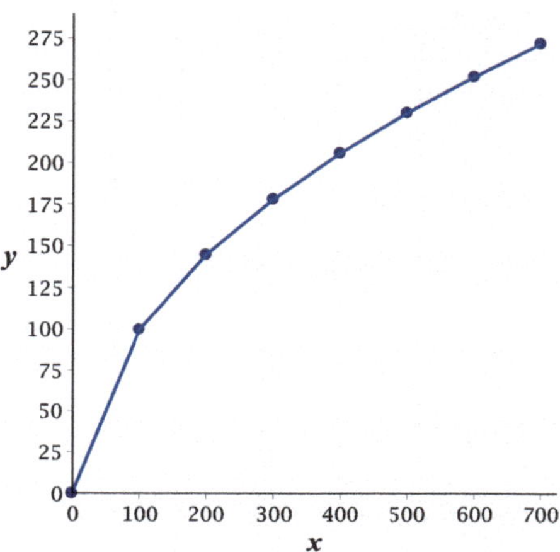

Fig. 4.1 An Euler polygonal curve approximation of the yacht's path

Now let us plot the points (x_n, y_n) for $n = 1, 2, \ldots, 7$ and then connect them with line segments. The result is the Euler polygonal curve shown in Fig. 4.1. We should not expect this polygonal curve to represent the actual path of the yacht all that well since each of these jumps, or ***steps***, are 100 feet long. The length of a step is called the ***step size***.

The Euler polygonal curve would better depict the actual path of the yacht if the step size were shortened. Let us illustrate this by reducing the step size to 50 feet. So now we need the slopes of tangent line segments every 50 feet instead of every 100 feet. Consequently, the first tangent line segment, $y = x$, will no longer be used as an approximation of the path all the way to $x = 100$ feet, but rather only to $x = 50$ feet. By (4.2.1), the slope of the second tangent line segment at $x = 50$ feet is

$$m_1 = \left.\frac{dy}{dx}\right|_{x=50} = \left.\frac{5}{\sqrt{x+25}}\right|_{x=50} = \frac{1}{\sqrt{3}}.$$

Thus the equation of the second line segment parallel to the direction arrow at the point $(x_1, y_1) = (50, 50)$ is

$$y - y_1 = m_1(x - x_1) \quad \Rightarrow \quad y - 50 = \frac{\sqrt{3}}{3}(x - 50),$$

which is also tangent to the solution curve passing through the point $(50, 50)$. Letting $x = 100$, we obtain the following approximation of the y-coordinate at $x_2 = 100$:

$$y_2 = 50 + \frac{\sqrt{3}}{3}(100 - 50) = 50 + \frac{50\sqrt{3}}{3} \approx 79 \text{ ft}.$$

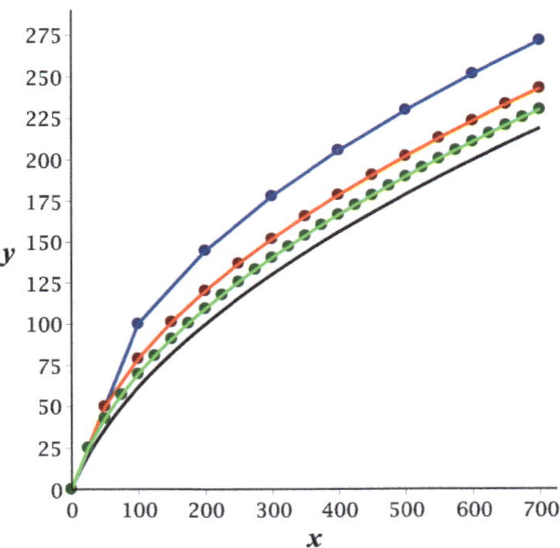

Fig. 4.2 Three Euler polygonal curve approximations and the actual path

By repeating these calculations until $x = 700$ feet is reached, we will have the points (x_n, y_n) for $n = 1, 2, \ldots, 14$ that are needed to construct an Euler polygonal curve with 14 connected tangent line segments instead of just 7. In Fig. 4.2, this curve is directly below the one consisting of 7 tangent line segments.

For comparison's sake, let us reduce the step size to 25 feet. Now the polygonal curve consists of 28 ($700 \div 25$) connected tangent line segments. Once again refer to Fig. 4.2 to view three Euler polygonal curves together with the actual path, which is the bottom curve in the figure. As expected, Euler polygonal curves look more and more like the actual path every time the step size is reduced. ♦

4.3 Euler's Method

The actual path of the yacht depicted in Fig. 4.2 (bottom curve) is the graph of the solution of the initial value problem

$$\frac{dy}{dx} = f(x), \quad y(0) = 0, \tag{4.3.1}$$

where $f(x) = 5/\sqrt{x + 25}$. The right-hand side of this particular differential equation only involves the independent variable x. However, generally speaking, the right-hand side of first-order equations expressed in this form will involve the dependent variable y as well. Let us examine each of the steps of the geometric procedure that we introduced in the previous section so that we can come up with a numerical algorithm to handle any initial value problem of the form

$$\frac{dy}{dx} = f(x, y), \quad y(x_0) = y_0 \tag{4.3.2}$$

without having to explicitly construct tangent lines as we did for equation (4.2.1). In the ensuing discussions, let us assume that the function $f(x, y)$ on the right-hand side of the differential equation satisfies conditions guaranteeing not only the existence of a solution of the initial value problem (4.3.2) but also its uniqueness.

In the example of the yacht, we found approximate values of the solution $y(x)$ of the initial value problem (4.3.1) at preselected evenly spaced points along the x-axis. These points are known as ***mesh points***. The distance between consecutive mesh points is called the ***step size***. Usually we will use the letter h to denote its value, although at times we will also use the symbol Δx.

Now let us turn to the general initial value problem (4.3.2). Suppose we need approximate values of the solution of (4.3.2) at a number of evenly spaced mesh points up to and including some point b to the right of x_0. One way to begin is to first assign the step size h some appropriate value. Then the goal will be to find approximate values of the solution at the mesh points $x_0 + h$, $x_0 + 2h$, $x_0 + 3h$, ... until the point b is reached. Most likely, however, b will not be a mesh point; so we need to keep approximating until we reach the first mesh point past b. This will allow us to approximate the value of the solution at b by interpolation.

Another way we could begin is to choose the number of approximations that we would like to have over the interval $[x_0, b]$. With N representing this number, the mesh points are found by first dividing the interval into N subintervals of equal length h, where

$$h = \frac{b - x_0}{N}.$$

Then the mesh points are $x_0, x_1, x_2, x_3, \ldots, x_N$, where

$$\begin{aligned}
x_1 &= x_0 + h \\
x_2 &= x_1 + h = x_0 + 2h \\
x_3 &= x_2 + h = x_0 + 3h \\
&\vdots \qquad \vdots \qquad \vdots
\end{aligned}$$

From the pattern and Fig. 4.3, we see that $x_n = x_0 + nh$ for $n = 0, 1, 2, \ldots, N$. Note that N is the *number of steps* that it takes to reach the point b:

$$x_N = x_0 + Nh = x_0 + N\left(\frac{b - x_0}{N}\right) = b.$$

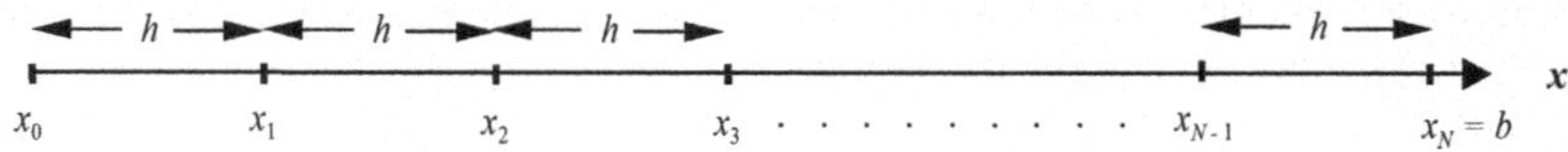

Fig. 4.3 The mesh points $x_0, x_1, x_2, \ldots, x_N$

We are now at a place where we can explain how to approximate the solution curve of (4.3.2) with an Euler polygonal curve and at the same time calculate approximations of the solution at the mesh points. Let us begin by approximating a portion of the solution curve with the line segment that starts at the initial point (x_0, y_0) and ends at a point (x_1, y_1) and which has the same slope as that of the direction arrow at (x_0, y_0). Recall from the previous chapter that the slope of this direction arrow is equal to the slope of the tangent line to the solution curve at (x_0, y_0). Although we do not know the value of y_1 yet, it can be found as follows: First, the slope of the line segment is equal to $y'(x_0)$, namely, the derivative of the solution at x_0. It follows from the differential equation (4.3.2) that

$$y'(x_0) = f(x_0, y_0).$$

Second, the slope of the line segment is also equal to

$$\frac{y_1 - y_0}{x_1 - x_0}.$$

Third, by equating these two expressions for the slope, we have

$$\frac{y_1 - y_0}{x_1 - x_0} = f(x_0, y_0).$$

Thus,

$$y_1 = y_0 + hf(x_0, y_0),$$

where $h = x_1 - x_0$. We will use this value to approximate $y(x_1)$, which denotes the actual value of the solution of (4.3.2) at x_1. The line segment with end points (x_0, y_0) and (x_1, y_1) is the first line segment constituting the Euler polygonal curve. This is illustrated in Fig. 4.4, where the curve through the point (0, 2) is the solution curve of the initial value problem

$$\frac{dy}{dx} = \frac{x-1}{y}, \quad y(0) = 2. \tag{4.3.3}$$

The Euler polygonal curve in the figure was generated using $h = 1$. Since $x_0 = 0$, $y_0 = 2$, and $f(x, y) = y^{-1}(x - 1)$,

$$y_1 = y_0 + hf(x_0, y_0) = y_0 + h \cdot \frac{x_0 - 1}{y_0} = 2 + \frac{0-1}{2} = 1.5.$$

Thus, the left endpoint of the first line segment of this curve is $(x_0, y_0) = (0, 2)$ and its right endpoint is $(x_1, y_1) = (1, 1.5)$.

Next, let us determine the second line segment of the Euler polygonal curve. It begins at (x_1, y_1) and ends at a point (x_2, y_2), where y_2 denotes the approximation of $y(x_2)$, which is the actual value of the solution of (4.3.2) at x_2. Again we turn to the differential equation to compute the slope of the line segment at $x = x_1$:

$$y'(x_1) = f(x_1, y_1).$$

Consequently, this line segment is parallel to the direction arrow located at the point (x_1, y_1), making it tangent to the solution curve that passes through (x_1, y_1). Since (x_1, y_1) and (x_2, y_2) are the endpoints of the second line segment, it follows that

$$\frac{y_2 - y_1}{x_2 - x_1} = f(x_1, y_1).$$

Thus the value of y_2 is given by

$$y_2 = y_1 + hf(x_1, y_1).$$

For example, consider the initial value problem (4.3.3) again . The y-coordinate of the right endpoint of the second line segment of the Euler polygonal curve in Fig. 4.4 is

$$y_2 = y_1 + hf(x_1, y_1) = y_1 + h \cdot \frac{x_1 - 1}{y_1} = 1.5 + \frac{1 - 1}{1.5} = 1.5.$$

So the right endpoint of the second line segment of this curve is $(x_2, y_2) = (2, 1.5)$. The middle curve is the solution curve of (4.3.3) through its left endpoint $(1, 1.5)$. Note that the second line segment is horizontal and tangent to the middle solution curve at this point. It is left as an exercise to show that the right endpoint of the third line segment of the Euler polygonal curve is $(x_3, y_3) = (3, 13/6)$.

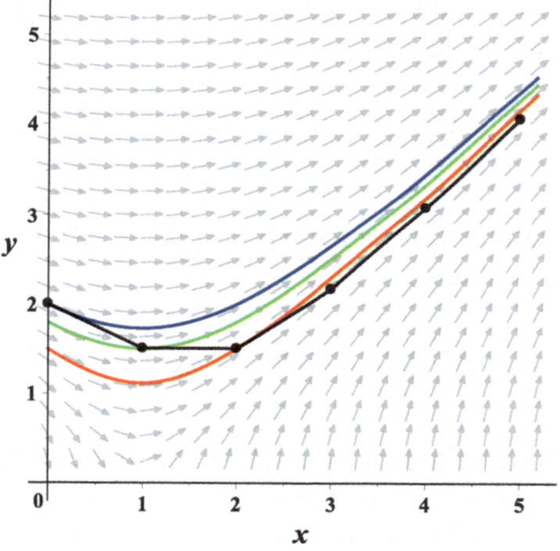

Fig. 4.4 Euler polygonal curve approximation of the solution curve of (4.3.3)

Continuing as before, we find that the third approximation is

$$y_3 = y_2 + hf(x_2, y_2)$$

while the fourth is

$$y_4 = y_3 + hf(x_3, y_3).$$

So, clearly

$$y_{n+1} = y_n + hf(x_n, y_n)$$

for $n = 0, 1, 2, 3, \ldots$. These calculations are repeated over and over again in order to obtain approximations of the actual values of the solution of the initial value problem (4.3.2) at the equally spaced points $x_1, x_2, x_3, x_4, \ldots$ until a desired stopping point on the x-axis is reached.

The process that we have just described amounts to little more than a finite number of repetitive calculations, whereby the coordinates of a point (x_{n+1}, y_{n+1}) are calculated using the coordinates of the preceding point (x_n, y_n) by means of the formulas:

$$x_{n+1} = x_n + h \quad \text{and} \quad y_{n+1} = y_n + hf(x_n, y_n).$$

In particular, the second of these formulas is known as ***Euler's recursive formula***. A recursive formula, such as this one, is a special type of formula that is used repeatedly for the purpose of calculating the value of a quantity in terms of its previous value. The following algorithm is known as ***Euler's method***.

Euler's Method

To approximate the values of the solution $y(x)$ of the initial value problem

$$\frac{dy}{dx} = f(x, y), \quad y(x_0) = y_0 \tag{4.3.4}$$

at equally spaced mesh points $x_0, x_1, x_2, \ldots$, execute the following steps:

1. Choose an appropriate step size h.
2. Let $y_0 = y(x_0)$. Then find y_{n+1} from y_n for $n = 0, 1, 2, \ldots$ as follows:

   (a) Calculate the slope $f(x_n, y_n)$.
   (b) Then calculate x_{n+1} and y_{n+1} using the recursive formulas

$$x_{n+1} = x_n + h \tag{4.3.5}$$

(continued)

$$y_{n+1} = y_n + hf(x_n, y_n). \tag{4.3.6}$$

3. Stop the calculations in Step 2 when the desired stopping point x_N on the x-axis is reached.

Remark Formula (4.3.6) is also known as Euler's formula. However, this is also the name of the formula $e^{it} = \cos t + i \sin t$ (see Chaps. 10 and 12). In order to avoid any confusion, we will always refer to (4.3.6) as Euler's recursive formula.

Example 4.3.1 Apply Euler's method to the initial value problem

$$\frac{dy}{dx} + 2xy = 1, \quad y(0) = 0. \tag{4.3.7}$$

Solution In order to use this method, we have to rewrite the differential equation in the form $y' = f(x, y)$. That is,

$$y' = 1 - 2xy.$$

Since $f(x, y) = 1 - 2xy$, it follows from (4.3.6) that Euler's recursive formula is

$$y_{n+1} = y_n + h(1 - 2x_n y_n).$$

From the initial condition $y(0) = 0$, we have $x_0 = 0$ and $y_0 = 0$. Suppose we choose $h = 0.1$, then the recursive formulas are

$$x_{n+1} = x_n + 0.1 \tag{4.3.8}$$

$$y_{n+1} = y_n + 0.1(1 - 2x_n y_n). \tag{4.3.9}$$

This pair of formulas will be used over and over until the desired stopping point on the x-axis is reached. For example, suppose we want to approximate the value of the solution at $x = 0.3$. Since $x_0 = 0$ and $h = 0.1$, this will entail using Euler's recursive formula three times in order to reach this point. Setting $n = 0$, we have

$$x_1 = x_0 + 0.1 = 0 + 0.1 = 0.1$$

$$y_1 = y_0 + 0.1(1 - 2x_0 y_0) = 0 + 0.1(1 - 2(0)(0)) = 0.1.$$

Now set $n = 1$:

$$x_2 = x_1 + 0.1 = 0.1 + 0.1 = 0.2$$

$$y_2 = y_1 + 0.1(1 - 2x_1 y_1) = 0.1 + 0.1(1 - 2(0.1)(0.1)) = 0.198.$$

Finally, set $n = 2$:

$$\begin{aligned} x_3 &= x_2 + 0.1 = 0.2 + 0.1 = 0.3 \\ y_3 &= y_2 + 0.1(1 - 2x_2y_2) = 0.198 + 0.1(1 - 2(0.2)(0.198)) = 0.29008. \end{aligned}$$

Hence, for the step size $h = 0.1$,

$$y(0.3) \approx 0.29008.$$

In Chap. 5, we will discuss a method that can be used to solve the initial value problem (4.3.7) analytically. In fact, for the initial value problem (5.2.36) in that chapter, we show that the solution is

$$y(x) = e^{-x^2} \int_0^x e^{u^2}\, du.$$

It is left as an exercise to verify that this function truly solves (4.3.7). Consequently, the actual value of $y(0.3)$ to 4 decimal places is

$$y(0.3) = e^{-(0.3)^2} \int_0^{0.3} e^{u^2}\, du = 0.2826. \qquad \blacklozenge$$

We used only one step size, namely $h = 0.1$, to calculate approximations of the solution of the initial value problem in the foregoing example. In the next example, we will consider another initial problem; however, we will choose several different step sizes in order to compare the resulting approximations.

Example 4.3.2 For the initial value problem

$$y' = 2x + y - 2, \quad y(0) = 1, \tag{4.3.10}$$

determine the appropriate Euler recursive formulas to use in order to approximate the values of the solution at $x = 0.1, 0.2, \ldots, 0.9, 1.0$. Choose the following three values for the step size h: $0.1, 0.01, 0.001$. Compare the results.

Solution Since $f(x, y) = 2x + y - 2$,

$$f(x_n, y_n) = 2x_n + y_n - 2.$$

So the recursive formulas are

$$x_{n+1} = x_n + h \tag{4.3.11}$$

$$y_{n+1} = y_n + hf(x_n, y_n) = y_n + h(2x_n + y_n - 2), \tag{4.3.12}$$

where $x_0 = 0$ and $y_0 = 1$.

Let us begin by choosing the value of the step size to be $h = 0.1$. Then the recursive formulas become

$$x_{n+1} = x_n + 0.1$$
$$y_{n+1} = y_n + 0.1(2x_n + y_n - 2).$$

Setting $n = 0$, we obtain an approximation of $y(0.1)$ as follows:

$$x_1 = x_0 + 0.1 = 0 + 0.1 = 0.1$$
$$y_1 = y_0 + 0.1(2x_0 + y_0 - 2) = 1 + 0.1(2(0) + 1 - 2) = 0.9.$$

Continuing with $n = 1$, we get

$$x_2 = x_1 + 0.1 = 0.1 + 0.1 = 0.2$$
$$y_2 = y_1 + 0.1(2x_1 + y_1 - 2) = 0.9 + 0.1(2(0.1) + 0.9 - 2) = 0.81.$$

And so $y(0.2) \approx 0.81$. With $n = 2$, we have

$$x_3 = x_2 + 0.1 = 0.2 + 0.1 = 0.3$$
$$y_3 = y_2 + 0.1(2x_2 + y_2 - 2) = 0.81 + 0.1(2(0.2) + 0.81 - 2) = 0.731.$$

Since all of these calculations are too laborious to do by hand, the rest of them were obtained with a computer. The results are shown in the second column in Table 4.2.

With $h = 0.01$, the recursive formulas are

$$x_{n+1} = x_n + 0.01$$
$$y_{n+1} = y_n + 0.01(2x_n + y_n - 2).$$

Table 4.2 Comparison of approximations of (4.3.10) using Euler's method

x	$h = 0.1$	$h = 0.01$	$h = 0.001$	$h = 0.0001$	**Actual value**
0.1	0.900000	0.904622	0.905116	0.905165	0.905171
0.2	0.810000	0.820190	0.821281	0.821391	0.821403
0.3	0.731000	0.747849	0.749656	0.749839	0.749859
0.4	0.664100	0.688864	0.691527	0.691795	0.691825
0.5	0.610510	0.644632	0.648309	0.648680	0.648721
0.6	0.571561	0.616697	0.621573	0.622064	0.622119
0.7	0.548717	0.606763	0.613048	0.613682	0.613753
0.8	0.543589	0.616715	0.624651	0.625452	0.625541
0.9	0.557948	0.648633	0.658497	0.659492	0.659603
1.0	0.593742	0.704814	0.716924	0.718146	0.718282

Set $n = 0$:

$$x_1 = x_0 + 0.01 = 0 + 0.01 = 0.01$$

$$y_1 = y_0 + 0.01(2x_0 + y_0 - 2) = 1 + 0.01(2(0) + 1 - 2) = 0.99.$$

Set $n = 1$:

$$x_2 = x_1 + 0.01 = 0.01 + 0.01 = 0.02$$

$$y_2 = y_1 + 0.01(2x_1 + y_1 - 2) = 0.99 + 0.01(2(0.01) + 0.99 - 2) = 0.9801.$$

It will take 8 additional evaluations of (4.3.12) with $h = 0.01$ in order to find y_{10}, the Euler approximation of $y(0.1)$. The result of doing so is

$$y_{10} = 0.904622.$$

In all, it takes 100 evaluations like these to find that an approximation of $y(1)$ is

$$y_{100} = 0.704814.$$

Obviously, calculating these by hand would be far too tedious and time-consuming. Consequently, the calculations were carried out on a computer. The approximations of $y(x)$ for $x = 0.1, 0.2, \ldots, 1.0$ are listed in Table 4.2 in the third ($h = 0.01$) column.

Now let us compare approximations of $y(1)$ using even smaller step sizes. The fourth column in Table 4.2 lists the approximations of the solution $y(x)$ at the points $x = 0.1, 0.2, \ldots 1.0$ when $h = 0.001$. With this step size, we see in the last row of this column that the approximate value of $y(1)$ after 1000 evaluations using Euler's recursive formula is

$$y_{1,000} = 0.716924.$$

Finally, the next-to-last column lists approximations of the solution using the step size $h = 0.0001$. With this step size, the approximate value of $y(1)$ is

$$y_{10,000} = 0.718146.$$

The approximations in the last row of the table indicate that the value of $y(1)$ to two decimal places is

$$y(1) = 0.72.$$

Fortunately, for this particular example, we can compare this approximation to its actual value because (4.3.10) has an analytical solution, which is

$$y(x) = e^x - 2x.$$

It is left as an exercise to verify that this function satisfies both the differential equation and the initial condition. So the actual value of the solution at $x = 1$ is

$$y(1) = e - 2 = 0.7182818\dots. \tag{4.3.13}$$

As one would expect, a perusal of Table 4.2 and a comparison of the approximations of the solution at a specified value of x to its actual value (last column) shows that the approximations increase in accuracy as the step size h is reduced. In order to get a better feel for this, let us see how reducing the step size affects the accuracy of the approximations in this particular example by calculating the *relative error* of each approximation. This will also give us the means to compare Euler's method with two other well-known numerical methods that we will introduce and discuss in the next two sections. The ***relative error*** in approximating the actual value $y(x_n)$ with the approximation y_n is defined to be

$$\frac{|y(x_n) - y_n|}{|y(x_n)|}, \tag{4.3.14}$$

provided that $y(x_n) \neq 0$. (If the actual value of the solution at x_n is zero, then the relative error at that point is undefined.) Let us calculate the relative errors in the approximations of the solution at $x = 1$ for each of the step sizes listed in Table 4.2. With $h = 0.1$, the relative error in approximating $y(1)$ with y_{10} is

$$\frac{|y(1) - y_{10}|}{|y(1)|} = \frac{|e - 2 - 0.593742|}{|e - 2|} = 0.1734.$$

Said another way, the percentage error is 17.34%. When the step size is reduced by a factor of 10 to $h = 0.01$, then the relative error in approximating $y(1)$ with y_{100} decreases to

$$\frac{|y(1) - y_{100}|}{|y(1)|} = \frac{|e - 2 - 0.704814|}{|e - 2|} = 0.0188.$$

Or, expressed as a percentage, the error is 1.88%. As for the step sizes $h = 0.001$ and $h = 0.0001$, similar calculations show that the errors in approximating $y(1)$ with $y_{1,000}$ and $y_{10,000}$ decrease even more to 0.189% and 0.0189%, respectively. ♦

4.4 Improved Euler's Method

The purpose of the rest of this chapter is to introduce a couple of well-known numerical methods that do a much better job of approximating solutions than does Euler's method. We begin by presenting another way of looking at Euler's method that is more amenable to improvements than is the tangent line formulation. As in the previous section, let $y(x)$ denote the actual solution of the initial value problem (4.3.2). In other words, on some interval I containing the point x_0,

$$\frac{d}{dx}y(x) = f(x, y(x)) \tag{4.4.1}$$

for all $x \in I$ and the value of $y(x)$ at x_0 is precisely y_0. Thus $y(x_n)$ denotes the actual value of the solution at the mesh point x_n. Keep in mind, for the remainder of this chapter, that we are supposing the actual value of $y(x_n)$, generally speaking, cannot be determined (either because we do not know how to find the actual solution $y(x)$ or there is no way of doing so). Our task then is to find a way of approximating it. As before, y_n denotes an approximation of $y(x_n)$.

Let $x_0, x_1, x_2, \ldots, x_N$ be equally spaced mesh points in I. Integrating (4.4.1) from $x = x_n$ to $x = x_{n+1}$, we obtain the formula

$$y(x_{n+1}) - y(x_n) = \int_{x_n}^{x_{n+1}} f(x, y(x))\,dx.$$

As $h = x_{n+1} - x_n$, we can rewrite this as

$$y(x_{n+1}) = y(x_n) + \int_{x_n}^{x_n+h} f(x, y(x))\,dx. \tag{4.4.2}$$

Recall that we are supposing that we cannot find a formula for $y(x)$; so we do not know $y(x)$ on the interval $[x_n, x_n + h]$. If, however, we did know the actual value of $y(x)$ at the left endpoint x_n, then at least we could approximate the integral in (4.4.2) by replacing the integrand $f(x, y(x))$ with its value at $x = x_n$:

$$\int_{x_n}^{x_n+h} f(x, y(x))\,dx \approx \int_{x_n}^{x_n+h} f(x_n, y(x_n))\,dx = hf(x_n, y(x_n)).$$

This has a simple interpretation: If $f(x, y(x)) \geq 0$ for all $x \in [x_n, x_n + h]$, then the area under the graph of $f(x, y(x))$ over this interval is being approximated with the area of a rectangle with dimensions h and $f(x_n, y(x_n))$. See Fig. 4.5. Using this integral approximation, (4.4.2) becomes

$$y(x_{n+1}) \approx y(x_n) + hf(x_n, y(x_n)).$$

The problem with this, as we pointed out earlier, is that we generally do not know the actual value of $y(x_n)$. Even so, let us forge on by supposing that we have a way of finding an approximation y_n for it. Then, replacing $y(x_n)$ in the previous formula with y_n, we get the formula

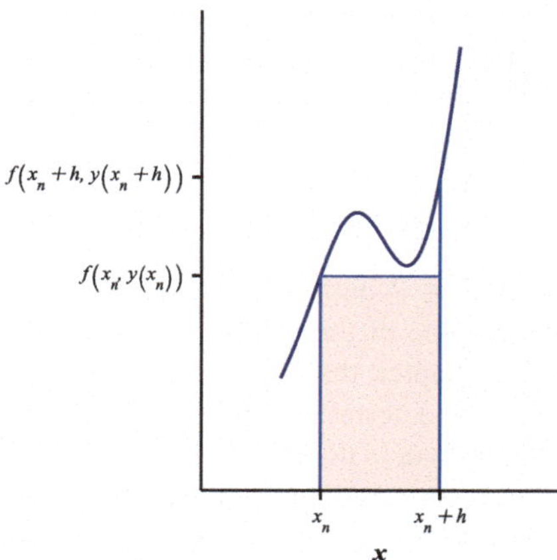

Fig. 4.5 Approximating $\int_{x_n}^{x_n+h} f(x, y(x))\,dx$ with the area of a rectangle

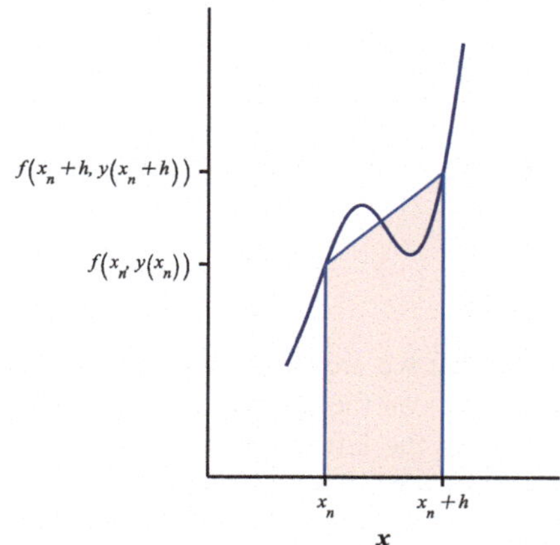

Fig. 4.6 Approximating $\int_{x_n}^{x_n+h} f(x, y(x))\,dx$ with the area of a trapezoid

$$y(x_{n+1}) \approx y_n + hf(x_n, y_n).$$

Letting y_{n+1} represent the value of the right-hand side, we obtain the recursive formula

$$y_{n+1} = y_n + hf(x_n, y_n) \qquad (n = 0, 1, 2, \dots). \tag{4.4.3}$$

Of course, this is the Euler recursive formula (4.3.6) that we had derived earlier in Sect. 4.3 using tangent lines. Nevertheless, this alternate derivation of Euler's method will allow us to make adjustments to improve the approximations given by (4.4.3), as we will see shortly.

The area of the trapezoid shown in Fig. 4.6 clearly approximates the integral better than does the area of the rectangle in Fig. 4.5. This is the key to finding a more accurate method for approximating the solution of the initial value problem (4.3.2) at the mesh points. Recall that the area of a trapezoid is the product of its height and the average length of its bases. Since the height of the trapezoid in Fig. 4.6 is h and the lengths of its bases are $f(x_n, y(x_n))$ and $f(x_n + h, y(x_n + h))$, its area is

$$h \cdot \frac{f(x_n, y(x_n)) + f(x_n + h, y(x_n + h))}{2}.$$

Thus, an approximation of the integral is given by

$$\int_{x_n}^{x_n+h} f(x, y(x))\, dx \approx \frac{h}{2}\left[f(x_n, y(x_n)) + f(x_n + h, y(x_n + h))\right]. \qquad (4.4.4)$$

However, there is an issue with computing the right-hand side of (4.4.4). Even if we have an approximation y_n for $y(x_n)$ from a previous calculation, there is still $y(x_n + h)$. How are we to approximate $y(x_n + h)$ if it is also part of the approximating formula? The remedy is to temporarily avoid this issue by first approximating $y(x_n + h)$ using Euler's recursive formula. Then we can improve this approximation in a subsequent step. Let us use the notation y_{n+1}^E to denote this initial Euler approximation. In other words,

$$y_{n+1}^E := y_n + hf(x_n, y_n).$$

Then replace $y(x_n)$ and $y(x_n + h)$ in (4.4.4) with their respective approximations y_n and y_{n+1}^E, thereby obtaining

$$\int_{x_n}^{x_n+h} f(x, y(x))\, dx \approx \frac{h}{2}\big[f(x_n, y_n) + f(x_n + h, y_n + hf(x_n, y_n))\big].$$

This approximation of the integral in (4.4.2) results in the recursive formula

$$y_{n+1} = y_n + \frac{h}{2}\big[f(x_n, y_n) + f(x_n + h, y_n + hf(x_n, y_n))\big],$$

which is called the ***improved Euler's formula***. Its implementation yields the following algorithm.

Improved Euler's Method
To approximate the values of the solution $y(x)$ of the initial value problem

$$\frac{dy}{dx} = f(x, y), \quad y(x_0) = y_0 \qquad (4.4.5)$$

(continued)

at equally spaced mesh points $x_0, x_1, x_2, \ldots$, execute the following steps:

1. Choose an appropriate step size h.
2. Let $y_0 = y(x_0)$. Then find y_{n+1} from y_n for $n = 0, 1, 2, \ldots$ as follows:

   (a) Calculate

$$x_{n+1} = x_n + h \tag{4.4.6}$$

$$y^E_{n+1} = y_n + hf(x_n, y_n). \tag{4.4.7}$$

   (b) Then use (4.4.6) and (4.4.7) to calculate

$$y_{n+1} = y_n + \frac{h}{2}\left[f(x_n, y_n) + f(x_{n+1}, y^E_{n+1})\right]. \tag{4.4.8}$$

3. Stop the calculations in Step 2 when the desired stopping point x_N on the x-axis is reached.

This is also known as ***Heun's method***, named after the German mathematician Karl Heun (1859–1929). (The correct pronunciation of his surname rhymes with "coin.") However, there also happens to be a more accurate method using a more complicated formula than (4.4.8) that some authors call Heun's method. To avoid any confusion, we will not use this name.

Example 4.4.1 Use the improved Euler's method to approximate the value of the solution of

$$y' = 2x + y - 2, \quad y(0) = 1 \tag{4.3.10}$$

at $x = 1$ using the values 0.1, 0.05, 0.01, and 0.005 for the step size h.

Solution This initial value problem is the same as the one in Example 4.3.2, where we used Euler's method to approximate solutions. With $f(x, y) = 2x + y - 2$ in (4.4.5), the recursive formulas (4.4.6)–(4.4.8) are

$$\begin{aligned} x_{n+1} &= x_n + h, \\ y^E_{n+1} &= y_n + h(2x_n + y_n - 2), \\ y_{n+1} &= y_n + \frac{h}{2}\left(2x_n + y_n - 2 + 2x_{n+1} + y^E_{n+1} - 2\right) \\ &= y_n + \frac{h}{2}\left(2x_n + 2x_{n+1} + y_n + y^E_{n+1} - 4\right). \end{aligned} \tag{4.4.9}$$

Let $h = 0.1$. To calculate x_1 and y_1^E, set $n = 0$. Moreover, due to the initial condition, let $x_0 = 0$ and $y_0 = 1$. With these values, we get

$$x_1 = x_0 + h = 0.1$$

$$y_1^E = y_0 + 0.1(2x_0 + y_0 - 2) = 1 + 0.1(2(0) + 1 - 2) = 0.9.$$

Thus the approximate value of the solution $y(0.1)$ with $h = 0.1$ is

$$\begin{aligned} y_1 &= y_0 + \frac{0.1}{2}\left(2x_0 + 2x_1 + y_0 + y_1^E - 4\right) \\ &= 1 + 0.05\left[2(0) + 2(0.1) + 1 + 0.9 - 4\right] = 0.905. \end{aligned}$$

Next, setting $n = 1$, we have

$$x_2 = x_1 + h = 0.2$$

$$y_2^E = y_1 + 0.1(2x_1 + y_1 - 2) = 0.905 + 0.1\left(2(0.1) + 0.905 - 2\right) = 0.8155.$$

Thus, an approximate value of the solution $y(0.2)$ is

$$\begin{aligned} y_2 &= y_1 + \frac{0.1}{2}\left(2x_1 + 2x_2 + y_1 + y_2^E - 4\right) \\ &= 0.905 + 0.05\left[2(0.1) + 2(0.2) + 0.905 + 0.8155 - 4\right] = 0.821025. \end{aligned}$$

Since carrying out these sets of calculations by hand eight more times in order to find the approximation y_{10} for $y(1)$ is something very few of us would relish doing, it would be best to either write a computer program for this method or to use one that has already been written. The rest of the approximations were obtained in this way and are listed in Table 4.3 in the $h = 0.1$ column, from which we see that $y_{10} = 0.714081$. In other words, according to the improved Euler's method with a

Table 4.3 Approximations of (4.3.10) using the improved Euler's method

x	$h = 0.1$	$h = 0.05$	$h = 0.01$	$h = 0.005$	**Actual value**
0.1	0.905000	0.905127	0.905169	0.905170	0.905171
0.2	0.821025	0.821305	0.821399	0.821402	0.821403
0.3	0.749233	0.749696	0.749852	0.749857	0.749859
0.4	0.690902	0.691585	0.691815	0.691822	0.691825
0.5	0.647447	0.648390	0.648708	0.648718	0.648721
0.6	0.620429	0.621680	0.622101	0.622114	0.622119
0.7	0.611574	0.613187	0.613729	0.613747	0.613753
0.8	0.622789	0.624826	0.625511	0.625534	0.625541
0.9	0.656182	0.658715	0.659566	0.659594	0.659603
1.0	0.714081	0.717191	0.718237	0.718271	0.718282

step size of $h = 0.1$, the approximate value of the solution at $x = 1$ is

$$y(1) \approx 0.714081.$$

There is a price to pay when this method is used. Every time the improved Euler's recursive formula (4.4.8) is called upon to do an approximation, the cost is two function evaluations of $f(x, y)$, whereas the cost of using Euler's recursive formula (4.3.6) is just one function evaluation. Nonetheless, the reward is great. Let us explain. We see from (4.3.13) that the actual value of the solution of this initial value problem at $x = 1$ is $e - 2$. So it follows from (4.3.14) that the relative error of the approximation y_{10} is

$$\frac{|y(1) - y_{10}|}{y(1)} = \frac{|e - 2 - 0.714081|}{|e - 2|} = 0.0058.$$

What is noteworthy is that this is less than the relative error of 0.0188 using Euler's method with $h = 0.01$ (see the second calculation after (4.3.14)). Moreover, Euler's method with this step size requires

$$\frac{1}{h} = \frac{1}{0.01} = 100$$

function evaluations to reach $x = 1$ from $x = 0$, whereas the improved Euler's method with the larger step size $h = 0.1$ requires only

$$2 \times \frac{1}{h} = 2 \times \frac{1}{0.1} = 20$$

function evaluations.

Taking a smaller step size, we see from Table 4.3 that with $h = 0.01$ the improved Euler's method yields the approximation 0.718237 for $y(1)$, where the relative error is

$$\frac{|y(1) - 0.718237|}{|y(1)|} = \frac{|e - 2 - 0.718237|}{|e - 2|} = 0.0000624.$$

This is much better than the relative error of 0.000189 using Euler's method with $h = 0.0001$. Moreover, Euler's method requires

$$\frac{1}{h} = \frac{1}{0.0001} = 10,000$$

function evaluations. By contrast, the improved Euler's method requires just

$$2 \times \frac{1}{h} = 2 \times \frac{1}{0.01} \times 2 = 200$$

function evaluations. So, with roughly one-fiftieth of the calculations, the improved Euler's approximation of $y(1)$ is still more accurate than Euler's method. ♦

4.5 Classical Runge-Kutta Method

When we consider the number of function evaluations and compare the relative errors that are needed to obtain good approximations of the values of the solution of the initial value problem

$$y' = f(x, y), \quad y(x_0) = y_0 \tag{4.4.5}$$

at equally spaced mesh points, it becomes apparent that the improved Euler's method is superior to Euler's method. In this section, we introduce an even more accurate and efficacious numerical method called the ***classical Runge-Kutta method***, which is also known as the ***Runge-Kutta method of order four*** or ***RK4 method***. Even though there are other Runge-Kutta methods, we will usually drop the word "classical" in the remainder of this discussion for the sake of brevity. The classical Runge-Kutta method originated with two German professors of mathematics: Carl Runge (1856–1927), who not only did research in mathematics but also conducted experiments in spectroscopy, and Wilhelm Kutta (1867–1944), who made important contributions to both mathematics and aerodynamics. In 1895, a paper by Carl Runge was published that presented several numerical methods, now known as Runge-Kutta methods, for solving a particular type of differential equation having to do with atomic spectra. Wilhelm Kutta, inspired by this paper, published a paper in 1901 (see [51]) that generalized the formulas in Runge's paper, which resulted in the classical Runge-Kutta method that we present in this section. Despite the fact that this method originated in the late nineteenth century, it is still widely used today: engineering and physics faculty refer to it now and then in their classes; professional engineers use it to model the motion of mechanical systems, simulate chemical processes, model the behavior of electrical grids, simulate the behavior of buildings and bridges under various conditions; physicists use it to model rigid body dynamics and missile trajectories; chemists use it to model the concentrations of reactants and products; meteorologists use it in numerical weather prediction models; and the list goes on.

Typically, an explanation of the classical Runge-Kutta method and the derivation of the recursive formulas that constitute this method involves the use of Taylor's formula for functions of two variables, which is how it is presented in most numerical analysis textbooks (cf. [17] or [69]). Since the first exposure to this formula usually takes place in an advanced calculus or numerical analysis course, it seems more appropriate to introduce the Runge-Kutta method with a more intuitive and less rigorous explanation.

We begin by letting $y(x)$ denote the solution of the initial value problem (4.4.5). As we have been doing all along, let $x_0, x_1, x_2, x_3, \ldots$ be equally spaced mesh points, where h is the step size, namely, the distance between consecutive mesh points. Recall that the derivation of the improved Euler's formula started with (4.4.2), which can also be written as

$$y(x_{n+1}) = y(x_n) + \int_{x_n}^{x_n+h} y'(x)\,dx \tag{4.5.1}$$

since $f(x, y(x)) = y'(x)$. After that, the integral in (4.5.1) was approximated with what turned out to be the (noncomposite) Trapezoidal rule. The classical Runge-Kutta method improves upon this trapezoidal approximation by replacing it with a more accurate approximation that is based on Simpson's rule, which is also known as Simpson's 1/3 rule.

Recall from calculus that Simpson's rule for approximating the integral of a function $g(x)$ over an interval $[a, b]$ using the step size $\bar{h} = (b - a)/2$ is

$$\int_a^b g(x)\,dx \approx \frac{\bar{h}}{3}\left[g(a) + 4g\left(\frac{a+b}{2}\right) + g(b)\right].$$

With $a = x_n$, $b = x_n + h$, and $g(x) = y'(x)$, this becomes

$$\int_{x_n}^{x_n+h} y'(x)\,dx \approx \frac{h}{6}\left[y'(x_n) + 4y'(x_n + \tfrac{h}{2}) + y'(x_n + h)\right]. \tag{4.5.2}$$

The Runge-Kutta method will turn out to be the result of approximating the integral in (4.5.1) with the right-hand side of (4.5.2), where the three derivatives in the bracketed quantity are approximated in the way that we will explain next.

Since $y(x)$ denotes the solution of the initial value problem (4.4.5),

$$y'(x_n) = f(x_n, y(x_n)).$$

Replacing $y(x_n)$ with its approximation y_n, we have

$$y'(x_n) \approx f(x_n, y_n),$$

which we rewrite as

$$y'(x_n) \approx k_{n1}, \tag{4.5.3}$$

where k_{n1} is defined by

$$k_{n1} := f(x_n, y_n).$$

In other words, k_{n1} is an approximation of the first term in the bracketed quantity on the right-hand side of (4.5.2). Note that k_{n1} is the slope of the solution curve $y(x)$ at the left endpoint of the interval $[x_n, x_n + h]$.

Now that we have an approximation for the first term, let us find one for the middle term $4y'(x_n + \frac{h}{2})$. Let us split this term up as follows:

$$2y'(x_n + \tfrac{h}{2}) + 2y'(x_n + \tfrac{h}{2}).$$

Then we will find two different approximations for $y'(x_n + \frac{h}{2})$, one of which we will use as an approximation of the first term and the other as an approximation of the second term. Using Euler's recursive formula to approximate $y(x)$ at the midpoint $x = x_n + \frac{h}{2}$ of the interval $[x_n, x_n + h]$, we get

$$y(x_n + \tfrac{h}{2}) \approx y(x_n) + \tfrac{h}{2} f(x_n, y_n) = y_n + \tfrac{h}{2} k_{n1}.$$

Then, as

$$y'(x_n + \tfrac{h}{2}) = f\big(x_n + \tfrac{h}{2}, y(x_n + \tfrac{h}{2})\big),$$

we have the approximation

$$y'(x_n + \tfrac{h}{2}) \approx f\big(x_n + \tfrac{h}{2}, y_n + \tfrac{h}{2} k_{n1}\big) \tag{4.5.4}$$

or

$$y'(x_n + \tfrac{h}{2}) \approx k_{n2}, \tag{4.5.5}$$

where k_{n2} is defined by

$$k_{n2} := f\big(x_n + \tfrac{h}{2}, y_n + \tfrac{h}{2} k_{n1}\big).$$

Replacing k_{n1} in (4.5.4) with k_{n2}, we obtain another approximation of the derivative $y'\big(x_n + \frac{h}{2}\big)$, namely,

$$y'\big(x_n + \tfrac{h}{2}\big) \approx f\big(x_n + \tfrac{h}{2}, y_n + \tfrac{h}{2} k_{n2}\big).$$

Let us rewrite this as

$$y'\big(x_n + \tfrac{h}{2}\big) \approx k_{n3} \tag{4.5.6}$$

where

$$k_{n3} := f\big(x_n + \tfrac{h}{2}, y_n + \tfrac{h}{2} k_{n2}\big).$$

Thus, we end up with the following approximation of the middle term in (4.5.2):

$$4y'(x_n + \tfrac{h}{2}) = 2y'(x_n + \tfrac{h}{2}) + 2y'(x_n + \tfrac{h}{2}) \approx 2k_{n2} + 2k_{n3}.$$

We still need an approximation of the third term $y'(x_n + h)$ in (4.5.2). By Euler's recursive formula, an approximation y_{n+1} of $y(x_n + h)$ is

$$y_{n+1} = y_n + hf(x_n, y_n) = y_n + hk_{n1}.$$

Let us replace k_{n1} with k_{n3}. This yields the approximation

$$y'(x_n + h) = f(x_n + h, y(x_n + h)) \approx f(x_n + h, y_n + hk_{n3}).$$

Defining $k_{n4} := f(x_n + h, y_n + hk_{n3})$, this becomes

$$y'(x_n + h) \approx k_{n4}. \tag{4.5.7}$$

Finally, let us replace the three terms on the right-hand side of (4.5.2) with the approximations given by (4.5.3), (4.5.5), (4.5.6), and (4.5.7) in order to get an approximation of the integral in (4.5.1). This results in the approximation

$$y(x_{n+1}) = y(x_n) + \int_{x_n}^{x_n+h} y'(x)\,dx \approx y_n + \frac{h}{6}\big(k_{n1} + 2k_{n2} + 2k_{n3} + k_{n4}\big).$$

As a result, we have the following numerical algorithm.

Classical Runge-Kutta Method
To approximate the values of the solution $y(x)$ of the initial value problem

$$y' = f(x, y), \quad y(x_0) = y_0 \tag{4.5.8}$$

at equally spaced mesh points $x_0, x_1, x_2, \ldots$, execute the following steps:

1. Choose an appropriate step size h.
2. Let $y_0 = y(x_0)$. Then find y_{n+1} from y_n for $n = 0, 1, 2, \ldots$ as follows:

   (a) First calculate

$$\begin{aligned} k_{n1} &= f(x_n, y_n) \\ k_{n2} &= f(x_n + \tfrac{h}{2}, y_n + \tfrac{h}{2}k_{n1}) \\ k_{n3} &= f(x_n + \tfrac{h}{2}, y_n + \tfrac{h}{2}k_{n2}) \\ k_{n4} &= f(x_n + h, y_n + hk_{n3}). \end{aligned} \tag{4.5.9}$$

   (b) Then use the values of the k_{ni}'s to calculate

(continued)

$$y_{n+1} = y_n + \frac{h}{6}\big(k_{n1} + 2k_{n2} + 2k_{n3} + k_{n4}\big). \tag{4.5.10}$$

3. Stop the calculations in Step 2 when the desired stopping point x_N on the x-axis is reached.

Example 4.5.1 Once again consider the initial value problem

$$y' = 2x + y - 2, \quad y(0) = 1. \tag{4.3.10}$$

Use the Runge-Kutta method with step size $h = 0.1$ to approximate the value of the solution at $x = 1$ to eight decimal places. Find the approximation when the step size is reduced to $h = 0.05$. Also, find the approximations using the improved Euler's method with $h = 0.01$ and $h = 0.001$. Compare the approximations of $y(1)$ to its actual value.

Solution Due to the initial condition $y(0) = 1$, take $x_0 = 0$ and $y_0 = 1$. Let us show how to calculate the Runge-Kutta approximations of $y(0.1)$ and $y(0.2)$ using the step size $h = 0.1$. First take $n = 0$ and calculate the four quantities in (4.5.9) as follows:

$$\begin{aligned}
k_{01} &= f(x_0, y_0) = f(0, 1) = -1.\\
k_{02} &= f\big(x_0 + \tfrac{h}{2}, y_0 + \tfrac{h}{2}k_{01}\big) = f\big(\tfrac{0.1}{2}, 1 + \tfrac{0.1}{2}(-1)\big) = -0.95.\\
k_{03} &= f\big(x_0 + \tfrac{h}{2}, y_0 + \tfrac{h}{2}k_{02}\big) = f\big(\tfrac{0.1}{2}, 1 + \tfrac{0.1}{2}(-0.95)\big) = -0.9475.\\
k_{04} &= f\big(x_0 + h, y_0 + hk_{03}\big) = f\big(0.1, 1 + 0.1(-0.9475)\big) = -0.89475.
\end{aligned}$$

Using these values and (4.5.10), we obtain

$$\begin{aligned}
y_1 &= y_0 + \frac{h}{6}\big(k_{01} + 2k_{02} + 2k_{03} + k_{04}\big) = 1 + \frac{0.1}{6}\big(k_{01} + 2k_{02} + 2k_{03} + k_{04}\big)\\
&= 1 + \frac{0.1}{6}\big(-1 + 2(-0.95) + 2(-0.9475) - 0.89475\big) = 0.905170833\ldots.
\end{aligned}$$

We conclude that with $h = 0.1$, the Runge-Kutta approximation of the value of $y(0.1)$ to eight decimal places is $y_1 = 0.90517083$. The actual value is (cf. Example 4.3.2):

$$y(0.1) = e^{0.1} - 2(0.1) = 0.905170918\ldots.$$

In order to approximate $y(0.2)$, let $n = 1$ and $x_1 = 0.1$. Then

$$k_{11} = f(x_1, y_1) = f(0.1, y_1) = -0.894892166\ldots$$
$$k_{12} = f\left(x_1 + \tfrac{h}{2}, y_1 + \tfrac{h}{2}k_{11}\right) = f\left(0.1 + \tfrac{0.1}{2}, y_1 + \tfrac{0.1}{2}k_{11}\right) = -0.839570625\ldots$$
$$k_{13} = f\left(x_1 + \tfrac{h}{2}, y_1 + \tfrac{h}{2}k_{12}\right) = f\left(0.1 + \tfrac{0.1}{2}, y_1 + \tfrac{0.1}{2}k_{12}\right) = -0.836807697\ldots$$
$$k_{14} = f\left(x_1 + h, y_1 + hk_{13}\right) = f\left(0.2, y_1 + 0.1k_{13}\right) = -0.778509936\ldots$$

Substituting these values into (4.5.10), we have

$$y_2 = y_1 + \frac{h}{6}\left(k_{11} + 2k_{12} + 2k_{13} + k_{14}\right) = 0.821402570\ldots.$$

Thus, with $h = 0.1$, the Runge-Kutta approximation of the value of $y(0.2)$ to eight decimal places is $y_2 = 0.82140257$. The actual value is

$$y(0.2) = e^{0.2} - 2(0.2) = 0.821402758\ldots.$$

The rest of the approximations in Table 4.4 were obtained with computer programs that implement the Runge-Kutta method and the improved Euler's method to solve differential equations. Note that the approximations obtained using the Runge-Kutta method with step size $h = 0.05$ are more accurate than those found using the improved Euler's method with step size $h = 0.001$. The actual value of the solution at $x = 1$ is

$$y(1) = e - 2 = 0.7182818284\ldots. \qquad \blacklozenge$$

Table 4.4 Comparison of results using the improved Euler's method and RK4 method

	Improved Euler's		Runge-Kutta (RK4)		
x	$h = 0.01$	$h = 0.001$	$h = 0.1$	$h = 0.05$	Actual value
0.1	0.90516909	0.90517090	0.90517083	0.90517091	0.90517092
0.2	0.82139872	0.82140272	0.82140257	0.82140275	0.82140276
0.3	0.74985211	0.74985874	0.74985850	0.74985879	0.74985881
0.4	0.69181483	0.69182460	0.69182424	0.69182467	0.69182470
0.5	0.64870763	0.64872113	0.64872064	0.64872123	0.64872127
0.6	0.62210072	0.62211862	0.62211796	0.62211875	0.62211880
0.7	0.61372939	0.61375247	0.61375163	0.61375264	0.61375271
0.8	0.62551148	0.62554063	0.62553956	0.62554084	0.62554093
0.9	0.65956649	0.65960274	0.65960141	0.65960300	0.65960311
1.0	0.71823686	0.71828138	0.71827974	0.71828169	0.71828183

Problems

No pain, no gain.

Old proverb

Euler's Method

In Problems 1 through 7, use Euler's method to approximate the solution of the initial value problem. Carry out the calculations in part (b) *of each problem by hand. If you use a programmable calculator, do not use any of its programming features for this part.*

1. $\dfrac{dy}{dx} = x - y^2, \quad y(0) = 1$
   (a) Find the recursive formulas for a step size of 0.5.
   (b) Approximate the solution at $x = 1$.
2. $y' = 2y - xy^2, \quad y(2) = 4$
   (a) Find the recursive formulas for a step size of 0.1.
   (b) Approximate the solution at $x = 2.3$.
3. $\dfrac{dy}{dx} = \frac{1}{2}y\sqrt{x}, \quad y(0) = 1$
   (a) Find the recursive formulas for a step size of 0.04.
   (b) Approximate the solution at $x = 0.08$.
4. $y' = xy - 2x, \quad y(0) = 0$
   (a) Find the recursive formulas for a step size of 0.1.
   (b) Approximate the solution at $x = 0.3$.
   (c) Find an analytical solution.
   (d) Use the result of (c) to find the exact value of the solution at $x = 0.3$. Compare the answer to the numerical approximation obtained in (b).
5. $y' = y - 2x, \quad y(0) = 3$
   (a) Find the recursive formulas for a step size of 0.2.
   (b) Approximate the solution at $x = 0.4$.
6. $y' = 2y\cos(\pi x), \quad y(1) = 3$
   (a) Find the recursive formulas for a step size of 0.1.
   (b) Approximate the solution at $x = 1.3$.
7. $\dfrac{dy}{dx} = -\dfrac{x}{y}, \quad y(3) = 4$
   (a) Find the recursive formulas for a step size of 0.5.
   (b) Approximate the solution at $x = 3.5$.
   (c) Find an analytical solution.
   (d) Use the result of (c) to find the exact value of the solution at $x = 3.5$. Compare the answer to the numerical approximation obtained in (b).

In Problems 8 and 9, use either a program or a CAS that employs Euler's method to approximate and plot the graphs of the solutions.

8. (a) Use Euler's method with step size $h = 0.1$ to approximate the solution of the initial value problem

$$\frac{dy}{dx} = y - xy^2, \quad y(1) = 2$$

on the interval $[1, 2]$. Print the approximations for $x = 1.0, 1.1, 1.2, \ldots 1.9, 2.0$ in tabular form.

(b) Use technology to draw the Euler polygonal curve corresponding to the approximations in part (a). Also, plot the solution of the initial value problem in part (a) using one of the computer algebra system's numeric solvers. In order to compare the two curves, draw both of them on the same set of axes.

(c) Approximate the solution at $x = 2.5$ using Euler's method.

9. (a) Use Euler's method with step size $h = 0.2$ to approximate the solution of the initial value problem

$$y' = x + \cos y \quad y(0) = -1$$

at the points $x = 0.2, 0.4, 0.6, 0.8$, and 1.0.

(b) Draw an Euler polygonal curve; then compare it to the solution curve drawn by a CAS.

10. (a) What is the value of the solution of the initial value problem

$$y' = y, \quad y(0) = 1$$

at the point $x = 1$?

(b) Show that Euler's approximation of the solution at $x = 1$ when the interval $[0, 1]$ is divided into m subintervals of equal length is

$$y_m = \left(1 + \frac{1}{m}\right)^m.$$

(c) Calculate y_m with $m = 2^n$ for $n = 0, 1, 2, 3, 4, 10, 20$.

(d) Find the limit of y_m as $m \to \infty$ and compare it to the answer for part (a).

Improved Euler's Method

In Problems 11 through 13, use the improved Euler's method to approximate the solution of the initial value problem at the specified point. If you use a programmable calculator, do not use any of its programming features for part (b) *of each problem.*

11. $\dfrac{dy}{dx} = 1 + \dfrac{y}{x} + \left(\dfrac{y}{x}\right)^2, \quad y(1) = 0$

(a) For the step size $h = 0.1$, determine the recursive formulas.

(b) Approximate to six decimal places the value of the solution at $x = 1.2$ using the recursive formulas in (a), pencil and paper, and a calculator.

(c) Use a CAS or write your own program for a computer or programmable calculator to approximate to six decimal places the value of the solution at $x = 2$ using the step size $h = 0.1$.

(d) Find the analytical solution. *Hint.* Change the dependent variable to $w := y/x$.

(e) Compare the approximate value of the solution to its actual value at $x = 2$.

12. $y' = 1 - 2xy, \quad y(0) = 0$

(a) For the step size $h = 0.25$, determine the recursive formulas.

(b) Approximate to six decimal places the value of the solution at $x = 0.5$ using the recursive formulas in (a), pencil and paper, and a calculator.

(c) Use a CAS or write your own program for a computer or programmable calculator to approximate to six decimal places the value of the solution at $x = 0.5$ using the step size $h = 0.05$. (As we shall see in Chap. 5, the solution of this initial value problem is a function called *Dawson's integral*. According to a table in reference [1, p. 319], the value of this function at $x = 0.5$ rounded to six decimal places is 0.424436.)

13. $\dfrac{dy}{dx} = \dfrac{2y\cos x}{y^2+1}, \quad y(-1) = 1$

(a) Approximate to eight decimal places the value of the solution at $x = -0.8$ using the improved Euler's method with step size $h = 0.1$, pencil and paper, and a calculator.

(b) Use a CAS or write your own program for a computer or programmable calculator to approximate to eight decimal places the value of the solution at $x = 1.5$ using the step size $h = 0.1$.

14. (a) Show that the improved Euler's approximation of the solution of

$$y' = y, \quad y(0) = 1$$

at $x = 1$ when the interval $[0, 1]$ is divided into m subintervals of equal length is

$$y_m = \left(1 + \frac{1}{m} + \frac{1}{2m^2}\right)^m.$$

(b) Calculate y_m with $m = 2^n$ for $n = 0, 1, 2, 3, 4, 10, 20$. What does the limit of y_m as $m \to \infty$ appear to be?

(c) Use l'Hôpital's rule to find the limit of y_m as $m \to \infty$.

Runge-Kutta Method

In Problems 15 through 17, use the Runge-Kutta method to approximate the solution of the initial value problem at the specified point. If you use a programmable calculator, do not use any of its programming features for part (b) *of each problem.*

15. $\dfrac{dy}{dx} = 1 + \dfrac{y}{x} + \left(\dfrac{y}{x}\right)^2, \quad y(1) = 0$

(a) Approximate to eight decimal places the value of the solution at $x = 1.2$ using the recursive formulas in (a), pencil and paper, and a calculator.

(b) Use a CAS or write your own program for a computer or programmable calculator to approximate to eight decimal places the value of the solution at $x = 2$ using the step size $h = 0.1$.

(c) Find the analytical solution. *Hint*. Change the dependent variable to $w := y/x$.

(d) Compare the approximate value of the solution to its actual value at $x = 2$.

16. $y' = 1 - 2xy, \quad y(0) = 0$

(a) Approximate to eight decimal places the value of the solution at $x = 0.2$ using the Runge-Kutta recursive formulas with step size $h = 0.1$, pencil and paper, and a calculator.

(b) Use a CAS or write your own program for a computer or programmable calculator to approximate to eight decimal places the value of the solution at $x = 1$ using the step size $h = 0.1$.

17. $\dfrac{dy}{dx} = \dfrac{2y\cos x}{y^2+1}$, $\quad y(-1) = 1$

(a) Approximate to eight decimal places the value of the solution at $x = -0.8$ using the Runge-Kutta recursive formulas with step size $h = 0.1$, pencil and paper, and a calculator.

(b) Use a CAS or write your own program for a computer or programmable calculator to approximate to eight decimal places the value of the solution at $x = 1.5$ using the step size $h = 0.1$.

Chapter 5
First-Order Linear and Related Equations

Mathematics is the tool specially suited for dealing with abstract concepts of any kind and there is no limit to its power in this field. For this reason a book on the new physics, if not purely descriptive of experimental work, must be essentially mathematical.

P. A. M. Dirac[1] [26, p. viii]

Abstract This chapter begins with an introduction to *integrating factors* and how they are used to find solutions of the standard form

$$\frac{dx}{dt} + p(t)x = q(t) \tag{1}$$

of the first-order linear equation $a_1(t)x' + a_0(t)x = b(t)$. Furthermore, an existence and uniqueness theorem for the initial value problem

$$\frac{dx}{dt} + p(t)x = q(t), \quad x(t_0) = x_0 \tag{2}$$

is proved by employing an integrating factor.

[1] Paul Adriene Maurice Dirac (1902–1984) was one of the greatest theoretical physicists of all time. One of his many achievements was to come up with a relativistic theory of the electron, which culminated in the Dirac wave equation. From the negative-energy solutions of this equation, he inferred the existence of a particle with the same mass as an electron but of opposite charge. So based solely on mathematics, Dirac predicted the existence of the anti-electron in 1931. It did not take much time for someone to discover that this particle did in fact exist. In 1932 an experimental physicist at Caltech, Carl Anderson, discovered anti-electrons (now called positrons) in cloud chamber experiments. Anderson received a Nobel prize in 1936. Dirac received a Nobel prize in 1933.

© The Author(s), under exclusive license to Springer Nature Switzerland AG 2026

L. C. Becker, *Ordinary Differential Equations: Concepts, Methods, and Models*,
https://doi.org/10.1007/978-3-032-15150-6_5

Gronwall's inequality, a result that yields a bound for an unknown nonnegative continuous function f satisfying the inequality

$$f(t) \leq K + \left| \int_{t_0}^{t} p(s) f(s) \, ds \right|,$$

where $K > 0$ is a constant and p is a known nonnegative continuous function, is derived with the help of an integrating factor. Gronwall's inequality is used to determine the extent to which two solutions of (1) can differ from each other over a given interval. It is instrumental in establishing one of the existence and uniqueness theorems for first-order equations in Chap. 9.

An alternative to finding the solution of the initial value problem (2) is the *variation of parameters method*. Its introduction here for first-order linear equations paves the way for explaining how to use this method to solve second-order nonhomogeneous linear equations (see Chap. 10). Furthermore, the variation of parameters formula

$$x(t) = z(t, t_0)x_0 + \int_{t_0}^{t} z(t, s)q(s) \, ds \tag{3}$$

is derived, where $z(t, s) = \exp\left(-\int_s^t p(u) \, du\right)$. A matrix analog of (3) for nonhomogeneous linear systems is one of the topics in Chap. 13.

It is pointed out that solutions of differential equations as simple as (1) may have to be expressed in terms of functions that are defined in terms of definite integrals, such as *Dawson's integral* and the *sine integral function*.

The last two sections of this chapter deal with two nonlinear first-order differential equations and their solutions: the *Bernoulli equation*

$$\frac{dx}{dt} + p(t)x = q(t)x^n$$

and the *Riccati equation*

$$\frac{dx}{dt} = a(t)x^2 + b(t)x + c(t),$$

both of which are important in engineering, physics, and other fields.

5.1 Standard Form

There are many people, besides mathematicians, who rely on mathematics in their respective disciplines: climatologists, econometricians, engineers, physical scientists, operations research analysts, just to name a few. It is a tool that enables people to model, predict, and to better understand reality. Reality involves incessant change. The linchpin of disciplines dealing with change—astronomy, biomathematics, classical and quantum mechanics, electrical engineering, economics, fluid mechanics, linear control systems, physical chemistry, and the like—is calculus and differential equations.

Some simple phenomena can be modeled by first-order differential equations of the form

$$a_1(t)\frac{dx}{dt} + a_0(t)x = b(t),$$

where $a_0(t)$, $a_1(t)$, and $b(t)$ denote known functions. As the notation indicates, each of these functions depends on the independent variable t but in no way on the dependent variable x. The term "linear" is used to describe first-order equations of this type.

First-Order Linear Equation

Definition 5.1.1 A first-order differential equation is said to be ***linear*** if it can be expressed in the form

$$a_1(t)\frac{dx}{dt} + a_0(t)x = b(t). \tag{5.1.1}$$

If b is the zero function ($b(t) \equiv 0$), then this equation is said to be ***homogeneous***. Otherwise, it is said to be ***nonhomogeneous***.

If a first-order differential equation does not look exactly like (5.1.1), it could still be linear. But we may not see that unless we perform some algebraic manipulations. For instance, consider

$$t^2\frac{dx}{dt} - \sin t = 4x + 9\frac{dx}{dt} + 1.$$

It is not expressed in the form of (5.1.1); however, we can rewrite it as

$$(t^2 - 9)\frac{dx}{dt} - 4x = 1 + \sin t. \tag{5.1.2}$$

Now it does look like (5.1.1). Thus it is a linear equation, where

$$a_1(t) = t^2 - 9, \quad a_0(t) = -4, \quad b(t) = 1 + \sin t.$$

On the other hand, the nearest we can make

$$tx\frac{dx}{dt} - \sin t = 4x + 9\frac{dx}{dt} + 1$$

look like (5.1.1) is

$$(tx - 9)\frac{dx}{dt} - 4x = 1 + \sin t.$$

Consequently, this equation is ***nonlinear*** (not linear) because the coefficient of dx/dt depends not only on t but on x as well. That is, the coefficient is a function of both t and x, which we indicate by naming it a_1 and writing

$$a_1(t, x) = tx - 9.$$

If the leading coefficient $a_1(t)$ of (5.1.1) is never zero on the interval over which we seek solutions, then (5.1.1) can be rewritten as

$$\frac{dx}{dt} + p(t)x = q(t), \tag{5.1.3}$$

where

$$p(t) := \frac{a_0(t)}{a_1(t)} \quad \text{and} \quad q(t) := \frac{b(t)}{a_1(t)}.$$

This is called the ***standard form*** of (5.1.1). Take, for example, Eq. (5.1.2). Its leading coefficient, namely $t^2 - 9$, is never zero on intervals that exclude $t = \pm 3$. Hence on intervals like $(-\infty, -3)$, $(-3, 3)$, and $(3, \infty)$, the equation can be rewritten in the standard form

$$\frac{dx}{dt} - \frac{4}{t^2 - 9}x = \frac{1 + \sin t}{t^2 - 9}.$$

Note that $p(t)$ is equal to $-4/(t^2 - 9)$, not $4/(t^2 - 9)$.

5.2 Solutions

We will now look for a way to solve first-order linear equations that are already in standard form. Since we have not faced these kinds of equations before, it is hard to know where to begin. A trick of the trade that mathematicians use to get a feel for new kinds of problems is to first consider examples and look at special cases. Let us try that with Eq. (5.1.3) by letting $p(t) \equiv r$ and $q(t) \equiv 0$, where r is a constant.

Then it becomes

$$x' + rx = 0. \tag{5.2.1}$$

Although we can easily solve (5.2.1) by separating its variables, doing so misses the point because (5.1.3) is not separable unless

- Either $p(t) \equiv 0$ or $q(t) \equiv 0$.
- Both $p(t)$ and $q(t)$ are constant functions.
- The functions $p(t)$ and $q(t)$ are not constant functions but are constant multiples of each other.

The upshot is that we will have to devise alternate means of handling linear equations.

We begin by multiplying both sides of (5.2.1) by e^{rt}. Then

$$e^{rt}x' + re^{rt}x = 0. \tag{5.2.2}$$

Now the left-hand side looks like the result of differentiating a product of two functions. In fact, it is the derivative of the product $e^{rt}x$. Hence (5.2.2) becomes

$$\frac{d}{dt}(e^{rt}x) = 0. \tag{5.2.3}$$

This says that the rate of change of the quantity $e^{rt}x$ with respect to t is always zero. Keep in mind that $e^{rt}x$ is a function of t even though it may seem at first to be a function of x as well. However, x is not an independent variable but rather a function of t; so we should think $e^{rt}x(t)$ when we see $e^{rt}x$. Now recall that a function is constant when its rate of change (derivative) is always zero. Thus, $e^{rt}x(t) = C$, for some constant C. Alternatively, we can integrate both sides of (5.2.3) with respect to t as follows:

$$\int \frac{d}{dt}(e^{rt}x)\,dt = \int 0\,dt + C,$$

which also yields $e^{rt}x = C$. Either way, we end up with the solutions

$$x(t) = Ce^{-rt}. \tag{5.2.4}$$

Since Eq. (5.2.3) can be rewritten as

$$e^{rt}(x' + rx) = 0$$

and the factor e^{rt} is never zero, Eqs. (5.2.1) and (5.2.3) have the same solutions. In other words, all of the functions given by (5.2.4) make up the complete set of the solutions of (5.2.1). But we already know this from studying the exponential function e^{rt} in calculus. Moreover, (5.2.4) is what we would get were we to apply the method of separation of variables to (5.2.1).

In short, we have shown that there is an alternative to separating variables to solve (5.2.1). Multiplication by the factor e^{rt} is the crucial step. This is ingenious in that it provides the means by which to obtain (5.2.4) by merely integrating (5.2.3). It is for this reason that e^{rt} is known as an ***integrating factor*** for (5.2.1).

Suppose we alter (5.2.1) a bit by replacing its right-hand side with an arbitrary continuous function $q(t)$, which is not identically zero. Then we have the linear equation

$$x' + rx = q(t) \tag{5.2.5}$$

whose variables cannot be separated unless $q(t)$ is a constant function. However, once again it can be solved by employing the function e^{rt}. When we multiply both sides of (5.2.5) by e^{rt} and use the product rule, we get

$$\frac{d}{dt}(e^{rt}x) = e^{rt}q(t). \tag{5.2.6}$$

Since this is

$$e^{rt}\left(\frac{dx}{dt} + rx\right) = e^{rt}q(t)$$

and $e^{rt} > 0$ for all values of t, Eq. (5.2.6) has precisely the same solutions as (5.2.5). And it is easier to deal with than (5.2.5) due to its left-hand side being the derivative of a quantity. Integrating both sides of this equation with respect to t, we obtain

$$e^{rt}x = \int e^{rt}q(t)\,dt + C.$$

Now the solutions of (5.2.5) can be obtained by carrying out the integration and solving for the dependent variable x.

To further clarify the previous discussion, let $r = 5$ and $q(t) = 4e^{3t}$. Then the differential equation (5.2.5) becomes

$$x' + 5x = 4e^{3t}. \tag{5.2.7}$$

Multiplying this by the integrating factor e^{5t}, we get

$$e^{5t}x' + 5e^{5t}x = 4e^{8t}.$$

Note that the left-hand side is equal to the derivative of the product of the "integrating factor e^{5t} and the dependent variable x." Consequently,

$$\frac{d}{dt}(e^{5t}x) = 4e^{8t}. \tag{5.2.8}$$

Integrating both sides with respect to t, we obtain

$$\int \left(\frac{d}{dt}(e^{5t}x)\right) dt = \int 4e^{8t}\, dt + C \quad \Rightarrow \quad e^{5t}x = \frac{1}{2}e^{8t} + C.$$

Solving for x, we get

$$x = \frac{1}{2}e^{3t} + Ce^{-5t}. \tag{5.2.9}$$

These functions are solutions of (5.2.8). Moreover, they are also the solutions of the original equation (5.2.7). Let us verify this. Substituting (5.2.9) and its derivative into the left-hand side of (5.2.7), we have

$$\frac{dx}{dt} + 5x = \frac{3}{2}e^{3t} - 5Ce^{-5t} + 5\left(\frac{1}{2}e^{3t} + Ce^{-5t}\right) = 4e^{3t},$$

which is the right-hand side of (5.2.7). This is the case regardless of the value of C. Therefore, all solutions of (5.2.7) are given by (5.2.9). For this reason, (5.2.9) is called the ***general solution*** of (5.2.7).

5.2.1 Integrating Factors

Now that we have managed to find solutions of Eqs. (5.2.1) and (5.2.5) when the coefficient of x is constant, let us consider a more general problem and ask if the integrating factor approach is applicable to any first-order linear equation expressed in the standard form

$$\frac{dx}{dt} + p(t)x = q(t), \tag{5.2.10}$$

where the functions $p(t)$ and $q(t)$ are continuous on some common interval. The objective is to find an integrating factor when $p(t)$ actually varies with t and is not constant. If we were to take a guess at this point, we might come up with $e^{tp(t)}$. That this could be an integrating factor is plausible because it reduces to e^{rt} when $p(t) \equiv r$. But alas, it turns out not to be an integrating factor (see Problem 41). So how do we go about finding an integrating factor?

Suppose we have somehow managed to come up with an integrating factor for (5.2.10). Whatever it may turn out to be, let us call it $\mu(t)$ and imitate the way in which we solved Eqs. (5.2.1) and (5.2.5). Multiplying both sides of (5.2.10) by $\mu(t)$, we have

$$\mu(t)\frac{dx}{dt} + p(t)\mu(t)x = \mu(t)q(t). \tag{5.2.11}$$

In view of how we handled (5.2.1) and (5.2.5), this will work if the left-hand side of (5.2.11) can be rewritten as the derivative of the product of the "integrating factor $\mu(t)$ and the dependent variable x," that is, if

$$\mu(t)\frac{dx}{dt} + p(t)\mu(t)x = \frac{d}{dt}\big[\mu(t)x\big]$$

or

$$\mu(t)\frac{dx}{dt} + p(t)\mu(t)x = \mu(t)\frac{dx}{dt} + x\frac{d}{dt}\mu(t).$$

It follows from this that if a function $\mu(t)$ is an integrating factor for equation (5.2.10), then it has to satisfy the condition

$$\frac{d}{dt}\mu(t) = p(t)\mu(t). \tag{5.2.12}$$

Fortunately, this is a separable equation and easy to solve. Separating the variables and integrating, we have

$$\frac{1}{\mu}d\mu = p(t)\,dt \quad \Rightarrow \quad \ln|\mu| = \int p(t)\,dt + K.$$

Exponentiating both sides, we end up with the formula

$$\mu(t) = k\exp\left(\int p(t)\,dt\right), \tag{5.2.13}$$

where k is an arbitrary constant.

As we stated above, for (5.2.13) to be an integrating factor, it must be a solution of (5.2.12). Let us verify this by differentiating it. Recalling that $\int p(t)\,dt$ denotes any function whose derivative is $p(t)$, we have

$$\frac{d}{dt}\mu(t) = \frac{d}{dt}\left(ke^{\int p(t)\,dt}\right) = ke^{\int p(t)\,dt}\frac{d}{dt}\int p(t)\,dt = \mu(t)p(t).$$

So $\mu(t)$ is an integrating factor irrespective of the value of k. So simply let $k = 1$.

We have just established that

$$\mu(t) = e^{\int p(t)\,dt} \tag{5.2.14}$$

is an integrating factor for (5.2.10). Note that it is not "the" integrating factor but "an" integrating factor. And why is that? First, we set $k = 1$ in (5.2.13) to get (5.2.14) but any other value for k would do just as well. Second, $\int p(t)\,dt$ denotes any antiderivative of $p(t)$, of which there are infinitely many from which to choose. So even though we might encounter the statement that (5.2.14) is the integrating factor for Eq. (5.2.10), it is by no means the only one. Finally, note that in the case of $p(t) \equiv r$, formula (5.2.14) reduces to e^{rt}, as we already had discovered in working with (5.2.1).

With (5.2.14) chosen as the integrating factor, (5.2.11) can be replaced by

$$\frac{d}{dt}[\mu(t)x(t)] = \mu(t)q(t). \tag{5.2.15}$$

Integrating this equation, we obtain

$$\mu(t)x(t) = \int \mu(t)q(t)\,dt + C.$$

Then multiplication by

$$\frac{1}{\mu(t)} = e^{-\int p(t)\,dt}$$

yields

$$x(t) = \frac{1}{\mu(t)}\left(\int \mu(t)q(t)\,dt + C\right). \tag{5.2.16}$$

That is,

$$x(t) = Ce^{-\int p(t)\,dt} + e^{-\int p(t)\,dt}\left(\int e^{\int p(t)\,dt} q(t)\,dt\right). \tag{5.2.17}$$

So if a function $x(t)$ is a solution of (5.2.10), then it will look like (5.2.17).

In fact, every member of the set of functions (5.2.16) is a solution of (5.2.10), irrespective of the value of C or what antiderivative of $p(t)$ is used. We can easily verify this with (5.2.12) and the product rule. Differentiating (5.2.16), we have

$$\begin{aligned}\frac{d}{dt}x(t) &= \frac{1}{\mu(t)}\cdot\frac{d}{dt}\left(\int \mu(t)\,q(t)\,dt + C\right) - \frac{\mu'(t)}{[\mu(t)]^2}\left(\int \mu(t)\,q(t)\,dt + C\right)\\ &= \frac{1}{\mu(t)}[\mu(t)q(t)] - \frac{p(t)\mu(t)}{[\mu(t)]^2}\left(\int \mu(t)\,q(t)\,dt + C\right).\end{aligned}$$

Simplifying, we have

$$\frac{d}{dt}x(t) = q(t) - p(t)\left[\frac{1}{\mu(t)}\left(\int \mu(t)\,q(t)\,dt + C\right)\right] = q(t) - p(t)x(t)$$

or

$$x'(t) + p(t)x(t) = q(t).$$

In summary, we have proved that all of the functions defined by (5.2.16) are solutions of (5.1.3). Moreover, these are its only solutions.

Since (5.2.17) represents the complete set of solutions of (5.2.10), it is said to be the ***general solution*** of (5.2.10). We could find solutions of first-order linear equations by committing this formula to memory and then carrying out the indicated integrations. However, it is not easily remembered for any length of time; so don't even bother! In truth, it is not needed—instead all we have to do is to memorize the integrating factor formula (5.2.14) by using it to solve several first-order linear equations following the steps enumerated below in the "Integrating Factor Method." Referring to these steps will eventually become unnecessary.

Integrating Factor Method

Follow these steps to find the general solution of $a_1(t)\dfrac{dx}{dt} + a_0(t)x = b(t)$:

1. If the leading coefficient $a_1(t)$ is not already equal to 1, multiply both sides of the equation by the reciprocal of $a_1(t)$ to rewrite it in the standard form

$$\frac{dx}{dt} + p(t)x = q(t).$$

2. Find an integrating factor $\mu(t)$ by exponentiating the integral of the coefficient of the dependent variable x in Step 1. That is,

$$\mu(t) = e^{\int p(t)\,dt}.$$

3. Multiply the standard form of the equation by $\mu(t)$:

$$\mu(t)\frac{dx}{dt} + p(t)\mu(t)x = \mu(t)q(t).$$

4. Replace the left-hand side of the equation in Step 3 by the derivative of the product of the integrating factor and the dependent variable:

$$\frac{d}{dt}(\mu(t)x) = \mu(t)q(t).$$

5. Verify the correctness of the integrating factor found in Step 2 by using the product rule to check that the derivative of $\mu(t)x$ is really equal to

$$\mu(t)\frac{dx}{dt} + p(t)\mu(t)x.$$

6. Integrate both sides of the equation in Step 4 with respect to t and then solve for the dependent variable x.

Example 5.2.1 Find the general solution of

$$\frac{dx}{dt} + 2x = 9e^t. \tag{5.2.18}$$

Solution Since the equation is already in standard form, we bypass Step 1 and go on to Step 2. An antiderivative of the coefficient of the dependent variable is

$$\int p(t)\,dt = \int 2\,dt = 2t.$$

So an integrating factor is

$$\mu(t) = e^{\int p(t)\,dt} = e^{2t}.$$

Moving on to Step 3, multiply (5.2.18) by e^{2t} obtaining

$$e^{2t}\frac{dx}{dt} + 2e^{2t}x = 9e^{3t}. \tag{5.2.19}$$

Doing as Step 4 says, we have

$$\frac{d}{dt}(e^{2t}x) = 9e^{3t}. \tag{5.2.20}$$

Step 5 is to verify that e^{2t} is truly an integrating factor. And it is since the derivative of the left-hand side of (5.2.20) is

$$\frac{d}{dt}(e^{2t}x) = e^{2t}\frac{dx}{dt} + x\frac{d}{dt}e^{2t} = e^{2t}\frac{dx}{dt} + 2e^{2t}x,$$

which is identical to the left-hand side of (5.2.19).

The general solution of Eq. (5.2.18) is obtained by executing Step 6. Carrying out the integrations

$$\int \frac{d}{dt}(e^{2t}x)\,dt = \int 9e^{3t}\,dt + C,$$

we have

$$e^{2t}x = 3e^{3t} + C.$$

Solving for x, we obtain the general solution

$$x = 3e^t + Ce^{-2t}.$$

♦

Example 5.2.2 Find the general solution of

$$\frac{1}{t}\frac{dx}{dt} = 10 - 2x \tag{5.2.21}$$

on the interval $(0, \infty)$.

Solution Unlike the equation in the previous example, this equation is not in standard form. So we must first multiply it by t and then place the terms involving the dependent variable x and its derivative $\dot{x}$ on the left-hand side as follows:

$$\frac{dx}{dt} + 2tx = 10t. \tag{5.2.22}$$

Then, as $p(t) = 2t$, an integrating factor is

$$\mu(t) = e^{\int 2t\,dt} = e^{t^2}.$$

Now multiply both sides of (5.2.22) by $\exp(t^2)$ obtaining

$$e^{t^2}\frac{dx}{dt} + 2te^{t^2}x = 10te^{t^2}.$$

Since the left-hand side of the equation is equal to the derivative of the product of the integrating factor $\exp(t^2)$ and the dependent variable x, we have

$$\frac{d}{dt}\left(e^{t^2}x\right) = 10te^{t^2}.$$

Integrating both sides of this equation with respect to t, we obtain

$$e^{t^2}x = 5e^{t^2} + C.$$

Solving for x, we obtain the general solution

$$x = 5 + Ce^{-t^2}. \tag{5.2.23}$$

Each value given to C results in a particular solution. In Fig. 5.1, a CAS was used to graph six solutions corresponding to $C = \pm 5, \pm 20, \pm 45$. From their graphs, we see that these solutions approach the constant solution $x(t) \equiv 5$. Observe this follows from the solution (5.2.23) itself since $Ce^{-t^2} \to 0$ as $t \to \infty$ if $C \neq 0$. ♦

Example 5.2.3 Find the general solution of

$$t\frac{dx}{dt} - x = t^2 \cos t. \tag{5.2.24}$$

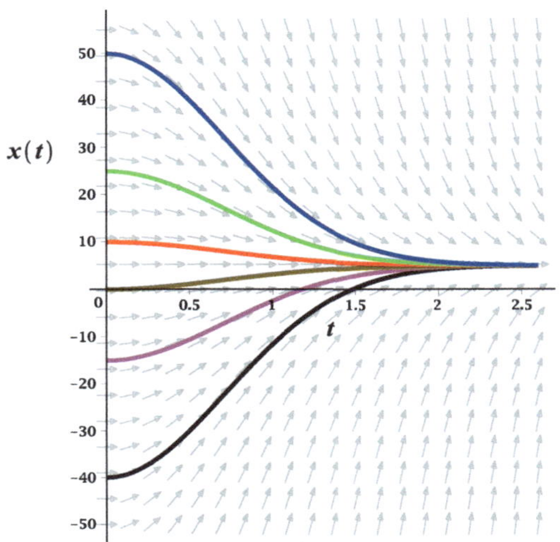

Fig. 5.1 Graphs of six solutions of $t^{-1}\dfrac{dx}{dt} = 10 - 2x$

Solution If $t \neq 0$, we can rewrite (5.2.24) in standard form by multiplying both sides by t^{-1} to get

$$\frac{dx}{dt} - \frac{1}{t}x = t\cos t. \tag{5.2.25}$$

Let us look for solutions on an interval that excludes $t = 0$. For example, consider the interval $(0, \infty)$. By Step 2 of the Integrating Factor Method,

$$\mu(t) = e^{-\int \frac{1}{t}\,dt} = e^{-\ln t} = \frac{1}{e^{\ln t}} = \frac{1}{t}. \tag{5.2.26}$$

Multiplying (5.2.25) by $\mu(t)$, we get

$$\frac{1}{t}\frac{dx}{dt} - \frac{1}{t^2}x = \cos t \tag{5.2.27}$$

or

$$\frac{d}{dt}\left(\frac{x}{t}\right) = \cos t.$$

Integrating this equation with respect to t, we have

$$\frac{x}{t} = \int \cos t\,dt + C.$$

Therefore, the general solution of Eq. (5.2.24) for $t > 0$ is

$$x = t \sin t + Ct. \tag{5.2.28}$$

Do we have to change anything if we require solutions on an interval where the t-values are all negative? For example, let us find solutions on the interval $(-\infty, 0)$. Then an integrating factor is

$$\mu(t) = e^{-\int \frac{1}{t}\,dt} = e^{-\ln|t|} = \frac{1}{e^{\ln|t|}} = \frac{1}{|t|} = -\frac{1}{t}.$$

Multiplying both sides of (5.2.25) by $-t^{-1}$, we get

$$-\frac{1}{t}\frac{dx}{dt} + \frac{x}{t^2} = -\cos t.$$

But note that this becomes (5.2.27) again when we multiply both sides of this equation by -1. We conclude (5.2.28) is also the general solution on $(-\infty, 0)$. In fact, it is the general solution of Eq. (5.2.24) on the entire interval $(-\infty, \infty)$. This is easy to verify by showing that (5.2.24) is satisfied when $t \sin t + Ct$ is substituted for x, irrespective of C's value. ♦

Observation. This example shows that if we employ the Integrating Factor Method to find solutions of the equation

$$\frac{dx}{dt} - \frac{1}{t}x = q(t),$$

we can ignore the absolute value bars that result from using Step 2. In other words, with $p(t) = -1/t$, Step 2 yields

$$\mu(t) = e^{\int p(t)\,dt} = e^{-\int \frac{1}{t}\,dt} = e^{-\ln|t|} = \frac{1}{|t|}.$$

Now simply drop the absolute value bars and use

$$\mu(t) = \frac{1}{t}$$

as an integrating factor for this equation. In this regard, see Problem 42.

5.2.2 *Initial Value Problems*

In order to model how an observable (physical quantity or property) will evolve over time using a differential equation, such as the model of the vertical position of a rock dropped from the top of a tall cliff (see Example 1.4.8 in Chap. 1), the value of the observable at some starting time must also be included. Suppose we use the notation $x(t)$ to denote the value of the observable at time t. Then if its value at a starting time $t = t_0$ is x_0, we can express this succinctly by writing $x(t_0) = x_0$. For example, instead of stating that the vertical position of the rock at time t_0 is x_0, we merely write $x(t_0) = x_0$. As we have already discussed in Chap. 2, $x(t_0) = x_0$ is known as an ***initial condition***. A differential equation together with an initial condition is called an ***initial value problem***. In this section, we investigate the solutions of first-order linear differential equations subject to various initial conditions.

Example 5.2.4 Find the solution of the differential equation

$$\frac{dx}{dt} + \frac{1}{t}x = \frac{1}{t^2 - 2t}, \tag{5.2.29}$$

such that $x(1) = 5$. It other words, find the function $x(t)$ that satisfies this equation on an interval that includes the point $t = 1$ so that its value is 5 when $t = 1$.

We will show how to solve this initial value problem by approaching it in two slightly different ways. Both ways will rely on finding an integrating factor. The first way is to proceed exactly as we have been doing in the examples up to this point. First, we will find the general solution of the differential equation (5.2.29). After that, we will still have to determine the value of the constant of integration in the general solution to obtain the solution that satisfies the initial condition.

Solution Observe that Eq. (5.2.29) is already in the standard form given in Step 1 of the Integrating Factor Method, where

$$p(t) = \frac{1}{t} \quad \text{and} \quad q(t) = \frac{1}{t^2 - 2t} = \frac{1}{t(t-2)}.$$

Not only must we find a function $x(t)$ that satisfies (5.2.29) on some interval but its value at $t = 1$ must also be 5. Since the interval has to contain the point $t = 1$ and because $p(t)$ is undefined at $t = 0$ while $q(t)$ is undefined at both $t = 0$ and $t = 2$, we will look for a solution on the interval $(0, 2)$.

First let us find an integrating factor. Since

$$e^{\int t^{-1}\,dt} = e^{\ln|t|} = |t| = t,$$

on $(0, 2)$, a suitable integrating factor is $\mu(t) = t$. Multiplying (5.2.29) by $\mu(t)$ and using the product rule, we obtain

$$\frac{d}{dt}(tx) = \frac{1}{t-2}. \tag{5.2.30}$$

Integration yields

$$tx = \ln|t-2| + C = \ln(2-t) + C.$$

The value of C is determined by the initial condition. Since it specifies that $x = 5$ when $t = 1$, we have

$$1 \cdot 5 = \ln 1 + C \quad \Rightarrow \quad C = 5.$$

Therefore, the solution of the initial value on the interval (0, 2) is

$$x(t) = \frac{\ln(2-t)+5}{t}.$$

♦

Since the solution of the initial problem (5.2.29) was obtained using indefinite integration, we could not use the initial condition until the very end. There is the alternative of using definite integration, which allows us to incorporate the initial condition right from the very start. Let us illustrate this by once again solving (5.2.29).

Alternative Solution First integrate both sides of (5.2.30) from 1 to t. The lower limit of integration is 1 due to the initial condition $x(1) = 5$. Since it would not make sense to use the letter t for both the upper limit and the variable of integration, let us change the name of the variable of integration from t to s (other choices would do just as well). Thus,

$$\int_{s=1}^{s=t} \frac{d}{ds}(sx(s))\,ds = \int_{s=1}^{s=t} \frac{1}{s-2}\,ds.$$

Integration yields

$$sx(s)\Big|_1^t = \ln|s-2|\Big|_1^t \quad \Rightarrow \quad tx(t) - 5 = \ln(2-t) - \ln 1.$$

As before, we get the solution $x(t) = t^{-1}(\ln(2-t)+5)$. ♦

5.2.3 Existence and Uniqueness

We were able to find the solution of Eq. (5.2.29) satisfying the initial condition $x(1) = 5$ because the functions $p(t) = t^{-1}$ and $q(t) = (t^2 - 2t)^{-1}$ are continuous on the interval (0, 2). This example invites us to consider the general initial value problem

$$\frac{dx}{dt} + p(t)x = q(t), \quad x(t_0) = x_0, \tag{5.2.31}$$

where p and q denote a pair of functions that are continuous on some interval (a, b). Although the initial value problem in Example 5.2.4 has a unique solution, will this generally be the case? That is, we would like to have answers to the following: If the functions p and q are continuous on an interval (a, b) and if x_0 and t_0 are real numbers with $t_0 \in (a, b)$, does (5.2.31) have a solution on this interval? And if so, is it unique?

As it turns out, we will be able to answer these questions if we can first answer this: If, for a pair of continuous functions p and q, a solution $x(t)$ of (5.2.31) is known to exist, what would it be? To answer this, we will employ an integrating factor as in Example 5.2.4. But now we are faced with a slight problem: the formula for integrating factors that we have been using is the one given in Step 2 of the Integrating Factor Method; however, unlike the previous examples where p and q are known functions, here they are not known. So how can we use this integrating factor formula, which involves an indefinite integral, to imitate the alternative solution method that we used to solve the initial value problem in Example 5.2.4, which involves definite integrals, to solve the general initial value problem (5.2.31)? The answer is to select the integrating factor

$$\mu(t) = e^{\int_{t_0}^{t} p(s)\,ds} \tag{5.2.32}$$

from all the possible integrating factors $e^{\int p(t)\,dt}$ given in Step 2. Note that this truly is an integrating factor for (5.2.31) since the integral in the exponent is an antiderivative of the function $p(t)$. With this choice, $\mu(t_0) = e^0 = 1$. Consequently, the differential equation can be rewritten (see (5.2.15)) as follows:

$$\frac{d}{dt}\left[\mu(t)x(t)\right] = \mu(t)q(t).$$

Integrating this equation from t_0 to t, where $a < t < b$, we have

$$\mu(t)x(t) - \mu(t_0)x(t_0) = \int_{t_0}^{t} \mu(s)q(s)\,ds.$$

Since $\mu(t_0) = 1$ and $x(t_0) = x_0$,

$$x(t) = \frac{1}{\mu(t)}\left(\int_{t_0}^{t} \mu(s)q(s)\,ds + x_0\right). \tag{5.2.33}$$

As a result, we conclude that if the initial value problem (5.2.31) has a solution, it would have to be the function

$$x(t) = e^{-\int_{t_0}^{t} p(s)\,ds} x_0 + \int_{t_0}^{t} e^{-\int_{s}^{t} p(v)\,dv} q(s)\,ds. \tag{5.2.34}$$

This rules out all other possibilities. It is left as an exercise (see Problem 43) to verify that (5.2.34) follows from (5.2.32) and (5.2.33).

So far we have ascertained that if a function solves the general initial value problem, then it would have to be (5.2.34). This brings up the question: Is (5.2.34) really a solution? In order to answer this, it will facilitate the following work if we first express (5.2.34) in terms of the function

$$P(t) := \int_{t_0}^{t} p(s)\,ds.$$

Since

$$e^{-\int_s^t p(v)\,dv} = e^{-\left[\int_{t_0}^t p(v)\,dv + \int_s^{t_0} p(v)\,dv\right]} = e^{-P(t)} \cdot e^{P(s)},$$

we can rewrite (5.2.34) as

$$x(t) = e^{-P(t)}\left[x_0 + \int_{t_0}^{t} e^{P(s)} q(s)\,ds\right]. \tag{5.2.35}$$

This function is differentiable on the entire interval (a, b) since by the Fundamental Theorem of Calculus,

$$\frac{d}{dt}P(t) = p(t) \quad \text{and} \quad \frac{d}{dt}\int_{t_0}^{t} e^{P(s)} q(s)\,ds = e^{P(t)} q(t)$$

for $a < t < b$. Using these two derivatives and the product and chain rules, we have

$$\begin{aligned} x'(t) &= e^{-P(t)}\frac{d}{dt}\left[x_0 + \int_{t_0}^{t} e^{P(s)} q(s)\,ds\right] + \left[x_0 + \int_{t_0}^{t} e^{P(s)} q(s)\,ds\right]\frac{d}{dt}e^{-P(t)} \\ &= e^{-P(t)} \cdot e^{P(t)} q(t) - p(t) e^{-P(t)}\left[x_0 + \int_{t_0}^{t} e^{P(s)} q(s)\,ds\right] \end{aligned}$$

for $a < t < b$. This simplifies to

$$x'(t) = q(t) - p(t)x(t).$$

This establishes that the function given by (5.2.34) is a solution of the general initial value problem (5.2.31) on (a, b) since its value at $t = t_0$ is x_0. Furthermore, because we had already narrowed down all the possible solutions to this particular function, we have also established its uniqueness. As a result, we have the following theorem.

Existence and Uniqueness of Solutions

Theorem 5.2.1 *Let t_0 and x_0 be real numbers. If $p(t)$ and $q(t)$ are continuous functions on an interval (a, b) and if $t_0 \in (a, b)$, then the initial value problem*

$$\frac{dx}{dt} + p(t)x = q(t), \quad x(t_0) = x_0$$

has a unique solution on (a, b). It is given by (5.2.34) *or* (5.2.35).

Example 5.2.5 Solve the initial value problem

$$\frac{dx}{dt} = 1 - 2tx, \quad x(0) = 0. \tag{5.2.36}$$

Solution We begin by rewriting the differential equation in the standard form

$$\frac{dx}{dt} + 2tx = 1. \tag{5.2.37}$$

Since $p(t) = 2t$ and $q(t) = 1$ are continuous on $(-\infty, \infty)$, Theorem 5.2.1 ensures the existence of a unique solution of (5.2.36) on $(-\infty, \infty)$.

An integrating factor for Eq. (5.2.37) is

$$\mu(t) = e^{\int p(t)\,dt} = e^{\int 2t\,dt} = e^{t^2}.$$

Multiplying both sides of (5.2.37) by $\mu(t)$ and using the product rule, we get

$$\frac{d}{dt}\left(e^{t^2}x\right) = e^{t^2}. \tag{5.2.38}$$

An integration yields

$$e^{t^2}x = \int e^{t^2}\,dt + C.$$

Now, however, it appears we have reached an impasse since we can go no further in evaluating the integral. The reason for this is that the integrand $\exp(t^2)$ does not have an elementary antiderivative. This means there is no elementary function[2]

[2] These are the familiar functions of algebra, trigonometry, and elementary calculus. More precisely, an ***elementary function*** is a polynomial, power function, exponential function, logarithmic function, trigonometric function, or inverse trigonometric function—or the result of

whose derivative is $\exp(t^2)$. The term ***nonelementary integral*** is used to designate such an integral.

There is a way of handling nonelementary integrals when they come up in solving first-order linear equations—and that is to express the solution in terms of a definite integral. Returning to where we left off, let us integrate both sides of (5.2.38) from 0 to t:

$$\int_0^t \frac{d}{ds}\left(e^{s^2}x(s)\right)\,ds = \int_0^t e^{s^2}\,ds.$$

Thus,

$$e^{t^2}x(t) - x(0) = \int_0^t e^{s^2}\,ds.$$

Using the initial condition $x(0) = 0$ and solving for $x(t)$, we obtain the unique solution

$$x(t) = e^{-t^2}\int_0^t e^{s^2}\,ds. \tag{5.2.39}$$

Even though this is expressed in terms of an integral, this is as far as we can simplify (5.2.39) since the integrand has no elementary antiderivative.

We can also get the solution directly from (5.2.34) or from (5.2.35). If we use the latter formula, we first evaluate $P(t)$. Since $t_0 = 0$ and $p(t) = 2t$,

$$P(t) = \int_0^t p(s)\,ds = \int_0^t 2s\,ds = t^2.$$

Thus, as $x_0 = 0$ and $q(t) = 1$, we once again obtain the solution

$$x(t) = e^{-P(t)}\left[x_0 + \int_{t_0}^t e^{P(s)}q(s)\,ds\right] = e^{-t^2}\int_0^t e^{s^2}\,ds.$$

Even though (5.2.34) or (5.2.35) can be used to solve initial value problems for first-order linear equations, we prefer the Integrating Factor Method instead of using these formulas. This method is easy to remember after it has been used several times whereas both formulas are readily misremembered or forgotten. ♦

combining these functions a finite number of times using the operations of addition, subtraction, multiplication, division, raising to powers, extraction of roots, and function composition.

5.3 Functions Defined by Integrals

Someone fresh out of a first-year calculus course will most likely be uncomfortable with functions that are defined in terms of definite integrals, such as (5.2.39). Even though it may not look like any of the garden-variety functions that are typically dealt with in undergraduate courses, it does satisfy the definition of a function: for each value of t, there is one, and only one, output $x(t)$. As it turns out, (5.2.39) is an important function because of its involvement in mathematical models of some physical systems. It is known as ***Dawson's integral***. From now on, we will denote it by $\mathcal{D}(t)$. That is,

$$\mathcal{D}(t) := e^{-t^2} \int_0^t e^{s^2}\, ds. \tag{5.3.1}$$

Now that we know that $\mathcal{D}(t)$ is a well-known function, we can say that the solution of the initial value problem (5.2.36) is $x(t) = \mathcal{D}(t)$.

Values of $\mathcal{D}(t)$ can be found in some handbooks, such as the classic *Handbook of Mathematical Functions* by Abramowitz and Stegun [1]. Even better is a revision of this handbook entitled *NIST Digital Library of Mathematical Functions*, which is published by the National Institute of Standards and Technology. It can be found in electronic form at the website address: https://dlmf.nist.gov. Functions that are defined by integrals can also be evaluated with a CAS. For example, we can evaluate

$$\mathcal{D}(1) = e^{-1} \int_0^1 e^{s^2}\, ds$$

with *Maple* using the command "dawson(1.0)", which yields 0.5380795069. Another option is to use a symbolic calculator, that is, one with CAS capabilities. For instance, entering

$$e\verb|^|(\text{-}1) \int(e\verb|^|(t\verb|^|2), t, 0, 1)$$

into a TI-89 calculator yields 0.53807951.

Example 5.3.1 Find the intervals on which the function $\mathcal{D}(t)$ is increasing and on which it is decreasing and determine if it has any local or absolute extrema. Also, find the intervals on which the graph of $\mathcal{D}(t)$ is concave up or concave down and determine if there are any points of inflection. If there are any extreme points or inflection points, compute their coordinates using either a CAS to evaluate $\mathcal{D}(t)$ or a handbook listing its values.

Solution The domain of $\mathcal{D}(t)$ is $(-\infty, \infty)$ since (5.3.1) is defined at all values of t. As e^{-t^2} is positive for all values of t and as the integral part of $\mathcal{D}(t)$ is positive for $t > 0$ and negative for $t < 0$, it follows that $\mathcal{D}(t) < 0$ on $(-\infty, 0)$ and $\mathcal{D}(t) > 0$ on $(0, \infty)$. And, of course, $\mathcal{D}(0) = 0$.

$\mathcal{D}(t)$ is an odd function. That is, $\mathcal{D}(-t) = -\mathcal{D}(t)$. This can be seen as follows:

$$\mathcal{D}(-t) = e^{-(-t)^2} \int_0^{-t} e^{s^2}\, ds = e^{-t^2} \int_0^{-t} e^{s^2}\, ds.$$

Letting $u = -s$, we have

$$\mathcal{D}(-t) = -e^{-t^2} \int_0^{t} e^{(-u)^2}\, du = -e^{-t^2} \int_0^{t} e^{u^2}\, du = -\mathcal{D}(t).$$

Consequently, the graph of $\mathcal{D}(t)$ is symmetric about the origin.

The graph of the function $\mathcal{D}(t)$ has no vertical asymptotes since $\mathcal{D}(t)$ is defined at every value of t. To ascertain whether there are any horizontal asymptotes, let us find the limits of $\mathcal{D}(t)$ as $t \to \pm\infty$. Employing l'Hôpital's rule, we have

$$\lim_{t\to\infty} \mathcal{D}(t) = \lim_{t\to\infty} e^{-t^2} \int_0^{t} e^{s^2}\, ds = \lim_{t\to\infty} \frac{\int_0^t e^{s^2}\, ds}{e^{t^2}} = \lim_{t\to\infty} \frac{\frac{d}{dt}\int_0^t e^{s^2}\, ds}{\frac{d}{dt} e^{t^2}}$$
$$= \lim_{t\to\infty} \frac{e^{t^2}}{2te^{t^2}} = \lim_{t\to\infty} \frac{1}{2t} = 0.$$

Thus, the t-axis is a horizontal asymptote of the graph of the function $y = \mathcal{D}(t)$. Moreover, as $\mathcal{D}(t) > 0$ for $t > 0$,

$$\lim_{t\to\infty} \mathcal{D}(t) = 0^+.$$

And as $\mathcal{D}(t)$ is symmetric about the origin,

$$\lim_{t\to-\infty} \mathcal{D}(t) = 0^-.$$

In order to determine where $\mathcal{D}(t)$ is increasing and decreasing, consider the sign of the first derivative $\mathcal{D}'(t)$, where

$$\mathcal{D}'(t) = \frac{d}{dt}\left[e^{-t^2} \int_0^t e^{s^2}\, ds\right] = e^{-t^2} \cdot e^{t^2} - 2te^{-t^2} \int_0^t e^{s^2}\, ds = 1 - 2t\mathcal{D}(t).$$

(Note that this, together with $\mathcal{D}(0) = 0$, confirms that $\mathcal{D}(t)$ is a solution of the initial value problem (5.2.36).) Since $\mathcal{D}'(t)$ is a continuous function, any changes in its sign can only occur where $\mathcal{D}'(t)$ is equal to zero; so let us see if there are values of t for which

$$1 - 2t\mathcal{D}(t) = 0. \tag{5.3.2}$$

Multiplying both sides of this equation by $\exp(t^2)$, we have

$$e^{t^2}[1 - 2t\mathcal{D}(t)] = 0 \quad \Rightarrow \quad e^{t^2} - 2t \int_0^t e^{s^2}\, ds = 0.$$

If there is a solution of this equation, then it would also be a solution of (5.3.2) and thereby a critical point of $\mathcal{D}(t)$. In other words, critical points of $\mathcal{D}(t)$, if any, are precisely the zeros of the function

$$F(t) := e^{t^2} - 2t \int_0^t e^{s^2}\, ds.$$

Since its derivative

$$F'(t) = 2te^{t^2} - 2te^{t^2} - 2\int_0^t e^{s^2}\, ds = -2\int_0^t e^{s^2}\, ds$$

is negative for $t > 0$, the function $F(t)$ is strictly decreasing on the interval $(0, \infty)$. Clearly $F(0) = 1$. Using a CAS, such as *Maple*, or some other computational tool, such as *Wolfram|Alpha*, we find that

$$F(2) = -11.2123\ldots .$$

Because $F(0) > 0$, $F(2) < 0$, and $F(t)$ is strictly decreasing on $(0, \infty)$, the function $F(t)$ has exactly one zero on $(0, \infty)$. Again with the help of a CAS, we find that it is at

$$t_1 = 0.9241388730\ldots . \tag{5.3.3}$$

Thus, $F(t) > 0$ for $0 < t < t_1$, $F(t_1) = 0$, and $F(t) < 0$ for $t > t_1$. Moreover, as

$$F(t) = e^{t^2}[1 - 2t\mathcal{D}(t)] = e^{t^2}\mathcal{D}'(t),$$

we also have

$$\mathcal{D}'(t) > 0 \text{ for } 0 < t < t_1, \quad \mathcal{D}'(t_1) = 0, \quad \mathcal{D}'(t) < 0 \text{ for } t > t_1.$$

Therefore, $\mathcal{D}(t)$ strictly increases on the interval $(0, t_1)$ and strictly decreases on (t_1, ∞). Furthermore, the symmetry of the graph of $\mathcal{D}(t)$ about the origin implies that $-t_1$ is also a critical point and that $\mathcal{D}(t)$ strictly decreases on $(-\infty, -t_1)$ and strictly increases on $(-t_1, 0)$. From this analysis, we conclude that the absolute maximum value of $\mathcal{D}(t)$ is $\mathcal{D}(t_1)$ and its absolute minimum value is $-\mathcal{D}(t_1)$, where

$$\mathcal{D}(t_1) = 0.5410442246\ldots .$$

Since the concavity of the graph of a twice-differentiable function can be determined from the sign of its second derivative, consider

$$\begin{aligned}\mathcal{D}''(t) &= \frac{d}{dt}\left[1 - 2t\mathcal{D}(t)\right] = -2t\mathcal{D}'(t) - 2\mathcal{D}(t) \\ &= -2\left[t(1 - 2t\mathcal{D}(t)) + \mathcal{D}(t)\right] = 2\Big[(2t^2 - 1)\mathcal{D}(t) - t\Big].\end{aligned}$$

Since $\mathcal{D}''(t)$ is a continuous function, it can only change its sign at points where $\mathcal{D}''(t) = 0$. Accordingly, let us find all the solutions of

$$\mathcal{D}(t) = \frac{t}{2t^2 - 1}. \tag{5.3.4}$$

Clearly, $t = 0$ is a solution. For $t > 0$, a CAS yields the solution (see Problem 65)

$$t_2 = 1.5019752682\ldots, \tag{5.3.5}$$

where $\mathcal{D}(t_2) = 0.4276866160\ldots$. But could there possibly be solutions to the right of the point t_2 that a CAS may fail to find? This can be answered by defining the function $G(t) := e^{t^2}\mathcal{D}''(t)$ (see Problem 44). With this function it can be shown that t_2 is the only solution of $\mathcal{D}''(t) = 0$ on $(0, \infty)$ much in the same way that $F(t) = e^{t^2}\mathcal{D}'(t)$ was used earlier to establish the uniqueness of the solution t_1 of $\mathcal{D}'(t) = 0$ on $(0, \infty)$. Furthermore, due to the relationship between $G(t)$ and $\mathcal{D}''(t)$, it can be shown that

$$\mathcal{D}''(t) < 0 \text{ for } 0 < t < t_2 \quad \text{and} \quad \mathcal{D}''(t) > 0 \text{ for } t > t_2.$$

Consequently, the graph of $\mathcal{D}(t)$ is concave down on $(0, t_2)$ and concave upward on (t_2, ∞). Additionally, the symmetry of the graph of $\mathcal{D}(t)$ about the origin implies that its graph is concave down on $(-\infty, -t_2)$ and concave up on $(-t_2, 0)$. Thus, we conclude that there are three points of inflection, to wit:

$$(-t_2, -\mathcal{D}(t_2)) \approx (-1.502, -0.4277), \quad (0, 0), \quad (t_2, \mathcal{D}(t_2)) \approx (1.502, 0.4277).$$

All of the important aspects of the graph of Dawson's integral that we have fleshed out with calculus are readily seen in Fig. 5.2. ♦

There are other functions defined in terms of definite integrals that are indispensable to the mathematical sciences and engineering. One of them is the ***error function***, which is denoted by $\text{erf}(t)$ and defined by

$$\text{erf}(t) := \frac{2}{\sqrt{\pi}} \int_0^t e^{-u^2}\, du. \tag{5.3.6}$$

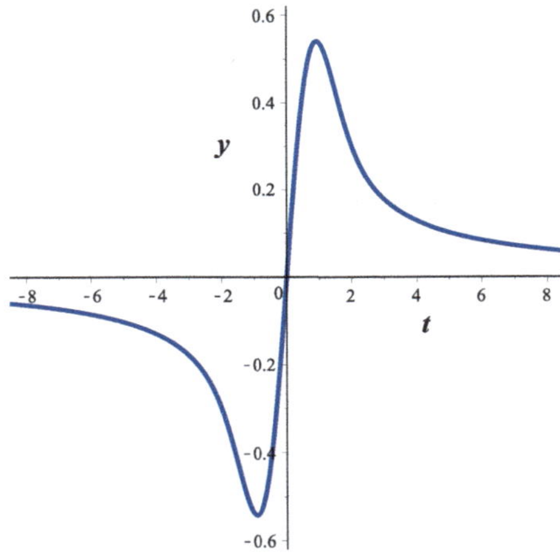

Fig. 5.2 Graph of Dawson's integral

It occurs in probability, statistics, and partial differential equations. Another important function is the ***sine integral function***, which is needed in certain areas of optics and electrical engineering. It is denoted by $\mathrm{Si}(t)$ and defined by the integral

$$\mathrm{Si}(t) := \int_0^t \frac{\sin u}{u}\, du. \tag{5.3.7}$$

5.4 Gronwall's Inequality

The integrating factor method for solving linear equations can also be applied to the linear integral inequality

$$f(t) \leq K + \int_{t_0}^{t} p(s) f(s)\, ds$$

for $t \geq t_0$, where K is a nonnegative constant and $p(t)$ is a given nonnegative, continuous function. As we will see in the proof of Theorem 5.4.1 below, the integrating factor method will enable us to find an upper bound for the unknown function $f(t)$. For certain types of differential and integral equations, this theorem often provides a way

- to obtain a bound on the difference between two solutions
- and to establish the uniqueness of solutions.

It can also be applied to integral equations as we shall see later in Example 5.4.2. The above inequality, as well as other similar forms, is known as ***Gronwall's inequality***. One of the earliest incipient forms of Gronwall's inequality can be traced back to a paper written by T.H. Gronwall [40] in 1919—hence its name. Gronwall's inequality has been modified and generalized a number of times by others; as a result, it has acquired other names too, chief among them are *Reid's inequality*, *Bellman's lemma*, and the *fundamental inequality*.[3] Gronwall's inequality also refers to the following theorem

Gronwall's Inequality

Theorem 5.4.1 *Let K be a given nonnegative constant. Let $p(t)$ be a given nonnegative, continuous function on an interval I and t_0 a point in I. If $f(t)$ is a nonnegative, continuous function on I and*

$$f(t) \le K + \left| \int_{t_0}^{t} p(s) f(s)\, ds \right| \tag{5.4.1}$$

for all $t \in I$, then

$$f(t) \le K e^{\left| \int_{t_0}^{t} p(s)\, ds \right|} \tag{5.4.2}$$

for all $t \in I$.

Proof If $t = t_0$, then (5.4.2) follows immediately from (5.4.1). Now suppose $t > t_0$, where $t \in I$. Then, as $p(s)f(s) \ge 0$ for all $s \in I$,

$$\left| \int_{t_0}^{t} p(s) f(s)\, ds \right| = \int_{t_0}^{t} p(s) f(s)\, ds.$$

Let $x(t)$ denote the right-hand side of (5.4.1). So, as $t > t_0$,

$$x(t) = K + \int_{t_0}^{t} p(s) f(s)\, ds. \tag{5.4.3}$$

Then (5.4.1) becomes $f(t) \le x(t)$. As $p(t) \ge 0$,

$$p(t) f(t) \le p(t) x(t). \tag{5.4.4}$$

[3] More information and applications of Gronwall's inequality can be found in Burton [18] and Driver [30]. Applications to integro-differential equations can be found in Becker [8].

Differentiating $x(t)$, we have

$$x'(t) = \frac{d}{dt}\left[K + \int_{t_0}^{t} p(s)f(s)\,ds\right] = p(t)f(t)$$

by the Fundamental Theorem of Calculus. Consequently it follows from (5.4.4) that

$$x'(t) - p(t)x(t) \le 0. \tag{5.4.5}$$

Now let

$$\mu(t) := e^{-\int_{t_0}^{t} p(s)\,ds}$$

and then multiply both sides of (5.4.5) by this function obtaining

$$\mu(t)x'(t) - \mu(t)p(t)x(t) \le 0.$$

Since $\mu'(t) = -\mu(t)p(t)$, this can be rewritten as

$$\frac{d}{dt}\big[\mu(t)x(t)\big] \le 0.$$

Integrating both sides from t_0 to t, we find that

$$\mu(t)x(t) \le \mu(t_0)x(t_0).$$

Since $\mu(t_0) = 1$ and $x(t_0) = K$, this becomes

$$\mu(t)x(t) \le K.$$

Therefore, we have

$$f(t) \le x(t) \le K\big[\mu(t)\big]^{-1} = Ke^{\int_{t_0}^{t} p(s)\,ds} = Ke^{\left|\int_{t_0}^{t} p(s)\,ds\right|},$$

which is (5.4.2).

The other case to consider is $t < t_0$, where $t \in I$. Then,

$$\left|\int_{t_0}^{t} p(s)f(s)\,ds\right| = -\int_{t_0}^{t} p(s)f(s)\,ds.$$

Hence (5.4.1) becomes

$$f(t) \le K - \int_{t_0}^{t} p(s)f(s)\,ds.$$

The completion of the proof for this case is left to Problem 50. ■

In the first of two examples illustrating the use of Gronwall's inequality, we will determine the extent to which two solutions of a first-order linear equation can differ from each other over an interval. In the second example, we prove that if the solution of a certain type of integral equation exists, then it must be unique.

Example 5.4.1 Let $x(t)$ and $\tilde{x}(t)$ be solutions of a first-order linear equation

$$\frac{dx}{dt} + p(t)x = q(t), \tag{5.4.6}$$

where $p(t)$ and $q(t)$ are continuous functions on an interval (a, b). Use Gronwall's inequality to show that if the respective values of $x(t)$ and $\tilde{x}(t)$ at a given point t_0 in (a, b) are x_0 and $\tilde{x_0}$, then

$$|x(t) - \tilde{x}(t)| \leq |x_0 - \tilde{x}_0| e^{\left|\int_{t_0}^{t} |p(s)|\, ds\right|} \tag{5.4.7}$$

for all $t \in (a, b)$.

Solution Since $p(t)$ and $q(t)$ are continuous on (a, b), both $x(t)$ and $\tilde{x}(t)$ satisfy equation (5.4.6) for all $t \in (a, b)$ (cf. Theorem 5.2.1). As a result, we have

$$x'(t) + p(t)x(t) - \left(\tilde{x}'(t) + p(t)\tilde{x}(t)\right) = q(t) - q(t) = 0.$$

That is,

$$(x(t) - \tilde{x}(t))' + p(t)\,(x(t) - \tilde{x}(t)) = 0$$

for $a < t < b$. Letting $u(t) := x(t) - \tilde{x}(t)$, this simplifies to

$$u'(t) + p(t)u(t) = 0.$$

An integration of both sides of this equation from t_0 to any $t \in (a, b)$ yields

$$u(t) = u(t_0) - \int_{t_0}^{t} p(s)u(s)\, ds.$$

Taking the absolute value and then using the triangle inequality, we get

$$|u(t)| = \left| u(t_0) - \int_{t_0}^{t} p(s)u(s)\, ds \right| \leq |u(t_0)| + \left| \int_{t_0}^{t} p(s)u(s)\, ds \right|.$$

It follows from Problem 49 that

$$\left| \int_{t_0}^{t} p(s)u(s)\, ds \right| \leq \left| \int_{t_0}^{t} |p(s)u(s)|\, ds \right|.$$

Thus,

$$|u(t)| \le |u(t_0)| + \left| \int_{t_0}^{t} |p(s)||u(s)|\,ds \right|.$$

Now we can apply Gronwall's inequality. Let $|u(t)|$, $|u(t_0)|$, and $|p(t)|$ play the roles of $f(t)$, K, and $p(t)$ in (5.4.1), respectively. Then from (5.4.2) we have

$$|u(t)| \le |u(t_0)| e^{\left| \int_{t_0}^{t} |p(s)|\,ds \right|},$$

which is (5.4.7) as $u(t) = x(t) - \tilde{x}(t)$ and $u(t_0) = x_0 - \tilde{x}_0$. ♦

Remarks If $\tilde{x}_0 = x_0$, then it follows from (5.4.7) that $\tilde{x}(t) = x(t)$ for all $t \in (a, b)$. This confirms the statement in Theorem 5.2.1 that a solution of the initial value problem (5.2.31) is unique.

A function $f(t)$ is said to be ***bounded*** on an interval I if a positive constant M exists such that $|f(t)| \le M$ for all $t \in I$. It is left as an exercise to show that if $p(t)$ is bounded by a positive constant M on (a, b), then

$$|x(t) - \tilde{x}(t)| \le |x_0 - \tilde{x}_0| e^{M|t-t_0|} \tag{5.4.8}$$

for all $t \in (a, b)$.

Example 5.4.2 Let $f(t, x)$ be a function that is defined and continuous for all $t, x \in \mathbb{R}$ and which has the following property: a constant $L > 0$ exists such that

$$|f(t, x) - f(t, y)| \le L|x - y| \tag{5.4.9}$$

for all $t \in [t_0, T)$ and $x, y \in \mathbb{R}$. (This is known as a ***Lipschitz condition.***) Prove that if there is a continuous function $x(t)$ which satisfies the integral equation

$$x(t) = x_0 + \int_{t_0}^{t} f(u, x(u))\,du \tag{5.4.10}$$

for all $t \in [t_0, T)$, then it is unique.

Solution Suppose there is another function $\tilde{x}(t)$ that is continuous on $[t_0, T)$ and which also satisfies (5.4.10) on this interval. Then

$$\begin{aligned} |x(t) - \tilde{x}(t)| &= \left| x_0 + \int_{t_0}^{t} f(u, x(u))\,du - x_0 - \int_{t_0}^{t} f(u, \tilde{x}(u))\,du \right| \\ &= \left| \int_{t_0}^{t} \big(f(u, x(u)) - f(u, \tilde{x}(u)) \big)\,du \right| \\ &\le \left| \int_{t_0}^{t} \big| f(u, x(u)) - f(u, \tilde{x}(u)) \big|\,du \right| \le \left| \int_{t_0}^{t} L|x(u) - \tilde{x}(u)|\,du \right|, \end{aligned}$$

where we have used Problem 49 and the Lipschitz condition in order to obtain the last two inequalities. Letting $f(t) := |x(t) - \tilde{x}(t)|$, we have

$$f(t) \le \left| \int_{t_0}^{t} Lf(u)\,du \right|.$$

This is (5.4.1) with $K = 0$ and $p(u) \equiv L$. It then follows from (5.4.2) that

$$f(t) \le Ke^{|\int_{t_0}^{t} p(s)\,ds|} = Ke^{L|t-t_0|}$$

for all $t \in [t_0, T)$. Thus, $f(t) \le 0$ because $K = 0$. And so

$$|x(t) - \tilde{x}(t)| \le 0$$

for all $t \in [t_0, T)$. This implies that $\tilde{x}(t) \equiv x(t)$ on $[t_0, T)$. ♦

5.5 Variation of Parameters Formula

There is another well-known method for solving the nonhomogeneous first-order linear equation

$$\frac{dx}{dt} + p(t)x = q(t) \tag{5.5.1}$$

that employs a formula called the *variation of parameters formula*,[4] which can be used in lieu of the integrating factor method. The idea behind this method comes from wondering whether the solutions of the associated homogeneous equation, namely, the equation

$$\frac{dx}{dt} + p(t)x = 0,$$

can somehow be used to solve (5.5.1). Let us tinker with this idea in the following example by first finding the general solution of the homogeneous equation associated with the nonhomogeneous equation (5.5.2), after which we can go on from there.

Example 5.5.1 Find the general solution of

$$t\frac{dx}{dt} + 2x = 3t. \tag{5.5.2}$$

[4] There is a comparable formula for nonautonomous linear systems, which is the topic of Sect. 13.6 in Chap. 13.

Solution We begin by finding a general solution of the associated homogeneous equation

$$t\frac{dx}{dt} + 2x = 0. \tag{5.5.3}$$

Multiplication by t yields the equation

$$t^2\frac{dx}{dt} + 2tx = 0 \quad \text{or} \quad \frac{d}{dt}\left(t^2x\right) = 0.$$

Consequently, $x = Ct^{-2}$. There is also the option of separating the variables of (5.5.3) and integrating as follows:

$$\int \frac{dx}{x} = -2\int \frac{dt}{t} + K \quad \Rightarrow \quad \ln|x| = -2\ln|t| + K.$$

Solving explicitly for x, we once again obtain

$$x = Ct^{-2}. \tag{5.5.4}$$

This is a solution of the homogeneous equation (5.5.3) for any value of C and on any interval that excludes $t = 0$.

Now let us try out the following idea: Perhaps there are solutions of the nonhomogeneous equation (5.5.2) of the form

$$x = vt^{-2}, \tag{5.5.5}$$

where v denotes a function of t, which of course is unknown an this point. Note that we have replaced the constant C in (5.5.4) with a function $v(t)$. The goal is to find v so that (5.5.5) is a solution of (5.5.2). Substituting (5.5.5) and its derivative into the left-hand side of (5.5.2), we obtain

$$t\frac{dx}{dt} + 2x = t(-2vt^{-3} + v't^{-2}) + 2vt^{-2} = v't^{-1}.$$

Thus, v is a solution of (5.5.2) if

$$v't^{-1} = 3t \quad \text{or} \quad v' = 3t^2.$$

Thus,

$$v = t^3 + C.$$

As a result, we have

$$x = (t^3 + C)t^{-2}.$$

Therefore, a solution of (5.5.2) is

$$x = t + Ct^{-2}. \tag{5.5.6}$$

This is the general solution since the solution of (5.5.2) subject to a given initial condition can always be expressed in this form for some value of C. For instance, suppose the initial condition is $x(1) = -9$. Substituting $t = 1$ and $x = -9$ into (5.5.6), we get $C = -10$. Therefore,

$$x = t - \frac{10}{t^2} \tag{5.5.7}$$

is the solution on $(0, \infty)$ that satisfies the initial condition. ♦

In light of Example 5.5.1, we can now explain how to go about finding the solution of the general linear equation (5.5.1) so that $x(t_0) = x_0$. As before, assume that p and q are continuous on an interval (a, b) and that t_0 belongs to this interval. We will try to figure out how to utilize the solutions of the associated homogeneous equation in order to obtain the solution of the initial value problem

$$\frac{dx}{dt} + p(t)x = q(t), \quad x(t_0) = x_0.$$

Let us start out with the solution of the associated homogeneous equation

$$\frac{dx}{dt} + p(t)x = 0$$

satisfying the initial condition $x(t_0) = 1$. That solution is

$$x(t) = e^{-\int_{t_0}^{t} p(s)\,ds} \tag{5.5.8}$$

since its derivative is $x'(t) = -p(t)x(t)$ and

$$x(t_0) = e^{-\int_{t_0}^{t_0} p(s)\,ds} = e^0 = 1.$$

Alternatively, we can obtain (5.5.8) from (5.2.34) by setting $x_0 = 1$ and letting $q(t) \equiv 0$. It follows from Theorem 5.2.1 that this solution exists on the entire interval (a, b). Furthermore, it is the only solution that is equal to 1 at $t = t_0$. This particular solution is known as the *principal solution* of the associated homogeneous equation. From now on, we will denote it by $z(t, t_0)$ in order to distinguish it from all other solutions of this equation. As a result, we have the following definition.

Principal Solution

Definition 5.5.1 Let $p(t)$ be a continuous function on an interval (a, b) and let $t_0 \in (a, b)$. The ***principal solution*** of the homogeneous linear equation

$$\frac{dx}{dt} + p(t)x = 0 \tag{5.5.9}$$

at the point $t = t_0$ is

$$z(t, t_0) := e^{-\int_{t_0}^{t} p(s)\,ds} \tag{5.5.10}$$

for $a < t < b$.

For given values of C and $t_0 \in (a, b)$, the function

$$x(t) = Ce^{-\int_{t_0}^{t} p(s)\,ds} \tag{5.5.11}$$

is a solution of (5.5.9) for $a < t < b$ since

$$\frac{d}{dt}x(t) = C\frac{d}{dt}e^{-\int_{t_0}^{t} p(s)\,ds} = Ce^{-\int_{t_0}^{t} p(s)\,ds}(-p(t)) = -p(t)x(t).$$

Theorem 5.2.1 tells us that it is the only solution of (5.5.9) with $x(t_0) = C$. Expressed in terms of the principal solution at $t = t_0$, solution (5.5.11) is

$$x(t) = Cz(t, t_0). \tag{5.5.12}$$

Recall that the goal is to express the solution of the initial value problem

$$\frac{dx}{dt} + p(t)x = q(t), \quad x(t_0) = x_0$$

in the form

$$x(t) = v(t) \cdot z(t, t_0) \tag{5.5.13}$$

for some $v(t)$. To find $v(t)$, substitute (5.5.13) and its derivative $x' = vz' + zv'$ into the left-hand side of the differential equation:

$$x' + px = vz' + zv' + pvz = v(z' + pz) + zv' = v \cdot 0 + zv' = zv'.$$

From this, we see that $x(t)$ given by (5.5.13) is a solution of the nonhomogeneous equation

$$\frac{dx}{dt} + p(t)x = q(t)$$

provided $zv' = q$. Division by z and integration yields

$$\int_{t_0}^{t} v'(u)\,du = \int_{t_0}^{t} \frac{q(u)}{z(u,t_0)}\,du \quad \Rightarrow \quad v(t) = v(t_0) + \int_{t_0}^{t} \frac{q(u)}{z(u,t_0)}\,du.$$

Setting $t = t_0$ in (5.5.13), we have

$$x(t_0) = v(t_0) \cdot z(t_0, t_0) = v(t_0) \cdot 1 = v(t_0).$$

And so $v(t_0) = x_0$. Thus,

$$v(t) = x_0 + \int_{t_0}^{t} \frac{q(u)}{z(u,t_0)}\,du.$$

From (5.5.13), we have

$$x(t) = z(t,t_0) \cdot v(t) = z(t,t_0)x_0 + \int_{t_0}^{t} \frac{z(t,t_0)}{z(u,t_0)} q(u)\,du.$$

Since

$$\frac{z(t,t_0)}{z(u,t_0)} = \frac{e^{-\int_{t_0}^{t} p(s)\,ds}}{e^{-\int_{t_0}^{u} p(s)\,ds}} = e^{-\int_{u}^{t} p(s)\,ds} = z(t,u),$$

we end up with the formula

$$x(t) = z(t,t_0)x_0 + \int_{t_0}^{t} z(t,u)q(u)\,du,$$

which is the solution of the nonhomogeneous first-order linear equation (5.5.1) that satisfies the initial condition $x(t_0) = x_0$. It is known as the ***variation of parameters formula*** for (5.5.1). In summary, we have the following result.

Variation of Parameters Formula

Theorem 5.5.2 *Let $p(t)$ and $q(t)$ be continuous functions on an interval (a,b). Let t_0 and x_0 be real numbers with $t_0 \in (a,b)$. The solution of the initial value problem*

$$\frac{dx}{dt} + p(t)x = q(t), \quad x(t_0) = x_0 \tag{5.5.14}$$

(continued)

is

$$x(t) = z(t, t_0)x_0 + \int_{t_0}^{t} z(t, s)q(s)\,ds \qquad (5.5.15)$$

for $a < t < b$, *where* $z(t, s)$ *denotes the principal solution of the associated homogeneous equation* (5.5.9) *at* $t = s$, *to wit:*

$$z(t, s) = e^{-\int_s^t p(u)\,du}. \qquad (5.5.16)$$

We derived (5.5.15) by replacing the constant C in (5.5.12) with the function $v(t)$, which could be thought of as "varying" this constant. Depending on whether C is viewed as a parameter or as a constant, (5.5.15) is called the ***variation of parameters formula*** or the ***variation of constants formula***. Finally, observe that formula (5.5.15) is exactly the same as (5.2.34).

Example 5.5.2 Solve the initial value problem

$$t\frac{dx}{dt} + 2x = 3t, \quad x(1) = -9. \qquad (5.5.17)$$

Solution Written in standard form, (5.5.17) is

$$\frac{dx}{dt} + \frac{2}{t}x = 3.$$

Using the notation of (5.5.14), $p(t) = 2/t$, $q(t) = 3$, $t_0 = 1$, and $x_0 = -9$. The principal solution of the associated homogeneous equation at $t = s$ is

$$z(t, s) = e^{-\int_s^t p(u)\,du} = e^{-\int_s^t 2/u\,du} = e^{-(\ln t^2 - \ln s^2)} = \frac{s^2}{t^2}.$$

It follows from the variation of parameters formula (5.5.15) that

$$\begin{aligned} x(t) &= z(t, t_0)x_0 + \int_{t_0}^{t} z(t, s)q(s)\,ds = -9z(t, 1) + 3\int_1^t z(t, s)\,ds \\ &= -9 \cdot \frac{1}{t^2} + 3\int_1^t \frac{s^2}{t^2}\,ds = -\frac{9}{t^2} + \frac{3}{t^2}\left(\frac{s^3}{3}\right)\Big|_1^t \\ &= -\frac{9}{t^2} + \frac{3}{t^2}\left(\frac{t^3}{3} - \frac{1}{3}\right) = t - \frac{10}{t^2} \end{aligned}$$

for $t > 0$. Note that this agrees with (5.5.7). ♦

Example 5.5.3 Find the solution of

$$\frac{dx}{dt} + (\cos t)x = e^{-\sin t}\cos t \tag{5.5.18}$$

satisfying the initial condition $x(\pi) = 10$.

Solution 1 Since $\cos t$ and the right-hand side of (5.5.18) are continuous everywhere, this initial value problem has a unique continuous solution on the interval $(-\infty, \infty)$ (see Theorem 5.2.1). To find it, we begin with the associated homogeneous equation

$$\frac{dx}{dt} + (\cos t)x = 0.$$

Its general solution is

$$x(t) = Ce^{-\int \cos t\,dt} = Ce^{-\sin t}.$$

Replacing C with $v(t)$, we have

$$x(t) = v(t)e^{-\sin t} \tag{5.5.19}$$

Substituting this and its derivative

$$x' = ve^{-\sin t}(-\cos t) + v'e^{-\sin t}$$

into (5.5.18) and simplifying, we get

$$v' = \cos t.$$

Thus,

$$v = \int \cos t\,dt + C = \sin t + C.$$

Consequently, (5.5.19) is

$$x(t) = (\sin t + C)e^{-\sin t}.$$

Finally, we have to determine the value of C so that the initial condition is satisfied. Letting $t = \pi$ and $x(\pi) = 10$ and then solving for C, we get $C = 10$. Therefore,

$$x(t) = (\sin t + 10)e^{-\sin t}$$

is the solution of the initial value problem for $-\infty < t < \infty$. ♦

Solution 2 Since $p(t) = \cos t$, the principal solution of the associated homogeneous equation at $t = s$ is

$$z(t, s) = e^{-\int_s^t p(u)\,du} = e^{-\int_s^t \cos u\,du} = e^{-(\sin t - \sin s)}.$$

As $t_0 = \pi$,

$$z(t, t_0) = z(t, \pi) = e^{-\sin t}.$$

Letting $x_0 = 10$ and $q(s) = e^{-\sin s}\cos s$ in the variation of parameters formula (5.5.15), we get

$$\begin{aligned} x(t) &= z(t, \pi) \cdot 10 + \int_\pi^t e^{-(\sin t - \sin s)} \cdot e^{-\sin s} \cos s\,ds \\ &= 10e^{-\sin t} + e^{-\sin t} \int_\pi^t \cos s\,ds = 10e^{-\sin t} + e^{-\sin t}(\sin t - \sin \pi). \end{aligned}$$

Therefore, as before, the solution is

$$x(t) = (10 + \sin t)e^{-\sin t}.$$

♦

5.6 Bernoulli's Equation

In this section we study ***Bernoulli's equation***, which is the name of a nonlinear first-order differential equation that can be written in the form

$$\frac{dx}{dt} + p(t)x = q(t)x^n, \tag{5.6.1}$$

where n is a real number. Observe that it is the factor x^n that makes this equation nonlinear—unless $n = 0$ or $n = 1$. If $n = 0$, then it is the linear equation

$$\frac{dx}{dt} + p(t)x = q(t) \tag{5.6.2}$$

that we have been studying in this chapter. If $n = 1$, then it is the linear homogeneous equation

$$\frac{dx}{dt} + (p(t) - q(t))x = 0. \tag{5.6.3}$$

Bernoulli's equation can be solved by transforming it into a linear equation. To explain this, let us suppose—in view of the above observation—that n has any value except for 0 and 1. The basic step in the transformation is to "get rid" of the x^n factor on the right-hand side of (5.6.1) by multiplying both sides of this equation by x^{-n}, thereby obtaining

$$x^{-n}\frac{dx}{dt} + p(t)x^{1-n} = q(t). \tag{5.6.4}$$

Observe that the first term of this equation is almost the derivative of the power function x^{1-n} with respect to t. So that it is, let us multiply both sides of (5.6.4) by $1-n$ in the hope that this somehow will lead to finding the solutions of (5.6.1). So now we have

$$(1-n)x^{-n}\frac{dx}{dt} + (1-n)p(t)x^{1-n} = (1-n)q(t).$$

At this juncture, note:

> *The first term is the derivative of the power function x^{1-n} in the second term with respect to t.*

Consequently, the previous equation can be written as

$$\frac{d}{dt}x^{1-n} + (1-n)p(t)x^{1-n} = (1-n)q(t). \tag{5.6.5}$$

Now introduce a new variable, say z, to replace x^{1-n}. As a result, we end up with the linear equation

$$\frac{dz}{dt} + P(t)z = Q(t),$$

where

$$P(t) := (1-n)p(t) \quad \text{and} \quad Q(t) := (1-n)q(t).$$

This equation can be solved by either using the integrating factor method or employing the variation of parameters formula.

In conclusion, all we really have to remember when solving a Bernoulli differential equation is to "get rid" of the x^n factor on the right-hand side of the equation because that will transform it into a linear equation. With some practice, it will become obvious on how to proceed after that. Let us take a look at some examples.

Example 5.6.1 Find the solutions of

$$\frac{dx}{dt} - \frac{2}{t}x = -3x^2 \tag{5.6.6}$$

on the interval $(-\infty, 0)$.

Solution To get rid of the x^2 factor on the right-hand side, divide each side of the equation by x^2 obtaining

$$x^{-2}\frac{dx}{dt} - \frac{2}{t}x^{-1} = -3.$$

Aside from "-1", the first term is the derivative of x^{-1} with respect to t. Multiplying both sides of the equation by -1, we get

$$-x^{-2}\frac{dx}{dt} + \frac{2}{t}x^{-1} = 3$$

or

$$\frac{d}{dt}x^{-1} + \frac{2}{t}x^{-1} = 3.$$

This last equation is now linear in x^{-1}, which becomes all the more apparent if we replace x^{-1} with z:

$$\frac{dz}{dt} + \frac{2}{t}z = 3. \tag{5.6.7}$$

We leave it as an exercise to verify that t^2 is an integrating factor for (5.6.7), which can then be used to obtain the solution

$$z = \frac{t^3 + C}{t^2},$$

where C is an arbitrary constant. As $x = 1/z$,

$$x = \frac{t^2}{t^3 + C}. \tag{5.6.8}$$

is a solution of (5.6.6) on $(-\infty, 0)$ for every value of C.

At first, it may seem that every solution of Eq. (5.6.6) could be obtained from (5.6.8) by assigning C an appropriate value. Note, however, that the *zero function* ($x(t) = 0$ for all $t < 0$) is also a solution of (5.6.6); but it is not included in the family of solutions (5.6.8) since no real value of C gives this ***zero solution***.[5] Since we divided both sides of (5.6.6) by x^2 (implicitly assuming $x \neq 0$), this additional solution should come as no surprise. It is called a ***singular solution***. ♦

Example 5.6.2 (Falling Chain) A chain with mass m and length l is placed in a coiled heap on a smooth table as close to its edge as possible. The height of the table exceeds the length of the chain. One end of the chain is pulled so that a piece of length y_0 hangs over the edge of the table but is temporarily held in place preventing it from falling. When it is released at time $t = 0$, assume that the hanging piece is long enough so that it begins to fall. Find a model for the velocity v of the hanging piece (whose length keeps increasing) up to the time that the last link of the chain leaves the table.

Solution Differential equations modeling the motion of the body in Example 1.4.8 were obtained by assuming that the only force acting on the body is due to gravity. In other words, we assumed that all other forces, such as air resistance, are negligible in comparison to gravity. This was done in order to obtain equations that we could solve with our current knowledge. In this example, we will also make such assumptions in order to obtain an equation of motion (differential equation) that we will actually be able to solve.

First of all, let us ignore all resistive forces, such as friction between the table top and the moving links of the chain. Secondly, let us assume that the mass of chain falling off the table changes continuously, even though in reality the change occurs in a discrete increment every time a link falls over the edge. In other words, with regard to mass, view the chain as a flexible, uniform metal cable with linear density $\rho = m/l$. Thirdly, assume that each link lying on the table does not move until it leaves the table.

Let $y(t)$ denote the length of the chain hanging over the edge of the table at time t. As stated above, the length of chain hanging over the edge at $t = 0$ is $y(0) = y_0$. Let $v(t)$ denote the velocity of the hanging piece at time t. Thus,

$$v(t) = \dot{y}(t).$$

Since the hanging piece is held in place until it is released, $v(0) = 0$. And as the mass m of the hanging piece at time t is

$$m(t) = \rho y(t),$$

its momentum p is

$$p(t) = m(t)v(t) = \rho y(t)\dot{y}(t).$$

[5] Or, ***trivial solution***.

The force due to gravity acting downward on this piece at time t is

$$F(t) = m(t)g = \rho y(t)g.$$

From Newton's second law of motion (see (1.4.9)), we have the equation of motion

$$\frac{d}{dt}[\rho y(t)\dot{y}(t)] = \rho y(t)g.$$

Since ρ is a constant, this simplifies to $y\ddot{y} + \dot{y}^2 = gy$ or

$$\ddot{y} + \frac{1}{y}\dot{y}^2 = g. \tag{5.6.9}$$

This is a second-order differential equation. But we can change it into a first-order equation by viewing v as a function of y instead of t. By the chain rule,

$$\ddot{y} = \frac{d}{dt}\dot{y} = \frac{dv}{dt} = \frac{dv}{dy}\frac{dy}{dt} = v\frac{dv}{dy}.$$

Replacing $\ddot{y}$ with $v(dv/dy)$ and $\dot{y}$ with v, Eq. (5.6.9) becomes

$$v\frac{dv}{dy} + \frac{1}{y}v^2 = g. \tag{5.6.10}$$

This is a Bernoulli equation since it can be written in the form of (5.6.1) with $n = -1$.

Since $v = 0$ when $y = y_0$, the model that we have derived for the velocity of the hanging piece of the chain—up to the moment it hits the floor—is given by the initial value problem

$$v\frac{dv}{dy} + \frac{1}{y}v^2 = g, \quad v(y_0) = 0. \tag{5.6.11}$$

The solution of (5.6.11) is left to Problem 64. ♦

Example 5.6.3 Find the solution of the initial value problem

$$\frac{dx}{dt} - tx = x^{-1}, \quad x(0) = 2. \tag{5.6.12}$$

Solution Multiplication of the differential equation by $2x$ yields

$$2x\frac{dx}{dt} - 2tx^2 = 2.$$

In terms of $z := x^2$, this is

$$\frac{dz}{dt} - 2tz = 2.$$

Multiplying by the integrating factor $\exp(-t^2)$, we have

$$\frac{d}{dt}\left[e^{-t^2}z(t)\right] = 2e^{-t^2}.$$

The result of integrating both sides of this equation from 0 to t is

$$e^{-t^2}z(t) - z(0) = 2\int_0^t e^{-s^2}\,ds.$$

Due to the initial condition, $z(0) = [x(0)]^2 = 4$. Consequently,

$$z(t) = e^{t^2}\left[4 + 2\int_0^t e^{-s^2}\,ds\right].$$

Let us rewrite this in terms of the *error function* (see (5.3.6)). Since

$$2\int_0^t e^{-s^2}\,ds = \sqrt{\pi}\,\mathrm{erf}(t),$$

$$z(t) = e^{t^2}\left[4 + \sqrt{\pi}\,\mathrm{erf}(t)\right].$$

Finally, as $z = x^2$ and $x(0) = 2$, we obtain

$$x(t) = e^{t^2/2}\sqrt{4 + \sqrt{\pi}\,\mathrm{erf}(t)},$$

which is the solution of (5.6.12) on the entire interval $(-\infty, \infty)$. ♦

5.7 Riccati's Equation

The nonlinear first-order differential equation

$$\frac{dx}{dt} = a(t)x^2 + b(t)x + c(t), \tag{5.7.1}$$

where $a(t)$, $b(t)$, and $c(t)$ are given functions, is called a ***generalized Riccati equation***. Analytical solutions of this equation are usually quite complicated in that they are difficult to obtain and cannot be expressed in terms of elementary functions.

However, the proof of the next theorem shows that if there happens to be a way of coming up with a particular solution of (5.7.1), then it can be transformed into a Bernoulli equation.

Generalized Riccati Equation

Theorem 5.7.1 *If $x = \varphi(t)$ is a particular solution of a given generalized Riccati equation, then the change of variable $y = x - \varphi(t)$ transforms it into the Bernoulli equation*

$$\frac{dy}{dt} = \left[2a(t)\varphi(t) + b(t)\right]y + a(t)y^2. \tag{5.7.2}$$

Proof Substituting

$$x = y + \varphi(t) \quad \text{and} \quad x' = y' + \varphi'(t)$$

into (5.7.1), we get

$$y' + \varphi'(t) = a(t)\left[y + \varphi(t)\right]^2 + b(t)\left[y + \varphi(t)\right] + c(t)$$

or

$$y' + \varphi'(t) = a(t)\left[y^2 + 2\varphi(t)y\right] + b(t)y + a(t)(\varphi(t))^2 + b(t)\varphi(t) + c(t).$$

Since $\varphi(t)$ is a solution of (5.7.1), the last three terms simplify to $\varphi'(t)$. Thus, we have

$$y' + \varphi'(t) = a(t)\left[y^2 + 2\varphi(t)y\right] + b(t)y + \varphi'(t),$$

which simplifies to (5.7.2). ■

Example 5.7.1 Find the solutions of

$$\frac{dx}{dt} = 4x^2 + 6x + 2. \tag{5.7.3}$$

Solution Observe that a constant function $x(t) = k$ is a solution if k satisfies the quadratic equation $4k^2 + 6k + 2 = 0$. Factoring, we see from

$$2(2k + 1)(k + 1) = 0$$

that (5.7.3) has two constant solutions, namely, $x(t) = -1$ and $x(t) = -1/2$ for $t \in \mathbb{R}$. In light of the theorem, let $\varphi(t) = -1$.[6] ($\varphi(t) = -1/2$ is also an option.) Changing variables, we have $y = x - \varphi(t) = x + 1$. Then, as $x = y - 1$ and $x' = y'$, Eq. (5.7.3) becomes

$$y' = 4(y-1)^2 + 6(y-1) + 2 = 4y^2 - 2y$$

or

$$y' + 2y = 4y^2.$$

This is a Bernoulli equation, which can also be obtained directly from (5.7.2) by setting $a(t) = 4$, $b(t) = 6$, $c(t) = 2$, and $\varphi(t) = -1$. Its solutions are

$$y(t) = \frac{1}{Ce^{2t} + 2},$$

where C is an arbitrary constant, as well as the singular solution $y(t) = 0$. Therefore, solutions of Eq. (5.7.3) are $x(t) = -1$ and

$$x(t) = \frac{1}{Ce^{2t} + 2} - 1$$

for all $t \in \mathbb{R}$. Note the constant solution $x(t) = -1/2$ is obtained by taking $C = 0$. ♦

Example 5.7.2 Find the solution of

$$\frac{dx}{dt} + x^2 = 1 + t^2 \tag{5.7.4}$$

satisfying the initial condition $x(0) = 2$.

Solution This is a Riccati equation of the form (5.7.1) with

$$a(t) \equiv -1, \ b(t) \equiv 0, \ \text{and } c(t) \equiv 1 + t^2.$$

Clearly $x = t$ is a solution; so let $y = x - t$. By either substituting $x = y + t$ and $x' = y' + 1$ into (5.7.4) or using (5.7.2) with $\varphi(t) = t$, we obtain the Bernoulli equation

$$\frac{dy}{dt} + 2ty = -y^2.$$

[6] Solving the quadratic equation is unnecessary if one can come up with $\varphi(t) = -1$ by inspection.

The change of variable $z := y^{-1}$ transforms this equation into the linear equation

$$\frac{dz}{dt} - 2tz = 1.$$

Multiplying both sides of this equation by the integrating factor e^{-t^2}, we get

$$\frac{d}{dt}\left[e^{-t^2}z\right] = e^{-t^2}.$$

The integration

$$\int_0^t \frac{d}{du}\left[e^{-u^2}z(u)\right] du = \int_0^t e^{-u^2}\, du$$

yields

$$e^{-t^2}z(t) - z(0) = \int_0^t e^{-u^2}\, du.$$

As a result,

$$z(t) = e^{t^2}\left[z(0) + \int_0^t e^{-u^2}\, du\right].$$

Since

$$z(t) = \frac{1}{y(t)} = \frac{1}{x(t) - t},$$

we have

$$x(t) - t = \frac{e^{-t^2}}{z(0) + \int_0^t e^{-u^2}\, du},$$

where $z(0) = 1/2$. Consequently,

$$x(t) = t + \frac{2e^{-t^2}}{1 + 2\int_0^t e^{-u^2}\, du}.$$

Expressed in terms of the error function (5.3.6), the solution is

$$x(t) = t + \frac{2e^{-t^2}}{1 + \sqrt{\pi}\,\mathrm{erf}(t)}.$$

♦

Problems

> The mere knowledge of a fact is pale; but when you come to *realize* your fact, it takes on color. It is all the difference between hearing of a man being stabbed to the heart, and seeing it done.
>
> Mark Twain, *A Connecticut Yankee in King Arthur's Court*, ch. VI

Linear, Separable, Both, or Neither

In Problems 1 through 10, determine whether the differential equation is linear, separable, both, or neither.

1. $\frac{dx}{dt} - 3x = 9e^{4t}$
2. $\frac{dx}{dt} + 6t^2 x = 5\sin(3t)$
3. $\frac{dy}{dx} - 10x = 5xy$
4. $\frac{dp}{ds} + 4p = -1$
5. $tx' + 3tx = \sin x$
6. $xx' + t = \cos x$
7. $\frac{dy}{dx} - y = y\tan x$
8. $(x^2+4)y' + 5xy = -2e^{-x}$
9. $x' + (8+t)x + 32 = 0$
10. $xy' + 5xy = 2\cos y$

Linear Equations

In Problems 11–22, find the general solution of the equation. Use either the integrating factor method or the variation of parameters method.

11. $\frac{dx}{dt} + x = e^t$
12. $\frac{dx}{dt} + x = e^{-4t}$
13. $x' - 2x = 3t$
14. $\cos t\,\frac{dx}{dt} + x\sin t = 1$
15. $\frac{dy}{dx} + 2x^2 = \frac{2y}{x}$
16. $x' + 2x = 5e^{-t}$
17. $xy' + 3y = 2x$
18. $\dot{x} = -6tx$
19. $x\frac{dy}{dx} = x^2 + y$
20. $\frac{dy}{dx} - \frac{y}{x} = \frac{x}{\sqrt{1-x^2}}$
21. $\frac{dp}{dt} + p\tan t = 2\sin t - \cos t$
22. $e^{x^2}\frac{dz}{dx} = \csc x - 2xe^{x^2}z$

Separable and Linear Equations

In Problems 23 through 32, first classify the differential equation as to its type: "separable" or "linear" or "neither of these." Solve all of the linear and separable equations. If it turns out that an equation is both separable and linear, solve it as you see fit. Find explicit solutions whenever possible.

23. $\frac{dx}{dt} = \frac{4tx}{2x^2+1}$
24. $t\frac{dx}{dt} = 8t^2 - x$
25. $(x+e^y)\frac{dy}{dx} = 2x - y$
26. $(1+x^2)\frac{dy}{dx} = 4xe^{2y}$
27. $\frac{dx}{dt} = 3 - 2t^{-1}x$
28. $\frac{1}{x^2}\frac{dy}{dx} = 5x^2\sec y$
29. $\frac{dy}{dx} = ye^{x^2 y}$
30. $(t^2-4)\frac{dx}{dt} = tx$
31. $(t+3)\frac{dx}{dt} = x$
32. $\frac{dy}{dt} = 3t\sin t^2$

Initial Value Problems

In Problems 33 through 39, solve the initial value problem.

33. $\frac{dy}{dx} - \frac{y}{x} = x\cos x, \quad y(\pi) = 1$
34. $\frac{dx}{dt} + 2tx = 5, \quad x(0) = 0$
35. $y' - 2ty = 4, \quad y(0) = 3$
36. $\frac{dx}{dt} + 2tx = 1, \quad x(0) = 1$
37. $\frac{dx}{dt} + \frac{2}{t}x = \frac{\sin t}{t^2}, \quad x(\pi) = 1$

38. $\dfrac{dx}{dt} = \dfrac{2tx^2}{t^2+4}, \quad x(-2) = 1$

39. $\dfrac{dv}{dt} + (8+t)v = -32, \quad v(0) = 0$

Integrating Factors

40. Find the solution of $x' - 2x = f(t)$ that satisfies the condition $x(0) = 5$, where f denotes any integrable function of t.

41. Explain why $e^{tp(t)}$ is not an integrating factor for the linear equation

$$x' + p(t)x = q(t).$$

For more context, see the paragraph containing Eq. (5.2.10).

42. If $\mu(t)$ is an integrating factor for (5.2.10), prove that $k\mu(t)$, for any nonzero constant k, is also an integrating factor.

43. Use the integrating factor (5.2.32) and Eq. (5.2.33) to derive (5.2.34).

Functions Defined by Integrals

Problems 44–46 pertain to Dawson's integral, namely, the function that is defined by (5.3.1) *and which is denoted by* $\mathcal{D}(t)$.

44. The purpose of this problem is to complete the details left out in Example 5.3.1 regarding the concavity and inflection points of the graph of $\mathcal{D}(t)$. Complete parts (a)–(f) to prove that the equation

$$\mathcal{D}''(t) = 0$$

has one, and only one, solution on the interval $(0, \infty)$, namely t_2, where the value of t_2 is given by (5.3.5).

(a) Let $G(t) := e^{t^2}\mathcal{D}''(t)$. Show its derivative is

$$G'(t) = -4e^{t^2}\mathcal{D}'(t).$$

(b) Use (a) to show that $G(t)$ is increasing on the interval (t_1, ∞), where the value of t_1 is given by (5.3.3).

(c) Use (b) to argue that t_2 is the only zero of the function $G(t)$ on the interval (t_1, ∞). *Hint.* Consider $G(t_1)$.

(d) What is the value of $G(0)$? Show that $G'(t) < 0$ on $(0, t_1)$.

(e) Explain why $G(t) < 0$ on $(0, t_2)$ and $G(t) > 0$ on (t_2, ∞).

(f) Use (e) to conclude that $(t_2, \mathcal{D}(t_2))$ is the only inflection point of the function $G(t)$ on the interval (t_2, ∞).

45. (a) Show that the derivatives of $\mathcal{D}(t)$ satisfy the recursive formula:

$$\mathcal{D}^{(n+1)}(t) + 2t\mathcal{D}^{(n)}(t) + 2n\mathcal{D}^{(n-1)}(t) = 0$$

for $n \geq 1$.

(b) Verify that the Maclaurin series expansion for $\mathcal{D}(t)$ is

$$\sum_{n=0}^{\infty} \frac{(-1)^n 2^{2n} n!}{(2n+1)!} t^{2n+1}.$$

(c) Use the Ratio Test to determine the interval of convergence of the series.

46. (a) Show

$$\int_a^b e^{s^2}\, ds = e^{b^2}\mathcal{D}(b) - e^{a^2}\mathcal{D}(a).$$

(b) For $t_0 > 0$ and any x_0, solve the initial value problem

$$tx' + (2t^2 - 1)x = t^2, \; x(t_0) = x_0$$

on the interval $(0, \infty)$. Use (a) to express the solution in terms of Dawson's integral.

(c) Use the solution found in (b) to obtain the solution $x(t)$ of

$$tx' + (2t^2 - 1)x = t^2, \; x(t_0) = 0$$

on $(0, \infty)$. Find $\lim_{t_0 \to 0^+} x(t)$.

(d) Part (c) suggests that the solution of

$$tx' + (2t^2 - 1)x = t^2, \; x(0) = 0$$

is the function $x(t) = t\mathcal{D}(t)$. Check that this is the solution on the entire interval $(-\infty, \infty)$.

(e) It was determined in Example 5.3.1 that $\mathcal{D}''(t)$ is positive on (t_2, ∞), where the value of t_2 is given by (5.3.5) Use this to prove that $x(t) = t\mathcal{D}(t)$ is strictly decreasing (t_2, ∞). Moreover, show that $t\mathcal{D}(t) \to 1/2$ as $t \to \infty$.

47. Find the solution of

$$t^2 \frac{dx}{dt} + tx = \sin t, \quad x(2) = 1$$

and express it in terms of the sine integral function defined by (5.3.7).

48. (a) Show that

$$\int_a^b e^{-u^2}\, du = \frac{1}{2}\sqrt{\pi}\big[\operatorname{erf}(b) - \operatorname{erf}(a)\big],$$

where $\operatorname{erf}(t)$ is the error function (5.3.6).

(b) Verify directly that $x = e^{t^2}\operatorname{erf}(t)$ is a solution of

$$\frac{dx}{dt} - 2tx = \frac{2}{\sqrt{\pi}}.$$

(c) Solve the initial value problem

$$\frac{dx}{dt} - 2tx = 8t^3, \quad x(0) = 2.$$

Gronwall's Inequality

49. A property of the definite integral that is used in the proof of Example 5.4.1 is that if f is integrable on an interval I, then so is $|f|$ and

$$\left|\int_a^b f(t)\, dt\right| \le \int_a^b |f(t)|\, dt$$

for all $a, b \in I$ if $b \ge a$. Use this to prove that

$$\left|\int_a^b f(t)\, dt\right| \le \left|\int_a^b |f(t)|\, dt\right|.$$

for all $a, b \in I$. That is, show the second inequality includes the case $b < a$ as well.

50. Complete the proof of Theorem 5.4.1 for the case $t < t_0$.

51. Let f be a continuous, nonnegative function on an interval $[\alpha, \beta]$ satisfying the inequality

$$f(t) \le A + B\int_\alpha^t f(u)\, du$$

where A and B are positive constants. Explain why

$$f(t) \le Ae^{B(t-\alpha)}$$

for $\alpha \le t \le \beta$.

52. Let a, t_0, x_0, $\tilde{x}_0$ be real numbers and let $q(t)$ be a continuous function on the interval $(-\infty, \infty)$. Find solutions $x(t)$ and $\tilde{x}(t)$ of

$$\frac{dx}{dt} + ax = q(t)$$

such that $x(t_0) = x_0$ and $\tilde{x}(t_0) = \tilde{x}_0$. Use these solutions to find $|x(t) - \tilde{x}(t)|$. Reconcile this result with (5.4.8).

53. Find all functions f that are continuous and nonnegative and which satisfy the inequality

$$f(t) \le \int_0^t f(s)\, ds$$

on the interval $0 \le t \le 1$.

A Kinematic Formula

54. (a) Suppose a body is moving along the x-axis with constant acceleration a. A well-known physics formula relating the velocity v of the body to its position x is

$$v^2 = v_0^2 + 2a(x - x_0),$$

where v_0 denotes the velocity of the body at $x = x_0$. Rather than replicating the algebraic derivation that is given in elementary physics textbooks, use the chain rule of calculus to derive the formula directly from $\dot{v} = a$.

(b) If the speed of a SUV traveling due east increases uniformly from

40 mph to 65 mph over a distance of 600 feet, what is the magnitude of the constant acceleration?

An Integral Equation

55. Find all functions $x(t)$ with continuous derivatives satisfying the integral equation

$$(x(t))^2 = \int_0^t \left[(x(s))^2 + (x'(s))^2\right] ds + 100.$$

Bernoulli's Equation

Find solutions of each equation in Problems 56 through 58. In Problem 59, find the solution satisfying the given initial condition.

56. $\dfrac{dx}{dt} + 6x = 5e^{-3t}x^2$
57. $6t^2\dfrac{dx}{dt} = tx + 2x^4 \quad (t > 0)$
58. $x' = ax - bx^3 \ (a > 0,\ b > 0)$
59. $x' - 3t^2x = 6t^2x^{-1},\ x(0) = -1$

Riccati's Equation

In Problems 60 through 63, find solutions of the generalized Riccati equation.

60. $\dfrac{dx}{dt} = 6 + x - x^2$
61. $x' = x^2 \sin t - 2 \sec t \tan t$, given that $\varphi(t) = -\sec t$ is a particular solution.
62. $x' = (x - t)^2 + 1$
63. $\dfrac{dx}{dt} = \dfrac{1}{t^2} - \dfrac{x}{t} - x^2$, given that $\varphi(t) = 1/t$ is a particular solution.

Falling Chain

64. (a) Solve the initial value problem (5.6.11) that was derived in Example 5.6.2.
    (b) Find the acceleration a of the falling chain. Show $a \approx g/3$ for $y \gg y_0$.

Computer Algebra System Problems

Use a CAS, such as Maple *or* Mathematica*, to work the following problems.*

65. (a) Graph $\mathcal{D}(t)$ (Dawson's integral) and the function g defined by

    $$g(t) := \frac{t}{2t^2 - 1}$$

    on the same set of axes. Find the absolute maximum of $\mathcal{D}(t)$ on $[0, \infty)$.
    (b) The graphs of $\mathcal{D}(t)$ and $g(t)$ intersect at a single point in the interval $(1, 2)$. Call this point t_2. Determine both of its coordinates.
    (c) From the graphs in (a), it appears that $\mathcal{D}(t) > g(t)$ on the interval (t_2, τ), where $\tau \approx 2.5$. But for $t > \tau$, it is unclear which function is larger. Use calculus, as in Example 5.3.1, to show $\mathcal{D}(t) > g(t)$ for all $t > t_2$.
66. Graph the error function

    $$\mathrm{erf}(t) = \frac{2}{\sqrt{\pi}} \int_0^t e^{-u^2}\, du.$$

    Evaluate erf(t) at several points, such as $t = 0.1, 0.5, 1.0, 1.5$, and 2.0.
67. Solve the initial value problem

    $$\frac{dv}{dt} + (8 + t)v = -32, \qquad v(0) = 0$$

    using a CAS. Then solve it by hand (see Problem 39). For context, see (3.3.9) in Chap. 3.
68. Verify the solution obtained in Problem 58 using a CAS.
69. Verify the solution obtained in Problem 59 using a CAS.
70. Graph some solutions of the generalized Riccati equation in Problem 60. Include the graphs of both constant solutions and at least five nonconstant solutions.

Chapter 6
Modeling with First-Order Equations

No subject better illustrates the divide between the two cultures—arts and sciences—than mathematics. To the outsider, mathematics is a strange, abstract world of horrendous technicality, full of weird symbols and complicated procedures, an impenetrable language and a black art. To the scientist, mathematics is the guarantor of precision and objectivity. It is also, astonishingly, the language of nature itself. No one who is closed off from mathematics can ever grasp the full significance of the natural order that is woven so deeply into the fabric of physical reality.

Paul Davies[1]

Abstract This chapter describes several everyday situations and real-world phenomena that are modeled with first-order differential equations. The first one is a practical application that takes place in an orange juice processing plant. Briefly, the problem is to determine the percentage by weight of soluble solids (sugars, acids, etc.) in a large tank of orange juice concentrate at any given time if water is pumped into the tank while the concentrate is pumped out of the tank at the same rate. This is referred to as the *orange juice problem* in this chapter. It serves as an introduction to a modeling technique known as *compartmental analysis*. A *one-compartment system* consists of a single compartment that contains a particular type of matter or some species of animal or plant, where the matter or organism enters and leaves the compartment at known rates. Some examples are: farm-raised catfish in a man-made pond, E. coli bacteria growing in a flask of nutrient broth, carbon monoxide in a house, and money in a savings account.

The orange juice problem is worked out in detail since it will serve as a paradigm for real-world phenomena that can be viewed as one-compartment systems. One of them is the *radioactive decay law*, which is derived using the hypothesis advanced

[1] See [24, ch. 4]. Paul Davies (born 1946) is an English mathematical physicist and professor at Arizona State University and the prolific author of a number of popular and technical books on physics.

© The Author(s), under exclusive license to Springer Nature Switzerland AG 2026

L. C. Becker, *Ordinary Differential Equations: Concepts, Methods, and Models*,
https://doi.org/10.1007/978-3-032-15150-6_6

by two Austrian physicists in the early 1900s that radioactive-decay processes are statistical in nature. Two other models that are also derived using compartmental analysis are the *Malthusian and logistic models of population growth.* These are well-known models from mathematical biology for predicting the populations of certain species of animals, such as fish in a pond, and other organisms, such as E. coli bacteria.

The chapter concludes with a detailed explanation of how to compute the balance of a deposit account, such as a savings or money market account, where the accrued interest is credited to the account periodically. This will prepare the way for one of the topics in Chapter 7, where it is shown how to set up a one-compartment model of a loan, such as a car loan or a home mortgage, in order to model the amortization of the loan using a differential equation and the *Dirac delta function*, which is also introduced in Chap. 7.

6.1 Mathematical Models

We present some real-world phenomena (natural and those caused by humans) in this chapter that have been investigated in the past and translated into the language of mathematics in order to obtain representations of these phenomena that embody their essence with as much realism as possible. These mathematical representations are called ***mathematical models***. Of course, there are both good and bad models. A good model represents reality as closely as possible, which aids in understanding the essential characteristics of the phenomenon being studied and provides the means to make accurate predictions about future outcomes. Since this book is about differential equations, all of the models that will be presented here and in later chapters involve ordinary differential equations. One of the models that we will consider is the *radioactive decay law*, which we will derive using the hypothesis advanced by two Austrian physicists in the early 1900s that radioactive-decay processes are statistical in nature. We will also present a couple of well-known models from mathematical biology for predicting the populations of certain species of animals, such as catfish in a pond, and other organisms, such as E. coli bacteria in a flask of nutrient broth. We will also show how to compute balances and annual percentage yields for deposit accounts, such as savings accounts and certificates of deposit where interest is usually compounded discretely, in preparation for showing how to model the amortization of a loan with a differential equation in Chap. 7. But before presenting these and other related models, we will start off with a practical application having to do with processing oranges to make orange juice, which will actually help us to understand the aforementioned models.

The situation we are about to describe in the next section takes place in a Florida processing plant, which makes frozen orange juice concentrate from oranges hauled in from local orchards. Briefly, the problem is to determine the percentage by weight of soluble solids (sugars, acids, etc.) in a large tank of orange juice at any given time. We refer to this as the ***orange juice problem***. One possible model of the orange juice problem, as we will see shortly, turns out to be a simple linear first-order differential equation—as will all of the models in this chapter—whose solution will allow us to calculate the percentage of soluble solids (by weight) in a tank of orange juice at any given time. The orange juice model will also serve as an introduction to a modeling technique known as ***compartmental analysis***. The type of ***compartment*** used in a model depends on the circumstance. Besides a tank in an orange juice processing plant, the compartment could be a catfish pond, a flask of nutrient broth containing some type of bacteria, a house, or a savings account. A ***one-compartment system*** consists of a single compartment that contains a particular type of matter or perhaps some species of animal or plant. Some examples of one-compartment systems are: orange juice in a tank, farm-raised catfish in a man-made Mississippi pond, E. coli bacteria in a flask of nutrient broth, carbon monoxide in a house, and money in a savings account. The matter or species enters and leaves the compartment at known rates. The topic of compartmental analysis is important to a wide spectrum of mathematical disciplines, such as biophysics, demography, economics, engineering, and so on. The following example is worked out in detail because it will serve as a paradigm for real-world phenomena that can be modeled using one-compartment systems.

6.2 Orange Juice Problem

6.2.1 Description

Consider the following scenario at a Florida processing plant that makes both frozen orange juice concentrate and ready-to-drink orange juice. Suppose after one stage of the process, orange juice concentrate is produced that consists of 42% soluble solids by weight while the balance is water. We call this particular mixture "42% concentrate." A portion of the 42% concentrate is used to make the canned, frozen orange juice concentrate sold in grocery stores, while the rest is pumped into large tanks. Water is added to the tanks to reduce the amount of soluble solids in order to make a ready-to-drink orange juice. The consistency of the orange juice is kept uniform throughout the tanks by continually stirring them with large blades. After the right amount of water has been added to the tanks, their contents are emptied to fill various containers, such as 1-gallon jugs and 52 fl. oz. cartons, that will be shipped to distributors.

Suppose one of the tanks contains 50,000 kg of orange juice, of which 10% by weight (actually mass) consists of soluble solids. However, an industry standard is that ready-to-drink orange juice consists of approximately 12% soluble solids by weight. Consequently a quality control engineer decides to raise the weight percentage of the soluble solids in the tank from 10% to 12% by pumping 42% concentrate into the tank at the rate of 100 kg per minute while at the same time pumping concentrate out of the tank at the same rate. The problem then is to determine how long it will take to reach the 12% target.

6.2.2 Mathematical Model

Let $x(t)$ be the percentage by weight of soluble solids in the tank after t minutes, where t is measured from the moment the pumps are turned on. Since the initial percentage is 10%, the amount of soluble solids in the tank at time $t = 0$ is

$$(50,000\,\text{kg})x(0) = (50,000\,\text{kg})(.10) = 5000\,\text{kg}.$$

This will eventually increase to (50,000 kg)(.12) = 6000 kg of soluble solids when the target of 12% is reached. Until that happens, the amount of soluble solids in the tank at time t is 50,000 $x(t)$ kg.

Our objective is to derive a differential equation with which we can determine the weight percentage $x(t)$ of soluble solids in the tank at any time t. We begin by equating the amount 50,000 $x(t+\Delta t)$ kg of soluble solids in the tank at time $t+\Delta t$ to the amount 50,000 $x(t)$ kg of soluble solids that was in the tank at time t plus the net amount of soluble solids that was added to the tank during the intervening Δt minutes. If Δt is much smaller than t, this net amount is about

$$(100\,\text{kg/min})(.42)(\Delta t\text{ min}) - (100\,\text{kg/min})\,x(t)\,(\Delta t\text{ min}).$$

The first term is the amount of soluble solids that was pumped into the tank during the time interval Δt. The second term approximates the amount of soluble solids that was pumped out during this time. This is of course an approximation because the amount actually varies from $100x(t)\Delta t$ to $100x(t+\Delta t)\Delta t$ as the time changes from t to $t+\Delta t$. However, the closer the value of Δt is to 0, the better the approximation. The result is the *recursive formula*

$$50{,}000x(t+\Delta t) \approx 50{,}000x(t) + 100(.42)(\Delta t) - 100x(t)\Delta t. \tag{6.2.1}$$

By factoring and dividing by Δt, we can rewrite (6.2.1) as

$$50{,}000\left(\frac{x(t+\Delta t)-x(t)}{\Delta t}\right) \approx 42 - 100x(t). \tag{6.2.2}$$

Since $x(t)$ is defined to be the percentage of soluble solids in the tank, the left-hand side represents the average rate of change in the amount of the soluble solids in the tank over a span of Δt minutes. As Δt approaches 0, the difference quotient $\Delta x/\Delta t$ approaches the derivative $x'(t)$. This together with the approximation (6.2.2) becoming increasingly better as $\Delta t \to 0$ yields the differential equation

$$50,000\,\frac{dx}{dt} = 42 - 100x(t).$$

Note that each term of this equation represents a rate of change in the amount of soluble solids (in kg/min) as indicated below:

$$\underbrace{\frac{d}{dt}[50{,}000x(t)]}_{\text{Rate of change in the amount of soluble solids in the tank}} = \underbrace{42}_{\text{Rate at which soluble solids are pumped into the tank}} - \underbrace{100x(t)}_{\text{Rate at which soluble solids are pumped out of the tank.}} \tag{6.2.3}$$

Equation (6.2.3) is a first-order linear equation, which in standard form is

$$\frac{dx}{dt} + 0.002x = 0.00084. \tag{6.2.4}$$

Since the initial condition is $x(0) = 0.10$, the mathematical model of the orange juice problem ends up being the initial value problem:

$$\frac{dx}{dt} + 0.002x = 0.00084, \quad x(0) = 0.10. \tag{6.2.5}$$

Its solution predicts the percentage by weight of soluble solids in the tank as a function of the time t. It is left as an exercise to verify that

$$x(t) = 0.42 - 0.32e^{-.002t}$$

is the solution of this IVP. Its graph is shown in Fig. 6.1.

To determine when the percentage of soluble solids in the tank will reach 12%, let us solve the equation

$$0.42 - 0.32e^{-.002t} = 0.12$$

for t. The result is

$$t = -500\ln\frac{30}{32} \approx 32.3\,\text{minutes}.$$

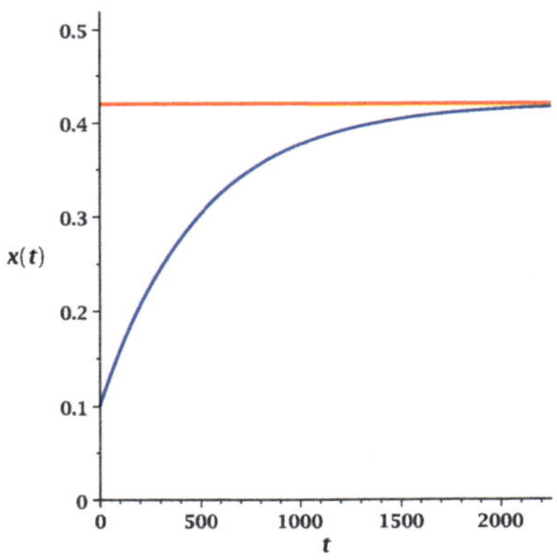

Fig. 6.1 Percentage by weight of soluble solids in the tank vs. time

Finally, it is worth noting that

$$\lim_{t \to \infty} x(t) = 0.42.$$

This says that eventually 42% of the tank's contents will consist of soluble solids, which is clearly depicted in Fig. 6.1.

6.3 One-Compartment Systems

The following observations will aid us in modeling other situations and phenomena that can be viewed as one compartment systems without having to go through all the details that were used in deriving Eq. (6.2.3). Note that the quantity in brackets on the left-hand side of (6.2.3) represents the total amount of soluble solids in the tank at time t; so the left-hand side is the rate at which this amount changes. As for the right-hand side, the first term is the rate at which soluble solids are pumped into the tank and the second term is the rate at which they are pumped out. Thus we have the following statement: The rate at which the amount of soluble solids in a tank changes is equal to the rate at which the soluble solids are pumped into the tank minus the rate at which they are pumped out of the tank. In a nutshell, we have the *word equation*:

[Rate of change in the amount of soluble solids with respect to time]

= [Inflow rate of soluble solids] − [Outflow rate of soluble solids].

Recall that in the language of differential calculus, finding the "rate at which a quantity changes" is the same as finding the "derivative of that quantity." Let us replace *tank* and *soluble solids* with the generic terms *compartment* and *matter*, respectively, since our goal, stated at the onset of the orange juice model, is to present a paradigm for all one-compartment systems. A schematic representation of a generic one-compartment system is shown in Fig. 6.2, where $A(t)$ denotes the amount of matter in the compartment at time t.

Since the word equation, which came about by analyzing the orange juice problem, accounts for all losses and gains of matter in the compartment, it can be viewed as a *conservation equation*. The differential equation (6.3.1) summarizes what we have learned about modeling one-compartment systems.

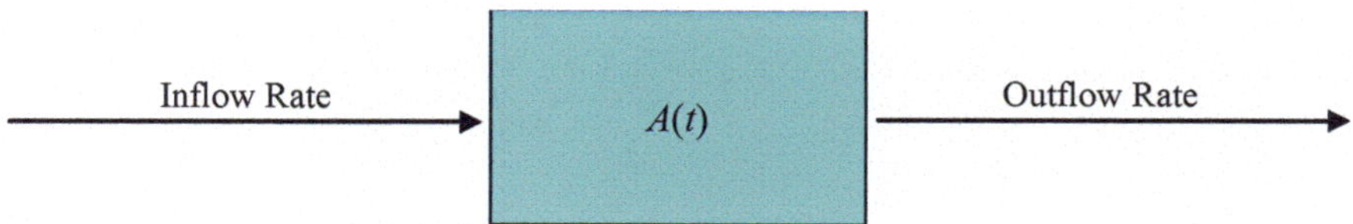

Fig. 6.2 Schematic of a one-compartment system

Conservation Equation for One-Compartment Systems
Let $A(t)$ be the amount of matter in a compartment at time t. Then,

$$\frac{dA}{dt} = [\text{Inflow rate}] - [\text{Outflow rate}]. \tag{6.3.1}$$

In other words,

[Rate of change in the amount of matter in the compartment with respect to t]
= [Rate of flow of matter into the compartment]
− [Rate of flow of matter out of the compartment].

Setting up a compartment schematic and then using the conservation equation is a practical and systematic way of finding a differential equation model for any situation or phenomenon that can be viewed as a one-compartment system. We illustrate this in the next example.

Example 6.3.1 At a water treatment facility, one of the treatment tanks for removing pollutants from city water is undergoing routine maintenance. After completion of the maintenance, the tank is filled with 25,000 gallons of pure water. The tank's pumps are then turned on and polluted water containing 10 ounces of pollutants per gallon is pumped into the tank at the rate of 1000 gallons per minute while the treated water is pumped out at the same rate. The circulation inside the tank keeps the pollutants thoroughly mixed with water. Ordinarily a series of filters inside the tank removes most of the pollutants from the water. However, a city sanitation engineer failed to replace the old filters with new ones. Find a formula to estimate the total number of ounces of pollutants in the treatment tank after the pumps have been turned on.

Solution Let $p(t)$ be the total number of ounces of pollutants in the treatment tank at time t (in minutes), where $t = 0$ denotes the time at which the pumps are turned on. The rate at which pollutants flow into the tank is the product of the rate at which the polluted water is pumped into the tank and the concentration of the pollutants in the inflowing water:

$$\text{Inflow rate} = \left(1000\ \frac{\text{gal}}{\text{min}}\right)\left(10\ \frac{\text{oz}}{\text{gal}}\right) = 10{,}000\ \frac{\text{oz}}{\text{min}}$$

Similarly, the rate at which the pollutants flow out of the tank is the product of the rate at which the polluted water in the tank is pumped out and the concentration of the pollutants in the outflowing water. Keep in mind that the concentration of the pollutants in the tank water is not the same as that of the polluted water entering

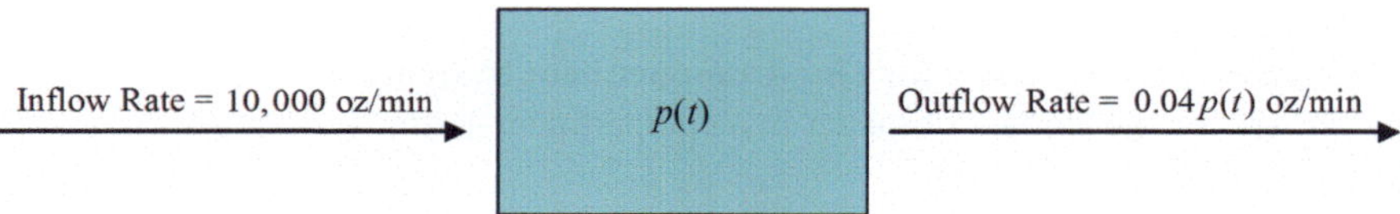

Fig. 6.3 Compartment model of the pollutants in a treatment tank

the tank because the pollutants are also mixed with whatever pure water remains in the tank.[2] Instead it is equal to the total number of ounces of pollutants in the tank, namely $p(t)$, divided by 25,000 gallons, the total volume of pollutants and water in the tank. Thus, the rate at which pollutants flow out of the tank is

$$\text{Outflow rate} = \left(1000\ \frac{\text{gal}}{\text{min}}\right)\left(\frac{p(t)\ \text{oz}}{25{,}000\ \text{gal}}\right) = 0.04p(t)\ \frac{\text{oz}}{\text{min}}.$$

The inflow and outflow rates are depicted in Fig. 6.3, where the rectangle represents the treatment tank. It follows from (6.3.1) that the rate of change in the number of ounces of pollutants in the treatment tank is

$$\frac{d}{dt}p(t) = [\text{Inflow rate}] - [\text{Outflow rate}] \quad \Rightarrow \quad \frac{d}{dt}p(t) = 10{,}000 - 0.04p(t).$$

Since the tank originally contained pure water, the initial condition is $p(0) = 0$. Thus, the formula for modeling the number of ounces of pollutants in the tank is the solution of the initial value problem

$$\frac{dp}{dt} + 0.04p(t) = 10{,}000, \qquad p(0) = 0.$$

It is left as an exercise to show that this solution is

$$p(t) = 250{,}000(1 - e^{-.04t}). \qquad \blacklozenge$$

Although we went into quite a bit of detail to derive a differential equation modeling the weight percentage of soluble solids in orange juice, it was well worth it because we were able to use it to come up with a conservation equation that can be used for any situation or phenomenon that may be viewed as a one-compartment system problem. We will illustrate this further by looking at

- the decay of radioactive elements,
- populations of animals or other organisms, and

[2] For the sake of simplicity, assume that pollutants and water mix instantaneously so that the concentration of pollutants is always uniform throughout the tank.

- balances of deposit accounts, such as regular savings accounts and certificate of deposit (CD) accounts.

6.3.1 Radioactive Decay

Let $N(t)$ be the number of nuclei of some radioactive element or one of its isotopes[3] at time t. The nuclei of radioactive atoms are unstable, some of which will eventually decay into other elements by emitting either α-particles (helium nuclei), β-particles (electrons), or γ-rays (high energy photons). For example, a nucleus of the radioactive isotope uranium-238 may at some moment spontaneously decay into a nucleus of thorium-234 by emitting an α-particle. The alpha decay of uranium-238 is represented using the shorthand notation

$$^{238}\mathrm{U} \to {}^{234}\mathrm{Th} + {}^{4}\mathrm{He}.$$

The ^{238}U nucleus is called the *parent nucleus* and the ^{234}Th nucleus the *daughter nucleus*. The emitted α-particle is a helium nucleus, which consists of two protons and two neutrons. It is impossible to predict if and when a given ^{238}U nucleus will decay. But it is known that a one-gram sample of uranium metal contains about 2.5×10^{21} ^{238}U nuclei; and during any given second, roughly 12,000 of these nuclei will decay (see Problem 13). Thus, the probability P that a nucleus will decay during a time interval of one second is roughly

$$P \approx \frac{12000}{2.5 \times 10^{21}} \approx 5 \times 10^{-18}.$$

Let us now derive the *radioactive decay law*. In the early 1900s, two Austrian physicists, Egon Schweidler and Stefan Meyer, advanced the hypothesis[4] that every nucleus of some given radioactive isotope has the same probability of decaying during a short time interval Δt and that the remaining undecayed nuclei will also have this same probability of decaying during the next interval Δt and during any subsequent Δt and that the probability that a particular nucleus will decay during the time interval Δt is roughly proportional to Δt. In other words, for a short time interval Δt, Schweidler and Meyer assumed that the probability P is a function of just Δt and that

$$P(\Delta t) \approx \lambda \Delta t, \tag{6.3.2}$$

[3] Nuclei of different ***isotopes*** of a chemical element have the same number of protons but a different number of neutrons. The most common naturally occurring isotopes of uranium are uranium-238 and uranium-235. Both of these isotopes have 92 protons. The nucleus of a uranium-238 atom has $238 - 92 = 146$ neutrons, whereas the nucleus of a uranium-235 atom has $235 - 92 = 143$ neutrons.

[4] See [52, p. 514 f.] and [80].

where $\lambda > 0$. Moreover, this approximation improves as Δt gets smaller and smaller so that

$$\lim_{\Delta t \to 0} \frac{P(\Delta t)}{\Delta t} = \lambda. \tag{6.3.3}$$

It then follows that the number of nuclei expected to decay during any short time interval Δt is equal to the product $N(t) \cdot P(\Delta t)$, where $N(t)$ denotes the number of undecayed nuclei at time t. Why? To answer this, think of an undecayed nucleus as an unbalanced coin that is tossed. If it lands "heads up" it has decayed; if it lands "tails up" it has not. Then the problem becomes mathematically equivalent to tossing N identical unbalanced coins. Suppose that the probability of a single coin landing heads up is P. Then if all N coins are tossed, the number of coins that we expect to see landing heads up is approximately NP.[5] So by (6.3.2), we have

$$NP = N(t)P(\Delta t) \approx N(t)\lambda\Delta t.$$

Since this is the approximate decrease in the number of nuclei during a short time interval Δt, it follows that

$$N(t + \Delta t) \approx N(t) - N(t)P(\Delta t).$$

Thus,

$$\frac{N(t + \Delta t) - N(t)}{\Delta t} \approx -N(t)\frac{P(\Delta t)}{\Delta t}. \tag{6.3.4}$$

As Δt approaches 0, this approximation becomes better and better. As a result and because of (6.3.3), we conclude

$$\lim_{\Delta t \to 0} \frac{N(t + \Delta t) - N(t)}{\Delta t} = -\lambda N(t). \tag{6.3.5}$$

In summary, the Schweidler-Meyer hypothesis has led to the differential equation

$$\frac{dN}{dt} = -\lambda N. \tag{6.3.6}$$

However, one could raise the following objection to (6.3.6): Changes in the number N of nuclei are discrete—in fact, integer-valued—but the above derivation proceeded as if N were a continuous function, and in fact a differentiable one. This objection can be circumvented somewhat by the following viewpoint. Any mathematical model is judged by how well it is in accord with experiments. Any

[5] In probability theory, tossing N identical coins is an example of a so-called binomial experiment. The expected number of successes (heads up) is NP.

experiment to test (6.3.6) would involve an unimaginably large number of nuclei. Since the amount of an isotope is proportional to the number N of nuclei, integer-valued changes in N translate generally to fractional changes in the amount. It seems reasonable to suppose that the amount of an isotope can be modeled nearly perfectly with a continuous function $A(t)$. And let us go even so far as to presume that $A(t)$ is differentiable. Then, as $A(t) = kN(t)$ for some $k > 0$, the result of multiplying both sides of (6.3.6) by k is

$$\frac{dA}{dt} = -\lambda A. \tag{6.3.7}$$

Its solution is $A(t) = A_0 e^{-\lambda t}$, where A_0 is the initial amount of isotope. Indeed, the correctness of this has been experimentally verified.

In the context of radioactivity, (6.3.7) is known as the ***radioactive decay law***. The constant of proportionality λ is called the ***decay constant***. It is a parameter whose value depends on the type of isotope. We could regard a sample of a given radioactive isotope in a laboratory or natural setting as a one-compartment system, where nothing is being added to the sample but the amount A of the radioactive isotope in the sample is decreasing at a rate λA, where λ is the decay constant of the isotope. This is depicted in Fig. 6.4.

6.3.2 Malthusian Population Model

Now let us take a look at the simplest and most well-known mathematical model for population growth. We will use the notation $p(t)$ to denote the size of the population of some organism or species of animal at time t. Of course, population size is integer-valued; nevertheless, in the same spirit as with modeling the number of radioactive nuclei, let us assume that $p(t)$ is sufficiently large so that it can be treated as if it were a differentiable function of t.

Imagine an experiment in which a small population of E. coli bacteria (Escherichia coli) is introduced into an Erlenmeyer flask containing a nutrient broth, such as Luria-Bertani broth, which is a nutritionally rich medium for growing bacteria. Since the initial population is small, some of the bacteria may perish but not from lack of nutrient broth or adequate space for survival during the early stages of the experiment. In fact, let us assume that virtually none of the bacteria die during

Fig. 6.4 Compartment model of radioactive decay

the course of the experiment. E. coli bacteria are single-celled microorganisms, which reproduce asexually through a process called ***binary fission***: a single *parent cell* divides into two new *daughter cells*. From the observer's viewpoint, it is impossible to predict which of the cells will divide at any given instant. Let us assume that a cell is just as likely to divide within a given time interval Δt as any other cell; and that even if the cell does not divide, the probability of it dividing during any future interval Δt remains constant. With these assumptions, reproduction by cell fission is similar to the decay of radioactive nuclei—the difference being that parent nuclei decay into daughter nuclei belonging to a different species, whereas parent cells divide into daughter cells belonging to the same species. And so the decay of a nucleus of some radioactive isotope results in a net loss of one nucleus of that isotope, whereas the fission of one E. coli cell results in a net gain of one E. coli cell. So now we have growth instead of decay. This analogy with radioactive decay suggests that the differential equation modeling population growth is

$$\frac{dp}{dt} = bp, \tag{6.3.8}$$

where $b > 0$. In words, the rate of population growth in this model is assumed to be (directly) proportional to the population. While the derivative $\dot{p}$ is the ***growth rate*** of the population, the ratio

$$\frac{\dot{p}}{p} = \frac{1}{p}\frac{dp}{dt} \tag{6.3.9}$$

is called the ***per capita growth rate*** of the population. In this particular model, the per capita growth rate is equal to the constant of proportionality b. The solution of (6.3.8) is

$$p(t) = p_0 e^{bt}, \tag{6.3.10}$$

where $p_0 = p(0)$, namely, the number of E. coli bacteria in the flask at time $t = 0$.

This mathematical model accords with laboratory experiments where it has been verified that the population of E. coli bacteria in a nutrient broth initially increases at an exponential rate for about an hour (cf. J. Maynard Smith [64, p. 17]). Even allowing for cell death, the model can still be used if the mortality rate of the cells is proportional to the population size. In other words, the mortality rate at time t is equal to $mp(t)$, where m is the constant of proportionality. A schematic for this is shown in Fig. 6.5. Consequently, the differential equation model is

$$\frac{dp}{dt} = bp - mp = (b - m)p = rp, \quad p(0) = p_0 \tag{6.3.11}$$

where $r = b - m$. Note that the differential equation in (6.3.11) is the same as (6.3.8), save for the change in the proportionality constant. If $b > m$ so that $r >$

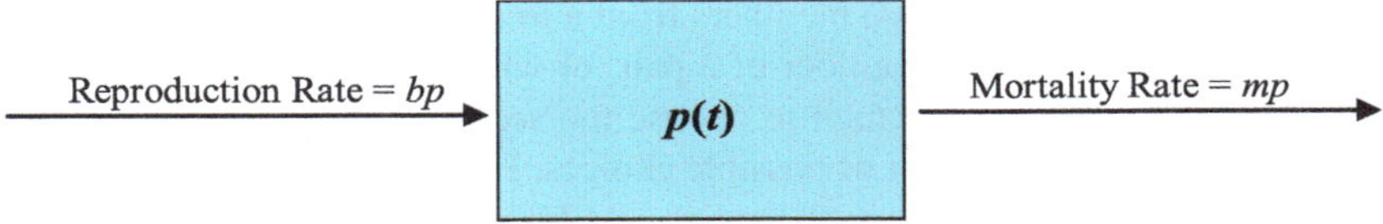

Fig. 6.5 Compartment model of E. coli bacteria in a flask of nutrient broth

0, then the population still increases exponentially. This particular mathematical model, which hypothesizes that the per capita growth rate r is a positive constant, is known as the ***Malthusian model of population growth***. It was named after Thomas Robert Malthus (1766–1834), an English economist and cleric, who published a very influential book in 1798 titled *An Essay on the Principle of Population as It Affects the Future Improvement of Society*. In it, one finds various statements and predictions about the growth of human populations, such as the following:

> In the United States of America, where the means of subsistence have been more ample, the manners of the people more pure, and consequently the checks to early marriages fewer than in any of the modern states of Europe, the population has been found to double itself in twenty-five years.
>
> This ratio of increase, though short of the utmost power of population, yet as the result of actual experience, we will take as our rule, and say, that population, when unchecked, goes on doubling itself every twenty-five years or increases in a geometric ratio.

Even though there are no differential equations in Malthus's essay, his statement that population, when unchecked, increases in a geometric ratio can be derived from Eq. (6.3.11). On this point, see Problem 15.

6.3.3 *Logistic Population Model*

As the population size of an organism in a closed ecosystem increases, competition for food and living space comes into play. Moreover, there are other factors adversely affecting the population, such as diseases and the build-up of waste products, just to name a few. A way of incorporating these factors is to modify the Malthusian model by introducing a "competition term" as follows:

$$\frac{dp}{dt} = rp - [\text{rate of competition}]. \tag{6.3.12}$$

Regardless of how this term is modeled, it should have a restraining effect on the growth rate of the population—hence, the minus sign. One common assumption is that the rate of competition is proportional to the total number of possible paired encounters between members of the species. Suppose at a given moment there are p members. Then the total number of possible paired encounters is equal to the number of ways of choosing two members from a total of p members. To calculate this number, choose the first member of a pair, of which there are p choices. That leaves $p - 1$ members from which to choose the second one. Hence, the product $p(p-1)$ gives the total number of possible choices. However, the pair "Hamlet and Claudius" is the same as the pair "Claudius and Hamlet." That means the number $p(p - 1)$ is too big by a factor of two. So the total number of possible pairs is $p(p-1)/2$. Or, viewing this as a combinations problem, this is equivalent to finding the number of combinations of p members taken two at a time, which is

$$\binom{p}{2} = \frac{p!}{2!(p-2)!} = \frac{p(p-1)}{2}.$$

As a result of assuming that the competition term is proportional to the total number of possible paired encounters, Eq. (6.3.12) becomes

$$\frac{dp}{dt} = rp - k\,\frac{p(p-1)}{2} = ap - bp^2,$$

where

$$a = r + \frac{k}{2} \quad \text{and} \quad b = \frac{k}{2}.$$

So, according to this model, at any given time t the reproduction rate is proportional to the population $p(t)$ while the mortality rate is proportional to $(p(t))^2$.

In the context of modeling the size of a population, the differential equation

$$\frac{dp}{dt} = ap - bp^2 \tag{6.3.13}$$

is known as the ***logistic model of population growth*** or simply the ***logistic equation***. The parameters a and b are known as the ***vital coefficients*** of the population. The first published appearance of this equation as a model for population growth was in a paper written in 1838 by the Belgian mathematician Pierre-François Verhulst entitled *Notice sur la loi que la population suit dans son accroissement* [75, p. 115].[6]

Observe that the logistic equation has two constant solutions, namely, $p(t) \equiv 0$ and $p(t) \equiv a/b$. The nontrivial solutions of this equation can be found by separating its variables and then integrating. There is also the alternative of transforming it into a linear equation since it is a Bernoulli equation (see Problem 23). Choosing the former, we have

$$\int \frac{dp}{ap - bp^2} = \int dt + K,$$

[6] Translation: Notice on the law that a population follows in its growth.

where $p \neq 0$ and $p \neq a/b$. In order to find the integral on the left-hand side, let us start with the partial fraction expansion

$$\frac{1}{p(a-bp)} = \frac{A}{p} + \frac{B}{a-bp}.$$

Multiplying both sides of this equation by $p(a-bp)$ to clear out the denominators and then combining like terms, we get

$$A(a-bp) + Bp = 1.$$

Now we can determine A and B by setting p equal to a/b and then to 0 as follows:

$$p = \frac{a}{b} \quad \Rightarrow \quad B\left(\frac{a}{b}\right) = 1 \quad \Rightarrow \quad B = \frac{b}{a}$$

$$p = 0 \quad \Rightarrow \quad Aa = 1 \quad \Rightarrow \quad A = \frac{1}{a}.$$

Hence,

$$\frac{1}{a}\int \frac{1}{p}\,dp - \frac{1}{a}\int \frac{-b}{a-bp}\,dp = t + K \quad \Rightarrow \quad \frac{1}{a}\ln|p| - \frac{1}{a}\ln|a-bp| = t + K.$$

Multiplying both sides by a, using a property of logarithms, and then exponentiating, we obtain

$$\ln\left|\frac{p}{a-bp}\right| = at + aK \quad \Rightarrow \quad \left|\frac{p}{a-bp}\right| = e^{aK}\cdot e^{at}.$$

Thus,

$$\frac{p}{a-bp} = \pm e^{aK}\cdot e^{at} \quad \Rightarrow \quad \frac{p}{a-bp} = Ce^{at},$$

where we have replaced $\pm\, e^{aK}$ with C. If the population is p_0 when $t = 0$, then

$$C = \frac{p_0}{a - bp_0}$$

provided $p_0 \neq a/b$. Consequently,

$$\frac{p}{a-bp} = \frac{p_0 e^{at}}{a-bp_0} \quad \Rightarrow \quad p(a-bp_0) = p_0(a-bp)e^{at}.$$

Solving for p, we obtain the function

$$p(t) = \frac{ap_0}{bp_0 + (a - bp_0)e^{-at}}, \tag{6.3.14}$$

namely, the solution of the logistic equation (6.3.13) satisfying the initial condition $p(0) = p_0$. Verhulst named this solution the ***logistic function*** (see [60]). The graph of a logistic function is called a ***logistic curve*** (see Problem 21).

Finally, note that

$$\lim_{t\to\infty} p(t) = \frac{a}{b}. \tag{6.3.15}$$

If $p_0 = a/b$, then $p(t) = a/b$ for all $t \geq 0$. If $p_0 \neq a/b$, the logistic function predicts that the population size will approach a/b irrespective of the initial population (ignoring the trivial case $p_0 = 0$). This limit is known as the ***carrying capacity of the environment***.

6.3.4 Deposit Accounts

Continuous Compounding of Interest The remainder of this chapter deals with the mathematics associated with ***deposit accounts***, to wit: savings, checking, and money market accounts, and certificates of deposit (CDs). For these types of accounts, we will look at how to compute interest and determine the balance of an account using the *compound interest formula*. Furthermore, we will learn how to decide which of two savings accounts advertised by competing banks is the better deal if the only differences between the accounts are the interest rates and the frequency with which interest is compounded. For example, later on in Example 6.3.5, we will determine whether "$5\frac{1}{4}$% compounded daily" or "$5\frac{1}{2}$% compounded quarterly" is better for the consumer.

The mathematics of personal finance is usually covered in finite math textbooks but rarely in differential equations books. However, as we shall see shortly, the balance of an account with a stated interest rate that is *compounded continuously* can be modeled with a differential equation. Moreover, we will use the compound interest formula and the "equivalence of annual rates" result that is stated at the end of this chapter to obtain a differential equation in Chap. 7 that we will employ to derive a formula for computing the monthly payments that are needed to pay off a loan, such as a car loan, given the time in which it is to be paid off and the interest rate.

We introduce *continuous compounding of interest* with the following scenario. Suppose a recent college graduate, Lexi Thérèse, finds her dream job and manages to save some money. Lexi decides it would be prudent to open an emergency savings account at her bank. She makes a one-time deposit of B_0 dollars into the account, which is worth 3 months of her living expenses, and resolves not to withdraw any money from this account unless an emergency should arise, such as an unexpected car repair bill. Let us imagine that the bank starts paying Lexi interest from the moment she makes the deposit (at time $t = 0$) by letting money flow continuously from its reserves into the account at a rate that is directly proportional to the balance in the account at time t. Let $B(t)$ denote this balance, where the time t is expressed in years (or parts of years). $B(t)$ consists of the initial deposit of $B(0) = B_0$ dollars plus the interest that has accrued up to time t. Of course, the bank does not actually transfer tangible forms of currency to some physical compartment called "Lexi's account." But if an emergency should arise at some point in the future and Lexi has to withdraw all of her money from the account, there is a well-known mathematical formula that the bank can use to determine the balance as if continuous transfers had been taking place. Let us find this formula.

The scenario we have just described can be handled by regarding it as a one-compartment problem, where the compartment represents the savings account containing $B(t)$ dollars at time t. The situation is depicted in Fig. 6.6. It shows money flowing into the the compartment (account) at a rate of $rB(t)$ dollars per year, where r is a positive constant. But no money is flowing out of the compartment since there will be no withdrawals from the account in the foreseeable future. So this particular compartment schematic involves an input but no output, which is reversed from the way it was with radioactive decay. From the conservation equation for one-compartment systems, we have the separable differential equation

$$\frac{d}{dt}B(t) = rB(t), \tag{6.3.16}$$

where r is the constant of proportionality. Since we have run into this equation many times before, we already know its solution to be

$$B(t) = B_0 e^{rt}. \tag{6.3.17}$$

This is the balance in the savings account at any time $t \geq 0$, provided no withdrawals are made and the value of r does not change.

Since Eqs. (6.3.16) and (6.3.17) have to do with the amount of money in a deposit account, the constant r is an interest rate, which is expressed as a percent in verbal statements. Depending on the financial institution, it goes by a number of different names, some of which are ***annual percentage rate (APR)***, ***annual***

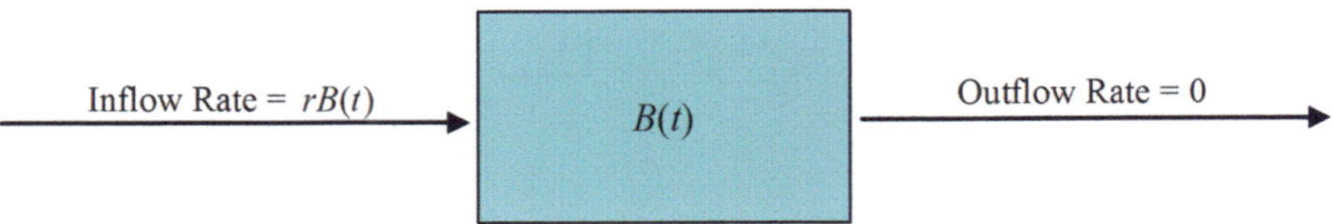

Fig. 6.6 Compartment model of continuous compounding of interest

(interest) rate, and ***nominal (interest) rate***. The parentheses indicate that the word *interest* is oftentimes omitted. If the balance of a deposit account or some other type of investment is given by Eq. (6.3.17), the interest is said to be accruing at an ***annual rate of r percent compounded continuously***. Although the annual percentage rate is expressed as a percent, it must be changed to a decimal whenever it is involved in calculations. For example, if a bank offers a savings account that pays interest at an annual rate of 10% compounded continuously, then $r = 10\%$. However, whenever formula (6.3.17) is used to calculate a balance, then $r = 0.10$.

Example 6.3.2 A deposit account pays interest at an annual rate of 10% compounded continuously. If a one-time deposit of \$1000 is made to this account and no money is ever withdrawn from it, what will be the balance in the account after 2 years (assuming the interest rate does not change)? After 10 years?

Solution It follows from (6.3.17) that the balance in the account after t years is

$$B(t) = 1000e^{0.10t}.$$

So, after 2 years, the balance in the account is

$$B(2) = 1000e^{(0.10)2} = \$1221.40.$$

Hence, the accrued interest is \$221.40. After 10 years, the balance is

$$B(10) = 1000e = \$2718.28.$$

A graph of $B(t)$ over a 10-year time period is shown in Fig. 6.7. ♦

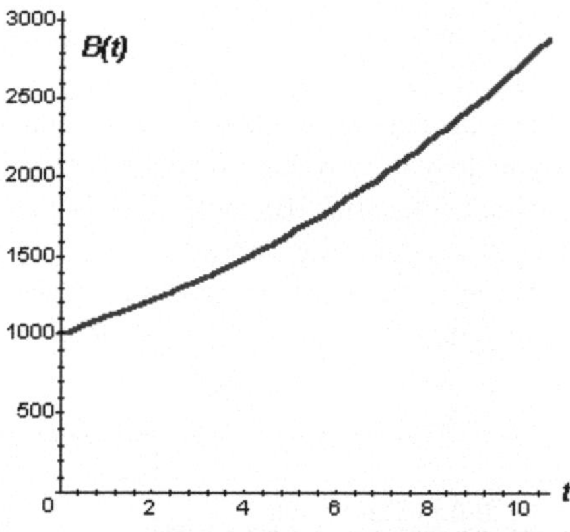

Fig. 6.7 Growth of \$1000 at 10% compounded continuously

The percent change in the balance after a year is called the ***annual percentage yield* (APY)** or ***effective annual rate***.[7] If B_0 denotes the balance at some given point in time and $B(1)$ the balance one year later, the interest earned during the course of the year is the difference $B(1) - B_0$. Thus the annual percentage yield is given by the formula

$$\text{APY} = \frac{\text{Interest over a year's time}}{\text{Beginning Balance}} = \frac{B(1) - B_0}{B_0}. \tag{6.3.18}$$

Example 6.3.3 What is the annual percentage yield of an investment drawing interest at an annual percentage rate of 10% compounded continuously?

Solution The value of $B(1)$, the balance after one year, is found by setting $r = 0.10$ and $t = 1$ in (6.3.17). Then it follows from (6.3.18) that the annual percentage yield is

$$\text{APY} = \frac{B(1) - B_0}{B_0} = \frac{B_0 e^{0.10} - B_0}{B_0} = \frac{B_0\left(e^{0.10} - 1\right)}{B_0} = e^{0.10} - 1 = 0.1051709\ldots.$$

So, for an APR of 10% compounded continuously, the APY to three decimal places is 10.517%. We can see from the above calculation that the APY does not depend on the value of B_0 and that

$$\text{APY} = e^r - 1 \tag{6.3.19}$$

irrespective of the value r of the annual percentage rate. ♦

Discrete Compounding of Interest To fully understand and appreciate the notion of *continuous compounding of interest*, we need to know how interest is usually computed. In standard business practices, interest on deposit accounts, loans, investments, and the like, do not accrue continuously; rather, they accrue in discrete amounts and are credited to these accounts periodically, such as quarterly (every three months), monthly, or daily. One of the oldest methods of computing interest is the ***simple interest method***. When this method is used, interest is earned only on the ***principal***, namely, the sum of money that has been deposited in an account or the amount of money borrowed from a lending institution. This interest is computed using the ***simple interest formula***:

$$\text{Interest} = \text{Principal} \times \text{rate} \times \text{time}.$$

[7] There appears to be nearly as many variations of these two terms in use by financial institutions as there are ways to combine and permute the words *annual*, *effective*, *rate*, and *yield*. Besides these, there are also *annual yield*, *effective rate*, *annual effective rate*, *annual effective yield*, and *true interest rate*.

With I denoting the interest, P the principal, r the annual simple interest rate (written as a decimal), and t the time (expressed in years), this formula becomes

$$I = Prt. \tag{6.3.20}$$

For instance, a principal of \$1000 invested at 10% simple interest will earn \$300 in 3 years since

$$I = 1000 \times 0.10 \times 3 = \$300.$$

At the end of the first year, the earned interest is $\$1000 \times 0.10 = \$100$. Likewise, the interest earned over the course of the second year is \$100. Finally, the third year adds another \$100, bringing the total accrued interest for the 3 years to \$300.

The standard for computing interest for most deposit accounts, investments, and loans is the ***compound interest method***, where interest is earned not only on the principal but also on the total of all previously accrued interest. Using this method, a year is divided into m equal parts called ***compounding periods*** or simply ***periods***. For $m = 1$, the period is a year and the interest is said to be ***compounded annually***. For $m = 4$, the period is a quarter of a year and the interest is said to be ***compounded quarterly***. If $m = 12$, then the period is one month and the interest is said to be ***compounded monthly***, and so forth.

So how is the interest and balance computed for different compounding periods? Before we can answer that, let us take a closer look at the some of the terminology associated with compounding of interest. One is the compound interest rate, which is stated as an annual percentage regardless of the number of compounding periods there are in a year. Depending on the financial institution, it is called the *annual rate* or *nominal rate* or *APR* (*annual percentage rate*). As with continuous compounding of interest, we will generally use the latter term. Suppose the stated APR is 10%. For one kind of investment, this could mean that the interest is compounded quarterly; for another, compounded monthly; yet for another, compounded daily. Of course, there other possibilities as well, such as annually and semiannually.[8] The point is that the APR is only one component of a compound interest rate; the other is the number of compounding periods per year.[9] For an APR of r with m periods per year, the interest that will accrue during any given period is computed using the simple interest formula (6.3.20):

[8] Even though the "true interest rates" for these cases are different, the annual percentage rate is still 10%. The true interest rate of an investment is given by the annual percentage yield, which we were introduced to earlier (see (6.3.18)). We will show later on how to compare two different compound interest rates by computing their respective annual percentage yields.

[9] Advertising the annual percentage rate to consumers without also stating the number of compounding periods in a year is disingenuous. Oftentimes an advertising brochure may state the APR in large print but omit the frequency with which the interest is compounded or bury that information somewhere in fine print.

$$I = Prt = \text{Balance} \times r \times \frac{1}{m}, \tag{6.3.21}$$

where P is the balance at the beginning of the period, r is the APR written as a decimal, and t is the time expressed in years. Because a year is divided into m periods, $t = 1/m$. That is, the interest for a given period is computed by multiplying the balance at the beginning of that period by r/m. Then the balance at the end of this period is obtained by adding this interest to the previous balance. Interest that is computed in this way is said to accrue at an ***annual percentage rate of r compounded m times per year***.

Now that we know how to compute the interest that accrues during any given period, let us turn our attention to how the balance is computed at the end of a period. Consider the following example.

Example 6.3.4 If \$1000 is invested at 10% compounded quarterly, what will the balance be after one year?

Solution Of course, it will be more than \$1100, as it would be in the case of simple interest, since interest is earned not only on the principal but also on any previous accrued interest—moreover, the crediting of interest occurs four times during the year. The meaning of this will become clearer as we carry out the following computations. Since the interest in this example is compounded quarterly, the year is divided into four 3-month periods called quarters. The interest that will accrue during any quarter is found by using the simple interest formula (6.3.21) with $r = 0.10$ and $m = 4$:

$$I = P \times 0.10 \times \frac{1}{4}. \tag{6.3.22}$$

Since $P = 1000$ dollars, the interest earned during the first quarter will be

$$I(\text{1st qtr}) = 1000 \times 0.10 \times \frac{1}{4} = \$25. \tag{6.3.23}$$

Consequently, the balance at the end of the first quarter will be

$$B(\text{1st qtr}) = 1000 + 25 = \$1025. \tag{6.3.24}$$

Since the new balance is \$1025, the interest at the end of the second quarter is found by substituting this amount for P into (6.3.21), thereby obtaining

$$I(\text{2nd qtr}) = 1025 \times 0.10 \times \frac{1}{4} = \$25.625. \tag{6.3.25}$$

Therefore, at the end of the second quarter the balance will increase to

$$B(\text{2nd qtr}) = B(\text{1st qtr}) + \text{Interest} = 1025 + 25.625 = \$1050.625. \tag{6.3.26}$$

Continuing with the computations, the interest for the third quarter will be

$$I(\text{3rd qtr}) = 1050.625 \times 0.10 \times \frac{1}{4} = \$26.265625.$$

So the balance at the end of the third quarter will be

$$I(\text{3rd qtr}) = 1050.625 + 26.265625 = \$1076.890625.$$

Finally, the interest at the end of the fourth quarter will be

$$I(\text{4th qtr}) = 1076.890625 \times 0.10 \times \frac{1}{4} = \$26.922265625$$

resulting in a balance of \$1103.81289063. Note that we have kept all of the digits in the computations to avoid round-off error, but if the fourth-quarter balance were to be withdrawn, it would of course be rounded to \$1103.81. The effect of compounding, where interest is earned not only on the original balance but also on all previously accrued interest, is made apparent by the increases in interest from quarter to quarter—and by the final balance of \$1103.81, which is \$3.81 more than the simple interest balance of \$1100. ♦

Even though the example illustrates the compound interest method, a formula for the balance at the end of a period is not apparent. Maybe the thing to do is to look at the calculations again but not let the computed numbers get in the way. Sometimes a pattern will emerge when attention is shifted from the end result to the means by which it was obtained. With this in mind, let B_k, for $k = 1, 2, 3, \ldots$, denote the balance at the end of the kth quarter. Then from (6.3.23) and (6.3.24), the balance at the end of the first quarter is

$$B_1 = 1000 + 1000(0.10)\left(\frac{1}{4}\right) = 1000\left(1 + \frac{.10}{4}\right). \tag{6.3.27}$$

From (6.3.25) and (6.3.26), the balance at the end of the second quarter is

$$B_2 = B_1 + B_1\,(0.10)\left(\frac{1}{4}\right) = B_1\left(1 + \frac{.10}{4}\right).$$

Then replacing B_1 with the expression from (6.3.27), this becomes

$$B_2 = 1000\left(1 + \frac{.10}{4}\right)^2.$$

Continuing in this manner, the third-quarter balance is

$$B_3 = B_2 + B_2(0.10)\left(\frac{1}{4}\right) = B_2\left(1 + \frac{.10}{4}\right) = 1000\left(1 + \frac{.10}{4}\right)^3.$$

And so the balance at the end of the fourth quarter, or year, is

$$B_4 = B_3\left(1 + \frac{.10}{4}\right) = 1000\left(1 + \frac{.10}{4}\right)^4.$$

Using a calculator to compute this, we obtain a balance of \$1103.81, as before in Example 6.3.4. But the point to be made here is that a pattern has emerged. So what would the balance be after two years? Since there are 8 periods in two years, the balance would be

$$B_8 = 1000\left(1 + \frac{.10}{4}\right)^8.$$

In general, after k periods the balance is given by the formula

$$B_k = 1000\left(1 + \frac{.10}{4}\right)^k. \tag{6.3.28}$$

From this we can see how to compute the balance at the end of k periods for any principal and interest rate. Simply replace the principal "1000" with "P", the APR ".10" with "r", and the number of periods "4" with "m". Then (6.3.28) becomes

$$B_k = P\left(1 + \frac{r}{m}\right)^k \tag{6.3.29}$$

for $k = 0, 1, 2, 3, \ldots$, where P denotes the principal, r the APR, m the number of compounding periods per year, and B_k the balance at the end of k periods. Note that B_0 is the beginning balance, namely, the principal P. This formula is known as the ***compound interest formula***. The ratio r/m is called the ***periodic interest rate***. For example, if the APR is 10% compounded quarterly, as in Example 6.3.4, the periodic interest rate or ***quarterly interest rate*** is

$$\frac{r}{m} = \frac{0.10}{4} = 0.025 = 2.5\%.$$

If interest is compounded daily, then the periodic interest rate, or ***daily periodic rate***, is $r/365$ since there are 365 days in a calendar year.[10]

We can also express the compound interest formula in terms of the time. If there are m periods in a year, then the time t (in years) corresponding to a total of k periods is $t = k/m$. Replacing k with mt and B_k with $B(t)$, the compound interest formula (6.3.29) becomes

$$B(t) = P\left(1 + \frac{r}{m}\right)^{mt}. \tag{6.3.30}$$

[10] Sometimes a ***banker's year*** of 360 days is used for calculating interest for certain financial products, such as short-term loans.

Since this formula comes from (6.3.29), the values that t are allowed to have for a given value of m are k/m, where $k = 0, 1, 2, 3, \ldots$. However, let us remove this restriction and allow t to have any nonnegative value. Then (6.3.30) can be used to compute the total amount of money in a deposit account at any time t, which would include all of the interest that would have accrued up to that point in time. Typically, however, if all of the money is withdrawn from the account before the end of a compounding period, a financial institution will not include the interest that would have accrued up to the time of withdrawal. Interest that has accrued is only credited to the account at the end of each compounding period. In short, if the domain of the compound interest formula (6.3.30) consists of all nonnegative numbers, the actual balance in a deposit account at time $t \in [k/m, (k+1)/m)$, where k is a nonnegative integer, is

$$B(k/m) = P\left(1 + \frac{r}{m}\right)^k .$$

For instance, consider Example 6.3.4 again. The balance at the end of the first quarter is \$1025. It remains at this amount during the entire second quarter, namely, the time from the end of the first quarter until the start of third quarter, at which point the balance jumps from \$1025 to \$1050.63. See (6.3.26). The balance then remains constant again until it jumps to \$1076.89 at the end of the third quarter and so on. The graph of the balance is shown in Fig. 6.8. It resembles an ascending straight flight of stairs except for the missing risers (the vertical sections connecting steps). The jumps between steps occur at the times quarterly interest is credited to the account, namely, at $t = \frac{1}{4}, \frac{1}{2}, \frac{3}{4}, 1, \frac{5}{4}, \ldots$. Note that the jumps between steps get larger with the passage of time since the interest that accrues during each compounding period keeps increasing.

In Fig. 6.9 we see a comparison of the graphs for the discrete growth of \$1000 earning interest at an APR of 10% compounded quarterly and the continuous growth of \$1000 earning interest at an APR of 10% compounded continuously. The

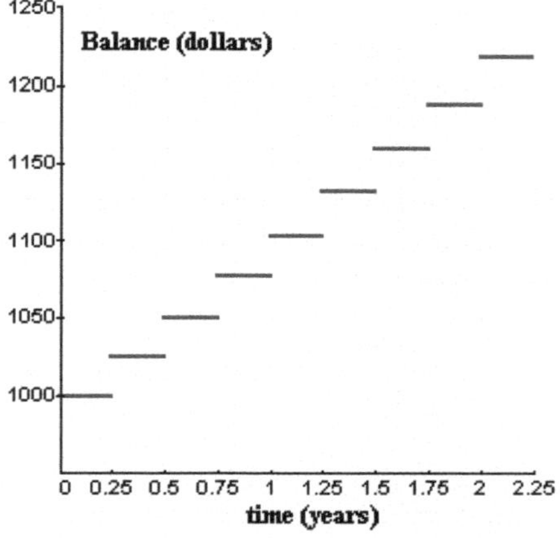

Fig. 6.8 Growth of \$1000 at 10% compounded quarterly

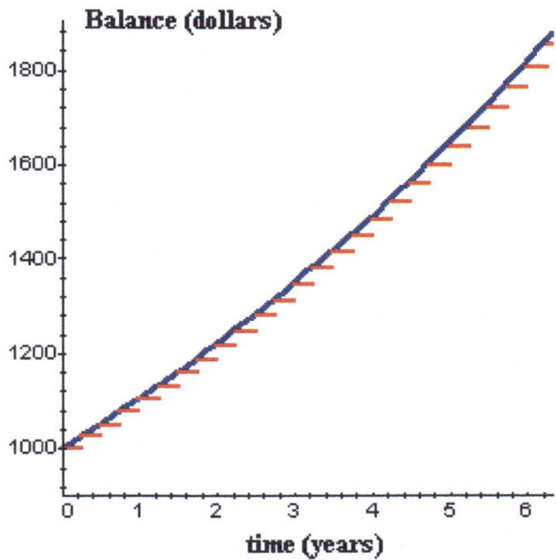

Fig. 6.9 Continuous vs. discrete growth of \$1000

continuous growth graph is always above the discrete growth graph even though it may appear that the balances are equal at $t = \frac{1}{4}, \frac{1}{2}, \frac{3}{4}, 1, \frac{5}{4}, \ldots$. But they are not. For instance, if the 10% interest rate is compounded quarterly the balance at the end of 5 years will be

$$B(5) = 1000\left(1 + \frac{0.10}{4}\right)^{(4)(5)} = \$1638.62.$$

If it is compounded continuously, then the balance will be

$$B(5) = 1000e^{(0.10)(5)} = \$1648.72.$$

This exceeds the balance for discrete growth by \$10.10.

Example 6.3.5 First Horizon Bank advertises a Certificate of Deposit (CD) for \$5000 at an APR of $5\frac{1}{4}\%$ compounded daily while First American National Bank advertises one at $5\frac{1}{2}\%$ compounded quarterly. Assuming that both CDs have the same maturity date, which bank offers the better deal with everything else being equal?

Solution Even though First Horizon's APR is lower by a quarter of a percentage point, perhaps the greater frequency with which First Horizon compounds its interest will offset First American's higher APR. But how can we determine whether or not this is the case? The way to choose the best interest rate, especially those with different compounding frequencies, is to compare the annual percentage yield (APY) of each of the CDs. Recall from (6.3.18) that the APY is the percent change in the balance after one year. So let us compute the APY of the CD advertised by

First Horizon:

$$\text{APY} = \frac{5000\left(1+\frac{.0525}{365}\right)^{365} - 5000}{5000} = \left(1+\frac{.0525}{365}\right)^{365} - 1 \approx 5.39\%.$$

Note that the \$5000 amounts canceled out in the computation, which goes to show that the APY is independent of the deposit or the beginning balance. Therefore, a general formula for the APY when the APR is r percent compounded m times per year is

$$\text{APY} = \left(1+\frac{r}{m}\right)^{m} - 1. \tag{6.3.31}$$

Now let us compute the APY of the CD advertised by First American:

$$\text{APY} = \left(1+\frac{.0550}{4}\right)^{4} - 1 \approx 5.61\%.$$

Consequently, First American offers the better deal. ♦

Consumers often find their mailboxes stuffed with pre-qualified credit card invitations. The APR is usually in plain sight, but it is sometimes near impossible to discern the frequency with which the interest on an outstanding balance is compounded. Generally speaking, one has to ferret out this information by reading the small print at the bottom of the invitation or somewhere on the reverse side. For instance, at the bottom of the cover letter of an invitation that this author once received was the statement: "The Annual Percentage Rate (APR) for Purchases, Bank, and ATM Cash Advances is 17.4%." However, as to the frequency of compounding, the cover letter provided no information. But the second page of the invitation disclosed a daily periodic rate of 0.0476%. Since by daily periodic rate, the ratio r/m is meant, where r is the APR and $m = 365$, this checks out because

$$\frac{r}{m} = \frac{0.174}{365} = .000476712\ldots \approx .04767\%.$$

In other words, this additional information reveals that the APR of 17.4% is compounded daily. As mentioned earlier, an APR is also referred to as a nominal rate. The word *nominal* means *in name only, not in fact*, as in nominal head of state, such as the present-day British monarch. So when the nominal rate (or APR) is 17.4% compounded daily, "17.4%" is not the true interest rate; instead it is given by the annual percentage yield, which is 19% according to formula (6.3.31):

$$\text{APY} = \left(1+\frac{.174}{365}\right)^{365} - 1 = 0.1900062\ldots .$$

Continuous Compounding vs. Discrete Compounding Recall that we introduced the notion of "continuous compounding of interest" before we ever said anything about "discrete compounding of interest." Had we introduced the latter first, we could have derived the continuous compounding of interest formula (6.3.17) without ever mentioning the differential equation (6.3.16). The way to do this is to start with the discrete compound interest formula:

$$B(t) = B_0 \left(1 + \frac{r}{m}\right)^{mt}. \tag{6.3.32}$$

When the value of m, the number of compounding periods per year, is 365, the interest is said to be compounded daily. Can we increase the frequency of compounding even more? Absolutely! There is nothing to prevent us from using (6.3.32) to compound hourly, or every minute, or even every second. Or every nanosecond (one billionth of a second)! Just set m equal to 3.1536×10^{16} as

$$m = 365\ \frac{\text{d}}{\text{yr}} \times 24\ \frac{\text{h}}{\text{d}} \times 3600\ \frac{\text{s}}{\text{h}} \times 10^9 \frac{\text{ns}}{\text{s}} = 3.1536 \times 10^{16}\ \frac{\text{ns}}{\text{yr}}.$$

And there is nothing to stop us from increasing m even more. So the question arises: What happens to the balance $B(t)$ as m increases without bound? The expression $(1 + r/m)^{mt}$ is related to the familiar expression $(1 + 1/m)^m$ from calculus. Recall that

$$\left(1 + \frac{1}{m}\right)^m \to e \quad \text{as} \quad m \to \infty.$$

So what is the limit of

$$\left(1 + \frac{r}{m}\right)^{mt} \quad \text{as} \quad m \to \infty?$$

The bothersome "r" in the numerator can be handled by replacing r/m with $1/n$. Then, as $m = nr$, we have

$$\left(1 + \frac{r}{m}\right)^{mt} = \left(1 + \frac{1}{n}\right)^{nrt}.$$

Since r is a nominal interest rate, which for a given problem is constant, $n \to \infty$ as $m \to \infty$. And so,

$$\left(1 + \frac{1}{n}\right)^{nrt} = \left[\left(1 + \frac{1}{n}\right)^n\right]^{rt} \to e^{rt}.$$

Thus, as $m \to \infty$,

$$B(t) = B_0 \left(1 + \frac{r}{m}\right)^{mt} \to B_0 e^{rt}.$$

In other words, the balance in the account after t years will be $B(t) = B_0 e^{rt}$.

The takeaway of this discussion is that *continuous compounding of interest* can be viewed in two different ways. One is that the balance is governed by the differential equation (6.3.16). The other is that the balance is the limit of the compound interest formula (6.3.32) as the number of compounding periods increases without bound.

We conclude this chapter with some observations. Note that the discrete compound interest formula (6.3.32) can be written as

$$B(t) = B_0 b^t \quad \text{where} \quad b := \left(1 + \frac{r}{m}\right)^m,$$

provided it is interpreted correctly in view of the fact that interest is not credited to an account until the end of each compounding period as we had explained earlier after rewriting (6.3.29) as (6.3.30). Now let us change the base from b to Euler's number so that $B(t) = B_0 e^{kt}$ for some constant k. Then k has to satisfy the equation $e^k = b$. Thus,

$$k = \ln b = m \ln\left(1 + \frac{r}{m}\right).$$

So it appears that the discrete compound interest formula (6.3.32) is equivalent to the continuous compounding interest formula

$$B(t) = B_0 e^{kt} \quad \text{where} \quad k = m \ln\left(1 + \frac{r}{m}\right). \tag{6.3.33}$$

But not quite! While the graph of (6.3.33) is continuous, the graph of (6.3.32) looks like a flight of stairs, but with the risers not shown,[11] lying under the graph of (6.3.33) except at the ends of the discrete compounding periods when the interest is credited to the account. At these particular times, the two balances are equal. Thus, we have:

Equivalent Annual Percentage Rates
An APR of "r compounded m times per year" is equivalent to an APR of

$$\text{“}m \ln\left(1 + \frac{r}{m}\right) \text{ compounded continuously”}$$

in the following sense: For an account earning interest at an APR of r compounded m times per year, the balance at the end of a compounding period, after the interest has been credited, is equal to the balance of an account earning interest at an APR of $m \ln(1 + r/m)$ compounded continuously.

[11] As in Fig. 6.8 with $B_0 = \$1000$, $r = 10\%$, and $m = 4$ quarters.

Example 6.3.6 An APR of 10% compounded quarterly is equivalent to an APR of

$$m \ln\left(1 + \frac{r}{m}\right) = 4 \ln\left(1 + \frac{.10}{4}\right) \approx 9.877\%$$

compounded continuously. For example, at the end of 2 years, with an initial deposit of \$1000, the balance can be computed in one of two ways: with the discrete compounding formula

$$B(2) = 1000\left(1 + \frac{.10}{4}\right)^8 = \$1,218.40$$

or with the continuous compounding formula

$$B(2) = 1000e^{(0.09877)(2)} = \$1,218.40.$$

♦

Problems

> Many of you no doubt have read Jeans[12] or Eddington. What they do when they want to explain the atom, or something of that sort, is to give you a description out of which you can make a mental picture. But then they warn you that this picture is not what the scientists actually believe. What the scientists believe is a mathematical formula. The pictures are there only to help you understand the formula. They are not really true in the way the formula is; they do not give you the real thing but only something more or less like it. They are only meant to help, and if they do not help you can drop them. The thing itself cannot be pictured, it can only be expressed mathematically.
>
> C. S. Lewis,[13] *Mere Christianity*, p. 58

One-Compartment Systems

1. Verify that the function $x(t) = 0.42 - 0.32e^{-.002t}$ is the solution of the initial value problem (6.2.5). When will the percentage of soluble solids in the tank reach 15%?
2. Initially an ice cooler contains 10 gallons of pure water. Salt is poured into the cooler at the rate of 1/2 pound per minute. At the moment the pouring begins, the cooler's drain valve is opened resulting in the salt water draining out of the cooler at the rate of 1 gallon per minute. However, pure water is poured into the cooler at a rate that always maintains 10 gallons of salt water in the cooler. The salt water is stirred continuously to keep its concentration uniform throughout the cooler. Find the amount of salt in the cooler after 5 minutes.
3. A 500-gallon tank is initially half full of a sugar water solution made by dissolving a total of 10 pounds of sugar in the water. A faucet located above the tank is turned on; at the same time, a plug at the bottom of the tank is unscrewed. Immediately, a sugar-water solution with

[12] That is, Sir James Jeans, who was quoted at the beginning of Chap. 1.

[13] Another excerpt from this book is found at the end of Sect. 2.2.

a concentration of 1 pound of sugar per gallon begins flowing into the tank at a rate of 2 gal/min while the solution in the tank flows out at the rate of 4 gal/min. Let t denote the time in minutes and $x(t)$ the concentration of the sugar, in pounds per gallon, in the tank at time t. Express your answers to the following parts in terms of these two variables.

(a) What is the initial concentration of the sugar in the tank?

(b) What expression represents the amount of sugar in the tank at time t?

(c) What derivative denotes the rate at which the concentration of the sugar in the tank changes?

(d) What derivative denotes the rate at which the amount of the sugar in the tank changes?

(e) What linear differential equation models the amount (or concentration) of the sugar in the tank?

(f) For what times is the differential equation in part (e) valid?

(g) Write the differential equation in part (e) in standard form. Then find an integrating factor, written as simply as possible. Use it to find a formula for the concentration of sugar in the tank at time t.

4. Initially an ice cooler contains 10 gallons of pure water. Salt water with a concentration of $1/2$ pound of salt per gallon is poured into the cooler at the rate of 1 gallon per minute. At the moment the pouring begins, the cooler's drain valve is opened and salt water drains out of the cooler at the rate of 2 gallons per minute. The salt water is stirred continuously in order to maintain a uniform concentration. Find a formula for the concentration of the salt in the cooler up to the time that the cooler becomes empty. Find the amount of salt after 3 minutes. When will the amount of salt in the tank reach a maximum value?

5. A 200-gallon tank initially contains 50 gallons of salt water that was made by dissolving 5 pounds of salt in water. Salt water with a concentration of 0.25 pounds of salt per gallon is pumped into the tank at the rate of 6 gallons per minute, while the salt water in the tank is pumped out at the rate of 4 gallons per minute. Assume that the salt water always remains a homogeneous mixture. Find a formula for the concentration of the salt in the tank up to the time that it begins overflowing.

6. A water treatment tank at a new city sanitation facility is designed to remove sediment from water that is pumped in from a nearby lake. Before the facility is set into operation, the tank will start off with 40,000 gallons of pure water. The lake water, averaging about 0.10 ounces of sediment per gallon, will be pumped in and out of the tank at the rate of 2000 gallons per minute. As pumps circulate the water-sediment mixture inside the tank, filters will remove 5% of the sediment per minute. What will be the concentration of the sediment still remaining in the water as it exits the tank? After several days of operation, estimate the total number of pounds of sediment mixed in with all 40,000 gallons of water in the tank at any given time.

7. The capacity of a car's radiator is 8 liters. The radiator is full of a dirty, two-year-old 50/50 solution of antifreeze and water. The owner, deciding that it is time to replace the old antifreeze with new antifreeze, starts the car, removes the radiator cap and the drain plug, and inserts a garden hose into the radiator. The faucet is turned up just enough so that the radiator remains full, without overflowing, while the solution runs out at an estimated rate of 4 liters per minute. Since the engine is kept idling, the solution in the radiator is continually stirred. Find the minimum time that is needed to ensure that at least 99% of the old antifreeze is removed.

8. Tom's water heater is located in his car garage, which is kept at a constant 80°F during the summer months. The water heater is anchored completely above a rectangular-shaped overflow pan with a base measuring 20 inches by 30 inches. One day Tom notices that the overflow pan contains water; so he looks and finds a leak in the water heater which he repairs. Although he estimates that there are still 3 inches of water in the pan to empty, he leaves that part of the job to evaporation. He hypothesizes that the water will evaporate at a rate

proportional to the volume of water in the pan. Although he could not know this, it turns out that the outflow pipe is almost completely stopped up; however, 2 cubic inches of water manage to drain from the pan per week. Tom observes that the height of the water in the pan after 5 days is 2 inches.

(a) Based on the given information and using Tom's hypothesis about the evaporation rate, model the above situation with an initial value problem whose solution can be used to predict the height of the water in the pan as a function of time.

(b) Find the solution of the model in part (a).

(c) Do you agree with the model in part (a)? If not, suggest a different one; then solve it and compare solutions.

9. After an operation, a glucose solution, with a concentration of b milligrams of glucose per milliliter of solution, is infused into a patient's blood at a constant rate of 80 milliliters per hour. The volume of blood in the patient is about 5 liters. Assume the patient's body eliminates the glucose from the blood at a rate proportional to the amount of glucose present in the blood and that the volume of blood remains constant.

(a) Write a differential equation for the concentration $x(t)$ of the glucose in the blood at time t hours.

(b) Let x_0 denote the concentration of glucose already in the blood at the time the intravenous feeding begins. Find a formula for the concentration $x(t)$ by solving the differential equation found in part (a). What happens to the concentration of the glucose as time increases?

Radioactive Decay

10. Verify that $A(t) = A_0 e^{-\lambda t}$ is the solution of the radioactive decay law (6.3.7).

11. A measure of the rate of decay of a radioactive isotope is its ***half-life***, defined as the time T required for half of a given amount of the isotope to decay. Express the half-life T in terms of the decay constant λ.

12. Write the solution of the radioactive decay law in terms of the exponential base 2 and its exponent in terms of the half-life T instead of the decay constant λ.

13. In 238 grams of naturally occurring uranium metal, there are approximately 6.02×10^{23} nuclei. Approximately 99.3% of uranium metal consists of the radioactive isotope uranium-238. The half-life of ^{238}U is 4.47 billion years.[14] Using these facts, answer the following questions:

(a) Find the decay constant λ of ^{238}U.

(b) Find the probability that a ^{238}U nucleus will decay during a one second interval.

(c) Find the approximate number of nuclei of ^{238}U in one gram of uranium metal.

(d) Find the approximate number of decays per second of ^{238}U in one gram of uranium metal.

14. A tank contains initially M grams of a radioactive isotope with a half-life of T days. Assume a process has been developed which can deliver additional isotope to the tank at a constant rate of m grams of isotope per day. Find a differential equation model for the mass of radioactive isotope in the tank as a function of time. Solve the differential equation. Predict what will happen in the long run.

---

[14] See Halliday-Resnick [43, p. 1082].

Malthusian Population Models

15. Use the solution of the Malthusian model (6.3.11) to find the factor by which the population increases during successive time intervals of equal duration, say T years. Show that the sequence formed by computing the population at the times $t = 0, T, 2T, 3T, \ldots$ is a geometric sequence.
16. The per capita growth rate of the world's human population in 2025 was approximately 0.85% per year. If this growth rate remains constant, how long will it take for the world's population to double according to the Malthusian model?
17. Thomas Malthus claimed that the population of the United States during the 18th century had been doubling itself every 25 years (see Sect. 6.3.2). Estimate the per capita growth rate using the Malthusian model of population growth.
18. Assume that the population of a certain species changes according to the Malthusian model $dp/dt = rp$, where r is the per capita birth rate minus the per capita death rate. Furthermore, assume that there is a ***migration*** (immigration minus emigration) of this species into the population at a constant rate m. Find the differential equation modeling the species' population. Find an explicit solution of the differential equation.
19. The ***annual growth rate*** of the population of a country is $\Delta p/p$, where Δp is the change in population during the course of a year and p is the population at the beginning of the year. If a certain country has a per capita growth rate of r (cf. (6.3.9)), what is its annual growth rate?
20. Predict the population of a certain city 10 years from now if its present population is 300,000 and growing at an annual rate of 2% per year. See Problem 19.

Logistic Population Models

21. Use curve-sketching techniques to sketch graphs of the logistic equation (6.3.13) for three different initial populations p_0. Choose values so that $0 < p_0 < \frac{a}{2b}$, $\frac{a}{2b} < p_0 < \frac{a}{b}$, and $p_0 > \frac{a}{b}$.
22. Suppose a biologist uses the logistic model

$$\frac{dp}{dt} = 0.08p - 0.00004p^2$$

to estimate the population $p(t)$ of smallmouth bass in a city lake, where t denotes the number of years since 2005.

(a) For what values of p is the population increasing?
(b) For what values of p is the population decreasing?
(c) What is the carrying capacity of the lake?
(d) For what value(s) of p is the population growing the fastest?
(e) The city council is considering a proposal to allow fishing in the lake. One plan would allow fishermen to harvest a total of $h(t) = 100t + 5t^2$ smallmouth bass from the lake after t years. In other words, the plan would allow a total of $h(1) = 105$ fish to be harvested after the first year, a total of $h(2) = 220$ fish after 2 years, etc. Modify the logistic model accordingly.

23. The logistic equation (6.3.13) is a Bernoulli equation. Use the method shown in Sect. 5.6 in the previous chapter to find the solution that satisfies the initial condition $p(t_0) = p_0$. Compare the result with (6.3.14).

24. Imagine that 5000 white-tailed prairie dogs inhabit a colony in a certain area of Colorado. In the absence of predators, the per capita growth rate of these prairie dogs is about 20% per year. However, each month predators, such as coyotes, hawks, and snakes, kill about 50 prairie dogs in the colony. Set up an initial-value problem to model the population of the colony. Then solve it to predict the future population of the colony. Use the solution to predict the population in 5 years.

Tumor Growth Models

25. The ***Gompertz model*** for predicting the volume of a tumor at time t is

$$\frac{dV}{dt} = r(t)V, \quad \frac{dr}{dt} = -br$$

where V denotes the volume and b is a parameter.

(a) Show that the Gompertz model can be written in one of the following two forms:

$$\frac{dV}{dt} = (a - b \ln V)V \quad \text{or} \quad \frac{dV}{dt} = ae^{-bt}V$$

where a is a constant.

(b) Find the solution of either equation in (a) satisfying the initial condition $V(0) = V_0$, where V_0 designates a given volume.

(c) Use (b) to express the maximum possible volume in terms of a, b, and V_0.

26. The ***von Bertalanffy model*** for predicting the volume V of a tumor at time t is

$$\frac{dV}{dt} = aV^{\gamma} - bV,$$

where a, b, and γ are constants. Assume $0 < \gamma < 1$.

(a) If the tumor has volume V_0 at $t = 0$, determine its volume for $t > 0$.

(b) Use (a) to express the maximum possible volume in terms of a, b, γ, and V_0.

Deposit Accounts

27. Find the balance at the end of a year if a one-time deposit of \$1000 is made at a bank that pays an interest rate of 5% compounded continuously. Assume that no withdrawals are made. What will the balance be after 5 years?

28. A one-time deposit of \$200 is made to an account that pays interest at an annual rate that is compounded continuously. If the balance of the account after 1 year is \$210, what will it be after 3 years? What is the annual interest rate?

29. Use the compound interest formula (6.3.30) to compare the balances at the end of a year if two deposits, each in the amount of \$1000, are made to separate savings accounts both of which pay interest at an annual rate of 5%; however, for one of the accounts interest is compounded quarterly whereas in the other one it is compounded daily. What will the balances be after 5 years if no withdrawals are made? Compare the balances with those of Problem 27.

30. Explain why formula (6.3.29) may be regarded as the result of applying Euler's Method to (6.3.16) with a step size of $1/m$.

31. The APY when interest is compounded m times a year is given by formula (6.3.31). When interest is compounded continuously, then the APY is given by formula (6.3.19). Explain why these two formulas are compatible with one another.
32. Find a continuous compound rate that is equivalent to the following annual percentage rates (after the interest has been credited at the end of a month):

    (a) 10% compounded monthly;

    (b) 10% compounded daily.

    Expand $m \ln\left(1 + \frac{r}{m}\right)$ in a Maclaurin series. Use this to contrast "discrete compounding of interest" with "continuous compounding of interest."
33. A deposit account is opened with a check for \$5000. The account pays interest at an annual rate of 4% compounded continuously. Six months after the opening of the account, a \$100 withdrawal is made. From then on, additional withdrawals, each in the amount of \$100, are made every six months. Find the balance in the account at the end of two years, after the fourth withdrawal of \$100.

Computer Algebra System Problems

34. Corroborate your pencil-and-paper graphs of the logistic equation (6.3.13) in Problem 21: set $a = 6$ and $b = 2$ and use a CAS to plot three graphs of (6.3.13), all on the same set of axes, for the initial values that you chose in Problem 21.

Chapter 7
Modeling Abrupt Changes

A discontinuity in the forcing term ... corresponds to an abrupt change in the nature of the external force .... A discontinuity in the coefficient corresponds to an abrupt change in the nature of the system as we go from one region to another. It turns out that it is very much simpler to portray these abrupt transitions by means of discontinuous functions than by means of rapidly varying functions ....

Bellman and Cooke[1]

Abstract This chapter introduces the *unit step function* (or *Heaviside function*) and the *Dirac delta function* so that students will be comfortable with these functions before the Laplace transform is introduced in Chap. 11. This is to avoid students having to grasp and assimilate new information while at the same time learning about Laplace transform methods. Moreover, it is a continuation of the modeling in Chap. 6 of the balance of a deposit account, such as a savings or money market account, where the accrued interest is credited to the account periodically. It is shown how to set up a one-compartment model of a loan, such as a car loan or a home mortgage, in order to model the amortization of the loan with a differential equation using the Dirac delta function.

The chapter begins with the narrative of a first-time gambler who has just arrived at a casino to try her luck at the slot machines. She has access to only \$100 while at the casino. A function called $x(t)$ is defined to be her "net worth" at time t, where $t = 0$ represents the moment she sits down at a slot machine to play. Thus, $x(0) = 100$ dollars. It is shown how to express $x(t)$ in terms of unit step functions to reflect the abrupt changes in her net worth due to any winnings and losses while gambling.

Other examples are also presented illustrating how unit step functions can be used to model abrupt changes, which are inherent features of sawtooth voltage pulses,

[1] See [14, ch. 2].

© The Author(s), under exclusive license to Springer Nature Switzerland AG 2026
L. C. Becker, *Ordinary Differential Equations: Concepts, Methods, and Models*,
https://doi.org/10.1007/978-3-032-15150-6_7

impulsive forces, rectangular wave functions, piecewise-defined functions, just to name a few.

The casino scenario is also used to introduce the Dirac delta function. This is achieved by formally differentiating the net worth function $x(t)$, which results in a differential equation with terms involving derivatives of the unit step function. It is shown that if the derivative of the unit step function is defined to have certain properties, such as the sifting property, then the net worth function can be recovered from the differential equation. This derivative is known as the Dirac delta function. Despite its name, the Dirac delta function is not really a function—at least not in the classical sense. Nonetheless, because of its unusual properties, it can be used to model situations involving abrupt changes, such as the impulsive force acting on a pitched baseball when it comes into contact with a bat and the sudden drops or increases in the population of an organism.

7.1 One-Armed Bandits

Imagine someone named Lily waking up from a nap one Saturday afternoon tired of being stuck in the same old routine and eager to do something totally different. Feeling adventuresome and confident enough to put Lady Luck to the test, she heads for Tunica, Mississippi,[2] a relatively short drive from Memphis with a total of \$100 in her wallet to try her hand at some casino games. After arriving, she finds herself not in the town of Tunica but rather in a sea of soybean and cotton fields incongruously interspersed with golf courses and islands of casinos and hotels. Driving around, she spots a casino whose colors and theme appeal to her feeling of wild abandon. After parking, Lily marches in confident of a big win. Awestruck by the throngs of people feeding their hard-earned money to a vast array of voracious slot machines, she decides to try her hand at doing the same thing. It looks simple enough. Feeling more comfortable with the old-fashioned mechanical slot machines rather than with the newfangled computer random-number-generator kind, she searches for a room in the casino that has some coin-operated machines with pull levers, pejoratively known as *one-armed bandits*. After purchasing \$100 worth of five-dollar tokens and dumping them into an oversized plastic cup, she spots her opponent—one promising easy money. She strides over to it and sits down to begin yet another struggle between woman and machine. Eyeing her mechanical opponent and wiggling into a comfortable playing position, she inserts a token into a slot and pulls the lever. Immediately, the combined action of the machine's mechanical mechanisms and electronic circuits set the reels of bars, bells, and fruit whirling. One by one each of the reels stop with the cherries lining up perfectly in a row. Lily waits with bated breath for something to happen. To her delight, 10 tokens tumble forth from the bowels of the machine!

[2] Tunica county, located south of Memphis, is the site of a number of casinos and hotels.

Enticed with the prospect of more and even bigger winnings, Lily continues to play. Suppose the time from her sitting down to inserting the token and pulling the lever was just shy of a minute—and the time it took the machine to release the 10 tokens, or \$50, was almost immediate. In other words, Lily won \$50 at approximately $t = 1$ minute, where $t = 0$ marks the instant she sat down to play. Let $x(t)$ denote her "net worth" at the casino after t minutes of play, which is the original \$100 plus any winnings minus any losses up to that time. Since she came to the casino with only \$100, $x(0) = 100$ dollars. In fact, her net worth was \$100 up to $t = 1$; but then shortly thereafter, it increased to $\$100 - \$5 + \$50 = \$145$. We can model this mathematically with the piecewise-defined function

$$x(t) = \begin{cases} 100, & 0 \leq t < 1 \\ 145, & t > 1. \end{cases} \tag{7.1.1}$$

This function states that Lily's net worth from $t = 0$ up to $t = 1$ was \$100; thereafter, it jumped to \$145, as indicated in Fig. 7.1, and will remain at that amount until she inserts some more tokens. Note, however, it does not say anything about her net worth exactly at $t = 1$. From the moment Lily inserted a token until the moment the machine released 10 tokens, her net worth was in a state of flux. Why? Because during those intervening moments, the machine's electronic circuits were determining whether to release tokens, and if so, how many. And so during this brief span of time, her net worth was indeterminable. All that can be said for certain is that it was at least \$95. The piecewise-defined function $x(t)$ defined by (7.1.1) is a mathematical model of the real situation. Its graph depicts an ***abrupt change*** in Lily's net worth that occurs at the time $t = 1$ minute: from \$100 to \$145. Strict inequalities are used in (7.1.1) because of the indeterminacy in her net worth at $t = 1$. The function $x(t)$ is continuous except for the ***jump discontinuity*** at $t = 1$.

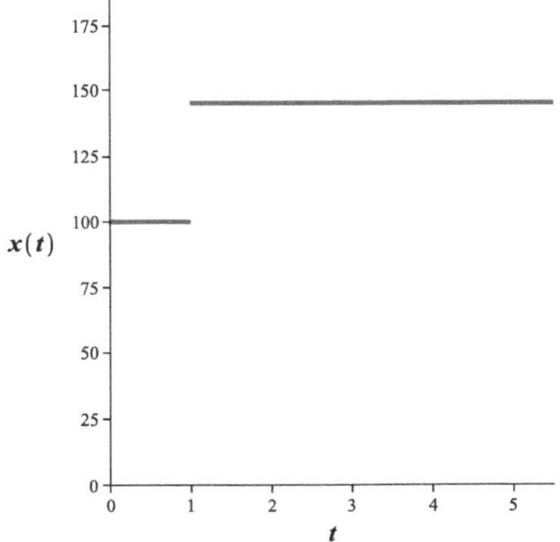

Fig. 7.1 Lily's net worth after one pull

7.2 Step Functions

In order to more effectively handle functions with jump discontinuities, such as the net worth function $x(t)$ in the previous section, we will learn how to rewrite them more concisely using so-called step functions. A *unit step function* is a function that is equal to 0 up to a certain point after which it then abruptly jumps to 1 and remains at this value forever thereafter. In engineering literature and in the computer algebra system *Maple*, a unit step function is also called a *Heaviside function*, after the noted English engineer Oliver Heaviside (1850–1925).[3]

Unit Step Function

Definition 7.2.1 The ***unit step function at*** T (or ***Heaviside function at*** T) is the piecewise-defined function that is denoted by $u(t - T)$ and defined as follows:

$$u(t - T) := \begin{cases} 0, & t < T \\ 1, & t > T. \end{cases} \tag{7.2.1}$$

Example 7.2.1 The unit step function at $T = 2$ is

$$u(t - 2) = \begin{cases} 0, & t < 2 \\ 1, & t > 2. \end{cases}$$

The unit step function at the origin ($T = 0$) is

$$u(t) = \begin{cases} 0, & t < 0 \\ 1, & t > 0. \end{cases} \tag{7.2.2}$$

A comparison of their graphs (see Fig. 7.2) shows that the graph of $u(t - 2)$ is the graph of $u(t)$ shifted 2 units to the right. This should come as no surprise. Recall that for a function f and a positive constant c, the graph of $f(t - c)$ is the graph of $f(t)$ shifted c units to the right. ♦

In some applications, a step function that jumps in value at a point $t = T$ is either not defined there or its value there is indeterminable, as in the casino example. That

[3] Oliver Heaviside was the main contributor and popularizer of a branch of mathematics that bears his name: Heaviside's operational calculus. He contributed significantly to the understanding and application of electromagnetic theory and predicted the existence of the ionosphere in the early 1900s.

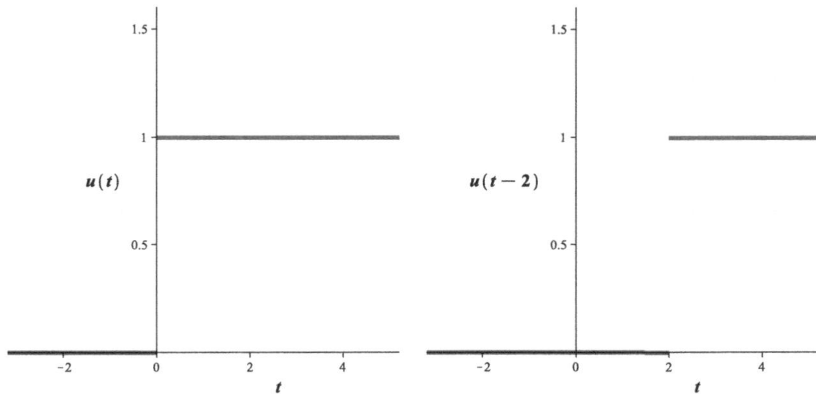

Fig. 7.2 Comparison of the unit step functions at $T = 0$ and $T = 2$

is our reason for not defining $u(t-T)$ at $t = T$. But then again, in other applications, we may need a step function that has a value at the location of the jump. For instance, in an example that we will consider shortly, we will use a unit step function that has the value 1 at the jump. In order to distinguish such a unit step function from $u(t-T)$, we employ the notation $H(t-T)$. That is, the function $H(t-T)$ is defined by

$$H(t-T) := \begin{cases} 0, & t < T \\ 1, & t \geq T. \end{cases} \tag{7.2.3}$$

Note that the only difference between these two functions is that $H(t-T) = 1$ at $t = T$ whereas $u(t-T)$ is undefined there. We should also point out that whether or not a unit step function is defined at a jump discontinuity varies from author to author. When it is, it is sometimes defined to be $1/2$ rather than 1 (see [1, p. 1020]). Some authors use the notation $u(t-T)$ rather than $H(t-T)$ to mean the right-hand side of (7.2.3). But the command "Heaviside($t - T$)" in *Maple* is the same as our $u(t-T)$.

Once again consider the casino example. With Definition 7.2.1, we can express Lily's net worth, given by (7.1.1), succinctly in terms of the unit step function $u(t-1)$. For $t \geq 0$,

$$x(t) = 100 + 45u(t-1). \tag{7.2.4}$$

Any function with a sequence of jump discontinuities varying in height can also be written in terms of step functions. To demonstrate this, let us suppose that Lily decides to insert more than 1 token. Many slot machines accept up to 3 tokens at a time: the more inserted increases the size of the winnings. Suppose 4 minutes after

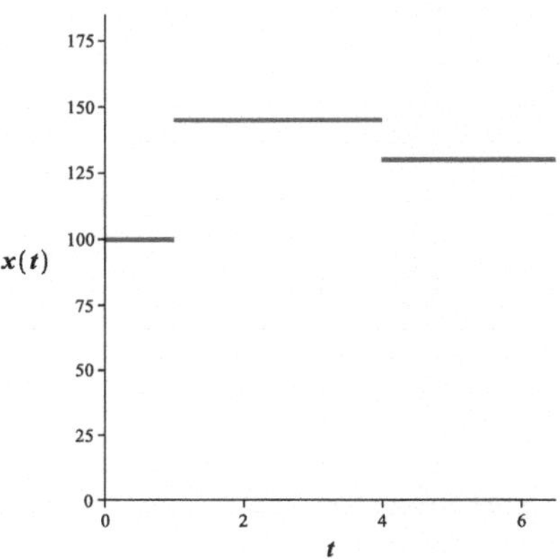

Fig. 7.3 Lily's net worth after two pulls

Lily sat down to play, she inserts 3 tokens but loses this time, thereby losing \$15. The graph of her net worth after 4 minutes of play is shown in Fig. 7.3, which shows it has decreased to \$130.In order to express $x(t)$ in terms of unit step functions, let us first write it out in detail as follows, noting the location and size of the jumps in net worth:

$$x(t) = 100 + \begin{Bmatrix} 0, & t < 1 \\ 45, & t > 1 \end{Bmatrix} - \begin{Bmatrix} 0, & t < 4 \\ 15, & t > 4 \end{Bmatrix}. \tag{7.2.5}$$

Now observe (7.2.5) is the same as

$$x(t) = 100 + 45 \cdot \begin{Bmatrix} 0, & t < 1 \\ 1, & t > 1 \end{Bmatrix} - 15 \cdot \begin{Bmatrix} 0, & t < 4 \\ 1, & t > 4 \end{Bmatrix}. \tag{7.2.6}$$

Written in terms of unit step functions, this becomes

$$x(t) = 100 + 45 \cdot u(t - 1) - 15 \cdot u(t - 4).$$

Next, let us take a look at a number of examples in other contexts.

Example 7.2.2 When a baseball player smacks a home run or hits a line drive, the bat exerts a very large force on the baseball dramatically changing its motion. However the time that the bat is in actual contact with the ball is extremely short. This is an example of an ***impulsive force***, which by definition is "a very large force acting on a body for a very short time." Such a force appreciably alters the body's motion. A graph illustrating the magnitude of a time-varying impulsive force $F(t)$ is shown in Fig. 7.4. Since it is usually impossible to ascertain the magnitude of an impulsive force at every instant, we compute its average value instead. This is then used to model the force and its effects on the subsequent motion of a body.

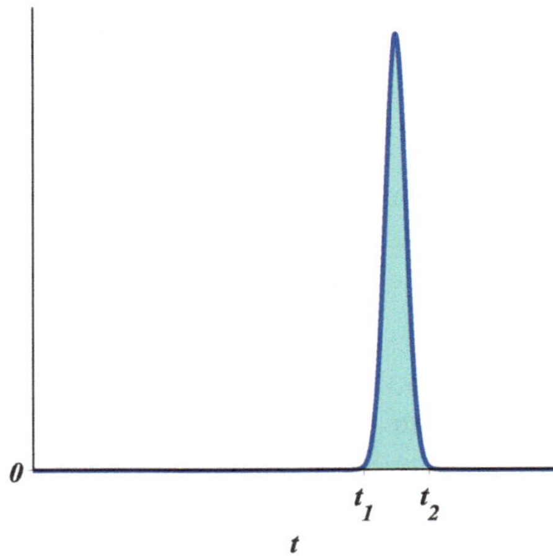

Fig. 7.4 An impulsive force $F(t)$

To simplify matters, let us suppose that an impulsive force is the only force acting on a body and that it always points in one direction. Recall that a typical problem in calculus is to compute the average value of a function over an interval, which is done by integrating the function over the interval and dividing by the interval's length. For an impulsive force $F(t)$ in contact with a body between the times t_1 and t_2, its average value F_{avg} over this time interval is

$$F_{avg} = \frac{1}{t_2 - t_1} \int_{t_1}^{t_2} F(t)\, dt. \tag{7.2.7}$$

The integral itself is known as the ***impulse*** of the force $F(t)$ and is denoted by J in many physics textbooks. But how can we compute F_{avg} using (7.2.7) if $F(t)$ is not known at every instant? The answer is that it can be found indirectly by means of Newton's second law of motion. For a body of constant mass m,

$$F(t) = mv'(t) = \frac{d}{dt}\big[mv(t)\big]$$

where $v(t)$ is the velocity of the body at time t. Hence, the impulse J is

$$J = \int_{t_1}^{t_2} F(t)\, dt = \int_{t_1}^{t_2} \frac{d}{dt}\big[mv(t)\big]\, dt = mv(t_2) - mv(t_1). \tag{7.2.8}$$

Since the product of the mass of a body and its velocity is the ***momentum*** of the body, we see that the impulse is the change in the body's momentum.

The upshot is that the average value of an impulsive force $F(t)$ cannot be computed using formula (7.2.7) alone, unless for some reason the force is known at every instant. But it can be computed if (7.2.7) is used in conjunction with (7.2.8).

Combining both formulas, we have

$$F_{avg} = \frac{mv(t_2) - mv(t_1)}{t_2 - t_1} = \frac{\Delta p}{\Delta t},$$

where Δp denotes the change in momentum $mv(t_2) - mv(t_1)$ over the time interval $\Delta t = t_2 - t_1$. Moreover, the impulse is

$$J = F_{avg} \cdot \Delta t = \Delta p.$$

In other words, an impulsive force, such as the one imagined in Fig. 7.4, can be modeled with a rectangular pulse of height F_{avg} and width Δt, which is depicted in Fig. 7.5. This model has the same impulse as that of the actual force: the area of the rectangular pulse is the same as the area of the shaded region in Fig. 7.4. Let us find the function whose graph is this rectangular pulse. It follows from Fig. 7.5 that

$$\begin{aligned} F(t) &= \begin{Bmatrix} 0, & t < t_1 \\ F_{avg}, & t > t_1 \end{Bmatrix} - \begin{Bmatrix} 0, & t < t_2 \\ F_{avg}, & t > t_2 \end{Bmatrix} \\ &= F_{avg} \cdot \begin{Bmatrix} 0, & t < t_1 \\ 1, & t > t_1 \end{Bmatrix} - F_{avg} \cdot \begin{Bmatrix} 0, & t < t_2 \\ 1, & t > t_2 \end{Bmatrix}. \end{aligned}$$

Writing this in terms of unit step functions, we have

$$F(t) = F_{avg} \cdot u(t - t_1) - F_{avg} \cdot u(t - t_2) = F_{avg}\big[u(t - t_1) - u(t - t_2)\big].$$

The bracketed quantity is called, among other things, a *box function*. This particular function is handy when it comes to dealing with periodic functions. More will be said about this in Chap. 11, where we will learn how to use Laplace transforms to solve linear ordinary differential equations with constant coefficients. ♦

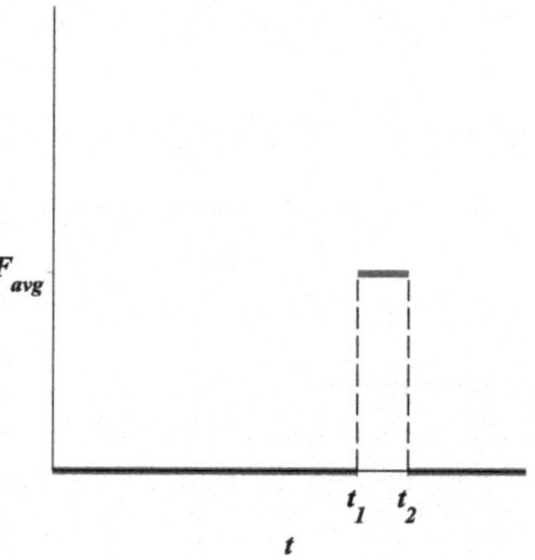

Fig. 7.5 Rectangular pulse model of the impulsive force $F(t)$

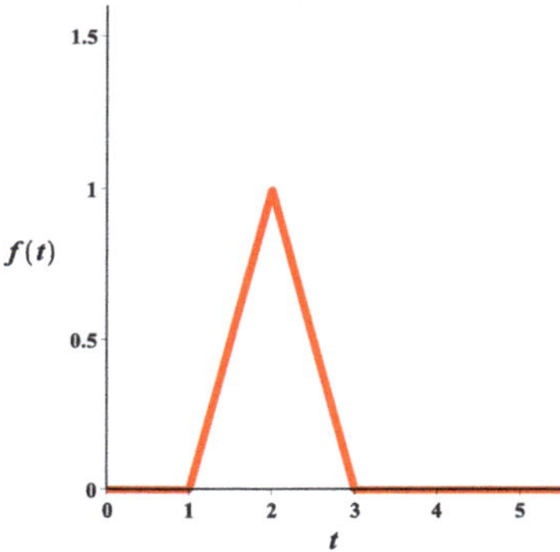

Fig. 7.6 A triangular pulse

Example 7.2.3 The graph of the triangular pulse in Fig. 7.6 defines a function $f(t)$ that is defined for all $t \geq 0$. Find the formula for f and express it in terms of the appropriate unit step functions.

Solution It follows from the graph of f that

$$f(t) = \begin{cases} 0, & \text{if } 0 \leq t < 1 \\ t-1, & \text{if } 1 \leq t < 2 \\ 3-t, & \text{if } 2 \leq t < 3 \\ 0, & \text{if } t \geq 3\,. \end{cases}$$

So as to better see how to express f in terms of unit step functions, let us rewrite it as follows:

$$\begin{aligned} f(t) &= 0 + \left\{ \begin{matrix} 0, & t<1 \\ t-1, & t \geq 1 \end{matrix} \right\} + \left\{ \begin{matrix} 0, & t<2 \\ -2t+4, & t \geq 2 \end{matrix} \right\} + \left\{ \begin{matrix} 0, & t<3 \\ t-3, & t \geq 3 \end{matrix} \right\} \\ &= (t-1)\cdot \left\{ \begin{matrix} 0, & t<1 \\ 1, & t \geq 1 \end{matrix} \right\} - 2(t-2)\cdot \left\{ \begin{matrix} 0, & t<2 \\ 1, & t \geq 2 \end{matrix} \right\} + (t-3)\cdot \left\{ \begin{matrix} 0, & t<3 \\ 1, & t \geq 3 \end{matrix} \right\}. \end{aligned}$$

Since $f(t)$ is actually defined at the points $t = 1, 2$, and 3, we use (7.2.3) instead of (7.2.1). Accordingly,

$$f(t) = (t-1)H(t-1) - 2(t-2)H(t-2) + (t-3)H(t-3). \qquad \blacklozenge$$

Example 7.2.4 *Sawtooth pulses* and *sawtooth waves*[4] are important when it comes to modeling the variations in voltages in certain types of electrical circuits. Consider the graph of the function $V(t)$ in Fig. 7.7, which consists of two successive sawtooth

[4] We see from Figs. 7.7 and 7.8 that the shapes of these graphs resemble the teeth of a handsaw.

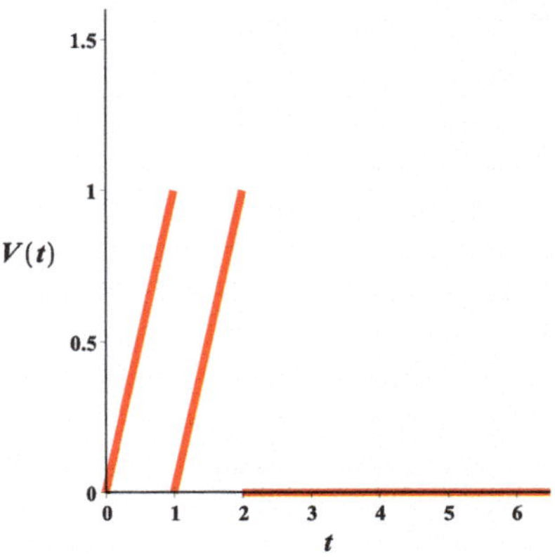

Fig. 7.7 Two sawtooth pulses

pulses from $t = 0$ to $t = 2$, after which its value drops to 0 for $t > 2$. In particular, $V(0) = 0$ and $V(t)$ is defined for all $t \geq 0$ except at $t = 1$ and $t = 2$. Express V in terms of appropriate unit step functions.

Solution From the graph of V in Fig. 7.7,[5] we see that

$$V(t) = \begin{cases} t, & \text{if } 0 \leq t < 1 \\ t - 1, & \text{if } 1 < t < 2 \\ 0, & \text{if } t > 2\,. \end{cases}$$

In order to express V in terms of the appropriate unit step functions, let us first write it as follows:

$$V(t) = t + \left\{ \begin{array}{cc} 0, & t < 1 \\ -1, & t > 1 \end{array} \right\} + \left\{ \begin{array}{cc} 0, & t < 2 \\ 1 - t, & t > 2 \end{array} \right\}$$

$$= t - \left\{ \begin{array}{cc} 0, & t < 1 \\ 1, & t > 1 \end{array} \right\} + (1 - t) \cdot \left\{ \begin{array}{cc} 0, & t < 2 \\ 1, & t > 2 \end{array} \right\}.$$

It follows that

$$V(t) = t - u(t - 1) + (1 - t) \cdot u(t - 2).$$

♦

[5] From now on, regard functions as undefined wherever jumps in values occur, unless otherwise noted.

Fig. 7.8 A sawtooth wave

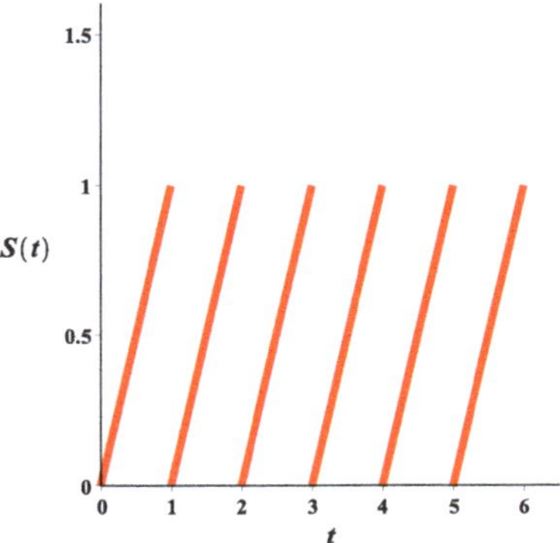

Example 7.2.5 The graph of the function V in the previous example consists of only two sawtooth pulses. Now consider the graph of the function S shown in Fig. 7.8, which consists of an infinite succession of sawtooth pulses. It is an example of a *sawtooth wave*. Find the formula for S. Express it in terms of appropriate unit step functions.

Solution The function defined by the sawtooth wave in Fig. 7.8 is

$$S(t) = \begin{cases} t, & \text{if } 0 \le t < 1 \\ t-1, & \text{if } 1 < t < 2 \\ t-2, & \text{if } 2 < t < 3 \\ t-3, & \text{if } 3 < t < 4 \\ \vdots & \vdots \end{cases}$$

$$= t + \left\{ \begin{array}{cc} 0, & t < 1 \\ -1, & t > 1 \end{array} \right\} + \left\{ \begin{array}{cc} 0, & t < 2 \\ -1, & t > 2 \end{array} \right\} + \left\{ \begin{array}{cc} 0, & t < 3 \\ -1, & t > 3 \end{array} \right\} + \cdots .$$

Therefore,

$$S(t) = t - u(t-1) - u(t-2) - u(t-3) - \cdots = t - \sum_{j=1}^{\infty} u(t-j)$$

for $t \ge 0$.

Remark In this example, $S(0) = 0$; so the first term in the series is t. If S were undefined at $t = 0$, then the first term of the series would be $tu(t)$. ♦

Example 7.2.6 Express the piecewise-defined function

$$g(t) = \begin{cases} \sin t, & \text{if } -\infty < t < \pi \\ \cos t, & \text{if } \pi < t < 4\pi \\ 5, & \text{if } t > 4\pi\ . \end{cases}$$

in terms of unit step functions.

Solution Since

$$g(t) = \sin t + \left\{ \begin{array}{ll} 0, & t < \pi \\ \cos t - \sin t, & t > \pi \end{array} \right\} + \left\{ \begin{array}{ll} 0, & t < 4\pi \\ 5 - \cos t, & t > 4\pi \end{array} \right\},$$

it follows that

$$g(t) = \sin t + (\cos t - \sin t)u(t - \pi) + (5 - \cos t)u(t - 4\pi).$$ ♦

Example 7.2.7 The graph of the rectangular wave function $r(t)$ shown in Fig. 7.9 consists of an infinite succession of positive and negative rectangular pulses for $t > 0$, where $r(t)$ is undefined at $t = 0, 1, 2, 3, \ldots$. Express $r(t)$ in terms of an infinite series of unit step functions.

Solution Clearly,

$$\begin{aligned} r(t) &= \left\{ \begin{array}{l} 0,\ t<0 \\ 2,\ t>0 \end{array} \right\} + \left\{ \begin{array}{l} 0,\ t<1 \\ -4,\ t>1 \end{array} \right\} + \left\{ \begin{array}{l} 0,\ t<2 \\ 4,\ t>2 \end{array} \right\} + \left\{ \begin{array}{l} 0,\ t<3 \\ -4,\ t>3 \end{array} \right\} + \cdots \\ &= 2u(t) - 4u(t-1) + 4u(t-2) - 4u(t-3) + \cdots \end{aligned}$$

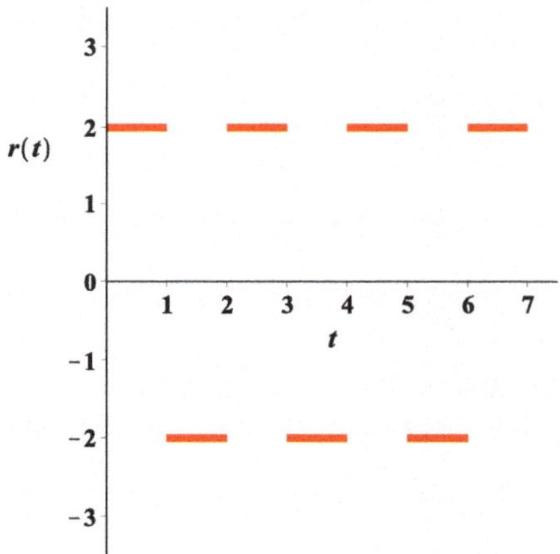

Fig. 7.9 A rectangular wave

for $t > 0$. Or, using summation notation,

$$r(t) = 2u(t) + 4\sum_{n=1}^{\infty}(-1)^n u(t-n).$$

Observe that

$$r(t+2) = r(t)$$

for $t \in (0, \infty)$, except at the points $t = 0, 1, 2, 3, \ldots$ where $r(t)$ is undefined. In other words, $r(t)$ is *periodic on the interval* $(0,\infty)$ *with period* 2. Furthermore, $r(t)$ is also *antiperiodic on* $(0,\infty)$ *with antiperiod* 1. That is to say,

$$r(t+1) = -r(t),$$

except at $t = 0, 1, 2, 3, \ldots$. ♦

7.3 The Dirac Delta Function

7.3.1 Introductory Example

Now that we have seen a number of examples showing how to express piecewise-defined functions in terms of unit step functions, let us return to the casino example and formally differentiate[6] both sides of the net worth function

$$x(t) = 100 + 45u(t-1) \tag{7.2.4}$$

with respect to t obtaining

$$\frac{dx}{dt} = 45u'(t-1). \tag{7.3.1}$$

Since the standard notation for $u'(t-1)$ turns out to be $\delta(t-1)$, where δ is the lower case Greek letter delta, let us rewrite (7.3.1) as follows:

$$\frac{dx}{dt} = 45\,\delta(t-1). \tag{7.3.2}$$

[6] The use of the word "formally" here means that the differentiation of (7.2.4) resulting in (7.3.1) is plausible, yet questionable and requires justification. Another example is obtaining a formula by "formally manipulating" symbols, such as canceling the dt's in the product $\frac{dy}{dt} \cdot \frac{dt}{dx}$ to obtain the Leibniz version of the chain rule. But that does not really prove it. Recall that a rigorous proof of this involves difference quotients.

However, having this differential equation is pointless unless we can use it to retrieve the net worth function. In other words, if we start with (7.3.2) and know nothing at all about x other than $x(0) = 100$, is there a way to solve this equation with its solution ending up being (7.2.4)? Recall that if the derivative of a quantity is a function of the independent variable alone, then that quantity can be found by integrating both sides of the differential equation with respect to the independent variable. Even though the expression on the right-hand side of (7.3.2) does not appear to be a function in the true sense of the word (which it is not as we shall soon see), it seems plausible that we could get (7.2.4) from (7.3.2) by formally integrating both sides of this equation from $t = 0$ to a later time $t = t_1$. Let us see where the integration

$$\int_0^{t_1} x'(t)\,dt = \int_0^{t_1} 45\,\delta(t-1)\,dt$$

leads us. Applying the Fundamental Theorem of Calculus to the left-hand side and using the initial condition $x(0) = 100$, we have

$$x(t_1) - 100 = 45\int_0^{t_1} \delta(t-1)\,dt.$$

We can make the argument of x look more like a variable instead of a constant by changing the variable of integration t to a different letter, such as s, and then replacing t_1 with t. As a result, we have

$$x(t) = 100 + 45\int_0^{t} \delta(s-1)\,ds. \tag{7.3.3}$$

For this to agree with (7.2.4), we see that the value of the definite integral has to be

$$\int_0^{t} \delta(s-1)\,ds = u(t-1) = \begin{Bmatrix} 0, & t < 1 \\ 1, & t > 1 \end{Bmatrix}. \tag{7.3.4}$$

Note there is nothing special about the time $t = 1$ or the amount of money won or lost in the casino. So the general form of (7.3.1) and (7.3.2) is

$$\frac{dx}{dt} = cu'(t-T) = c\,\delta(t-T), \tag{7.3.5}$$

where c denotes the amount of money won or lost at time $t = T$. With the same kind of reasoning as above, we have

$$\int_0^{t} \delta(s-T)\,ds = u(t-T) = \begin{Bmatrix} 0, & t < T \\ 1, & t > T \end{Bmatrix}. \tag{7.3.6}$$

The mathematical entity $\delta(t-T)$ is an oddity, the likes of which we have not encountered before. It is zero everywhere, except at the single point $t=T$ where it is undefined, because

$$\delta(t-T) = u'(t-T) = \left\{\begin{matrix} \dfrac{d}{dt}(0), & t<T \\ \dfrac{d}{dt}(1), & t>T \end{matrix}\right\} = 0 \quad \text{if } t \neq T. \tag{7.3.7}$$

And yet it follows from (7.3.6) that

$$\int_0^t \delta(s-T)\,ds = 1 \quad \text{if } t > T. \tag{7.3.8}$$

That is, its integral equals 1 and not 0 for all $t > T$. How can this be? This is not possible by virtue of the fact that the definite integral of a function that is equal to 0 at all points between its limits of integration is equal to 0. And this is also the case for a function that is 0 everywhere except at a single point, where it is either undefined or has some nonzero value. So we see that $\delta(t-T)$ is not a function in the classical sense—nevertheless, it will turn out to be quite useful in modeling real-world phenomena that involve sudden or abrupt changes.

So far, we have constructed a new mathematical entity, denoted by $\delta(t-T)$, which has the unusual properties:

$$\delta(t-T) = \begin{cases} 0, & t \neq T \\ \text{undefined}, & t = T \end{cases} \quad \text{and} \quad \int_0^t \delta(s-T)\,ds = \begin{cases} 0, & t<T \\ 1, & t>T. \end{cases}$$

Abrupt Changes To investigate the mathematical construct $\delta(t-T)$ further, let us suppose that the amount of matter A in a compartment at time t, such as the amount of money in a savings account or the population of E. coli bacteria in a flask, is given by

$$A(t) = g(t) + k_1 u(t-T_1) - k_2 u(t-T_2), \tag{7.3.9}$$

where $g(t)$ is a differentiable function and k_1, k_2, T_1, T_2 are positive constants with $T_2 > T_1$. Observe that $A(0) = g(0)$. Since

$$A'(t) = g'(t) + k_1\delta(t-T_1) - k_2\delta(t-T_2), \tag{7.3.10}$$

we see that the rate of flow of matter into the compartment is $g'(t) + k_1\delta(t-T_1)$ and the rate of flow out of the compartment is $k_2\delta(t-T_2)$. In other words, if an amount k_1 of matter is "instantaneously" added to an existing amount of matter at time $t = T_1$, then include the term $k_1\delta(t-T_1)$ in the expression for $A'(t)$. Similarly, if an amount k_2 of matter is "instantaneously" removed from an existing amount of matter at time $t = T_2$, then include the term $-k_2\delta(t - T_2)$ in the expression for $A'(t)$.

Now suppose instead of being given the amount $A(t)$, we are given that

$$\frac{dA}{dt} = g'(t) + k_1\delta(t - T_1) - k_2\delta(t - T_2), \quad A(0) = g(0). \tag{7.3.11}$$

Can we solve this initial value problem and will we end up with (7.3.9)? Integrating the differential equation from 0 to t, we have

$$\int_0^t A'(s)\,ds = \int_0^t g'(s)\,ds + k_1\int_0^t \delta(s - T_1)\,ds - k_2\int_0^t \delta(s - T_2)\,ds.$$

It follows from (7.3.6) that

$$A(t) - A(0) = g(t) - g(0) + k_1u(t - T_1) - k_2u(t - T_2).$$

This simplifies to (7.3.9) since $A(0) = g(0)$.

Sifting Property Let us revisit the casino example again so that we can introduce another important property of $\delta(t - T)$. Imagine that the protagonist, Lily, keeps on playing the slot machines. Unfortunately, she eventually loses all of the money she has won including the original \$100. Although she still dreams of hitting the jackpot, she decides that gambling is far too risky and expensive. However, she cannot get over all of those people who are more than willing to gamble away their hard-earned money for hours on end. So she decides to turn over a new leaf. Instead of gambling in a casino, she will invest in the stock of a company that owns and operates casinos. This decision marks the beginning of a new monetary chapter in her life. For at that very moment, she begins setting aside money so that in a year's time she will be able to purchase stock in MGM Resorts International.

Suppose one year later, Lily purchases N shares of MGM stock for \$1000. Let the function $f(t)$ denote the value of those N shares at time t, measured in years from the time Lily began saving, where $t = 0$ designates that moment. Since she bought N shares a year later, namely, at time $t = 1$, $f(1) = 1000$ dollars. Of course, Lily will not know the value of $f(t)$ at a time $t > 1$ until that time actually arrives. As for $0 \le t < 1$, she could check via the Internet to see what N shares of MGM stock had been worth during those times. Now let us define a second function called $y(t)$ that represents the value of Lily's MGM stock portfolio at time t. For $0 \le t < 1$, $y(t) \equiv 0$ since she did not own any MGM stock during this time. For $t > 1$, the two functions f and y are identical. Thus,

$$y(t) = \begin{cases} 0, & t < 1 \\ f(t), & t > 1 \end{cases} \tag{7.3.12}$$

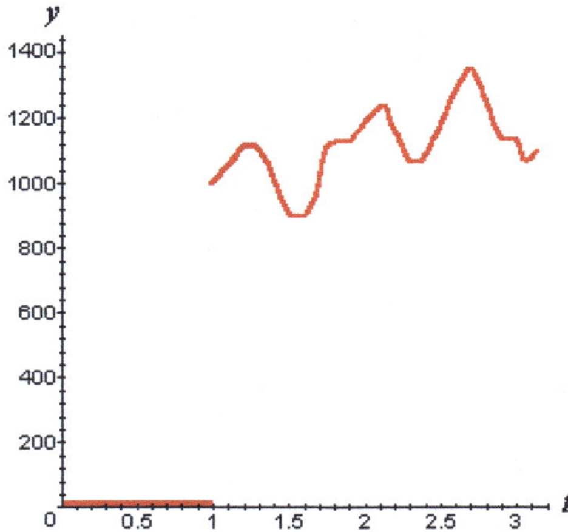

Fig. 7.10 Value of Lily's MGM stock portfolio

for $t \geq 0$. Even though $f(1) = 1000$, the value of $y(1)$ is ambiguous. Did Lily or MGM or did someone else own the N shares of stock at precisely $t = 1$? This is why there are strict inequalities in (7.3.12). The graph of $y(t)$ might look something like the one shown in Fig. 7.10. It depicts the value of the N shares fluctuating a lot for $t > 1$, which is what happens in reality since the price of a share of stock in any company changes frequently and sometimes sharply. However, for the sake of what we are about to show, let us assume that f is differentiable everywhere, even though this is not realistic due to the sudden drops or rises in the price of a share.Written in terms of a step function, (7.3.12) is

$$y(t) = f(t) \cdot \begin{Bmatrix} 0, & t < 1 \\ 1, & t > 1 \end{Bmatrix} = f(t) \cdot u(t-1) \tag{7.3.13}$$

for $t \geq 0$.

Now that we have provided some context for why an equation that looks like (7.3.13) might be useful, consider

$$y(t) = f(t) \cdot u(t-T), \tag{7.3.14}$$

where we have replaced the constant 1 with an arbitrary positive constant T. Formally taking the derivative of (7.3.14) with respect to t, we obtain

$$\frac{dy}{dt} = f(t) \cdot \delta(t-T) + f'(t) \cdot u(t-T), \tag{7.3.15}$$

where $\delta(t-T) = u'(t-T)$. As we did with (7.3.6), let us reverse our steps in an attempt to recover (7.3.14) from the differential equation (7.3.15). Integrating both sides of this equation from 0 to t, we have

$$\int_0^t y'(s)\,ds = \int_0^t f(s)\cdot\delta(s-T)\,ds + \int_0^t f'(s)\cdot u(s-T)\,ds.$$

If $y(0) = 0$, this becomes

$$y(t) = \int_0^t f'(s)\cdot u(s-T)\,ds + \int_0^t f(s)\cdot\delta(s-T)\,ds. \tag{7.3.16}$$

When $0 \le t < T$, the step function $u(s-T) = 0$, since $0 \le s \le t < T$; thus the first integral on the right-hand side of (7.3.16) is zero. On the other hand, when $t > T$, it becomes

$$\int_0^t f'(s)\cdot u(s-T)\,ds = \lim_{a\to T^-}\int_0^a f'(s)\cdot 0\,ds + \lim_{b\to T^+}\int_b^t f'(s)\cdot 1\,ds = f(t)-f(T)$$

since $u(s-T) = 1$ for $s > T$. Thus (7.3.16) simplifies to

$$y(t) = \begin{cases} \int_0^t f(s)\cdot\delta(s-T)\,ds, & t < T \\ f(t) - f(T) + \int_0^t f(s)\cdot\delta(s-T)\,ds, & t > T. \end{cases}$$

For this to agree with (7.3.14),

$$\int_0^t f(s)\cdot\delta(s-T)\,ds = \left\{\begin{matrix} 0, & t < T \\ f(T), & t > T \end{matrix}\right\} = f(T)u(t-T). \tag{7.3.17}$$

This result is known as the ***sifting property*** of $\delta(t-T)$ since it sifts out the value $f(T)$ from the rest of values of f on $[0,\infty)$ when $t > T$.

7.3.2 Difference Quotient and Geometric Viewpoint

The symbol $\delta(t-T)$ denotes $u'(t-T)$. However, as we pointed out earlier, if $\delta(t-T)$ were a classical function, (7.3.6) and (7.3.7) would be incompatible. Nonetheless, from a utilitarian point of view, there are practical benefits in treating $\delta(t-T)$ as an ordinary derivative, particularly when it comes to solving differential equations modeling abrupt changes. Let us see what the implications are if we apply the definition of the derivative to $u(t-T)$.

Consider the difference quotient of $u(t-T)$ with increment h:

$$\frac{u(t-T+h) - u(t-T)}{h}. \tag{7.3.18}$$

Of course, h can be either positive or negative. But for what we are about to do, let us assign it negative values. That is, let $h = -\tau$, where $\tau > 0$. Then (7.3.18) becomes

$$\frac{u(t-T-\tau)-u(t-T)}{-\tau} = \frac{u(t-T)-u(t-(T+\tau))}{\tau}. \tag{7.3.19}$$

Let $\delta_\tau\,(t-T)$ denote this difference quotient. This notation expresses its connection to $\delta(t-T)$ and indicates its dependence on τ. The meaning of $\delta_\tau(t-T)$ becomes clearer if we rewrite (7.3.19) as

$$\delta_\tau(t-T) = \begin{cases} \dfrac{1}{\tau}, & \text{if } T < t < T+\tau \\ 0, & \text{if } t < T \text{ or } t > T+\tau. \end{cases} \tag{7.3.20}$$

If $T = 1$ and $\tau = 1$, then $\delta_\tau(t-T) = 1$ for $1 < t < 2$; otherwise it is equal to 0. The graph of $\delta_\tau(t-1)$ with $\tau = 1$ is shown in Fig. 7.11.

If we draw a vertical line segment from the t-axis to the graph at $t = 1$ and another one at $t = 2$, we obtain a rectangular pulse of width $\tau = 1$ and height $1/\tau = 1$ enclosing an area of 1 square unit. See Fig. 7.12. If the width τ of the rectangular pulse is decreased, its height increases in accordance with (7.3.20). Figure 7.13 shows graphs of $\delta_\tau(t-1)$ for values of τ equal to 1, 0.25, 0.10, and 0.05. Note that regardless of the value of τ, the area of the corresponding rectangular pulse is always 1 square unit.

Since the symbol $\delta(t-T)$ denotes the derivative of $u(t-T)$ with respect to t, it follows that

$$\delta(t-T) = \lim_{\tau\to 0} \delta_\tau(t-T), \tag{7.3.21}$$

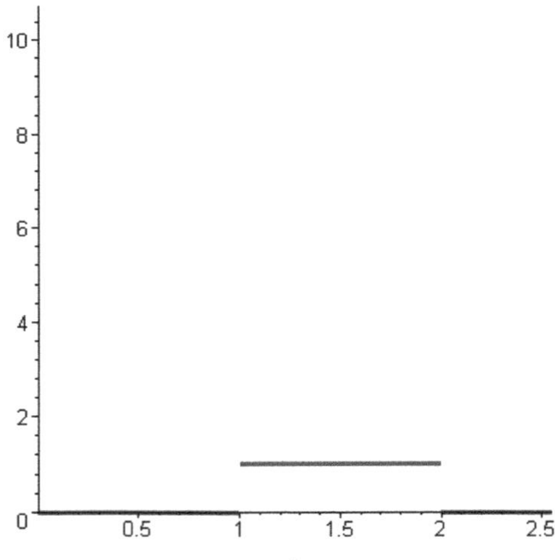

Fig. 7.11 Graph of $\delta_\tau(t-1)$ with $\tau = 1$

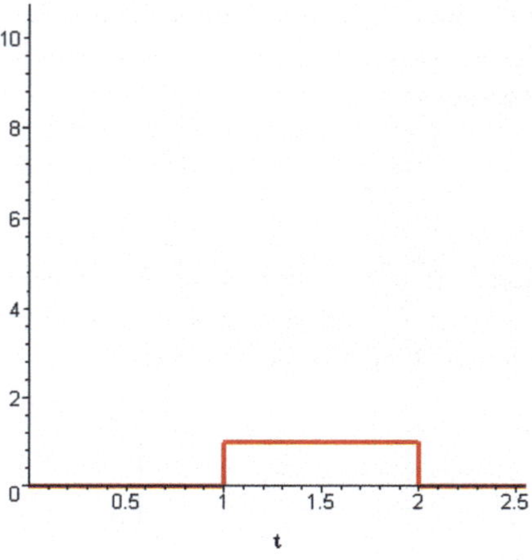

Fig. 7.12 Rectangular pulse at $\tau = 1$

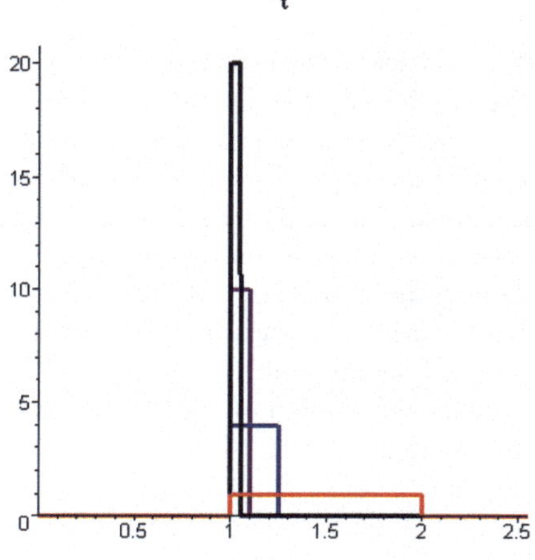

Fig. 7.13 Graphs of $\delta_\tau(t-1)$ with τ =1, 0.25, 0.10, 0.05

where $\delta_\tau(t-T)$ is given by (7.3.20), which is the difference quotient (7.3.19). We can see from (7.3.20) that the interval on which $\delta_\tau(t-T)$ is nonzero is $(T, T+\tau)$. Since any fixed $t \neq T$ lies outside this interval for sufficiently small values of τ, it follows that $\delta(t-T) = 0$ for all $t \neq T$, which we had already determined in the previous subsection.

Finally, note that the quantity $1/\tau$ in (7.3.20) does not approach a limit; rather it keeps increasing without bound. In other words, the value of $1/\tau$ becomes larger and larger as τ gets closer and closer to 0. We say that $1/\tau$ becomes infinite as τ approaches 0, which we indicate by the symbol ∞. Thus,

$$\delta(t-T) = \lim_{\tau \to 0} \delta_\tau(t-T) = \begin{cases} 0, & t \neq T \\ \infty, & t = T. \end{cases} \tag{7.3.22}$$

Depending on what is being modeled, we can view $\delta(t-T)$ as being either "infinite" or "undefined" at $t = T$.

7.3.3 *Properties*

As we have seen, some of the properties of $\delta(t - T)$ are incompatible with the definition and properties of an ordinary function. For instance, we see from (7.3.22) that

$$\delta(t - T) = 0$$

except at the single point $t = T$. But then (7.3.8) tells us that

$$\int_0^t \delta(s - T)\, ds = 1$$

instead of 0 when $t > T$. The way to resolve these conflicts is to view $\delta(t - T)$ not as an ordinary function, a rule that assigns a number to each value of t, but as an *operator*. An ***operator*** is also a rule but one that acts on a set of functions. That is, the set on which it acts or operates, known as the ***domain*** of the operator, consists of functions rather than numbers. It assigns to each function in its domain a function belonging to another set of functions. The set of assigned functions is called the ***range***. In short, an operator maps functions to functions. It is similar to a function except it is a rule dealing with functions instead of numbers. For example, the derivative is an operator in that the derivative of a function (when it exists) is another function. Other common operators involve integrals. Suppose we define an integral operator $\mathcal{I}[\,\cdot\,]$, where the "dot" is a placeholder for an integrable function, by

$$\mathcal{I}[f](t) := \int_0^t f(u)\, du.$$

Applying this operator to $f(u) = 2u$ and $g(u) = \sin u$, we obtain

$$\mathcal{I}[f](t) = \int_0^t 2u\, du = t^2; \quad \mathcal{I}[g](t) = \int_0^t \sin u\, du = 1 - \cos t.$$

Another important integral operator that we will study later on in Chap. 11 is the Laplace transform. We denote it with an uppercase script letter L, to wit: $\mathcal{L}$. The Laplace transform of a function f is defined as the improper integral

$$\mathcal{L}\{f\}(s) := \int_0^\infty e^{-st} f(t)\, dt,$$

provided this integral converges. For example, for $f(t) = \cos t$,

$$\mathcal{L}\{f\}(s) := \int_0^\infty e^{-st} \cos t \, dt = \frac{s}{s^2 + 1}.$$

In view of the previous discussions, it is clear that $\delta(t - T)$ is not a function in the classical sense. Although misleading, the notation

$$\int_0^t f(s)\delta(s - T)\, ds$$

is the way to specify that $\delta(t - T)$ operates on a function f. Even though this looks like a Riemann integral, it is not—rather, it is defined by (7.3.17):

$$\int_0^t f(s)\delta(s - T)\, ds := f(T)u(t - T).$$

Although $\delta(t - T)$ is not an ordinary function, it is for many people, such as physicists, engineers, and mathematicians, useful and indispensable. It belongs to a special class of mathematical constructs called *distributions* or *generalized functions*. The study of distributions is a specialized subject requiring an understanding of graduate-level mathematics. It is dealt with in monographs, such as the one by Zemanian [81]. The use of $\delta(t - T)$ dates back to the nineteenth century where it was applied to practical problems by some engineers, even though they did not fully understand it at the time, at least not in the rigorous sense of distributions. The British theoretical physicist and Nobel laureate Paul A. M. Dirac[7] [26, p. 58 ff.] used it in quantum mechanics in the 1930s with considerable success. Despite not being a (classical) function, it is known as the ***Dirac delta function***. The French mathematician Laurent Schwartz thoroughly justified the Dirac delta function mathematically with his systematic exposition of the theory of distributions, for which he was awarded the Fields medal in 1950.

The casino scenario, together with the geometric interpretation of the Dirac delta function, have led to the following properties: (7.3.6), (7.3.7), (7.3.17), and (7.3.22). Let us regard these as defining the Dirac delta function. However, a truly correct definition requires much more mathematical knowledge than we have at this point. Nevertheless, physicists, engineers, and system theory analysts find the following properties adequate for most of their needs.

[7] In 1928, Dirac set about to reconcile Einstein's theory of special relativity with the new theory of quantum mechanics. His mathematical results implied the existence of so-called antiparticles, which at that time no one even knew existed. In 1932, the American physicist C. D. Anderson discovered the positron, the antiparticle of the electron, from studying cosmic rays.

Dirac Delta Function

Despite its name, the Dirac delta function is not a function in the usual sense, but rather a distribution. Nevertheless, in the context of solving differential equations, it can be viewed as a function with the following unusual characteristics:

- ***Derivative.*** The Dirac delta function $\delta(t - T)$, where $T \geq 0$, models the rate of change of $u(t - T)$ with respect to t.[8] That is,

$$\delta(t - T) = u'(t - T). \tag{7.3.23}$$

At $t = T$, $\delta(t - T)$ is undefined; or in some contexts, it may be more appropriate to view its value there to be "infinitely large." Thus,

$$\delta(t - T) = \begin{cases} 0, & t \neq T \\ \text{“undefined” or } \infty, & t = T. \end{cases}$$

- ***Sifting Property.*** If a function $f(t)$ is continuous on $[0, \infty)$, then

$$\int_0^t f(s)\delta(s - T)\,ds = f(T)u(t - T) = \begin{cases} 0, & t < T \\ f(T), & t > T. \end{cases} \tag{7.3.24}$$

In particular, for $f(t) \equiv 1$,

$$\int_0^t \delta(s - T)\,ds = u(t - T) = \begin{cases} 0, & t < T \\ 1, & t > T. \end{cases}$$

7.4 Abrupt Population Changes and Amortized Loans

In this section, we use differential equations and the Dirac delta function to model real-world phenomena involving abrupt changes. In the first example, we derive a

[8] We can solve differential equations involving the Dirac delta function if we regard $\delta(t - T)$ as the ordinary derivative of $u(t - T)$. The fact of the matter, however, is that this is not really correct: $\delta(t - T)$ is the so-called *distributional derivative* of $u(t - T)$. A precise formulation of the Dirac delta function can be found in monographs dealing with the theory of distributions, such as in Zemanian [81, p. 47].

differential equation model of the population of E. coli bacteria in a flask when some of the bacteria are removed from the flask at regular intervals.

Example 7.4.1 The per capita growth rate of a population of E. coli bacteria in a flask of nutrient broth is r bacteria per hour per bacterium, where r is a constant. Let p_0 denote the number of bacteria in the flask at time $t = 0$. At time $t = T$ hours, a sample containing b bacteria is removed from the flask. Then later on, another sample of b bacteria is removed at time $t = 2T$ hours. Assume that the rate of growth of the bacteria is unaffected by these removals. Find the differential equation modeling the number $p(t)$ of bacteria in the flask at time t. Determine the number of bacteria remaining in the flask immediately after the removal of the first and second samples of bacteria.

Solution Since the per capita growth rate $\dot{p}/p$ is equal to r, the rate at which the bacteria are reproducing is rp. The number $n(t)$ of bacteria removed at time t is

$$n(t) = \left\{ \begin{array}{ll} 0, & \text{if } 0 \leq t < T \\ b, & \text{if } T < t < 2T \\ 2b, & \text{if } t > 2T\,. \end{array} \right\} = bu(t-T) + bu(t-2T).$$

Thus,

$$n'(t) = b\delta(t-T) + b\delta(t-2T).$$

A compartment model of this example, where $p(t)$ denotes the number of bacteria in the compartment (flask) at time t, is shown in Fig. 7.14. It follows from the conservation equation for one-compartment systems that

$$\frac{dp}{dt} = rp - b\delta(t-T) - b\delta(t-2T)$$

or

$$\frac{dp}{dt} - rp = -b\delta(t-T) - b\delta(t-2T).$$

Multiplying both sides of this equation by the integrating factor e^{-rt}, we get

$$e^{-rt}\frac{dp}{dt} - re^{-rt}p = -be^{-rt}\delta(t-T) - be^{-rt}\delta(t-2T).$$

Fig. 7.14 Compartment model of the population of E. coli bacteria

Thus,

$$\frac{d}{dt}\left[e^{-rt}p(t)\right] = -be^{-rt}\delta(t-T) - be^{-rt}\delta(t-2T).$$

Integrating both sides of this equation from 0 to t, we obtain

$$\int_0^t \frac{d}{ds}\left[e^{-rs}p(s)\right]ds = -b\int_0^t e^{-rs}\delta(s-T)\,ds - b\int_0^t e^{-rs}\delta(s-2T)\,ds.$$

It follows from the sifting property (7.3.24) that

$$e^{-rt}p(t) - p(0) = -be^{-rT}u(t-T) - be^{-2rT}u(t-2T).$$

Multiplying both sides of this equation by e^{rt} and using that $p(0) = p_0$, we obtain the following formula for the number of bacteria at time t:

$$p(t) = p_0e^{rt} - be^{r(t-T)}u(t-T) - be^{r(t-2T)}u(t-2T).$$

In other words,

$$p(t) = \begin{cases} p_0e^{rt}, & 0 \le t < T \\ (p_0 - be^{-rT})e^{rt}, & T < t < 2T \\ (p_0 - be^{-rT} - be^{-2rT})e^{rt}, & t > 2T. \end{cases} \tag{7.4.1}$$

Now let us determine the number of bacteria remaining in the flask immediately after b of them have been removed at time $t = T$. Because of the strict inequality at $t = T$, the population immediately before the removal of the bacteria at time T is given by the left-hand limit

$$\lim_{t\to T^-} p(t) = \lim_{t\to T^-} p_0e^{rt} = p_0e^{rT}.$$

After the removal of b bacteria, the number of bacteria in the flask drops down to $p_0e^{rT} - b$. This can also be obtained from the right-hand limit

$$\lim_{t\to T^+} p(t) = \lim_{t\to T^+} (p_0 - be^{-rT})e^{rt} = p_0e^{rT} - b.$$

Notice that this result could also have been found without the use of the Dirac delta function. Since the per capita growth rate is r, the population is governed by the Malthusian model. So the number of bacteria at time t, before any are removed, is given by the exponential function: $p(t) = p_0e^{rt}$. It follows then that there are p_0e^{rT} bacteria in the flask at time T before any of them are removed. Immediately afterwards, when b of them are removed, there are $p_0e^{rT} - b$ bacteria left in the flask.

It is left as an exercise to show that the population (7.4.1) can be found without using the Dirac delta function. See Problem 16.

Finally, the number of bacteria remaining in the flask after the second batch of b bacteria is removed at time $t = 2T$ is

$$\lim_{t\to 2T^+} p(t) = \lim_{t\to 2T^+} (p_0 - be^{-rT} - be^{-2rT})e^{rt} = p_0e^{2rT} - be^{rT} - b. \quad \blacklozenge$$

Example 7.4.2 In Example 7.2.2, we introduced the notion of an impulsive force, which is a very large force acting on a body for a very short time. Recall that if such a force $F(t)$ acts on a body of mass m from the time t_1 to the time t_2, the impulse acting on the body is

$$\text{Impulse} = \int_{t_1}^{t_2} F(t)\,dt = \Delta p, \tag{7.4.2}$$

where Δp is the change in momentum of the body. Let T denote any time between t_1 and t_2 ($t_1 < T < t_2$). Then it follows from the sifting property that

$$\int_{t_1}^{t_2} \Delta p\,\delta(t-T)\,dt = \Delta p. \tag{7.4.3}$$

Comparing (7.4.2) and (7.4.3), we see that the force $F(t)$ and $\Delta p\,\delta(t-T)$ have the same impulse. Consequently, if $F(t)$ is unknown or simply cannot be determined, we can at least approximate it with $\Delta p\,\delta(t-T)$. ♦

The following examples illustrate how to incorporate the Dirac delta function into a differential equation in order to model the balance of a deposit account when there are periodic withdrawals of equal amounts from the account.

Example 7.4.3 A deposit account is opened with a check for \$5000. The account pays interest at an annual rate of 4% compounded continuously. Six months after the opening of the account, a \$100 withdrawal is made. From then on, additional withdrawals, each in the amount of \$100, are made every six months. Find the balance in the account at the end of two years, after the fourth withdrawal of \$100.

Solution Let $B(t)$ denote the balance after t years. Since we are to determine the balance after two years, we shall restrict our attention to $0 \le t < 2.5$, that is, from the time of the deposit up to the time of the fifth withdrawal. Because interest is accruing at the rate of 4% compounded continuously, the balance at time t is increasing at the rate of $0.04B(t)$—except at $t = j/2$, for $j = 1, 2, 3, 4$, the times at which withdrawals of \$100 are made. In other words, as soon as t reaches $\frac{1}{2}$, 1, $\frac{3}{2}$, and 2 years, the balance immediately decreases by \$100. These abrupt drops in the balance and the times at which they occur are modeled by the step functions: $100u(t-\frac{j}{2})$ for $j = 1, 2, 3, 4$. Consequently, a model of the rate at which the balance is decreasing at each of these times is $100\delta(t-\frac{j}{2})$. The compartment model

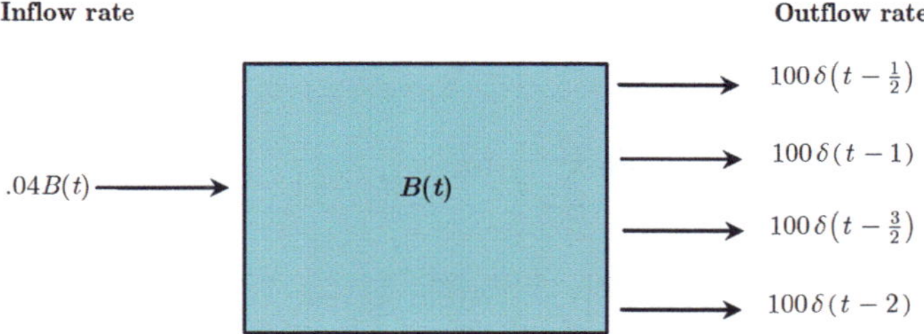

Fig. 7.15 Compartment model of the balance of the account in Example 7.4.3

shown in Fig. 7.15 depicts the dynamics of the account for $0 \le t < 2.5$. By the conservation equation, the rate of change in the balance $B(t)$, for $0 \le t < 2.5$, is modeled by the differential equation

$$\frac{d}{dt}B(t) = 0.04B(t) - 100\delta(t-\tfrac{1}{2}) - 100\delta(t-1) - 100\delta(t-\tfrac{3}{2}) - 100\delta(t-2). \tag{7.4.4}$$

To solve this equation for $B(t)$, first multiply it by the integrating factor $e^{-0.04t}$, thereby obtaining

$$\begin{aligned}\frac{d}{dt}\big[e^{-0.04t}B(t)\big] &= -100e^{-0.04t}\delta(t-\tfrac{1}{2}) - 100e^{-0.04t}\delta(t-1)\\ &\quad - 100e^{-0.04t}\delta(t-\tfrac{3}{2}) - 100e^{-0.04t}\delta(t-2).\end{aligned}$$

Integration from 0 to t yields

$$\begin{aligned}&\int_0^t \frac{d}{ds}\big[e^{-0.04s}B(s)\big]\,ds\\ &\quad = -100\int_0^t e^{-0.04s}\delta\big(s-\tfrac{1}{2}\big)\,ds - 100\int_0^t e^{-0.04s}\delta(t-1)\,ds\\ &\quad\quad - 100\int_0^t e^{-0.04s}\delta\big(s-\tfrac{3}{2}\big)\,ds - 100\int_0^t e^{-0.04s}\delta(s-2)\,ds\end{aligned}$$

or

$$e^{-0.04t}B(t) - B(0) = -100\sum_{j=1}^{4}\int_0^t e^{-0.04s}\delta\big(s-\tfrac{j}{2}\big)\,ds. \tag{7.4.5}$$

It then follows from the sifting property (7.3.24) that

$$\int_0^t e^{-0.04s}\delta\big(s-\tfrac{j}{2}\big)\,ds = e^{-0.04(j/2)}u\big(t-\tfrac{j}{2}\big)$$

for $j = 1, 2, 3, 4$. Finally, setting $B(0) = 5000$ and solving for $B(t)$ results in the following formula for the balance for $0 \le t < 2.5$:

$$\begin{aligned} B(t) &= 5000e^{0.04t} - 100e^{0.04t} \sum_{j=1}^{4} e^{-0.02j} u\left(t - \tfrac{j}{2}\right) \qquad (7.4.6) \\ &= 5000e^{0.04t} - 100e^{0.04t}\left[e^{-0.02}u\left(t - \tfrac{1}{2}\right) + e^{-0.04}u(t-1) \right. \\ &\qquad \left. + e^{-0.06}u\left(t - \tfrac{3}{2}\right) + e^{-0.08}u(t-2)\right]. \end{aligned}$$

Thus, for $1.5 < t < 2$, the balance is

$$B(t) = 5000e^{0.04t} - 100e^{0.04t}\left(e^{-0.02} + e^{-0.04} + e^{-0.06}\right). \qquad (7.4.7)$$

Just before the fourth withdrawal is made, the balance is given by the limit of (7.4.7) as t approaches 2 from the left:

$$\begin{aligned} \lim_{t \to 2^-} B(t) &= 5000e^{0.08} - 100e^{0.08}\left(e^{-0.02} + e^{-0.04} + e^{-0.06}\right) \\ &= 5000e^{0.08} - 100\left(e^{0.06} + e^{0.04} + e^{0.02}\right) = 5104.15. \end{aligned}$$

Therefore, the balance immediately after the fourth withdrawal of \$100 is \$5004.15.

We can just as easily find this balance using the right-hand limit at $t = 2$. From (7.4.6), the balance for $2 < t < 2.5$ is

$$B(t) = 5000e^{0.04t} - 100e^{0.04t}\left(e^{-0.02} + e^{-0.04} + e^{-0.06} + e^{-0.08}\right). \qquad (7.4.8)$$

So the balance after the fourth withdrawal is given by the limit of (7.4.8) as t approaches 2 from the right:

$$\begin{aligned} \lim_{t \to 2^+} B(t) &= 5000e^{0.08} - 100e^{0.08}\left(e^{-0.02} + e^{-0.04} + e^{-0.06} + e^{-0.08}\right) \\ &= 5000e^{0.08} - 100\left(e^{0.06} + e^{0.04} + e^{0.02} + 1\right) = 5004.15. \end{aligned}$$

In sum, according to either limit, the balance in the account immediately after the fourth withdrawal is \$5004.15. ♦

Example 7.4.4 We can generalize formula (7.4.6) for the balance in the account by replacing the numerical values in Example 7.4.3 with variables: a deposit of \$5000 with a deposit of B_0 dollars, withdrawals of \$100 with withdrawals of P dollars, the annual interest rate of 4% compounded continuously with the annual interest rate of k percent compounded continuously, and the balance after 2 years with the balance after n years. Let us also change the frequency of withdrawals from 2 per year to m per year, where m is a positive integer. Typical values for m are 1, 2, 4, 12, and 365, which correspond to annual, semiannual, quarterly, monthly, and daily withdrawals,

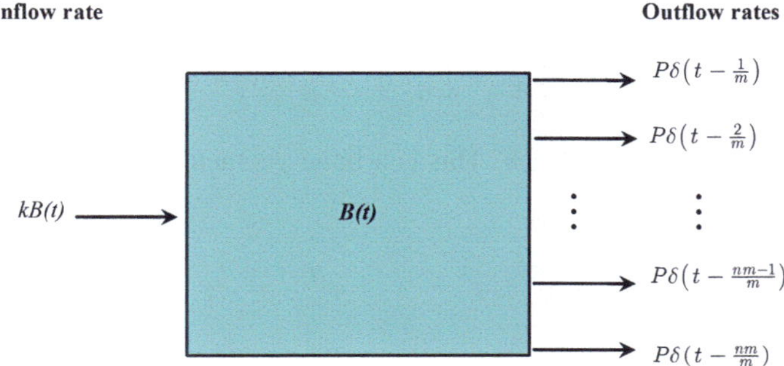

Fig. 7.16 Compartment model of the balance of the account in Example 7.4.4

respectively. After n years, there are a total of nm withdrawals. Accordingly, the times of the withdrawals change from $t = j/2$ years for $j = 1, 2, 3, 4$ to $t = j/m$ years for $j = 1, 2, 3, \ldots, nm$. Our goal is to determine the balance in the account after n years, immediately after the nm-st withdrawal is made.

It follows from the compartment model shown in Fig. 7.16 that the differential equation for the balance is

$$\frac{d}{dt}B(t) = kB(t) - P\sum_{j=1}^{nm}\delta\left(t - \frac{j}{m}\right).$$

This equation corresponds to (7.4.4) in Example 7.4.3. It gives the rate of change of the account balance $B(t)$ with respect to t for $0 \le t < n + \frac{1}{m}$, the time from the deposit of B_0 dollars up to the $(nm + 1)$-st withdrawal of P dollars. Employing the integrating factor e^{-kt}, we find that

$$e^{-kt}B(t) - B(0) = -P\sum_{j=1}^{nm}\int_0^t e^{-ks}\delta\left(s - \frac{j}{m}\right)\,ds,$$

which corresponds to (7.4.5). Let $B(0) = B_0$. Using the sifting property (7.3.24) and solving for $B(t)$, we obtain the formula

$$B(t) = B_0e^{kt} - Pe^{kt}\sum_{j=1}^{nm}e^{-k(j/m)}u\left(t - \frac{j}{m}\right),$$

for the balance in the account for $0 \le t < n + \frac{1}{m}$ (cf. (7.4.6)). Consequently, the balance for $n < t < n + \frac{1}{m}$ is

$$B(t) = B_0e^{kt} - Pe^{kt}\left(e^{-k/m} + e^{-2(k/m)} + e^{-3(k/m)} + \cdots + e^{-nm(k/m)}\right),$$

which generalizes formula (7.4.8). The quantity in parenthesis has the form

$$q + q^2 + q^3 + \cdots + q^N,$$

where $q = e^{-k/m}$ and $N = nm$. This is a finite geometric series. From the well-known formula

$$1 + q + q^2 + q^3 + \cdots + q^{N-1} = \frac{1 - q^N}{1 - q},$$

we obtain

$$\begin{aligned} q + q^2 + q^3 + \cdots + q^N &= q\left(1 + q + q^2 + \cdots + q^{N-1}\right) \\ &= e^{-k/m}\frac{1 - (e^{-k/m})^{nm}}{1 - e^{-k/m}} = \frac{1 - e^{-kn}}{e^{k/m} - 1}. \end{aligned}$$

This considerably simplifies the formula for the balance for $n < t < n + \frac{1}{m}$ to

$$B(t) = B_0 e^{kt} - P e^{kt}\frac{1 - e^{-kn}}{e^{k/m} - 1}. \qquad (7.4.9)$$

Thus the balance after n years, after the withdrawal is made, is the limit of (7.4.9) as t approaches n from the right, namely,

$$\lim_{t \to n^+} B(t) = B_0 e^{kn} - P\frac{e^{kn} - 1}{e^{k/m} - 1}. \qquad (7.4.10)$$

Finally, let us corroborate the validity of this formula by applying it to Example 7.4.3. Substituting $k = 0.04$, $m = 2$, $B_0 = 5000$, $P = 100$, and $n = 2$ into (7.4.10), we get

$$\lim_{t \to 2^+} B(t) = 5000e^{0.08} - 100\frac{e^{0.08} - 1}{e^{0.02} - 1} = \$5004.15,$$

which is what we obtained in Example 7.4.3. ♦

In our final example, we show how to use (7.4.10) to derive a well-known formula for calculating the equal periodic payments that are needed to ***amortize*** a loan in a specified length of time, that is, to pay off both the original amount of money borrowed from a lender and all of the interest that has accrued during the life of the loan.

Example 7.4.5 Find a formula to calculate the monthly payments that are required to pay off a loan of A dollars in N years when the interest on the loan accrues at an annual rate of r compounded monthly. Use it to calculate the monthly payments on a home mortgage of \$139,850 at an annual interest rate of 7.875% for 30 years.

Solution The formula that we will derive to determine the monthly payments applies to loans such as student loans, car loans, personal loans, and home mortgages. Let us assume that the monthly payments will always be the same throughout the life of the loan: let P dollars denote the amount of each payment. From the point of view of a lending institution, a loan resembles a savings account earning interest at an APR of r percent compounded monthly. The loan amount is analogous to a one-time deposit and the monthly payments to the withdrawals. As a result, the compartment model used in Example 7.4.4 applies to this loan situation; however, the compartment now represents a loan account rather than a savings account. But there is a fly in the ointment: the interest in Example 7.4.4 is compounded continuously, whereas the interest here is compounded monthly. Even so, there is a way of getting around this. Recall from the "Equivalent Annual Percentage Rates" statement before Example 6.3.6 in the previous chapter that an APR of r compounded m times per year is equivalent[9] to an APR of

$$m \ln \left(1 + \frac{r}{m}\right)$$

compounded continuously. Since the payments are monthly, $m = 12$. So all that we have to do in order to obtain a formula for the balance of the loan after n years is to modify (7.4.10) by letting $m = 12$, $k = 12 \ln(1 + r/12)$, and $B_0 = A$. Then, as

$$e^k = \left(1 + \frac{r}{12}\right)^{12},$$

the limit (7.4.10) becomes

$$\begin{aligned} \lim_{t \to n^+} B(t) &= A(e^k)^n - P \frac{(e^k)^n - 1}{e^{k/12} - 1} \qquad (7.4.11) \\ &= A\left(1 + \frac{r}{12}\right)^{12n} - P \frac{\left(1 + \frac{r}{12}\right)^{12n} - 1}{\frac{r}{12}}. \end{aligned}$$

After $n = N$ years, the balance will be 0 dollars since the loan is to be repaid by then. So by setting (7.4.11) equal to 0 and solving for P , we obtain the following monthly repayment formula or ***loan amortization formula***:[10]

[9] That is, the two balances are equal at the end of each discrete compounding period.

[10] Typically in differential equations textbooks, the loan amortization formula is derived by solving difference equations. ***Difference equations*** involve discrete changes (or differences) in the dependent variable in contrast to differential equations which involve instantaneous rates of change (derivatives) of the dependent variable. See Braun [15, ch. 1.12] or Becker [11, p. 191].

$$P = A\left[\frac{\frac{r}{12}}{1-\left(1+\frac{r}{12}\right)^{-12N}}\right]. \tag{7.4.12}$$

Now let us employ the amortization formula to determine the monthly payments that are needed to pay off a \$139,850 mortgage at 7.875% compounded monthly in 30 years. With $r = 0.07875$, $N = 30$, and $12N = 360$, the formula yields

$$P = \$139,850\left[\frac{\frac{0.07875}{12}}{1-\left(1+\frac{0.07875}{12}\right)^{-360}}\right] = \$1014.01.$$

Note that the actual cost of purchasing this home on credit, if it is kept and paid off in 30 years, is

$$\$1014.01 \text{ per month} \times 360 \text{ months} = \$365{,}043.60.$$

This amount does not include the down payment and the closing costs, such as appraisal fees, origination fees, and title insurance. ♦

Problems

"The best thing for being sad," replied Merlyn, beginning to puff and blow, "is to learn something. That is the only thing that never fails. You may grow old and trembling in your anatomies, you may lay awake at night listening to the disorder in your veins, you may miss your only love, you may see the world about you devastated by evil lunatics, or know your honour trampled in the sewers of baser minds. There is only one thing for it then—to learn. Learn why the world wags and what wags it. That is the only thing which the mind can never exhaust, never alienate, never be tortured by, never fear or distrust, and never dream of regretting. Learning is the thing for you. Look at what a lot of things there are to learn—pure science, the only purity there is. You can learn astronomy in a lifetime, natural history in three, literature in six. And then, after you have exhausted a milliard lifetimes in biology and medicine and theocriticism and geography and history and economics—why you can start to make a cartwheel out of the appropriate wood, or spend fifty years learning to begin to learn to beat your adversary at fencing. After that you can start again on mathematics, until it is time to learn to plough."

T. H. White[11] [78, p. 183]

[11] Terence H. White (1906–1964) was a British novelist. The quote is from his book *The Once and Future King*, which is a version of the epic tale of King Arthur and the knights of the Round Table. It is based on Sir Thomas Mallory's *Le Morte d'Arthur*. The 1960 Broadway musical *Camelot* and the 1967 movie of the same name were based on these books.

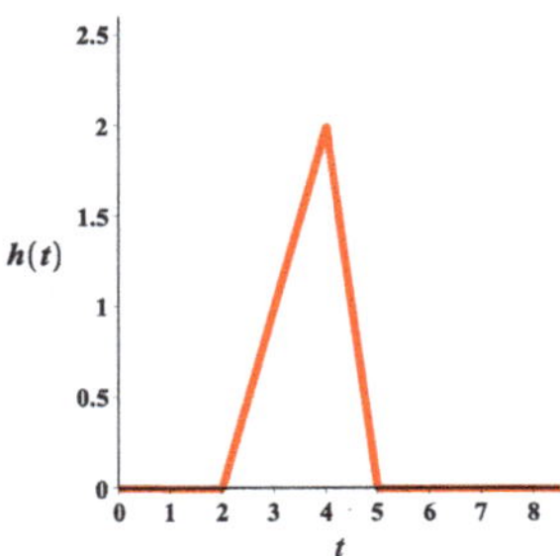

Fig. 7.17 Graph of the triangular pulse function $h(t)$ in Problem 3

Expressing Functions in Terms of Unit Step Functions

In Problems 1–12, express each function in terms of the appropriate unit step functions.

1. $f(t) = \begin{cases} 0, & 0 \le t < 2 \\ 1, & 2 < t < 4 \\ -5, & t > 4 \end{cases}$

2. $g(t) = \begin{cases} -1, & 0 \le t < 1 \\ 2, & 1 < t < 3 \\ 1, & t > 3 \end{cases}$

3. The function $h(t)$ with domain $[0, \infty)$ whose graph is the triangular pulse shown in Fig. 7.17.

4. $c(t) = \begin{cases} 1, & t < 2 \\ \cos t, & t > 2 \end{cases}$

5. $\varphi(t) = \begin{cases} e^{2t}, & t < 5 \\ 0, & t > 5 \end{cases}$

6. $p(t) = \begin{cases} 0, & 0 \le t < 2 \\ 4 - 2t, & 2 < t < 3 \\ t - 5, & t > 3 \end{cases}$

7. $x(t) = \begin{cases} \sin t, & 0 \le t < \pi/2 \\ 0, & \pi/2 < t < 2 \\ t - 2, & 2 < t < 3 \\ 1, & t > 3 \end{cases}$

8. The ***ramp function*** $r(t)$ whose graph is shown in Fig. 7.18. The function is defined for all $t \ge 0$ and $r(3) = 2$.

9. The function $s(t)$ whose graph is the square wave shown in Fig. 7.19 and which is defined at all points $t \ge 0$, except for the jump discontinuities at $t = 1, 2, 3, \ldots$.

10. The function $y(t)$ with domain $0 \le t < \infty$ that is defined by the half-rectified sine wave shown in Fig. 7.20. (A ***half-rectified sine wave*** is obtained from a sine wave by replacing all negative values with 0.)

11. The ***unit staircase function*** $h(t)$ whose graph is shown in Fig. 7.21 and which is defined at all points $t \ge 0$, except for the jump discontinuities at $t = 1, 2, 3, \ldots$.

Fig. 7.18 Graph of the ramp function $r(t)$ in Problem 8

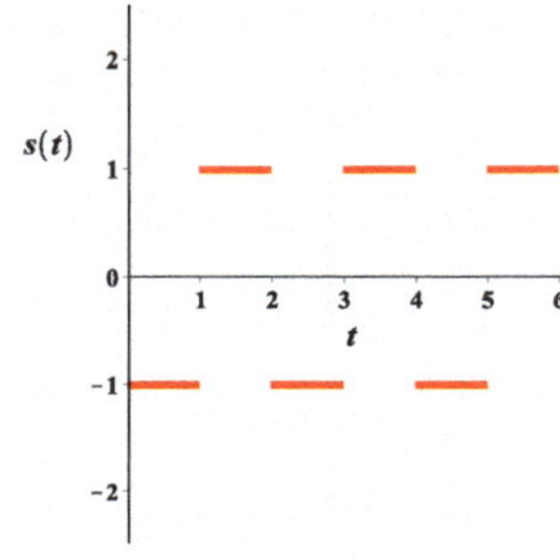

Fig. 7.19 Graph of the square wave function $s(t)$ in Problem 9

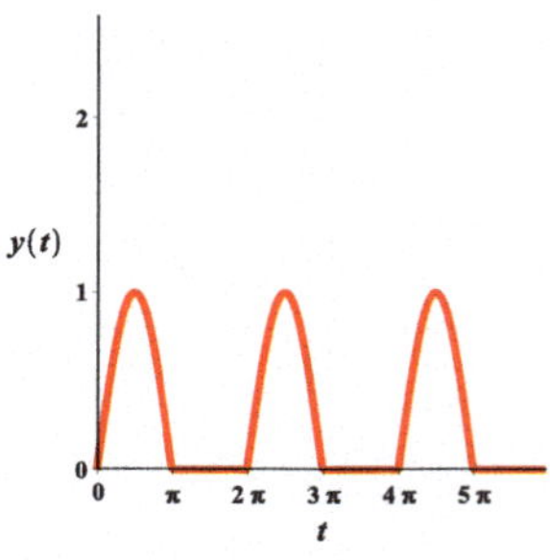

Fig. 7.20 Graph of the half-rectified sine wave $y(t)$ in Problem 10

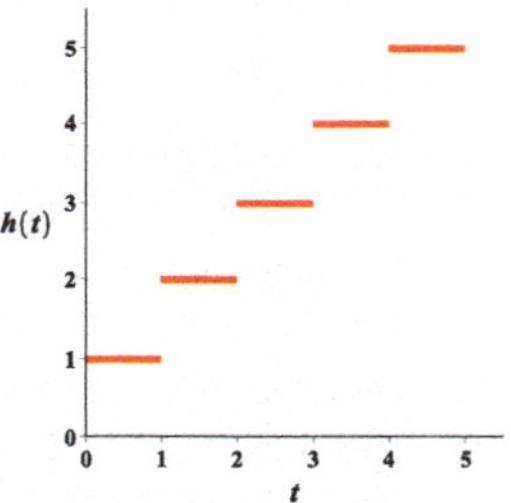

Fig. 7.21 Graph of the unit staircase function $h(t)$ in Problem 11

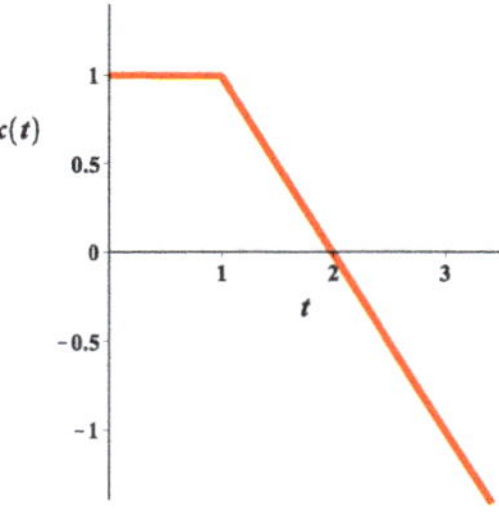

Fig. 7.22 Graph of the function $x(t)$ in Problem 12

12. The function $x(t)$ whose graph is shown in Fig. 7.22. Also, formally take the derivative of $x(t)$ and interpret the result.

Applications: Unit Step Functions

13. What is the maximum that a letter can weigh for it to be considered first-class mail? Use unit step functions to express the cost of mailing a letter as a function of its weight. A complete domestic postage rate chart for first-class mail can be found at the website of the U.S. Postal Service.

14. An underground storage tank initially contains 1000 gallons of gasoline. At time $t = 0$ minutes, a pump is turned on and gasoline is pumped into the storage tank at a constant rate of 50 gallons per minute. The pump is equipped with a shutoff valve that will automatically turn it off when the tank reaches its capacity of 6000 gallons. Unfortunately, a crack in the tank develops 4 minutes after the pump is turned on. Suppose the number of gallons that has leaked out of the tank at time t minutes can be modeled by the function: $G(t) = t^2 - 8t + 17$ for $t \geq 4$.

    (a) Express the number of gallons that has leaked out at time t in terms of a step function.

    (b) Express the number of gallons of gasoline in the tank in terms of t: first as a piecewise-defined function and then as a step function.

    (c) For what times are the functions in parts (a) and (b) valid?

    (d) Will the pump ever shut off automatically? If not, when will the tank become empty?

15. A librarian picks up a differential equations book lying on the floor. The force exerted on the book by the librarian is given by the function

$$F_l(y) = \begin{cases} mg + \epsilon, & 0 \leq y < \epsilon \\ mg, & \epsilon \leq y < h - \epsilon \\ mg - \epsilon, & h - \epsilon \leq y < h \\ mg, & y \geq h \end{cases}$$

where y is the vertical position of the book relative to the floor, m is the book's mass, g is the acceleration due to gravity, and ϵ and h are positive constants.

    (a) Express the force $F_l(y)$ in terms of the Heaviside function as defined by (7.2.3).

    (b) Find the acceleration of the book.

    (c) Find the velocity of the book.

    (d) The ***kinetic energy*** K of a body with mass m moving with speed v is $K = \frac{1}{2}mv^2$. What is the kinetic energy of the book when $0 \leq y \leq \epsilon$? When $y = h$?

(e) Recall from a calculus or physics course that if the displacement of a body lies along the same straight line as a force acting on the body, say the y-axis, and if the magnitude of the force depends on the body's position, then the ***work*** done on the body by the force when the body moves from $y = a$ to $y = b$ is $W = \int_a^b F(y)\,dy$. How much work is done on the book by the force $F_l(y)$ that the librarian exerts on the book if the librarian puts the book on a shelf that is located h feet above the floor? Compare that with the work done on the book by earth's gravity.

Applications: Dirac Delta Function

16. Verify (7.4.1) by reworking Example 7.4.1 without using the Dirac delta function.
17. Solve the initial value problem: $\dfrac{dx}{dt} + x = \delta(t-2),\ \ x(0) = 5$. Find the value of the solution $x(t)$ at $t = 10$.
18. Solve the initial-value problem: $\dfrac{dx}{dt} - 2x = 3\delta(t-1),\ \ x(0) = 4$. Find the value of the solution $x(t)$ at $t = 2$.
19. An underground storage tank initially contains 1000 gallons of gasoline. At time $t = 0$ minutes, a pump is turned on and gasoline is pumped into the storage tank at a constant rate of 50 gallons per minute. The pump is equipped with a shutoff valve that will automatically turn it off when the tank reaches its capacity of 6000 gallons. Unfortunately, a crack in the tank develops 4 minutes after the pump is turned on. As a result, gasoline leaks out of the tank at a rate that can be modeled by the function: $(2t-8)u(t-4) + (t^2 - 8t + 17)\delta(t-4)$ gallons per minute from $t = 0$ until the moment the tank becomes empty.

    (a) Express the rate of change of the number of gallons in the tank with respect to t in terms of a unit step function and the Dirac delta function. What is the initial condition?

    (b) Use the differential equation and initial condition from part (a) to find a formula for the number of gallons of gasoline in the tank. Compare this formula to the one obtained in Problem 14.

    (c) For what times is the differential equation in part (a) valid?

20. Express the acceleration of the book in Problem 15 in terms of the Heaviside function as defined by (7.2.3). A sudden change in the acceleration of a body is called a *jerk*. To be precise, ***jerk*** is the derivative of acceleration with respect to time. Express the jerk of the book in Problem 15 in terms of the Dirac delta function.
21. A certain isolated ranch in New Mexico has 1000 mustangs. The per capita growth rate of the mustang population remains nearly constant at 10%, regardless of the size of the population. The owners decide to reduce the population to 500 mustangs over the course of the next 2 years. This is to be done by selling R of the mustangs a year from now and then R more of them at the end of the second year.

    (a) Determine the value of R.

    (b) Rework the problem if the reduction in the mustang population is to take place over 10 years, instead of over 2 years, and the same number of mustangs is sold every year at the same time.

22. A Texas rancher owns 1600 longhorn cattle and estimates that their number increases at an annual per capita rate of approximately 8%, regardless of the size of the herd. Deciding that the herd is less profitable than her Hereford herd, she decides to sell off the entire herd over the course of the next 5 years. Her plan is to sell the same number of longhorns every 6 months.

(a) If she waits 6 months to sell the first lot of longhorns, determine the number in a lot so that the entire herd is sold 5 years from the day of her decision to sell.

(b) Suppose that the rancher begins selling longhorns right away instead of holding off for 6 months. How does that change the number of longhorns in a lot?

Deposit Accounts and Amortized Loans

23. A $95,000 home is purchased with a 10% down payment. The rest of the purchase price plus $3200 in closing costs is financed with a mortgage at 8.25% for 30 years. Find the monthly payment. Find the monthly payment if the home mortgage is for 15 years instead of 30 years.

24. In the *Commercial Appeal*, a Memphis newspaper, a car dealer advertises 5.9% APR financing for up to 48 months on a Mercury Grand Marquis with 10% down. The ad states the monthly payments will be $23.44 per $1000 financed if the car is financed for 48 months. Is this correct? What will the monthly payments be per $1000 financed if the car is financed for 60 months?

25. Find the APR of a car loan in the amount of $21,250 that is to be paid off in 5 years with monthly payments of $385.

26. Generalize the loan amortization formula (7.4.12) by determining the periodic payment needed to pay off a loan of A dollars in n periods, where the interest on the loan accrues at an annual rate of r percent compounded m times per year.

27. An account is opened with a deposit of A dollars. The account pays interest at an annual rate of r percent compounded quarterly. Three months after the opening of the account, an additional deposit of P dollars is made. From then on, additional deposits, each in the amount of P dollars are made every three months. Determine the total amount that can be withdrawn from the account after N years.

Computer Algebra System Problems

Use a CAS to work the following problems.

28. Problem 11. Replicate the graph shown in Fig. 7.21.

29. Problem 21.

30. Problem 22.

Chapter 8
Exact and Related Equations

Yet at about the age of sixteen, I was offered a choice which, in retrospect, I can see I was not mature enough, at the time, to make wisely. This choice was between starting on the calculus and, alternately, giving up mathematics altogether and spending the time saved from it on reading Latin and Greek literature more widely. I chose to give up mathematics and I have lived to regret this keenly .... The calculus, even a taste of it, would have given me an important and illuminating additional outlook on the Universe .... I was not good at mathematics; I did not like the stuff.... One ought, after all, to be initiated into the life of the world in which one is going to have to live. I was going to have to live in the Western World at its transition from the modern to the post-modern chapter of its history; and the calculus, like the full-rigged sailing ship, is ... one of the characteristic expressions of the modern Western genius.

Arnold Toynbee[1]

Abstract This chapter is an introduction to *exact differential equations*. It begins with the observation that every differential equation $y' = f(x, y)$ can always be written in the form

$$M(x, y) + N(x, y)\frac{dy}{dx} = 0 \tag{1}$$

and that sometimes there is a function F such that

$$\frac{d}{dx}F(x, y) = M(x, y) + N(x, y)\frac{dy}{dx} \tag{2}$$

[1] See [74, p. 12 f.]. Arnold J. Toynbee (1889–1975) was a British historian, a professor of modern Greek studies at King's College, and research professor of international history at the London School of Economics. One of his most influential works was a 12-volume series entitled *A Study of History*, which was a study of civilizations and presented a cyclical theory for their rise and fall.

© The Author(s), under exclusive license to Springer Nature Switzerland AG 2026

L. C. Becker, *Ordinary Differential Equations: Concepts, Methods, and Models*,
https://doi.org/10.1007/978-3-032-15150-6_8

at every point (x, y) belonging to some open rectangular region $\mathcal{R}$. When this is the case, solutions of (1) are given implicitly by

$$F(x, y) = C, \tag{3}$$

where C is an arbitrary constant. It is shown that the criterion for the existence of such a function is that

$$\frac{\partial M}{\partial y} = \frac{\partial N}{\partial x} \tag{4}$$

at each point of $\mathcal{R}$.

Employing (4) to ascertain whether or not a given Eq. (1) is exact and then finding a function F if it is exact requires a basic understanding of *first- and second-order partial derivatives*. However, students with only two semesters of calculus under their belt probably have never heard of a partial derivative, much less know what to do with one. Since this book is written for such students, whatever they need to know about partial derivatives in order to understand this chapter is gently introduced in Chap. 1 and then covered more extensively in the first four sections of this chapter.

Two types of *non-exact differential equations* that can be converted into exact equations by means of *integrating factors* are covered in the last section.

Exact differential equations are important for enrollees in engineering and physical science courses. For instance, in electrostatics, these equations define equipotential surfaces. In thermodynamics, equations of state for ideal and real gases involve exact and non-exact differentials and integrating factors. In a multivariable calculus course, *Green's theorem* can be used to prove that a line integral $\int_C M\,dx + N\,dy$ is equal to zero if C is a closed path in an open simply-connected region and M and N satisfy condition (4) in that region.

8.1 Initial Observations

There is a type of first-order differential equation called an *exact differential equation*, which can be transformed into an equivalent form that makes it particularly simple to solve. Exact differential equations have a way of cropping up now and then in various physical science and engineering applications. For example, in thermodynamics, the equation of state of a real gas can be derived from an exact differential equation if certain rates of change are known (cf. [7, ch. 3].) In electrostatics, finding surfaces in space with the same electrical potential involve exact equations (see Problem 53). Before we formally define what is meant by an *exact differential equation*, let us take a look at several differential equations that turn out to be exact equations.

Consider the equation

$$x\frac{dy}{dx} + y = 0. \tag{8.1.1}$$

Since this equation is linear, we can solve it using either the integrating factor method or the variation of parameters formula. But note that it is also separable; so we could also solve it by integrating the separated terms. Unfortunately, we cannot rely on these methods to solve the other equations in this chapter. Besides (8.1.1) being both a linear and separable equation, it turns out to be a so-called exact differential equation (which we will define in the next section). Our goal is to find another way to solve (8.1.1). Hopefully, this will lead to a method for solving exact equations that are neither separable nor linear.

A method for solving (8.1.1), besides separating variables or finding an integrating factor, is based on the observation that the left-hand side of (8.1.1) is the derivative of the product "xy" with respect to x. This means that we can rewrite (8.1.1) as

$$\frac{d}{dx}(xy) = 0. \tag{8.1.2}$$

The product xy can be regarded as a function of x alone by virtue of y being a function of x, albeit an unknown one at this point. A basic result of differential calculus is that a function whose derivative is zero at all points of an interval must be constant over that interval. Applying that result here, we conclude that xy is constant. Consequently, solutions of (8.1.1) satisfy the equation

$$xy = C, \tag{8.1.3}$$

where C is a constant. Solving for y, we find

$$y = \frac{C}{x}. \tag{8.1.4}$$

For any value of C, (8.1.4) is an explicit solution of (8.1.1) on any interval that excludes $x = 0$, such as $(0, \infty)$. The reader should verify that this same result can be obtained by separating variables and by finding an integrating factor.

Whatever an exact equation might be, another example of one is

$$x\frac{dy}{dx} + y - 2x = 0. \tag{8.1.5}$$

Obviously the difference between (8.1.1) and (8.1.5) is the term "$-2x$". Even though (8.1.5) is no longer separable, it is still linear. Accordingly, we could solve it by finding an integrating factor. However, at the moment, our objective is to discover some other way of solving (8.1.5). So, let us try an approach similar to the last one.

Thus, the question before us is whether a function $F(x, y)$ can be found whose derivative with respect to x is the left-hand side of (8.1.5). The existence of one would allow us to replace (8.1.5) with the equivalent equation

$$\frac{d}{dx}F(x, y) = 0, \tag{8.1.6}$$

just as we replaced (8.1.1) with (8.1.3). This in turn would imply that $F(x, y) = C$. It only takes a comparison of the left-hand sides of (8.1.1), (8.1.2), and (8.1.5) to come up with the function $F(x, y) = xy - x^2$. Therefore, the equation

$$xy - x^2 = C$$

represents the family of implicit solutions of (8.1.5). Solving for y, we obtain the explicit solutions

$$y = x + \frac{C}{x}.$$

A third example of an exact equation is

$$(x + e^y)\frac{dy}{dx} + y - 2x = 0. \tag{8.1.7}$$

Aside from the term $e^y y'$, this equation looks like (8.1.5). However, (8.1.7) is neither linear nor separable! So, the methods that we learned for solving these types of equations are of no use here. We have no choice but to try something else. Let us see if this scheme of finding a function $F(x, y)$ works here too. Even though finding one with its derivative equal to the left-hand side of the equation, in this case (8.1.7), is more difficult than it was in the two previous examples, it soon becomes apparent that

$$F(x, y) = xy - x^2 + e^y$$

works. Therefore, solutions of (8.1.7) are given implicitly by

$$xy - x^2 + e^y = C. \tag{8.1.8}$$

By this we mean that solutions of the differential equation (8.1.7) satisfy equation (8.1.8), and conversely.

Unlike the two previous examples, we cannot express the solutions of (8.1.7) solely in terms of x because (8.1.8) cannot be solved for y. That is to say, even though (8.1.8) has a solution, it is impossible to express it in terms of elementary functions (see the end of Sect. 5.2). However, it is possible to express the solution in terms of x alone using a non-elementary function called the *Lambert W function*.

The ***Lambert W function***, denoted by the letter W, is defined to be the inverse of the function

$$f(x) = xe^x$$

when the domain of f is restricted to $[-1, \infty)$.[2] In terms of this function, the solutions of (8.1.8) and (8.1.7) are

$$y = x + \frac{C}{x} - W\left(\frac{e^{x+Cx^{-1}}}{x}\right). \tag{8.1.9}$$

Since this function and most other non-elementary functions are unfamiliar to most readers, let us stick with implicit solutions when it is impossible or too difficult to obtain explicit solutions.

8.2 Exact Equations

It is always possible to rewrite any first-order equation $y' = g(x, y)$ in the form

$$M(x, y) + N(x, y)\frac{dy}{dx} = 0. \tag{8.2.1}$$

For example, we can rewrite

$$\frac{dy}{dx} = \frac{e^{xy}\sin y}{3y - x^2}$$

as

$$e^{xy}\sin y + (x^2 - 3y)\frac{dy}{dx} = 0.$$

But there are other possibilities too, such as

$$\frac{e^{xy}\sin y}{x^2 - 3y} + \frac{dy}{dx} = 0.$$

[2] It follows that $W(xe^x) = x$ for $x \geq -1$. Because of this, the solutions of a variety of equations can be expressed in terms of the Lambert W function, such as $xe^x = a$, $a^x + x = b$, and (8.1.8). For example, the solution of $xe^x = 2$ is $x = W(2)$. Using a CAS or an online Lambert W function calculator, $W(2) = 0.8526\ldots$.

The three differential equations considered in the previous section—(8.1.1), (8.1.5), and (8.1.7)—are written in the form (8.2.1). From now on, we shall assume that $M(x, y)$ and $N(x, y)$ are continuous throughout some *open rectangular region*

$$\mathcal{R} = \{(x, y) : a < x < b,\ c < y < d\}$$

in the xy-plane, where $-\infty \le a < b \le \infty$ and $-\infty \le c < d \le \infty$. A half-plane, such as all points to the right of the y-axis, and the xy-plane are open rectangular regions.

In the previous section we claimed that the differential equations (8.1.1), (8.1.5), and (8.1.7) are exact. The reason for this is due to the following definition.

Exact Equation

Definition 8.2.1 Equation (8.2.1) is said to be ***exact*** in an open rectangular region $\mathcal{R}$ if there is a function $F(x, y)$ whose derivative with respect to x is equal to the left-hand side of (8.2.1) at each point of $\mathcal{R}$. That is, (8.2.1) is exact if a function $F(x, y)$ exists such that

$$\frac{d}{dx}F(x, y) = M(x, y) + N(x, y)\frac{dy}{dx} \tag{8.2.2}$$

holds throughout $\mathcal{R}$.

Now suppose one is faced with the task of finding solutions of an exact equation (8.2.1), where M and N are given continuous functions. The goal then is to find differentiable functions that will satisfy this equation if they are substituted for the unknown y. Let φ denote such a function. Substituting $\varphi(x)$ for y, we obtain

$$M(x, \varphi(x)) + N(x, \varphi(x))\frac{d}{dx}\varphi(x) = 0.$$

Due to the existence of the function F in (8.2.2), this equation may be rewritten as

$$\frac{d}{dx}F(x, \varphi(x)) = 0. \tag{8.2.3}$$

So we can see that finding solutions of an exact equation is essentially over once such a function is found since (8.2.3) implies that $F(x, \varphi(x)) = C$ for some constant C. Therefore, solutions of (8.2.1) are given implicitly by

$$F(x, y) = C. \tag{8.2.4}$$

Sometimes explicit solutions of (8.2.1) can be obtained by solving (8.2.4) for y. However, more often than not, this may not be possible. Then the only way of expressing the solutions of (8.2.1) analytically is through the implicit representation given by (8.2.4).

8.3 Total Derivatives and Differentials

The left-hand side of each of Eqs. (8.1.1), (8.1.5), and (8.1.7) was contrived so as to be the derivative of a function $F(x, y)$ with respect to x. Part of the contrivance was making it relatively easy to come up with such a function. In practice, however, we must first determine if such a function exists for a given equation; and if it does, by what method can it be found. This is especially important for more complicated equations where coming up with functions without having a systematic way of finding them would be difficult. Before we begin our search for a method, we need to know what is meant by the *total derivative* of a function.

Let us first provide some insight by way of an example. Consider the function

$$F(x, y) = 2x^5y^3 + 5x - \frac{2}{3}y - 7.$$

First, let us compute the first-order partial derivatives of F:

$$\frac{\partial}{\partial x}F(x, y) = \frac{\partial}{\partial x}\left(2x^5y^3 + 5x - \frac{2}{3}y - 7\right) = 10x^4y^3 + 5 \tag{8.3.1a}$$

$$\frac{\partial}{\partial y}F(x, y) = \frac{\partial}{\partial y}\left(2x^5y^3 + 5x - \frac{2}{3}y - 7\right) = 6x^5y^2 - \frac{2}{3}. \tag{8.3.1b}$$

Generally speaking, y is independent of x. However, if y is known to be a function of x, then the ordinary derivative of F with respect to x is

$$\begin{aligned}\frac{dF}{dx} &= 2x^5 \cdot 3y^2y' + y^3 \cdot 10x^4 + 5 - \frac{2}{3}y' \\ &= \left(6x^5y^2 - \frac{2}{3}\right)\frac{dy}{dx} + \left(10x^4y^3 + 5\right).\end{aligned} \tag{8.3.2}$$

Note the use of the chain rule in differentiating the powers of y. Observe from the computed partials in (8.3.1) that the coefficient of dy/dx in the first term is F_y and the second term is F_x. The derivative in (8.3.2) is called the *total derivative of F*. This name allows us to distinguish this type of derivative from the partial derivatives given in (8.3.1). This example motivates the following definition of the *total derivative* of a function of two variables.

Total Derivative

Definition 8.3.1 Let $F(x, y)$ be a function with continuous first-order partial derivatives. Let y be a differentiable function of x. The ***total derivative*** of F, denoted by dF/dx, is

$$\frac{dF}{dx} := F_x(x, y) + F_y(x, y)\frac{dy}{dx} = \frac{\partial F}{\partial x} + \frac{\partial F}{\partial y} \cdot \frac{dy}{dx}. \tag{8.3.3}$$

A more symmetric form that is easier to remember is the *total differential* version of (8.3.3), which is defined next.

Total Differential

Definition 8.3.2 The ***total differential*** of F, denoted by dF, is

$$dF := \frac{\partial F}{\partial x}\, dx + \frac{\partial F}{\partial y}\, dy. \tag{8.3.4}$$

Equating the expressions in (8.2.2) and (8.3.3) for the total derivative of F, we have

$$\frac{\partial}{\partial x} F(x, y) + \frac{\partial}{\partial y} F(x, y)\frac{dy}{dx} = M(x, y) + N(x, y)\frac{dy}{dx}.$$

This suggests the following alternate, yet equivalent, definition of an exact equation.

Exact Equation (Alternate Definition)

Definition 8.3.3 The differential equation

$$M(x, y)\, dx + N(x, y)\, dy = 0$$

is said to be ***exact*** in an open rectangular region $\mathcal{R}$ of the xy-plane if a function $F(x, y)$ exists such that

$$\frac{\partial}{\partial x} F(x, y) = M(x, y) \quad \text{and} \quad \frac{\partial}{\partial y} F(x, y) = N(x, y)$$

at every point (x, y) in $\mathcal{R}$.

That is, if $M\,dx + N\,dy = 0$ is exact, then it can be written as

$$\frac{\partial F}{\partial x}\,dx + \frac{\partial F}{\partial y}\,dy = 0,$$

for some function F. Or, as

$$\frac{\partial F}{\partial x} + \frac{\partial F}{\partial y} \cdot \frac{dy}{dx} = 0.$$

At any rate,

$$\frac{d}{dx}F(x, y) = 0,$$

from which we conclude $F(x, y) = C$ for some constant C.

8.4 Second-Order Partial Derivatives

Let $f(x, y)$, as the notation indicates, denote a function of both x and y. Suppose both first-order partial derivatives $f_x(x, y)$ and $f_y(x, y)$ exist. If these functions can also be differentiated with respect to x and y, then $f(x, y)$ will have four *second-order partial derivatives*. Below is a description of each of them. They are counterparts of the second derivative of a function of a single variable.

1. $\dfrac{\partial^2 f}{\partial x^2} := \dfrac{\partial}{\partial x}\left(\dfrac{\partial f}{\partial x}\right)$

Its value at a point (x_0, y_0) is indicated either by writing $\dfrac{\partial^2 f}{\partial x^2}(x_0, y_0)$ or $f_{xx}(x_0, y_0)$.

2. $\dfrac{\partial^2 f}{\partial y \partial x} := \dfrac{\partial}{\partial y}\left(\dfrac{\partial f}{\partial x}\right)$

This notation indicates that the order of differentiation is first with respect to x and then with respect to y. This is a ***mixed second-order partial derivative***, which is also denoted by f_{xy}.

3. $\dfrac{\partial^2 f}{\partial x \partial y} := \dfrac{\partial}{\partial x}\left(\dfrac{\partial f}{\partial y}\right)$

Or we can write f_{yx}. Note that the order of differentiation of this partial derivative is the reverse of the previous one. The notation indicates that the order of differentiation is first with respect to y, then with respect to x.

4. $\dfrac{\partial^2 f}{\partial y^2} := \dfrac{\partial}{\partial y}\left(\dfrac{\partial f}{\partial y}\right)$

Not surprisingly an alternate notation is f_{yy}.

Example 8.4.1 Let $F(x, y) = 5x^2y$. The first-order partial derivatives of F are $F_x(x, y) = 10xy$ and $F_y(x, y) = 5x^2$. The second-order partial derivatives of F are

1. $F_{xx}(x, y) = \dfrac{\partial}{\partial x}F_x(x, y) = \dfrac{\partial}{\partial x}(10xy) = 10y,$
2. $F_{xy}(x, y) = \dfrac{\partial}{\partial y}F_x(x, y) = \dfrac{\partial}{\partial y}(10xy) = 10x,$
3. $F_{yx}(x, y) = \dfrac{\partial}{\partial x}F_y(x, y) = \dfrac{\partial}{\partial x}(5x^2) = 10x,$
4. $F_{yy}(x, y) = \dfrac{\partial}{\partial y}F_y(x, y) = \dfrac{\partial}{\partial y}(5x^2) = 0.$

Observe that $F_{xy}(x, y) = F_{yx}(x, y)$. ♦

Example 8.4.2 Compute the mixed second-order partial derivatives of

$$g(x, y) = e^{2x}\sin(3y) - \frac{x}{y}.$$

Solution Since $\dfrac{\partial g}{\partial x} = 2e^{2x}\sin(3y) - \dfrac{1}{y}$,

$$\frac{\partial^2 g}{\partial y \partial x} = \frac{\partial}{\partial y}\left(2e^{2x}\sin(3y) - \frac{1}{y}\right) = 6e^{2x}\cos(3y) + \frac{1}{y^2}.$$

Similarly, as $\dfrac{\partial g}{\partial y} = 3e^{2x}\cos(3y) + \dfrac{x}{y^2}$,

$$\frac{\partial^2 g}{\partial x \partial y} = \frac{\partial}{\partial x}\left(3e^{2x}\cos(3y) + \frac{x}{y^2}\right) = 6e^{2x}\cos(3y) + \frac{1}{y^2}.$$

Once again the mixed partial derivatives are equal. ♦

It is no accident that the mixed second-order partial derivatives in the previous examples are equal. Generally speaking, mixed partial derivatives of any of the functions that one normally encounters in the applied sciences, economics, and engineering are equal. There are exceptions, pathological functions which mathematicians have been able to construct, but we have no need for them in this book. The next theorem, known as ***Clairaut's Theorem***, gives sufficient conditions that will ensure the equality of mixed partial derivatives. However, this name is not the only one in use: it is also called the ***Mixed Partials Theorem*** or ***Euler's reciprocity***

relation. A proof of this theorem[3] is best left to a more advanced course, such as multivariable calculus or real analysis.

Clairaut's Theorem

Theorem 8.4.1 *Let $f(x, y)$ be a function that is defined on an open rectangular region $\mathcal{R}$. If the partial derivatives f_x, f_y, and f_{xy} exist and are continuous on $\mathcal{R}$, then f_{yx} also exists and*

$$f_{yx}(x_0, y_0) = f_{xy}(x_0, y_0),$$

for all $(x_0, y_0) \in \mathcal{R}$.

8.5 Solutions of Exact Equations

We have given several examples of equations that look like (8.2.1) whose solutions were obtained by inspection[4] and from the form of these solutions we were able to determine that the equations were exact. But that is backwards from it should be. What we ought to have is (i) a test that would disclose whether a differential equation is exact, and if it is, (ii) a method for solving this equation. Looking for such a test and method is the raison d'être of this and the next section. For example, consider the equation

$$\left(2\cos x - 2\cos(2x)\cos y + \frac{1}{x^2+4}\right) dx + \sin(2x)\sin y\, dy = 0. \tag{8.5.1}$$

Is there a way of determining whether this complicated-looking equation is exact or not? And if it is exact, how do we go about solving it?

Rather than dealing with (8.5.1) directly, let us start with the general form of this equation, namely,

$$M(x, y)\, dx + N(x, y)\, dy = 0, \tag{8.5.2}$$

where we assume $M(x, y)$ and $N(x, y)$ are continuous and have continuous first-order partial derivatives in an open rectangular region $\mathcal{R}$. Let us consider the implications of such an equation being exact. If it is, then it follows from

[3] See *Advanced Calculus* by Friedman [36, Thm. 1, p. 215].

[4] Recall we obtained solutions of the form $F(x, y) = C$ with virtually no pencil-and-paper work and without the aid of any formal methods. We say the solutions were obtained ***by inspection***.

Definition 8.3.3 that a function $F(x, y)$ exists whose first-order partial derivatives exist at all points of $\mathcal{R}$; and at these points,

$$\frac{\partial}{\partial x}F(x, y) = M(x, y) \quad \text{and} \quad \frac{\partial}{\partial y}F(x, y) = N(x, y).$$

Consequently,

$$\frac{\partial}{\partial y}\left(\frac{\partial}{\partial x}F(x, y)\right) = M_y(x, y) \quad \text{and} \quad \frac{\partial}{\partial x}\left(\frac{\partial}{\partial y}F(x, y)\right) = N_x(x, y).$$

Since by assumption M_y and N_x are continuous in $\mathcal{R}$, both of these mixed second-order partial derivatives are continuous in $\mathcal{R}$. As a result, it follows from Clairaut's Theorem that

$$\frac{\partial}{\partial y}\left(\frac{\partial F}{\partial x}\right) = \frac{\partial}{\partial x}\left(\frac{\partial F}{\partial y}\right). \tag{8.5.3}$$

Replacing F_x with M and F_y with N, we obtain the following necessary condition for a differential equation to be exact: If (8.5.2) is exact on an open rectangular region $\mathcal{R}$, then

$$\frac{\partial}{\partial y}M(x, y) = \frac{\partial}{\partial x}N(x, y) \tag{8.5.4}$$

at all points $(x, y) \in \mathcal{R}$. Thus, we have established the truth of the following statement:

$$\text{If } M\,dx + N\,dy = 0 \text{ is exact, then } M_y = N_x. \tag{8.5.5}$$

Logicians call this a ***conditional proposition***. It is actually a compound statement since it is composed of two components which themselves are statements. The first component

"$M\,dx + N\,dy = 0$ is exact"

is called the ***antecedent*** while the second component

"$M_y = N_x$"

is the ***consequent***. There are other ways to express (8.5.5), such as:

- $M\,dx + N\,dy = 0$ is exact only if $M_y = N_x$.
- $M\,dx + N\,dy = 0$ is exact implies that $M_y = N_x$.
- The condition $M_y = N_x$ is necessary for $M\,dx + N\,dy = 0$ to be exact.

Another conditional proposition that is logically equivalent to (8.5.5) is obtained by negating both the antecedent and the consequent and then interchanging them. The result is the following proposition known as the ***contrapositive*** of (8.5.5):

$$\text{If } M_y \neq N_x, \text{ then } M\,dx + N\,dy = 0 \text{ is not exact.} \tag{8.5.6}$$

At this point, it makes sense to ask if the converse of (8.5.5) is true. The ***converse*** of a proposition is obtained by interchanging its antecedent and consequent. So the converse of (8.5.5) is the proposition:

$$\text{If } M_y = N_x, \text{ then } M\,dx + N\,dy = 0 \text{ is exact.} \tag{8.5.7}$$

This may very well be true and we will look into this in the next section. However, it would be a mistake to assume the converse of some statement is true without further analysis. For instance, suppose someone were to exclaim: "It's raining cats and dogs!" By that, we would know that there are clouds overhead. Clearly, the statement "If it's raining cats and dogs, then clouds are overhead" is true. But if we were to switch "it's raining cats and dogs" with "clouds are overhead", then we would get the preposterous statement: "If clouds are overhead, then it's raining cats and dogs."

Before attempting to ascertain the truth or falsity of (8.5.7), let us look for a way to solve equations we already know to be exact, such as (8.1.7). Its differential form is

$$(y - 2x)\,dx + (x + e^y)\,dy = 0. \tag{8.5.8}$$

Let us begin with Definition 8.3.3 and take note of the steps that we use to solve it. Since it is exact, there is a function $F(x, y)$ whose first-order partial derivatives are

$$\frac{\partial F}{\partial x} = M(x, y) = y - 2x \quad \text{and} \quad \frac{\partial F}{\partial y} = N(x, y) = x + e^y. \tag{8.5.9}$$

Now it is rather obvious how to proceed: first find F from one of its partial derivatives by integrating either M with respect to x or N with respect to y. If we choose the former, then

$$F(x, y) = \int (y - 2x)\,dx + g(y) = xy - x^2 + g(y), \tag{8.5.10}$$

where $g(y)$ denotes a function of y alone. The reason here for $g(y)$ instead of the usual "C" is that the integration is with respect to x; so even though y is a variable, it is treated as if it were a constant at this intermediate step. Thus we have to allow for the possibility that the so-called "constant of integration" may not be just a constant but could involve y. Now we still have to find $g(y)$ but that is easy. Taking the partial derivative of (8.5.10) with respect to y and then setting it equal to $\partial F/\partial y$ in (8.5.9), we get

$$x + g'(y) = x + e^y.$$

Thus $g'(y) = e^y$. An integration yields $g(y) = \int e^y \, dy = e^y$.[5] Note that we do not need to add a C to e^y since all we need is any function whose partial derivatives are given by (8.5.9). Thus $F(x, y) = xy - x^2 + e^y$. We conclude that solutions of (8.5.8), or (8.1.7), are given implicitly by

$$xy - x^2 + e^y = C,$$

which agrees with (8.1.8).

Recapitulating the steps that were used to solve Eq. (8.5.8), we end up with the following method for solving exact differential equations.

Method I for Solving Exact Equations

To find implicit solutions of an exact equation $M(x, y)\, dx + N(x, y)\, dy = 0$, execute the following steps:

1. Find a function $F(x, y)$ so that $\partial F/\partial x = M$ by integrating $M(x, y)$ with respect to x while holding y constant, thereby obtaining

$$F(x, y) = \int M(x, y)\, dx + g(y). \tag{8.5.11}$$

   The integral represents any function of x and y whose partial derivative with respect x is $M(x, y)$. The function $g(y)$ denotes a differentiable function of y alone. At this point, $g(y)$ is unknown; so $F(x, y)$ is not fully known.
2. Now find $g(y)$ by taking the partial derivative of both sides of (8.5.11) with respect to y and setting the result equal to $N(x, y)$. Solve for $g'(y)$. Then integrate it with respect to y and let $g(y)$ be any one of the antiderivatives.
3. Substitute $g(y)$ into (8.5.11) at which point $F(x, y)$ becomes fully known.
4. Solutions of equation $M(x, y)\, dx + N(x, y)\, dy = 0$ are given implicitly by $F(x, y) = C$, where C is an arbitrary constant.

Alternatively, start by integrating $N(x, y)$ with respect to y. See Method II below.

[5] Recall the "Indefinite Integral Convention" in Sect. 2.1.

Example 8.5.1 Solve the differential equation

$$(x - 3y) + (y - 3x)\frac{dy}{dx} = 0$$

given that it is exact in the xy-plane.

Solution Since it is stated that

$$(x - 3y)\,dx + (y - 3x)\,dy = 0$$

is exact everywhere, there is a function $F(x, y)$ such that

$$F_x(x, y) = x - 3y \quad \text{and} \quad F_y(x, y) = y - 3x$$

holds at all points (x, y) in the xy-plane. Integrating F_x with respect to x, we get

$$F(x, y) = \int (x - 3y)\,dx + g(y) = \frac{1}{2}x^2 - 3xy + g(y).$$

Consequently,

$$F_y(x, y) = \frac{\partial}{\partial y}\left[\frac{1}{2}x^2 - 3xy + g(y)\right] = -3x + g'(y).$$

Equating both of the expressions for $F_y(x, y)$, we have

$$-3x + g'(y) = y - 3x,$$

which simplifies to $g'(y) = y$. Since the antiderivatives of $g'(y)$ are $y^2/2 + C$, let $g(y) = y^2/2$. Thus,

$$F(x, y) = \frac{1}{2}x^2 - 3xy + \frac{y^2}{2}.$$

Note that the claim that the differential equation is exact is correct (cf. Definition 8.3.3) since

$$\frac{\partial}{\partial x}F(x, y) = \frac{\partial}{\partial x}\left[\frac{1}{2}x^2 - 3xy + \frac{y^2}{2}\right] = x - 3y$$

and

$$\frac{\partial}{\partial y}F(x, y) = \frac{\partial}{\partial y}\left[\frac{1}{2}x^2 - 3xy + \frac{y^2}{2}\right] = -3x + y.$$

Therefore, solutions of the differential equation are given implicitly by

$$\frac{1}{2}x^2 - 3xy + \frac{y^2}{2} = C.$$

Or, by multiplying both sides of this equation by 2 and then "absorbing" it into the constant C, we can also say the solutions are given implicitly by

$$x^2 - 6xy + y^2 = C.$$

♦

We can also solve an exact differential equation by integrating with respect to y first. For some differential equations, this may be a way to avoid complicated integrations. With this in mind, consider the following alternative method for solving exact differential equations.

Method II for Solving Exact Equations
To find implicit solutions of an exact equation $M(x, y)\,dx + N(x, y)\,dy = 0$, execute the following steps:

1. Find a function $F(x, y)$ so that $\partial F/\partial y = N$ by integrating $N(x, y)$ with respect to y while holding x constant, thereby obtaining

$$F(x, y) = \int N(x, y)\,dy + h(x). \tag{8.5.12}$$

2. Find $h(x)$ by taking the partial derivative of both sides of (8.5.12) with respect to x and setting the result equal to $M(x, y)$. Solve for $h'(x)$. Then integrate it with respect to x. Let $h(x)$ be any antiderivative of $h'(x)$.
3. Substitute $h(x)$ into (8.5.12) to fully determine $F(x, y)$.
4. Solutions of equation $M(x, y)\,dx + N(x, y)\,dy = 0$ are given implicitly by $F(x, y) = C$, where C is an arbitrary constant.

Alternatively, first integrate $M(x, y)$ with respect to x. See Method I.

Remark The function $F(x, y)$ is known as a ***first integral*** of the differential equation $M\,dx + N\,dy = 0$. This is discussed in more detail in Sect. 8.7.

Example 8.5.2 Find a first integral of the differential equation

$$\left(2xy + \frac{1}{1+x^2}\right) dx + \left(x^2 - \sin y\right) dy = 0 \tag{8.5.13}$$

and its implicit solutions.

Solution Although we do not know if (8.5.13) is exact in some open rectangular region $\mathcal{R}$, let us assume for now that it is. If that is the case, then there is a function F whose first-order partial derivatives are

$$F_x(x, y) = M(x, y) = 2xy + \frac{1}{1+x^2} \tag{8.5.14}$$

and

$$F_y(x, y) = N(x, y) = x^2 - \sin y \tag{8.5.15}$$

for $(x, y) \in \mathcal{R}$. Now let us illustrate Method II by first integrating $N(x, y)$ with respect to y to find F. By (8.5.12), we have

$$F(x, y) = \int (x^2 - \sin y)\, dy + h(x) = x^2 y + \cos y + h(x). \tag{8.5.16}$$

The partial derivative of (8.5.16) with respect to x is

$$F_x(x, y) = 2xy + h'(x).$$

Setting this equal to F_x in (8.5.14) and solving for $h'(x)$ gives

$$h'(x) = \frac{1}{1+x^2}.$$

An integration of $h'(x)$ with respect to x yields the antiderivatives $\tan^{-1} x + C$. So let $h(x) = \tan^{-1} x$. Thus,

$$F(x, y) = x^2 y + \cos y + \tan^{-1} x,$$

which is a first integral of (8.5.13). However, this result for F is predicated on (8.5.13) being an exact differential equation. It is left to the reader to verify that it is in fact an exact equation by showing that the total differential of F is equal to the left-hand side of (8.5.13). Therefore, solutions of (8.5.13) are given implicitly by

$$x^2 y + \cos y + \tan^{-1} x = C.$$

♦

8.6 Criterion for Exactness

As in the previous section, let the functions $M(x, y)$ and $N(x, y)$ be continuous and have continuous first-order partial derivatives in an open rectangular region $\mathcal{R}$. In that section, we determined that $M\, dx + N\, dy = 0$ is exact only if $M_y = N_x$. Now the question arises whether or not the converse of this is true, which is statement (8.5.7). So let start off by supposing

$$\frac{\partial}{\partial y}M(x, y) = \frac{\partial}{\partial x}N(x, y)$$

at all points $(x, y) \in \mathcal{R}$ and that this implies the consequent of (8.5.7) is true. Then by Definition 8.3.3 a function $F(x, y)$ exists such that

$$F_x(x, y) = M(x, y) \quad \text{and} \quad F_y(x, y) = N(x, y)$$

for all $(x, y) \in \mathcal{R}$. Is there such a function? Finding a function F so that $F_x = M$ presents no problem because we can simply let

$$F(x, y) = \int M(x, y)\, dx + g(y), \tag{8.6.1}$$

where $g(y)$ denotes any differentiable function of y. Now we have to find a $g(y)$, if there is one, so that $F_y = N$. That would establish that (8.5.7), the converse of (8.5.5), is a true statement.

The partial derivative of (8.6.1) with respect to y is

$$F_y(x, y) = \frac{\partial}{\partial y}\int M(x, y)\, dx + g'(y). \tag{8.6.2}$$

Since we are seeking a function $F(x, y)$ whose partial derivative F_y is equal to N, let us solve this equation for $g'(y)$ and replace F_y with N:

$$g'(y) = N(x, y) - \frac{\partial}{\partial y}\int M(x, y)\, dx. \tag{8.6.3}$$

However, this equation seems self-contradictory. According to the notation on the left-hand side, g depends only on y. However, the right-hand side suggests that g depends on both x and y. But in fact, it does not—it is a function of y alone. This is seen by taking the partial derivative of the right-hand side with respect to x. Since $M_y = N_x$,

$$\begin{aligned}
&\frac{\partial}{\partial x}\left(N(x, y) - \frac{\partial}{\partial y}\int M(x, y)\, dx\right) \\
&= \frac{\partial}{\partial x}N(x, y) - \frac{\partial}{\partial x}\left(\frac{\partial}{\partial y}\int M(x, y)\, dx\right) = N_x(x, y) - \frac{\partial}{\partial y}\left(\frac{\partial}{\partial x}\int M(x, y)\, dx\right) \\
&= N_x(x, y) - \frac{\partial}{\partial y}M(x, y) = N_x(x, y) - N_x(x, y) = 0.
\end{aligned}$$

Note that the change in the order of differentiation is valid due to the continuity of $\int M(x, y)\,dx$ and its first-order partial derivatives in $\mathcal{R}$, which is inherited from the continuity of $M(x, y)$ and its first-order partial derivatives (see Clairaut's Theorem).

As a result, for $g(y)$ we can choose any antiderivative of the right-hand side of (8.6.3). It follows from (8.6.1) that

$$F(x, y) = \int M(x, y)\,dx + \int \left(N(x, y) - \frac{\partial}{\partial y} \int M(x, y)\,dx \right) dy.$$

Finally, let us verify that this has the correct partial derivatives. Since the second term depends only on y, $\partial F/\partial x = M(x, y)$. Moreover,

$$\frac{\partial}{\partial y} F(x, y) = \frac{\partial}{\partial y} \int M(x, y)\,dx + N(x, y) - \frac{\partial}{\partial y} \int M(x, y)\,dx = N(x, y).$$

This establishes that the converse (8.5.7) is a true statement. As a result of this and the discussion in the previous section, we have the following important criterion that allows us to ascertain whether or not an equation is exact simply by computing and comparing two partial derivatives.

Criterion for Exactness

Theorem 8.6.1 *If the functions $M(x, y)$ and $N(x, y)$ are continuous and if they have continuous first-order partial derivatives in an open rectangular region $\mathcal{R}$, then*

$$M(x, y)\,dx + N(x, y)\,dy = 0 \textit{ is an exact differential equation in } \mathcal{R}$$

if and only if

$$\frac{\partial}{\partial y} M(x, y) = \frac{\partial}{\partial x} N(x, y) \textit{ at every point } (x, y) \in \mathcal{R}.$$

In light of this criterion, we now have a way of testing whether equation (8.5.1) in Sect. 8.5 is exact or not. Recall that at the time that it was too difficult to determine this by inspection.

Example 8.6.1 Use the Criterion for Exactness (Theorem 8.6.1) to show that

$$\left(2\cos x - 2\cos(2x)\cos y + \frac{1}{x^2 + 4} \right) dx + \sin(2x)\sin y\,dy = 0 \qquad (8.5.1)$$

is an exact differential equation. Then solve it.

Solution Using the notation of Theorem 8.6.1,

$$M(x, y) = 2\cos x - 2\cos(2x)\cos y + \frac{1}{x^2+4}; \quad N(x, y) = \sin(2x)\sin y.$$

Since

$$\frac{\partial M}{\partial y} = 2\cos(2x)\sin y = \frac{\partial N}{\partial x}$$

holds everywhere, the equation is exact in the entire xy-plane. Thus, there exists a function $F(x, y)$ such that

$$\frac{\partial F}{\partial x} = 2\cos x - 2\cos(2x)\cos y + \frac{1}{x^2+4}$$

and

$$\frac{\partial F}{\partial y} = \sin(2x)\sin y$$

at all points in the xy-plane. To find F, let us integrate $\partial F/\partial y$, or N, with respect to y (see Method II in Sect. 8.5):

$$F(x, y) = \int \sin(2x)\sin y\,dy + h(x) = -\sin(2x)\cos y + h(x). \tag{8.6.4}$$

To find a function $h(x)$ so that $F(x, y)$ has the above partial derivatives, let us take the partial derivative of (8.6.4) with respect to x and then set the result equal to M:

$$-2\cos(2x)\cos y + h'(x) = 2\cos x - 2\cos(2x)\cos y + \frac{1}{x^2+4}.$$

And so

$$h'(x) = 2\cos x + \frac{1}{x^2+4}.$$

Integrating this with respect to x, we have

$$h(x) = 2\sin x + \frac{1}{2}\tan^{-1}\frac{x}{2}.$$

Substituting h into (8.6.4), we obtain

$$F(x, y) = -\sin(2x)\cos y + 2\sin x + \frac{1}{2}\tan^{-1}\frac{x}{2}.$$

Therefore, the solutions of the differential equation are given implicitly by

$$2 \sin x - \sin(2x) \cos y + \frac{1}{2} \tan^{-1} \frac{x}{2} = C, \tag{8.6.5}$$

which concludes this example. ♦

Example 8.6.2 Solve the differential equation

$$\frac{dy}{dx} = \frac{\cos x - y^2}{2xy}. \tag{8.6.6}$$

Solution First note that (8.6.6) is neither a linear nor a separable equation. Hopefully, it is an exact equation. To ascertain whether this is the case, let us first rewrite the equation in the differential form $M(x, y)\, dx + N(x, y)\, dy = 0$:

$$(\cos x - y^2)\, dx - 2xy\, dy = 0. \tag{8.6.7}$$

Thus,

$$M(x, y) = \cos x - y^2 \quad \text{and} \quad N(x, y) = -2xy.$$

Computing the partial derivatives M_x and N_y in order to ascertain whether the equation is exact or not, we get

$$\frac{\partial M}{\partial y} = \frac{\partial}{\partial y}(\cos x - y^2) = -2y; \quad \frac{\partial N}{\partial x} = \frac{\partial}{\partial x}(-2xy) = -2y.$$

Thus we see that

$$M_x(x, y) = -2y = N_y(x, y)$$

at all points in the xy-plane. Hence, according to the Criterion for Exactness, Eq. (8.6.7) is exact everywhere. Consequently, a function $F(x, y)$ exists with the first-order partial derivatives:

$$\frac{\partial F}{\partial x} = \cos x - y^2 \quad \text{and} \quad \frac{\partial F}{\partial y} = -2xy. \tag{8.6.8}$$

Integrating $\partial F/\partial x$ with respect to x (see Method I in Sect. 8.5), we get

$$F(x, y) = \int (\cos x - y^2)\, dx + g(y) = \sin x - xy^2 + g(y), \tag{8.6.9}$$

where g denotes a differentiable function of y. To find g, let us take the partial derivative of (8.6.9) with respect to y:

$$\frac{\partial F}{\partial y} = \frac{\partial}{\partial y}[\sin x - xy^2 + g(y)] = -2xy + g'(y).$$

It follows from (8.6.8) that

$$-2xy + g'(y) = -2xy.$$

Thus, $g'(y) = 0$. A function with this derivative is $g(y) \equiv 0$. But, of course, we could also choose $g(y) \equiv k$, where k is any real number. Regardless of the value chosen for k, we conclude that the solutions of (8.6.6) are given implicitly by

$$\sin x - xy^2 = C.$$ ♦

8.7 First Integrals and Integral Curves

We introduced the term "first integral" in the remark before Example 8.5.2 to denote a function $F(x, y)$ whose total differential is equal to the left-hand side of the differential equation $M(x, y)\,dx + N(x, y)\,dy = 0$. Let us examine this term more closely as it pertains to a given first-order differential equation[6] expressed in the form

$$y' = f(x, y), \tag{8.7.1}$$

where f is continuous on an open set U in the xy-plane.[7]

First Integral

Definition 8.7.1 Let a nonconstant function $F(x, y)$ and its first-order partial derivatives exist and be continuous on an open rectangular region $\mathcal{R} \subseteq U$, where U is an open set in the xy-plane. The function F is called a ***first integral*** of (8.7.1) if

$$F_x(x, y) + F_y(x, y)f(x, y) = 0 \tag{8.7.2}$$

for all $(x, y) \in \mathcal{R}$.

[6] See [44, p. 115].

[7] A set U in the xy-plane is said to be ***open*** if each point $(x, y) \in U$ is the center of a disk of positive radius that lies entirely in U.

Remark A first integral is not unique. If F is a first integral of (8.7.1), then so are $F + k$ and kF, where k is a nonzero constant, since they also satisfy (8.7.2).

Example 8.7.1 Use Definition 8.7.1 to show that

$$F(x, y) = x^3 y + \frac{x^2}{y}$$

is a first integral of the differential equation

$$\frac{dy}{dx} = \frac{3xy^3 + 2y}{x - x^2 y^2}.$$

Solution Let $\mathcal{R}$ be any open rectangular region that excludes the x- and y-axes and the curve $xy^2 = 1$. Letting $f(x, y)$ designate the right-hand side of the differential equation, we have

$$\begin{aligned} F_x(x, y) + F_y(x, y) \cdot f(x, y) &= 3x^2 y + \frac{2x}{y} + \left(x^3 - \frac{x^2}{y^2}\right) \cdot \frac{3xy^3 + 2y}{x - x^2 y^2} \\ &= 3x^2 y + \frac{2x}{y} - \frac{x^2(1 - xy^2)}{y^2} \cdot \frac{3xy^3 + 2y}{x(1 - xy^2)} \\ &= 3x^2 y + \frac{2x}{y} - x\left(3xy + \frac{2}{y}\right) = 0. \end{aligned}$$

Thus $F(x, y)$ is a first integral of the differential equation on $\mathcal{R}$. ♦

Theorem 8.7.2 *A first integral of* (8.7.1) *is constant along each of its solutions.*

Proof Let $f(x, y)$ be a function that is continuous on an open set U in the xy-plane. Suppose a function $F(x, y)$ exists that possesses the properties stated in Definition 8.7.1 and which is a first integral of the differential equation

$$\frac{dy}{dx} = f(x, y).$$

Then it follows from Definition 8.7.1 that

$$F_x(x, y) + F_y(x, y) f(x, y) = 0$$

for all $(x, y) \in \mathcal{R}$, which can be rewritten as

$$\frac{d}{dx} F(x, y) = 0.$$

Consequently, if $y = \varphi(x)$ is a solution of Eq. (8.7.1) on an interval I, where $\{(x, \varphi(x)) : x \in I\} \subset \mathcal{R}$, then

$$\frac{d}{dx}F(x, \varphi(x)) = 0$$

for $x \in I$. Thus $F(x, \varphi(x)) = k$ for some constant k. Choose any point $x_0 \in I$. Then $k = F(x_0, \varphi(x_0))$. Therefore,

$$F(x, \varphi(x)) = F(x_0, \varphi(x_0))$$

for all $x \in I$. ■

If $F(x, y)$ is a first integral of a differential equation $y' = f(x, y)$, then the graph of $F(x, y) = C$ for a given value of C is called an ***integral curve*** of the differential equation. However, unlike what is usually meant by a curve, an integral curve may consist of disconnected pieces as we will see in the next example.

Example 8.7.2 Find the integral curves of the differential equation

$$\frac{dy}{dx} = \frac{x}{y}. \tag{8.7.3}$$

Determine which of the integral curves pass through the point $(-2, -1)$.

Solution Separating the variables of (8.7.3) and integrating, we get

$$y^2 - x^2 = C. \tag{8.7.4}$$

Since the first-order partial derivatives of the function $F(x, y) = y^2 - x^2$ satisfies Eq. (8.7.2) with $f(x, y) = x/y$ on any region that excludes the x-axis, it is a first integral of the differential equation (8.7.3). For a given value of the constant C, Eq. (8.7.4) defines an integral curve of (8.7.3). For example, let $C = 3$. Then (8.7.4) becomes

$$y^2 - x^2 = 3. \tag{8.7.5}$$

The graph of this equation is the hyperbola shown in Fig. 8.1, which consists of two branches: one branch is concave up with vertex $(0, \sqrt{3})$ and the other one is concave down with vertex $(0, -\sqrt{3})$. In other words, the hyperbola is an integral curve: it consists of two solution curves, to wit: the graphs of the solutions

$$y = \sqrt{x^2 + 3} \quad \text{and} \quad y = -\sqrt{x^2 + 3}$$

of the differential equation (8.7.3).

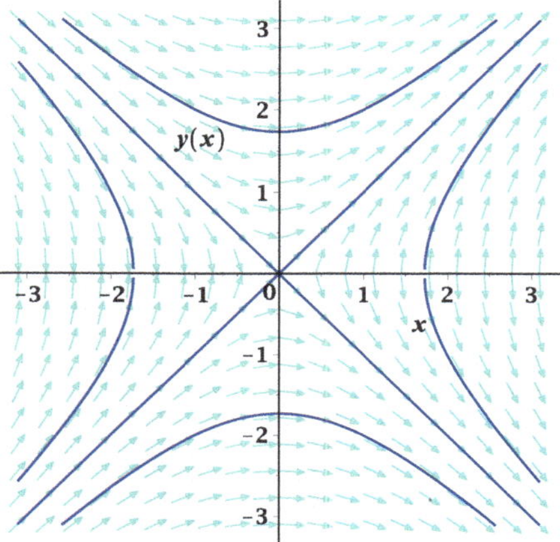

Fig. 8.1 Graphs of some integral curves of $y' = x/y$

To find the integral curve containing the point $(-2, -1)$, set $x = -2$ and $y = -1$ in (8.7.4), thereby obtaining $C = (-1)^2 - (-2)^2 = -3$. Thus,

$$y^2 - x^2 = -3 \quad \text{or} \quad x^2 - y^2 = 3.$$

Its graph is the other hyperbola shown in Fig. 8.1, which consists of a left branch with vertex $(-\sqrt{3}, 0)$ and a right branch with vertex $(\sqrt{3}, 0)$. This integral curve is composed of four solution curves. Two of them are located above the x-axis: the graphs of the solutions $y = \sqrt{x^2 - 3}$ on $(\sqrt{3}, \infty)$ and $y = \sqrt{x^2 - 3}$ on $(-\infty, -\sqrt{3})$. The other two solution curves are located below the x-axis: the graph of $y = -\sqrt{x^2 - 3}$ on $(\sqrt{3}, \infty)$ and also on $(-\infty, -\sqrt{3})$. The latter is the one whose graph includes the point $(-2, -1)$.

It is patently clear from the direction field shown in Fig. 8.1 that all integral curves of (8.7.3) are hyperbolas, except for the two lines. Setting $C = 0$, we get the integral curve $y^2 - x^2 = 0$, which yields the equations $y = \pm x$ of these lines (solution curves). Note that they are also the slant asymptotes of the hyperbolas. ♦

Before we look at the next example, let us introduce a new term that we will need. Suppose we are dealing with a function $z = F(x, y)$, regardless of whether or not it is associated with a differential equation. Its graph is a surface in three-dimensional space, such as the one shown in Fig. 8.3. For a constant C whose value lies in the range of F, the set of all points in the xy-plane satisfying the equation $F(x, y) = C$ is known as a ***level curve*** of the function F. If F also happens to be a first integral of the differential equation (8.7.1), then the level curve $F(x, y) = C$ is an integral curve of this differential equation.

Generally speaking, integral curves are more complicated than lines, hyperbolas, circles, parabolas, or any of the other familiar curves of analytic geometry. For example, consider the integral curves of the differential equation

$$\left(2\cos x - 2\cos(2x)\cos y + \frac{1}{x^2+4}\right)dx + \sin(2x)\sin y\,dy = 0 \qquad (8.5.1)$$

which we determined in Example 8.6.1 to be

$$2\sin x - \sin(2x)\cos y + \frac{1}{2}\tan^{-1}\frac{x}{2} = C. \qquad (8.6.5)$$

Without the assistance of technology, who could imagine what these integral curves look like? The result of using a CAS to graph some of them is shown in Fig. 8.2. The smallest closed curve shown in the interior of Quadrant I is the integral curve through the point $(1, 3)$. It corresponds to the value of C given by

$$C = 2\sin(1) - \sin(2)\cdot\cos(3) + \frac{1}{2}\tan^{-1}\frac{1}{2} = 2.81496\ldots.$$

Another example is the integral curve corresponding to $C = 0$. It consists of the set of points satisfying the equation

$$2\sin x - \sin(2x)\cos y + \frac{1}{2}\tan^{-1}\frac{x}{2} = 0.$$

Note that a point $(0, y)$ for any value of y satisfies this equation. Thus, the integral curve corresponding to $C = 0$ is the vertical line $x = 0$.

The graph of the function

$$z = 2\sin x - \sin(2x)\cos y + \frac{1}{2}\tan^{-1}\frac{x}{2} \qquad (8.7.6)$$

is the surface shown in Fig. 8.3. The intersection of a plane that is parallel to the xy-plane with the surface $z = F(x, y)$ results in a curve. This curve is called a ***trace***

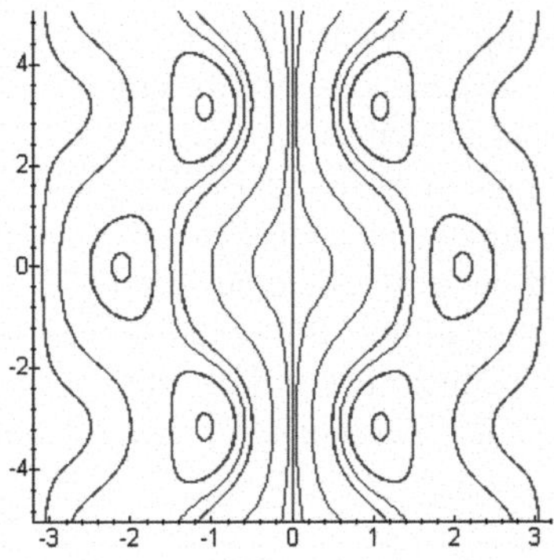

Fig. 8.2 Integral curves of (8.5.1)

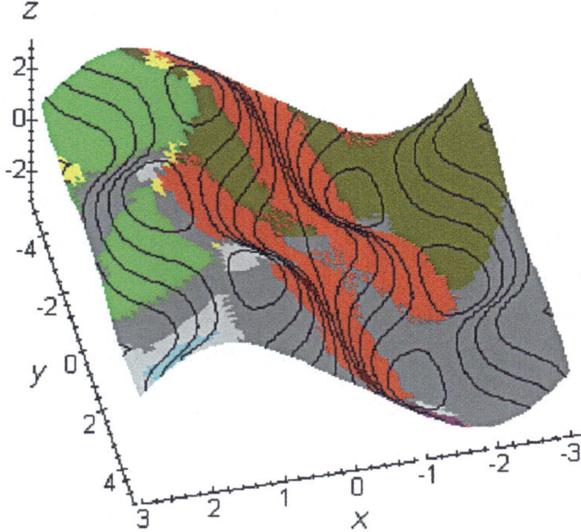

Fig. 8.3 Graph of the function (8.7.6)

(or cross-section) of the surface. That is, for a given value of the constant C, the ***trace*** is the graph of the curve $F(x, y) = C$ that lies in the $z = C$ plane. So what is the difference between a trace and a level curve? The answer is that the graph of a trace lies in the intersecting plane, whereas the graph of the corresponding level curve lies in the xy-plane. In other words, the projection of a trace onto the xy-plane is the corresponding level curve. The traces corresponding to the level curves (8.6.5) in Fig. 8.2 can be easily seen on the surface defined by (8.7.6) in Fig. 8.3.

Integral curves of a differential equation

$$M(x, y)\,dx + N(x, y)\,dy = 0, \tag{8.7.7}$$

irrespective of whether or not it is exact, can be graphed with a CAS. A way to do this is to first rewrite (8.7.7) as

$$\frac{dy}{dx} = -\frac{M(x, y)}{N(x, y)}. \tag{8.7.8}$$

Then introduce a parameter t in order to express x and y as functions of this parameter. By the chain rule,

$$\frac{dy}{dx} \cdot \frac{dx}{dt} = \frac{dy}{dt}.$$

Using this to replace the derivative dy/dx in (8.7.8), we have

$$\frac{dy/dt}{dx/dt} = -\frac{M(x, y)}{N(x, y)}. \tag{8.7.9}$$

Note that a pair of functions $x(t)$ and $y(t)$ satisfying both of the equations of the system

$$\frac{dx}{dt} = N(x, y) \tag{8.7.10a}$$

$$\frac{dy}{dt} = -M(x, y). \tag{8.7.10b}$$

on an interval $a < t < b$ would also satisfy Eq. (8.7.9) on this interval, and conversely. The parametric plot of $(x(t), y(t))$ in the xy-plane is an integral curve of (8.7.7). In order to obtain an integral curve containing a particular point (x_0, y_0), use the initial conditions

$$x(0) = x_0, \quad y(0) = y_0.$$

This is how the integral curves of (8.5.1) depicted in Fig. 8.2 were obtained. That is, the differential equation (8.5.1) was first rewritten as the system of equations:

$$\frac{dx}{dt} = N(x, y) = \sin(2x)\sin y$$

$$\frac{dy}{dt} = -M(x, y) = 2\cos(2x)\cos y - 2\cos x - \frac{1}{x^2+4}.$$

Then, after specifying some desired initial conditions, the computer algebra system *Maple* was used to graph the integral curves. For example, the integral curve through the point (1, 3) was graphed by specifying that $x(0) = 1$ and $y(0) = 3$. *Maple* was also used to draw the surface and traces shown in Fig. 8.3. Note that the projections of these traces onto the xy-plane are the integral curves depicted in Fig. 8.2.

8.8 Special Integrating Factors

Up to now, we have dealt solely with exact equations: how to recognize and solve them. But most differential equations are not exact; so other methods will have to be employed to find solutions of these equations. On the other hand, as we will see shortly, even though a differential equation may not be exact, there is still the possibility that it could be converted into one by multiplying the equation by an integrating factor. In Chap. 5 we showed how to solve the first-order linear equation

$$\frac{dy}{dx} + p(x)y = q(x) \tag{8.8.1}$$

by means of an integrating factor $\mu(x)$. Recall that $\mu(x)$ is

$$\mu(x) = e^{\int p(x)\,dx}, \tag{8.8.2}$$

where the exponent $\int p(x)\,dx$ denotes any antiderivative of $p(x)$. It is clear from the differential form

$$\big[p(x)y - q(x)\big]\,dx + dy = 0 \tag{8.8.3}$$

of (8.8.1) that it is not an exact differential equation because

$$\frac{\partial}{\partial y}\big[p(x)y - q(x)\big] \neq \frac{\partial}{\partial x}1.$$

In other words, it fails to satisfy the Criterion for Exactness. Even so, could we still make use of $\mu(x)$ to find solutions of (8.8.3) without having to revert to the standard form (8.8.1)? Let us attempt to answer this by multiplying (8.8.3) by $\mu(x)$ in order to apply the Criterion for Exactness to

$$\big[\mu(x)p(x)y - \mu(x)q(x)\big]\,dx + \mu(x)\,dy = 0 \tag{8.8.4}$$

to see where it might lead. Taking the appropriate partial derivatives of the coefficients of the differentials dx and dy, we get

$$\frac{\partial}{\partial y}\big[\mu(x)p(x)y - \mu(x)q(x)\big] = \mu(x)p(x)$$

and

$$\frac{\partial}{\partial x}\mu(x) = \frac{d}{dx}e^{\int p(x)\,dx} = e^{\int p(x)\,dx}\,p(x) = \mu(x)p(x).$$

Consequently,

$$\frac{\partial}{\partial y}\big[\mu(x)p(x)y - \mu(x)q(x)\big] = \frac{\partial}{\partial x}\mu(x)$$

holds everywhere. It follows from the Criterion for Exactness that Eq. (8.8.4) is exact throughout the entire xy-plane. We conclude that the method for solving exact equations can be used to solve (8.8.3) provided we first multiply it by the integrating factor given by (8.8.2) in order to convert it into an exact equation. We illustrate this with the next example.

Example 8.8.1 Find the general solution of

$$\frac{dy}{dx} + 2y = 9e^x. \tag{8.8.5}$$

Solution The differential form of this equation is

$$(2y - 9e^x)\,dx + dy = 0. \tag{8.8.6}$$

An integrating factor for (8.8.6) is given by (8.8.2) with $p(x) \equiv 2$. Thus,

$$\mu(x) = e^{\int p(x)\,dx} = e^{\int 2\,dx} = e^{2x}.$$

Multiplying both sides of (8.8.6) by $\mu(x)$, we get

$$(2e^{2x}y - 9e^{3x})\,dx + e^{2x}dy = 0. \tag{8.8.7}$$

Before continuing, it would be prudent to verify that this equation is exact in case a computation was carried out incorrectly. Since

$$\frac{\partial}{\partial y}\left(2e^{2x}y - 9e^{3x}\right) = \frac{\partial}{\partial x}e^{2x},$$

it follows from the Criterion for Exactness that it is indeed an exact equation. Hence, a function $F(x, y)$ exists whose first-order partial derivatives are

$$F_x(x, y) = 2e^{2x}y - 9e^{3x} \quad \text{and} \quad F_y(x, y) = e^{2x}.$$

Integrating the latter with respect to y, we obtain

$$F(x, y) = \int F_y(x, y)\,dy = \int e^{2x}\,dy = e^{2x}y + h(x),$$

where h designates a differentiable function of x. From this, we have

$$F_x(x, y) = \frac{\partial}{\partial x}[e^{2x}y + h(x)] = 2e^{2x}y + h'(x).$$

Equating both expressions for F_x, we get

$$2e^{2x}y + h'(x) = 2e^{2x}y - 9e^{3x}.$$

Thus $h'(x) = -9e^{3x}$. So let $h(x) = -3e^{3x}$. Thus, a first integral of Eq. (8.8.6) is

$$F(x, y) = e^{2x}y - 3e^{3x}.$$

Therefore, solutions are given by $e^{2x}y - 3e^{3x} = C$, where C is a constant. Solving for y, we find that the general solution of Eq. (8.8.5) is

$$y = 3e^{x} + Ce^{-2x} \tag{8.8.8}$$

for $-\infty < x < \infty$. ♦

Example 8.8.1 is not meant to suggest supplanting the integrating factor method for finding solutions of first-order linear equations that was described in Chap. 5.

The solution (8.8.8) is precisely what we would have obtained had we kept (8.8.5) in standard form rather than changing it to its differential form (8.8.6); in fact, working with the standard form requires less work as can be seen in Example 5.2.1. The real purpose of this example is to point out that some differential equations that are non-exact can be converted into equations that are exact.

Before continuing with more examples, let us define what we mean by the term *integrating factor* as it pertains to a non-exact equation $M(x, y)\,dx + N(x, y)\,dy = 0$.

Integrating Factor

Definition 8.8.1 If the differential equation

$$M(x, y)\,dx + N(x, y)\,dy = 0 \tag{8.8.9}$$

is not exact but there is a nonzero function $\mu(x, y)$ such that

$$\mu(x, y)M(x, y)\,dx + \mu(x, y)N(x, y)\,dy = 0 \tag{8.8.10}$$

is exact on some open rectangular region, then $\mu(x, y)$ is called an ***integrating factor*** for Eq. (8.8.9).

Example 8.8.2 The function $\mu(x) = e^{2x}$ is an integrating factor for (8.8.6). ♦

Example 8.8.3 Show that $\mu(x, y) = xy^2$ is an integrating factor for

$$(3xy^2 - 5y^{-2}\sin x^2)\,dx + 4x^2y\,dy = 0 \tag{8.8.11}$$

and then use it to solve this differential equation. Is there a solution that satisfies the initial condition $y(\sqrt{\pi}) = 1$?

Solution It follows from the Criterion for Exactness that (8.8.11) is non-exact because

$$\frac{\partial}{\partial y}\left(3xy^2 - 5y^{-2}\sin x^2\right) \neq \frac{\partial}{\partial x}\left(4x^2y\right).$$

Multiplying (8.8.11) by $\mu(x, y) = xy^2$, we get

$$(3x^2y^4 - 5x\sin x^2)\,dx + 4x^3y^3\,dy = 0. \tag{8.8.12}$$

Since

$$\frac{\partial}{\partial y}\left(3x^2y^4 - 5x\sin x^2\right) = \frac{\partial}{\partial x}\left(4x^3y^3\right),$$

Equation (8.8.12) is exact. Thus, $\mu(x, y) = xy^2$ is an integrating factor for (8.8.11).

Let us use $\mu(x, y)$ to find solutions of (8.8.12); but before we do, note how its solutions and those of the original equation (8.8.11) are related. If an implicit solution of Eq. (8.8.12) defines an explicit solution $y = \varphi(x)$ on some interval I, then

$$\left(3x^2(\varphi(x))^4 - 5x\sin x^2\right)dx + 4x^3(\varphi(x))^3\,\varphi'(x)\,dx = 0$$

for $x \in I$. Factoring out $x(\varphi(x))^2$, this becomes

$$x(\varphi(x))^2\left[\left(3x(\varphi(x))^2 - 5(\varphi(x))^{-2}\sin x^2\right)dx + 4x^2\varphi(x)\,\varphi'(x)\,dx\right] = 0.$$

Suppose $0 \notin I$ and that $\varphi(x) \neq 0$ for all $x \in I$. Then as $x(\varphi(x))^2 \neq 0$, the bracketed quantity is equal to 0 for all $x \in I$. In other words, the solution $y = \varphi(x)$ of the exact equation (8.8.12) is also a solution of the original equation (8.8.11).

Now let us continue with solving (8.8.12). Since it is an exact equation, there is a function $F(x, y)$ with the first-order partial derivatives

$$F_x(x, y) = 3x^2y^4 - 5x\sin x^2 \quad \text{and} \quad F_y(x, y) = 4x^3y^3.$$

To find $F(x, y)$, let us integrate the second term (since it has only one term) with respect to y (cf. Method II in Sect. 8.5):

$$F(x, y) = \int 4x^3y^3\,dy + h(x) = x^3y^4 + h(x).$$

Its partial derivative with respect to x is

$$F_x(x, y) = 3x^2y^4 + h'(x).$$

It follows from comparing both expressions for F_x that

$$h'(x) = -5x\sin x^2.$$

Thus,

$$h(x) = -5\int x\sin x^2\,dx = \frac{5}{2}\cos x^2$$

and so

$$F(x, y) = x^3y^4 + \frac{5}{2}\cos x^2.$$

Consequently, solutions of Eq. (8.8.12) are given implicitly by

$$2x^3y^4 + 5\cos x^2 = C. \tag{8.8.13}$$

To answer the question as to whether there is a solution that satisfies the initial condition $y(\sqrt{\pi}) = 1$, set $x = \sqrt{\pi}$ and $y = 1$ in (8.8.13):

$$C = 2(\sqrt{\pi})^3 + 5\cos\pi = 2\pi\sqrt{\pi} - 5.$$

For this value of C, Eq. (8.8.13) becomes

$$2x^3y^4 + 5\cos x^2 = 2\pi\sqrt{\pi} - 5.$$

Solving this equation for y, we get

$$y = \sqrt[4]{\frac{2\pi\sqrt{\pi} - 5(1+\cos x^2)}{2x^3}}. \tag{8.8.14}$$

Note that

$$2\pi\sqrt{\pi} - 5(1+\cos x^2) > 0$$

since $2\pi\sqrt{\pi} > 11$ and $5(1+\cos x^2) \leq 10$. It follows that this function is defined for all $x > 0$.

We conclude that the function (8.8.14) satisfies the exact equation (8.8.12) on the interval $(0, \infty)$ as well as the initial condition $y(\sqrt{\pi}) = 1$. Moreover, since for this function $xy^2 > 0$ for all $x > 0$, it follows from what we had noted earlier that this function is also the solution of the original equation (8.8.11) on $(0, \infty)$ which satisfies the initial condition $y(\sqrt{\pi}) = 1$. ♦

The integrating factor $\mu(x, y) = xy^2$ in the example just popped up out of the blue with no explanation as to where it came from. But had it not been given, is there some way to find it? The answer to this lies with the Criterion for Exactness. In order for a function $\mu(x, y)$ to be an integrating factor for (8.8.9), it has to satisfy the condition

$$\frac{\partial}{\partial y}\left[\mu(x, y)M(x, y)\right] = \frac{\partial}{\partial x}\left[\mu(x, y)N(x, y)\right],$$

which can be rewritten as

$$\mu\frac{\partial M}{\partial y} + M\frac{\partial \mu}{\partial y} = \mu\frac{\partial N}{\partial x} + N\frac{\partial \mu}{\partial x}.$$

In other words, an integrating factor $\mu(x, y)$ must satisfy the condition

$$\mu\left(\frac{\partial M}{\partial y}-\frac{\partial N}{\partial x}\right)=N\frac{\partial\mu}{\partial x}-M\frac{\partial\mu}{\partial y}. \tag{8.8.15}$$

Unfortunately, (8.8.15) is a partial differential equation that is too difficult to be of any use here unless some simplifying assumptions are made. We will consider a few of the most well-known. One is to assume that an integrating factor for (8.8.9) has the special form $\mu(x, y) = x^m y^n$. However, because of (8.8.15), it is not an integrating factor unless

$$x^m y^n\left(\frac{\partial M}{\partial y}-\frac{\partial N}{\partial x}\right)=N\frac{\partial}{\partial x}\left(x^m y^n\right)-M\frac{\partial}{\partial y}\left(x^m y^n\right),$$

which simplifies to

$$M_y - N_x = m\left(\frac{N}{x}\right)-n\left(\frac{M}{y}\right). \tag{8.8.16}$$

Example 8.8.4 Use (8.8.16) to find an integrating factor for

$$(3xy^2 - 5y^{-2}\sin x^2)\,dx + 4x^2y\,dy = 0. \tag{8.8.11}$$

Solution The function $\mu(x, y) = x^m y^n$ is an integrating factor for (8.8.11) if there are values for m and n satisfying (8.8.16) with $M = 3xy^2 - 5y^{-2}\sin x^2$ and $N = 4x^2y$. For this M and N, condition (8.8.16) is

$$6xy + 10y^{-3}\sin x^2 - 8xy = m\left(\frac{4x^2y}{x}\right)-n\left(\frac{3xy^2-5y^{-2}\sin x^2}{y}\right).$$

This simplifies to

$$-2xy + 10y^{-3}\sin x^2 = xy(4m-3n) + 5ny^{-3}\sin x^2.$$

This is an identity if

$$4m - 3n = -2 \quad\text{and}\quad 5n = 10.$$

Since $m = 1$ and $n = 2$ solves this system, we conclude that xy^2 is an integrating factor for (8.8.11). This explains where it came from in Example 8.8.3. ♦

Condition (8.8.16) resulted from looking for an integrating factor of the form $x^m y^n$ for Eq. (8.8.9). If this type of integrating factor does not work for some particular equation that we are trying to solve, perhaps there is an alternative. Let us determine the condition that would have to be satisfied in order to have an integrating factor that is a function of x alone. Let $\mu(x)$ denote this integrating factor. Then, as $\partial\mu/\partial y = 0$ and $\partial\mu/\partial x = d\mu/dx$, condition (8.8.15) becomes

$$\mu\left(\frac{\partial M}{\partial y} - \frac{\partial N}{\partial x}\right) = N\frac{d\mu}{dx}$$

or

$$\frac{1}{\mu}\frac{d\mu}{dx} = \frac{M_y - N_x}{N}.$$

Since the left-hand side depends on x alone, the same would have to be true of the right-hand side. Let $p(x)$ denote the right-hand side. Thus we have the condition

$$\frac{1}{\mu}\frac{d\mu}{dx} = p(x).$$

If $p(x)$ is continuous on an interval, then $\exp\left(\int p(x)\,dx\right)$ is a solution of this equation on the interval. Consequently, it is an integrating factor for (8.8.9).

Special Integrating Factor $\mu(x)$

Theorem 8.8.2 *Suppose*

$$M(x, y)\,dx + N(x, y)\,dy = 0 \tag{8.8.9}$$

is not an exact equation. If p defined by

$$p := \frac{M_y - N_x}{N} \tag{8.8.17}$$

is a continuous function of x alone, then

$$\mu(x) = e^{\int p(x)\,dx}. \tag{8.8.18}$$

is an integrating factor for (8.8.9).

Example 8.8.5 Solve the differential equation

$$\left(\frac{x}{y^2} - x^2\right)\frac{dy}{dx} = 3xy + \frac{2}{y}. \tag{8.8.19}$$

Find solutions satisfying the initial conditions $y(-1) = -1$ and $y(1) = 1$.

Solution Observe that Eq. (8.8.19) is not defined at points on the coordinate axes. In order to solve the equation, we begin by writing it in the differential form

$$\left(3xy+\frac{2}{y}\right)dx+\left(x^2-\frac{x}{y^2}\right)dy=0. \tag{8.8.20}$$

Let

$$M(x,y)=3xy+\frac{2}{y} \quad \text{and} \quad N(x,y)=x^2-\frac{x}{y^2}.$$

Since

$$\frac{\partial M}{\partial y}=3x-\frac{2}{y^2} \quad \text{and} \quad \frac{\partial N}{\partial x}=2x-\frac{1}{y^2},$$

the equation is not exact as $M_y \neq N_x$. However, because

$$p=\frac{M_y-N_x}{N}=\frac{3x-\dfrac{2}{y^2}-\left(2x-\dfrac{1}{y^2}\right)}{x^2-\dfrac{x}{y^2}}=\frac{x-\dfrac{1}{y^2}}{x\left(x-\dfrac{1}{y^2}\right)}=\frac{1}{x}$$

is a continuous function of x alone on any interval that excludes $x=0$, the equation has an integrating factor. From (8.8.18), we have

$$e^{\int p(x)\,dx}=e^{\int \frac{1}{x}\,dx}=e^{\ln|x|}=|x|.$$

So either x or $-x$ can be used as an integrating factor for (8.8.20). Choosing x and multiplying the equation by it, we get

$$\left(3x^2y+\frac{2x}{y}\right)dx+\left(x^3-\frac{x^2}{y^2}\right)dy=0. \tag{8.8.21}$$

This is an exact equation on any open rectangular region $\mathcal{R}$ that excludes both axes. Hence, there is a function $F(x,y)$ on $\mathcal{R}$ such that

$$F_x(x,y)=3x^2y+\frac{2x}{y} \quad \text{and} \quad F_y(x,y)=x^3-\frac{x^2}{y^2}.$$

It is easy to see that a function with these partial derivatives is $F(x,y)=x^3y+x^2y^{-1}$ (cf. Example 8.7.1). Consequently, solutions of (8.8.21) are defined implicitly by

$$x^3y+\frac{x^2}{y}=C. \tag{8.8.22}$$

This is also an implicit solution of (8.8.20) since the only difference between the two differential equations is the factor x.

To find solution satisfying the initial condition $y(-1) = -1$, set $x = -1$ and $y = -1$ in (8.8.22). This yields $C = 0$. Thus,

$$x^3 y + \frac{x^2}{y} = 0 \quad \text{or} \quad x^2 \left(xy + \frac{1}{y} \right) = 0.$$

It follows that an explicit solution of (8.8.20) (or (8.8.19)) such that $y(-1) = -1$ must satisfy the equation

$$xy + \frac{1}{y} = 0.$$

This defines two functions $y = \pm 1/\sqrt{-x}$ on the interval $(-\infty, 0)$. Because of the initial condition, the solution of (8.8.19) with $y(-1) = -1$ is

$$y = -\frac{1}{\sqrt{-x}}$$

for $-\infty < x < 0$. Its graph is the upper curve in Quadrant III in Fig. 8.4.

As for the initial condition $y(1) = 1$, it is satisfied if $C = 2$ in (8.8.22). Thus, a solution must satisfy the equation

$$x^3 y + \frac{x^2}{y} = 2.$$

Its graph is the integral curve shown in Fig. 8.4, which consists of four disconnected continuous curves. Since $x^3 y^2 - 2y + x^2 = 0$ is quadratic in y, the quadratic formula

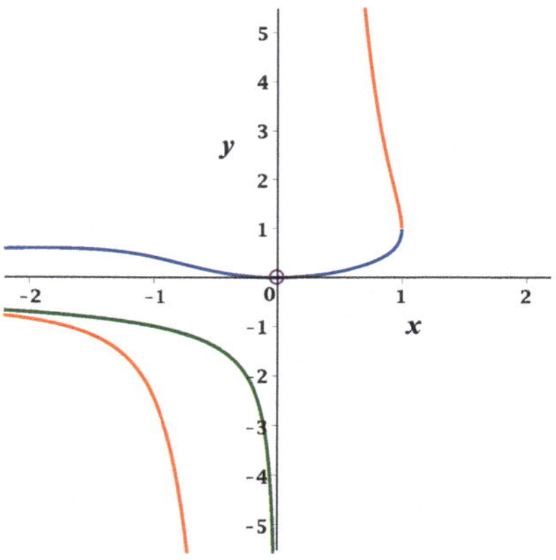

Fig. 8.4 Five solution curves of $(xy^{-2} - x^2)y' = 3xy + 2y^{-1}$

yields

$$y = \frac{2 \pm \sqrt{4 - 4x^5}}{2x^3} = \frac{1 \pm \sqrt{1 - x^5}}{x^3}.$$

Both of these functions are solutions of (8.8.19) on the interval (0, 1] and both satisfy the initial condition. Their graphs are the two curves joined at the point (1, 1) in Quadrant I, where

$$\frac{1 - \sqrt{1 - x^5}}{x^3} \to 0 \quad \text{and} \quad \frac{1 + \sqrt{1 - x^5}}{x^3} \to \infty$$

as $x \to 0^+$. ♦

Just as $M(x, y)\,dx + N(x, y)\,dy = 0$ may have an integrating factor that depends on x alone, another equation may have an integrating factor that depends only on y. The proof of the next theorem is similar to the proof of Theorem 8.8.2 and is left as an exercise (see Problem 49).

Special Integrating Factor $\mu(y)$

Theorem 8.8.3 *Suppose*

$$M(x, y)\,dx + N(x, y)\,dy = 0 \tag{8.8.9}$$

is not an exact equation. If q defined by

$$q := \frac{M_y - N_x}{M} \tag{8.8.23}$$

is a continuous function of y alone, then

$$\mu(y) = e^{-\int q(y)\,dy} \tag{8.8.24}$$

is an integrating factor for (8.8.9).

Example 8.8.6 Find a general solution of

$$2y\,dx + (4x + y)\,dy = 0. \tag{8.8.25}$$

Solution Let $M(x, y) = 2y$ and $N(x, y) = 4x + y$. Then

$$\frac{\partial M}{\partial y} = 2 \quad \text{and} \quad \frac{\partial N}{\partial x} = 4.$$

Since $M_y \neq N_x$, the equation is not exact. However,

$$q = \frac{M_y - N_x}{M} = \frac{2-4}{2y} = -\frac{1}{y}$$

is a continuous function of y alone for $y \neq 0$. Since

$$e^{-\int q(y)\,dy} = e^{\int \frac{1}{y}\,dy} = e^{\ln|y|} = |y|,$$

an integrating factor is $\mu(y) = y$. Multiplying equation (8.8.25) by y, we obtain the exact equation

$$2y^2\,dx + (4xy + y^2)\,dy = 0.$$

Consequently, a function $F(x, y)$ exists with the first-order partial derivatives:

$$F_x(x, y) = 2y^2 \quad \text{and} \quad F_y(x, y) = 4xy + y^2.$$

It is left as an exercise to show that $F(x, y) = 6xy^2 + y^3$ is a first integral of (8.8.25). Therefore, solutions are defined implicitly by

$$6xy^2 + y^3 = C.$$ ♦

Problems

God gives every bird his worm, but He does not throw it into the nest.

P. D. James[8]

Initial Observations

In Problems 1–3, there is a function $F(x, y)$ whose total derivative is equal to the left-hand side of the differential equation. Find a general solution of the differential equation by finding $F(x, y)$ by inspection (as in Sect. 8.1).

1. $e^{2x}\dfrac{dy}{dx} + 2e^{2x}y = 0$
2. $2x^2 \cos 2y\dfrac{dy}{dx} + 2x \sin 2y = 0$
3. $(x^2 - 1)\dfrac{dy}{dx} + (2xy + 3) = 0$

[8] See [46, p. 325]. P. D. James, a British mystery novelist born in 1920, is best known for her books involving Adam Dalgliesh, a fictional detective at New Scotland Yard. *Devices and Desires* was originally published in 1989 and aired in 1991 on the television PBS series “Mystery!”.

Total Derivative and Differential

Answer the questions in Problems 4–7 using (8.2.3) *and* (8.2.4) *and Definitions 8.2.1 and 8.3.3.*

4. Find the total derivative of the function $F(x, y) = xy - x^2 + e^y$. Compare the result with Eq. (8.1.7)? What is a general solution of (8.1.7)?
5. Find the total differential of $G(x, y) = e^y \cos x + x^2 + 3y^{1/3}$. What is a general solution of
$$(2x - e^y \sin x)\,dx + \left(e^y \cos x + y^{-2/3}\right) dy = 0\,?$$
6. Find the total differential of $g(x, y) = y + x \ln y$, where $y > 0$. What is a general solution of
$$\ln y\,dx + \left(1 + \frac{x}{y}\right) dy = 0\,?$$
7. Find the total differential of $H(x, y) = x^2 + y^2 - \tan^{-1}(xy)$. What is a general solution of
$$\left(2x - \frac{y}{1 + x^2y^2}\right) dx + \left(2y - \frac{x}{1 + x^2y^2}\right) dy = 0\,?$$

Second-Order Partial Derivatives

8. Find all second-order partial derivatives of $g(x, y) = \ln(x + y)$. Verify that the mixed partial derivatives are equal.
9. Find all second-order partial derivatives of $f(x, y) = 3x^4y^2 - \cos(xy) + 5x \tan y$. Verify that the mixed partial derivatives are equal.

Solutions of Exact Equations

10. Use Method I in Sect. 8.5 to solve (see (8.1.7))
$$(y - 2x)\,dx + \left(x + e^y\right)\,dy = 0.$$
Also, use Method II to solve the equation. Compare the results.
11. Prove that every separable equation is also exact. That is, prove $y' = g(x)h(y)$ is an exact differential equation.

Criterion for Exactness

In Problems 12 through 20, use the Criterion for Exactness to verify that the differential equation is exact. Then solve it.

12. $(2x - y)\,dx + (2y - x)\,dy = 0$
13. $(y - 3x^2)\,dx + (x - 1)\,dy = 0$
14. $(ye^x + 1)\,dx + (e^x - 2)\,dy = 0$
15. $(\tan y - 2x)\,dx + \left(x \sec^2 y + \dfrac{1}{y}\right) dy = 0$
16. $(x + 2y)\,dx + (2x + y)\,dy = 0$
17. $\dfrac{dy}{dx} = \dfrac{3x^2 + y}{e^{2y} - x}$
18. $\dfrac{dy}{dx} = \dfrac{x + y^2}{\cos(2y) - 2xy}$
19. $(\sec^2 x - 2y)\,dx + (3 - 2x + y)\,dy = 0$

20. $15x^2y\,dx + (5x^3 - 2y)\,dy = 0$

Classifying and Solving Equations

In Problems 21 through 38, state whether the differential equation is separable, linear, or exact. If it can be classified in more than one way, identify each of the types to which it belongs. Then solve it using any method discussed up to this point in the book. If however the equation belongs to none of the three types, do not attempt to solve it; but simply write the phrase "none of these types".

21. $\dfrac{dy}{dx} + x\sin y = x$
22. $\dfrac{1}{x^2}\dfrac{dy}{dx} = x^2\sec y$
23. $(5y\sin x + xy^2)\,dx + (x^2y - 5\cos x)\,dy = 0$
24. $(x^2 + y)\,dx - x\,dy = 0$
25. $\dfrac{dy}{dx} - \dfrac{y}{x^2} = \dfrac{1}{yx^2}$
26. $(3x^2 + 4xy^2)\,dx + 4x^2y\,dy = 0$
27. $\dfrac{dy}{dx} = \dfrac{1-y}{x+2y+1}$
28. $\sin y\dfrac{dy}{dx} - x\cos y = x$
29. $(x+y+1)\,\dfrac{dy}{dx} = 2 - y$
30. $(xy + y)\dfrac{dy}{dx} + y^2 = 3y\tan x$
31. $\dfrac{dy}{dx} = x + \dfrac{y}{x}$
32. $x^2y' = \sqrt{y} - y'$
33. $\dfrac{dy}{dx} = 2e^{x-0.5y}$
34. $\dfrac{dy}{dx} = \dfrac{10xy + y\sin x - 2}{\cos x - 5x^2}$
35. $(1 + e^y)\,dx - (e^y + x^2e^y)\,dy = 0$
36. $\dfrac{dy}{dx} = ye^{x^2y}$
37. $\dfrac{dy}{dx} = \dfrac{2x + \sin y + \cos y}{x(\sin y - \cos y)}$
38. $\dfrac{dy}{dx} = \dfrac{e^x - y\cos x}{\sin x}$

Cobb-Douglas Production Function

39. The Cobb-Douglas production function $P = bL^{\alpha}K^{1-\alpha}$ was introduced in Example 2.3.6 in Chap. 2. Suppose a company uses it to model the production P of one of its products, where L and K are the corresponding amounts of labor and capital investment, respectively, and $\alpha \in (0, 1)$ and $b > 0$ are constants.

    (a) Show that the total differential of P is

$$dP = \frac{\alpha P}{L}\,dL + \frac{(1-\alpha)P}{K}\,dK.$$

(b) Suppose the company decides the most that can be spent on labor and capital investment is D dollars. If a unit of labor costs the company m and a unit of capital investment costs n, then production of the product is subject to the constraint $mL + nK = D$. Find the values of L and K that would maximize the production P of the product. Use these values to show that the maximum production is

$$P = bD\left(\frac{\alpha}{m}\right)^{\alpha}\left(\frac{1-\alpha}{n}\right)^{1-\alpha}.$$

Integral Curves

40. Find the integral curves of the differential equation $yy' = -x$. Find the solution curve passing through the point $(-2, -3)$ and the interval of existence of the solution.
41. Find the integral of the equation $y' = y/x$. Find the solution curve passing through the point $(1, 2)$ and the interval of existence of the solution.
42. Find the integral curves of the differential equation

$$\frac{dy}{dx} = \frac{2y\cos x}{1+y^2}.$$

See Problem 56.

43. Find the integral curves of the differential equation

$$\frac{dy}{dx} = \frac{x+y^2}{\cos y - 2xy}.$$

Special Integrating Factors

For each equation in Problems 44 through 48, find an integrating factor. If Theorems 8.8.2 and 8.8.3 are inapplicable, try formula (8.8.16). *After finding a suitable integrating factor, solve the equation.*

44. $(3xy + 3y^2 - 1)\,dx + (x^2 + 3xy)\,dy = 0$
45. $(xy - x + y)\,dx + x\,dy = 0$
46. $\dfrac{dy}{dx} = \dfrac{y^2 + 4xy}{2(x^2 - y)}$
47. $(x^2y^5 + y^3)\,dx + (x^3y^4 + x)\,dy = 0$
48. $2xy\sin y\,dx + (x^2y\cos y + 3x^2\sin y)\,dy = 0$
49. Prove Theorem 8.8.3.
50. (a) Suppose the equation $M(x, y)\,dx + N(x, y)\,dy = 0$ is not exact. Show that if the expression

$$r := \frac{M_y - N_x}{yN - xM}$$

depends only on the product xy and is continuous on some interval, then the equation has an integrating factor of the form $\mu(z)$, where $z = xy$. Find a formula for $\mu(xy)$.

(b) Use the result of part (a) to solve the equation

$$(3xy + 2y)\,dx + (2x^2 + 2x)\,dy = 0.$$

Bored Bugs

51. Four juvenile zyzzyvas[9] are spending the afternoon just hanging out. Bored to tears, Moe convinces the others to play "Follow the Leader." However, unable to decide on a leader, they compromise by positioning themselves at the corners of a square. They agree that Moe will start at (10, 10) and will always crawl toward Larry, who will start at (−10, 10). Likewise, Larry will always crawl toward Curly, who will start out at (−10, −10). And Curly will always crawl toward Shemp, who will start at (10, −10) and always crawl toward Moe. If they begin at time $t = 0$ and crawl at the same constant speed, find a differential equation that describes Moe's path and then solve it. If this type of differential equation is unfamiliar to you, then research how it can be solved by hand.

Computer Algebra System Problems

Use a CAS to work the following problems.

52. Plot some integral curves of $(2x + y)\,dx + (x - 2y)\,dy = 0$. Is this equation exact? If so, solve it by hand.
53. Let $F(x, y) = y^2 - x^2$. Take the total differential of both sides of $F(x, y) = C$, where C denotes a constant, to find a differential equation with integral curves that are level curves of F. Now find another differential equation that has integral curves orthogonal to the integral curves of the first differential equation.[10] Graph at least five integral curves for each of these differential equations on the same set of axes.

Remark The procedure in this problem can be used whenever a family of curves is to be found which intersects another family of curves orthogonally at each point. For example, we can use it to find the equipotential surfaces of an electric field since they are orthogonal to the lines of force of the electric field (cf. Halliday and Resnick [43, Ch. 24]).

54. Solve the differential equation in Problem 51 that describes Moe's path. Graph the paths of each of the zyzzyvas.
55. Find a general solution of the differential equation in Problem 14.
56. Graph the integral curve of the equation in Problem 42 that passes through the point $(\pi, 1)$.

[9] Zyzzyvas are a genus of tropical weevils native to South America, the last entry in some dictionaries, and the ultimate Scrabble word.

[10] Two curves are ***orthogonal*** if at the points where they meet their respective tangent lines are perpendicular.

Chapter 9
Existence-Uniqueness Theorems

Mathematicians are machines for turning coffee into theorems.
Paul Erdős[1]

Abstract This chapter introduces *Picard's method of successive approximations*. This method is sometimes used to approximate the solution of the initial value problem

$$y' = f(x, y), \quad y(x_0) = y_0 \tag{1}$$

for a given function f, assuming it has one, when (1) does not have a closed-form solution or it is difficult to find.

A number of examples using Picard's method are presented, such as calculating several *Picard approximations* of the solution of

$$y' = 2\sin x - y, \quad y(0) = -1. \tag{2}$$

Eventually, it becomes clear from these approximations that the solution of (2) is $y(x) = \sin x - \cos x$, which is what the *integrating factor method* or the *variation of parameters formula* in Chap. 5 would yield.

The real purpose of examples like (2) is to get students comfortable with Picard's method before it is used to prove an *existence and uniqueness theorem* that guarantees a unique solution of (1) exists when the function f is continuous and satisfies a *Lipschitz condition* with respect to y on a closed rectangular region $\mathcal{R}$. In another version of this theorem, it is assumed that both f and the partial derivative $\partial f/\partial y$ are continuous on $\mathcal{R}$.

[1] Paul Erdős (1913–1996) authored or co-authored more mathematical papers than anyone else in history with more that 1520 papers according to "The Erdös Number Project" website. His mathematical genius and eccentric personality inspired many humorous anecdotes about his life. These are told in a captivating, amusing way in *The Man Who Loved Only Numbers* by Paul Hoffman [45].

© The Author(s), under exclusive license to Springer Nature Switzerland AG 2026
L. C. Becker, *Ordinary Differential Equations: Concepts, Methods, and Models*,
https://doi.org/10.1007/978-3-032-15150-6_9

Another existence and uniqueness theorem is stated for initial value problems involving the second-order differential equation

$$y'' = f(x, y, y'). \tag{3}$$

Examples are provided demonstrating how to use this theorem to establish the existence and uniqueness of the solution of (3) for a given function f and set of initial conditions. In one of these examples, the theorem is used to prove that

$$y'' = xy, \quad y(0) = 0, \quad y'(0) = 1 \tag{4}$$

has a unique solution. Then Picard's method is used to obtain this solution, which turns out to be a convergent series that cannot be expressed in terms of the elementary functions of calculus. The differential equation in (3), called *Airy's equation*, is important in many fields, such as optics and fluid dynamics.

9.1 Introduction

We begin this chapter with *Picard's method of successive approximations*, which is sometimes used to approximate solutions of a differential equation when analytical methods fail or one is unaware of any method that could be used to solve the equation. After describing Picard's method in the next section, we present some examples. For the differential equations in the first two of these examples, we will see that this method will yield successive approximations from which it will be apparent what the analytical solutions of both equations are. In the third example, however, it will be hard to discern an analytical solution even though the method yields very good approximations of a relatively simple-looking differential equation.

Picard's method can also be used to prove theorems that tell us when and on what intervals solutions of first-order differential equations exist and are unique. Two of these existence and uniqueness theorems are stated and proved in Sect. 9.3. Furthermore, we will employ Picard's method to obtain an existence and uniqueness theorem for second-order equations in Sect. 9.4.

9.2 Method of Successive Approximations

We describe a well-known method that can be used to approximate the solution of the initial value problem

$$\frac{dy}{dx} = f(x, y), \quad y(x_0) = y_0 \tag{9.2.1}$$

when the differential equation is not one of those typical first-order equations for which we have well-known analytical methods, such as the methods for solving separable, linear, or exact equations. As we will see in the next section, this method can also be used to show that solutions of (9.2.1) exist and are unique provided the function f satisfies certain conditions.

Let us suppose, for some particular initial value problem of the form (9.2.1), that somehow we know that it has a unique solution $y(x)$ on an interval I containing the initial point x_0.[2] The objective is to approximate this solution with a well-known method called ***Picard's method of successive approximations***. Implementation of this method begins with an ***initial approximation*** of the solution that we denote by $y_0(x)$ and define to be the function

$$y_0(x) \equiv y_0 \tag{9.2.2}$$

on the interval I. That is, $y_0(x)$ is a constant function that is equal to the initial value y_0 for all $x \in I$. At $x = x_0$, it agrees exactly with the actual solution $y(x)$ since both are equal to y_0. For values of x "very close" to x_0, it approximates the solution $y(x)$ rather well since $y(x)$ (being a solution and hence differentiable) is continuous on I. In particular, as $x_0 \in I$, $y(x)$ is continuous at x_0, which means

$$\lim_{x \to x_0} y(x) = y(x_0).$$

Consequently, it follows from the initial condition and (9.2.2) that

$$\lim_{x \to x_0} y(x) = y_0(x).$$

This says that $y(x) \approx y_0(x)$ when x is not too far from x_0. However, for values of x further away from x_0, we cannot expect $y_0(x)$ to approximate $y(x)$ well.

Obviously, the immediate goal is to improve upon the approximation $y_0(x)$. This is done by replacing the unknown y on the right-hand side of the differential equation in (9.2.1) with $y_0(x)$ and then calculating the solution of the initial value problem

$$y'(x) = f(x, y_0(x)), \quad y(x_0) = y_0. \tag{9.2.3}$$

Changing the name of the variable from x to u and then integrating both sides of the differential equation from $u = x_0$ to $u = x$, we obtain

$$\int_{x_0}^{x} y'(u)\, du = \int_{x_0}^{x} f(u, y_0(u))\, du,$$

[2] For now we are not concerned about how this is known.

which simplifies to

$$y(x) - y(x_0) = \int_{x_0}^{x} f(u, y_0(u))\,du.$$

Solving for $y(x)$, renaming it $y_1(x)$, and using the initial condition and (9.2.2), we get

$$y_1(x) = y_0 + \int_{x_0}^{x} f(u, y_0)\,du. \tag{9.2.4}$$

Hopefully, $y_1(x)$ will approximate the solution of (9.2.1) on the interval I better than does $y_0(x)$. Assuming this to be the case, let us improve upon this approximation by finding the solution of the initial value problem

$$y'(x) = f(x, y_1(x)), \quad y(x_0) = y_0.$$

Integrating from x_0 to x as before, we obtain

$$y(x) - y(x_0) = \int_{x_0}^{x} f(u, y_1(u))\,du.$$

Solving for $y(x)$ and renaming it $y_2(x)$, we obtain the formula

$$y_2(x) = y_0 + \int_{x_0}^{x} f(u, y_1(u))\,du.$$

Continuing with these calculations, with the hope that each new approximation improves upon the previous one, we end up with the recursive formula

$$y_n(x) = y_0 + \int_{x_0}^{x} f(u, y_{n-1}(u))\,du \tag{9.2.5}$$

for $n = 1, 2, 3, \ldots$, which can be used to generate the successive approximating functions: $y_1(x)$, $y_2(x)$, $y_3(x)$, ... This formula is called ***Picard's (recursive) formula***. The function $y_n(x)$ denotes the nth ***Picard approximation*** of the solution of the initial value problem (9.2.1).

Example 9.2.1 Apply Picard's method of successive approximations to the initial value problem

$$\frac{dy}{dx} = ay, \quad y(0) = 1 \tag{9.2.6}$$

where a is a constant.

Solution Using the notation of (9.2.1), $f(x, y) = ay$, $x_0 = 0$, and $y_0 = 1$. Thus, as $f(u, y_n(u)) = ay_n(u)$, the Picard formula (9.2.5) applied to (9.2.6) becomes

$$y_n(x) = 1 + \int_0^x ay_{n-1}(u)\, du \tag{9.2.7}$$

where $y_0(x) \equiv 1$. Setting $n = 1$ in (9.2.7), we obtain the first Picard approximation of the solution of (9.2.6):

$$y_1(x) = 1 + \int_0^x ay_0(u)\, du = 1 + \int_0^x a\, du = 1 + ax.$$

To calculate the next Picard approximation, set $n = 2$ in (9.2.7) and use the result of the previous calculation:

$$y_2(x) = 1 + \int_0^x ay_1(u)\, du = 1 + \int_0^x a(1 + au)\, du = 1 + ax + \frac{1}{2}a^2x^2.$$

In the same way, the third approximation is

$$\begin{aligned} y_3(x) &= 1 + \int_0^x ay_2(u)\, du = 1 + \int_0^x a\left(1 + au + \frac{1}{2}a^2u^2\right) du \\ &= 1 + ax + \frac{1}{2}a^2x^2 + \frac{1}{6}a^3x^3 \end{aligned}$$

and the fourth approximation is

$$\begin{aligned} y_4(x) &= 1 + \int_0^x ay_3(u)\, du = 1 + \int_0^x a\left(1 + au + \frac{1}{2}a^2u^2 + \frac{1}{6}a^3u^3\right) du \\ &= 1 + ax + \frac{1}{2}a^2x^2 + \frac{1}{6}a^3x^3 + \frac{1}{24}a^4x^4. \end{aligned}$$

Recognizing that $2 = 2!$, $6 = 3!$, and $24 = 4!$, we see that these Picard approximations are given by the formula

$$y_n(x) = 1 + ax + \frac{a^2x^2}{2!} + \cdots + \frac{a^nx^n}{n!} \tag{9.2.8}$$

for $n = 1, 2, 3$, and 4. Moreover, (9.2.8) is also valid for $n = 0$ as $0! = 1$.

Now suppose that (9.2.8) is also true for some integer $n > 4$. Then it follows from (9.2.7) that

$$y_{n+1}(x) = 1 + \int_0^x ay_n(u)\, du = 1 + \int_0^x a\left(1 + au + \frac{a^2u^2}{2!} + \cdots + \frac{a^nu^n}{n!}\right) du.$$

Carrying out the integration and using that $(n+1)n! = (n+1)!$, this simplifies to

$$y_{n+1}(x) = 1 + ax + \frac{a^2x^2}{2!} + \cdots + \frac{a^{n+1}x^{n+1}}{(n+1)!}. \tag{9.2.9}$$

This says that if (9.2.8) is the nth Picard approximation, then the very next one, namely the $(n+1)$th Picard approximation, must be (9.2.9). This along with (9.2.8) being valid for $n = 0, 1, 2, 3,$ and 4 implies that (9.2.8) is in fact valid for all positive integers n. This method of proving that a formula is valid for all positive integers is known as ***mathematical induction***.

It gets even better when we realize that (9.2.8) is the Taylor polynomial of degree n for e^{ax} expanded about $x = 0$ and that

$$\lim_{n\to\infty} y_n(x) = \lim_{n\to\infty} \sum_{k=0}^{n} \frac{(ax)^k}{k!} = e^{ax}. \tag{9.2.10}$$

This suggests that $y(x) = e^{ax}$ is the solution of the initial value problem (9.2.6). But of course we already knew this from the outset. The point is that now have a way of approximating solutions of more difficult initial value problems—and possibly even obtaining their precise solutions. ♦

Example 9.2.2 Solve the initial value problem

$$y' + y = 2\sin x, \quad y(0) = -1 \tag{9.2.11}$$

using the method of successive approximations.

Solution Since this is a first-order linear equation, we could use the integrating factor method to find the solution. But as in the previous example, the point right now is to illustrate Picard's method. The integration

$$\int_0^x y'(u)\,du = \int_0^x (2\sin u - y(u))\,du$$

together with the initial condition yields

$$y(x) = -1 + \int_0^x (2\sin u - y(u))\,du.$$

Now we can get Picard's formula for (9.2.11) by merely affixing the subscript n to $y(x)$ on the left-hand side of the integral equation and the subscript $n-1$ to $y(u)$ on the right-hand side. Thus,

$$y_n(x) = -1 + \int_0^x (2\sin u - y_{n-1}(u))\,du \tag{9.2.12}$$

for $n = 1, 2, 3, \ldots$, where $y_0(x) \equiv -1$.

To find the first Picard approximation, let $n = 1$. Then,

$$\begin{aligned} y_1(x) &= -1 + \int_0^x (2\sin u - y_0(u))\,du = -1 + \int_0^x (2\sin u + 1)\,du \\ &= -1 + [x - 2\cos x + 2] = 1 + x - 2\cos x. \end{aligned}$$

The second and third Picard approximations are

$$\begin{aligned} y_2(x) &= -1 + \int_0^x (2\sin u - y_1(u))\,du = -1 + \int_0^x (2\sin u - 1 - u + 2\cos u)\,du \\ &= 1 - x - \frac{1}{2}x^2 - 2\cos x + 2\sin x. \end{aligned}$$

$$\begin{aligned} y_3(x) &= -1 + \int_0^x (2\sin u - y_2(u))\,du \\ &= -1 + \int_0^x \left(2\sin u - 1 + u + \frac{1}{2}u^2 + 2\cos u - 2\sin u\right) du \\ &= -1 + \int_0^x \left(-1 + u + \frac{1}{2}u^2 + 2\cos u\right) du = -1 - x + \frac{1}{2}x^2 + \frac{1}{6}x^3 + 2\sin x. \end{aligned}$$

Continuing, we have

$$\begin{aligned} y_4(x) &= -1 + \int_0^x (2\sin u - y_3(u))\,du \\ &= -1 + \int_0^x \left(2\sin u + 1 + u - \frac{1}{2}u^2 - \frac{1}{6}u^3 - 2\sin u\right) du \\ &= -1 + \int_0^x \left(1 + u - \frac{1}{2}u^2 - \frac{1}{6}u^3\right) du = -1 + x + \frac{1}{2}x^2 - \frac{1}{6}x^3 - \frac{1}{24}x^4. \end{aligned}$$

$$\begin{aligned} y_5(x) &= -1 + \int_0^x (2\sin u - y_4(u))\,du \\ &= -1 + \int_0^x \left(2\sin u + 1 - u - \frac{1}{2}u^2 + \frac{1}{6}u^3 + \frac{1}{24}u^4\right) du \\ &= -2\cos x + 1 + x - \frac{1}{2}x^2 - \frac{1}{6}x^3 + \frac{1}{24}x^4 + \frac{1}{120}x^5. \end{aligned}$$

Taylor polynomials in these approximations become recognizable when terms are rearranged as follows:

$$y_3(x) = -\left(x - \frac{x^3}{3!}\right) - \left(1 - \frac{x^2}{2!}\right) + 2\sin x,$$

$$y_4(x) = \left(x - \frac{x^3}{3!}\right) - \left(1 - \frac{x^2}{2!} + \frac{x^4}{4!}\right),$$

$$y_5(x) = -2\cos x + \left(x - \frac{x^3}{3!} + \frac{x^5}{5!}\right) + \left(1 - \frac{x^2}{2!} + \frac{x^4}{4!}\right).$$

It follows from the Maclaurin series expansions

$$\cos x = \sum_{n=0}^{\infty}(-1)^n \frac{x^{2n}}{(2n)!} = 1 - \frac{x^2}{2!} + \frac{x^4}{4!} - \frac{x^6}{6!} + \cdots \tag{9.2.13}$$

$$\sin x = \sum_{n=0}^{\infty}(-1)^n \frac{x^{2n+1}}{(2n+1)!} = x - \frac{x^3}{3!} + \frac{x^5}{5!} - \frac{x^7}{7!} + \cdots \tag{9.2.14}$$

that

$$\begin{aligned} y_3(x) &= -\left(x - \frac{x^3}{3!}\right) - \left(1 - \frac{x^2}{2!}\right) + 2\sin x \\ &\approx -\sin x - \cos x + 2\sin x = \sin x - \cos x \end{aligned}$$

if x is not too far away from 0. Of course, the closer x is to 0, the better will be the approximation. Furthermore,

$$y_4(x) = \left(x - \frac{x^3}{3!}\right) - \left(1 - \frac{x^2}{2!} + \frac{x^4}{4!}\right) \approx \sin x - \cos x$$

and

$$y_5(x) \approx -2\cos x + \sin x + \cos x = \sin x - \cos x.$$

These approximations suggest that the sequence $y_1(x), y_2(x), y_3(x), \ldots$ converges to the function

$$y(x) = \sin x - \cos x$$

and that it could quite possibly be the solution of the initial value problem (9.2.11) for all $x \in \mathbb{R}$. And indeed it is as $y(0) = \sin 0 - \cos 0 = -1$ and

$$y'(x) + y(x) = \frac{d}{dx}(\sin x - \cos x) + (\sin x - \cos x) = 2\sin x.$$

♦

Example 9.2.3 Compute the fourth Picard approximation of the solution of

$$\frac{dx}{dt} = 1 - 2tx, \quad x(0) = 0. \tag{9.2.15}$$

What is the value of this approximation at $t = 0.5$?

Solution Since $f(t, x) = 1 - 2tx$ and $t_0 = 0$, Picard's iteration formula for (9.2.15) is

$$x_n(t) = x_0 + \int_{t_0}^{t} f(u, x_{n-1}(u))\, du = \int_0^t \left[1 - 2ux_{n-1}(u)\right] du \tag{9.2.16}$$

where $x_0(t) \equiv 0$. Letting $n = 1$, the first Picard approximation is

$$x_1(t) = \int_0^t \left[1 - 2ux_0(u)\right] du = \int_0^t \left[1 - 2u \cdot 0\right] du = t.$$

The next three approximations are

$$x_2(t) = \int_0^t \left[1 - 2ux_1(u)\right] du = \int_0^t \left[1 - 2u^2\right] du = t - \frac{2}{3}t^3,$$

$$x_3(t) = \int_0^t \Big[1 - 2ux_2(u)\Big] du = \int_0^t \Big[1 - 2u\Big(u - \frac{2}{3}u^3\Big)\Big] du = t - \frac{2}{3}t^3 + \frac{4}{15}t^5,$$

and

$$\begin{aligned} x_4(t) &= \int_0^t \Big[1 - 2ux_3(u)\Big] du = \int_0^t \Big[1 - 2u\Big(u - \frac{2}{3}u^3 + \frac{4}{15}u^5\Big)\Big] du \\ &= t - \frac{2}{3}t^3 + \frac{4}{15}t^5 - \frac{8}{105}t^7. \end{aligned}$$

Thus, the value of the fourth Picard approximation at $t = 0.5$ is

$$x_4(0.5) = \frac{1}{2} - \frac{2}{3}\left(\frac{1}{2}\right)^3 + \frac{4}{15}\left(\frac{1}{2}\right)^5 - \frac{8}{105}\left(\frac{1}{2}\right)^7 = 0.42440476\ldots.$$

It is shown in Chap. 5 (see Example 5.2.5) that the solution of the initial value problem (9.2.15) is $x(t) = \mathcal{D}(t)$, where $\mathcal{D}(t)$ denotes Dawson's integral, namely, the function

$$\mathcal{D}(t) = e^{-t^2} \int_0^t e^{s^2}\, ds. \tag{9.2.17}$$

The actual value of $\mathcal{D}(0.5)$ is $0.42443638\ldots$, which can be found in a table of values of Dawson's integral, such as in [1], or by evaluating (9.2.17) at $t = 0.5$ with a CAS or an appropriate calculator. Also, it can be shown (see Problem 45, Chap. 5) that

$$\mathcal{D}(t) = \sum_{n=0}^{\infty} \frac{(-1)^n 2^{2n} n!}{(2n+1)!} t^{2n+1} = t - \frac{2}{3}t^3 + \frac{4}{15}t^5 - \frac{8}{105}t^7 + \frac{16}{945}t^9 - \ldots$$

and that this series converges for all $t \in \mathbb{R}$. ♦

9.3 First-Order Equations

Now that we have gained some experience with applying Picard's method, we will use it to prove Theorem 9.3.2 after the following definition and examples. It concerns the existence and uniqueness of solutions of the differential equation $y' = f(x, y)$ when the function f is continuous and satisfies a so-called *Lipschitz condition*.

Lipschitz Condition

Definition 9.3.1 Let $\mathcal{S}$ be a given set in the xy-plane. A function $f(x, y)$ is said to satisfy a ***Lipschitz condition*** with respect to y on $\mathcal{S}$ if there is a constant $L > 0$ (called a ***Lipschitz constant***) such that

$$|f(x, y) - f(x, z)| \leq L|y - z| \tag{9.3.1}$$

for all (x, y) and (x, z) in $\mathcal{S}$.

Example 9.3.1 The function $f(x, y) = 5y \cos x$ satisfies a Lipschitz condition with respect to y on the entire xy-plane with Lipschitz constant $L = 5$ since

$$|f(x, y) - f(x, z)| = |5y \cos x - 5z \cos x)| = 5|\cos x||y - z| \leq 5|y - z|$$

for all points (x, y) and (x, z) in the plane. ♦

Example 9.3.2 Show that the function $f(x, y) = x + y^2$ satisfies a Lipschitz condition with respect to y on the closed rectangular region $\mathcal{R} = \{(x, y) : |x| \leq a, |y| \leq b\}$, where a, b are a pair of arbitrary positive numbers.

Solution Let (x, y) and (x, z) be points in $\mathcal{R}$. Then

$$\begin{aligned}|f(x, y) - f(x, z)| &= |(x + y^2) - (x + z^2)| = |y^2 - z^2| \\ &= |y + z||y - z| \leq 2b|y - z|.\end{aligned}$$

Thus, f satisfies a Lipschitz condition on $\mathcal{R}$ with Lipschitz constant $L = 2b$. ♦

The following theorem is known as an *existence and uniqueness theorem.*

Existence of a Unique Solution I

Theorem 9.3.2 *Suppose a function $f(x, y)$ is continuous on a closed rectangular region*

$$\mathcal{R} = \{\, (x, y) : |x - x_0| \leq a, |y - y_0| \leq b \,\}$$

where $a, b, x_0, y_0 \in \mathbb{R}$ and $a, b > 0$. Let M be the maximum value that $|f(x, y)|$ attains in $\mathcal{R}$. Also, suppose $f(x, y)$ satisfies a Lipschitz condition with respect to y on $\mathcal{R}$. Then there exists a unique solution $y(x)$ of the initial value problem

$$\frac{dy}{dx} = f(x, y), \quad y(x_0) = y_0 \tag{9.3.2}$$

on the interval

$$[x_0 - h, x_0 + h] \quad \text{where} \quad h = \min\left\{a, \frac{b}{M}\right\}. \tag{9.3.3}$$

Furthermore, the sequence $\{y_n(x)\}$ of Picard approximations, where $y_0(x) \equiv y_0$ and $y_n(x)$ is given by (9.2.5) for $n = 1, 2, 3, \ldots$, converges to the solution $y(x)$ on $[x_0 - h, x_0 + h]$.

Proof In order to get some idea how this theorem could be proved, let us first suppose that there is a function $y(x)$ that is a solution of the initial value problem (9.3.2) on an interval $I \subset [x_0 - a, x_0 + a]$ such that $|y(x) - y_0| \leq b$ for every $x \in I$. That is, $y(x_0) = y_0$ and

$$y'(x) = f(x, y(x)) \tag{9.3.4}$$

for $x \in I$. Since, by hypothesis, $f(x, y)$ is continuous on the rectangular region $\mathcal{R}$, it follows that $f(x, y(x))$ is continuous on the interval I. Thus, $y'(x)$ is continuous on I. Consequently, we can integrate both sides of (9.3.4) from x_0 to an arbitrary

$x \in I$:

$$\int_{x_0}^{x} y'(u)\,du = \int_{x_0}^{x} f(u, y(u))\,du,$$

which, as $y(x_0) = y_0$, simplifies to

$$y(x) = y_0 + \int_{x_0}^{x} f(u, y(u))\,du. \tag{9.3.5}$$

It turns out that this integral equation will be instrumental in proving this theorem and that Picard's method of successive approximations will play a big role in the proof.

Now let us begin the proof by supposing there is a function $y(x)$ that is continuous on an interval $I \subset [x_0 - a, x_0 + a]$ and that $|y(x) - y_0| \leq b$ for every $x \in I$ and which satisfies the integral equation (9.3.5). Then it will also solve the initial value problem (9.3.2) on I. To see this, first note that $f(x, y(x))$ is continuous on I since, by hypothesis, $f(x, y)$ is continuous on the rectangular region $\mathcal{R}$. Thus, the right-hand side of (9.3.5) is differentiable on I, which means that $y(x)$ also differentiable on I. Taking the derivative of both sides of (9.3.5) and using the Fundamental Theorem of Calculus, we have

$$\frac{d}{dx}y(x) = \frac{d}{dx}y_0 + \frac{d}{dx}\int_{x_0}^{x} f(u, y(u))\,du = f(x, y(x))$$

for all $x \in I$. This says that $y(x)$ is a solution of the differential equation $y' = f(x, y)$ on the interval I. Setting $x = x_0$ in (9.3.5), we obtain $y(x_0) = y_0$, which shows that $y(x)$ also satisfies the initial condition.

In sum, we have established that if a function solves the integral equation (9.3.5) on an interval I, then this same function solves the initial value problem (9.3.2) on I. So if we can prove that the integral equation (9.3.5) has a unique continuous solution $y(x)$, then we will also have proven that $y(x)$ is the unique solution of the initial value problem (9.3.2).

This is where Picard's method of successive approximations comes into play. We will show that the sequence of Picard approximations $\{y_n(x)\}_{n=0}^{\infty}$, where $y_n(x)$ is given by (9.2.5), converges to a continuous function that is a solution of the integral equation (9.3.5). To that end, we will first show that every Picard approximation is continuous.

Because the region $\mathcal{R}$ is closed and bounded, $|f(x, y)|$ attains an absolute maximum value at least once in $\mathcal{R}$. (For more information, consult any calculus book dealing with functions of two or more variables.) Let M denote this maximum value. This will enable us to prove that each Picard approximation is continuous on the interval $[x_0 - h, x_0 + h]$, where h is defined by

$$h := \min\left\{a, \frac{b}{M}\right\}.$$

Note that $[x_0 - h, x_0 + h]$ is a subset of $[x_0 - a, x_0 + a]$ as $h \le a$.

Consider the Picard approximation $y_1(x)$. Picard's recursive formula (9.2.5) with $n = 1$ is

$$y_1(x) = y_0 + \int_{x_0}^{x} f(u, y_0(u))\, du = y_0 + \int_{x_0}^{x} f(u, y_0)\, du$$

as $y_0(u) \equiv y_0$. Because $f(x, y_0)$ is a continuous function of x on $[x_0 - h, x_0 + h]$ (so integrable), the integral $\int_{x_0}^{x} f(u, y_0)\, du$ is likewise a continuous function of x on this interval. Consequently, the right-hand side is continuous, making $y_1(x)$ continuous on the interval. Furthermore,

$$\begin{aligned} |y_1(x) - y_0| &\le \left| \int_{x_0}^{x} f(u, y_0)\, du \right| \le \left| \int_{x_0}^{x} |f(u, y_0)|\, du \right| \le \left| \int_{x_0}^{x} M\, du \right| \\ &\le M|x - x_0| \le Mh \le b \end{aligned} \tag{9.3.6}$$

since $h \le b/M$. This shows that for $x \in [x_0 - h, x_0 + h]$ all values of $y_1(x)$ belong to the interval $[y_0 - b, y_0 + b]$.

Now let us show that all of the Picard approximations are continuous with the following inductive reasoning. If, at each $x \in [x_0 - h, x_0 + h]$, the nth Picard approximation y_n is continuous and $|y_n(x) - y_0| \le b$, then it would follow from (9.2.5) that the next Picard approximation y_{n+1} is also continuous at each of these x's and that

$$\begin{aligned} |y_{n+1}(x) - y_0| &\le \left| \int_{x_0}^{x} f_n(u, y_n(u))\, du \right| \le \left| \int_{x_0}^{x} |f(u, y_n(u))|\, du \right| \le \left| \int_{x_0}^{x} M\, du \right| \\ &\le M|x - x_0| \le Mh \le M\frac{b}{M} = b. \end{aligned}$$

In other words, if for all $x \in [x_0 - h, x_0 + h]$ all values of $y_n(x)$ are in $[y_0 - b, y_0 + b]$, then the same is true of $y_{n+1}(x)$. Recall that we have already established that $y_1(x)$ is continuous on $[x_0 - h, x_0 + h]$ and that its values are in $[y_0 - b, y_0 + b]$. Therefore, it follows from the principle of mathematical induction that every Picard approximation is continuous on $[x_0 - h, x_0 + h]$ and that all values fall in the interval $[y_0 - b, y_0 + b]$.

The trick for proving that the sequence $\{y_n(x)\}_{n=0}^{\infty}$ converges to a continuous solution of the integral equation (9.3.5) is to express $y_n(x)$ in terms of a telescoping sum as follows:

$$\begin{aligned} y_n(x) &= y_0(x) + [y_1(x) - y_0(x)] + [y_2(x) - y_1(x)] + \cdots + [y_n(x) - y_{n-1}(x)] \\ &= y_0(x) + \sum_{k=1}^{n} [y_k(x) - y_{k-1}(x)]. \end{aligned} \tag{9.3.7}$$

This can be viewed as the nth partial sum of the infinite series

$$y_0(x) + \sum_{n=1}^{\infty} [y_n(x) - y_{n-1}(x)]. \tag{9.3.8}$$

If we can prove that this series converges at each point x of $[x_0 - h, x_0 + h]$, thereby defining a function $y(x)$ on this interval, then we will have succeeded in proving that the sequence $\{y_n(x)\}_{n=0}^{\infty}$ converges to $y(x)$.

Clearly convergence of (9.3.8) depends on the differences $y_n(x) - y_{n-1}(x)$. Subtracting $y_1(x)$ from

$$y_2(x) = y_0 + \int_{x_0}^{x} f(u, y_1(u))\, du$$

and taking the absolute value, we have

$$\begin{aligned} |y_2(x) - y_1(x)| &= \left| y_0 + \int_{x_0}^{x} f(u, y_1(u))\, du - y_0 - \int_{x_0}^{x} f(u, y_0)\, du \right| \\ &\le \left| \int_{x_0}^{x} \big(f(u, y_1(u)) - f(u, y_0)\big)\, du \right| \le \left| \int_{x_0}^{x} |f(u, y_1(u)) - f(u, y_0)|\, du \right|. \end{aligned}$$

We can take this a bit further due to the Lipschitz condition, namely, that there is a constant $L > 0$ such that $f(x, y)$ satisfies (9.3.1). This along with

$$|y_1(x) - y_0| \le M|x - x_0|$$

from (9.3.6)) implies

$$\begin{aligned} |y_2(x) - y_1(x)| &\le \left| \int_{x_0}^{x} L|y_1(u) - y_0|\, du \right| \\ &\le \left| \int_{x_0}^{x} LM|u - x_0|\, du \right| \le LM \left| \int_{x_0}^{x} |u - x_0|\, du \right|. \end{aligned}$$

It is left as an exercise to show that

$$\left| \int_{x_0}^{x} |u - x_0|\, du \right| = \frac{|x - x_0|^2}{2}.$$

As a result,

$$|y_2(x) - y_1(x)| \le LM \frac{|x - x_0|^2}{2} \le LM \frac{h^2}{2} = \frac{M}{L} \frac{(Lh)^2}{2!}. \tag{9.3.9}$$

Next, consider $|y_3(x) - y_2(x)|$. Using (9.2.5), (9.3.1), and (9.3.9), we have

$$
\begin{aligned}
|y_3(x) - y_2(x)| &= \left| y_0 + \int_{x_0}^{x} f(u, y_2(u))\,du - y_0 - \int_{x_0}^{x} f(u, y_1(u))\,du \right| \\
&\le \left| \int_{x_0}^{x} |f(u, y_2(u)) - f(u, y_1(u))|\,du \right| \le \left| \int_{x_0}^{x} L|y_2(u) - y_1(u)|\,du \right| \\
&\le \left| \int_{x_0}^{x} L^2 M \frac{|u - x_0|^2}{2}\,du \right| = L^2 M \frac{|x - x_0|^3}{2 \cdot 3} \le L^2 M \frac{h^3}{3!} = \frac{M}{L}\frac{(Lh)^3}{3!}.
\end{aligned}
$$

The previous calculations suggest that

$$
|y_n(x) - y_{n-1}(x)| \le \frac{M}{L}\frac{(Lh)^n}{n!} \tag{9.3.10}
$$

for $n = 1, 2, 3, \ldots$ and $|x - x_0| \le h$. It is left as an exercise to prove this using mathematical induction.

We are now in a position to apply a well-known convergence theorem for a series of functions known as the **Weierstrass M-test**. Unfortunately, however, we cannot discuss this test here because it requires knowledge of upper-level mathematics that is found in either an advanced calculus or a real analysis course. Suffice it to say here that because of this test, it can be argued that the series in (9.3.8) converges and the reason for this is that the series

$$
\sum_{n=1}^{\infty} \frac{M}{L}\frac{(Lh)^n}{n!} = \frac{M}{L}\left[Lh + \frac{(Lh)^2}{2!} + \frac{(Lh)^3}{3!} + \ldots \right],
$$

whose terms are the upper bounds from (9.3.10), converges. In fact, we can see from (9.2.10) that this series converges to $(M/L)(e^{Lh} - 1)$. As a result, it follows from (9.3.7) that

$$
\begin{aligned}
\lim_{n\to\infty} y_n(x) &= \lim_{n\to\infty} \left(y_0(x) + \sum_{k=1}^{n} [y_k(x) - y_{k-1}(x)] \right) \\
&= y_0(x) + \sum_{k=1}^{\infty} [y_k(x) - y_{k-1}(x)]
\end{aligned}
$$

converges for all $x \in [x_0 - h, x_0 + h]$—and, as a result, defines a function $y(x)$ on this interval. Moreover, the Weierstrass M-test tells us even more: the series converges not only to the function $y(x)$ but also *uniformly*. Again what this means precisely is best left to more advanced courses.

The following items are consequences of the uniform convergence of the series and the Lipschitz condition that the function $f(x, y)$ satisfies.

(i) Since each $y_n(x)$ is continuous on $[x_0 - h, x_0 + h]$, so is the limit function $y(x)$.
(ii) It follows from the Lipschitz condition that

$$|f(x, y_n(x)) - f(x, y(x))| \leq L|y_n(x) - y(x)|.$$

Since $y_n(x) \to y(x)$ as $n \to \infty$,

$$\lim_{n\to\infty} f(x, y_n(x)) = f(x, y(x)).$$

(iii) Because $\{y_n(x)\}$ converges uniformly to $y(x)$ on $[x_0 - h, x_0 + h]$, the order in which the limit and integral is taken can be interchanged. Therefore, taking the limit of both sides of (9.2.5), we have

$$\begin{aligned}\lim_{n\to\infty} y_n(x) &= \lim_{n\to\infty} y_0 + \lim_{n\to\infty} \int_{x_0}^{x} f(u, y_{n-1}(u))\, du \\ &= y_0 + \int_{x_0}^{x} \lim_{n\to\infty} f(u, y_{n-1}(u))\, du.\end{aligned}$$

It follows from (ii) that

$$y(x) = y_0 + \int_{x_0}^{x} f(u, y(u))\, du,$$

which is (9.3.5).

Since $y(x)$ is continuous on $[x_0 - h, x_0 + h]$, we conclude that this function is a continuous solution of the integral equation (9.3.5) on $[x_0 - h, x_0 + h]$ and thereby of the initial value problem (9.3.2).

Finally, let us prove that $y(x)$ is the unique solution of (9.3.2) on $[x_0 - h, x_0 + h]$ by proving that it is the only continuous solution of the integral equation (9.3.5). Suppose that $\tilde{y}(x)$ is another continuous solution. Then

$$\begin{aligned}|y(x) - \tilde{y}(x)| &= \left|y_0 + \int_{x_0}^{x} f(u, y(u))\, du - y_0 - \int_{x_0}^{x} f(u, \tilde{y}(u))\, du\right| \\ &= \left|\int_{x_0}^{x} \left[f(u, y(u)) - f(u, \tilde{y}(u))\right] du\right| \\ &\leq \left|\int_{x_0}^{x} \left|f(u, y(u)) - f(u, \tilde{y}(u))\right| du\right| \leq \left|\int_{x_0}^{x} L|y(u) - \tilde{y}(u)|\, du\right|\end{aligned}$$

for all $x \in [x_0 - h, x_0 + h]$. Letting $r(x) = |y(x) - \tilde{y}(x)|$, this becomes

$$r(x) \leq K + \left|\int_{x_0}^{x} p(u) r(u)\, du\right|,$$

where $K = 0$ and $p(u) = L$. By Gronwall's inequality (see Theorem 5.4.1),

$$r(x) \leq K e^{\left|\int_{x_0}^{x} p(u)\,du\right|} = K e^{L|x-x_0|}$$

for $x \in [x_0 - h, x_0 + h]$. So as $K = 0$, we have

$$0 \leq r(x) \leq 0.$$

Therefore, $r(x) = 0$ for all $x \in [x_0 - h, x_0 + h]$, which implies $\tilde{y}(x) \equiv y(x)$. ■

Example 9.3.3 Show that Theorem 9.3.2 guarantees that the initial value problem

$$\frac{dy}{dx} = x + y^2, \quad y(0) = 0 \tag{9.3.11}$$

has a unique solution.

Solution Let $x_0 = 0$ and $y_0 = 0$ due to the initial condition. Theorem 9.3.2 does apply here because the function $f(x, y) = x + y^2$ is continuous everywhere in the xy-plane; and, for any a and b, it satisfies a Lipschitz condition with respect to y on the rectangular region

$$\mathcal{R} = \{\, (x, y) : |x| \leq a, |y| \leq b \,\}$$

with Lipschitz constant $2b$ (see Example 9.3.2). The maximum value that $|f(x, y)|$ attains on $\mathcal{R}$ is

$$M = \max\{|f(x, y)| : (x, y) \in \mathcal{R}\} = \max\{|x + y^2| : |x| \leq a, |y| \leq b\} = a + b^2.$$

Since we merely want to establish that there is a unique solution through the origin, we can choose any positive values for a and b. With $a = 1$ and $b = 1$, $M = 2$. Thus,

$$h = \min\left\{a, \frac{b}{M}\right\} = \min\left\{1, \frac{1}{2}\right\} = \frac{1}{2}.$$

Therefore, Theorem 9.3.2 guarantees that a unique solution of the initial value problem (9.3.11) exists on the interval $[-0.5, 0.5]$. ♦

Remark Generally speaking, the sizes of the intervals obtained with this and other existence and uniqueness theorems are conservative. The actual sizes are usually much larger. For example, see the graph of the solution of the initial value problem (9.3.11) that is shown in Fig. 9.1.

The Lipschitz condition in Theorem 9.3.2 is automatically satisfied if the partial derivative $\partial f/\partial y$ exists and is continuous on the closed rectangular region $\mathcal{R}$. To

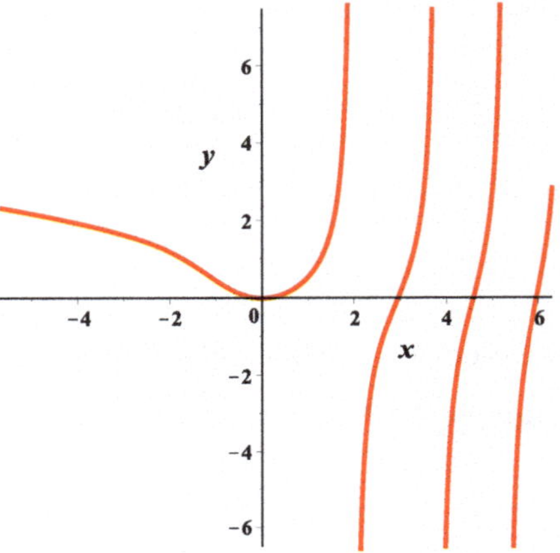

Fig. 9.1 Graph of the solution of the initial value problem (9.3.11)

prove this, apply the Mean Value Theorem: for (x, y) and (x, z) in $\mathcal{R}$, there exists a number ξ between y and z such that

$$f(x, y) - f(x, z) = f_y(x, \xi)(y - z).$$

Since $f_y(x, y)$ is continuous on $\mathcal{R}$, a constant L exists such that $|f_y(x, y)| \leq L$ for all $(x, y) \in \mathcal{R}$. As a result,

$$|f(x, y) - f(x, z)| = |f_y(x, \xi)||(y - z)| \leq L|y - z|$$

for all $(x, y) \in \mathcal{R}$. Because of this, we have the following existence and uniqueness result that involves a partial derivative rather than a Lipschitz condition.

Existence of a Unique Solution II

Theorem 9.3.3 *Suppose a function $f(x, y)$ and its partial derivative $f_y(x, y)$ are continuous on a closed rectangular region*

$$\mathcal{R} = \{(x, y) : |x - x_0| \leq a, |y - y_0| \leq b\}$$

where $a, b, x_0, y_0 \in \mathbb{R}$ and $a, b > 0$. Let $M = \max\{|f(x, y)| : (x, y) \in \mathcal{R}\}$. Then there exists a unique solution $y(x)$ of the initial value problem (9.3.2) *on the interval*

$$[x_0 - h, x_0 + h] \quad \textit{where} \quad h = \min\left\{a, \frac{b}{M}\right\}.$$

Moreover, the sequence $\{y_n(x)\}$ of Picard approximations, where $y_0(x) \equiv y_0$ and $y_n(x)$ is given by (9.2.5) *for $n = 1, 2, \ldots,$ converges to $y(x)$ on $[x_0 - h, x_0 + h]$.*

Example 9.3.4 Use the theorem to show that the initial value problem

$$\frac{dy}{dx} = \frac{2y\cos x}{y^2+1}, \quad y(\pi) = 1 \tag{9.3.12}$$

has a unique solution.

Solution First, note that Theorem 9.3.3 is applicable to this initial value problem because $f(x, y) = 2y\cos x/(y^2+1)$ and its partial derivative $f_y(x, y)$ are continuous on the rectangular region

$$\mathcal{R} = \{\,(x, y) : |x-\pi| \le a,\, |y-1| \le b\,\},$$

for any given positive values of a and b. Then the value of M in the statement of the theorem is

$$\begin{aligned} M &= \max\{|f(x, y)| : (x, y) \in \mathcal{R}\} = \max\left\{\left|\frac{2y\cos x}{y^2+1}\right| : (x, y) \in \mathcal{R}\right\} \\ &= \max\left\{2|\cos x|\frac{|y|}{y^2+1} : |x-\pi| \le a,\, |y-1| \le b\right\}. \end{aligned}$$

Since the maximum values of $|\cos x|$ and $|y|/(y^2+1)$ for all $x, y \in \mathbb{R}$ are 1 and $\frac{1}{2}$, respectively, we can choose $M = 1$. With $b = a$,

$$h = \min\{a, b/M\} = \min\{a, a\} = a.$$

So, regardless of the value of a, it follows from the theorem that (9.3.12) has a unique solution on the interval $[\pi - a, \pi + a]$. This implies that there is a unique solution of (9.3.12) on the entire interval $(-\infty, \infty)$. ◆

Example 9.3.5 Use Theorem 9.3.3 to show that the initial value problem

$$y' = 1 + y^2, \quad y(0) = 0 \tag{9.3.13}$$

has a unique solution. Compute the first three or four terms of the Picard sequence of approximations for (9.3.13). Find the analytical solution of this initial value problem.

Solution The function $f(x, y) = 1 + y^2$ and the partial derivative $f_y(x, y) = 2y$ are continuous everywhere in the xy-plane. In particular, for any given $a, b > 0$, they are continuous on the closed rectangular region

$$\mathcal{R} = \{\,(x, y) : |x| \le a,\, |y| \le b\,\}.$$

And so

$$M = \max\{|f(x, y)| : (x, y) \in \mathcal{R}\} = \max\{1 + y^2 : |x| \le a, |y| \le b\} = 1 + b^2.$$

With $a = 1$ and $b = 1$, we have

$$h = \min\left\{a, \frac{b}{M}\right\} = \min\left\{a, \frac{b}{1+b^2}\right\} = \frac{1}{2}.$$

Therefore, a solution of (9.3.13) exists and is unique on the interval $[-\frac{1}{2}, \frac{1}{2}]$.

Picard's formula for (9.3.13) is

$$y_n(x) = \int_0^x \left[1 + y_{n-1}^2(u)\right] du,$$

where $y_0(x) \equiv 0$. The first three Picard approximations are

$$y_1(x) = \int_0^x \left[1 + y_0^2(u)\right] du = \int_0^x du = x,$$

$$y_2(x) = \int_0^x \left[1 + y_1^2(u)\right] du = \int_0^x \left[1 + u^2\right] du = x + \frac{1}{3}x^3,$$

and

$$y_3(x) = \int_0^x \left[1 + y_2^2(u)\right] du = \int_0^x \left[1 + \left(u + \frac{1}{3}u^3\right)^2\right] du$$
$$= x + \frac{1}{3}x^3 + \frac{2}{15}x^5 + \frac{1}{63}x^7.$$

According to the last statement of Theorem 9.3.3, the sequence $y_1(x), y_2(x), y_3(x), \ldots$ converges to the solution of (9.3.13) on the interval $[-\frac{1}{2}, \frac{1}{2}]$.

We can easily find the analytical solution by separating variables and integrating the differential equation, thereby obtaining

$$\tan^{-1} y = x + C.$$

It follows from the initial condition that $C = 0$. Thus, $y = \tan x$ is the solution that exists and is unique on $[-\frac{1}{2}, \frac{1}{2}]$. In fact, $y = \tan x$ is a solution on the open interval $(-\frac{\pi}{2}, \frac{\pi}{2})$, which is more than three times longer than $[-\frac{1}{2}, \frac{1}{2}]$. There is a lesson that we should take away from this example and Example 9.3.3.

The actual interval of existence of the solution is usually much larger than the interval given by any of the existence and uniqueness theorems.

Since every convergent power series is the Taylor series of its sum (see Theorem 9.3.4 below), it appears from the Picard approximations that

$$\tan x = \lim_{n\to\infty} y_n(x) = x + \frac{1}{3}x^3 + \frac{2}{15}x^5 + \frac{1}{63}x^7 + \ldots.$$

However, the Maclaurin series for $\tan x$ is

$$\tan x = x + \frac{1}{3}x^3 + \frac{2}{15}x^5 + \frac{17}{315}x^7 + \ldots,$$

which is discussed in most calculus textbooks, such as [68, Ch. 11]. It seems something is amiss here. But what? The answer is revealed in the fourth Picard approximation:

$$y_4(x) = x + \frac{1}{3}x^3 + \frac{2}{15}x^5 + \frac{17}{315}x^7 + \frac{38}{2835}x^9 + \ldots.$$

So another lesson this example teaches us is to make sure that the coefficient of a given power of x is correct by continuing to compute Picard approximations until the coefficient does not change. ♦

In the example, we referred to an important result about power series, namely, that every convergent power series is the Taylor series of its sum. Before continuing, let us review some results for power series that pertain to the rest of this chapter. These results can be found in the "infinite sequences and series" chapter of most calculus textbooks (e.g., [68, Sec. 11.9]).

Power Series

Theorem 9.3.4 *If a power series*

$$\sum_{n=0}^{\infty} a_n(x-x_0)^n = a_0 + a_1(x-x_0) + a_2(x-x_0)^2 + a_3(x-x_0)^3 + \ldots$$

converges on an interval $I = (x_0-R, x_0+R)$, where $R > 0$, or on the interval $I = (-\infty, \infty)$, then it defines a function f on I, where $f(x)$ represents the ***sum of the series*** *at x. Furthermore:*

(i) *The function f is continuous and has (continuous) derivatives of all orders on I. Moreover, f' can be computed by differentiating f term-by-term:*

$$f'(x) = a_1 + 2a_2(x-x_0) + 3a_3(x-x_0)^2 + \cdots = \sum_{n=1}^{\infty} na_n(x-x_0)^{n-1}.$$

(continued)

Likewise, f'' can be computed by differentiating f' term-by-term:

$$f''(x) = 2a_2 + 6a_3(x - x_0) + \cdots = \sum_{n=2}^{\infty} n(n-1)a_n(x - x_0)^{n-2}$$

and so on.

(ii) *The coefficients of the power series are given by the formula:*

$$a_n = \frac{f^n(x_0)}{n!} \quad \text{for} \quad n = 0, 1, 2, \ldots.$$

In other words, the power series is the Taylor series of its sum $f(x)$.

Remark The maximum value of R for which the power series converges at all x in $(x_0 - R, x_0 + R)$ is called the ***radius of convergence*** of the series. For instance, in Example 9.3.5, $R = \pi/2$, not $1/2$. If the series converges on $(-\infty, \infty)$, then we say $R = \infty$. If the series converges only at $x = x_0$, then $R = 0$.

9.4 Second-Order Equations

Picard's method of successive approximations can also be used to obtain existence and uniqueness theorems for higher-order ordinary differential equations. The following theorem pertains to second-order equations.

Existence of a Unique Solution III

Theorem 9.4.1 *If a function $f(x, y, z)$ is continuous on a closed box*

$$\mathcal{B} = \{\, (x, y, z) : |x - x_0| \leq a, |y - y_0| \leq b, |z - z_0| \leq b \,\},$$

where $a, b, x_0, y_0 \in \mathbb{R}$ and $a, b > 0$, and if there is a Lipschitz constant $L > 0$ such that

$$|f(x, y, z) - f(x, u, v)| \leq L\big(|y - u| + |z - v|\big) \tag{9.4.1}$$

for all $(x, y, z), (x, u, v) \in \mathcal{B}$, then there exists a unique solution of the second-order equation

(continued)

$$y'' = f(x, y, y') \tag{9.4.2}$$

satisfying the initial conditions

$$y(x_0) = y_0, \; y'(x_0) = z_0 \tag{9.4.3}$$

on an interval $[x_0 - h, x_0 + h]$, *provided* h *is sufficiently small.*

Proof The proof begins by replacing (9.4.2) with an equivalent system of two first-order differential equations. After that, the rest of the proof proceeds very much like the proof of Theorem 9.3.2. However, since we have not yet considered systems of equations, we omit the proof. The interested reader can find a proof in Corduneanu [23, Theorem 2.3]. ■

Example 9.4.1 Use Theorem 9.4.1 to show that the initial value problem

$$\frac{d^2y}{dx^2} - y = 0, \quad y(0) = 0, \quad y'(0) = 1 \tag{9.4.4}$$

has a unique solution. Use Picard's method to find approximations of the solution.

Solution Comparing (9.4.2) to the differential equation $y'' - y = 0$, we see for this example that $f(x, y, y') = y$. From the initial conditions in (9.4.4), we have $x_0 = 0$, $y_0 = 0$, and $z_0 = 1$. Let a and b be any fixed pair of positive constants. Note that one of the conditions of the theorem is satisfied since $f(x, y, z) = y$ is continuous on the box

$$\mathcal{B} := \{ (x, y, z) : |x| \leq a, |y| \leq b, |z - 1| \leq b \}.$$

As for the Lipschitz condition (9.4.1),

$$|f(x, y, z) - f(x, u, v)| = |y - u|$$

for all $(x, y, z), (x, u, v) \in \mathcal{B}$. So the Lipschitz condition is met with $L = 1$. Since all of the conditions of Theorem 9.4.1 are met, we conclude that there is a unique solution $y(x)$ of the initial value problem (9.4.4) on an interval $[-h, h]$ if h is sufficiently small.

In order for us to apply Picard's method to the second-order equation in (9.4.4), let us rewrite it so that it looks like a first-order equation, such as in Examples 9.2.1–9.2.3. This can be accomplished by defining w by $w = y'$. Then,

$$y(x) = \int_0^x w(u)\, du \tag{9.4.5}$$

due to one of the initial conditions, namely, $y(0) = 0$. Because of the differential equation in (9.4.4),

$$w' = y'' = y = \int_0^x w(u)\,du.$$

Moreover, it follows from the other initial condition that $w(0) = y'(0) = 1$. Thus, we have transformed (9.4.4) to the following integro-differential equation and initial condition:

$$w'(x) = \int_0^x w(u)\,du, \qquad w(0) = 1. \tag{9.4.6}$$

Now suppose we can somehow find a function $w(x)$ that is a solution of the initial value problem (9.4.6). Then the function $y(x)$ defined by (9.4.5) is a solution of the differential equation in (9.4.4) because

$$y''(x) = w'(x) = \int_0^x w(u)\,du = y(x).$$

Moreover, the initial conditions in (9.4.4) are satisfied because $y'(0) = w(0) = 1$ and $y(0) = 0$ due to (9.4.5). In other words, we have shown that if there is a function $w(x)$ that is a solution of (9.4.6), then the function $y(x)$ defined by (9.4.5) is a solution of (9.4.4). Note that $w(x)$ would be the only solution of (9.4.6) since (9.4.4) has one and only one solution. It is left as an exercise to explain why the existence and uniqueness of a solution of (9.4.4) implies that (9.4.6) does in fact have a unique solution.

Now let us find Picard's formula for (9.4.6) so that we can use it to find approximations of the solution of (9.4.4). Integrating (9.4.6) as follows

$$\int_0^x w'(u)\,du = \int_0^x \left(\int_0^u w(v)\,dv\right) du,$$

we get

$$w(x) = 1 + \int_0^x \left(\int_0^u w(v)\,dv\right) du$$

since $w(0) = 1$. Consequently, Picard's formula is

$$w_n(x) = 1 + \int_0^x \left(\int_0^u w_{n-1}(v)\,dv\right) du \tag{9.4.7}$$

for $n = 1, 2, 3, \ldots$, where $w_0(x) \equiv 1$. Setting $n = 1$, we get

$$w_1(x) = 1 + \int_0^x \left(\int_0^u w_0(v)\,dv\right) du = 1 + \int_0^x \left(\int_0^u 1\,dv\right) du = 1 + \frac{x^2}{2}.$$

From (9.4.5), we also have the approximation

$$y_1(x) = \int_0^x w_1(u)\,du = \int_0^x \left(1 + \frac{u^2}{2}\right) du = x + \frac{x^3}{6}.$$

With $n = 2$,

$$\begin{aligned} w_2(x) &= 1 + \int_0^x \left(\int_0^u w_1(v)\,dv\right) du = 1 + \int_0^x \left(\int_0^u \left(1 + \frac{v^2}{2}\right) dv\right) du \\ &= 1 + \int_0^x \left(u + \frac{u^3}{3!}\right) du = 1 + \frac{x^2}{2!} + \frac{x^4}{4!} \end{aligned}$$

and

$$y_2(x) = \int_0^x w_2(u)\,du = \int_0^x \left(1 + \frac{u^2}{2!} + \frac{u^4}{4!}\right) du = x + \frac{x^3}{3!} + \frac{x^5}{5!}.$$

For an integer $n > 2$, the results of these calculations suggest that $w_n(x)$ and $y_n(x)$ are given by the sums

$$w_n(x) = \sum_{k=0}^{n} \frac{x^{2k}}{(2k)!} \quad \text{and} \quad y_n(x) = \sum_{k=0}^{n} \frac{x^{2k+1}}{(2k+1)!}. \tag{9.4.8}$$

If these sums are correct, then it would follow that $w_{n+1}(x)$ and $y_{n+1}(x)$ can be obtained from (9.4.8) by merely replacing n with $n + 1$. This is truly the case since it follows from (9.4.7) that

$$\begin{aligned} w_{n+1}(x) &= 1 + \int_0^x \left(\int_0^u w_n(v)\,dv\right) du = 1 + \int_0^x \left(\int_0^u \sum_{k=0}^{n} \frac{v^{2k}}{(2k)!}\,dv\right) du \\ &= 1 + \int_0^x \sum_{k=0}^{n} \frac{u^{2k+1}}{(2k+1)!}\,du = 1 + \sum_{k=0}^{n} \frac{x^{2k+2}}{(2k+2)!} \\ &= 1 + \frac{x^2}{2!} + \frac{x^4}{4!} + \cdots + \frac{x^{2n+2}}{(2n+2)!} = \sum_{k=0}^{n+1} \frac{x^{2k}}{(2k)!} \end{aligned}$$

and from (9.4.5) that

$$y_{n+1}(x) = \int_0^x w_{n+1}(u)\,du = \int_0^x \sum_{k=0}^{n+1} \frac{u^{2k}}{(2k)!}\,du = \sum_{k=0}^{n+1} \frac{x^{2k+1}}{(2k+1)!}.$$

Since we already have established that (9.4.8) is correct for $n = 1$, it follows from the principle of mathematical induction that $w_n(x)$ and $y_n(x)$ are given by the sums (9.4.8) for all integers $n \geq 1$. Therefore, the Picard approximations of the solution $y(x)$ of the initial value problem (9.4.4) are

$$y_n(x) = \sum_{k=0}^{n} \frac{x^{2k+1}}{(2k+1)!} = x + \frac{x^3}{3!} + \frac{x^5}{5!} + \cdots + \frac{x^{2n+1}}{(2n+1)!}$$

for $n = 0, 1, 2, \ldots$.

The approximations suggest that the infinite series

$$x + \frac{x^3}{3!} + \frac{x^5}{5!} + \frac{x^7}{7!} + \cdots = \sum_{n=0}^{\infty} \frac{x^{2n+1}}{(2n+1)!}$$

could be the solution $y(x)$ on an interval $[-h, h]$ if h is small enough (or quite possibly even on a larger interval containing this one). But does this series converge? Letting $a_n = x^{2n+1}/(2n+1)!$ and using the well-known Ratio Test for determining the convergence of series, we have

$$\ell := \lim_{n\to\infty} \left| \frac{a_{n+1}}{a_n} \right| = \lim_{n\to\infty} \left| \frac{x^{2n+3}}{(2n+3)!} \cdot \frac{(2n+1)!}{x^{2n+1}} \right| = \lim_{n\to\infty} \frac{x^2}{(2n+3)(2n+2)} = 0.$$

Since $\ell < 1$, the series converges for all values of x. Therefore, it defines a function on $\mathbb{R}$ that we call $y(x)$. That is,

$$y(x) := \sum_{n=0}^{\infty} \frac{x^{2n+1}}{(2n+1)!} = x + \frac{x^3}{3!} + \frac{x^5}{5!} + \frac{x^7}{7!} + \cdots \tag{9.4.9}$$

for all $x \in \mathbb{R}$.

Now the question is this: Does the function $y(x)$ defined by the series (9.4.9) solve the initial value problem (9.4.4)? To answer this, let us differentiate $y(x)$. It follows from Theorem 9.3.4 that

$$y'(x) = \frac{d}{dx} \sum_{n=0}^{\infty} \frac{x^{2n+1}}{(2n+1)!} = \sum_{n=0}^{\infty} \frac{x^{2n}}{(2n)!} = 1 + \frac{x^2}{2!} + \frac{x^4}{4!} + \frac{x^6}{6!} + \ldots$$

and

$$y''(x) = \frac{d}{dx} \sum_{n=0}^{\infty} \frac{x^{2n}}{(2n)!} = \sum_{n=1}^{\infty} \frac{x^{2n-1}}{(2n-1)!} = x + \frac{x^3}{3!} + \frac{x^5}{5!} + \cdots = \sum_{n=0}^{\infty} \frac{x^{2n+1}}{(2n+1)!}$$

for $x \in \mathbb{R}$. Thus, $y''(x) = y(x)$. That is, $y(x)$ is a solution of $y'' - y = 0$ on $\mathbb{R}$. Moreover, $y(0) = 0$ and $y'(0) = 1$. So $y(x)$ satisfies the initial conditions as well.

Even though everything called for in this example has been done, let us point out one more thing. Oftentimes, one has to be content with a series solution, but sometimes the series can be expressed in terms of elementary or familiar nonelementary functions. Such is the case with the series solution $y(x)$. Observe that it resembles the Maclaurin series expansion of e^x except that the even powers of x with their coefficients are missing. Adding in those terms and interspersing them with the odd powers of x, we have

$$y(x) = x + \frac{x^3}{3!} + \frac{x^5}{5!} + \frac{x^7}{7!} + \cdots = \frac{1}{2}\left[\left(1 + x + \frac{x^2}{2!} + \frac{x^3}{3!} + \frac{x^4}{4!} + \frac{x^5}{5!} + \ldots\right)\right.$$
$$\left.- \left(1 - x + \frac{x^2}{2!} - \frac{x^3}{3!} + \frac{x^4}{4!} - \frac{x^5}{5!} + \ldots\right)\right].$$

Since the second series enclosed by the brackets is the Maclaurin series for e^{-x}, the solution of the initial value problem (9.4.4) turns out to be

$$y(x) = \frac{e^x - e^{-x}}{2}$$

for $-\infty < x < \infty$. This particular linear combination[3] of the exponential functions e^x and e^{-x} defines the function $\sinh x$, which is called the ***hyperbolic sine function***. As a result, we can also write the solution as

$$y(x) = \sinh x.$$

♦

Remark In the next chapter, it is shown how to obtain this solution and other solutions of the second-order differential equation in (9.4.4) very easily and quickly using "characteristic roots" (see Theorem 10.3.1). Also, see Example 10.9.5.

Example 9.4.2 The differential equation $y'' - xy = 0$ is known as ***Airy's equation***, which is named after the British astronomer and mathematician George B. Airy, who found a solution of this equation, called the Airy integral, which he used to explain rainbow phenomena. It is an important equation in many fields, such as optics, quantum mechanics, electromagnetism, and fluid dynamics. Show that the initial value problem

$$y'' - xy = 0, \quad y(0) = 0, \quad y'(0) = 1 \tag{9.4.10}$$

has a unique solution. Apply Picard's method to find approximations of the solution. Then express the unique solution in terms of a series.

---

[3] The expression $c_1 f_1(x) + c_2 f_2(x)$, where c_1, c_2 are constants, is called a ***linear combination*** of the functions f_1 and f_2.

Solution Airy's equation is of the form $y'' = f(x, y, y')$, where $f(x, y, y') = xy$. Let $a, b > 0$. Due to the initial conditions, $x_0 = 0$, $y_0 = 0$, and $z_0 = 1$. Notice that $f(x, y, z) = xy$ is continuous on the closed box

$$\mathcal{B} := \{ (x, y, z) : |x| \leq a, |y| \leq b, |z - 1| \leq b \} .$$

Moreover,

$$\begin{aligned} |f(x, y, z) - f(x, u, v)| &= |xy - xu| = |x||y - u| \\ &\leq a|y - u| \leq a\big(|y - u| + |z - v|\big) \end{aligned}$$

for $(x, y, z), (x, u, v) \in \mathcal{B}$. Thus, f satisfies the Lipschitz condition (9.4.1) on $\mathcal{B}$ with Lipschitz constant $L = a$. Since all of the conditions of Theorem 9.4.1 are met, we conclude that there is a unique solution of the initial value problem (9.4.10) on an interval $[-h, h]$ if h is sufficiently small.

To find Picard's formula, let us rewrite Airy's equation in terms of $w = y'$. Since $y = 0$ when $x = 0$,

$$y = \int_0^x w(u)\, du. \tag{9.4.11}$$

This, together with Airy's equation, yields

$$w' = y'' = xy = x \int_0^x w(u)\, du.$$

Also, $w(0) = 1$ since $w(0) = y'(0) = 1$. Therefore, the initial value problem (9.4.10) expressed in terms of w is

$$w'(x) = x \int_0^x w(u)\, du, \quad w(0) = 1. \tag{9.4.12}$$

An integration yields

$$w(x) = 1 + \int_0^x \left(u \int_0^u w(v)\, dv \right) du$$

since $w(0) = 1$. Thus, Picard's formula is

$$w_n(x) = 1 + \int_0^x \left(u \int_0^u w_{n-1}(v)\, dv \right) du \tag{9.4.13}$$

for $n = 1, 2, 3, \ldots$, where $w_0(x) \equiv 1$.

With $n = 1$, Picard's formula (9.4.13) yields

$$w_1(x) = 1 + \int_0^x \left(u \int_0^u w_0(v)\, dv \right) du = 1 + \int_0^x u^2\, du = 1 + \frac{x^3}{3}.$$

It then follows from (9.4.11) that the first Picard approximation of the solution of the initial value problem (9.4.10) is

$$y_1(x) = \int_0^x w_1(u)\, du = \int_0^x \left(1 + \frac{u^3}{3} \right) du = x + \frac{x^4}{3 \cdot 4}.$$

Now let $n = 2$. Then,

$$\begin{aligned} w_2(x) &= 1 + \int_0^x \left(u \int_0^u w_1(v)\, dv \right) du = 1 + \int_0^x u \left(\int_0^u \left[1 + \frac{v^3}{3} \right] dv \right) du \\ &= 1 + \int_0^x u \left(u + \frac{u^4}{3 \cdot 4} \right) du = 1 + \int_0^x \left(u^2 + \frac{u^5}{3 \cdot 4} \right) du \\ &= 1 + \frac{x^3}{3} + \frac{x^6}{3 \cdot 4 \cdot 6}. \end{aligned}$$

Thus, the second Picard approximation of the solution of (9.4.10) is

$$y_2(x) = \int_0^x w_2(u)\, du = \int_0^x \left(1 + \frac{u^3}{3} + \frac{u^6}{3 \cdot 4 \cdot 6} \right) du = x + \frac{x^4}{3 \cdot 4} + \frac{x^7}{3 \cdot 4 \cdot 6 \cdot 7}.$$

It is left as an exercise to verify that

$$w_3(x) = 1 + \int_0^x \left(u \int_0^u w_2(v)\, dv \right) du = 1 + \frac{x^3}{3} + \frac{x^6}{3 \cdot 4 \cdot 6} + \frac{x^9}{3 \cdot 4 \cdot 6 \cdot 7 \cdot 9}$$

and

$$y_3(x) = \int_0^x w_3(u)\, du = x + \frac{x^4}{3 \cdot 4} + \frac{x^7}{3 \cdot 4 \cdot 6 \cdot 7} + \frac{x^{10}}{3 \cdot 4 \cdot 6 \cdot 7 \cdot 9 \cdot 10}.$$

Another way of expressing $y_3(x)$ is

$$y_3(x) = x + \frac{2}{4!}x^4 + \frac{2 \cdot 5}{7!}x^7 + \frac{2 \cdot 5 \cdot 8}{10!}x^{10}.$$

The pattern suggests that the nth Picard approximation is

$$y_n(x) = x + \frac{2}{4!}x^4 + \frac{2 \cdot 5}{7!}x^7 + \cdots + \frac{2 \cdot 5 \cdots (3n-1)}{(3n+1)!}x^{3n+1}.$$

It is left as an exercise to prove the validity of this formula for all integers $n \geq 1$ using a mathematical induction argument as in Example 9.4.1.

Since it has already been established that the initial value problem (9.4.10) has a unique solution on a sufficiently small interval, let us look into the possibility that the infinite series

$$x + \frac{2}{4!}x^4 + \frac{2 \cdot 5}{7!}x^7 + \frac{2 \cdot 5 \cdot 8}{10!}x^{10} + \cdots = x + \sum_{n=1}^{\infty} \frac{2 \cdot 5 \cdots (3n-1)}{(3n+1)!}x^{3n+1}$$

is that unique solution. Of course, it has to converge, at least on some interval containing $x = 0$, in order to define a function on this interval. Turning to the Ratio Test, we have

$$\begin{aligned} \ell &:= \lim_{n\to\infty} \left| \frac{2 \cdot 5 \cdots (3n-1)(3n+2)x^{3n+4}}{(3n+4)!} \cdot \frac{(3n+1)!}{2 \cdot 5 \cdots (3n-1)x^{3n+1}} \right| \\ &= \lim_{n\to\infty} \frac{|x|^3}{(3n+3)(3n+4)} = 0. \end{aligned}$$

Therefore, the series converges for all $x \in \mathbb{R}$. As a result, the series defines the function

$$y(x) := x + \sum_{n=1}^{\infty} \frac{2 \cdot 5 \cdots (3n-1)}{(3n+1)!}x^{3n+1} \tag{9.4.14}$$

on $\mathbb{R}$.

Now let us show that it a solution. Differentiating $y(x)$ term-by-term, we have

$$y'(x) = 1 + \sum_{n=1}^{\infty} \frac{2 \cdot 5 \cdots (3n-1)}{(3n)!}x^{3n}. \tag{9.4.15}$$

Likewise, differentiating $y'(x)$ term-by-term yields

$$\begin{aligned} y''(x) &= \sum_{n=1}^{\infty} \frac{2 \cdot 5 \cdots (3n-1)}{(3n-1)!}x^{3n-1} = x^2 + \sum_{n=2}^{\infty} \frac{2 \cdot 5 \cdots (3n-4)}{(3n-2)!}x^{3n-1} \\ &= x^2 + \sum_{n=1}^{\infty} \frac{2 \cdot 5 \cdots (3n-1)}{(3n+1)!}x^{3n+2}. \end{aligned}$$

It is obvious from (9.4.14) that $y''(x)$ is equal to $xy(x)$. Therefore, $y(x)$ is a solution of Airy's equation for $-\infty < x < \infty$. Moreover, it follows from (9.4.14) that $y(0) = 0$ and from (9.4.15) that $y'(0) = 1$.

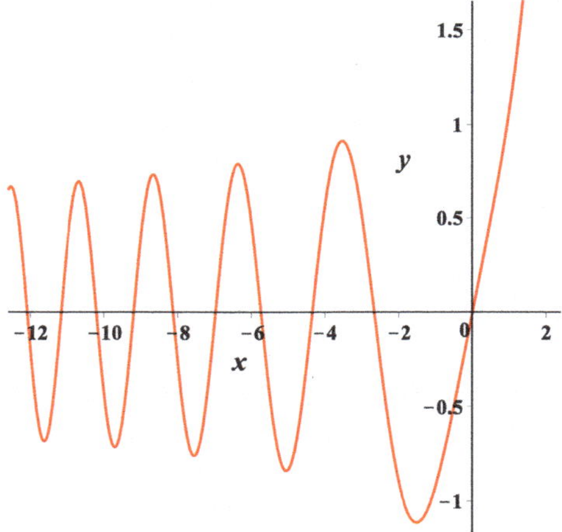

Fig. 9.2 Graph of the solution of the initial value problem (9.4.10)

In summary, we have established that the series $y(x)$ defined by (9.4.14) is the solution of the initial value problem (9.4.10). Its graph is shown in Fig. 9.2.

Remarks A couple of remarks are in order. The computer algebra system *Maple* was used to create the graph in Fig. 9.2. However, to do so, the numerator in the general formula for the terms of the series in (9.4.14), which has missing factors indicated by the ellipsis, was replaced with an equivalent expression that *Maple* could recognize and use. Even though we can clearly see from the pattern what the missing factors are, *Maple* needs something more explicit. Fortunately, a way to handle this is to rewrite the series (9.4.14) using the ***Pochhammer function*** (also known as the ***rising factorial***) since the command for this function, namely, pochhammer(x,n), is built into *Maple*. The mathematical notation for the Pochhammer function is not standard. One of the commonly-used symbols is $x^{(n)}$. It is defined as follows: For a nonnegative integer n,

$$x^{(n)} = x(x+1)(x+2)\cdots(x+n-1), \text{ where } x^{(0)} = 1.$$

It is a generalization of the ordinary factorial of a nonnegative integer:

$$1^{(n)} = 1 \cdot 2 \cdot 3 \cdots (n-1) \cdot n = n!$$

Expressed in terms of the Pochhammer function, (9.4.14) is

$$y(x) = \sum_{n=0}^{\infty} 3^n \left(\frac{2}{3}\right)^{(n)} \frac{x^{3n+1}}{(3n+1)!}.$$

Let us verify this that this is correct by writing out the first several terms:

$$\begin{aligned}
y(x) &= 3^0\left(\frac{2}{3}\right)^{(0)}\frac{x}{1!} + 3^1\left(\frac{2}{3}\right)^{(1)}\frac{x^4}{4!} + 3^2\left(\frac{2}{3}\right)^{(2)}\frac{x^7}{7!} + 3^3\left(\frac{2}{3}\right)^{(3)}\frac{x^{10}}{10!} + \cdots \\
&= x + 3\left(\frac{2}{3}\right)\frac{x^4}{4!} + 3^2\left(\frac{2}{3}\right)\left(\frac{5}{3}\right)\frac{x^7}{7!} + 3^3\left(\frac{2}{3}\right)\left(\frac{5}{3}\right)\left(\frac{8}{3}\right)\frac{x^{10}}{10!} + \cdots \\
&= x + \frac{2}{4!}x^4 + \frac{2\cdot 5}{7!}x^7 + \frac{2\cdot 5\cdot 8}{10!}x^{10} + \cdots .
\end{aligned}$$

The second remark concerns the decay in the oscillation of $y(x)$ when $x < 0$ and the contrasting rapid growth of $y(x)$ when $x > 0$. These features of the solution could not have been anticipated from merely viewing Airy's equation $y'' - xy = 0$ or even its series solution (9.4.14). In hindsight, however, these features make sense when we compare Airy's equation with the initial conditions in (9.4.10) to the following two initial value problems. Suppose we replace the variable x in Airy's equation with the constant -1 when $x \le 0$ and with 1 when $x \ge 0$. In other words, replace (9.4.10) with

$$y'' + y = 0, \quad y(0) = 0, \quad y'(0) = 1 \tag{9.4.16}$$

when $x \le 0$ and with

$$y'' - y = 0, \quad y(0) = 0, \quad y'(0) = 1 \tag{9.4.17}$$

when $x \ge 0$. By inspection, the solution of (9.4.16) is $y = \sin x$. And in Example 9.4.1, we determined that the solution of (9.4.16) is $y = \sinh x$. Splicing these two solutions together at $x = 0$, we obtain the piecewise-defined function

$$y = \begin{cases} \sin x, & x \le 0 \\ \sinh x, & x > 0. \end{cases} \tag{9.4.18}$$

Likewise, replacing x with -4 when $x \le 0$ and with 4 when $x \ge 0$, solving the resulting initial value problems, and splicing the solutions together yields the function

$$y = \begin{cases} \frac{1}{2}\sin 2x, & x \le 0 \\ \frac{1}{2}\sinh 2x, & x > 0. \end{cases} \tag{9.4.19}$$

The graphs of both piecewise-defined functions are shown in Fig. 9.3. It is evident from comparing the graphs in Figs. 9.2 and 9.3 that the variable coefficient x of the dependent variable y in Airy's equation is the reason that the oscillation in Fig. 9.2 decays in amplitude and increases in frequency as the distance from the

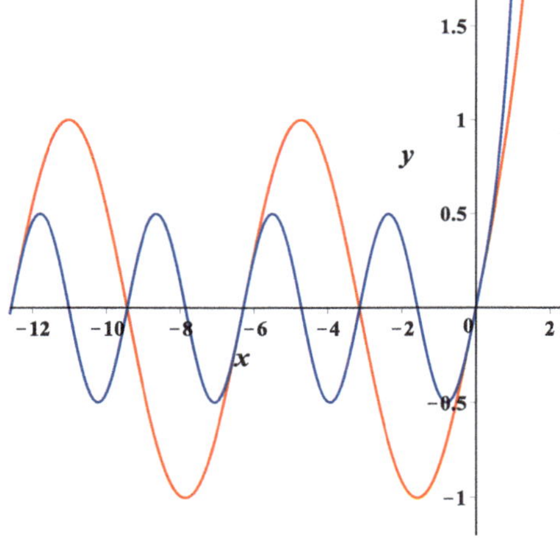

Fig. 9.3 Graphs of the functions (9.4.18) and (9.4.19)

origin increases; and why, to the right of the origin, the graph increases so rapidly, which is explained by the fact that

$$\sinh kx = \frac{e^{kx} - e^{-kx}}{2} \to \infty$$

when $k > 0$.

♦

Problems

It is quite a three-pipe problem, and I beg that you won't speak to me for fifty minutes.
(Mr. Sherlock Holmes to Dr. Watson)
Sir Arthur Conan Doyle, *The Red-Headed League* (1891) [28, p. 428]

Picard Approximations

1. Compute the third Picard approximation $y_3(x)$ of the solution of

$$\frac{dy}{dx} = 2xy, \quad y(0) = 4.$$

Solve this initial value problem to find its analytical solution. Find the Maclaurin polynomial of degree 6 that approximates the solution and compare it to $y_3(x)$.

2. Compute the third Picard approximation $y_3(x)$ of the solution of

$$\frac{dy}{dx} = 1 + 2xy, \quad y(1) = 4.$$

Show that the analytical solution is

$$y(x) = \frac{\sqrt{\pi}}{2} e^{x^2}\left[\operatorname{erf}(x) - \operatorname{erf}(1)\right] + 4e^{x^2-1}.$$

Compare the approximation $y_3(x)$ to the solution $y(x)$ at $x = 0.8$.

3. Compute the third Picard approximation of the solution of

$$\frac{dy}{dx} = x + y^2, \quad y(0) = 0.$$

Use it to approximate the solution at $x = 3/4$.

4. It is shown in Example 9.3.4 that the initial value problem

$$\frac{dy}{dx} = \frac{2y\cos x}{y^2+1}, \quad y(\pi) = 1$$

has a unique solution $y(x)$ on $(-\infty, \infty)$.

(a) Find the first Picard approximation $y_1(x)$ of the solution $y(x)$. Find the Taylor series for $y_1(x)$ about $x = \pi$.

(b) Show that an implicit solution of the differential equation is

$$y^2 + \ln y^2 = 1 + 4\sin x.$$

Use this and the method of implicit differentiation to compute the cubic Taylor polynomial $T_3(x)$ about π that approximates the solution $y(x)$. Compare $T_3(x)$ with the Taylor series for $y_1(x)$.

5. Compute at least five Picard approximations of the solution of

$$y' = x^2 y + 2x, \quad y(0) = 0.$$

Show that the nth Picard approximation is

$$y_n(x) = x^2 + \sum_{k=1}^{n-1} \frac{x^{3k+2}}{\prod_{j=1}^{k}(3j+2)},$$

for $n = 1, 2, 3, \ldots$, where $y_0(x) \equiv 0$. Note that $y_1(x) = x^2$ since $n = 1$ results in an empty sum (cf. (11.7.18)). Just as the uppercase Greek letter sigma is used to denote a sum, the uppercase Greek letter pi denotes the following product:

$$\prod_{j=1}^{k} b_j := (b_1)(b_2)\cdots(b_{k-1})(b_k).$$

For example,

$$\prod_{j=1}^{4}(3j+2) = (5)(8)(11)(14) = 6160.$$

Existence and Uniqueness of Solutions

6. (a) Apply Theorem 9.3.3 to Problem 5. With $a = 1$, show that a unique solution of the initial value problem exists on the interval $[-h, h]$, where $h = b/(b+2)$. Then explain why the solution exists at least on the open interval $(-1, 1)$.

   (b) Find an analytical solution of the initial value problem in Problem 5 using the integrating factor method (cf. Example 5.2.5). Explain why

$$e^{x^3/3} \int_0^x u e^{-u^3/3}\, du = \frac{1}{2}\left(x^2 + \sum_{k=1}^{\infty} \frac{x^{3k+2}}{\prod_{j=1}^{k}(3j+2)}\right)$$

   holds at least on the interval $(-1, 1)$.

7. Use one of the existence and uniqueness theorems in Sect. 9.3 to establish that

$$y' = -y^2, \quad y(0) = 1$$

   has a unique solution. Find an interval over which the solution is guaranteed to exist by the theorem. Compute several Picard approximations of the solution. Does this sequence suggest what the solution could be? Find it by the method of separation of variables. What, in fact, is the largest interval on which this solution exists and satisfies the initial value problem?

8. Show by using an appropriate existence and uniqueness theorem that the initial value problem

$$\frac{dy}{dx} = \cos^2 y, \quad y(0) = 0$$

   has a unique solution. Find the largest interval of existence of the solution. Compute the second Picard approximation of the solution. Use it to estimate the value of the solution at $x = \pi/8$. Solve this initial value problem analytically and then find the exact value of the solution at $x = \pi/8$.

Second-Order Equations

9. In Example 9.4.1, we determined that the solution of the initial value problem

$$y'' - y = 0, \quad y(0) = 0, \quad y'(0) = 1$$

   is $y = \sinh x$ for all $x \in \mathbb{R}$, where

$$\sinh x := \frac{e^x - e^{-x}}{2},$$

   which is called the ***hyperbolic sine function***. A related function is the ***hyperbolic cosine function***, which is defined by

$$\cosh x := \frac{e^x + e^{-x}}{2}.$$

   Using these definitions, show that

$$\frac{d}{dx}\cosh x = \sinh x \quad \text{and} \quad \frac{d}{dx}\sinh x = \cosh x.$$

   Verify $y = \sinh x$ is the solution of this initial value problem via these differentiation formulas.

10. Use Picard's method to find the solution of the initial value problem

$$y'' - y = 0, \quad y(0) = 1, \quad y'(0) = 0.$$

Hint. See Example 9.4.1.

11. Use Picard's method to find the solution of the initial value problem

$$y'' - xy = 0, \quad y(0) = 1, \quad y'(0) = 0.$$

Hint. See Example 9.4.2.

12. Use Picard's method to find the solution of the initial value problem

$$y'' + xy = 0, \quad y(0) = 0, \quad y'(0) = 1.$$

Chapter 10
Second-Order Linear Equations

Pedagogically speaking, a good share of physics and mathematics was—and is—writing differential equations on a blackboard and showing students how to solve them. Differential equations represent reality as a continuum, changing smoothly from place to place and from time to time, not broken in discrete grid points or time steps.

James Gleick[1]

Abstract This chapter explores how to find solutions of *second-order linear differential equations with constant coefficients*, namely equations of the form

$$a\frac{d^2x}{dt^2} + b\frac{dx}{dt} + cx = g(t), \tag{1}$$

where the constant $a \neq 0$ and g represents a given function.

The chapter begins with a description of the physical setup of a body sitting on a virtually frictionless air track and attached to a so-called *ideal spring* whose mass is negligible compared to that of the body. In one scenario, it is assumed that the only forces acting on the body are the restoring force of the spring and a damping force that is proportional to the velocity of the body. Newton's second law and Hooke's law are used to derive a differential equation that models the position x of the body at time t when it is displaced from its resting position and then released. This equation is of the form

$$ax'' + bx' + cx = 0, \tag{2}$$

[1] James Gleick (born 1954) helped popularize the mathematics and science of chaos through his newspaper column in the *New York Times* and in his best-selling nonfiction book *Chaos: Making a New Science* [39], where the above quotation is found on page 67. After that, the general public began hearing of chaos elsewhere, even in the movies, such as Spielberg's 1993 film *Jurassic Park* in which Jeff Goldblum played the role of a mathematician whose expertise is chaos theory.

© The Author(s), under exclusive license to Springer Nature Switzerland AG 2026
L. C. Becker, *Ordinary Differential Equations: Concepts, Methods, and Models*,
https://doi.org/10.1007/978-3-032-15150-6_10

which is (1) with $g(t) \equiv 0$. This is the launching point for explaining how to find a general solution of a given *homogeneous equation* (2) when the associated *characteristic equation*

$$a\lambda^2 + b\lambda + c = 0 \tag{3}$$

has distinct real roots or a repeated real root.

The case of complex roots begins with the simple equation

$$x'' + \omega^2 x = 0 \quad (\omega > 0). \tag{4}$$

The reason for using this particular equation is that it models the periodic motion of a body attached to an ideal spring when friction and other damping forces are totally absent and because the roots of the associated characteristic equation are $\pm i\omega$. This suggests solutions of (4) are linear combinations of $\sin \omega t$ and $\cos \omega t$ and that there is a connection between complex characteristic roots and the sine and cosine functions. After this is shown to be the case, attention turns to finding a general solution of a given equation (2) when the characteristic equation for it has complex roots with nonzero real parts. Details on how all of this unfolds are presented in Sect. 10.3.

Section 10.4 introduces *Euler's formula*, which defines the complex exponential function e^{it}. Instead of the usual definition involving the Maclaurin series for e^t and extending it to complex numbers, Euler's formula is justified by explaining how to define e^{it} so that it yields the solutions $\cos t$ and $\sin t$ of two initial value problems involving (4) with $\omega = 1$.

Among the other topics covered in this chapter are *Cauchy-Euler equations* in Sect. 10.5; *linear homogeneous equations of higher order* in Sect. 10.6; and an introduction to *second-order nonhomogeneous linear equations* in Sect. 10.7, that is, equations of the form (1) where $g(t)$ is not identically zero. Sections 10.8 and 10.9 cover the methods of *undetermined coefficients* and *variation of parameters*.

10.1 Horizontal Mass-Spring System

Consider the simple apparatus depicted in Fig. 10.1. A body of mass m sits on an air track and is attached to a spring, such as can be found in most college physics labs. The left end of the spring is fastened to a stationary support. This is an example of a ***horizontal mass-spring system.***[2] Sometimes we will use the letter m to refer to the body itself and at other times to the mass of the body; it should be clear from the context in which sense m is used. Our objective is to model the back-and-forth motion of the body m when it is pulled to the right or pushed to the left and then

[2] The word "mass" in "horizontal mass-spring system" refers to the body attached to the spring and not to the quantity of matter that makes up the body.

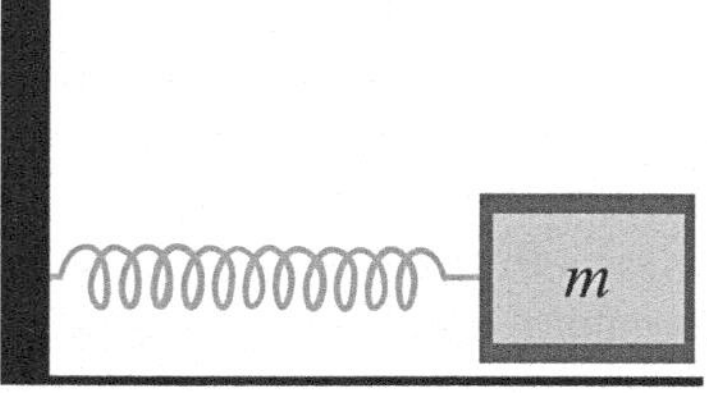

Fig. 10.1 Horizontal mass-spring system with no damping

released. Suppose the mass-spring system has the following features:

1. The mass of the spring is negligible compared to the mass of m. In essence, the spring can be viewed as massless.
2. The track on which m is free to slide is like the surface of an air hockey table. There are many small perforations in the track through which air is pumped to significantly reduce the frictional force exerted on m by the track. Consequently, friction will have a negligible effect on the motion of m.
3. When the natural length of the spring is changed by stretching or compressing it, a force acts to restore it back to its natural length. Assume that the spring obeys ***Hooke's law***, namely, that the magnitude of the restoring force at any given instant is proportional to the difference between the length of the spring at that instant and its natural length. This is the case for many common springs provided the elastic limit of the spring is not exceeded. Such springs are said to be ***linear*** (or ***ideal***) springs.

To specify the horizontal position of m at time t, let

$$x(t) := [\text{length of the spring at time } t] - [\text{natural length of the spring}].$$

Thus, when $x > 0$, the spring is stretched and when $x < 0$, it is compressed. See Fig. 10.2, where P marks the right end of the spring that is attached to m.

Obviously, when the spring is stretched, the restoring force is directed to the left. So in this case the force is a negative quantity. It follows from the statement of Hooke's law that this restoring force F is given by the formula

$$F = -kx \tag{10.1.1}$$

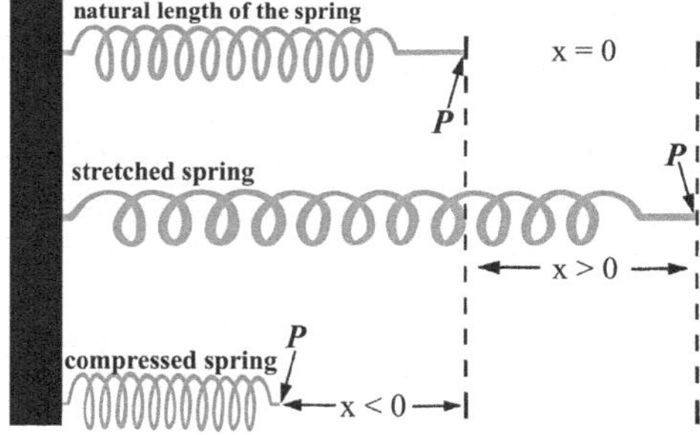

Fig. 10.2 Comparison of a deformed spring to its natural length

where $k > 0$. The negative sign is included so that the constant of proportionality k, called the ***spring constant*** or ***stiffness of the spring***, is positive. When the spring is compressed, the restoring force is directed to the right: so $F > 0$. Note that this is the direction given by Eq. (10.1.1) since $-kx > 0$ when $x < 0$. Moreover, it also follows from (10.1.1) that $F = 0$ when $x = 0$, as it should be when the spring is in its natural state. We conclude that (10.1.1) is the correct formulation of Hooke's law.

Since the force $-kx$ is the only horizontal force acting on m, Newton's second law of motion yields the differential equation

$$m\ddot{x} = -kx. \tag{10.1.2}$$

With $\omega := \sqrt{k/m}$, this can be rewritten as

$$\ddot{x} + \omega^2 x = 0. \tag{10.1.3}$$

The parameter ω is positive since both k and m are positive.

A physical system with a time-dependent quantity that can be modeled by Eq. (10.1.3), such as the position $x(t)$ of the mass (body) in a mass-spring system, is known as a ***simple*** (or ***linear***) ***harmonic oscillator***. Besides the mass-spring system, there are other physical systems with quantities that can be modeled by (10.1.3), such as the current in an LC circuit: an electrical circuit consisting of an inductor and a charged capacitor but no voltage source and which has no resistance. Of course, in a real LC circuit, there is always some resistance just as there is always some friction present in a mechanical system.

The horizontal mass-spring system that we have described above is an ideal system because in reality the frictional force between the track and the body m cannot be totally ignored. Even if we could somehow pull off an engineering feat by eliminating friction altogether there is still a ***damping*** (***retarding***) ***force*** acting on m as it moves through the air. This damping force would be even more pronounced if the mass moved through a more viscous medium, such as oil. In fact, the suspension system for each wheel of a motor vehicle consists of a mass-spring system connected to a shock absorber, which damps or smooths out the constant up-and-down motion of the vehicle's frame. A simple shock absorber consists of a cylinder, with a sliding piston inside it, that is filled with a fluid, such as oil or pressurized air.

Experimental evidence indicates the damping force described above is more or less proportional to the velocity of the body m; that is, the force is $-r\dot{x}$, where $r > 0$ is the constant of proportionality. The negative sign arises from the opposing directions of the damping force and the velocity of m. If we still ignore friction but now add the damping force, (10.1.2) becomes

$$m\ddot{x} = -r\dot{x} - kx \tag{10.1.4}$$

or

$$\ddot{x} + b\dot{x} + \omega^2 x = 0, \tag{10.1.5}$$

where $b = r/m$ and $\omega = \sqrt{k/m}$.

Equations (10.1.2)–(10.1.5) are members of the family of differential equations known as *second-order homogeneous linear differential equations*. The objective of this chapter is to learn how to solve these kind of equations.

10.2 Second-Order Homogeneous Linear Equations

Equations (10.1.3) and (10.1.5) are examples of second-order homogeneous linear equations with constant coefficients. The order of a differential equation is two when the equation has a second-order derivative but no derivatives of higher order. Members of this family are easy to recognize. Any equation that looks like (10.2.1) below, or which can be rewritten like it, is a member of this family.

Second-Order Homogeneous Linear Equations

Definition 10.2.1 A differential equation expressible in the form

$$a\frac{d^2x}{dt^2} + b\frac{dx}{dt} + cx = 0, \tag{10.2.1}$$

where a, b, c are constants and $a \neq 0$ is called a ***second-order homogeneous linear differential equation with constant coefficients***.

It goes without saying that x'' must be present in order for (10.2.1) to be considered a second-order equation, hence, the proviso $a \neq 0$. On the other hand, the other terms may or may not be present. That depends on the values of b and c. For example, for $a = 1$, $b = -1$, and $c = 0$, (10.2.1) is

$$\frac{d^2x}{dt^2} - \frac{dx}{dt} = 0. \tag{10.2.2}$$

The term ***homogeneous*** means the right-hand side of (10.2.1) is always equal to zero, regardless of the value of the independent variable t. If we replace "0" with a function g that is not identically equal to zero, then (10.2.1) becomes

$$a\frac{d^2x}{dt^2} + b\frac{dx}{dt} + cx = g(t)$$

and it is then called a ***nonhomogeneous***[3] ***second-order differential equation***.

[3] Or, ***inhomogeneous***.

10.3 Characteristic Roots and Solutions

We begin our search for a method to solve equations of the form (10.2.1). First, consider the following simple examples which will help us get some experience with dealing with these equations and a feel for the nature of their solutions. Unless it is otherwise indicated, t is the independent variable and the prime symbol ($'$) denotes differentiation with respect to t.

Example 10.3.1 Find solutions of the first-order equation $x' - x = 0$.

Solution By inspection (or by separating variables or the integrating factor method),

$$x(t) = ce^t$$

is a solution on the interval $(-\infty, \infty)$ for any constant c. ♦

Example 10.3.2 Find solutions of the second-order equation $x'' - x' = 0$.

Solution Solutions of this equation are easy to find due to the missing x term: simply rewrite the equation as

$$(x')' - x' = 0$$

and then replace x' with y. In this way, the second-order equation reduces to the first-order equation

$$y' - y = 0.$$

By the previous example, it has the general solution $y(t) = ce^t$. And so $x'(t) = ce^t$. Integrating, we see that any function of the form

$$x(t) = ce^t + k,$$

where c and k are constants, is a solution on $(-\infty, \infty)$. Notice that we could view this solution as a combination of two exponential functions

$$x(t) = ce^t + ke^{0t}.$$ ♦

Example 10.3.3 Verify that e^{2t} and e^{-3t} are solutions of $x'' + x' - 6x = 0$.

Solution For $x(t) = e^{2t}$, $x'(t) = 2e^{2t}$, and $x''(t) = 4e^{2t}$. Thus,

$$x''(t) + x'(t) - 6x(t) = 4e^{2t} + 2e^{2t} - 6e^{2t} = 0,$$

which shows that $x(t) = e^{2t}$ is a solution. It is left to the reader to show $x(t) = e^{-3t}$ is also a solution. In fact,

$$x(t) = c_1 e^{2t} + c_2 e^{-3t}$$

is a solution on $(-\infty, \infty)$ for any values of the constants c_1 and c_2. ♦

The second-order equations in the previous examples have solutions of the form $e^{\lambda t}$, where λ is a constant. So it seems reasonable for us to conjecture that all second-order equations of the form (10.2.1) have exponential solutions. Since we are concerned only with these kinds of equations from now on, it will be understood that $a \neq 0$ throughout the rest of this chapter.

To find out if our conjecture is correct, let us substitute $e^{\lambda t}$ for x in (10.2.1) to see if it is a solution for some value of λ:

$$a\frac{d^2}{dt^2}e^{\lambda t} + b\frac{d}{dt}e^{\lambda t} + ce^{\lambda t} = (a\lambda^2 + b\lambda + c)e^{\lambda t} = 0.$$

It follows from this that $e^{\lambda t}$ is a solution if the factor $a\lambda^2 + b\lambda + c$ is equal to zero. Moreover, it is a solution only if $a\lambda^2 + b\lambda + c = 0$ since $e^{\lambda t}$ is never zero regardless of the value of λ. In other words, $e^{\lambda t}$ is a solution of (10.2.1) if λ is a solution of the quadratic equation

$$a\lambda^2 + b\lambda + c = 0, \tag{10.3.1}$$

but not otherwise. In the context of solving a differential equation, this quadratic equation is known as a ***characteristic equation***. So (10.3.1) is the characteristic equation for the differential equation (10.2.1). The polynomial

$$P(\lambda) := a\lambda^2 + b\lambda + c \tag{10.3.2}$$

is called the ***characteristic polynomial***. By the quadratic formula, Eq. (10.3.1) has the roots

$$\lambda = \frac{-b \pm \sqrt{b^2 - 4ac}}{2a}. \tag{10.3.3}$$

Roots of a characteristic equation are called ***characteristic roots***. Let us label them λ_1 and λ_2 as follows:

$$\lambda_1 = \frac{-b - \sqrt{b^2 - 4ac}}{2a} \quad \text{and} \quad \lambda_2 = \frac{-b + \sqrt{b^2 - 4ac}}{2a}. \tag{10.3.4}$$

Observe that the characteristic roots are unequal, or distinct, if the discriminant of the quadratic equation $(b^2 - 4ac)$ is nonzero. If $b^2 - 4ac > 0$, these distinct roots are real numbers. However, if $b^2 - 4ac < 0$, then they clearly are not real numbers, but rather complex (or imaginary) numbers. There is also the case: $b^2 - 4ac = 0$.

Then

$$\lambda_1 = \lambda_2 = -\frac{b}{2a}.$$

For this case, we could say, as in an algebra course, that there is only one root. However, it will be advantageous, as we shall see shortly, to view all characteristic equations as having two roots. With this point of view, there are still two characteristic roots when $b^2 - 4ac = 0$, namely, the real and equal roots

$$\lambda_1 = -\frac{b}{2a} \quad \text{and} \quad \lambda_2 = -\frac{b}{2a}.$$

We say that $-b/2a$ is a ***repeated root*** or a ***double root*** or a ***root of multiplicity 2***.

Up to this point, we have discovered that

Characteristic Roots and Exponential Solutions

$$e^{\lambda t} \text{ is a solution of } ax'' + bx' + cx = 0 \tag{10.3.5a}$$

if and only if

$$\lambda \text{ is a root of } a\lambda^2 + b\lambda + c = 0. \tag{10.3.5b}$$

It is clear that a multiple of $e^{\lambda}t$ is also a solution of (10.2.1). That brings up the question as to whether there are other solutions that we are unaware of. We will answer this next by searching for a general solution of (10.2.1)—the form in which all of its solutions can be written. Our approach will be to rewrite (10.2.1) in terms of its characteristic roots and then reduce it to a first-order equation similar to what was done in Example 10.3.2. To this end, suppose $x(t)$ is a solution of equation (10.2.1), which means of course that it satisfies (10.2.1) on some interval J. And so

$$ax''(t) + bx'(t) + cx(t) = 0 \tag{10.3.6}$$

for $t \in J$. Whether the characteristic roots are distinct or repeated, their respective sum and product are

$$\lambda_1 + \lambda_2 = -\frac{b}{a} \quad \text{and} \quad \lambda_1\lambda_2 = \frac{c}{a}. \tag{10.3.7}$$

This is easily shown by adding and multiplying the values of λ_1 and λ_2 given by (10.3.4). These two equations make it possible to rewrite (10.3.6) in terms of its characteristic roots. To see this, first divide (10.3.6) by the nonzero coefficient a:

$$x''(t) + \frac{b}{a}x'(t) + \frac{c}{a}x(t) = 0.$$

Now replace the coefficients of $x'(t)$ and $x(t)$ with their respective counterparts from (10.3.7):

$$x''(t) - (\lambda_1 + \lambda_2)x'(t) + \lambda_1\lambda_2 x(t) = 0. \tag{10.3.8}$$

By rewriting this as

$$\big[x'(t) - \lambda_2 x(t)\big]' - \lambda_1\big[x'(t) - \lambda_2 x(t)\big] = 0, \tag{10.3.9}$$

we can simplify it by defining $y(t)$ to be

$$y(t) := x'(t) - \lambda_2 x(t). \tag{10.3.10}$$

As a result, the second-order equation (10.3.8), which we do not know yet how to solve directly, has been changed to the first-order equation

$$y'(t) - \lambda_1 y(t) = 0, \tag{10.3.11}$$

which we do know how to solve. All of its solutions are of the form $y(t) = k_1 e^{\lambda_1 t}$, where k_1 denotes a constant. Consequently, by (10.3.10),

$$x'(t) - \lambda_2 x(t) = k_1 e^{\lambda_1 t}.$$

Multiplying by the integrating factor $e^{-\lambda_2 t}$, we get

$$\frac{d}{dt}\big[e^{-\lambda_2 t}x(t)\big] = k_1 e^{(\lambda_1 - \lambda_2)t}. \tag{10.3.12}$$

If the roots are distinct ($\lambda_2 \neq \lambda_1$), we can integrate (10.3.12) and solve for $x(t)$ obtaining

$$x(t) = \frac{k_1}{\lambda_1 - \lambda_2}e^{\lambda_1 t} + k_2 e^{\lambda_2 t}, \tag{10.3.13}$$

where k_2 is a constant. We conclude that if a function $x(t)$ is a solution of (10.2.1) and the characteristic roots are the distinct numbers λ_1 and λ_2, then $x(t)$ is given by (10.3.13) for some k_1 and k_2.

On second thought we need to rethink the very last statement. It is certainly correct if λ_1 and λ_2 are real numbers. But what if they are complex numbers? Then there is a problem because $e^{\lambda_1 t}$ and $e^{\lambda_2 t}$ would make no sense to someone who has not dealt with complex powers before. In point of fact, it does make sense but the explanation of this will have to wait until Sect. 10.4. As for now, we will only deal with real characteristic roots.

Note that (10.3.13) is undefined when $\lambda_2 = \lambda_1$. We will investigate what to do when this happens after we have disposed of the case of distinct real roots.

10.3.1 Distinct Real Characteristic Roots

Let us replace the constants $k_1/(\lambda_1 - \lambda_2)$ and k_2 in (10.3.13) by c_1 and c_2, respectively. Then (10.3.13) simplifies to

$$x(t) = c_1 e^{\lambda_1 t} + c_2 e^{\lambda_2 t}. \tag{10.3.14}$$

We discovered in the previous section that if the characteristic roots λ_1 and λ_2 are real and distinct, then any solution $x(t)$ of (10.2.1) can be expressed in the form (10.3.14) for some c_1 and c_2. Conversely, any function of the form (10.3.14) is a solution of (10.2.1) if λ_1 and λ_2 are roots of (10.3.1). This is easy to verify: take the first and second derivatives of (10.3.14), substitute them and $x(t)$ into the left-hand side of (10.2.1), and use that λ_1 and λ_2 are solutions of (10.3.1). The result is zero. We conclude that (10.3.14) is the form in which all solutions of (10.2.1) can be expressed when the characteristic roots are real and distinct—for that reason it is known as a ***general solution*** of (10.2.1). Finally, we note that (10.3.14) is a solution of (10.2.1) on the entire interval $(-\infty, \infty)$.

Assigning actual values to the coefficients c_1 and c_2 results in a ***particular solution*** of (10.2.1). For example, setting $c_1 = 2$ and $c_2 = -3/7$ gives the particular solution

$$x(t) = 2e^{\lambda_1 t} - \frac{3}{7}e^{\lambda_2 t}.$$

The particular solution

$$x(t) = e^{\lambda_1 t}$$

is obtained by setting $c_1 = 1$ and $c_2 = 0$. Similarly, setting $c_1 = 0$ and $c_2 = 1$ gives the solution

$$x(t) = e^{\lambda_2 t}.$$

An expression of the form $c_1 e^{\lambda_1 t} + c_2 e^{\lambda_2 t}$, where c_1 and c_2 are constants, is called a ***linear combination*** of the functions $e^{\lambda_1 t}$ and $e^{\lambda_2 t}$. Accordingly, we see from (10.3.14) that every solution of equation (10.2.1) is some linear combination of the solutions $e^{\lambda_1 t}$ and $e^{\lambda_2 t}$ whenever λ_1 and λ_2 are distinct real characteristic roots.

For this reason, the pair of functions $e^{\lambda_1 t}$ and $e^{\lambda_2 t}$ are said to be a ***fundamental set of solutions*** of (10.2.1). Furthermore, they are completely different functions in the sense of not being proportional to each other. The mathematical term for this is *linear independence.* That is to say, two functions are said to be ***linearly independent*** on an interval J if neither is a constant multiple of the other. As this is the case for $e^{\lambda_1 t}$ and $e^{\lambda_2 t}$ for $-\infty < t < \infty$, they are linearly independent solutions of (10.2.1) on $(-\infty, \infty)$.

Theorem 10.3.1 summarizes the case of distinct real roots.

Distinct Real Characteristic Roots and General Solution

Theorem 10.3.1 *If the characteristic equation for*

$$ax'' + bx' + cx = 0 \tag{10.3.15}$$

has the distinct real roots λ_1 and λ_2, then $e^{\lambda_1 t}$ and $e^{\lambda_2 t}$ are linearly independent solutions of (10.3.15) *on the interval* $(-\infty, \infty)$. *Furthermore,*

$$x(t) = c_1 e^{\lambda_1 t} + c_2 e^{\lambda_2 t}, \tag{10.3.16}$$

where c_1 and c_2 are arbitrary constants, is a general solution on $(-\infty, \infty)$.

A mnemonic for coming up with a general solution of (10.3.15) is to associate the functions $e^{\lambda_1 t}$ and $e^{\lambda_2 t}$ with the characteristic roots λ_1 and λ_2 as follows:

$$\lambda_1 \rightarrow e^{\lambda_1 t}, \quad \lambda_2 \rightarrow e^{\lambda_2 t}.$$

Their linear combination is the general solution (10.3.16).

Example 10.3.4 Find a general solution of $3x'' - 13x' - 10x = 0$.

Solution The characteristic equation is $3\lambda^2 - 13\lambda - 10 = 0$. By the quadratic formula, the characteristic roots are

$$\lambda = \frac{-(-13) \pm \sqrt{(-13)^2 - 4(3)(-10)}}{2(3)} = \frac{13 \pm \sqrt{289}}{6}$$

$$= \frac{13 \pm 17}{6} = \begin{cases} -\dfrac{4}{6} = -\dfrac{2}{3} \\ \dfrac{30}{6} = 5. \end{cases}$$

But of course these roots could have been obtained by merely factoring the characteristic equation as follows:

$$(3\lambda + 2)(\lambda - 5) = 0.$$

At any rate, the particular solutions associated with these roots are

$$\lambda_1 = -\frac{2}{3} \;\rightarrow\; e^{-2t/3}, \quad \lambda_2 = 5 \;\rightarrow\; e^{5t}.$$

By Theorem 10.3.1, the linear combination

$$x(t) = c_1 e^{-2t/3} + c_2 e^{5t}.$$

is a general solution of the differential equation on $(-\infty, \infty)$.

10.3.2 Repeated Real Characteristic Roots

Theorem 10.3.1 pertains to real characteristic roots λ_1 and λ_2 that are not equal. However, if it also were true when a characteristic root λ is repeated, then, as $\lambda_1 = \lambda_2 = \lambda$, the theorem would state that a general solution of (10.3.15) is

$$x(t) = c_1 e^{\lambda t} + c_2 e^{\lambda t} = (c_1 + c_2)e^{\lambda t}.$$

This would say that every solution of (10.3.15) is merely a multiple of $e^{\lambda t}$. But this flies in the face of the distinct roots case where two linearly independent solutions are needed to express all solutions. The discrepancy is resolved when we come to realize Theorem 10.3.1 does not apply to repeated roots because the integration yielding (10.3.13) is not valid when $\lambda_2 = \lambda_1$. This means we have to backtrack to (10.3.12) when λ_1 and λ_2 have the same value λ. In that case,

$$\frac{d}{dt}\left[e^{-\lambda t}x(t)\right] = k_1.$$

Changing the name of the constant k_1 to c_2 and integrating, we get

$$e^{-\lambda t}x(t) = c_2 t + c_1.$$

So,

$$x(t) = (c_1 + c_2 t)e^{\lambda t}.$$

In sum, if the characteristic equation for (10.3.15) has the repeated root λ, then any solution of (10.3.15) can be expressed in the form

$$x(t) = c_1 e^{\lambda t} + c_2 t e^{\lambda t} \tag{10.3.17}$$

for some c_1 and c_2. Conversely, it is left as an exercise (see Problem 2) to show that if λ is a repeated root of the characteristic equation for (10.3.15), then any function of the form (10.3.17) is a solution of (10.3.15). Clearly, it is a solution on $(-\infty, \infty)$.

Setting $c_1 = 1$ and $c_2 = 0$, we obtain the particular solution $x(t) = e^{\lambda t}$. Similarly, setting $c_1 = 0$ and $c_2 = 1$ gives the particular solution

$$x(t) = te^{\lambda t}.$$

Observe that $te^{\lambda t}$ is not a constant multiple of $e^{\lambda t}$ since t is a variable, not a constant. Or, in the new terminology, $e^{\lambda t}$ and $te^{\lambda t}$ are linearly independent solutions of (10.3.15). Our findings are summarized in the next theorem.

Repeated Characteristic Roots and General Solution

Theorem 10.3.2 *If the characteristic equation for*

$$ax'' + bx' + cx = 0 \tag{10.3.18}$$

has the real repeated root $\lambda = -b/2a$, *then* $e^{\lambda t}$ *and* $te^{\lambda t}$ *are linearly independent solutions of* (10.3.18) *on the interval* $(-\infty, \infty)$. *Furthermore,*

$$x(t) = c_1 e^{\lambda t} + c_2 t e^{\lambda t}, \tag{10.3.19}$$

where c_1 *and* c_2 *are arbitrary constants, is a general solution on* $(-\infty, \infty)$.

This theorem is easy to apply if we merely remember that two linearly independent solutions associated with the repeated root λ are

$$\lambda \rightarrow \begin{cases} e^{\lambda t} \\ te^{\lambda t} \end{cases}$$

and that their linear combination is a general solution.

Example 10.3.5 Find a general solution of $9x'' + 24x' + 16x = 0$.

Solution The characteristic equation is $9\lambda^2 + 24\lambda + 16 = 0$. Note that its left-hand side is a perfect square trinomial, which can be factored as $(3\lambda + 4)^2$. If we rewrite the equation as $(3\lambda + 4)(3\lambda + 4) = 0$, we see that $\lambda = -4/3$ is a repeated root. From

$$\lambda = -\frac{4}{3} \rightarrow \begin{cases} e^{-4t/3} \\ te^{-4t/3}, \end{cases}$$

we obtain the general solution

$$x(t) = c_1 e^{-4t/3} + c_2 t e^{-4t/3}$$

for $t \in (-\infty, \infty)$. ♦

10.3.3 Differential Operators and the Superposition Principle

It follows from Theorems 10.3.1 and 10.3.2 that each linear combination of the two fundamental solutions of

$$a\frac{d^2x}{dt^2} + b\frac{dx}{dt} + cx = 0 \tag{10.2.1}$$

is also a solution of this equation. This suggests that all linear combinations of solutions, regardless of how they are obtained, are solutions. Let us examine this more closely.

Our investigation will proceed more smoothly if we first introduce some new terms and notation. We will focus on functions $f(t)$ that are defined on an open interval $J = (\alpha, \beta)$ and whose second derivatives $f''(t)$ exist at each $t \in J$. The reason is that these are the kinds of functions that we must deal with when searching for solutions of equation (10.2.1). Such functions are said to be ***twice differentiable*** on J. We will assign to each f the output $af'' + bf' + cf$, which is also a function. A concise way to express this is as follows:

$$f \mapsto af'' + bf' + cf. \tag{10.3.20}$$

Notice this too is a rule—similar to what is typically called a function. However, its domain consists of a set of twice-differentiable functions rather than a set of numbers. So it would be far too confusing to also refer to this as a function since the domain of a function consists of real numbers. Instead this is called a ***differential operator***. We denote it with the symbol $L[\,\cdot\,]$, where the "dot" is a placeholder for a twice-differential function. $L[f]$ denotes the output of applying the differential operator L to f. That is,

$$L[f] = af'' + bf' + cf. \tag{10.3.21}$$

It is common to shorten this to

$$L = aD^2 + bD + c,$$

where D and D^2 denote the operations of taking the first and second derivatives, respectively, of functions placed after them. In other words, D is an alternate symbol for the derivative operator d/dt. Likewise, D^2 is an alternate symbol for d^2/dt^2. So the result of applying them to a twice-differentiable function f is

$$Df(t) = \frac{d}{dt}f(t) = f'(t) \quad \text{and} \quad D^2 f(t) = \frac{d^2}{dt^2}f(t) = f''(t).$$

The way to write that L operates on f is

$$L[f] = \big(aD^2 + bD + c\big)[f] = aD^2 f + bDf + cf = a\frac{d^2 f}{dt^2} + b\frac{df}{dt} + cf.$$

Note that the result is (10.3.21). The point to be made is that we can now represent the homogeneous differential equation (10.2.1) in the abbreviated form

$$L[x] = 0.$$

The operator L has the important property of ***linearity***. What this means is that for any two functions f and g, both of which are twice differentiable on an interval J, and for any two constants γ and λ, the result of L operating on the sum $\gamma f + \lambda g$ is

$$L[\gamma f + \lambda g] = \gamma L[f] + \lambda L[g], \tag{10.3.22}$$

where equality holds on the interval J. The sum $\gamma f + \lambda g$ is known as a ***linear combination*** of the functions f and g. Property (10.3.22) follows directly from the definition of L:

$$\begin{aligned} L[\gamma f + \gamma g] &= a\frac{d^2}{dt^2}(\gamma f + \gamma g) + b\frac{d}{dt}(\gamma f + \gamma g) + c(\gamma f + \gamma g) \\ &= \gamma(af'' + bf' + cf) + \gamma(ag'' + bg' + cg) = \gamma L[f] + \gamma L[g]. \end{aligned}$$

Because L is expressed in terms of the derivative operators D and D^2 and has the linearity property (10.3.22), it is said to be a ***linear differential operator***.

It is common to see (10.3.22) recast as

$$L[\gamma f] = \gamma L[f] \tag{10.3.23}$$

and

$$L[f+g] = L[f] + L[g]. \tag{10.3.24}$$

The reader should verify that (10.3.23) and (10.3.24) implies (10.3.22), and conversely. Note that the linear differential operator L becomes the derivative operator D when $a = c = 0$ and $b = 1$. Then (10.3.23) and (10.3.24) are the familiar statements that "the derivative of a constant times a function is the constant times the derivative of the function" and "the derivative of the sum is the sum of the derivatives," respectively.

Besides the operators D and L, there are other well-known operators that are used extensively in the sciences and engineering, such as the Laplace transform (see Chap. 11). Linear operators by definition have properties (10.3.23) and (10.3.24). If an operator fails to satisfy either one of them, or both, then it is not a linear operator, but rather a ***nonlinear operator***.

Finally we are ready to prove that all linear combinations of solutions of equation (10.2.1) are also solutions.

Superposition Principle for Homogeneous Equations

Theorem 10.3.3 *Let $x_1(t)$ and $x_2(t)$ be solutions of the homogeneous linear differential equation*

$$ax'' + bx' + cx = 0$$

on $(-\infty, \infty)$. If c_1 and c_2 are constants, then the linear combination

$$c_1x_1(t) + c_2x_2(t)$$

is also a solution of the equation on $(-\infty, \infty)$.

Proof Let $x_1(t)$ and $x_2(t)$ be any two solutions of $L[x] = 0$ on $(-\infty, \infty)$, where $L = aD^2 + bD + c$. This means that

$$L[x_1(t)] = 0 \quad \text{and} \quad L[x_2(t)] = 0$$

for all $-\infty < t < \infty$. Let c_1 and c_2 be constants. Then it follows from the linearity property (10.3.22) that

$$L[c_1x_1(t) + c_2x_2(t)] = c_1L[x_1(t)] + c_2L[x_2(t)] = c_1 \cdot 0 + c_2 \cdot 0 = 0$$

for all $-\infty < t < \infty$. ■

10.3.4 Complex Characteristic Roots

10.3.4.1 Introduction to Complex Numbers

Now let us determine the solutions of second-order linear equations when their characteristic equations have complex roots. This occurs when $b^2 - 4ac < 0$. To see this, let us rewrite the characteristic roots given by (10.3.3) as

$$\lambda = -\frac{b}{2a} \pm \frac{\sqrt{b^2 - 4ac}}{2a} = -\frac{b}{2a} \pm \frac{\sqrt{-1} \cdot \sqrt{4ac - b^2}}{2a}$$

and then introduce some new notation. Since $4ac - b^2 > 0$, $\sqrt{4ac - b^2}$ is a positive real number. But $\sqrt{-1}$ is different: it is not even a real number. Why? The answer has to do with the meaning of $\sqrt{x}$. It is defined as the nonnegative real number whose square is x. According to this then, $\sqrt{-1}$ should represent a nonnegative real number whose square is -1. But this cannot be: the square of any real number is nonnegative. Nonetheless we can still say it is a number, just not a real number; instead it is known as a *pure imaginary number*[4] and is usually denoted by the letter i.[5] Thus, $i = \sqrt{-1}$ and $i^2 = -1$. By defining the real numbers α and ω by

$$\alpha := -\frac{b}{2a} \quad \text{and} \quad \omega := \frac{\sqrt{4ac - b^2}}{2a}, \tag{10.3.25}$$

the expressions for the two characteristic roots simplify to $\alpha + i\omega$ and $\alpha - i\omega$, thereby making them less tedious to deal with in calculations. Since a denotes any nonzero real number and $\sqrt{4ac - b^2} > 0$, it follows that ω is a nonzero real number.

Numbers that are not real can always be expressed in the form $a + ib$, where a and $b \neq 0$ are real numbers. These numbers are known as ***complex numbers***.[6] It is easy to show that complex numbers can be handled algebraically much in the same way as real numbers. It follows from (10.3.3) that one characteristic root cannot be a complex number and the other one a real number. Either both roots are complex or both are real. Since characteristic roots always occur as an $\alpha \pm i\omega$ pair, they are said to be ***complex conjugates*** of each other. The word *conjugate* used in this sense means *joined together in a pair*. Ordinarily, a bar is placed over a complex number

---

[4] ***Pure imaginary numbers*** are numbers of the form bi, where $b \neq 0$ is a real number. The derivation of the term "imaginary" goes back to the 17th century when such numbers were first being discovered but were not fully understood. The term is a misnomer because there is nothing unreal about imaginary numbers if they are interpreted correctly.

[5] In some disciplines, such as electrical engineering, j is used instead of i because i is reserved for designating electrical current.

[6] Every pure imaginary number bi is a complex number since it can be written as $0 + bi$. Every real number a can also be regarded as a complex number since it can be written as $a + 0i$. Of course, not every complex number is a real number. It comes down to this: The set of all complex numbers is larger than the set of real numbers in the sense that it includes all of the real numbers.

to denote its conjugate. With this notation, the two characteristic roots are

$$\lambda = \alpha + i\omega \quad \text{and} \quad \overline{\lambda} = \alpha - i\omega. \tag{10.3.26}$$

The real number α is called the ***real part*** of the complex number λ. It is convenient to write the real part of λ as follows:

$$\text{Re}\,\lambda = \alpha.$$

The real number ω after i is called the ***imaginary part*** of λ and we write

$$\text{Im}\,\lambda = \omega.$$

Note that the imaginary part does not include i. Thus, $\text{Re}\,\overline{\lambda} = \alpha$ and $\text{Im}\,\overline{\lambda} = -\omega$.

Remark Observe that we could replace ω with $-\omega$ in (10.3.26) and nothing really changes: we still have the same pair of complex numbers.

10.3.4.2 Simple Harmonic Oscillator and Solutions

When the characteristic roots are complex numbers, we expect the mathematics to be more involved than when they are real numbers, because somehow we will have to figure out how to use the complex roots to obtain real-valued solutions of equation (10.2.1). At this point, we have no idea where to even start. So instead of tackling the general version of Eq. (10.2.1) head on, let us consider a specific example: let $a = 1$, $b = 0$, $c = w^2$, where ω denotes any nonzero real number. With these values, (10.2.1) simplifies to

$$x'' + \omega^2 x = 0. \tag{10.3.27}$$

The roots of (10.3.27) are complex numbers since $b^2 - 4ac = -4\omega^2 < 0$. Recall that (10.3.27) is the equation of motion of a simple harmonic oscillator if the independent variable t denotes time and $\omega > 0$ (cf. Sect. 10.1).

We will prove a number of results for Eq. (10.3.27) because, as it turns out, these results will eventually allow us to find a general solution of any second-order homogeneous linear equation (10.2.1) that has complex characteristic roots.

Lemma 10.3.1 *Let t_0 and ω be given real numbers where $\omega \neq 0$. The initial value problem*

$$x'' + \omega^2 x = 0 \tag{10.3.28a}$$

(continued)

Lemma 10.3.1 (continued)

$$x(t_0) = 0, \; x'(t_0) = 0 \tag{10.3.28b}$$

has the unique solution $x(t) \equiv 0$ *on* $(-\infty, \infty)$.

Proof Obviously the function $x(t) = 0$ solves the differential equation (10.3.28a) for $-\infty < t < \infty$. It also satisfies the initial conditions (10.3.28b). So we only have to establish that there is no other such function. To that end, suppose a function $u(t)$ solves (10.3.28a) for $-\infty < t < \infty$ and also satisfies (10.3.28b). In other words, let us suppose that

$$u''(t) + \omega^2 u(t) = 0 \tag{10.3.29}$$

and

$$u(t_0) = 0, \; u'(t_0) = 0. \tag{10.3.30}$$

Now multiply (10.3.29) by $2u'(t)$ and then integrate the result from t_0 to t:

$$\int_{t_0}^{t} 2u'(s)\big[u''(s) + \omega^2 u(s)\big]\, ds = \int_{t_0}^{t} 0\, ds. \tag{10.3.31}$$

By the Fundamental Theorem of Calculus and (10.3.30),

$$\begin{aligned}
\int_{t_0}^{t} \big[2u'(s)u''(s) + 2\omega^2 u(s)u'(s)\big]\, ds &= \Big\{ [u'(s)]^2 + \omega^2 [u(s)]^2 \Big\}\Big|_{t_0}^{t} \\
&= [u'(t)]^2 + \omega^2 [u(t)]^2 - [u'(t_0)]^2 \\
&\quad - \omega^2 [u(t_0]^2 \\
&= [u'(t)]^2 + \omega^2 [u(t)]^2.
\end{aligned}$$

Since the right-hand side of (10.3.31) is equal to 0, we have

$$[u'(t)]^2 + \omega^2 [u(t)]^2 = 0. \tag{10.3.32}$$

Since both terms on the left-hand side of (10.3.32) are nonnegative, each must in fact be equal to 0. Finally, as $\omega^2 > 0$, we are forced to conclude that $u(t) \equiv 0$. ∎

Remark Lemma 10.3.1 has an obvious physical interpretation. If at time $t = t_0$, the mass of a horizontal mass-spring system is located at the origin (the position where the spring is neither stretched nor compressed) and it has no velocity, then it will continue to remain there unless a force acts on it at some point in the future.

Lemma 10.3.2 *Let t_0, ζ, η, ω be given real numbers where $\omega \neq 0$. If*

$$x'' + \omega^2 x = 0$$

$$x(t_0) = \zeta, \; x'(t_0) = \eta$$

has a solution on $(-\infty, \infty)$, then it is unique.

Proof Suppose $x(t)$ and $\tilde{x}(t)$ are solutions of this initial value problem on the interval $(-\infty, \infty)$. Consider the difference $u(t) := x(t) - \tilde{x}(t)$. By the Superposition Principle, the function $u(t)$ is also a solution of $x'' + \omega^2 x = 0$. Moreover,

$$u(t_0) = x(t_0) - \tilde{x}(t_0) = \zeta - \zeta = 0$$

and

$$u'(t_0) = x'(t_0) - \tilde{x}'(t_0) = \eta - \eta = 0.$$

So $u(t)$ is also a solution of the initial value problem in Lemma 10.3.1. Consequently, $u(t) \equiv 0$. Therefore, $x(t) \equiv \tilde{x}(t)$ on $(-\infty, \infty)$. ■

We have just established that there can be at most one solution of the initial value problem in Lemma 10.3.2. However, aside from the zero solution in Lemma 10.3.1, where $\zeta = 0$ and $\eta = 0$, we have not proved that it actually has a solution. Does it? It does in fact, regardless of the values of ζ and η, as we will now prove. But first let us point out that in both of the lemmas, ω represents any nonzero real number. However, in the next theorem, we state that ω is positive. As it turns out, there is no loss of generality in doing so as we will explain in the remark following the proof of the theorem.

Simple Harmonic Oscillator and Solutions I

Theorem 10.3.4 *Let t_0, ζ, η, ω be real numbers where $\omega > 0$. There exists a unique solution of*

$$x'' + \omega^2 x = 0 \tag{10.3.33}$$

on $(-\infty, \infty)$ satisfying the initial conditions

$$x(t_0) = \zeta, \; x'(t_0) = \eta. \tag{10.3.34}$$

(continued)

This solution can be expressed in the form

$$x(t) = c_1 \cos \omega t + c_2 \sin \omega t, \tag{10.3.35}$$

where the constants c_1 and c_2 are uniquely determined by (10.3.34).

Proof Recall that Eq. (10.3.33) models the motion of the mass of an undamped mass-spring system when $\omega > 0$. Because of the total absence of any damping, our intuition suggests that the mass will move back and forth in the same repetitious manner forever. This suggests that sine and cosine functions may be solutions. Let us confirm this by substituting $\cos \omega t$ for x:

$$x'' + \omega^2 x = \frac{d^2}{dt^2} \cos \omega t + \omega^2 \cos \omega t = -\omega^2 \cos \omega t + \omega^2 \cos \omega t = 0.$$

Likewise, $x = \sin \omega t$ is also a solution of (10.3.33). It follows from the Superposition Principle that all linear combinations of $\cos \omega t$ and $\sin \omega t$ are solutions.

So far we have established that all functions of the form (10.3.35) are solutions of equation (10.3.33). However, this does not rule out the possibility that the equation may have other kinds of solutions that we have not thought of. So let us find out if there is always a solution of (10.3.33) that satisfies the initial conditions (10.3.34) irrespective of the values of ζ and η. And if it does, can it always be expressed in the form (10.3.35)? If so, then this would rule out the existence of other solutions.

Since we have already determined that (10.3.35) is a solution of (10.3.33) for all values of c_1 and c_2, let us see if a pair of values for c_1 and c_2 exist that will also satisfy the initial conditions (10.3.34) for any given pair ζ and η. It boils down to this: Does the system of equations

$$c_1 \cos \omega t_0 + c_2 \sin \omega t_0 = \zeta$$

$$-c_1 \omega \sin \omega t_0 + c_2 \omega \cos \omega t_0 = \eta$$

have a solution? The answer is yes. This can be seen by multiplying the first equation by $\omega \cos \omega t_0$, the second one by $-\sin \omega t_0$, and then adding the results. For c_1, we obtain the value

$$c_1 = \zeta \cos \omega t_0 - \frac{\eta}{\omega} \sin \omega t_0. \tag{10.3.36}$$

Similarly,

$$c_2 = \zeta \sin \omega t_0 + \frac{\eta}{\omega} \cos \omega t_0. \tag{10.3.37}$$

Therefore,

$$x(t) = c_1 \cos \omega t + c_2 \sin \omega t,$$

where c_1 and c_2 are given by (10.3.36) and (10.3.37), solves the differential equation (10.3.33). Moreover, it also satisfies the initial conditions (10.3.34). Finally, Lemma 10.3.2 asserts this is the only solution! ■

Remark In Theorem 10.3.4, we state that $\omega > 0$. But what if $\omega < 0$? Then $-\omega > 0$. In that case, by (10.3.35), a general solution of (10.3.33) is

$$x(t) = c_1 \cos(-\omega t) + c_2 \sin(-\omega t).$$

Recall that the cosine and sine functions are even and odd functions, respectively. Thus, $\cos(-\omega t) = \cos \omega t$ and $\sin(-\omega t) = -\sin \omega t$. Letting $k_1 := c_1$ and $k_2 := -c_1$, we have

$$x(t) = c_1 \cos \omega t - c_2 \sin \omega t = k_1 \cos \omega t + k_2 \sin \omega t.$$

The point to be made is that we still obtain the same general solution. So from now on we will always choose $\omega > 0$ to avoid the hassle of dealing with a negative sign.

Example 10.3.6 Find the solution of the initial value problem

$$\frac{d^2x}{dt^2} + 4x = 0; \quad x(0) = 2, \; x'(0) = 4\sqrt{3}.$$

Solution Since $\omega^2 = 4$, let $\omega = 2$ (see the preceding remark). It follows from (10.3.35) that the solution is of the form

$$x(t) = c_1 \cos 2t + c_2 \sin 2t.$$

Now we must find values of c_1 and c_2 so as to satisfy both initial conditions. To that end, with $t = 0$, set $x(t)$ and its derivative

$$x'(t) = -2c_1 \sin 2t + 2c_2 \cos 2t$$

to 2 and $4\sqrt{3}$, respectively. The result is the system

$$c_1 \cos 0 + c_2 \sin 0 = 2$$

$$-2c_1 \sin 0 + 2c_2 \cos 0 = 4\sqrt{3},$$

which has the solution $c_1 = 2$ and $c_2 = 2\sqrt{3}$. Therefore, the solution of the initial value problem is

$$x(t) = 2 \cos 2t + 2\sqrt{3} \sin 2t.$$

♦

There are other ways of expressing solutions of the undamped mass-spring system that are more suited for graphing and determining the frequency of the oscillation of the mass m. Referring back to (10.3.35), define the constant A by

$$A := \sqrt{c_1^2 + c_2^2} \qquad (10.3.38)$$

and the angle φ by

$$A\cos\varphi = c_1 \quad \text{and} \quad A\sin\varphi = c_2. \qquad (10.3.39)$$

Then, using one of the addition formulas of trigonometry, (10.3.35) can be written as

$$x(t) = A\cos\varphi\cos\omega t + A\sin\varphi\sin\omega t = A\cos(\omega t - \varphi).$$

By (10.3.38), $A \geq 0$. As a result, we have the following alternative form of a general solution of a simple harmonic oscillator.

Simple Harmonic Oscillator and Solutions II

Theorem 10.3.5 *Let $\omega > 0$. A general solution of*

$$x'' + \omega^2 x = 0$$

on $(-\infty, \infty)$ is

$$x(t) = A\cos(\omega t - \varphi), \qquad (10.3.40)$$

where $A \geq 0$ and φ are arbitrary constants.

Remark If we define the angle φ by

$$A\sin\varphi = c_1 \quad \text{and} \quad A\cos\varphi = c_2$$

rather than by (10.3.39), then we obtain

$$x(t) = A\sin(\omega t + \varphi),$$

which is another alternative form that can be used. Moreover,

$$A\sin(\omega t - \varphi) \quad \text{and} \quad A\cos(\omega t + \varphi)$$

are also general solutions. The one that is used is a matter of personal preference.

Example 10.3.7 Find the solution of the initial value problem

$$x'' + 4x = 0; \quad x(0) = 2,\ x'(0) = 4\sqrt{3}.$$

Solution According to (10.3.40), a general solution of the differential equation is

$$x(t) = A\cos(2t - \varphi).$$

Its derivative is

$$x'(t) = -2A\sin(2t - \varphi).$$

Due to the initial conditions,

$$A\cos(-\varphi) = A\cos\varphi = 2 \quad \text{and} \quad -2A\sin(-\varphi) = 2A\sin\varphi = 4\sqrt{3}.$$

Consequently,

$$\cos\varphi = \frac{2}{A} \quad \text{and} \quad \sin\varphi = \frac{2\sqrt{3}}{A}.$$

Since $\cos^2\varphi + \sin^2\varphi = 1$,

$$\left(\frac{2}{A}\right)^2 + \left(\frac{2\sqrt{3}}{A}\right)^2 = 1.$$

Thus $A = 4$. As a result,

$$\cos\varphi = \frac{1}{2} \quad \text{and} \quad \sin\varphi = \frac{\sqrt{3}}{2}.$$

So $\varphi = \pi/3$. We conclude that the solution of the initial value problem is

$$x(t) = 4\cos\left(2t - \frac{\pi}{3}\right). \qquad \blacklozenge$$

Before looking at the next example, let us introduce some of the terminology associated with the simple harmonic oscillator and the significance of the constants A, φ, and ω in (10.3.40).

Since the cosine function takes on values only from -1 to 1, the minimum and maximum values of (10.3.40) are $-A$ and A, respectively. So $|A|$ is the maximum displacement of m from its equilibrium position and is called the ***amplitude*** of the motion. As for φ, let us compare the graphs of $x(t) = A\cos(\omega t - \varphi)$ and $A\cos\omega t$. Since

$$A\cos(\omega t - \varphi) = A\cos(\omega(t - \varphi/\omega)),$$

we see that the graph of $x(t)$ can be obtained from the graph of $A \cos \omega t$ by shifting the latter along the t-axis $|\varphi|/\omega$ units (to the right if $\varphi > 0$ and to the left if $\varphi < 0$). The constant φ is called the ***phase angle*** and φ/ω is the ***phase shift***.

Since the period of the cosine function is 2π, we see from

$$\begin{aligned} x(t) &= A \cos(\omega t - \varphi) = A \cos(\omega t - \varphi + 2\pi) \\ &= A \cos\left[\omega\left(t + \frac{2\pi}{\omega}\right) - \varphi\right] = x\left(t + \frac{2\pi}{\omega}\right). \end{aligned}$$

that

$$T := \frac{2\pi}{\omega}$$

is the ***period*** of the motion, the time it takes for m to make one complete oscillation. So, the ***frequency*** f of motion, the number of oscillations per unit of time, is

$$f = \frac{1}{T} = \frac{\omega}{2\pi} = \frac{1}{2\pi}\sqrt{\frac{k}{m}}. \tag{10.3.41}$$

The constant ω is called the ***angular frequency*** of the motion since it is related to the frequency f by $\omega = 2\pi f$. The special type of periodic motion that is given by (10.3.40) is known as ***simple*** (or ***linear***) ***harmonic motion***. ♦

Example 10.3.8 Suppose the mass-spring system in Fig. 10.1 is modified by attaching a second massless spring as depicted in Fig. 10.3. If the spring constants of the left and right springs are k_1 and k_2, respectively, determine the frequency of oscillation of m when it is in motion.

Solution Suppose at some point in time m is located x units to the right of its equilibrium position (the position where both springs are neither stretched nor compressed). Then the left spring is stretched beyond its natural length, say by x_1 units. As a result, the right spring is stretched by x_2 units, where $x_2 = x - x_1$. By Hooke's law, the right spring exerts a force

$$F = -k_2 x_2$$

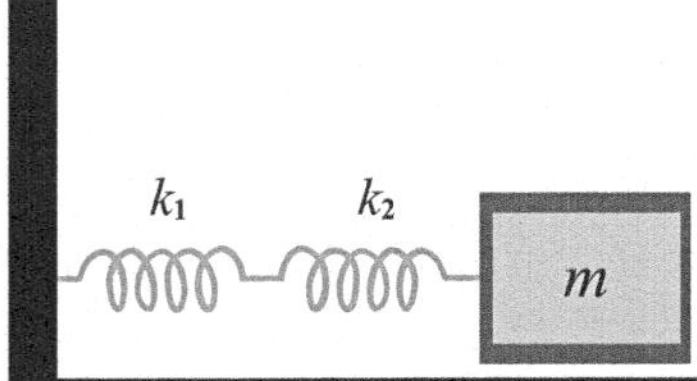

Fig. 10.3 Mass and two springs in series

on m. It also exerts a force $F_2 = k_2x_2$ on the left spring. Since the left spring is stretched by x_1 units, it exerts a force $F_1 = -k_1x_1$ on the right spring. It follows from Newton's third law of motion that $F_2 = -F_1$.[7] Thus, $k_2x_2 = k_1x_1$. And so

$$F = -k_2x_2 = -k_1x_1.$$

Consequently,

$$x = x_1 + x_2 = -\frac{F}{k_1} - \frac{F}{k_2} = -F\left(\frac{k_2 + k_1}{k_1k_2}\right).$$

Solving for F, we get

$$F = -\left(\frac{k_1k_2}{k_1 + k_2}\right)x. \tag{10.3.42}$$

It is left to the reader to show that this equation is also valid when m is located x units to the left of its equilibrium position.

For a given set of initial conditions, Eq. (10.3.42) implies that the motion of m attached to two springs connected in series (that is, end-to-end) with spring constants k_1 and k_2 is identical to the motion of m attached to a single spring with spring constant

$$k := \frac{k_1k_2}{k_1 + k_2}.$$

Consequently, the frequency of motion (see (10.3.41)) is

$$f = \frac{1}{2\pi}\sqrt{\frac{k}{m}} = \frac{1}{2\pi}\sqrt{\frac{k_1k_2}{m(k_1 + k_2)}}.$$

Furthermore, the position x of m at time t is given by (10.3.40), where the angular frequency is

$$\omega = \frac{2\pi}{T} = 2\pi f$$

and the amplitude A and phase angle φ are determined by the initial position and velocity of m. ♦

---

[7] Newton's three laws of motion appeared in 1687 in the three-volume work titled *Philosophiæ Naturalis Principia Mathematica* (Latin for "mathematical principles of natural philosophy"), which was written by the great English physicist Isaac Newton (1643–1727). A typical translation of the third law is: "To every action there is always opposed an equal reaction." Action and reaction refer to the pair of forces that two interacting bodies exert on each other. In today's vernacular, Newton's third law of motion states that if a body A exerts a force on a body B, then body B exerts an equal but oppositely directed force on body A.

10.3.4.3 Complex Roots: General Case

Besides telling us how to find the solutions of a simple harmonic oscillator, Theorem 10.3.4 will enable us to find a general solution of

$$ax'' + bx' + cx = 0 \tag{10.3.43}$$

when the characteristic roots are complex numbers. To this end, let $\lambda = \alpha + i\omega$ and $\overline{\lambda} = \alpha - i\omega$ denote these roots, where α and ω are defined by (10.3.25). We may suppose that $\omega > 0$ in view of the remark preceding Example 10.3.6. Let us replace the coefficients a, b, c in (10.3.43) with α and ω. To do this, first divide (10.3.43) by a:

$$x'' + \frac{b}{a}x' + \frac{c}{a}x = 0. \tag{10.3.44}$$

Then, due to (10.3.7), the coefficients of (10.3.44) can be written in terms of the constants α and ω as follows:

$$\frac{b}{a} = -(\lambda + \overline{\lambda}) = -2\alpha$$

and

$$\frac{c}{a} = \lambda \cdot \overline{\lambda} = (\alpha + i\omega)(\alpha - i\omega) = \alpha^2 + \omega^2.$$

Consequently, Eq. (10.3.44) becomes

$$x'' - 2\alpha x' + (\alpha^2 + \omega^2)x = 0. \tag{10.3.45}$$

In summary, if the characteristic equation for (10.3.43) has the complex roots $\alpha \pm i\omega$, then equations (10.3.43) and (10.3.45) are equivalent in that they have the same solutions. It follows that a general solution of (10.3.43) is a general solution of (10.3.45), and conversely.

Now let us find a general solution of (10.3.45). First, notice that the characteristic equation for

$$x'' - 2\alpha x' + \alpha^2 x = 0 \tag{10.3.46}$$

has the repeated real root α. Then, it follows from Theorem 10.3.2 that a general solution of (10.3.46) is

$$x(t) = c_1 e^{\alpha t} + c_2 t e^{\alpha t}.$$

Thus, when $\alpha \neq 0$, $e^{\alpha t}$ is a solution of (10.3.46) but not of (10.3.45). This, however, suggests that something may come of trying to express a solution $x(t)$ of (10.3.45) in the form

$$x(t) = e^{\alpha t} y(t)$$

for some twice-differentiable function $y(t)$. Replacing x on the left-hand side of (10.3.45) with $e^{\alpha t} y$, we obtain

$$\begin{aligned} x'' - 2\alpha x' + (\alpha^2 + \omega^2)x &= (\alpha^2 + \omega^2)(e^{\alpha t} y) - 2\alpha \frac{d}{dt}(e^{\alpha t} y) + \frac{d^2}{dt^2}(e^{\alpha t} y) \\ &= \alpha^2 e^{\alpha t} y + \omega^2 e^{\alpha t} y - 2\alpha(e^{\alpha t} y' + \alpha e^{\alpha t} y) + e^{\alpha t} y'' + 2\alpha e^{\alpha t} y' + \alpha^2 e^{\alpha t} y \\ &= e^{\alpha t}(y'' + \omega^2 y). \end{aligned}$$

As a result, we have shown the following:

If a function $y(t)$ is a solution of

$$y'' + \omega^2 y = 0, \tag{10.3.47}$$

then $x(t) = e^{\alpha t} y(t)$ is a solution of (10.3.45). And conversely, if $x(t)$ is a solution of (10.3.45), then $y(t) = e^{-\alpha t} x(t)$ is a solution of (10.3.47).

It then follows from Theorem 10.3.4 that a general solution of (10.3.47) is

$$y(t) = c_1 \cos \omega t + c_2 \sin \omega t.$$

Consequently, as (10.3.45) is equivalent to (10.3.43), we conclude:

If a function $x(t)$ is a solution of (10.3.43), then

$$x(t) = e^{\alpha t}(c_1 \cos \omega t + c_2 \sin \omega t)$$

for some constants c_1 and c_2.

By setting $c_1 = 1$ and $c_2 = 0$, we obtain the particular solution $e^{\alpha t} \cos \omega t$. Likewise, $c_1 = 0$ and $c_2 = 1$ yields the particular solution $e^{\alpha t} \sin \omega t$. They are clearly linearly independent solutions, that is, not constant multiples of each other.

Finally, we see from Theorem 10.3.5 and the previous discussion that a general solution of (10.3.43) can also be written in the form

$$x(t) = Ae^{\alpha t} \cos(\omega t - \varphi),$$

where $A \geq 0$ and φ are arbitrary constants.

In conclusion, we have found two different ways of expressing a general solution of (10.3.45), and thereby of (10.3.43). Theorem 10.3.6 is a summary of the foregoing results.

Complex Characteristic Roots and General Solution

Theorem 10.3.6 *If the characteristic equation for*

$$ax'' + bx' + cx = 0 \tag{10.3.48}$$

has the complex roots $\alpha \pm i\omega$, *where* $\omega > 0$, *then* $e^{\alpha t}\cos\omega t$ *and* $e^{\alpha t}\sin\omega t$ *are linearly independent solutions of* (10.3.48) *on* $(-\infty, \infty)$. *A general solution on* $(-\infty, \infty)$ *is*

$$x(t) = c_1 e^{\alpha t}\cos\omega t + c_2 e^{\alpha t}\sin\omega t, \tag{10.3.49}$$

where c_1 *and* c_2 *are arbitrary constants. An alternative form is*

$$x(t) = Ae^{\alpha t}\cos(\omega t - \varphi), \tag{10.3.50}$$

where $A \geq 0$ *and* φ *are arbitrary constants.*

A mnemonic for remembering this is to associate the particular solutions $e^{\alpha t}\cos\omega t$ and $e^{\alpha t}\sin\omega t$ with the characteristic roots $\lambda = \alpha + i\omega$ and $\overline{\lambda} = \alpha - i\omega$:

$$\alpha \pm i\omega \;\to\; \begin{cases} e^{\alpha t}\cos\omega t \\ e^{\alpha t}\sin\omega t. \end{cases}$$

A general solution of (10.3.48) is a linear combination of these two solutions.

Example 10.3.9 Find a general solution of $2x'' - 2x' + 5x = 0$.

Solution The characteristic equation is $2\lambda^2 - 2\lambda + 5 = 0$. By the quadratic formula,

$$\lambda = \frac{2 \pm \sqrt{(-2)^2 - 4(2)(5)}}{2(2)} = \frac{2 \pm \sqrt{-36}}{4} = \frac{2 \pm 6i}{4} = \frac{1}{2} \pm \frac{3}{2}i.$$

An alternative to the quadratic formula is to divide the characteristic equation by the coefficient of λ^2 and then to complete the square as follows:

$$\lambda^2 - \lambda + \frac{1}{4} = -\frac{5}{2} + \frac{1}{4} \quad\Rightarrow\quad \left(\lambda - \frac{1}{2}\right)^2 = -\frac{9}{4}.$$

Since $\lambda - \frac{1}{2} = \pm\frac{3}{2}i$, the characteristic roots are $\lambda = \frac{1}{2} \pm \frac{3}{2}i$. Then it follows from

$$\frac{1}{2} \pm \frac{3}{2}i \;\to\; \begin{cases} e^{t/2}\cos\dfrac{3t}{2} \\ e^{t/2}\sin\dfrac{3t}{2} \end{cases}$$

that a general solution is

$$x(t) = c_1 e^{t/2} \cos\frac{3t}{2} + c_2 e^{t/2} \sin\frac{3t}{2}.$$

Alternatively, we have

$$x(t) = Ae^{t/2} \cos\left(\frac{3t}{2} - \varphi\right)$$

from (10.3.50). ♦

Example 10.3.10 Find the solution of the initial value problem

$$x'' - 2x' + 10x = 0; \quad x(0) = 0,\ x'(0) = 6.$$

Solution The characteristic equation is $\lambda^2 - 2\lambda + 10 = 0$, which can be written as $(\lambda - 1)^2 = -9$ by completing the square. Consequently, the characteristic roots are

$$\lambda = 1 \pm 3i.$$

By (10.3.49), a general solution is

$$x(t) = c_1 e^t \cos 3t + c_2 e^t \sin 3t.$$

As $x(0) = c_1$, it follows that $c_1 = 0$ from the initial condition $x(0) = 0$. Thus,

$$x(t) = c_2 e^t \sin 3t.$$

Differentiation of $x(t)$ with respect to t yields

$$x'(t) = 3c_2 e^t \cos 3t + c_2 e^t \sin 3t.$$

As $x'(0) = 3c_2$ and $x'(0) = 6$, $c_2 = 2$. Therefore,

$$x(t) = 2e^t \sin 3t. \tag{10.3.51}$$

It is left as an exercise to show that (10.3.50) gives the alternative form

$$x(t) = 2e^t \cos\left(3t - \frac{\pi}{2}\right),$$

which by one of the trigonometric cofunction identities is equal to (10.3.51). ♦

10.4 Complex Exponential Functions

If the characteristic equation for a second-order homogeneous linear equation has a real root λ, one of the solutions of the equation is the exponential function $e^{\lambda t}$. What if λ is a complex number? Is it still correct to say $e^{\lambda t}$ is a solution? But there is a more basic question: Does it make sense to raise Euler's number to a complex power? To answer this, we consider the following two examples.

Example 10.4.1 Find the solution of the initial value problem

$$x'' + x = 0; \quad x(0) = 1, \ x'(0) = 0. \tag{10.4.1}$$

Solution The characteristic equation is

$$\lambda^2 + 1 = 0.$$

Thus, $\lambda = \pm i$ are the characteristic roots. Now for a new idea—well new to us but really an old one. Suppose we try to express solutions in terms of these complex characteristic roots just as we did with real roots. If this could be justified, then it seems plausible that a general solution of the differential equation in (10.4.1) could be written as

$$x(t) = c_1 e^{it} + c_2 e^{-it}. \tag{10.4.2}$$

But how do we define e^{it} and e^{-it} so that (10.4.2) makes sense and is really a solution? To answer this, let us begin by defining the derivatives of e^{it} and e^{-it} by the formulas

$$\frac{d}{dt}e^{it} = ie^{it} \quad \text{and} \quad \frac{d}{dt}e^{-it} = -ie^{-it}$$

in order to be consonant with the differentiation formula

$$\frac{d}{dt}e^{\lambda t} = \lambda e^{\lambda t}$$

when λ is a real number. Then the derivative of (10.4.2) is

$$x'(t) = ic_1 e^{it} - ic_2 e^{-it}. \tag{10.4.3}$$

From (10.4.2), (10.4.3), and the initial conditions, we obtain the system

$$\begin{aligned} c_1 + c_2 &= 1 \\ ic_1 - ic_2 &= 0. \end{aligned} \tag{10.4.4}$$

Its solution is $c_1 = c_2 = 1/2$. With these values (10.4.2) becomes

$$x(t) = \frac{1}{2}e^{it} + \frac{1}{2}e^{-it}. \tag{10.4.5}$$

But is (10.4.5) a legitimate way of expressing the solution? And if it is, how do we interpret it? The answer is found by comparing (10.4.5) to what we truly know is the solution. By inspection, $x(t) = \cos t$ is a solution of (10.4.1). Moreover, by Theorem 10.3.4, it is the only solution of (10.4.1)! It follows that if we want to view (10.4.5) as another way of expressing this solution, we have to say that it is equal to $\cos t$. In short, any definition of e^{it} would have to be compatible with the formula

$$\cos t = \frac{e^{it} + e^{-it}}{2}. \tag{10.4.6}$$

♦

Example 10.4.2 Find the solution of the initial value problem

$$x'' + x = 0; \quad x(0) = 0, \ x'(0) = 1. \tag{10.4.7}$$

Solution The only differences between (10.4.1) and (10.4.7) are the initial values. Changing (10.4.4) to reflect these new values, we have

$$\begin{aligned} c_1 + c_2 &= 0 \\ ic_1 - ic_2 &= 1, \end{aligned} \tag{10.4.8}$$

which has the solution $c_1 = 1/2i$, $c_2 = -1/2i$. Thus,

$$x(t) = \frac{1}{2i}e^{it} - \frac{1}{2i}e^{-it}. \tag{10.4.9}$$

Again, as was the case with (10.4.5), how is this to be interpreted? By inspection and from Theorem 10.3.4, we see that $x(t) = \sin t$ is the unique solution of (10.4.7). So if we want to be able to say that (10.4.9) is another way of expressing this solution, we have to interpret it as being equal to $\sin t$, to wit:

$$\sin t = \frac{e^{it} - e^{-it}}{2i}. \tag{10.4.10}$$

♦

Now consider the linear combination $\cos t + i \sin t$. If we replace $\cos t$ and $\sin t$ with the right-hand sides of (10.4.6) and (10.4.10), respectively, we get

$$\cos t + i \sin t = \frac{e^{it} + e^{-it}}{2} + i \frac{e^{it} - e^{-it}}{2i} = e^{it}.$$

This is how we define e^{it} since it is compatible with both (10.4.6) and (10.4.10). It is a well-known and very useful formula named after Leonhard Euler.

Euler's Formula

Definition 10.4.1 For any real number t, the complex exponential function e^{it} is defined by

$$e^{it} := \cos t + i \sin t. \tag{10.4.11}$$

Several noteworthy observations are:

1. Replacing t with $-t$, we get $e^{i(-t)} = \cos(-t) + i \sin(-t)$. Since $\cos t$ is an even function and $\sin t$ is an odd function, this simplifies to

$$e^{-it} = \cos t - i \sin t. \tag{10.4.12}$$

2. Differentiating (10.4.11) and (10.4.12) with respect to t, we obtain the formulas (see Problem 34)

$$\frac{d}{dt} e^{it} = i e^{it} \quad \text{and} \quad \frac{d}{dt} e^{-it} = -i e^{-it}. \tag{10.4.13}$$

3. Setting $t = \pi$, Euler's formula yields $e^{i\pi} = \cos \pi + i \sin \pi = -1 + i0$, which simplifies to $e^{i\pi} = -1$ or

$$e^{i\pi} + 1 = 0. \tag{10.4.14}$$

This equation, known as ***Euler's identity***, is remarkable in that it relates five basic mathematical constants inherent to several branches of mathematics: the numbers 0 and 1 from arithmetic, π from geometry, Euler's number e from calculus and real analysis, and i from complex analysis and abstract algebra.

4. Let us extend the laws of exponents for real numbers to complex numbers so that

$$e^{a \pm ib} = e^{a} \cdot e^{\pm ib}.$$

As a result, we obtain the following generalizations of (10.4.11) and (10.4.12):

$$e^{a \pm ib} := e^{a}(\cos b \pm i \sin b). \tag{10.4.15}$$

Now that we have managed to give meaning to complex exponential functions, a review of the work in Sect. 10.3 that led to (10.3.13) and (10.3.14) and which appeared at the time to be valid only for real characteristic roots is in fact valid for complex characteristic roots as well. So allowing for complex-valued solutions, we have the following extension of Theorem 10.3.1.

Distinct Characteristic Roots & General Solution

Theorem 10.4.1 *If the characteristic equation for*

$$ax'' + bx' + cx = 0 \tag{10.4.16}$$

has the distinct (real or complex) roots λ_1 and λ_2, then

$$x(t) = c_1 e^{\lambda_1 t} + c_2 e^{\lambda_2 t}, \tag{10.4.17}$$

where c_1 and c_2 are arbitrary constants, is a general (complex-valued) solution on $(-\infty, \infty)$.

Example 10.4.3 Use Theorem 10.4.1 to find the solution of the initial value problem

$$x'' + 4x = 0; \quad x(0) = 1, \; x'(0) = 2i.$$

Solution The characteristic roots are $\lambda = \pm 2i$. It follows from (10.4.17) that a general solution of the differential equation is

$$x(t) = c_1 e^{2it} + c_2 e^{-2it}.$$

Differentiating this with respect to t, we get

$$x'(t) = 2ic_1 e^{2it} - 2ic_2 e^{-2it}.$$

From $x(0)$, $x'(0)$, and the initial conditions, we obtain the system of equations

$$c_1 + c_2 = 1$$
$$2ic_1 - 2ic_2 = 2i.$$

By inspection, $c_1 = 1$ and $c_2 = 0$. Therefore, the solution of the initial value problem is the complex-valued function

$$x(t) = e^{2it} = \cos 2t + i \sin 2t.$$ ♦

Example 10.4.4 Find the solution of the initial value problem

$$x'' - 2x' + 10x = 0; \quad x(0) = 0, \ x'(0) = 6.$$

Solution The characteristic roots are $\lambda = 1 \pm 3i$. By (10.4.17), a general solution of the differential equation is

$$x(t) = c_1 e^{(1+3i)t} + c_2 e^{(1-3i)t}.$$

Differentiation of $x(t)$ with respect to t yields

$$x'(t) = (1+3i)c_1 e^{(1+3i)t} + (1-3i)c_2 e^{(1-3i)t}.$$

Setting $t = 0$ and using the initial conditions, we get

$$\begin{aligned} c_1 + c_2 &= 0 \\ (1+3i)c_1 + (1-3i)c_2 &= 6, \end{aligned}$$

which has the solution $c_1 = -i$ and $c_2 = i$. Thus,

$$x(t) = -ie^{(1+3i)t} + ie^{(1-3i)t} = -ie^t(e^{3it} - e^{-3it}).$$

At first, this solution appears to be complex-valued; however, we see from (10.4.10) that it simplifies to

$$x(t) = -ie^t \left(\frac{e^{3it} - e^{-3it}}{2i} \right)(2i) = 2e^t \sin 3t.$$

Note that we obtained this very same result in Example 10.3.10. ♦

10.5 Cauchy-Euler Equations

Even though the procedure for finding solutions of a homogeneous linear equation

$$a\frac{d^2x}{dt^2} + b\frac{dx}{dt} + cx = 0$$

with constant coefficients a, b, c is relatively simple, it is quite another story if any of the coefficients vary with t. Unfortunately, the characteristic roots method does not work for

$$p(t)\frac{d^2x}{dt^2} + q(t)\frac{dx}{dt} + r(t)x = 0 \tag{10.5.1}$$

if any of p, q, r vary with t. Generally speaking, its solutions cannot be expressed in terms of finitely many elementary functions. Other means, such as power series methods or numerical techniques, are usually needed to handle such equations. However, if the coefficients of (10.5.1) have the special form

$$p(t) = at^2, \ q(t) = bt, \text{ and } r(t) = c,$$

where a, b, and c are constants, then its solutions are relatively simple and can easily be obtained, which we shall see in the examples at the end of this section. Names given to equations with these particular coefficients are: *Cauchy-Euler equations*, *Euler-Cauchy equations*, or *Euler equations*. Of these, let us use the former.

Homogeneous Cauchy-Euler Equation

Definition 10.5.1 A ***second-order homogeneous Cauchy-Euler equation*** is an equation of the form

$$at^2\frac{d^2x}{dt^2} + bt\frac{dx}{dt} + cx = 0 \tag{10.5.2}$$

where a, b, c are constants.

Fortunately, we can transform equation (10.5.2) to a homogeneous linear equation with constant coefficients. A way to achieve this is through a clever change of variable: from t to another variable, u say, so that t and t^2 will drop out of (10.5.2). If we are seeking a solution on an interval where t is always positive, we can define u by $u = \ln t$, or, equivalently, by

$$t = e^u. \tag{10.5.3}$$

First, let us find the relation between the derivatives $x'(t)$ and $x'(u)$. By the chain rule,

$$\frac{dx}{du} = \frac{dx}{dt} \cdot \frac{dt}{du}. \tag{10.5.4}$$

Then as

$$\frac{dt}{du} = \frac{d}{du}e^u = e^u = t,$$

this simplifies to

$$t\frac{dx}{dt} = \frac{dx}{du}. \tag{10.5.5}$$

Next, let us find the relation between the second-order derivatives $x''(t)$ and $x''(u)$. By (10.5.5) and the product and chain rules, we obtain

$$\frac{d^2x}{du^2} = \frac{d}{du}\left(t\frac{dx}{dt}\right) = t\frac{d}{du}\left(\frac{dx}{dt}\right) + \frac{dx}{dt}\cdot\frac{dt}{du} = t\frac{d}{dt}\left(\frac{dx}{dt}\right)\cdot\frac{dt}{du} + \frac{dx}{du}.$$

As $dt/du = t$, this simplifies to

$$\frac{d^2x}{du^2} = t^2\frac{d^2x}{dt^2} + \frac{dx}{du}$$

and so

$$t^2\frac{d^2x}{dt^2} = \frac{d^2x}{du^2} - \frac{dx}{du}. \tag{10.5.6}$$

Note that with the change of variable we have managed to transform the variable coefficient terms to constant coefficient terms. Substituting (10.5.5) and (10.5.6) into (10.5.2), we obtain the equation

$$a\left(\frac{d^2x}{du^2} - \frac{dx}{du}\right) + b\frac{dx}{du} + cx = 0,$$

which simplifies to

$$a\frac{d^2x}{du^2} + (b-a)\frac{dx}{du} + cx = 0. \tag{10.5.7}$$

The characteristic equation for (10.5.7) is

$$a\lambda^2 + (b-a)\lambda + c = 0. \tag{10.5.8}$$

Because of its connection to the Cauchy-Euler equation, let us refer to (10.5.8) as the ***Cauchy-Euler characteristic equation.***

Now we can explain how to find a general solution of (10.5.2). Since the characteristic roots λ_1 and λ_2 of (10.5.8) may be real and distinct numbers, real and equal numbers, or complex numbers, there are three different cases to consider.

Case 1. The roots λ_1, λ_2 are real and $\lambda_1 \neq \lambda_2$.

The general solution of (10.5.7) is

$$x(u) = c_1e^{\lambda_1 u} + c_2e^{\lambda_2 u}.$$

Use (10.5.3) to revert back to the original variable t: $e^{\lambda_1 u} = (e^u)^{\lambda_1} = t^{\lambda_1}$. Similarly, $e^{\lambda_2 u} = t^{\lambda_2}$. Thus, a general solution of (10.5.2) for this case is

$$x(t) = c_1 t^{\lambda_1} + c_2 t^{\lambda_2}. \tag{10.5.9}$$

Case 2. The roots λ_1, λ_2 are real and $\lambda_1 = \lambda_2$.

This is the repeated roots case. Let $\lambda := \lambda_i$ for $i = 1, 2$. A general solution of (10.5.7) is

$$x(u) = c_1 e^{\lambda u} + c_2 u e^{\lambda u},$$

From (10.5.3), $u = \ln t$ and so a general solution of (10.5.2) for this case is

$$x(t) = c_1 t^{\lambda} + c_2 t^{\lambda} \ln t. \tag{10.5.10}$$

Case 3. The roots λ_1, λ_2 are complex numbers.

Let $\lambda_1 = \alpha + i\omega$ and $\lambda_2 = \alpha - i\omega$, where $\omega > 0$, denote the two complex roots. Then a general solution of (10.5.7) is

$$x(u) = c_1 e^{\alpha u} \cos \omega u + c_2 e^{\alpha u} \sin \omega u.$$

Since $u = \ln t$, a general solution of (10.5.2) for the case of complex roots is

$$x(t) = c_1 t^{\alpha} \cos(\omega \ln t) + c_2 t^{\alpha} \sin(\omega \ln t). \tag{10.5.11}$$

If we seek solutions on intervals where t is always negative, then the definition of u given in (10.5.3) must be changed to $-t = e^u$. Accordingly,

$$\frac{dx}{du} = \frac{dx}{dt} \cdot \frac{dt}{du} = \frac{dx}{dt} \cdot \frac{d}{du}\left(-e^u\right) = t\frac{dx}{dt},$$

which is (10.5.5) again. Likewise (10.5.6) remains the same. Consequently the characteristic equation is still given by (10.5.8). The only change occurs in reverting back to the original variable t: $u = \ln(-t)$ rather than $u = \ln t$.

There is no need to memorize the general solutions (10.5.9)–(10.5.11) for the three different cases. All we need to remember is the following method.

Method for Solving Homogeneous Cauchy-Euler Equations

To find a general solution of

$$at^2\frac{d^2x}{dt^2}+bt\frac{dx}{dt}+cx=0 \tag{10.5.12}$$

on an interval that excludes $t=0$, complete the following steps:

1. Find a general solution of the associated equation

$$a\frac{d^2x}{du^2}+(b-a)\frac{dx}{du}+cx=0. \tag{10.5.13}$$

2. After completing Step 1, replace u with $\ln|t|$ and e^u with $|t|$.

Example 10.5.1 Find a general solution of

$$t^2x''+tx'-4x=0 \tag{10.5.14}$$

on the interval $(0,\infty)$.

Solution Comparing (10.5.14) to (10.5.12), we see that $a=1$, $b=1$, and $c=-4$. So, $b-a=0$ and the associated equation is

$$\frac{d^2x}{du^2}-4x=0,$$

where $u=\ln|t|=\ln t$. The characteristic roots are $\lambda=\pm 2$. A general solution of the associated equation is

$$x(u)=c_1e^{-2u}+c_2e^{2u}.$$

Since $e^u=t$,

$$e^{-2u}=(e^u)^{-2}=t^{-2}.$$

Likewise, $e^{2u}=t^2$. Therefore a general solution of (10.5.14) is

$$x(t)=\frac{c_1}{t^2}+c_2t^2.$$

♦

Example 10.5.2 Find the solution of the initial value problem

$$4t^2x''+x=0;\ x(1)=2,\ x'(1)=0. \tag{10.5.15}$$

Solution Since $a = 4$, $b = 0$, $c = 1$, and $b - a = -4$, the associated equation is

$$4\frac{d^2x}{du^2} - 4\frac{dx}{du} + x = 0.$$

The solution is defined on the interval $(0, \infty)$ since the initial conditions are given at $t = 1$. So the appropriate change of variable is $u = \ln t$. The characteristic equation for the preceding equation is $(2\lambda - 1)^2 = 0$. Since $\lambda = 1/2$ is a repeated root, a general solution of the associated equation is

$$x(u) = c_1 e^{u/2} + c_2 u e^{u/2}.$$

As $u = \ln t$ and $e^u = t$, a general solution of (10.5.15) is

$$x(t) = c_1\sqrt{t} + c_2\sqrt{t}\ln t.$$

Finally, we need values for c_1 and c_2 so that the solution satisfies the initial conditions. Since $x(1) = c_1$, it follows that $c_1 = 2$. Differentiating $x(t)$, we have

$$x'(t) = 2\frac{d}{dt}t^{1/2} + c_2\frac{d}{dt}(t^{1/2}\ln t) = \frac{1}{\sqrt{t}} + c_2\left(\frac{1}{\sqrt{t}} + \frac{\ln t}{2\sqrt{t}}\right).$$

Letting $t = 1$, we get $x'(1) = 1 + c_2 = 0$. Thus, $c_2 = -1$. Therefore, the solution of the initial value problem is

$$x(t) = 2\sqrt{t} - \sqrt{t}\ln t.$$ ♦

Example 10.5.3 Find a general solution of

$$t^2x'' + 3tx' + 2x = 0 \tag{10.5.16}$$

on the interval $(-\infty, 0)$.

Solution Since $a = 1$, $b = 3$, $c = 2$, and $b - a = 2$, the associated equation is

$$\frac{d^2x}{du^2} + 2\frac{dx}{du} + 2x = 0,$$

where $u = \ln(-t)$ since $t < 0$. The characteristic equation for the previous equation is $\lambda^2 + 2\lambda + 2 = 0$, which has the complex roots $\lambda = -1 \pm i$. A general solution of the associated equation is

$$x(u) = c_1 e^{-u}\cos u + c_2 e^{-u}\sin u.$$

In terms of t, the right-hand side is

$$c_1(e^u)^{-1}\cos u + c_2(e^u)^{-1}\sin u = c_1(-t)^{-1}\cos(\ln(-t)) + c_2(-t)^{-1}\sin(\ln(-t)).$$

Therefore, a general solution of equation (10.5.16) is

$$x(t) = \frac{k_1}{t}\cos(\ln(-t)) + \frac{k_2}{t}\sin(\ln(-t)),$$

where k_1 and k_2 are arbitrary constants. ♦

10.6 Homogeneous Linear Equations of Higher Order

A ***homogeneous linear differential equation of order n*** with constant coefficients is an equation of the form

$$a_0\frac{d^n x}{dt^n} + a_1\frac{d^{n-1}x}{dt^{n-1}} + \cdots + a_{n-1}\frac{dx}{dt} + a_n x = 0, \tag{10.6.1}$$

where $a_0, a_1, \ldots, a_n$ denote the constant coefficients and $a_0 \neq 0$. Now that we know how to find a general solution of the second-order homogeneous linear equation

$$a_0\frac{d^2x}{dt^2} + a_1\frac{dx}{dt} + a_2 x = 0, \tag{10.6.2}$$

we will show how to obtain a general solution of (10.6.1) when $n = 3$. In other words, we will find a general solution of the third-order homogeneous linear equation

$$a_0\frac{d^3x}{dt^3} + a_1\frac{d^2x}{dt^2} + a_2\frac{dx}{dt} + a_3 x = 0. \tag{10.6.3}$$

Let us begin by dividing both sides of (10.6.3) by a_0:

$$\frac{d^3x}{dt^3} + b_1\frac{d^2x}{dt^2} + b_2\frac{dx}{dt} + b_3 x = 0, \tag{10.6.4}$$

where $b_i = a_i/a_0$ for $i = 1, 2, 3$. It is easy to verify that $e^{\lambda t}$ is a solution of (10.6.4) if and only if λ is a root of

$$\lambda^3 + b_1\lambda^2 + b_2\lambda + b_3 = 0. \tag{10.6.5}$$

Of course, (10.6.5) is the ***characteristic equation*** for (10.6.4). It has exactly 3 roots as does any cubic polynomial when both real and complex roots are counted. Now recall the Remainder Theorem of algebra: If a polynomial $P(\lambda)$ is divided by $\lambda - c$, then the remainder is $P(c)$. It follows that $\lambda - c$ is a factor of $P(\lambda)$ if and only if $P(c) = 0$. Consequently, if $\lambda_1, \lambda_2, \lambda_3$ are the three roots of (10.6.5), then we can write it in the factored form

$$(\lambda - \lambda_1)(\lambda - \lambda_2)(\lambda - \lambda_3) = 0. \tag{10.6.6}$$

Multiplying out the left-hand side of (10.6.6), we obtain

$$\lambda^3 - (\lambda_1 + \lambda_2 + \lambda_3)\lambda^2 + (\lambda_1\lambda_2 + \lambda_1\lambda_3 + \lambda_2\lambda_3)\lambda - \lambda_1\lambda_2\lambda_3 = 0. \tag{10.6.7}$$

Comparing (10.6.7) to (10.6.5), we see that

$$\begin{aligned} b_1 &= -(\lambda_1 + \lambda_2 + \lambda_3) \\ b_2 &= \lambda_1\lambda_2 + \lambda_1\lambda_3 + \lambda_2\lambda_3 \\ b_3 &= -\lambda_1\lambda_2\lambda_3. \end{aligned}$$

Hence, written in terms of its characteristic roots, (10.6.4) is

$$x''' - (\lambda_1 + \lambda_2 + \lambda_3)x'' + (\lambda_1\lambda_2 + \lambda_1\lambda_3 + \lambda_2\lambda_3)x' - \lambda_1\lambda_2\lambda_3 x = 0.$$

Now rearrange the terms as follows:

$$(x' - \lambda_3 x)'' - \lambda_2(x' - \lambda_3 x)' - \lambda_1(x' - \lambda_3 x)' + \lambda_1\lambda_2(x' - \lambda_3 x) = 0. \tag{10.6.8}$$

Then by defining

$$y := x' - \lambda_3 x \tag{10.6.9}$$

Eq. (10.6.8) becomes

$$y'' - (\lambda_1 + \lambda_2)y' + \lambda_1\lambda_2 y = 0. \tag{10.6.10}$$

Since a general solution of (10.6.4) depends on the characteristic roots—whether they are all distinct or some are equal or whether they are all real or some are complex—let us take a look at all of the possibilities. We know from algebra that complex roots of polynomials must occur in conjugate pairs. That is, if $z = \alpha + i\omega$ is a complex root of a polynomial $P(z)$, then so is its conjugate $\overline{z} = \alpha - i\omega$. This implies that a polynomial cannot have an odd number of complex roots. Thus, the roots of (10.6.5) are either all real or two of them are complex conjugate pairs while the other is real. There are four distinct cases to consider.

Case 1. All three characteristic roots are real and distinct.

Since the characteristic roots associated with (10.6.10) are λ_1 and λ_2, its general solution is

$$y(t) = k_1 e^{\lambda_1 t} + k_2 e^{\lambda_2 t}, \tag{10.6.11}$$

where k_1 and k_2 are constants. An integrating factor for (10.6.9) is $e^{-\lambda_3 t}$. Hence,

$$\frac{d}{dt}\left(e^{-\lambda_3 t} x\right) = e^{-\lambda_3 t} y = k_1 e^{(\lambda_1 - \lambda_3)t} + k_2 e^{(\lambda_2 - \lambda_3)t}.$$

An integration yields

$$x(t) = \frac{k_1}{\lambda_1 - \lambda_3} e^{\lambda_1 t} + \frac{k_2}{\lambda_2 - \lambda_3} e^{\lambda_2 t} + k_3 e^{\lambda_3 t}, \tag{10.6.12}$$

where k_3 is another constant. This is a solution regardless of the values of k_1, k_2, and k_3. Hence, we can replace the coefficients in (10.6.12) with c_1, c_2, and c_3. In short, for this case, a general solution of equation (10.6.3) is

$$x(t) = c_1 e^{\lambda_1 t} + c_2 e^{\lambda_2 t} + c_3 e^{\lambda_3 t}. \tag{10.6.13}$$

Case 2. The characteristic roots are real and two of them are equal.

Let us label the roots as follows: $\lambda_1 = \lambda_2 = m_1$ and $\lambda_3 = m_2$, where $m_2 \neq m_1$. Then (10.6.10) becomes

$$y'' - 2m_1 y' + m_1^2 y = 0 \tag{10.6.14}$$

with a general solution

$$y(t) = k_1 e^{m_1 t} + k_2 t e^{m_1 t}. \tag{10.6.15}$$

From (10.6.9), we have

$$x' - m_2 x = k_1 e^{m_1 t} + k_2 t e^{m_1 t},$$

which has the solutions

$$x(t) = \frac{(m_1 - m_2)k_1 - k_2}{(m_1 - m_2)^2} e^{m_1 t} + \frac{k_2}{m_1 - m_2} t e^{m_1 t} + k_3 e^{m_2 t}.$$

Consequently, we conclude that a general solution of (10.6.3) is

$$x(t) = c_1 e^{m_1 t} + c_2 t e^{m_1 t} + c_3 e^{m_2 t}. \tag{10.6.16}$$

Case 3. The three characteristic roots are real and equal.

Let λ denote their common value. This makes λ a root of multiplicity 3. Then, as in Case 2 (cf. (10.6.15)),

$$y(t) = k_1 e^{\lambda t} + k_2 t e^{\lambda t}.$$

Hence,

$$x' - \lambda x = k_1 e^{\lambda t} + k_2 t e^{\lambda t}.$$

Integrating and renaming constants, we obtain the general solution

$$x(t) = c_1 e^{\lambda t} + c_2 t e^{\lambda t} + c_3 t^2 e^{\lambda t}. \tag{10.6.17}$$

Case 4. The roots consist of a complex conjugate pair $\alpha \pm i\omega$ and a real number λ.

Let $\lambda_1 = \alpha + i\omega$ and $\lambda_2 = \alpha - i\omega$ in (10.6.10). And in (10.6.9), let $\lambda_3 = \lambda$. Then, as $\lambda_1 + \lambda_2 = 2\alpha$ and $\lambda_1\lambda_2 = \alpha^2 + \omega^2$, Eq. (10.6.10) becomes

$$y'' - 2\alpha y' + (\alpha^2 + \omega^2)y = 0, \tag{10.6.18}$$

where

$$y = x' - \lambda x. \tag{10.6.19}$$

Because $\alpha \pm i\omega$ are the roots of the characteristic equation for (10.6.18), a general solution is

$$y(t) = k_1 e^{\alpha t} \cos \omega t + k_2 e^{\alpha t} \sin \omega t.$$

Consequently, if $x(t)$ is a solution of (10.6.4), it must be one of the solutions of

$$x' - \lambda x = k_1 e^{\alpha t} \cos \omega t + k_2 e^{\alpha t} \sin \omega t.$$

Employing the integrating factor $e^{-\lambda t}$, we find that

$$x(t) = e^{\lambda t} k_1 \int e^{(\alpha-\lambda)t} \cos \omega t \, dt + e^{\lambda t} k_2 \int e^{(\alpha-\lambda)t} \sin \omega t \, dt + k_3 e^{\lambda t}.$$

It is left as an exercise to verify that when these integrations are carried out that

$$x(t) = \frac{k_1(\alpha - \lambda) - k_2\omega}{(\alpha-\lambda)^2 + \omega^2} e^{\alpha t} \cos \omega t + \frac{k_1\omega + k_2(\alpha - \lambda)}{(\alpha-\lambda)^2 + \omega^2} e^{\alpha t} \sin \omega t + k_3 e^{\lambda t}.$$

Finally, after replacing the above coefficients with c_1, c_2, and c_3, we obtain

$$x(t) = c_1 e^{\alpha t}\cos\omega t + c_2 e^{\alpha t}\sin\omega t + c_3 e^{\lambda t}.$$

Example 10.6.1 Find a general solution of $\dfrac{d^3x}{dt^3} - 3\dfrac{dx}{dt} + 2x = 0.$

Solution Written in terms of the operator $D = d/dt$, the equation is

$$(D^3 - 3D + 2)x = 0.$$

The corresponding characteristic equation is

$$\lambda^3 - 3\lambda + 2 = 0.$$

By inspection, $\lambda = 1$ is a root. Consequently, $\lambda - 1$ is a factor of the characteristic polynomial $P(\lambda) = \lambda^3 - 3\lambda + 2$. To obtain another factor, divide $P(\lambda)$ by $\lambda - 1$. By the method of synthetic division:

$$\begin{array}{r|rrrr} 1 & 1 & 0 & -3 & 2 \\ & & 1 & 1 & -2 \\ \hline & 1 & 1 & -2 & \boxed{0} \end{array}$$

The numbers 1, 1, -2 in the last row represent the factor $\lambda^2 + \lambda - 2$. Thus,

$$\lambda^3 - 3\lambda + 2 = (\lambda - 1)(\lambda^2 + \lambda - 2) = (\lambda - 1)^2(\lambda + 2).$$

So the characteristic roots are $\lambda = -2$ and the double root $\lambda = 1$. It follows from Case 2 that a general solution of the differential equation is

$$x(t) = c_1 e^{-2t} + c_2 e^t + c_3 t e^t.$$ ♦

Example 10.6.2 Find the solution of

$$4x''' - 4x'' - 3x' + 5x = 0$$

satisfying the initial conditions

$$x(0) = -17,\ x'(0) = 0,\ x''(0) = 34.$$

Solution The corresponding characteristic equation is

$$4\lambda^3 - 4\lambda^2 - 3\lambda + 5 = 0.$$

It is easy to see that $\lambda = -1$ is a root; so a factor of the characteristic polynomial is $\lambda - (-1) = \lambda + 1$. Synthetic division yields

$$\begin{array}{r|rrrr} -1 & 4 & -4 & -3 & 5 \\ & & -4 & 8 & -5 \\ \hline & 4 & -8 & 5 & \boxed{0} \end{array}$$

Hence, $4\lambda^2 - 8\lambda + 5$ is another factor. Solutions of

$$4\lambda^2 - 8\lambda + 5 = 0$$

are $\lambda = 1 \pm \frac{1}{2}i$. Since the characteristic roots are $-1, 1 \pm \frac{1}{2}i$, it follows from Case 4 that a general solution is

$$x(t) = c_1 e^{-t} + c_2 e^t \sin\frac{t}{2} + c_3 e^t \cos\frac{t}{2}.$$

Now let us find values for c_1, c_2, c_3 so that $x(t)$ satisfies the initial conditions. The first and second derivatives of $x(t)$ are

$$x'(t) = -c_1 e^{-t} + c_2 e^t \left(\sin\frac{t}{2} + \frac{1}{2}\cos\frac{t}{2}\right) + c_3 e^t \left(\cos\frac{t}{2} - \frac{1}{2}\sin\frac{t}{2}\right)$$

$$x''(t) = c_1 e^{-t} + c_2 e^t \left(\frac{3}{4}\sin\frac{t}{2} + \cos\frac{t}{2}\right) + c_3 e^t \left(\frac{3}{4}\cos\frac{t}{2} - \sin\frac{t}{2}\right).$$

Setting $t = 0$ and using the initial conditions, we see that the constants must satisfy the system of equations

$$\begin{aligned} c_1 + c_3 &= -17 \\ -c_1 + \frac{1}{2}c_2 + c_3 &= 0 \\ c_1 + c_2 + \frac{3}{4}c_3 &= 34. \end{aligned}$$

The solution of this system is

$$c_1 = 3, \ c_2 = 46, \ c_3 = -20.$$

Therefore, the solution of the initial value problem is

$$x(t) = 3e^{-t} + 46e^t \sin\frac{t}{2} - 20e^t \cos\frac{t}{2}.$$ ♦

We could look for a general solution of a fourth-order homogeneous linear differential equation (namely, (10.6.1) with $n = 4$) and examine the various cases as we did with the third-order equation (10.6.3), but that would be tedious. There is a much better way but it requires knowledge of linear algebra, something not assumed in this book. So let us merely state how to find a general solution of an nth-order homogeneous linear differential equation without providing a proof.

Solutions of Homogeneous Linear Equations of nth Order

Theorem 10.6.1 *A general solution $x(t)$ of*

$$a_0 \frac{d^n x}{dt^n} + a_1 \frac{d^{n-1} x}{dt^{n-1}} + \cdots + a_{n-1} \frac{dx}{dt} + a_n x = 0 \tag{10.6.20}$$

is

$$x(t) = c_1 x_1(t) + c_2 x_2(t) + \cdots + c_n x_n(t) = \sum_{i=1}^{n} c_i x_i(t)$$

where the $x_i(t)$ denote the following functions:

(a) If λ is a real characteristic root of multiplicity j, then j of the $x_i(t)$ are: $e^{\lambda t}, te^{\lambda t}, t^2 e^{\lambda t}, \ldots, t^{j-1} e^{\lambda t}$.

(b) If $\alpha + i\omega$ is a characteristic root of multiplicity k, then $2k$ of the $x_i(t)$ are: $e^{\alpha t}\cos\omega t$, $e^{\alpha t}\sin\omega t$, $te^{\alpha t}\cos\omega t$, $te^{\alpha t}\sin\omega t$, $\ldots$, $t^{k-1}e^{\alpha t}\cos\omega t$, $t^{k-1}e^{\alpha t}\sin\omega t$.

Example 10.6.3 Find a general solution of the fourth-order linear equation

$$x^{(4)} + 8\ddot{x} + 16x = 0. \tag{10.6.21}$$

Solution Writing this equation in terms of the derivative operator D, it becomes

$$(D^4 + 8D^2 + 16)x = 0.$$

In this form, it is particularly easy to see that the characteristic equation is

$$\lambda^4 + 8\lambda^2 + 16 = 0,$$

which is a perfect square polynomial. Thus, we see from

$$(\lambda^2 + 4)^2 = 0$$

that $2i$ is a characteristic root of multiplicity 2. It then follows from Theorem 10.6.1 that a general solution of (10.6.21) is

$$x(t) = c_1 \cos 2t + c_2 \sin 2t + c_3 t \cos 2t + c_4 t \sin 2t. \qquad \blacklozenge$$

10.7 Nonhomogeneous Linear Equations

A ***second-order nonhomogeneous***[8] ***linear differential equation*** with constant coefficients is an equation of the form

$$a\frac{d^2x}{dt^2} + b\frac{dx}{dt} + cx = g(t), \tag{10.7.1}$$

where $a \neq 0$ and g is a function that is not identically zero. The function $g(t)$ is often referred to as a ***forcing function*** (or ***forcing term***), especially when this equation is used to model a physical system and $g(t)$ represents an external force. If $g(t) \equiv 0$, then (10.7.1) is no longer a nonhomogeneous equation but rather the homogeneous equation

$$a\frac{d^2x}{dt^2} + b\frac{dx}{dt} + cx = 0. \tag{10.7.2}$$

From this point on, we focus on nonhomogeneous equations and how to solve them.

An example of a second-order nonhomogeneous linear equation is

$$x'' + 2x' - 3x = -12.$$

This is (10.7.1) with $a = 1$, $b = 2$, $c = -3$, and $g(t) = -12$. It is clear, with no need for any pencil-and-paper work, that a solution of this equation is $x(t) \equiv 4$. A function that is a solution of a nonhomogeneous differential equation and which contains no arbitrary constants is known as a ***particular solution***. Thus, $x(t) \equiv 4$ is a particular solution. So are the functions

$$e^{-3t} + 4,\ 2e^{-3t} + 4,\ e^{-3t} + e^t + 4,\ e^t + 4,\ 5e^t + 4,$$

just to list a few. In fact, this equation has infinitely many solutions of the form

$$x(t) = c_1 e^{-3t} + c_2 e^t + 4,$$

where c_1 and c_2 are constants. A point to be made here is that we obtained the constant solution $x(t) \equiv 4$ *by inspection*; that is, it took a mere cursory glance to see that this function satisfies the differential equation on the interval $(-\infty, \infty)$.

[8] The term ***inhomogeneous*** is also used.

A second example of a nonhomogeneous linear differential equation is

$$4x'' + x = t$$

This equation results from setting $a = 4$, $b = 0$, $c = 1$, and $g(t) = t$ in (10.7.1). As with the first example, this equation has more than one particular solution. One that comes to mind with barely any thought is $x(t) = t$. Clearly, this function satisfies the equation on $(-\infty, \infty)$ since its second derivative is always zero.

The reason for wanting a particular solution of equation (10.7.1) is that its general solution, as we will prove next, can be expressed as the sum of any one of its particular solutions and a general solution of equation (10.7.2). This latter equation is called the ***homogeneous differential equation associated with*** Eq. (10.7.1). In the literature, a general solution of the homogeneous equation (10.7.2) is often called the ***complementary function*** for the nonhomogeneous equation (10.7.1).

Solutions of Second-Order Nonhomogeneous Equations

Theorem 10.7.1 *If a function $x_p(t)$ is a particular solution of*

$$ax'' + bx' + cx = g(t) \tag{10.7.3}$$

on an interval J, then

$$x(t) = x_h(t) + x_p(t) \tag{10.7.4}$$

is a general solution of (10.7.3) *on J, where $x_h(t)$ is a general solution of the associated homogeneous equation*

$$ax'' + bx' + cx = 0. \tag{10.7.5}$$

Proof Recall from Sect. 10.3.3 that the result of applying the linear differential operator

$$L = aD^2 + bD + c$$

to a twice-differentiable function x is

$$L[x] = (aD^2 + bD + c)x = aD^2x + bDx + cx = ax'' + bx' + cx.$$

Consequently, Eq. (10.7.3) can be written in the abbreviated form $L[x] = g(t)$.

Let $x_p(t)$ be any particular solution of the nonhomogeneous equation (10.7.3) on an interval J that we have somehow managed to find.[9] This means of course that

$$L[x_p(t)] = g(t)$$

for $t \in J$. Likewise, let $x(t)$ denote another solution of (10.7.3) on J. And so

$$L[x(t)] = g(t)$$

for $t \in J$. Now apply the operator L to the difference $x(t) - x_p(t)$. Since L is a linear operator,

$$L[x(t) - x_p(t)] = L[x(t)] - L[x_p(t)] = g(t) - g(t) = 0.$$

In other words, the function $x(t) - x_p(t)$ is a solution of $L[x] = 0$; that is, it is a solution of the associated homogeneous equation (10.7.5). So, depending on whether the roots of the characteristic equation for (10.7.5) are real or complex, $x(t) - x_p(t)$ is either (10.3.16) or (10.3.19) or (10.3.49) for some specific values of the constants c_1 and c_2. Therefore, as $x(t) - x_p(t)$ is one of the solutions belonging to the general solution $x_h(t)$, which represents the set of all solutions of (10.7.5), we see that all solutions $x(t)$ of the nonhomogeneous equation (10.7.3) are given by

$$x(t) = x_h(t) + x_p(t),$$

which is (10.7.4). ■

Example 10.7.1 Find a general solution of the nonhomogeneous equation

$$x'' + 2x' - 3x = -12. \tag{10.7.6}$$

Solution Recall that $x_p(t) \equiv 4$ is a particular solution of this equation on the interval $(-\infty, \infty)$. The associated homogeneous differential equation is

$$x'' + 2x' - 3x = 0, \tag{10.7.7}$$

which has the characteristic equation $\lambda^2 + 2\lambda - 3 = 0$. Since its factorization is

$$(\lambda + 3)(\lambda - 1) = 0,$$

the characteristic roots are $\lambda = -3$ and $\lambda = 1$. Therefore, a general solution of the homogeneous equation (10.7.7) is

$$x_h(t) = c_1 e^{-3t} + c_2 e^t.$$

[9] How to actually come up with such a solution is the subject of the next two sections.

It follows from Theorem 10.7.1 that a general solution of (10.7.6) is

$$x(t) = x_p(t) + x_h(t) = c_1 e^{-3t} + c_2 e^t + 4.$$

for all $t \in (-\infty, \infty)$. ♦

Example 10.7.2 Find a general solution of

$$4x'' + x = t. \tag{10.7.8}$$

Solution As we found earlier, $x_p(t) = t$ is a particular solution of (10.7.8) on the interval $(-\infty, \infty)$. The associated homogeneous differential equation is

$$4x'' + x = 0.$$

Since the characteristic equation is $4\lambda^2 + 1 = 0$, the characteristic roots are $\lambda = \pm i/2$. Thus, a general solution of the associated homogeneous equation is

$$x_h(t) = c_1 \cos \frac{t}{2} + c_2 \sin \frac{t}{2}.$$

By Theorem 10.7.1, a general solution of (10.7.8) is

$$x(t) = x_h(t) + x_p(t) = c_1 \cos \frac{t}{2} + c_2 \sin \frac{t}{2} + t.$$

for all $t \in (-\infty, \infty)$.

Example 10.7.3 Find a general solution of

$$x'' + 4x = 4t^{-1} + 2t^{-3} \tag{10.7.9}$$

on the interval $(0, \infty)$.

Solution It is easy to mentally verify that $x_p(t) = t^{-1}$ is a solution of (10.7.9) for $t > 0$. Since the characteristic roots are $\pm 2i$, a general solution of the associated homogeneous equation $x'' + 4x = 0$ is

$$x_h(t) = c_1 \cos 2t + c_2 \sin 2t.$$

Consequently, a general solution of (10.7.9) is

$$x(t) = c_1 \cos 2t + c_2 \sin 2t + t^{-1}.$$

for all $t \in (0, \infty)$. ♦

10.8 Method of Undetermined Coefficients

The examples in Sect. 10.7 may give the false impression that the only way of finding a particular solution is by making an educated guess and then keeping one's fingers crossed that the guess will actually turn out to be a solution. Unfortunately, particular solutions usually do not just come out of nowhere. In this section, we consider some differential equations with particular solutions that are not as obvious as they have been up to now. This necessitates developing a method whereby particular solutions can be found without completely relying on guesswork. There may still be some guesswork involved as to the form of a particular solution—however, unlike the previous equations, we will not usually be able to come up with a particular solution without first doing some pencil-and-paper work. Before we say anymore, let us look at examples of some more nonhomogeneous equations.

Example 10.8.1 Find a particular and then a general solution of

$$x'' + 4x = -20e^{6t}. \tag{10.8.1}$$

Solution The form of a particular solution of a nonhomogeneous equation obviously depends on the function on its right-hand side as we will explore here and in subsequent examples. For this equation that function is an exponential function. The fact that derivatives of exponential functions are also exponential functions suggests that a particular solution has the form

$$x_p(t) = Ae^{6t}. \tag{10.8.2}$$

Let us consider this possibility and call it a ***trial particular solution***. Now substitute it and its second derivative into (10.8.1) to see if the coefficient A can be assigned a value to really make it a particular solution. These substitutions give

$$x_p''(t) + 4x_p(t) = 36Ae^{6t} + 4Ae^{6t} = 40Ae^{6t}.$$

Clearly the right-hand side is equal to the right-hand side of (10.8.1) for $A = -1/2$. Therefore,

$$x_p(t) = -\frac{1}{2}e^{6t}$$

is a particular solution of (10.8.1). Since $x_h(t) = c_1 \cos 2t + c_2 \sin 2t$ is a general solution of the associated homogeneous differential equation, it follows from Theorem 10.7.1 that a general solution of (10.8.1) is

$$x(t) = x_h(t) + x_p(t) = c_1 \cos 2t + c_2 \sin 2t - \frac{1}{2}e^{6t}. \quad ♦$$

Before continuing with more examples, let us reflect on how we obtained the particular solution in Example 10.8.1. First, we managed to come up with the so-called trial particular solution (10.8.2) by asking what type of function would generate the right-hand side of the differential equation when the differential operator $L = D^2 + 4$ is applied to it. At that point, the value of the coefficient A was unknown. But then its value was found by substituting the trial particular solution (10.8.2) into the differential equation (10.8.1) to ascertain what the value of A would have to be in order for it to truly be a solution. Let us explore this approach further by considering more examples.

Example 10.8.2 Find a general solution of

$$x'' - 2x' - 3x = 2 - 9t^2. \tag{10.8.3}$$

Solution Let $L[x]$ denote the left-hand side of (10.8.3), where L is the differential operator $L = D^2 - 2D - 3$. The right-hand side of (10.8.3) is a quadratic function. Applying L to this quadratic function yields another quadratic function. This suggests that a particular solution of (10.8.3) has the form

$$x_p(t) = At^2 + Bt + C. \tag{10.8.4}$$

Even though the t-term is missing from the right-hand side of (10.8.3), it would be assuming too much not to expect such a term in $x_p(t)$. Applying L to $x_p(t)$, we have

$$\begin{aligned} L[x_p(t)] &= x_p''(t) - 2x_p'(t) - 3x_p(t) \\ &= 2A - 2(2At + B) - 3(At^2 + Bt + C) \\ &= -3At^2 + (-4A - 3B)t + (2A - 2B - 3C). \end{aligned}$$

Comparing this with the right-hand side of (10.8.3) reveals that $x_p(t)$ is a solution of (10.8.3) provided that

$$-3At^2 + (-4A - 3B)t + (2A - 2B - 3C) = -9t^2 + 2.$$

Equating coefficients of like terms, we obtain the following three simultaneous equations in terms of the three undetermined coefficients A, B, and C:

$$\begin{aligned} -3A &= -9 \\ -4A - 3B &= 0 \\ 2A - 2B - 3C &= 2. \end{aligned}$$

The solution of these simultaneous equations is: $A = 3$, $B = -4$, $C = 4$. Therefore, a particular solution is

$$x_p(t) = 3t^2 - 4t + 4. \tag{10.8.5}$$

The characteristic equation for the associated homogeneous equation

$$x'' - 2x' - 3x = 0,$$

is $\lambda^2 - 2\lambda - 3 = 0$. Since its roots are $\lambda = -1$ and $\lambda = 3$, a general solution of the associated homogeneous equation is

$$x_h(t) = c_1 e^{-t} + c_2 e^{3t}.$$

Then, by Theorem 10.7.1, a general solution of (10.8.3) is

$$x(t) = c_1 e^{-t} + c_2 e^{3t} + 3t^2 - 4t + 4.$$ ♦

Recapitulating, we found the particular solution (10.8.5) of the nonhomogeneous equation (10.8.3) by first coming up with (10.8.4) as a trial particular solution, which was based on the form of the right-hand side of (10.8.3). At that point, the coefficients A, B, C had *not yet been determined.* But then their values *were determined* by substituting the trial particular solution (10.8.4) into (10.8.3) to ascertain what values would truly make it a particular solution.

The procedure that we have been using to find particular solutions of the nonhomogeneous equations (10.8.1) and (10.8.3) is known as the ***method of undetermined coefficients***. Here are the essential steps of this method:

1. First, find a trial particular solution based on the form of the right-hand side of the nonhomogeneous equation.
2. Second, substitute the trial particular solution into the nonhomogeneous equation in order to ascertain the values of its coefficients.

The choice of a trial particular solution is not always so straightforward as it was in the two previous examples. Some complications may arise as we shall see shortly in the next set of examples.

10.8.1 Complications

The previous examples suggest that this method will work whenever the function $g(t)$ on the right-hand side of (10.7.3) is either an exponential function or a polynomial. Consider

$$ax'' + bx' + cx = ke^{\alpha t}, \tag{10.8.6}$$

where a, k, and α are nonzero constants. The obvious choice for a trial particular solution is

$$x_p(t) = Ae^{\alpha t} \tag{10.8.7}$$

since the result of $L = aD^2 + bD + c$ operating on it is

$$L[Ae^{\alpha t}] = aD^2(Ae^{\alpha t}) + bD(Ae^{\alpha t}) + cAe^{\alpha t} = (a\alpha^2 + b\alpha + c)Ae^{\alpha t}.$$

Comparing this with the right-hand side of (10.8.6), we see that (10.8.7) is a particular solution of (10.8.6) if

$$A = \frac{k}{a\alpha^2 + b\alpha + c}$$

provided $a\alpha^2 + b\alpha + c \neq 0$. However, if $a\alpha^2 + b\alpha + c = 0$, then (10.8.7) is not a particular solution of the nonhomogeneous equation (10.7.3), but rather a solution of the associated homogeneous equation (10.7.5). In other words, if α is a root of the characteristic equation

$$a\lambda^2 + b\lambda + c = 0.$$

then some sort of adjustment to the trial particular solution $x_p(t) = Ae^{\alpha t}$ will have to be made as the next example demonstrates.

Example 10.8.3 Find a general solution of

$$x'' - 2x' = e^{2t}. \tag{10.8.8}$$

Solution The function e^{2t} on the right-hand side suggests that a particular solution of (10.8.8) will be of the form

$$x_p(t) = Ae^{2t} \tag{10.8.9}$$

for some value of A. Unfortunately,

$$x_p''(t) - 2x_p'(t) = 4Ae^{2t} - 2(2Ae^{2t}) = 0.$$

This trial particular solution fails because we get 0 instead of a constant multiple of e^{2t}. It is in fact a solution of the corresponding homogeneous equation, to wit:

$$x'' - 2x' = 0. \tag{10.8.10}$$

This could have been ascertained earlier had we found a general solution of (10.8.10) first, which due to the characteristic roots being 0 and 2 is

$$x_h(t) = c_1 + c_2e^{2t}.$$

Notice that (10.8.9) is a member of this family of solutions: this is seen by setting $c_1 = 0$ and $c_2 = A$.

So what is a suitable form of a trial particular solution? The answer lies with the derivative

$$\frac{d}{dt}\left(te^{2t}\right) = 2te^{2t} + e^{2t}.$$

That is to say, we can still generate the term e^{2t} by differentiating te^{2t}. This suggests replacing (10.8.9) with the trial particular solution

$$x_p(t) = Ate^{2t}. \tag{10.8.11}$$

Differentiating, we get

$$x_p'(t) = 2Ate^{2t} + Ae^{2t}$$
$$x_p''(t) = 4Ate^{2t} + 2Ae^{2t} + 2Ae^{2t} = 4Ate^{2t} + 4Ae^{2t}.$$

Thus,

$$x_p''(t) - 2x_p'(t) = 4Ate^{2t} + 4Ae^{2t} - 2(2Ate^{2t} + Ae^{2t}) = 2Ae^{2t}.$$

Comparing this with the right-hand side of (10.8.8), we see that (10.8.11) is a particular solution if $A = 1/2$. Hence, by Theorem 10.7.1, a general solution of (10.8.8) is

$$x(t) = c_1 + c_2e^{2t} + \frac{1}{2}te^{2t}.$$ ♦

Next, we take a look at two examples of nonhomogeneous differential equations with forcing functions that include sine and cosine functions. The forcing function in the first of these examples involves a single cosine function.

Example 10.8.4 Find a particular solution of

$$x'' + 4x' + 4x = -3\cos t. \tag{10.8.12}$$

Then find its general solution.

Solution The cosine function on the right-hand side of (10.8.12) suggests that the function

$$x_p(t) = A\cos t \tag{10.8.13}$$

is a particular solution for some $A \neq 0$. So let us see if we can find a value for A so that it really is a solution of (10.8.12). Substituting $A\cos t$ for the dependent variable x, we get

$$x_p'' + 4x_p' + 4x_p = \frac{d^2}{dt^2}(A\cos t) + 4\frac{d}{dt}(A\cos t) + 4(A\cos t)$$
$$= -A\cos t - 4A\sin t + 4A\cos t = 3A\cos t - 4A\sin t.$$

Equating this to the right-hand side of (10.8.12), we have

$$3A\cos t - 4A\sin t = -3\cos t$$

or

$$\cos t = \frac{4A}{3A+3}\sin t.$$

However, as we well know from trigonometry, $\cos t$ and $\sin t$ are not constant multiples of each other. In other words, they are linearly independent (cf. Example 10.9.1). Consequently, the trial particular solution (10.8.13) fails to be a solution of (10.8.12).

The problem lies with choosing (10.8.13) as the trial particular solution since the result of

$$L = D^2 + 4D + 4$$

operating on $\cos t$ is $\sin t$ as well as $\cos t$. So it turns out that we were too restrictive in our choice. We should have allowed ourselves a little more latitude by including a sine term to complement the cosine term as follows:

$$x_p(t) = A\cos t + B\sin t. \tag{10.8.14}$$

So do values of A and B exist so that $x_p(t)$ is a particular solution of (10.8.12)? To answer this, let us differentiate $x_p(t)$ twice:

$$x_p'(t) = -A\sin t + B\cos t, \qquad x_p''(t) = -A\cos t - B\sin t.$$

Now multiply $x_p(t)$ and its derivatives by the appropriate coefficients of the operator $L = D^2 + 4D + 4$ and add the results as follows:

$$\begin{aligned} x_p''(t) &= -A\cos t - B\sin t \\ 4x_p'(t) &= 4B\cos t - 4A\sin t \\ +\ 4x_p(t) &= 4A\cos t + 4B\sin t \\ \hline L[x_p(t)] &= (3A+4B)\cos t + (3B-4A)\sin t. \end{aligned}$$

This layout using a column format to align like terms rather than stringing them out in a single row, as was previously done, safeguards against accidentally omitting any of the terms. The function $x_p(t)$ is a particular solution if $L[x_p(t)] = -3\cos t$. This means that A and B must be given values so that

$$\begin{aligned} 3A + 4B &= -3 \\ -4A + 3B &= 0. \end{aligned}$$

The solution of this linear system is

$$A = -\frac{9}{25}, \ B = -\frac{12}{25}.$$

Therefore, a particular solution of (10.8.12) is

$$x_p(t) = -\frac{9}{25}\cos t - \frac{12}{25}\sin t.$$

Since the characteristic equation $\lambda^2 + 4\lambda + 4 = 0$ has the repeated root $\lambda = -2$, a general solution of (10.8.12) is

$$x(t) = c_1 e^{-2t} + c_2 t e^{-2t} - \frac{9}{25}\cos t - \frac{12}{25}\sin t.$$ ♦

The lesson to be taken from Example 10.8.4 is this:

If a cosine function is used in a trial particular solution, then it must be paired it up with a sine function whose argument is the same as that of the cosine function, and vice versa.

Example 10.8.5 Find a general solution of

$$x'' + x = \sin t. \tag{10.8.15}$$

Solution Due to the presence of the sine function in (10.8.15), we will heed the lesson of Example 10.8.4, namely, that sine and cosine functions should always appear together in a trial particular solution. Accordingly, choose the trial particular solution

$$x_p(t) = A\sin t + B\cos t. \tag{10.8.16}$$

Substituting it into the left-hand side of (10.8.15), we obtain

$$\begin{aligned} x_p''(t) + x_p(t) &= \frac{d^2}{dt^2}(A\sin t + B\cos t) + (A\sin t + B\cos t) \\ &= -A\sin t - B\cos t + A\sin t + B\cos t = 0. \end{aligned}$$

Stating the obvious, this trial particular solution does not work. What we have unwittingly shown here is that it is a solution of the associated homogeneous equation

$$x'' + x = 0 \tag{10.8.17}$$

for all values of A and B.

There is a lesson to be learned here. The characteristic equation for (10.8.17) is $\lambda^2 + 1 = 0$; so its roots are $\lambda = \pm i$. Hence, a general solution of (10.8.17) is

$$x_h(t) = c_1 \cos t + c_2 \sin t.$$

Note this is precisely (10.8.16), aside from the letters used for the coefficients. Had we instead started with finding a general solution of the associated homogeneous equation (10.8.17) rather than finding a particular solution of the nonhomogeneous equation (10.8.15), we could have saved some time and effort by avoiding the foregoing work, which led nowhere. As a result of this example (see also Example 10.8.3), we have learned that a good rule of thumb to follow is this:

Before looking for a particular solution of a nonhomogeneous equation, first find a general solution of the associated homogeneous equation.

So, if (10.8.16) does not work, what does? The answer is that we need to come up with a different trial particular solution $x_p(t)$ so that

$$L[x_p(t)] = (D^2 + 1)x_p(t) = x_p''(t) + x_p(t)$$

includes a term that is a constant multiple of $\sin t$. In view of the product rule, this is the case for the function $t \cos t$. Again as sine and cosine functions should be paired together, let us try

$$x_p(t) = At \sin t + Bt \cos t.$$

By the product rule,

$$\begin{aligned} x_p'(t) &= At \cos t + A \sin t - Bt \sin t + B \cos t \\ x_p''(t) &= -At \sin t + A \cos t + A \cos t - Bt \cos t - B \sin t - B \sin t. \end{aligned}$$

Thus, we have

$$\begin{array}{rrl} & x_p''(t) = & -At \sin t - Bt \cos t + 2A \cos t - 2B \sin t \\ + & x_p(t) = & At \sin t + Bt \cos t \\ \hline & L[x_p(t)] = & \qquad\qquad\qquad\quad + 2A \cos t - 2B \sin t. \end{array}$$

From this we can see that $x_p(t)$ is a particular solution of equation (10.8.15) if A and B have values so that $L[x_p(t)] = \sin t$. This is the case for $A = 0$ and $B = -1/2$. Thus,

$$x_p(t) = -\frac{1}{2} t \cos t$$

is a particular solution. Therefore, a general solution of (10.8.15) is

$$x(t) = x_h(t) + x_p(t) = c_1 \cos t + c_2 \sin t - \frac{1}{2} t \cos t. \qquad \blacklozenge$$

Aside from the exceptions noted in some of the examples, Table 10.1 lists the form of a particular solution $x_p(t)$ in the right-hand column corresponding to the forcing function $g(t)$ in the left-hand column.

Example 10.8.6 Find a general solution of

$$x'' - 2x' = 8te^{2t}. \tag{10.8.18}$$

Solution Heeding the advice of the rule of thumb noted earlier, let us first find a general solution of the associated homogeneous differential equation

$$(D^2 - 2D)x = 0. \tag{10.8.19}$$

Since the characteristic equation is $\lambda(\lambda - 2) = 0$, the characteristic roots are $\lambda_1 = 0$ and $\lambda_2 = 2$. And so a general solution of (10.8.19) is

$$x_h(t) = c_1 + c_2 e^{2t}. \tag{10.8.20}$$

Table 10.1 Form of a particular solution of $L[x] = g(t)$

$g(t)$	$x_p(t)$
ae^{kt}	Ae^{kt}
$p_n(t)$[a]	$A_n t^n + A_{n-1} t^{n-1} + \cdots + A_1 t + A_0$
$a \cos \omega t + b \sin \omega t$	$A \cos \omega t + B \sin \omega t$ (always include both terms even if $a = 0$ or $b = 0$)
$e^{kt} p_n(t)$	$e^{kt}\left(A_n t^n + A_{n-1} t^{n-1} + \cdots + A_1 t + A_0\right)$
$e^{kt}(a \cos \omega t + b \sin \omega t)$	$e^{kt}\left(A \cos \omega t + B \sin \omega t\right)$
$p_n(t)(a \cos \omega t + b \sin \omega t)$	$\left(A_n t^n + A_{n-1} t^{n-1} + \cdots + A_1 t + A_0\right) \cos \omega t$ $+ \left(B_n t^n + B_{n-1} t^{n-1} + \cdots + B_1 t + B_0\right) \sin \omega t$

[a] $p_n(t)$ denotes a polynomial of degree n: $p_n(t) = a_n t^n + a_{n-1} t^{n-1} + \cdots + a_1 t + a_0$

Since the right-hand side of (10.8.18) is $8te^{2t}$, which in Table 10.1 is of the form

$$g(t) = e^{kt} p_n(t),$$

take $k = 2$ and $n = 1$. Then, according to the entry just to its right, a trial particular solution has the form

$$x_p(t) = e^{2t} (A_1 t + A_0).$$

So as not to have to contend with subscripts, let us rewrite this as

$$x_p(t) = (At + B)e^{2t} = Ate^{2t} + Be^{2t}.$$

However, no matter what values are assigned to A and B, this trial solution is not a solution of (10.8.18). It is left to the reader to confirm this. The reason is that the term Be^{2t} is a solution of the associated homogeneous equation (10.8.19), which we can see from (10.8.20). But this is easily rectifiable: simply multiply $x_p(t)$ by t. As a result, the new trial particular solution is

$$x_p(t) = At^2e^{2t} + Bte^{2t}. \tag{10.8.21}$$

Note that neither of its terms is a solution of (10.8.19).[10] We leave it as an exercise to verify that

$$x''_p - 2x'_p = 4Ate^{2t} + (2A + 2B)e^{2t}.$$

By matching its coefficients with those of (10.8.18), we get

$$4A = 8, \quad 2A + 2B = 0.$$

Thus, $A = 2$ and $B = -2$. So a particular solution of (10.8.18) is

$$x_p(t) = 2t^2e^{2t} - 2te^{2t} = (2t^2 - 2t)e^{2t}.$$

Therefore, a general solution of (10.8.18) is

$$x(t) = c_1 + c_2e^{2t} + 2(t^2 - t)e^{2t}.$$ ♦

The next theorem will enable us to build a trial particular solution using Table 10.1 when the forcing function of a given nonhomogeneous equation is a mix of different functions of the type listed in the left-hand column of the table.

---

[10] Had this not been the case, we would have to multiply again by t, that is, (10.8.21) by t.

Superposition Principle for Nonhomogeneous Equations

Theorem 10.8.1 *If $x_p(t)$ is a solution of*

$$ax'' + bx' + cx = g(t)$$

and $x_q(t)$ a solution of

$$ax'' + bx' + cx = h(t),$$

then $x_p(t) + x_q(t)$ is a solution of

$$ax'' + bx' + cx = g(t) + h(t).$$

Proof According to the hypotheses,

$$L[x_p(t)] = g(t) \quad \text{and} \quad L[x_q(t)] = h(t),$$

where $L[x] := ax'' + bx' + cx$. Thus,

$$L[x_p(t) + x_q(t)] = L[x_p(t)] + L[x_q(t)] = g(t) + h(t),$$

which proves the theorem since this says that $x_p(t) + x_q(t)$ is a solution of

$$L[x] = g(t) + h(t).$$

■

Example 10.8.7 Find a general solution of

$$x'' + x = 2t^2 - 1 + \sin t. \tag{10.8.22}$$

Solution We have already determined in Example 10.8.5 that a general solution of the associated homogeneous equation $x'' + x = 0$ is

$$x_h(t) = c_1 \cos t + c_2 \sin t.$$

According to Theorem 10.8.1, a particular solution of (10.8.22) is the sum of particular solutions of the following two nonhomogeneous equations:

(i) $x'' + x = 2t^2 - 1$
(ii) $x'' + x = \sin t$.

First, let us find a particular solution of equation (i). From Table 10.1, a trial particular solution is $x_p(t) = At^2 + Bt + C$. Substituting this into (i), we get

$$x_p'' + x_p = At^2 + Bt + (2A + C).$$

Setting the right-hand side equal to the right-hand side of (i), we have

$$At^2 + Bt + (2A + C) = 2t^2 - 1.$$

Equating coefficients of like terms yields

$$A = 2, \quad B = 0, \quad 2A + C = -1.$$

And so $A = 2$, $B = 0$, $C = -5$. Thus, a particular solution of (i) is

$$x_p(t) = 2t^2 - 5.$$

In Example 10.8.5, we found that a particular solution of (ii) is

$$x_q(t) = -\frac{1}{2}t\cos t.$$

It follows from the Superposition Principle that a particular solution of (10.8.22) is

$$x_p(t) + x_q(t) = 2t^2 - 5 - \frac{1}{2}t\cos t.$$

Therefore,

$$x(t) = x_h(t) + [x_p(t) + x_q(t)] = c_1\cos t + c_2\sin t + 2t^2 - 5 - \frac{1}{2}t\cos t$$

is a general solution of (10.8.22). ♦

Example 10.8.8 Find a general solution of

$$5x'' + 3x' - 2x = 14t^3 - 12e^{-t} + 6. \qquad (10.8.23)$$

Solution The roots of the characteristic equation for the associated homogeneous equation

$$5x'' + 3x' - 2x = 0 \qquad (10.8.24)$$

are $\lambda_1 = -1$ and $\lambda_2 = \frac{2}{5}$. Hence, a general solution of (10.8.24) is

$$x_h(t) = c_1e^{-t} + c_2e^{2t/5}. \qquad (10.8.25)$$

Now we are ready to find a particular solution of (10.8.23). The first step is to inspect the right-hand side of (10.8.23). We note that its terms can be rearranged as the sum of the cubic polynomial $14t^3 + 6$ and the exponential function $-12e^{-t}$. And so we have the sum of two of the types of functions that are listed in Table 10.1 under the column heading $g(t)$.

First, let us find a particular solution of

$$5x'' + 3x' - 2x = 14t^3 + 6. \tag{10.8.26}$$

Because of the cubic polynomial, Table 10.1 suggests the trial particular solution

$$x_p(t) = At^3 + Bt^2 + Ct + D. \tag{10.8.27}$$

Since none of its terms look like any of the terms of (10.8.25), this is the correct trial particular solution. The result of substituting it and its first and second derivatives into the left-hand side of (10.8.26) is

$$5x_p'' + 3x_p' - 2x_p = \\ -2At^3 + (9A - 2B)t^2 + (30A + 6B - 2C)t + (10B - 3C - 2D). \tag{10.8.28}$$

Equating the coefficients of like terms on the right-hand sides of (10.8.26) and (10.8.28), we obtain the system

$$\begin{aligned} -2A &= 14 \\ 9A - 2B &= 0 \\ 30A + 6B - 2C &= 0 \\ 10B + 3C - 2D &= 6. \end{aligned}$$

The solution of the system is

$$A = -7,\ B = -\frac{63}{2},\ C = -\frac{399}{2},\ D = -\frac{1839}{4}.$$

Hence, a particular solution of (10.8.26) is

$$x_p(t) = -7t^3 - \frac{63}{2}t^2 - \frac{399}{2}t - \frac{1839}{4}. \tag{10.8.29}$$

Next, let us find a particular solution of

$$5x'' + 3x' - 2x = -12e^{-t}. \tag{10.8.30}$$

Due to the exponential function on the right-hand side, $x_q(t) = Ee^{-t}$ appears to be a viable trial particular solution. But when we compare $x_q(t)$ with (10.8.25), we see that it is a solution of (10.8.24), irrespective of the value of E. So it cannot be a solution of (10.8.30) as well. A way out of this predicament is to modify $x_q(t)$ by multiplying it by the independent variable t and then checking to see if

$$x_q(t) = Ete^{-t}$$

is a solution of (10.8.24). Since (10.8.25) shows that it is not, let us substitute it into the left-hand side of (10.8.30). It is left as an exercise to verify that the result of this is

$$5x_q'' + 3x_q' - 2x_q = -7Ee^{-t}.$$

Thus, $x_q(t)$ is a solution of (10.8.30) if $-7E = -12$. It follows that

$$x_q(t) = \frac{12}{7}te^{-t} \tag{10.8.31}$$

is a particular solution of (10.8.30).

It follows from the Superposition Principle that a particular solution of equation (10.8.23) is the sum of the particular solutions x_p and x_q of equations (10.8.29) and (10.8.31), respectively. Adding this sum to the general solution (10.8.25) of the associated homogeneous equation (10.8.24), we obtain

$$x(t) = c_1e^{-t} + c_2e^{2t/5} - 7t^3 - \frac{63}{2}t^2 - \frac{399}{2}t - \frac{1839}{4} + \frac{12}{7}te^{-t},$$

which according to Theorem 10.7.1 is a general solution of (10.8.23). ♦

10.9 Method of Variation of Parameters

In this final section we present another well-known and widely-used method for finding particular solutions of nonhomogeneous differential equations. It is called the *method of variation of parameters*. The steps in this method are slightly more involved than those in the method of undetermined coefficients, but it can be used to solve certain types of nonhomogeneous equations when the latter method fails for some reason. For instance, in view of the derivatives of $\sec t$ and $\tan t$, the method of undetermined coefficients cannot be used to find a particular solution of the differential equation

$$x'' + x = \sec t$$

since no linear combination of trigonometric functions when summed with its second derivative equals $\sec t$ or a constant multiple of this function. However, the method of variation of parameters can be used to solve this equation, which we will show later on in Example 10.9.7. Furthermore, unlike the method of undetermined coefficients, the method of variation of parameters is not restricted to equations with constant coefficients. By and large, differential equations with variable coefficients do not have analytical solutions that can be expressed in terms of finitely many elementary functions. Even for the equations that do, no general method exists for obtaining these solutions. However, special methods exist for some types of equations with variable coefficients. For example, in Sect. 10.5, we presented a method for finding analytical solutions of homogeneous Cauchy-Euler equations. As a result, it turns out that the method of variation of parameters can be used to find particular solutions of nonhomogeneous Cauchy-Euler equations. In fact, this method will enable us to find a particular solution of a nonhomogeneous equation (with constant or with variable coefficients) if (i) a general analytical solution of its associated homogeneous equation exists and can be found and if (ii) the integrals resulting from applying the method can be evaluated.

In Chap. 5 we derived the variation of parameters formula for first-order nonhomogeneous linear equations. Likewise, we can derive a variation of parameters formula for second-order nonhomogeneous linear equations. But first let us review the method for first-order equations. Consider

$$\frac{dx}{dt} + 2tx = 5e^{-t^2}. \tag{10.9.1}$$

The method of variation of parameters presupposes we have (or can find) a solution of the associated homogeneous equation, which in this case is

$$\frac{dx}{dt} + 2tx = 0. \tag{10.9.2}$$

It is easy to show by separating variables or by using the integrating factor method that a general solution of the homogeneous equation (10.9.2) is

$$x_h(t) = ke^{-t^2},$$

where k is an arbitrary constant. Now let us replace k with a function $u(t)$ to look for a particular solution of (10.9.1) of the form

$$x_p(t) = u(t)e^{-t^2}. \tag{10.9.3}$$

This may be viewed as *varying the parameter k* (or *varying the constant k*). Substituting $x_p(t)$ and its first derivative into Eq. (10.9.1) yields

$$x_p'(t) + 2tx_p(t) = \left(-2tu(t)e^{-t^2} + u'(t)e^{-t^2}\right) + 2tu(t)e^{-t^2} = u'(t)e^{-t^2}.$$

On comparing its right-hand side to that of Eq. (10.9.1), it can be seen that (10.9.3) is a solution of (10.9.1) provided

$$u'(t) = 5.$$

One such function is $u(t) = 5t$. As a result,

$$x_p(t) = u(t)e^{-t^2} = 5te^{-t^2}$$

is a particular solution of (10.9.1) and

$$x(t) = x_h(t) + x_p(t) = (k + 5t)e^{-t^2}$$

is a general solution.

Now we can show how to extend this method to the second-order linear nonhomogeneous equation

$$a_2(t)\frac{d^2x}{dt^2} + a_1(t)\frac{dx}{dt} + a_0(t)x = b(t),$$

where a_0, a_1, a_2, and b denote functions that are continuous on an open interval J. Moreover, let us assume $a_2(t) \neq 0$ for all $t \in J$. As with first-order nonhomogeneous linear equations, finding a particular solution depends on having a general solution of the associated homogeneous equation

$$a_2(t)\frac{d^2x}{dt^2} + a_1(t)\frac{dx}{dt} + a_0(t)x = 0$$

for $t \in J$. Under the hypothesis of continuity of the coefficients, solutions of this equation exist. The result that allows us to make this claim is the following existence and uniqueness theorem. No proof is given here since doing so would take us too far afield from the present discussion.

Existence and Uniqueness of Solutions

Theorem 10.9.1 *Let $a_0(t)$, $a_1(t)$, $a_2(t)$, and $b(t)$ be continuous functions on an open interval J, where $a_2(t) \neq 0$ for all $t \in J$. For any $t_0 \in J$ and any real numbers ζ and η, a solution of the differential equation*

$$a_2(t)\frac{d^2x}{dt^2} + a_1(t)\frac{dx}{dt} + a_0(t)x = b(t) \tag{10.9.4}$$

(continued)

exists satisfying the initial conditions

$$x(t_0) = \zeta, \quad x'(t_0) = \eta. \tag{10.9.5}$$

Furthermore, there is only one such solution and it is defined and satisfies (10.9.4) *on all of* J.

As we shall see shortly, we can express all solutions of the homogeneous equation associated with (10.9.4) as a linear combination of two of its solutions. Moreover, any two will do as long as neither is equal to a constant times the other. The mathematical term for this is ***linear independence***.

Linear Independent and Dependent Functions

Definition 10.9.1 Two functions f and g are said to be ***linearly independent*** on an interval J provided neither is a constant multiple of the other on the interval. Otherwise, they are said to be ***linearly dependent*** on J. That is, f and g are ***linearly dependent*** on J if there is a constant k such that either

$$f(t) = kg(t) \quad \text{or} \quad g(t) = kf(t)$$

for all $t \in J$.

In the following examples, the letter a denotes either a real number or $-\infty$ while b denotes either a real number or $+\infty$.

Example 10.9.1 Consider the functions $f(t) = \sin t$ and $g(t) = \cos t$ on an interval $J = (a, b)$. Suppose a constant k exists such that $\sin t = k \cos t$ for $t \in J$. But this contradicts the fact that $\tan t$ is not constant on intervals. We conclude that no such k exists. Likewise, there does not exist a constant k such that $\cos t = k \sin t$ for $t \in J$. Thus, $\sin t$ and $\cos t$ are linearly independent on every interval (a, b). ♦

Example 10.9.2 Let $f(t) = \sin t$ and $g(t) \equiv 0$ on an interval (a, b). Certainly no constant k exists such that $f(t) = \sin t = kg(t)$ since $g(t) = 0$ for all $t \in (a, b)$. On the other hand,

$$g(t) = 0 \cdot \sin t = 0 \cdot f(t)$$

for all $t \in (a, b)$. Therefore f and g are linearly dependent on (a, b). ♦

Example 10.9.3 The functions $\sin 2t$ and $\sin t \cos t$ on linearly dependent on (a, b). This follows immediately from the double-angle formula: $\sin 2t = 2 \sin t \cos t$. ♦

There is a convenient way to test if any two given functions are linear independent. It involves a function that is known as the *Wronskian*, named after the Polish-born French mathematician Josef Hoëné de Wronski (1778–1853).

Wronskian

Definition 10.9.2 Let f and g be two differentiable functions. The function

$$W(t) := f(t)g'(t) - f'(t)g(t), \tag{10.9.6}$$

is called the ***Wronskian*** of the functions.

If we are dealing with more than two functions, say with f, g, and h, then we can indicate the Wronskian of two of them, say f and g, with the more explicit notation $W[f, g](t)$. A convenient way to remember the Wronskian is to express it as a *determinant*. A two by two ***determinant***—the determinant of a matrix with two rows and two columns—is computed by subtracting the product of the entries on the secondary diagonal from the product of the entries on the main diagonal. For example, the Wronskian of f and g is

$$W[f, g](t) = \begin{vmatrix} f(t) & g(t) \\ f'(t) & g'(t) \end{vmatrix} = f(t)g'(t) - f'(t)g(t).$$

Criterion for Linearly Independent Functions

Theorem 10.9.2 *Let f and g be a pair of functions that are differentiable on an interval J. If there is at least one point in J where the Wronskian*

$$W(t) = \begin{vmatrix} f(t) & g(t) \\ f'(t) & g'(t) \end{vmatrix} \tag{10.9.7}$$

is not zero, then f and g are linearly independent on J.

Proof Suppose there is a point $t_0 \in J$ where the Wronskian of f and g is not zero:

$$W(t_0) = \begin{vmatrix} f(t_0) & g(t_0) \\ f'(t_0) & g'(t_0) \end{vmatrix} = f(t_0)g'(t_0) - f'(t_0)g(t_0) \neq 0.$$

To show this implies f and g are linearly independent on J, suppose the contrary, namely, that there is a constant k such that either

$$f(t) = kg(t) \quad \text{or} \quad g(t) = kf(t)$$

for all $t \in J$. Without loss of generality, suppose that $f(t) = kg(t)$ (if otherwise, merely interchange the names of the functions). Differentiating, we get

$$f'(t) = kg'(t)$$

for all $t \in J$. In particular, at $t = t_0$, we have the pair of equations

$$f(t_0) - kg(t_0) = 0$$

$$f'(t_0) - kg'(t_0) = 0.$$

Multiplying the first equation by $g'(t_0)$ and the second one by $-g(t_0)$ and adding the results, we get

$$f(t_0)g'(t_0) - f'(t_0)g(t_0) = 0.$$

But this contradicts the supposition. Therefore, f and g must be linearly independent on the interval J. ■

Example 10.9.4 Compute the Wronkskian of the functions

$$f(t) = t \quad \text{and} \quad g(t) = t \ln t$$

on the interval $(0, \infty)$.

Solution From (10.9.7), we have

$$W[f, g](t) = \begin{vmatrix} f(t) & g(t) \\ f'(t) & g'(t) \end{vmatrix} = \begin{vmatrix} t & t\ln t \\ 1 & 1 + \ln t \end{vmatrix} = t(1 + \ln t) - t \ln t = t.$$

It is evident from Definition 10.9.1 that this pair of functions is linearly independent on the interval $(0, \infty)$. But note that this also follows from Theorem 10.9.2 since $W[f, g](t) \neq 0$ for all $t > 0$. ♦

We show next that if the functions $a_0(t)$, $a_1(t)$, and $a_2(t)$ of the nonhomogeneous equation (10.9.4) are continuous, then the associated homogeneous equation of this equation always has a pair of linearly independent solutions.

Linearly Independent Solutions of Homogeneous Equations

Theorem 10.9.3 *Let $a_0(t)$, $a_1(t)$, and $a_2(t)$ be continuous functions on an open interval J, where $a_2(t) \neq 0$ for all $t \in J$. Then two linearly independent solutions of the homogeneous linear equation*

$$a_2(t)\frac{d^2x}{dt^2} + a_1(t)\frac{dx}{dt} + a_0(t)x = 0 \tag{10.9.8}$$

exist on the interval J.

Proof Select any $t_0 \in J$. According to Theorem 10.9.1, there is a unique solution $x_1(t)$ of (10.9.8) satisfying the initial conditions: $x_1(t_0) = 1$, $x_1'(t_0) = 0$. Likewise, there is a unique solution $x_2(t)$ with $x_2(t_0) = 0$ and $x_2'(t_0) = 1$. The value of the Wronskian of these two solutions at $t = t_0$ is

$$W(t_0) = \begin{vmatrix} x_1(t_0) & x_2(t_0) \\ x_1'(t_0) & x_2'(t_0) \end{vmatrix} = \begin{vmatrix} 1 & 0 \\ 0 & 1 \end{vmatrix} = 1.$$

Since $W(t_0) \neq 0$, the solutions $x_1(t)$ and $x_2(t)$ are linearly independent on J. ■

It can be proven that a general solution of a homogeneous equation can be expressed as a linear combination of any two of its linearly independent solutions. This is the content of the next theorem.

General Solutions of Homogeneous Equations

Theorem 10.9.4 *Let $a_0(t)$, $a_1(t)$, and $a_2(t)$ be continuous functions on an open interval J, where $a_2(t) \neq 0$ for all $t \in J$. Let $x_1(t)$ and $x_2(t)$ be linearly independent solutions of* (10.9.8) *on J. If $x(t)$ is a solution of* (10.9.8) *on J, then constants c_1 and c_2 exist such that*

$$x_h(t) = c_1x_1(t) + c_2x_2(t). \tag{10.9.9}$$

for all $t \in J$.

Note the importance of Theorem 10.9.4. It states that all solutions of (10.9.8) become known once we have found two linearly independent solutions of it. In other words, if $x_1(t)$ and $x_2(t)$ are linearly independent solutions of (10.9.8), then every solution $x(t)$ of (10.9.8) can be expressed as a linear combination of these two solutions. For this reason, we say that (10.9.9) is a general solution of (10.9.8).

Example 10.9.5 Find a general solution of

$$x'' - x = 0. \tag{10.9.10}$$

Solution Clearly, $x_1(t) = e^t$ and $x_2(t) = e^{-t}$ are solutions for $t \in (-\infty, \infty)$. Since neither is a constant multiple of the other, they are linearly independent on $(-\infty, \infty)$. Thus, by Theorem 10.9.4, a general solution of (10.9.10) on $(-\infty, \infty)$ is

$$x(t) = c_1 e^t + c_2 e^{-t}. \tag{10.9.11}$$

Notice that Theorem 10.3.1 also yields this general solution. ♦

Engineers and physicists find it practical to express solutions of equations, such as (10.9.10), in terms of the hyperbolic functions. The three most common ones are the ***hyperbolic cosine***, ***hyperbolic sine***, and ***hyperbolic tangent*** functions, which are defined by

$$\cosh t := \frac{e^t + e^{-t}}{2}, \qquad \sinh t := \frac{e^t - e^{-t}}{2}, \qquad \tanh t := \frac{e^t - e^{-t}}{e^t + e^{-t}}, \tag{10.9.12}$$

respectively. It is left as an exercise to show that

$$\frac{d}{dt} \cosh t = \sinh t \quad \text{and} \quad \frac{d}{dt} \sinh t = \cosh t$$

and to verify that $x_1(t) = \cosh t$ and $x_2(t) = \sinh t$ are solutions of (10.9.10) for $t \in (-\infty, \infty)$. Since neither is a constant multiple of the other, $\cosh t$ and $\sinh t$ are linearly independent on $(-\infty, \infty)$. Therefore, it follows from Theorem 10.9.4 that

$$x(t) = c_1 \cosh t + c_2 \sinh t. \tag{10.9.13}$$

is also a general solution of (10.9.10) on $(-\infty, \infty)$.

To sum up, we have shown that both (10.9.11) and (10.9.13) are general solutions of (10.9.10). This means that any solution of (10.9.10) can be expressed either as a linear combination of the exponential functions e^t and e^{-t} or of the hyperbolic functions $\cosh t$ and $\sinh t$. That is why we use the phrase "a general solution" rather than "the general solution."

To find a particular solution of the nonhomogeneous equation (10.9.4), let us begin with two linearly independent solutions $x_1(t)$ and $x_2(t)$ of the associated homogeneous (10.9.8) on an interval J. Then $x_h(t)$ given by (10.9.9) is a general solution of this equation on J. As we have seen with first-order linear equations in Chap. 5, the idea behind the variation of parameters (or variation of constants) method is to think of c_1 and c_2 in (10.9.9) as parameters that can be varied, so to speak, that will allow us to somehow come up with a particular solution of equation (10.9.4). In other words, let us replace c_1 and c_2 in (10.9.9) with functions, say $u_1(t)$ and $u_2(t)$, thereby transforming (10.9.9) into

$$x_p(t) = u_1(t)x_1(t) + u_2(t)x_2(t), \tag{10.9.14}$$

which we can only hope at this point is the form of a particular solution of equation (10.9.4). Of course, not any two functions will do: somehow u_1 and u_2 have to be chosen so that (10.9.14) is actually a particular solution of (10.9.4). Now the question boils down to this: Do such functions exist—and if so, how do we go about finding them?

To facilitate answering these questions, we employ the differential operator

$$L = a_2(t)D^2 + a_1(t)D + a_0(t).$$

Then equations (10.9.4) and (10.9.8) can be written in the abbreviated forms

$$L[x(t)] = b(t) \quad \text{and} \quad L[x(t)] = 0.$$

To see if a function of the form (10.9.14) can be found that will satisfy the nonhomogeneous equation (10.9.4), let us substitute $x_p = u_1x_1 + u_2x_2$ and its derivatives

$$x_p' = (u_1x_1' + u_2x_2') + (u_1'x_1 + u_2'x_2)$$

$$x_p'' = (u_1x_1'' + u_2x_2'') + (u_1'x_1' + u_2'x_2') + \frac{d}{dt}(u_1'x_1 + u_2'x_2).$$

into the equation. For the sake of brevity, we have omitted writing the argument t of each of these functions. So as to simplify matters, let us eliminate some of the terms in both derivatives by imposing the following condition:

$$u_1'x_1 + u_2'x_2 = 0. \tag{10.9.15}$$

At this juncture, we do not know if this will work; nonetheless, let us forge on and see what will happen. Multiply x_p and its derivatives by their respective coefficients and then add the results as follows:

$$\begin{array}{rl} a_2(t)x_p'' &= a_2u_1x_1'' + a_2u_2x_2'' + a_2(u_1'x_1' + u_2'x_2') \\ a_1(t)x_p' &= a_1u_1x_1' + a_1u_2x_2' \\ + \ a_0(t)x_p &= a_0u_1x_1 + a_0u_2x_2 \\ \hline L[x_p] &= u_1L[x_1] + u_2L[x_2] + a_2(u_1'x_1' + u_2'x_2'). \end{array}$$

This simplifies to

$$L[x_p] = a_2(u_1'x_1' + u_2'x_2')$$

because $x_1(t)$ and $x_2(t)$ are solutions of the homogeneous equation $L[x] = 0$ (which is the abbreviated form of Eq. (10.9.8)). We conclude that (10.9.14) is a particular solution of (10.9.4) if there are functions $u_1(t)$ and $u_2(t)$ that satisfy the condition

$$a_2(u_1'x_1' + u_2'x_2') = b(t), \tag{10.9.16}$$

as well as the other condition that we imposed earlier, namely, (10.9.15).

So let us see if there are functions u_1 and u_2 whose derivatives are solutions of the system of equations

$$\begin{aligned} x_1u_1' + x_2u_2' &= 0 \\ x_1'u_1' + x_2'u_2' &= b(t)/a_2(t). \end{aligned}$$

First, let us solve for the unknown u_1' by eliminating u_2'. Multiplying the first equation by x_2' and the second one by $-x_2$ and then adding the results, we obtain

$$\begin{array}{rl} x_1x_2'u_1' + x_2x_2'u_2' &= 0 \\ + \ -x_1'x_2u_1' - x_2x_2'u_2' &= -\dfrac{b(t)x_2(t)}{a_2(t)} \\ \hline u_1'(x_1x_2' - x_1'x_2) &= -\dfrac{b(t)x_2(t)}{a_2(t)}. \end{array}$$

Likewise, eliminating u_1', we have

$$u_2'(x_1x_2' - x_1'x_2) = \frac{b(t)x_1(t)}{a_2(t)}.$$

It follows that the system has a solution on the interval J if the Wronskian

$$W(t) = x_1(t)x_2'(t) - x_1'(t)x_2(t) \neq 0 \tag{10.9.17}$$

for all $t \in J$. Suppose this to be true. Then solving for u_1' and u_2' and integrating, we obtain

$$u_1(t) = -\int \frac{b(t)x_2(t)}{a_2(t)W(t)}\,dt, \quad u_2(t) = \int \frac{b(t)x_1(t)}{a_2(t)W(t)}\,dt.$$

From this and (10.9.14), it follows that a particular solution of the nonhomogeneous equation (10.9.4) is given by the formula

$$x_p(t) = -x_1(t)\int \frac{b(t)x_2(t)}{a_2(t)W(t)}\,dt + x_2(t)\int \frac{b(t)x_1(t)}{a_2(t)W(t)}\,dt, \tag{10.9.18}$$

provided $W(t) \neq 0$ on the interval J. This is the case as it turns out because the solutions $x_1(t)$ and $x_2(t)$ of the homogeneous equation (10.9.8) are linearly independent on J. Although we do not prove this here, it is stated in the next theorem.

Wronskian and Linear Independence

Theorem 10.9.5 *Two solutions $x_1(t)$ and $x_2(t)$ of the homogeneous linear equation* (10.9.8) *on an interval J are linearly independent if and only if $W[x_1, x_2](t) \neq 0$ for all $t \in J$.*

The proof of this theorem is based on the existence-and-uniqueness theorem stated earlier in this section, namely, Theorem 10.9.1. Its proof is similar to the proofs of Theorems 12.5.23 and 12.5.25 in Chap. 12.

Due to the complexity of formula (10.9.18) and the likelihood of either forgetting or misplacing the minus sign, we suggest not committing it to memory. Instead it is easier to memorize the system of equations (10.9.15) and (10.9.16) due to their similarity. In summary, we have the following method for obtaining particular solutions of nonhomogeneous linear second-order equations.

Method of Variation of Parameters

Let $x_1(t)$ and $x_2(t)$ be linearly independent solutions of

$$a_2(t)x'' + a_1(t)x' + a_0(t)x = 0 \tag{10.9.19}$$

on an interval J that excludes points where $a_2(t) = 0$. A particular solution of

$$a_2(t)x'' + a_1(t)x' + a_0(t)x = b(t) \tag{10.9.20}$$

of the form

$$x_p(t) = u_1(t)x_1(t) + u_2(t)x_2(t) \tag{10.9.21}$$

can be obtained on the interval J by completing the following steps:

1. Solve the system of equations

$$\begin{aligned} u_1'x_1 + u_2'x_2 &= 0 \\ u_1'x_1' + u_2'x_2' &= b(t)/a_2(t) \end{aligned} \tag{10.9.22}$$

   for u_1' and u_2'.
2. Find the functions u_1 and u_2 by integrating u_1' and u_2' with respect to t.
3. Substitute u_1 and u_2 into (10.9.21) to obtain a particular solution of the nonhomogeneous equation (10.9.20) on J.

Example 10.9.6 Find a particular solution of the Cauchy-Euler equation

$$t^2x'' + tx' - 4x = 6t \tag{10.9.23}$$

on the interval $(0, \infty)$.

Solution Referring to Eq. (10.9.20), we see that

$$a_2(t) = t^2, \ a_1(t) = t, \ a_0(t) = -4, \ \text{and} \ b(t) = 6t.$$

In Example 10.5.1, it is shown that a general solution of the associated homogeneous equation

$$t^2x'' + tx' - 4x = 0,$$

on the interval $(0, \infty)$ is

$$x_h(t) = \frac{c_1}{t^2} + c_2 t^2.$$

With $x_1(t) = t^{-2}$ and $x_2(t) = t^2$, system (10.9.22) is

$$\begin{aligned} t^{-2}u_1' + t^2 u_2' &= 0 \\ -2t^{-3}u_1' + 2tu_2' &= \frac{6t}{t^2}. \end{aligned} \tag{10.9.24}$$

Now solve for $u_1'(t)$ and $u_2'(t)$. With the aim of eliminating u_2', let us multiply the first equation by -2 and the second one by t, thereby obtaining

$$\begin{aligned} -2t^{-2}u_1' - 2t^2u_2' &= 0 \\ -2t^{-2}u_1' + 2t^2u_2' &= 6. \end{aligned}$$

Adding the two equations yields

$$-4t^{-2}u_1' = 6 \quad \Rightarrow \quad u_1' = -\frac{3}{2}t^2.$$

Then integrating, we have

$$u_1 = -\frac{3}{2}\int t^2\,dt = -\frac{1}{2}t^3.$$

Substituting this into the first equation of system (10.9.24) gives

$$t^{-2}\left(-\tfrac{3}{2}t^2\right) + t^2u_2' = 0 \quad \Rightarrow \quad u_2' = \frac{3}{2t^2}.$$

Thus,

$$u_2 = \frac{3}{2}\int t^{-2}\,dt = -\frac{3}{2}t^{-1}.$$

Since

$$u_1(t)x_1(t) + u_2(t)x_2(t) = -\frac{1}{2}t^3 \cdot t^{-2} - \frac{3}{2}t^{-1} \cdot t^2 = -2t,$$

a particular solution of (10.9.23) is

$$x_p(t) = -2t. \qquad \blacklozenge$$

In the derivation of the formulas, the use of u_1 and u_2 made it easy to remember that u_1 is paired with x_1 and u_2 with x_2. Nonetheless, it is awkward dealing with these subscripts when actually solving equations. For this reason, we will use u and v rather than u_1 and u_2 from now on.

Example 10.9.7 Find a general solution of

$$x'' + x = \sec t \tag{10.9.25}$$

on the interval $[0, \pi/2)$.

Solution The associated homogeneous equation is

$$(D^2 + 1)x = 0.$$

Since its characteristic equation is $\lambda^2 + 1 = 0$, a general solution of this equation is

$$x_h(t) = c_1 \cos t + c_2 \sin t.$$

Now we look for a particular solution of equation (10.9.25) of the form

$$x_p(t) = u(t) \cos t + v(t) \sin t,$$

where the derivatives of u and v satisfy the system of equations

$$\begin{aligned} u' \cos t + v' \sin t &= 0 \\ -u' \sin t + v' \cos t &= \sec t. \end{aligned}$$

Solving for u and v using Cramer's rule,[11] we get

$$u' = \frac{\begin{vmatrix} 0 & \sin t \\ \sec t & \cos t \end{vmatrix}}{\begin{vmatrix} \cos t & \sin t \\ -\sin t & \cos t \end{vmatrix}} = -\frac{\sin t \cdot \sec t}{\cos^2 t + \sin^2 t} = -\frac{\sin t}{\cos t}$$

[11] See Problem 85.

and

$$v' = \frac{\begin{vmatrix} \cos t & 0 \\ -\sin t & \sec t \end{vmatrix}}{\begin{vmatrix} \cos t & \sin t \\ -\sin t & \cos t \end{vmatrix}} = \cos t \cdot \sec t = 1.$$

As a result,

$$u = -\int \frac{\sin t}{\cos t}\, dt = \ln|\cos t| = \ln(\cos t) \quad \text{and} \quad v = \int dt = t.$$

Note that we dropped the absolute bars because $\cos t > 0$ for $0 \le t < \pi/2$. Hence, a particular solution of (10.9.25) is

$$x_p(t) = \ln(\cos t) \cdot \cos t + t \sin t.$$

It follows from Theorem 10.7.1 that a general solution of (10.9.25) is

$$x(t) = c_1 \cos t + c_2 \sin t + \ln(\cos t) \cdot \cos t + t \sin t. \qquad \blacklozenge$$

Example 10.9.8 Find a general solution of

$$x'' - 6x' + 9x = t^{-1}e^{3t} \tag{10.9.26}$$

on the interval $(-\infty, 0)$.

Solution The associated homogeneous equation is

$$(D^2 - 6D + 9)x = 0. \tag{10.9.27}$$

Since its characteristic equation is

$$(\lambda^2 - 6\lambda + 9) = (\lambda - 3)^2 = 0,$$

$\lambda = 3$ is a repeated root. Consequently, (10.9.27) has the linearly independent solutions e^{3t} and te^{3t}, which means

$$x_h(t) = c_1 e^{3t} + c_2 t e^{3t}.$$

is a general solution. Now we look for a particular solution of (10.9.26) of the form

$$x_p(t) = u(t)e^{3t} + v(t)te^{3t},$$

where the derivatives of u and v satisfy the system of equations

$$u'e^{3t} + v'te^{3t} = 0$$
$$u'(3e^{3t}) + v'(3te^{3t} + e^{3t}) = \frac{t^{-1}e^{3t}}{1}.$$

Multiplying each of these equations by $\exp(-3t)$, we have

$$u' + tv' = 0$$
$$3u' + (3t+1)v' = t^{-1}.$$

Using Cramer's rule and then integrating, we obtain

$$u' = \frac{\begin{vmatrix} 0 & t \\ t^{-1} & 3t+1 \end{vmatrix}}{\begin{vmatrix} 1 & t \\ 3 & 3t+1 \end{vmatrix}} = \frac{0-1}{(3t+1)-3t} = -1 \quad \Rightarrow \quad u = -t$$

and

$$v' = \frac{\begin{vmatrix} 1 & 0 \\ 3 & t^{-1} \end{vmatrix}}{\begin{vmatrix} 1 & t \\ 3 & 3t+1 \end{vmatrix}} = t^{-1} \quad \Rightarrow \quad v = \int t^{-1}\,dt = \ln|t| = \ln(-t).$$

Therefore, a particular solution of (10.9.26) is

$$x_p(t) = ue^{3t} + vte^{3t} = -te^{3t} + \ln(-t)\cdot te^{3t} = [-t + t\ln(-t)]\,e^{3t}.$$

Consequently, a general solution of equation (10.9.26) on the interval $(-\infty, 0)$ is

$$x(t) = [c_1 + c_2 t - t + t\ln(-t)]\,e^{3t}.$$

Factoring out t from the middle two terms and then replacing $(c_2 - 1)$ with c_2, this simplifies to

$$x(t) = [c_1 + c_2 t + t\ln(-t)]\,e^{3t}. \qquad \blacklozenge$$

Problems

> Something else an academic education will do for you. If you go along with it any considerable distance, it'll begin to give you an idea what size mind you have. What it'll fit and, maybe, what it won't. After a while, you'll have an idea what kind of thoughts your particular size mind should be wearing. For one thing, it may save you an extraordinary amount of time trying on ideas that don't suit you, aren't becoming to you. You'll begin to know your true measurements and dress your mind accordingly.
>
> Mr. Antolini's advice to Holden Caulfield
>
> *The Catcher in the Rye* by J. D. Salinger (ch. 24)

General Solution

1. Let λ_1 and λ_2 denote the roots of the characteristic equation for (10.3.15). If they are real and distinct, verify that the function given by (10.3.16) is a solution of (10.3.15) (for all values of c_1 and c_2) by substituting it into the left-hand side of (10.3.15).
2. Suppose that $\lambda = -b/2a$ is a repeated root of the characteristic equation for the homogeneous equation (10.3.18). Verify that (10.3.19) is a solution of this equation (for all values of c_1 and c_2) by substituting it into the left-hand side of (10.3.18).

Second-Order Homogeneous Equations

In Problems 3–26, find a general solution of the differential equation. All derivatives are with respect to t, unless otherwise indicated.

3. $\dfrac{d^2x}{dt^2} - 9\dfrac{dx}{dt} = 0$
4. $x'' + 3x' - 10x = 0$
5. $9\dfrac{d^2p}{ds^2} + 4p = 0$
6. $15\dfrac{d^2y}{dx^2} + 44\dfrac{dy}{dx} - 36y = 0$
7. $2x'' = 3x - 5x'$
8. $2y'' - 7y' = 0$
9. $\dfrac{d^2y}{dt^2} - \dfrac{dy}{dt} + 7y = 0$
10. $20y'' - 33y' - 27y = 0$
11. $4\dfrac{d^2Q}{dt^2} - 12\dfrac{dQ}{dt} + 9Q = 0$
12. $\dfrac{d^2x}{dt^2} + 5x = 0$
13. $12\ddot{s} + \dot{s} - 35s = 0$
14. $3\dfrac{d^2x}{dt^2} + 7\dfrac{dx}{dt} = 4x$
15. $2y'' = -2y' - 3y$
16. $x'' + 4x' + 4x = 0$
17. $2x'' = 3x - x'$
18. $3x'' - 4x' + 12x = 0$
19. $x'' - 6x' + 13x = 0$
20. $x'' - 4x = 0$
21. $9\ddot{x} + 30\dot{x} + 25x = 0$
22. $9\dfrac{d^2y}{dx^2} + 18\dfrac{dy}{dx} + 10y = 0$
23. $4\dfrac{d^2y}{dx^2} - 20\dfrac{dy}{dx} + 25y = 0$
24. $\dfrac{d^2p}{ds^2} = 5\dfrac{dp}{ds}$
25. $2x'' = x' - 3x$
26. $3x'' + 7x' - 20x = 0$

Second-Order Initial Value Problems

In Problems 27–31, find the solution of the given initial value problem, where the independent variable is t.

27. $x'' + x' = 0; x(0) = 2,\ x'(0) = 1$
28. $x'' - 2x' + x = 0; x(0) = 0,\ x'(0) = -1$
29. $x'' - 2x' + 2x = 0; x(\pi) = 1,\ x'(\pi) = 0$
30. $y'' + 5y' - 14y = 0; y(0) = 5,\ y'(0) = 1$
31. $\ddot{x} + x = 0; x(\pi) = -5,\ x'(\pi) = 2$

Complex Exponential Functions

32. Use Euler's formula (10.4.11) to derive the double-angle formulas:

$$\sin 2t = 2\sin t\cos t$$

$$\cos 2t = \cos^2 t - \sin^2 t.$$

33. Use (10.4.6) and (10.4.10) to derive the trigonometric half-angle formulas:

$$\cos^2 t = \frac{1+\cos 2t}{2}$$

$$\sin^2 t = \frac{1-\cos 2t}{2}.$$

34. Verify the formulas in (10.4.13), where e^{it} and e^{-it} are defined by (10.4.11) and (10.4.12), respectively.

Homogeneous Cauchy-Euler Equations

In Problems 35–38, find a general solution of each equation on the interval $(0, \infty)$.

35. $t^2\dfrac{d^2x}{dt^2} + t\dfrac{dx}{dt} + x = 0$
36. $4t^2x'' - 3tx' - 2x = 0$
37. $5t^2x'' - 3tx' + 3x = 0$
38. $t^2\dfrac{d^2x}{dt^2} - 5t\dfrac{dx}{dt} + 9x = 0$
39. (a) Find a general solution of

$$t^2x'' - tx' + x = 0.$$

(b) Use the answer to part (a) to find the solution satisfying the initial conditions:

$$x(1) = 3,\ x'(1) = 1.$$

40. Solve the initial value problem:

$$t^2\frac{d^2x}{dt^2} + 3t\frac{dx}{dt} + 2x = 0;$$

$$x(1) = 2,\ x'(1) = 5.$$

41. Prove: If $x(t)$ is a solution of the Cauchy-Euler equation (10.5.2) for $t > 0$, then the function $y(t) := x(-t)$ is a solution for $t < 0$.

Homogeneous Equations of Higher Order

In Problems 42–47, find a general solution of the equation. All derivatives are with respect to t*, unless otherwise indicated.*

42. $x''' + x'' - 8x' - 12x = 0$
43. $\dfrac{d^4x}{dt^4} - 4\dfrac{d^3x}{dt^3} + 5\dfrac{d^2x}{dt^2} - 4\dfrac{dx}{dt} + 4x = 0$
44. $\dddot{x} + \ddot{x} + 4\dot{x} + 4x = 0$
45. $y''' - y'' + 4y' - 4y = 0 \ \left(y' = \dfrac{dy}{dx}\right)$
46. $\dfrac{d^3x}{dt^3} + 2\dfrac{d^2x}{dt^2} - \dfrac{dx}{dt} - 2x = 0$
47. $2x''' - 3x'' + 8x' + 5x = 0$

Higher-Order Differential Equations: Initial Value Problems

In Problems 48–50, find a solution of the initial value problem. All derivatives are with respect to the independent variable t.

48. $4x''' - 11x' + 10x = 0;\ x(0) = -37,\ x'(0) = 0,\ x''(0) = 74$
49. $x''' - x' = 0;\ x(0) = 6,\ x'(0) = 1,\ x''(0) = 2$
50. $2x^{(4)} - x''' - 9x'' + 4x' + 4x = 0;\ x(0) = 10,\ x'(0) = -12,\ x''(0) = 0,\ x'''(0) = 2$

Nonhomogeneous Equations and General Solutions

In Problems 51–53, find a general solution.

51. $3x'' - 5x' = 6 - 10t$, given $x_p(t) = t^2$ is a particular solution on the interval $(-\infty, \infty)$.
52. $\dfrac{d^2y}{dx^2} + y = 5e^{2x}$, given $y_p(x) = e^{2x}$ is a particular solution on the interval $(-\infty, \infty)$.
53. $x'' - 4x' + 4x = e^t$

Method of Undetermined Coefficients

In Problems 54–66, use the method of undetermined coefficients to find a particular solution of the given differential equation. Then find a general solution.

54. $x'' + 3x' + 2x = \sin t$
55. $x'' + 3x' = 12\cos 2t$
56. $x'' + 9x = 6t$
57. $x'' - 3x' + 2x = 5e^{2t} - \sin t$
58. $x'' - 4x' = 3e^{t/2} - 2e^{4t} + 1$
59. $15y'' - 6y' - 48y = 10x^2 - 13x$
60. $x'' + 9x = 4\sin 3t - \cos 3t - 27$
61. $4x'' + 4x' + x = 9te^t$
62. $6x'' + 3x' - 30x = 10e^t$
63. $x'' + 4x' + 7x = 14 - 7t$
64. $x'' + x' - 2x = -8e^t$
65. $x'' - 4x' + 4x = e^{2t}$
66. $4x'' - 12x = t\sin 2t$

Method of Variation of Parameters

In Problems 67–72, use the method of variation of parameters to find a particular solution of the given differential equation. Then find a general solution.

67. $x'' + x = \csc t$ for $0 < t \le \pi/2$

68. $x'' - 4x' + 4x = \dfrac{e^{2t}}{t^2}$ for $t > 0$

69. $t^2\dfrac{d^2x}{dt^2} + t\dfrac{dx}{dt} + x = t$ for $t > 0$

70. $t^2\dfrac{d^2x}{dt^2} + 6t\dfrac{dx}{dt} + 4x = 18t^2 + 5t - 12$ for $t > 0$

71. $x'' + 4x = 8\csc 2t$

72. $\ddot{x} + x = \tan t$ for $-\pi/2 < t < \pi/2$

Nonhomogeneous Cauchy-Euler Equations: Initial Value Problems

In Problems 73–76, solve the initial value problem for the given nonhomogeneous Cauchy-Euler equation.

73. $t^2x'' - 5tx' + 9x = -36;\ x(1) = 2,\ x'(1) = -5$

74. $t^2x'' - tx' - 3x = 4t + 3;\ x(2) = 1,\ x'(2) = -1$

75. $t^2x'' - tx' + x = 2\ln t;\ x(1) = 7,\ x'(1) = 6$

76. $t^2x''+tx'+4x = 8t^2; x(1) = 3,\ x'(1) = 4$

Choosing an Appropriate Method

In Problems 77–82, find a general solution of the differential equation. Use whatever method is most appropriate. However, there is no need to use a method if a particular solution can be obtained by inspection.

77. $t^2x'' - 3tx' + 3x = t \quad (t > 0)$

78. $t^2x'' - 5tx' + 9x = 8t - 8 \quad (t > 0)$

79. $x'' + x = \dfrac{1}{\sin t \cdot \cos^2 t}$ on $(0, \pi/2)$

80. $x'' + x' + x = \cos^2 t$, *Hint.* See Problem 33.

81. $x'' + x = 4 - t^2 + \tan t$ for $-\pi/2 < t < \pi/2$ *Hint.* Use the method of undetermined coefficients to find a particular solution of

$$x'' + x = 4 - t^2.$$

Use the method of variation of parameters to to find a particular solution of

$$x'' + x = \tan t.$$

Then use the superposition principle for nonhomogeneous equations.

82. $\dfrac{d^2x}{dt^2} - x = 10\sin^2 t$

Wronskians

83. Compute the Wronskians

$$W[x_1, x_2](t), \quad W[x_3, x_4](t)$$
$$W[x_2, x_3](t), \quad W[x_3, x_5](t)$$

of the following functions:

$$x_1(t) = \cos t,\ x_2(t) = \sin t,$$
$$x_3(t) = e^{2t},\ x_4(t) = te^{2t}, \text{ and}$$
$$x_5(t) = e^{5t}.$$

84. Prove the statement: The Wronskian of two functions that are linearly dependent on an interval is identically zero on that interval.

Cramer's Rule

85. Consider the system

$$ax + by = e$$
$$cx + dy = f,$$

where a, b, c, d, e, and f are given and x and y are to be determined. Does this system always have solutions? Explain why it has a unique solution if and only if

$$\begin{vmatrix} a & b \\ c & d \end{vmatrix} = ad - bc \neq 0.$$

Show that the solution can be expressed in terms of determinants as follows:

$$x = \frac{\begin{vmatrix} e & b \\ f & d \end{vmatrix}}{\begin{vmatrix} a & b \\ c & d \end{vmatrix}}, \quad y = \frac{\begin{vmatrix} a & e \\ c & f \end{vmatrix}}{\begin{vmatrix} a & b \\ c & d \end{vmatrix}}.$$

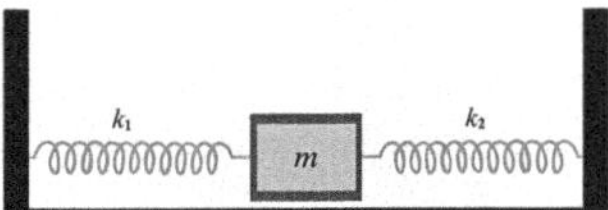

Fig. 10.4 Problem 86

This way of expressing the solution of a linear system of equations in terms of determinants is known as ***Cramer's rule***.

Mass Between Two Springs

86. Two springs with spring constants k_1 and k_2 are attached to a mass m and stationary supports as shown in Fig. 10.4. Assume that friction is negligible. If m is pulled x_0 units to the right of its equilibrium position ($x = 0$) and then released so that m has no initial velocity, determine the position x of m at time t and its frequency of oscillation.

Chapter 11
Laplace Transforms

A few worked examples should convince the reader that the Laplace transform furnishes a useful technique for solving linear differential equations. (However, the study of differential equations is no more contained in a table of Laplace transforms than the study of geometry in a set of trigonometric tables.)

T. W. Körner[1]

Sire, I had no need of that hypothesis.

Pierre-Simon de Laplace

Abstract This chapter introduces the *Laplace transform*—an integral transform often used by engineers and scientists in industry and academia to solve linear differential equations with constant coefficients, especially if those equations involve Heaviside and Dirac delta functions (see Chap. 7).

Section 11.1 begins with the initial value problem

$$\frac{dx}{dt} + 2x = 3e^{t}, \quad x(0) = 1, \tag{1}$$

where it is shown that the solution of (1) can be found by multiplying both sides of the differential equation by e^{-st} (without specifying a value for s) and then integrating from 0 to ∞ with respect to t . This leads to the definition of the Laplace transform, which is employed in Sect. 11.2 to obtain Laplace transforms of basic functions, such as e^{bt}, t^n, $\sin bt$, and the Heaviside and Dirac functions. These transforms are among those included in a short table titled "Basic Laplace Transforms." The examples in Sect. 11.3 illustrate how to use this table together

[1] Thomas William Körner (born 1946) is a Cambridge University mathematician who has written research papers and books on Fourier analysis, including a popular textbook for engineers, physicists, and mathematicians entitled *An Introduction to Fourier Analysis*. This particular quotation comes from another one of his books, *Fourier Analysis* [50, p. 378], which consists of essays on the ideas and results of Fourier analysis and their applications to other disciplines.

© The Author(s), under exclusive license to Springer Nature Switzerland AG 2026
L. C. Becker, *Ordinary Differential Equations: Concepts, Methods, and Models*,
https://doi.org/10.1007/978-3-032-15150-6_11

with the *linearity property* of the transform to solve initial value problems like (1) without having to resort to the definition.

Section 11.4 introduces the class of functions of *exponential order*. This sets the stage for deriving the formula for the *Laplace transform of the first derivative* and then shows how to use it and a version of *Lerch's theorem* to solve first-order initial value problems.

Piecewise continuous functions and the existence and long-term behavior of Laplace transforms are covered in Sect. 11.5. The *shift* (*translation*) *properties* are derived in Chap. 6 along with other important properties of the Laplace transform, such as the formulas involving *derivatives and integrals of Laplace transforms*. Formulas for computing transforms of *periodic functions* and the *convolution of two functions* are also derived. Examples abound illustrating how to use these properties to find the transforms of functions that are not listed in the aforementioned table of "Basic Laplace Transforms."

Although the notion of an *inverse Laplace transform* of a function is evident from solving the first-order initial value problems in Sect. 11.4, it is formally defined in Sect. 11.7 and illustrated with many examples. The shift properties are restated in a form more amenable for finding inverse transforms. This section includes a number of results stating the conditions under which inverse Laplace transforms of functions of the form

$$F(s) = \frac{G(s)}{1 - ae^{-bs}}, \tag{2}$$

are periodic, where a, b are constants with $a \neq 0$ and $b > 0$.

Section 11.8 is replete with examples illustrating how to use the transform formulas for first and second derivatives and the properties derived in the previous sections to solve second-order initial value problems with a variety of different forcing functions. Furthermore, it is shown in Sect. 11.9 that even a certain type of second-order differential equation with variable coefficients can be solved with Laplace transform methods.

11.1 Definition of the Laplace Transform

Engineers and scientists in academia and industry often employ a method to solve a linear differential equation with constant coefficients that involves an improper integral with an infinite upper limit of integration and an integrand that is the product of an exponential function and the dependent variable in the equation. This improper integral is known as a *Laplace transform*. Even though a first-order linear equation can be solved using the integrating factor method and a linear differential equation of any order with constant coefficients can be solved by finding the roots of the characteristic equation for the differential equation and then using those roots to obtain its solutions (see Chap. 10), it turns out that Laplace transforms are particularly well-suited for solving differential equations containing Heaviside and Dirac delta functions.[2] Although we will only explain how to use Laplace transforms to solve first- and second-order linear differential equations with constant coefficients, they can also be used to solve higher-order equations and linear systems of differential equations with constant coefficients. We will also show how to use Laplace transforms to solve a particular kind of differential equation with variable coefficients and certain types of integral and integro-differential equations.

The goal of this section is to introduce the idea behind the Laplace transform and then to formally define it. The Laplace transform was pioneered by Pierre-Simon de Laplace[3] in *Théorie Analytique des Probabilités*, a treatise on probability published in 1812. However, Laplace did not have the final word on this subject. Its application to differential equations was still being developed as late as the early 1930s by Gustav Doetsch [27], who is largely responsible for the present-day formulation of the Laplace transform and its applications.

Despite already having methods from earlier chapters for solving first-order linear differential equations, such as the integrating factor method, we will show that the Laplace transform is an alternative method for solving linear differential equations with constant coefficients. Moreover, we will see that the Laplace transform changes (or transforms) these equations into algebraic equations and that even the initial conditions can be incorporated into these algebraic equations. Furthermore, we will learn that it is particularly efficacious in handling differential equations involving the Heaviside and Dirac delta functions—and that there are certain linear equations with variable coefficients that are more easily solved with Laplace transform methods than with other methods. Hopefully, this will serve as motivation to continue our investigations of the Laplace transform for the purpose of solving higher-order linear equations and to develop a sense of knowing when it is practical and expedient to use it.

Let us begin with the initial value problem

$$\frac{dx}{dt} + 2x = 3e^t, \quad x(0) = 1. \tag{11.1.1}$$

---

[2] The Heaviside function, also known as the unit step function, and the Dirac delta function are introduced in Chap. 7.

[3] Pierre-Simon de Laplace (1749–1827) was a French mathematician and astronomer. All of his mathematical research in celestial mechanics, such as applying the laws of Newton to model the motions of the planets, were published from 1798 to 1825 in a five-volume treatise entitled *Traité de Mécanique Céleste*. Besides being an academic, he also held political offices and was made a count by Napoleon Bonaparte. The quotation from Laplace following the title of this chapter was in response to Napoleon's remark, "You have written this huge book on the system of the world without once mentioning the author of the universe." More on Laplace and other famous mathematicians can be found in the classic book *Men of Mathematics* by E. T. Bell [13, ch. 11], from which the quotation was taken.

Since the differential equation is already in standard form, its solutions are readily found by multiplying it by the integrating factor e^{2t} and then integrating the result:

$$\begin{aligned} e^{2t}\frac{dx}{dt} + 2e^{2t}x = 3e^{3t} \quad &\Rightarrow \quad \frac{d}{dt}\left(e^{2t}x\right) = 3e^{3t} \\ &\Rightarrow \quad \int \frac{d}{dt}\left(e^{2t}x\right)dt = \int 3e^{3t}\,dt + C \\ &\Rightarrow \quad e^{2t}x = e^{3t} + C. \end{aligned} \tag{11.1.2}$$

Solving for x, we obtain the general solution

$$x(t) = e^t + Ce^{-2t}.$$

Setting $t = 0$ and $x(0) = 1$, we get

$$x(0) = e^0 + Ce^0 = 1 + C = 1.$$

Thus, $C = 0$. Therefore, the solution of the initial value problem (11.1.1) is

$$x(t) = e^t. \tag{11.1.3}$$

As we have seen many times before, there is usually more than one way to attack a problem. Suppose we try a different approach, disregarding the fact that the left-hand side of the first equation in (11.1.2) can be written as the derivative of a product. Instead let us integrate each of its terms from $t = 0$ to $t = T$ to see if

$$\int_0^T e^{2t}x'(t)\,dt + \int_0^T 2e^{2t}x(t)\,dt = \int_0^T 3e^{3t}\,dt \tag{11.1.4}$$

will also lead to the solution (11.1.3). Let us integrate the first integral on the left-hand side by parts. Let $u = e^{2t}$ and $dv = x'(t)\,dt$. Then $du = 2e^{2t}\,dt$ and $v = x(t)$. Thus,

$$\int e^{2t}x'(t)\,dt = \int u\,dv = uv - \int v\,du = e^{2t}x(t) - \int 2e^{2t}x(t)\,dt$$

and so

$$\int_0^T e^{2t}x'(t)\,dt = \left[e^{2t}x(t)\right]_0^T - \int_0^T 2e^{2t}x(t)\,dt.$$

As a result, (11.1.4) becomes

$$e^{2T}x(T) - x(0) - 2\int_0^T e^{2t}x(t)\,dt + 2\int_0^T e^{2t}x(t)\,dt = e^{3T} - 1. \tag{11.1.5}$$

Since the initial condition is $x(0) = 1$, this simplifies to $x(T) = e^T$. We obtain the solution (11.1.3) by simply replacing T with t.

Now let us look at a variation of the preceding work that is more general in that we multiply both sides of the differential equation in (11.1.1) by the exponential function e^{-st} without assigning a value to s.[4] Even though e^{-st} is an integrating factor when $s = -2$, let us ignore this fact to see if we can still obtain the solution of (11.1.1). As before, let us integrate each term of the resulting equation from $t = 0$ to $t = T$:

$$\int_0^T e^{-st}x'(t)\,dt + \int_0^T 2e^{-st}x(t)\,dt = \int_0^T 3e^{-st}e^t\,dt. \tag{11.1.6}$$

Note that this simplifies to (11.1.4) when $s = -2$. Similar to our handling of (11.1.4), let us integrate the first integral on the left-hand side of (11.1.6) by parts by setting $u = e^{-st}$ and $dv = x'(t)\,dt$, thereby obtaining

$$\left[e^{-st}x(t)\right]_{t=0}^{t=T} - \int_{t=0}^{t=T} (-s)e^{-st}x(t)\,dt + \int_0^T 2e^{-st}x(t)\,dt = \int_0^T 3e^{-st}e^t\,dt,$$

which simplifies to

$$e^{-sT}x(T) - x(0) + (s+2)\int_0^T e^{-st}x(t)\,dt = -\frac{3}{s-1}\left[e^{-(s-1)T} - 1\right]. \tag{11.1.7}$$

Note that this simplifies to (11.1.5) when $s = -2$, which yields the solution. However, let us pretend that we did not notice this. And so we take a different path by letting T increase without bound to see if that might somehow lead to the solution. That is, let $T \to \infty$. If $s < 1$, the right-hand side of (11.1.7) diverges; but if s is restricted to values greater than 1, it converges to the function $3/(s-1)$ since $e^{-(s-1)T} \to 0$ as $T \to \infty$. So when we take the limit of both sides of (11.1.6) as $T \to \infty$ restricting s to $s > 1$, the right-hand side of (11.1.6) becomes

$$\begin{aligned}\lim_{T\to\infty}\int_0^T 3e^{-st}e^t\,dt &= 3\lim_{T\to\infty}\int_0^T e^{-(s-1)t}\,dt \\ &= -\frac{3}{s-1}\lim_{T\to\infty}\left[e^{-(s-1)T} - 1\right] = \frac{3}{s-1}.\end{aligned} \tag{11.1.8}$$

This is what is meant by

$$\int_0^\infty 3e^{-st}e^t\,dt = \frac{3}{s-1}.$$

---

[4] The reason for using e^{-st} instead of e^{st} is dictated by convention. We will be concerned shortly with the convergence of improper integrals of the form $\int_0^\infty e^{-st}x(t)\,dt$.

The integral $\int_0^\infty 3e^{-st}e^t\,dt$ is improper because the symbol ∞ does not represent a real number; as a result, it must be handled differently. Recall from calculus courses that

$$\int_a^\infty f(t)\,dt$$

is really mathematical shorthand for

$$\lim_{T\to\infty}\int_a^T f(t)\,dt.$$

Let us also assume that $e^{-sT}x(T) \to 0$ as $T \to \infty$. This will be justified shortly. So, as $T \to \infty$, (11.1.7) becomes

$$-x(0) + (s+2)\int_0^\infty e^{-st}x(t)\,dt = \frac{3}{s-1}. \tag{11.1.9}$$

Setting $x(0) = 1$ and solving for the improper integral, we get

$$\int_0^\infty e^{-st}x(t)\,dt = \frac{3}{(s-1)(s+2)} + \frac{1}{s+2},$$

which simplifies to

$$\int_0^\infty e^{-st}x(t)\,dt = \frac{1}{s-1}. \tag{11.1.10}$$

At first this result may seem interesting yet useless since we have not explicitly solved for $x(t)$. Instead we have what is known as an *integral transform* of $x(t)$. But then our hopes are rekindled when we recall that $1/(s-1)$ also appeared earlier in (11.1.8). It follows from this that

$$\int_0^\infty e^{-st}e^t\,dt = \frac{1}{s-1}.$$

Even though we do not know how to find solutions of the integral equation (11.1.10) directly, it does not really matter here, for it follows from (11.1.8) that the differentiable function $x(t) = e^t$ is a solution of this equation. Furthermore, the assumption that led to this result, namely, that

$$e^{-sT}x(T) \to 0 \quad \text{as } T \to \infty,$$

turns out to be valid since

$$e^{-sT}x(T) = e^{-sT}e^T = e^{-(s-1)T}$$

does approach 0 as $T \to \infty$ for $s > 1$. And so we have managed to obtain the solution $x(t) = e^t$ of the initial value problem (11.1.1) for a third time!

Truth be told, (11.1.1) was contrived so that the solution of the integral equation (11.1.10) just happened to be lying around, namely, in (11.1.8). Ordinarily, one is usually not so lucky. For instance, if the initial condition is changed from $x(0) = 1$ to $x(0) = 2$, then that changes (11.1.10) to

$$\int_0^\infty e^{-st}x(t)\,dt = \frac{3}{(s-1)(s+2)} + \frac{2}{s+2} = \frac{2s+1}{(s-1)(s+2)}.$$

So if this approach to solving linear differential equations is to be an effective alternative to the integrating factor method, we need some way of coming up with functions $x(t)$ that will satisfy the integral equation

$$\int_0^\infty e^{-st}x(t)\,dt = X(s) \tag{11.1.11}$$

when $X(s)$ is a known function of s. The function $X(s)$ is called an *integral transform* of $x(t)$. Actually, there are other well-known integral transforms[5]—this particular one is called the *Laplace transform*. And right now we have our first example: the Laplace transform of $x(t) = e^t$ is

$$X(s) = \frac{1}{s-1}$$

for $s > 1$. Before continuing, let us formally define what is meant by the Laplace transform of a function.

Laplace Transform

Definition 11.1.1 The ***Laplace transform*** of a function $f(t)$ that is defined for $t \geq 0$ is the function $F(s)$ defined by the improper integral

$$F(s) := \int_0^\infty e^{-st}f(t)\,dt, \tag{11.1.12}$$

provided that this integral converges. The set of all values of s for which it converges is the ***domain*** of $F(s)$. On the other hand, if the integral diverges for all values of s, then the Laplace transform of $f(t)$ does not exist.

---

[5] The *Fourier transform* is an important integral transform, which the reader may have encountered.

Remark A function does not have to be defined at all points in the interval $[0, \infty)$ in order for its Laplace transform to exist. A simple example of this is the step function $u(t - b)$ with $b \geq 0$, which is undefined at $t = b$. The integral (11.1.12) for this function (see Example 11.2.3) converges. Moreover, it can converge for even more complicated functions that are undefined at finitely many points. In fact, in later sections, we will be considering the Laplace transforms of a certain class of functions that are either undefined or discontinuous at infinitely many isolated points but which have only finitely many of these points in every finite interval $[0, T]$.

A word about notation: first note the use of the lowercase f and the uppercase F in (11.1.12). When Laplace transforms are involved, the convention is to use a lowercase letter for the original function and the uppercase form of the same letter to denote its Laplace transform.[6] For instance, just as we used X to denote the Laplace transform of the function $x(t) = e^t$, we would use F and G to denote the Laplace transforms of functions called f and g, respectively.

There is also another way to denote the Laplace transform of a function $f(t)$ and that is to write $\mathcal{L}\{f(t)\}(s)$. For example, either

$$\mathcal{L}\{e^t\}(s) = \frac{1}{s-1} \quad \text{or} \quad \mathcal{L}\{x(t)\}(s) = \frac{1}{s-1}$$

expresses the fact that the Laplace transform of $x(t) = e^t$ is $X(s) = 1/(s-1)$. As a matter of convenience, it is customary to shorten this to

$$\mathcal{L}\{e^t\} = \frac{1}{s-1}.$$

This is a mild abuse of notation: for just as we write

$$X(s) = \frac{1}{s-1} \quad \text{and not} \quad X = \frac{1}{s-1},$$

we should write

$$\mathcal{L}\{e^t\}(s) = \frac{1}{s-1} \quad \text{instead of} \quad \mathcal{L}\{e^t\} = \frac{1}{s-1}.$$

Nonetheless, since including the argument "s" after $\mathcal{L}\{f(t)\}$ every time it is written in working out a problem is tedious and time-consuming and as it is already understood that $\mathcal{L}\{f(t)\}$ is a function of s, there is no harm in omitting it.

When taking the Laplace transform of a function, the original function $f(t)$ is transformed into a new function called $\mathcal{L}\{f(t)\}$ or $F(s)$. We portray this symbolically by writing

[6] A caveat on notation: Some authors do just the opposite: the uppercase letter denotes the original function while the corresponding lowercase letter denotes its transform.

$$f(t) \xrightarrow{\mathcal{L}} F(s).$$

Looking at it like this, we can say that the Laplace transform is a rule $\mathcal{L}$ that assigns to the original function some other function—but only one other function. Recall that we have seen rules like this before: they are called *operators*.

The operator $\mathcal{L}$ not only transforms a function $f(t)$ into another function $F(s)$, it is also accompanied by a change in domain. For example, we just determined that the Laplace transform of $x(t) = e^t$ is the function

$$\mathcal{L}\{x(t)\} = \frac{1}{s-1} \quad \text{for} \quad s > 1.$$

Associated with this transform is a change from the *t-domain* $[0, \infty)$ for $f(t)$ to the *s-domain* $(1, \infty)$ for $F(s)$. In applications involving time with t denoting time, the *t-domain* is called the *time domain*.

The task of determining pairs of functions satisfying (11.1.12) that will be useful to us is not as daunting as it might first seem. The reason for this is that the aim of our investigation is not to find functions $f(t)$ satisfying (11.1.12) for every imaginable function $F(s)$: rather only for those $F(s)$ that correspond to functions $f(t)$ which are actual solutions of linear differential equations. From our prior experiences with linear equations, these are the exponential, sinusoidal, and algebraic functions, along with certain combinations of these functions. So our game plan will be to substitute these types of functions into (11.1.12) for $f(t)$ and then work out the resulting improper integral to determine its Laplace transform $F(s)$. In this way, we can first create a list of the Laplace transforms of the most frequently used functions and then see where to go on from there.

11.2 Laplace Transforms of Basic Functions

In the previous section we determined that $\mathcal{L}\{e^t\} = 1/(s-1)$, which is the succinct way to write that the Laplace transform of the function $f(t) = e^t$ is the function

$$F(s) = \frac{1}{s-1}.$$

In the next example, let us generalize this result by finding the Laplace transform of $f(t) = e^{bt}$, where b represents any real number. The purpose of this example and those to follow is to demonstrate how to use Definition 11.1.1 to find the Laplace transform of a given function.

Example 11.2.1 Find $\mathcal{L}\{e^{bt}\}$, where b is a constant.

Solution Using Definition 11.1.1, we have

$$\mathcal{L}\{e^{bt}\} = \int_0^\infty e^{-st} e^{bt}\, dt = \lim_{T\to\infty} \int_0^T e^{(b-s)t}\, dt$$

$$= \lim_{T\to\infty} \frac{1}{b-s} e^{(b-s)t}\Big|_{t=0}^{t=T} = \frac{1}{b-s} \lim_{T\to\infty} \left[e^{(b-s)T} - e^0\right].$$

Implicit in carrying out the integration is that $s \neq b$. As $T \to \infty$,

$$e^{(b-s)T} \to 0 \quad \text{if} \quad s > b,$$

whereas

$$e^{(b-s)T} \to \infty \quad \text{if} \quad s < b.$$

From this, we conclude that the integral converges to $1/(s-b)$ if $s > b$ and diverges if $s < b$. For the case $s = b$,

$$\mathcal{L}\{e^{bt}\} = \int_0^\infty e^{-bt} e^{bt}\, dt = \lim_{T\to\infty} \int_0^T 1\, dt = \lim_{T\to\infty} T = \infty.$$

Therefore,

$$\mathcal{L}\{e^{bt}\} = \begin{cases} \dfrac{1}{s-b}, & \text{for } s > b \\ \text{undefined}, & \text{for } s \leq b. \end{cases} \tag{11.2.1}$$

This is the same as saying that the Laplace transform of $f(t) = e^{bt}$ is the function $F(s) = 1/(s-b)$ with domain (b, ∞). ♦

Example 11.2.2 Show that $\mathcal{L}\{1\} = \dfrac{1}{s}$ for $s > 0$.

Solution As far as the values of t go, we are only concerned with the interval $[0, \infty)$. So the argument "1" of the operator $\mathcal{L}$ really denotes the function whose value at each $t \in [0, \infty)$ is 1. In other words, $\mathcal{L}\{1\}$ this is a succinct way of denoting the Laplace transform of the function f, where $f(t) \equiv 1$ for $t \geq 0$. By (11.1.12),

$$\mathcal{L}\{1\} = \int_0^\infty e^{-st} \cdot 1\, dt = \lim_{T\to\infty} \int_0^T e^{-st}\, dt$$

$$= \lim_{T\to\infty} -\frac{1}{s} e^{-st}\Big|_{t=0}^{t=T} = -\frac{1}{s} \lim_{T\to\infty} \left[e^{-sT} - 1\right]$$

$$= \begin{cases} \dfrac{1}{s}, & s > 0 \\ \text{undefined}, & s < 0. \end{cases}$$

If $s = 0$, the integral diverges. Therefore, for $s > 0$,

$$\mathcal{L}\{1\} = \frac{1}{s}.$$

Note this also follows from Example 11.2.1. Setting $b = 0$ in Eq. (11.2.1), we have

$$\mathcal{L}\{1\} = \left.\frac{1}{s-b}\right|_{b=0} = \frac{1}{s} \quad (s > 0).$$

♦

In Chap. 7 we introduced an important function whose value is always 0 up to a certain point but then jumps to 1 and thereafter remains at this value. This function is known either as the *unit step function* or the *Heaviside function*. If the jump discontinuity occurs at $t = b$ but is not defined there, we denote the function by $u(t - b)$. In other words,

$$u(t-b) := \begin{cases} 0, & t < b \\ 1, & t > b. \end{cases} \tag{11.2.2}$$

But if the function is defined at $t = b$ with its value there being 1, then we write $H(t - b)$:

$$H(t-b) := \begin{cases} 0, & t < b \\ 1, & t \geq b. \end{cases} \tag{11.2.3}$$

Since unit step functions are so important when it comes to modeling abrupt changes in engineering, physics, and other disciplines, let us find its Laplace transform.

Example 11.2.3 Find the Laplace transform of $u(t - b)$, where $b \geq 0$.

Solution Suppose $b > 0$. From (11.1.12), we have $\mathcal{L}\{u(t-b)\} = \int_0^\infty e^{-st} u(t-b)\, dt$. Since $u(t - b)$ is equal to 0 up to $t = b$ but then 1 afterward, we split the improper integral at $t = b$ into two integrals as follows:

$$\mathcal{L}\{u(t-b)\} = \int_0^b e^{-st} \cdot 0\, dt + \int_b^\infty e^{-st} \cdot 1\, dt = \lim_{T\to\infty} \int_b^T e^{-st}\, dt$$

$$= \lim_{T\to\infty} -\frac{1}{s} e^{-st}\Big|_{t=b}^{t=T} = -\frac{1}{s} \lim_{T\to\infty} \left[e^{-sT} - e^{-sb}\right].$$

Thus

$$\mathcal{L}\{u(t-b)\} = \frac{e^{-bs}}{s} \quad \text{if } s > 0.$$

Now observe that if we formally set $b = 0$ in this formula, we obtain

$$\mathcal{L}\{u(t)\} = \frac{1}{s}.$$

This is in fact correct. Since $u(t) = 1$ for $t > 0$,

$$\mathcal{L}\{u(t)\} = \int_0^\infty e^{-st}u(t)\,dt = \lim_{a\to 0^+}\int_a^\infty e^{-st}\cdot 1\,dt = \lim_{a\to 0^+}\frac{e^{-as}}{s} = \frac{1}{s}.$$

Since there is no difference between $u(t-b)$ and $H(t-b)$ except at $t = b$, we conclude that

$$\mathcal{L}\{u(t-b)\} = \mathcal{L}\{H(t-b)\} = \frac{e^{-bs}}{s} \qquad (s > 0).$$

♦

The next two examples demonstrate the indispensability of the method of integration by parts as a tool for finding Laplace transforms. The first one also demonstrates the importance of recognizing patterns.

Example 11.2.4 For $n = 1, 2, 3, \ldots$, show that

$$\mathcal{L}\{t^n\} = \frac{n!}{s^{n+1}} \qquad (s > 0).$$

Solution Referring to Definition 11.1.1 again, we have

$$\mathcal{L}\{t^n\} = \lim_{T\to\infty}\int_0^T e^{-st}t^n\,dt.$$

This is one of those classic integration-by-parts integrand. According to the *LIATE* mnemonic, let $u = t^n$ and $dv = e^{-st}\,dt$. Then $du = nt^{n-1}\,dt$ and $v = -\frac{1}{s}e^{-st}$. Thus,

$$\begin{aligned}\int_0^T e^{-st}t^n\,dt &= -\frac{1}{s}e^{-st}t^n\Big|_0^T - \int_0^T\left(-\frac{1}{s}e^{-st}\right)nt^{n-1}\,dt \\ &= -\frac{1}{s}e^{-sT}T^n + \frac{n}{s}\int_0^T e^{-st}t^{n-1}\,dt.\end{aligned}$$

Let us determine the long-term behavior of the first term on the right-hand side using the following Maclaurin series expansion of e^x:

$$1 + x + \frac{x^2}{2!} + \frac{x^3}{3!} + \frac{x^3}{4!} + \cdots + \frac{x^n}{n!} + \cdots .$$

Recall that this series converges to e^x regardless of the value of x. Setting $x = sT$, we have

$$\frac{T^n}{e^{sT}} = \frac{T^n}{1 + sT + \frac{s^2T^2}{2!} + \frac{s^3T^3}{3!} + \cdots + \frac{s^nT^n}{n!} + \frac{s^{n+1}T^{n+1}}{(n+1)!} + \frac{s^{n+2}T^{n+2}}{(n+2)!} + \cdots}$$

$$= \frac{1}{\frac{1}{T^n} + \frac{s}{T^{n-1}} + \frac{s^2}{2!T^{n-2}} + \cdots + \frac{s^{n-1}}{(n-1)!T} + \frac{s^n}{n!} + \frac{s^{n+1}T}{(n+1)!} + \frac{s^{n+2}T^2}{(n+2)!} + \cdots}.$$

Since $s^{n+k}T^k/(n+k)! \to \infty$ as $T \to \infty$ for $k = 1, 2, 3, \ldots$ and $s > 0$, we conclude that $T^n/e^{sT} \to 0$ as $T \to \infty$, regardless of the value of n.[7] So if the integral in the second term on the right-hand side converges, its limit would be $\mathcal{L}\{t^{n-1}\}$, which in turn would imply that the integral on the left-hand side converges to what would be $\mathcal{L}\{t^n\}$. The end result is the recursive formula

$$\mathcal{L}\{t^n\} = \frac{n}{s}\mathcal{L}\{t^{n-1}\} \quad (s > 0) \tag{11.2.4}$$

provided $\mathcal{L}\{t^{n-1}\}$ exists.

Substituting $n = 1$ into (11.2.4) gives

$$\mathcal{L}\{t^1\} = \frac{1}{s}\mathcal{L}\{t^0\} = \frac{1}{s}\mathcal{L}\{1\}.$$

Now $\mathcal{L}\{1\}$ does exist since we determined in Example 11.2.2 that $\mathcal{L}\{1\} = 1/s$. Its existence implies the validity of formula (11.2.4) for $n = 1$ and thereby the existence of $\mathcal{L}\{t\}$:

$$\mathcal{L}\{t\} = \frac{1}{s^2}.$$

[7] Another way to obtain this limit is with *L'Hôpital's Rule*.

The existence of $\mathcal{L}\{t\}$ then validates (11.2.4) for $n = 2$; therefore,

$$\mathcal{L}\{t^2\} = \frac{2}{s}\mathcal{L}\{t^1\} = \frac{2}{s} \cdot \frac{1}{s^2} = \frac{2}{s^3}.$$

Continuing in this manner, we obtain

$$\begin{aligned} n = 3: \quad & \mathcal{L}\{t^3\} = \frac{3}{s}\mathcal{L}\{t^2\} = \frac{3}{s} \cdot \frac{2}{s^3} = \frac{3 \cdot 2}{s^4} \\ n = 4: \quad & \mathcal{L}\{t^4\} = \frac{4}{s}\mathcal{L}\{t^3\} = \frac{4}{s} \cdot \frac{3 \cdot 2}{s^4} = \frac{4 \cdot 3 \cdot 2}{s^5} \\ n = 5: \quad & \mathcal{L}\{t^5\} = \frac{5}{s}\mathcal{L}\{t^4\} = \frac{5}{s} \cdot \frac{4 \cdot 3 \cdot 2}{s^5} = \frac{5 \cdot 4 \cdot 3 \cdot 2}{s^6} \end{aligned}$$

and so on. The pattern indicates that

$$\mathcal{L}\{t^n\} = \frac{n!}{s^{n+1}} \quad (s > 0) \tag{11.2.5}$$

for every positive integer n. Certainly, we have established that the formula is true for $n = 1, 2, 3, 4, 5$. Note that it is also true for $n = 0$ with the understanding that $0! = 1$. What about for $n > 5$? Suppose (11.2.4) were true for some positive integer k; that is,

$$\mathcal{L}\{t^k\} = \frac{k!}{s^{k+1}}. \tag{11.2.6}$$

The previous cases suggest that it then should also be true for $k + 1$. Let us check this. By the recursive formula (11.2.4),

$$\mathcal{L}\{t^{k+1}\} = \frac{k+1}{s}\mathcal{L}\{t^k\}.$$

This together with (11.2.6) yields

$$\mathcal{L}\{t^{k+1}\} = \frac{k+1}{s} \cdot \frac{k!}{s^{k+1}} = \frac{(k+1)!}{s^{k+2}}.$$

So we have established that it is impossible for (11.2.5) to be true for some positive integer but not for the next one. The validity of the formula for $n = 1, 2, 3, 4$, and 5 has already been established. Since it is valid for $n = 5$, it must be valid for $n = 6$. Then its validity for $n = 6$ implies its validity for $n = 7$, which in turn implies its validity for $n = 8$. This "domino effect" makes it valid for all positive integers. In short, we have established the validity of formula (11.2.5) for all positive integers by mathematical induction. ♦

Example 11.2.5 Find the Laplace transform of $\mathcal{L}\{\sin bt\}$, where b is a constant.

Solution The Laplace transform of $\sin bt$ with $b \neq 0$ is given by

$$\mathcal{L}\{\sin bt\} = \lim_{T\to\infty} \int_0^T e^{-st} \sin bt \, dt.$$

Applying the integration-by-parts formula twice to the integral, we find that

$$\left(1 + \frac{b^2}{s^2}\right) \int_0^T e^{-st} \sin bt \, dt = -\frac{1}{s} e^{-sT} \sin bT + \frac{b}{s^2} e^{-sT} \cos bT + \frac{b}{s^2}$$

for $s \neq 0$. For $s > 0$, $e^{-sT} \to 0$ as $T \to \infty$. Since

$$\left| e^{-sT} \sin bT \right| = e^{-sT} \left| \sin bT \right| \leq e^{-sT},$$

we see that $e^{-sT} \sin bT \to 0$ as $T \to \infty$. Likewise $e^{-sT} \cos bT \to 0$ as $T \to \infty$. We conclude that

$$\left(1 + \frac{b^2}{s^2}\right) \mathcal{L}\{\sin bt\} = \frac{b}{s^2}.$$

Solving for $\mathcal{L}\{\sin bt\}$ gives

$$\mathcal{L}\{\sin bt\} = \frac{b}{s^2 + b^2}$$

for $s > 0$. Note that this formula is also valid for $b = 0$; so we can remove the restriction that $b \neq 0$. ♦

The Dirac delta function $\delta(t - b)$ was introduced in Chap. 7. In the next example, we use the sifting property (7.3.24) to find its Laplace transform.

Example 11.2.6 Find the Laplace transform of $\delta(t - b)$, where $b \geq 0$.

Solution Resorting once again to (11.1.12), we have

$$\mathcal{L}\left\{\delta(t - b)\right\} = \int_0^\infty e^{-st} \delta(t - b) \, dt.$$

By the sifting property of the Dirac delta function,

$$\mathcal{L}\left\{\delta(t - b)\right\} = \lim_{T\to\infty} \int_0^T e^{-st} \delta(t - b) \, dt = \lim_{T\to\infty} e^{-sb} u(T - b).$$

Since $u(T - b) = 1$ for $T > b$,

$$\mathcal{L}\{\delta(t - b)\} = e^{-bs}.$$

In particular,

$$\mathcal{L}\{\delta(t)\} = 1.$$

♦

In the previous examples, we used Definition 11.1.1 to find the Laplace transforms of some of basic functions. However, not every function has a Laplace transform.

Example 11.2.7 Explain why $f(t) = e^{t^2}$ has no Laplace transform.

Solution In order for e^{t^2} to have a Laplace transform, the improper integral

$$\int_0^\infty e^{-st} \cdot e^{t^2}\, dt = \lim_{b\to\infty} \int_0^b e^{(t-s)t}\, dt \tag{11.2.7}$$

would have to converge for some values of s (see Definition 11.1.1). Accordingly, consider any value of $s < 0$. Let $b > 0$. Then, as $(t-s)t \geq 0$ for all $t \geq 0$, we have

$$\int_0^b e^{(t-s)t}\, dt \geq \int_0^b dt \to \infty$$

as $b \to \infty$. Thus the improper integral (11.2.7) diverges when $s < 0$. Now suppose $s \geq 0$. Then the exponent $(t-s)t \geq 0$ for all $t \geq s$. And so, for $b > s$,

$$\int_0^b e^{(t-s)t}\, dt \geq \int_s^b e^{(t-s)t}\, dt \geq \int_s^b dt = b - s \to \infty$$

as $b \to \infty$. Since (11.2.7) diverges for all values of s, the transform of $f(t) = e^{t^2}$ does not exist. ♦

Example 11.2.8 The function $f(t) = t^{-1}$ has no Laplace transform because

$$\int_0^\infty e^{-st} t^{-1}\, dt$$

diverges for all values of s. Obviously, it diverges when $s = 0$. So let us see what the situation is for any arbitrary value of $s > 0$. With the change of variable $x = st$, the improper integral is transformed into another improper integral, but one without s:

$$\int_0^\infty e^{-st} \frac{1}{t}\, dt = \int_0^\infty e^{-x} \left(\frac{s}{x}\right)\left(\frac{1}{s}\right) dx = \int_0^\infty e^{-x} x^{-1}\, dx.$$

It is left to the reader (see Problem 40) to show that the integral on the right "without the s" is bounded below by a simpler integral that is divergent. It is also left to the reader to analyze the case $s < 0$. ♦

Table 11.1 in the next section lists all of the Laplace transforms that have been found so far, including the transform of the cosine function whose derivation is very much like that of the sine function in Example 11.2.5. The table also includes the Laplace transforms of the hyperbolic cosine and sine functions. The former is derived later on in Example 11.3.3.

11.3 Linearity of the Laplace Transform

Laplace transforms can be computed with a CAS or looked up in mathematical tables and handbooks (or even in the appendices of many engineering and physics textbooks); even so, the transforms of some of the most basic functions should be memorized for the sake of efficiency. These are listed in Table 11.1.

Note that the Laplace transforms of $\sin bt$ and $\cos bt$ are the same except for their numerators. The statement

"some birds can swim"

is a way to remember that the numerator of the Laplace transform of $\sin bt$ is b, whereas for $\cos bt$ it is s. This mnemonic also applies to the Laplace transforms of the hyperbolic functions $\sinh bt$ and $\cosh bt$.

Can we use Table 11.1 to find the Laplace transform of a function that is expressed in terms of some of the functions listed in the table? It turns out that the answer is yes if that function can be expressed as a linear combination of those

Table 11.1 Basic Laplace transforms

$f(t)$	$F(s) = \mathcal{L}\{f(t)\}$
1	$\frac{1}{s} \quad (s > 0)$
t	$\frac{1}{s^2} \quad (s > 0)$
$t^n \ (n = 0, 1, 2, \ldots)$	$\frac{n!}{s^{n+1}} \quad (s > 0)$
e^{bt}	$\frac{1}{s-b} \quad (s > b)$
$\sin bt$	$\frac{b}{s^2+b^2} \quad (s > 0)$
$\cos bt$	$\frac{s}{s^2+b^2} \quad (s > 0)$
$\sinh bt$	$\frac{b}{s^2-b^2} \quad (s > \lvert b \rvert)$
$\cosh bt$	$\frac{s}{s^2-b^2} \quad (s > \lvert b \rvert)$
$u(t-b), H(t-b) \ (b \geq 0)$	$\frac{e^{-bs}}{s} \quad (s > 0)$
$\delta(t-b) \quad (b \geq 0)$	e^{-bs}

functions. A ***linear combination*** of the functions $f_1, f_2, \cdots, f_n$ is a function of the form

$$c_1 f_1 + c_2 f_2 + \cdots + c_n f_n,$$

where $c_1, c_2, \cdots, c_n$ are constants. In calculus we learned that the integration of a linear combination of integrable functions can be carried out by first integrating each of the functions and then by taking the same linear combination of the results; that is to say, in the case of $n = 2$:

$$\int_a^b (c_1 f_1(t) + c_2 f_2(t))\,dt = c_1 \int_a^b f_1(t)\,dt + c_2 \int_a^b f_2(t)\,dt.$$

This property of integration is known as ***linearity***. Since Laplace transforms are improper integrals, they inherit this linearity property.

Linearity of the Laplace Transform

Theorem 11.3.1 *Let f_1 and f_2 be functions whose Laplace transforms exist for $s > b_1$ and $s > b_2$, respectively. Let b be the larger of b_1 and b_2. Then for $s > b$,*

$$\mathcal{L}\{c_1 f_1(t) + c_2 f_2(t)\} = c_1 \mathcal{L}\{f_1(t)\} + c_2 \mathcal{L}\{f_2(t)\}, \tag{11.3.1}$$

where c_1 and c_2 are any constants.

Proof To show this, we merely use the linearity property of integration: For all $T \geq 0$,

$$\int_0^T e^{-st}\,[c_1 f_1(t) + c_2 f_2(t)]\,dt = c_1 \int_0^T e^{-st} f_1(t)\,dt + c_2 \int_0^T e^{-st} f_2(t)\,dt.$$

Then letting $T \to \infty$, we obtain (11.3.1). ■

An equivalent way to express (11.3.1) is with the following pair of properties:

(a) $\mathcal{L}\{f_1(t) + f_2(t)\} = \mathcal{L}\{f_1(t)\} + \mathcal{L}\{f_2(t)\}$
(b) $\mathcal{L}\{c_1 f_1(t)\} = c_1 \mathcal{L}\{f_1(t)\}$.

Obviously, (11.3.1) implies (a) (let $c_1 = c_2 = 1$) and it also implies (b) (let $c_2 = 0$). In other words, (11.3.1) implies both (a) and (b). It is left as an exercise to show the converse is also true, namely, that (a) and (b) taken together imply (11.3.1). Therefore, it follows that the pair (a) and (b) is equivalent to (11.3.1).

An operator $\mathcal{O}$, such as the Laplace transform $\mathcal{L}$ or the differential operator d/dt, with the property that

$$\mathcal{O}\{c_1 f_1(t) + c_2 f_2(t)\} = c_1 \mathcal{O}\{f_1(t)\} + c_2 \mathcal{O}\{f_2(t)\}$$

for all functions f_1 and f_2 in its domain and all constants c_1 and c_2 is called a ***linear operator***. So the statement that the Laplace transform $\mathcal{L}$ is a linear operator means that it satisfies property (11.3.1). It is left as an exercise to show that (11.3.1) can be extended to a finite linear combination of more than two functions:

$$\mathcal{L}\{c_1 f_1(t) + c_2 f_2(t) + \cdots + c_n f_n(t)\} = c_1 \mathcal{L}\{f_1(t)\} + c_2 \mathcal{L}\{f_2(t)\} + \cdots + c_n \mathcal{L}\{f_n(t)\}. \tag{11.3.2}$$

In the following examples, we demonstrate how to use the linearity property in conjunction with Table 11.1 to find the Laplace transforms of linear combinations of a variety of functions.

Example 11.3.1 Find $\mathcal{L}\{t^3 + 2e^{5t}\}$.

Solution By (11.3.1), or (11.3.2) with $n = 2$,

$$\mathcal{L}\{t^3 + 2e^{5t}\} = \mathcal{L}\{t^3\} + 2\mathcal{L}\{e^{5t}\}.$$

By Table 11.1, $\mathcal{L}\{t^3\} = 3!/s^4$ for $s > 0$ and $\mathcal{L}\{e^{5t}\} = 1/(s-5)$ for $s > 5$. Therefore,

$$\mathcal{L}\{t^3 + 2e^{5t}\} = \frac{3!}{s^4} + 2 \cdot \frac{1}{s-5} = \frac{6}{s^4} + \frac{2}{s-5}$$

for $s > 5$. ♦

Example 11.3.2 Find the Laplace transform of the function

$$f(t) = 5e^{-3t} - \frac{1}{2}\sin 6t + \pi.$$

Solution It follows from (11.3.2) with $n = 3$ and Table 11.1 that

$$\begin{aligned}\mathcal{L}\{f(t)\} &= 5\mathcal{L}\{e^{-3t}\} - \frac{1}{2}\mathcal{L}\{\sin 6t\} + \mathcal{L}\{\pi \cdot 1\} \\ &= 5 \cdot \frac{1}{s-(-3)} - \frac{1}{2} \cdot \frac{6}{s^2+6^2} + \pi\mathcal{L}\{1\} = \frac{5}{s+3} - \frac{3}{s^2+36} + \frac{\pi}{s}\end{aligned}$$

for $s > 0$. ♦

As the last example in this section, let us find the Laplace transform of one of the hyperbolic functions listed in Table 11.1.

Example 11.3.3 Find the Laplace transform of $\cosh bt$, where b is a constant.

Solution Recall that

$$\cosh bt := \frac{e^{bt} + e^{-bt}}{2}.$$

If $b \neq 0$, then it follows from the linearity of the Laplace transform that

$$\mathcal{L}\{\cosh bt\} = \frac{1}{2}\mathcal{L}\{e^{bt} + e^{-bt}\} = \frac{1}{2}\left(\mathcal{L}\{e^{bt}\} + \mathcal{L}\{e^{-bt}\}\right) = \frac{1}{2}\left(\frac{1}{s-b} + \frac{1}{s+b}\right)$$

for $s > b$ and $s > -b$. Consequently,

$$\mathcal{L}\{\cosh bt\} = \frac{s}{s^2 - b^2} \qquad (s > |b|) \tag{11.3.3}$$

If $b = 0$, then

$$\mathcal{L}\{\cosh bt\} = \mathcal{L}\{\cosh 0\} = \mathcal{L}\{1\} = \frac{1}{s}$$

for $s > 0$. Therefore, (11.3.3) is valid for all values of b. ♦

11.4 First-Order Linear Equations

Recall that the primary goal of this chapter is to develop a method for solving linear differential equations using the Laplace transform, so we must bear this in mind as we search for more of its properties. Recall that we began this chapter with the initial value problem (11.1.1), namely,

$$\frac{dx}{dt} + 2x = 3e^t, \quad x(0) = 1.$$

In our attempt to discover another way of solving (11.1.1), one that did not involve integrating factors, we stumbled upon the idea of the Laplace transform. Now that we have formally defined the Laplace transform, discovered that it is a linear operator, and found the transform of some commonly occurring functions, let us retrace the steps that led to the solution of (11.1.1) and tidy them up so as to streamline the procedure of using Laplace transforms to find solutions. We began by tinkering with the differential equation in (11.1.1). This eventually resulted in Eq. (11.1.6). Then we took the limit of both of its sides as $T \to \infty$. Admittedly all of this was done in a clumsy, roundabout way; however, it is really more straightforward than it seems. Observe that the process of obtaining (11.1.6) and then taking its limit amounts to nothing more than taking the Laplace transform of both sides of the differential equation, thereby obtaining

$$\mathcal{L}\{x' + 2x\} = \mathcal{L}\{3e^t\},$$

and then using the linearity property to get

$$\mathcal{L}\{x'\} + 2\mathcal{L}\{x\} = 3\mathcal{L}\{e^t\}.$$

From Table 11.1, $\mathcal{L}\{e^t\} = 1/(s-1)$. Thus,

$$\mathcal{L}\{x'\} + 2\mathcal{L}\{x\} = \frac{3}{s-1}. \tag{11.4.1}$$

If $\mathcal{L}\{x\}$ were the only unknown in this equation, we could solve for it, which would give us the transform of the solution. However, the presence of $\mathcal{L}\{x'\}$ in the first term prevents this. But perhaps there is a formula connecting $\mathcal{L}\{x'\}$ and $\mathcal{L}\{x\}$ in which case we could use it to convert (11.4.1) into one involving only $\mathcal{L}\{x\}$. So the utility of employing Laplace transforms to solve differential equations hinges on the existence of such a formula.

By Definition 11.1.1, the Laplace transform of the derivative of a function f is

$$\mathcal{L}\{f'(t)\} = \lim_{T\to\infty} \int_0^T e^{-st} f'(t)\,dt.$$

Integrating by parts with $u = e^{-st}$, $dv = f'(t)\,dt$, $du = -se^{-st}\,dt$, and $v = f(t)$, we obtain

$$\begin{aligned} \int_0^T e^{-st} f'(t)\,dt &= e^{-st} f(t)\Big|_0^T + s\int_0^T e^{-st} f(t)\,dt \qquad (11.4.2)\\ &= e^{-sT} f(T) - f(0) + s\int_0^T e^{-st} f(t)\,dt. \end{aligned}$$

For $\mathcal{L}\{f'\}$ to exist, the right-hand side must converge as $T \to \infty$. Now the last term converges to $s\mathcal{L}\{f\}$ if $\mathcal{L}\{f\}$ exists. There is no problem with the second term since it is a constant. But what about the first term? If you recall, we faced this same problem before in solving the initial value problem (11.1.1). There we assumed from the outset that its solution $x(t)$ would behave in the following manner: $e^{-st}x(t) \to 0$ as $t \to \infty$. After we found it, we verified that this was indeed the case for $s > 1$. This long-term behavior was also used in Example 11.2.4 to find the Laplace transform of $f(t) = t^n$; there we proved for any fixed $s > 0$ that $e^{-st} f(t) = e^{-st}t^n \to 0$ as $t \to \infty$. Let us define a class of functions whose long-term behavior for any fixed value $s > \gamma$ will turn out to be

$$e^{-st} f(t) \to 0 \quad \text{as} \quad t \to \infty, \tag{11.4.3}$$

where the value of γ depends on the particular function f.

Exponential Order

Definition 11.4.1 A function $f(t)$ is said to be of ***exponential order*** γ as $t \to \infty$ if there are constants $M > 0$ and $t_1 \geq 0$ such that

$$|f(t)| \leq Me^{\gamma t} \tag{11.4.4}$$

for $t \geq t_1$.

Many authors omit the constant γ and the phrase "as $t \to \infty$" and simply say that "$f(t)$ is of *exponential order*." Another way of expressing exactly the same thing is to say that "$f(t)$ is of the ***order of*** $e^{\gamma t}$ as $t \to \infty$," which is the meaning of the notation

$$f(t) = O(e^{\gamma t}). \tag{11.4.5}$$

The use of the upper case letter O in this way is known as "big O" notation, which comes from the expression "order of magnitude." So we can also say that "$f(t)$ is big O of $e^{\gamma t}$." In particular, when $\gamma = 0$, (11.4.5) becomes $f(t) = O(e^{0t}) = O(1)$. In other words, the notation

$$f(t) = O(1)$$

means that constants $M > 0$ and $t_1 \geq 0$ exist such that $|f(t)| \leq M$ for all $t \geq t_1$. This is just a concise way of expressing that $f(t)$ is bounded on the interval $[t_1, \infty)$ for some $t_1 \geq 0$.

Let us consider the following examples to better acquaint ourselves with the new notation and to become proficient at recognizing functions that are of exponential order. After that, we will search for a formula that expresses the Laplace transform of the derivative of a function in terms of the Laplace transform of the function itself.

Example 11.4.1 Let n be any nonnegative integer. Determine whether $f(t) = t^n$ with domain $[0, \infty)$ is of exponential order. If it is, find suitable values for the constants γ, M, and t_1 in Definition 11.4.1.

Solution From the Maclaurin series

$$e^t = 1 + t + \frac{t^2}{2!} + \frac{t^3}{3!} + \cdots + \frac{t^n}{n!} + \ldots,$$

it follows that $e^t > t^n/n!$ and so

$$|f(t)| = \left|t^n\right| = t^n \leq n!\,e^t \quad \text{for all } t \geq 0.$$

Therefore, (11.4.4) holds for this function with $M = n!$, $\gamma = 1$, and $t_1 = 0$. This shows that $f(t) = t^n$ is of exponential order 1, which we can also express by writing $t^n = O(e^t)$. In fact, $t^n = O(e^{\gamma t})$ for any $\gamma > 0$. In order to see this, let γ be any fixed positive number, no matter how small. Recall we determined as we worked through Example 11.2.4 that $e^{-\gamma t}t^n \to 0$ as $t \to \infty$. Since $e^{-\gamma t}t^n$ is continuous on $[0, \infty)$, it must be bounded on this interval. That is, a constant $K > 0$ exists such that $\left|e^{-\gamma t}t^n\right| \leq K$ for all $t \geq 0$. Thus,

$$\left|t^n\right| \leq Ke^{\gamma t}$$

for all $t \geq 0$. ♦

Example 11.4.2 Explain why $u(t-b) = O(1)$.

Solution Recall that $u(t-b)$ denotes the unit step function, which is defined by (11.2.2). Let t_1 denote any constant greater than b. Then, for all $t \geq t_1$,

$$|u(t-b)| = u(t-b) = 1 \leq 1 \cdot e^{0t}.$$

This is (11.4.4) with $M = 1$ and $\gamma = 0$. Thus, $u(t-b)$ is of exponential order 0 as $t \to \infty$. ♦

Example 11.4.3 Determine whether the function

$$g(t) = -6te^{3t/2}\sin t,$$

where $t \geq 0$, is of exponential order.

Solution Since $-1 \leq \sin t \leq 1$ for all t and as $t < e^t$ for all $t \geq 0$,

$$|g(t)| \leq 6te^{3t/2} \leq 6e^t e^{3t/2} = 6e^{5t/2}$$

for $t \geq 0$. Thus, g is of exponential order 2.5. ♦

Example 11.4.4 Observe that the function

$$p(t) = \frac{5}{\sqrt{t}}$$

is unbounded on the interval $(0, \infty)$ since $p(t) \to \infty$ as $t \to 0^+$. However, p is bounded on $[1, \infty)$ because it is a positive, decreasing function on this interval and $p(1) = 5$. As a result, $p(t) = O(1)$ as $t \to \infty$. ♦

We conclude with an example of a function that is *not* of exponential order.

Example 11.4.5 The function

$$h(t) = e^{t^2}$$

is not of exponential order. For if it were, then constants γ, M, and t_1 would exist such that $|h(t)| = e^{t^2} \leq Me^{\gamma t}$ for all $t \geq t_1$. Equivalently, $e^{t^2 - \gamma t} \leq M$ for all $t \geq t_1$. But this contradicts the fact that $e^{t(t-\gamma)} \to \infty$ as $t \to \infty$. ♦

Now that we know what functions of exponential order are and have some idea of what they are supposed to look like, let us show that these functions display the long-term behavior that was claimed earlier in (11.4.3). Suppose $f(t) = O\left(e^{\gamma t}\right)$ as $t \to \infty$. Then, according to (11.4.4), there are constants $M > 0$ and $t_1 \geq 0$ such that

$$-Me^{\gamma t} \leq f(t) \leq Me^{\gamma t} \quad \text{for all } t \geq t_1.$$

Multiplying both inequalities by e^{-st}, we have

$$-Me^{-(s-\gamma)t} \leq e^{-st} f(t) \leq Me^{-(s-\gamma)t}.$$

If $s > \gamma$, then $\pm Me^{-(s-\gamma)t} \to 0$ as $t \to \infty$. Since $e^{-st} f(t)$ is sandwiched between these two functions, it too must approach 0 as $t \to \infty$. Thus, we have the following result:

Long-Term Behavior of Functions of Exponential Order

Theorem 11.4.2 *If $f(t) = O\left(e^{\gamma t}\right)$ as $t \to \infty$, then for $s > \gamma$,*

$$\lim_{t\to\infty} e^{-st} f(t) = 0. \tag{11.4.6}$$

At last all of the necessary pieces are in place to derive a formula relating the Laplace transform of the derivative of a function to the transform of the function itself.

Laplace Transform of the First Derivative

Theorem 11.4.3 *Let $f(t)$ be a function that is differentiable on $[0, \infty)$ and of exponential order γ. If the Laplace transform of $f(t)$ exists and if the derivative $f'(t)$ is integrable on $[0, T]$ for every $T \geq 0$, then the Laplace transform of $f'(t)$ exists for all $s > \gamma$ and is given by the formula*

$$\mathcal{L}\{f'(t)\} = s\mathcal{L}\{f(t)\} - f(0). \tag{11.4.7}$$

Proof For every $T > 0$, the product $e^{-st} f'(t)$ is integrable on $[0, T]$ since both factors are integrable on $[0, T]$.[8] Integrating $e^{-st} f'(t)$ by parts (see (11.4.2)), we have

$$\int_0^T e^{-st} f'(t)\,dt = e^{-sT} f(T) - f(0) + s \int_0^T e^{-st} f(t)\,dt.$$

It follows from (11.4.6) that for $s > \gamma$, $e^{-sT} f(T) \to 0$ as $T \to \infty$. Furthermore, the hypothesis that the transform of f exists means $\int_0^T e^{-st} f(t)\,dt \to \mathcal{L}\{f(t)\}$. It follows that the integral on the left-hand side has a limit and that it is given by the right-hand side of (11.4.7). ■

Example 11.4.6 Find the Laplace transform of $\cos bt$, where b is a constant given that the Laplace transform of $\sin bt$ (see Example 11.2.5) is

$$\mathcal{L}\{\sin bt\} = \frac{b}{s^2 + b^2}.$$

Solution Let $f(t) = \sin bt$. Then, $f'(t) = b \cos bt$ and $f(0) = 0$. Then it follows from (11.4.7) that

$$\mathcal{L}\{b \cos bt\} = s\mathcal{L}\{\sin bt\} - 0 = s\left(\frac{b}{s^2 + b^2}\right).$$

By the linearity property of the transform,

$$\mathcal{L}\{\cos bt\} = \frac{s}{s^2 + b^2}.$$

♦

Example 11.4.7 Find $\mathcal{L}\{\sin^2 bt\}$.

Solution Use (11.4.7) letting $f(t) = \sin^2 bt$. Then as $f(0) = 0$, we have

$$\mathcal{L}\left\{\frac{d}{dt}\sin^2 bt\right\} = s\mathcal{L}\{\sin^2 bt\} - 0.$$

Carrying out the differentiation on the left-hand side, we get

$$\mathcal{L}\{2b(\sin bt)(\cos bt)\} = s\mathcal{L}\left\{\sin^2 bt\right\}.$$

[8] If f and g are integrable functions on an interval $[a, b]$, then their product fg is also integrable on $[a, b]$. For a proof of this, see any advanced calculus or real analysis book, such as [6, p. 254] and [37, p. 120].

By the double-angle identity $2(\sin bt)(\cos bt) = \sin 2bt$. Hence,

$$\mathcal{L}\left\{\sin^2 bt\right\} = \frac{1}{s}\,\mathcal{L}\left\{b\sin 2bt\right\} = \frac{b}{s}\cdot\frac{2b}{s^2+(2b)^2} = \frac{2b^2}{s\left(s^2+4b^2\right)}.$$

♦

Now that we have a formula relating the transform of a function to the transform of its derivative, namely (11.4.7), let us go back to (11.4.1) to continue retracing the steps that led to the solution of the initial value problem (11.1.1). Recall it is our intention to streamline the roundabout way in which we used Laplace transforms to solve (11.1.1). This will be accomplished through the $\mathcal{L}$ operator. As

$$\mathcal{L}\{x'(t)\} = s\mathcal{L}\{x(t)\} - x(0)$$

and $x(0) = 1$, (11.4.1) becomes

$$s\mathcal{L}\{x(t)\} - 1 + 2\mathcal{L}\{x(t)\} = \frac{3}{s-1}.$$

It is here that we should point out one of the advantages in using the Laplace transform to solve initial value problems.[9] Note that $\mathcal{L}$ has transformed (11.1.1), the differential equation and the initial condition, into an algebraic equation:

$$\left\{\begin{array}{l} x'+2x=3e^t \\ x(0)=1 \end{array}\right\} \xrightarrow{\ \mathcal{L}\ } \quad s\mathcal{L}\{x(t)\} - 1 + 2\mathcal{L}\{x(t)\} = \frac{3}{s-1}.$$

Solving the algebraic equation for the unknown $\mathcal{L}\{x(t)\}$, we get

$$(s+2)\mathcal{L}\{x(t)\} = \frac{3}{s-1} + 1 = \frac{s+2}{s-1}$$

and so

$$\mathcal{L}\{x(t)\} = \frac{1}{s-1}.$$

Recall this is (11.1.10). Of course, this is the transform of the solution of (11.1.1), not the solution. Let us look for a continuous function with this transform. From Table 11.1, we see that

$$\frac{1}{s-1} = \mathcal{L}\{e^t\}.$$

[9] Other advantages will be pointed out later, such as the efficient manner in which the transform can handle discontinuous functions.

Thus,

$$\mathcal{L}\{x(t)\} = \mathcal{L}\{e^t\}.$$

From this we surmise that

$$x(t) = e^t.$$

Recall from earlier discussions that this is indeed the unique solution of the initial value problem (11.1.1).

When employing Laplace transforms to solve an initial value problem, we will eventually arrive at the point where the transform of the solution $x(t)$ is expressed in terms of the transform of some other function, say $f(t)$. That is,

$$\mathcal{L}\{x(t)\} = \mathcal{L}\{f(t)\}.$$

(For example, we found $\mathcal{L}\{x(t)\} = \mathcal{L}\{e^t\}$ in the previous discussion.) So we ask if $x(t) = f(t)$? The answer to this is given by a special version of a result due to Mathias Lerch.[10] For continuous functions, it says that different continuous functions have different Laplace transforms. Therefore, there can only be one continuous function whose transform is equal to the transform of the solution.

Lerch's Theorem

Theorem 11.4.4 *If* $f(t)$ *and* $g(t)$ *are continuous on the entire interval* $[0, \infty)$ *and* $\mathcal{L}\{f(t)\} = \mathcal{L}\{g(t)\}$ *for* $s > \gamma$ *for some* $\gamma \geq 0$, *then* $f(t) = g(t)$ *for all* $t \geq 0$.

This version of Lerch's theorem is found in Marsden [58, p. 390; p. 398 f.]. We cannot prove it here because the proof depends on advanced results from real and complex analysis. A more general statement of Lerch's theorem and where a proof can be found is given in Churchill's *Operational Mathematics* [21, p. 14; p. 184].

Example 11.4.8 Let us try out the method that we just used to solve (11.1.1) on the initial value problem:

$$\frac{dx}{dt} - 2x = 4, \quad x(0) = 0. \tag{11.4.8}$$

[10] Mathias Lerch (1860–1922), born in Czechoslovakia, was a professor of mathematics at the University of Freiburg in Germany and at Masaryk University in Czechoslovakia. He published about 240 papers, mostly in analysis and number theory. Some of his results are important to the study of operators, such as the Laplace transform operator.

Solution First, take the Laplace transform of both sides of the differential equation:

$$\mathcal{L}\left\{x' - 2x\right\} = \mathcal{L}\{4\}.$$

Due to the linearity property of the Laplace transform, this becomes

$$\mathcal{L}\{x'\} - 2\mathcal{L}\{x\} = 4\mathcal{L}\{1\}.$$

It then follows from formula (11.4.7) and the first entry in Table 11.1 that

$$s\mathcal{L}\{x\} - x(0) - 2\mathcal{L}\{x\} = 4 \cdot \frac{1}{s}.$$

Since the initial condition is $x(0) = 0$, this simplifies to

$$s\mathcal{L}\{x\} - 2\mathcal{L}\{x\} = \frac{4}{s}.$$

Solving this equation for $\mathcal{L}\{x(t)\}$, we get

$$\mathcal{L}\{x(t)\} = \frac{4}{s(s-2)}. \tag{11.4.9}$$

Now that we have $\mathcal{L}\{x(t)\}$, we can find $x(t)$ by searching for a continuous function with this particular transform. But the right-hand side of (11.4.9) or anything resembling it does not appear in Table 11.1. We could look for it in an engineering or math handbook that has a more extensive table of Laplace transforms. There we would find an entry for $1/[(s-a)(s-b)]$. On the other hand, we do not need to do this since we can express (11.4.9) in terms of the functions listed in Table 11.1. Note that we would run into this same problem if we had to integrate (11.4.9): if that were the case, we would simply write the right-hand side of (11.4.9) as a sum of two rational functions and then integrate. In other words, we would first apply the method of partial fractions. Let us try that here. The form of the partial fraction expansion of (11.4.9) is

$$\frac{4}{s(s-2)} = \frac{A}{s-2} + \frac{B}{s}.$$

To find A and B, we first multiply by $s(s-2)$ to clear the fractions of their denominators obtaining

$$As + B(s-2) = 4.$$

Then set $s = 2$ to find A:

$$s = 2 \quad \Rightarrow \quad 2A = 4 \quad \Rightarrow \quad A = 2.$$

And $s = 0$ to find B:

$$s = 0 \quad \Rightarrow \quad -2B = 4 \quad \Rightarrow \quad B = -2.$$

Therefore,

$$\frac{4}{s(s-2)} = \frac{2}{s-2} - \frac{2}{s}$$

and so

$$\mathcal{L}\{x(t)\} = 2 \cdot \frac{1}{s-2} - 2 \cdot \frac{1}{s}.$$

From Table 11.1 and the linearity of the transform, we see that

$$2 \cdot \frac{1}{s-2} - 2 \cdot \frac{1}{s} = 2\mathcal{L}\{e^{2t}\} - 2\mathcal{L}\{1\} = \mathcal{L}\{2e^{2t} - 2\}.$$

Since $\mathcal{L}\{x(t)\} = \mathcal{L}\{2e^{2t} - 2\}$, we conclude from Lerch's theorem (Theorem 11.4.4) that the solution of the initial value problem (11.4.8) is

$$x(t) = 2e^{2t} - 2.$$

♦

An examination of how we solved the initial value problems (11.1.1) and (11.4.8) reveals the steps that are needed to solve initial value problems of the form

$$x' + ax = q(t), \quad x(0) = x_0$$

using Laplace transforms.

Method of Laplace Transforms
To solve an IVP using Laplace transforms, execute the following steps:

1. Take the Laplace transform of both sides of the differential equation.
2. Use the initial condition and properties of the Laplace transform to express the algebraic equation obtained in Step 1 in terms of the Laplace transform of the solution.
3. Solve the equation obtained in Step 2 for the transform of the solution.
4. Determine the solution by finding a continuous function with a Laplace transform the same as the one obtained in Step 3.

Before we consider any more examples, let us point out the assumptions which are unsaid but implicit when we employ the above method. In Step 1, we assume the existence of a function $x(t)$ that is continuous on $[0, \infty)$ and which satisfies the first-order differential equation and an accompanying initial condition. This is a valid assumption since for any constant a and continuous function q on $[0, \infty)$ there is a unique continuous solution $x(t)$ of

$$x' + ax = q(t), \quad x(0) = x_0$$

on $[0, \infty)$ (see Theorem 5.2.1). It follows from the continuity of the right-hand side of

$$x'(t) = q(t) - ax(t)$$

that $x'(t)$ is continuous on $[0, \infty)$. Consequently, $x'(t)$ is integrable on $[0, T]$ for every $T \geq 0$. Thus, some of the hypotheses of Theorem 11.4.3 are automatically fulfilled. This still leaves open the question of the existence of the Laplace transform of $x(t)$ and whether $x(t)$ is of exponential order. What we do in practice is to go ahead with the method of Laplace transforms assuming that all the hypotheses are satisfied. If Steps 3 and 4 can be carried out, then we can check if the result of Step 4 fulfills the hypotheses and is a solution.

In the next section, we will list conditions that guarantee the existence of the Laplace transform of a given function, thereby avoiding the tedium that comes from applying Definition 11.1.1. But first let us solve a couple of initial value problems in order to illustrate the above method.

Example 11.4.9 Solve the initial value problem

$$\frac{dx}{dt} - x = -t, \quad x(0) = 1. \tag{11.4.10}$$

Solution First, take the Laplace transform of both sides of the differential equation and apply the linearity property:

$$\mathcal{L}\{x'\} - \mathcal{L}\{x\} = -\mathcal{L}\{t\}.$$

Then use Theorem 11.4.3, the entry for t^n with $n = 1$ in Table 11.1, and the alternative notation $X(s)$ for $\mathcal{L}\{x(t)\}$ to get

$$sX(s) - x(0) - X(s) = -\frac{1}{s^2}.$$

Next, use the initial condition $x(0) = 1$ and solve this equation for $X(s)$:

$$X(s) = \frac{1}{s-1} - \frac{1}{s^2(s-1)}.$$

Factor out $1/(s-1)$ and simplify:

$$X(s) = \frac{1}{s-1}\left[1 - \frac{1}{s^2}\right] = \frac{1}{s-1}\left[\frac{s^2}{s^2-1}\right] = \frac{s+1}{s^2} = \frac{1}{s} + \frac{1}{s^2}.$$

Thus, the transform of the solution $x(t)$ of the initial value problem (11.4.10) is

$$\mathcal{L}\{x(t)\} = \frac{1}{s} + \frac{1}{s^2}.$$

Consulting Table 11.1, we see that

$$\frac{1}{s} + \frac{1}{s^2} = \mathcal{L}\{1\} + \mathcal{L}\{t\}.$$

Consequently,

$$\mathcal{L}\{x(t)\} = \mathcal{L}\{1\} + \mathcal{L}\{t\} = \mathcal{L}\{1+t\}.$$

Therefore, by Theorem 11.4.4,

$$x(t) = 1 + t.$$

♦

Example 11.4.10 Solve

$$x' + 4x = 2 - 25\cos 3t, \qquad x(0) = 2.$$

Solution Transforming the differential equation, we get

$$\mathcal{L}\{x'\} + 4\mathcal{L}\{x\} = 2\,\mathcal{L}\{1\} - 25\,\mathcal{L}\{\cos 3t\}.$$

As in the previous example, use (11.4.7) and consult Table 11.1. As a result,

$$sX(s) - x(0) + 4X(s) = 2\cdot\frac{1}{s} - 25\cdot\frac{s}{s^2+3^2},$$

where $X(s) = \mathcal{L}\{x(t)\}$. Solving for $X(s)$, we get

$$X(s) = \frac{2}{s+4} + \frac{2}{s(s+4)} - \frac{25s}{(s+4)(s^2+9)}.$$

Now we need to determine the solution by finding a continuous function with this same transform. Unfortunately, the second and third terms are not listed in

Table 11.1. However, we can still use the table by expressing these terms as sums of partial fractions. First let us expand the second term by expressing it as the sum

$$\frac{2}{s(s+4)} = \frac{A}{s} + \frac{B}{s+4}.$$

To obtain the values of the coefficients, we multiply both of its sides by $s(s+4)$:

$$A(s+4) + Bs = 2.$$

Now we can readily obtain the values of A and B from this equation by setting $s = 0$ and $s = -4$:

$$s = 0 \quad \Rightarrow \quad 4A = 2 \quad \Rightarrow \quad A = \frac{1}{2}$$

$$s = -4 \quad \Rightarrow \quad -4B = 2 \quad \Rightarrow \quad B = -\frac{1}{2}.$$

Thus,

$$\frac{2}{s(s+4)} = \frac{1}{2} \cdot \frac{1}{s} - \frac{1}{2} \cdot \frac{1}{s+4}. \tag{11.4.11}$$

The partial fraction expansion of the third term is

$$\frac{25s}{(s+4)(s^2+9)} = \frac{C}{s+4} + \frac{Ds+E}{s^2+9}.$$

Multiplying by $(s+4)(s^2+9)$ gives

$$C(s^2+9) + (Ds+E)(s+4) = 25s.$$

Setting s successively equal to -4, 0, and 1 gives

$$s = -4: \quad 25C = -100 \quad \Rightarrow \quad C = -4$$

$$s = 0: \quad 9C + 4E = 0 \quad \Rightarrow \quad E = -\frac{9}{4}C = 9$$

$$s = 1: \quad 10C + 5(D+E) = 25 \quad \Rightarrow \quad D = 5 - 2C - E = 5 - 2(-4) - 9 = 4.$$

Therefore,

$$\frac{25s}{(s+4)(s^2+9)} = -4 \cdot \frac{1}{s+4} + \frac{4s+9}{s^2+9} \tag{11.4.12}$$

$$= -4 \cdot \frac{1}{s+4} + 4 \cdot \frac{s}{s^2+9} + 3 \cdot \frac{3}{s^2+9}.$$

Now let us substitute the two partial fraction expansions, (11.4.11) and (11.4.12), for the second and third terms of $X(s)$, respectively, and then simplify:

$$X(s) = \frac{1}{2} \cdot \frac{1}{s} + \frac{11}{2} \cdot \frac{1}{s-(-4)} - 4 \cdot \frac{s}{s^2+3^2} - 3 \cdot \frac{3}{s^2+3^2}$$

$$= \frac{1}{2} \mathcal{L}\{1\} + \frac{11}{2} \mathcal{L}\{e^{-4t}\} - 4 \mathcal{L}\{\cos 3t\} - 3 \mathcal{L}\{\sin 3t\}.$$

This is the same as

$$\mathcal{L}\{x(t)\} = \mathcal{L}\left\{\frac{1}{2} + \frac{11}{2} e^{-4t} - 4\cos 3t - 3\sin 3t\right\}$$

by the linearity property. Finally, it follows from Lerch's theorem that the solution is

$$x(t) = \frac{1}{2} + \frac{11}{2} e^{-4t} - 4\cos 3t - 3\sin 3t.$$

♦

11.5 Existence of the Laplace Transform

In this section we state and prove some results guaranteeing the existence of Laplace transforms for certain types of functions. The following theorem is one of those results.

Existence of the Laplace Transform I

Theorem 11.5.1 *If a function $f(t)$ is bounded and integrable on every interval $[0, T]$ and of exponential order γ, then its Laplace transform exists for all $s > \gamma$.*

Proof To prove the existence of the Laplace transform of $f(t)$ for all $s > \gamma$, we have to establish that the improper integral $\int_0^\infty e^{-st} f(t)\,dt$ converges for every $s > \gamma$. That is, we must prove $\int_0^T e^{-st} f(t)\,dt$ exists for all $T > 0$ and $s > \gamma$ and that it approaches a finite limit as $T \to \infty$. The integral clearly exists on $[0, T]$ for every $T > 0$ since $e^{-st} f(t)$ is bounded and integrable on $[0, T]$.[11] And as $f(t)$ is of exponential order γ, constants M and t_1 exist such that $|f(t)| \leq Me^{\gamma t}$ for all $t \geq t_1$. In fact, since $f(t)$ is bounded on $[0, t_1]$, we can even say that

$$|f(t)| \leq Me^{\gamma t}$$

for all $t \geq 0$ by increasing the value of M if necessary. Now we can address the question of the convergence of $\int_0^\infty e^{-st} f(t)\, dt$ for $s > \gamma$. Observe that the integrand is dominated by the exponential function $Me^{-(s-\gamma)t}$ for $t \geq 0$ since

$$|e^{-st} f(t)| = e^{-st}|f(t)| \leq e^{-st} Me^{\gamma t} \leq Me^{-(s-\gamma)t}. \tag{11.5.1}$$

Then as $s > \gamma$,

$$\lim_{T\to\infty} \int_0^T Me^{-(s-\gamma)t}\, dt = \lim_{T\to\infty} \left[\frac{M}{-(s-\gamma)} e^{-(s-\gamma)t} \right]_0^T = \frac{M}{s-\gamma}. \tag{11.5.2}$$

By a well-known comparison test for improper integrals,[12] $\int_0^\infty e^{-st} f(t)\, dt$ converges. In other words, $\mathcal{L}\{f(t)\}$ exists for $s > \gamma$. ■

Example 11.5.1 In Example 11.4.3 we showed that

$$g(t) = -6te^{3t/2} \sin t$$

is of exponential order 2.5. Since g is continuous everywhere, it is bounded and integrable on $[0, T]$ for every $T > 0$. Therefore, the theorem allows us to conclude that $\mathcal{L}\{g(t)\}$ exists for $s > 2.5$ without having to resort to Definition 11.1.1 in order to determine whether or not the improper integral

$$\int_0^\infty e^{-st} g(t)\, dt$$

converges. Sometimes this is all that we need to know. ♦

Example 11.5.2 The function $tH(t-b)$, where $b > 0$, is continuous on $[0, \infty)$ except at $t = b$; so it is bounded and integrable on every interval $[0, T]$. Since for

[11] The product of integrable functions is integrable (cf. Ross [62, §33 Exercises] or Fulks [37, p. 120]).

[12] If g and h are integrable on $[a, T]$ for all $T > a$ and if $|g(t)| \leq h(t)$ for all $t \geq a$, then $\int_a^\infty g(t)\, dt$ converges if $\int_a^\infty h(t)\, dt$ converges. For more information, consult a real analysis textbook, such as Bartle [5, p. 262] or Fulks [37, p. 468, p. 472]. Weaker versions of the comparison test are stated in elementary calculus textbooks, such as Stewart [68, p. 533] and Thomas and Finney [72, p. 525].

$t \geq 0$,

$$|tH(t-b)| \leq t < e^t,$$

$tH(t-b)$ is of exponential order 1. Thus, its Laplace transform exists for $s > 1$. ♦

The proof of Theorem 11.5.1 suggests the following result.

Long-Term Behavior of Laplace Transforms

Corollary 11.5.2 *If a function $f(t)$ satisfies the hypotheses of Theorem 11.5.1, then*

$$\lim_{s\to\infty} \mathcal{L}\{f\}(s) = 0. \tag{11.5.3}$$

Proof It follows from (11.5.1) and (11.5.2) that

$$|\mathcal{L}\{f\}(s)| = \left|\int_0^{\infty} e^{-st} f(t)\, dt\right| \leq \int_0^{\infty} \left|e^{-st} f(t)\right| dt \leq \frac{M}{s-\gamma}$$

for $s > \gamma$. As a result,

$$0 \leq |\mathcal{L}\{f\}(s)| \leq \frac{M}{s-\gamma}.$$

This implies (11.5.3) since the lower bound is fixed at 0 while the upper bound approaches 0 as $t \to \infty$. ■

For instance, consider the function

$$F(s) = \frac{s}{s-1}.$$

It follows from the corollary that there is no bounded, integrable function of exponential order with $F(s)$ as its Laplace transform. That is because

$$\lim_{s\to\infty} F(s) = \lim_{s\to\infty} \frac{s}{s-1} = 1.$$

Corollary 11.5.2 will be put to good use later on in Sect. 11.9 to help solve differential equations with variable coefficients.

11.5.1 Piecewise Continuous Functions

The set of continuous functions suffice for mathematically modeling a lot of things. Nevertheless, there are physical phenomena and situations, such as the current in an electrical circuit that is switched on and off, that require a broader class of functions with which to model them. One of those consists of functions that are discontinuous at isolated points but whose Laplace transforms exist. In order to fully describe these functions, we need to know what is meant by a *jump discontinuity* and a *removable discontinuity*. Since their definitions involve the concept of continuity, let us first recall that a function $f(t)$ is said to be ***continuous*** at an interior point τ of a closed interval $[a, b]$ on which it is defined if

$$\lim_{t\to\tau} f(t) = f(\tau).$$

This is the succinct way of conveying that

(i) $f(\tau)$ is defined;
(ii) $\lim_{t\to\tau^-} f(t)$ exists and has a finite value;
(iii) $\lim_{t\to\tau^+} f(t)$ exists and has a finite value;

and that

(iv) $\lim_{t\to\tau^-} f(t) = \lim_{t\to\tau^+} f(t) = f(\tau)$.

As for the endpoints of $[a, b]$, $f(t)$ is said to be continuous at a if $\lim_{t\to a^+} f(t) = f(a)$. It is continuous at b if $\lim_{t\to b^-} f(t) = f(b)$.

Now we can define the terms: jump discontinuity and removable discontinuity. A function $f(t)$ is said to have a ***jump discontinuity*** at a point $\tau \in (a, b)$ if both $\lim_{t\to\tau^-} f(t)$ and $\lim_{t\to\tau^+} f(t)$ exist and are finite but

$$\lim_{t\to\tau^-} f(t) \neq \lim_{t\to\tau^+} f(t)$$

while $f(\tau)$ may or may not be defined. The difference

$$\lim_{t\to\tau^+} f(t) - \lim_{t\to\tau^-} f(t)$$

is called the ***jump of*** f ***at the point*** τ, which may be either positive or negative. The definition of a jump discontinuity at an endpoint of $[a, b]$ is very much like its definition at an interior point: $f(t)$ is said to have a jump discontinuity at a if $\lim_{t\to a^+} f(t)$ exists and is finite and if $f(a)$ is defined but

$$\lim_{t\to a^+} f(t) \neq f(a).$$

Similarly, we say that $f(t)$ has a jump discontinuity at b if $\lim_{t\to b^-} f(t)$ exists and is finite and if $f(b)$ is defined but $\lim_{t\to b^-} f(t) \neq f(b)$.

A function $f(t)$ has a ***removable discontinuity*** at a point $\tau \in (a, b)$ if $\lim_{t\to\tau} f(t)$ exists and is finite but either (i) $f(\tau)$ is not defined or (ii) $f(\tau)$ is defined but

$$\lim_{t\to\tau} f(t) \neq f(\tau).$$

This term is appropriate since $f(t)$ can be made continuous at τ by defining $f(\tau)$ if (i) is the case or redefining $f(\tau)$ if (ii) is the case so that

$$f(\tau) = \lim_{t\to\tau} f(t).$$

If $\lim_{t\to a^+} f(t)$ exists and is finite but $f(a)$ is not defined, then we say that $f(t)$ has a removable discontinuity at the left endpoint a. The definition of a removable discontinuity at a right endpoint is completely analogous to this definition and is left to the reader to formulate. Finally, we point out that jump continuities at endpoints are also removable but those at interior points cannot be removed.

Piecewise Continuous Functions

Definition 11.5.3

- A function f is said to be ***piecewise continuous on a finite interval*** $[a, b]$ if it is continuous at each point in $[a, b]$ except for at most a finite number of jump discontinuities and removable discontinuities.
- A function f is said to be ***piecewise continuous on the infinite interval*** $[0, \infty)$ if it is piecewise continuous on $[0, T]$ for every $T > 0$.

Example 11.5.3 For a given $b > 0$, the unit step function $u(t - b)$ is continuous on $[0, \infty)$ except at $t = b$ since it is not defined at this point but is defined and continuous at all other points of the interval by virtue of its definition, namely, that $u(t - b) = 0$ for $t < b$ and $u(t - b) = 1$ for $t > b$. Furthermore, the discontinuity at $t = b$ is a jump discontinuity since both one-sided limits at $t = b$ exist and are finite but are not equal:

$$\lim_{t\to b^-} u(t - b) = 0 \quad \text{whereas} \quad \lim_{t\to b^+} u(t - b) = 1.$$

It follows from Definition 11.5.3 that $u(t - b)$ is piecewise continuous on $[0, \infty)$. ♦

Example 11.5.4 The graph of the function

$$f(t) = \begin{cases} 3 + \cos t, & \text{if } -8 \le t < -4 \\ 4, & \text{if } -4 < t \le -2 \\ 5.5 + \dfrac{\sin t}{t}, & \text{if } -2 < t < 1 \\ 2, & \text{if } t = 1 \\ \dfrac{t^3}{10} - 4\sin 3t - 6, & \text{if } 1 < t \le 5 \\ 2 - t, & \text{if } 5 < t < 8 \\ -3, & \text{if } t = 8 \end{cases}$$

on the closed interval $[-8, 8]$ is shown in Fig. 11.1.

Note that $f(t)$ is undefined at $t = 0$, which is what the small circle on the graph at that location signifies. This discontinuity can be removed by defining $f(0)$ to be

$$f(0) = \lim_{t\to 0} f(t) = \lim_{t\to 0}\left(5.5 + \frac{\sin t}{t}\right) = 5.5 + \lim_{t\to 0}\frac{\sin t}{t} = 6.5.$$

The jump discontinuity at the right endpoint $t = 8$ can also be removed by changing its value there from -3 to

$$f(8) = \lim_{t\to 8^-} f(t) = \lim_{t\to 8^-}(2 - t) = -6.$$

The other four discontinuities at $t = -4, -2, 1$, and 5 are not removable because they are jump discontinuities that are located at interior points. As $f(t)$ is continuous at all other points, we conclude that $f(t)$ is piecewise continuous on $[-8, 8]$. ♦

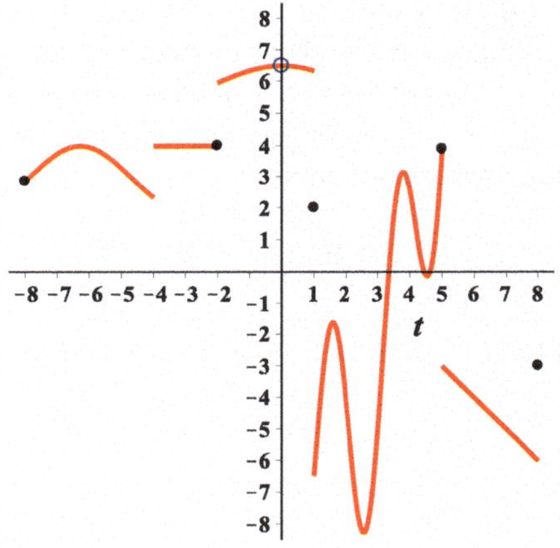

Fig. 11.1 The piecewise continuous function $f(t)$ in Example 11.5.4

Example 11.5.5 A continuous function on an interval is also considered piecewise continuous on that interval since it has neither jump discontinuities nor removable discontinuities. ♦

Existence of the Laplace Transform II

Corollary 11.5.4 *If a function $f(t)$ is piecewise continuous on $[0, \infty)$ and of exponential order γ, then its Laplace transform exists for $s > \gamma$.*

Proof It follows from the definition of piecewise continuity that $f(t)$ is bounded and has at most a finite number of isolated discontinuities on every finite interval $[0, T]$. This implies $f(t)$ is integrable on $[0, T]$.[13] Consequently, by Theorem 11.5.1, $\mathcal{L}\{f(t)\}$ exists for $s > \gamma$. ■

11.6 Properties of the Laplace Transform

We have already found two properties of the Laplace transform that are needed to solve initial value problems, namely,

- the linearity property:

$$\mathcal{L}\{c_1 f_1(t) + c_2 f_2(t)\} = c_1 \mathcal{L}\{f_1(t)\} + c_2 \mathcal{L}\{f_2(t)\}$$

- the Laplace transform of the first derivative of a function:

$$\mathcal{L}\{f'(t)\} = s\mathcal{L}\{f(t)\} - f(0).$$

Now we will derive some other very useful properties. The first two of these involve shifting (translating) along the s-axis and the t-axis.

11.6.1 Shift in s Property

Oftentimes we need the transform of a product of two functions, one of which is an exponential function while the other is a function whose Laplace transform we already have. In other words, the product has the form $e^{bt} f(t)$, where b is a constant and $\mathcal{L}\{f\}$ is known or can easily be found. This particular product is important; one

[13] See Bartle and Sherbert [6, p. 252], Ross [62, §33], or Gaughan [38, p. 167].

reason being that it may turn up as a solution of a linear equation. For example, you can check that $e^{5t}\sin t$ is a solution of $\dot{y} - 5y = e^{5t}\cos t$.

Now that we have established the importance of knowing the transform of $e^{bt}f(t)$, let us get down to the business of finding it. We begin by assuming that $\mathcal{L}\{f(t)\}$ exists and is equal to a function $F(s)$ for $s > a$. Then, by (11.1.12),

$$\mathcal{L}\{e^{bt}f(t)\} = \int_0^\infty e^{-st} \cdot e^{bt}\, dt = \int_0^\infty e^{-(s-b)t} f(t)\, dt.$$

Since

$$F(s) = \mathcal{L}\{f(t)\} = \int_0^\infty e^{-st} f(t)\, dt,$$

it follows that

$$\mathcal{L}\{e^{bt}f(t)\} = F(s-b)$$

for $s - b > a$. So when a function f is multiplied by e^{bt}, its Laplace transform can be obtained by merely replacing the argument s of F (the Laplace transform of f) with $s - b$. This shifts the graph of F along the s-axis either b units to the right or to the left depending on the sign of b. Accordingly, we refer to this property as the *shift in s property.*

Shift in s

Theorem 11.6.1 *If $\mathcal{L}\{f(t)\} = F(s)$ for $s > a$, then, for any constant b,*

$$\mathcal{L}\{e^{bt}f(t)\} = F(s-b) \tag{11.6.1}$$

for $s > a + b$.

Example 11.6.1 Find $\mathcal{L}\{e^{2t}\cos 3t\}$.

Solution We begin with

$$\mathcal{L}\{\cos 3t\} = \frac{s}{s^2 + 3^2} = F(s),$$

which holds for $s > 0$. Using the notation of Theorem 11.6.1, $a = 0$ and $b = 2$. Thus,

$$\mathcal{L}\{e^{2t}\cos 3t\} = F(s-2) = \frac{s-2}{(s-2)^2 + 3^2}$$

for $s > 2$. ♦

Example 11.6.2 Find $\mathcal{L}\{e^{-t}t^5\}$.

Solution Start with

$$\mathcal{L}\{t^5\} = \frac{5!}{s^6} = F(s) \qquad (s > 0).$$

Now set $b = -1$ in (11.6.1):

$$\mathcal{L}\{e^{-t}t^5\} = F(s+1) = \frac{5!}{(s+1)^6} \qquad (s > -1).$$

♦

The next example illustrates using the shift property in reverse.

Example 11.6.3 Find a function f with the Laplace transform

$$\mathcal{L}\{f(t)\} = \frac{1}{s^2 - s + 1}.$$

Solution Although $1/(s^2 - s + 1)$ is not listed in Table 11.1, it is related to the $\sin bt$ entry. To see this, complete the square of its denominator to write it in the form $b/(s^2 + b^2)$:

$$\frac{1}{s^2 - s + 1} = \frac{1}{(s - \frac{1}{2})^2 + \frac{3}{4}} = \frac{2}{\sqrt{3}} \cdot \frac{\frac{\sqrt{3}}{2}}{(s - \frac{1}{2})^2 + (\frac{\sqrt{3}}{2})^2}.$$

From Table 11.1, we see

$$\frac{\frac{\sqrt{3}}{2}}{s^2 + (\frac{\sqrt{3}}{2})^2} = \mathcal{L}\{\sin \tfrac{\sqrt{3}}{2} t\}.$$

By the shift property (11.6.1),

$$\frac{\frac{\sqrt{3}}{2}}{(s - \frac{1}{2})^2 + (\frac{\sqrt{3}}{2})^2} = \mathcal{L}\{e^{\frac{1}{2}t} \sin (\tfrac{\sqrt{3}}{2} t)\}.$$

Thus,

$$f(t) = \frac{2}{\sqrt{3}} e^{\frac{1}{2}t} \sin (\tfrac{\sqrt{3}}{2} t).$$

♦

11.6.2 Shift in t Property

Abrupt changes can occur in many different types of situations. Some of them that we describe in Chap. 7 involve

- culling a herd of horses,
- hitting a baseball with a bat, and
- depositing money into an account at regular intervals.

There are other examples which you may encounter sometime, such as determining the current in an electrical circuit when a switch is turned from off to on and vice versa.

A situation involving an abrupt change in the value of a quantity, such as those just cited, is oftentimes modeled with a linear nonhomogeneous differential equation, where the nonhomogeneous terms involve constant multiples of unit step functions. Recall that the Laplace transform of $u(t-b)$ is

$$\mathcal{L}\{u(t-b)\} = \frac{e^{-bs}}{s}$$

where $b \geq 0$. With this transform we can solve linear differential equations involving step functions. But how do we handle equations having terms involving abrupt changes that are modeled with slightly more complicated functions of the form

$$f(t)u(t-b) = \begin{cases} 0, & \text{if } t < b \\ f(t), & \text{if } t > b? \end{cases} \tag{11.6.2}$$

And what is the Laplace transform of such a function?

To answer this, let us start off with a function f that is defined for $t \geq 0$ and whose Laplace transform exists for $s > a$. That is, assume that

$$\mathcal{L}\{f(t)\}(s) = \int_0^\infty e^{-st} f(t)\,dt \tag{11.6.3}$$

converges for $s > a$. Applying the definition of the Laplace transform to (11.6.2), we have

$$\mathcal{L}\{f(t)u(t-b)\} = \int_0^\infty e^{-st} f(t)u(t-b)\,dt = \int_b^\infty e^{-st} f(t)\,dt,$$

which obviously converges for $s > a$ since our assumption is that (11.6.3) converges for $s > a$

With an appropriate change of variable, we can change the lower limit of integration b back to 0 so that the integral is a Laplace transform. All we have to do is to let $\tau := t - b$, for then $\tau = 0$ when $t = b$. As a result, we have

$$\mathcal{L}\{f(t)u(t-b)\} = \int_0^\infty e^{-s(\tau+b)} f(\tau+b)\,d\tau = e^{-sb}\int_0^\infty e^{-st} f(t+b)\,dt$$

Now the integral is the transform of the function $f(t+b)$. Thus, we conclude that

$$\mathcal{L}\{f(t)u(t-b)\} = e^{-bs}\mathcal{L}\{f(t+b)\}.$$

This result involves changing the argument of f which causes the graph of f to shift along the t-axis. For that reason, let us call this result the *shift in t property*. Thus, we have just proved the following result.

Shift in *t*

Theorem 11.6.2 *Let f be a function whose Laplace transform exists for $s > a$ and let b be any positive constant. Then*

$$\mathcal{L}\{f(t)u(t-b)\} = e^{-bs}\mathcal{L}\{f(t+b)\} \tag{11.6.4}$$

for $s > a$.

Observe that (11.6.4) is trivially true if $b = 0$ since $f(t)u(t) = f(t)$ for $t > 0$.

Example 11.6.4 Find the Laplace transform of $e^{2t}u(t-1)$.

Solution Using the notation of (11.6.4), $f(t) = e^{2t}$ and $b = 1$. Thus,

$$f(t+b) = f(t+1) = e^{2(t+1)} = e^2 \cdot e^{2t},$$

and so

$$\mathcal{L}\{f(t+1)\} = e^2 \cdot \mathcal{L}\{e^{2t}\} = \frac{e^2}{s-2}.$$

Therefore, by (11.6.4), we have

$$\mathcal{L}\{e^{2t}u(t-1)\} = e^{-s} \cdot \frac{e^2}{s-2} = \frac{e^{2-s}}{s-2}.$$

♦

Example 11.6.5 Find the Laplace transform of

$$g(t) = \begin{cases} 0, & \text{if } t < \pi \\ \sin(0.5t), & \text{if } t > \pi. \end{cases}$$

Solution The function g has a jump discontinuity at $t = \pi$. Written in terms of the unit step function at π,

$$g(t) = \sin(0.5t) \cdot u(t - \pi).$$

In the notation of (11.6.4), $f(t) = \sin(0.5t)$ and $b = \pi$. And so

$$f(t + \pi) = \sin[0.5(t + \pi)] = \sin\left(\frac{t}{2} + \frac{\pi}{2}\right) = \cos\frac{t}{2}.$$

Hence,

$$\mathcal{L}\{\sin(0.5t) \cdot u(t - \pi)\} = e^{-\pi s}\mathcal{L}\left\{\cos\frac{t}{2}\right\} = e^{-\pi s} \cdot \frac{s}{s^2 + \frac{1}{4}}.$$

♦

The shift in t property is perfect for finding Laplace transforms of functions of the form (11.6.2), but less so for functions of the form

$$g(t) = \begin{cases} 0, & \text{if } t < b \\ f(t - b), & \text{if } t > b, \end{cases} \tag{11.6.5}$$

despite their similarity. There is the matter of the unit step function $u(t - b)$ being used here because as it is now we may not be able to rewrite (11.6.5) as $f(t - b)u(t - b)$. And that is because $f(t - b)$ is not defined for $t < b$ if f is defined only on the interval $[0, \infty)$. Up to now, it has been irrelevant as to whether f is defined for $t < 0$ because only the values of f on the interval $[0, \infty)$ determine its Laplace transform. For this reason we can assign any values to f on $(-\infty, 0)$ that we deem appropriate. In fact, we can use this here to our advantage: if we define (or redefine) f to be identically equal to zero on $(-\infty, 0)$, then we can rewrite (11.6.5) as

$$g(t) = f(t - b)u(t - b). \tag{11.6.6}$$

Consequently, the Laplace transform of (11.6.5) can be found by carrying out the following integration:

$$\mathcal{L}\{g(t)\} = \mathcal{L}\{f(t-b)u(t-b)\} = \int_0^\infty e^{-st} f(t-b)u(t-b)\, dt = \int_b^\infty e^{-st} f(t-b)\, dt.$$

The rest of the details are similar to the proof of Theorem 11.6.2 and are left as an exercise. The end result is Theorem 11.6.3 below.

Shift in t (Alternate Form)

Theorem 11.6.3 *Let f be a function whose Laplace transform exists for $s > a$. If $b > 0$, then*

$$\mathcal{L}\{f(t-b)u(t-b)\} = e^{-bs}\mathcal{L}\{f(t)\} \tag{11.6.7}$$

for $s > a$.

Example 11.6.6 Find the solution of

$$\dot{x} + 9x = 2\delta(t-3)$$

satisfying the initial condition $x(0) = -5$.

Solution Taking the Laplace transform of the equation and using the linearity property as well as the transform of the first derivative property, we get

$$sX(s) - x(0) + 9X(s) = 2e^{-3s},$$

where as usual we use the alternate notation $X(s)$ in lieu of $\mathcal{L}\{x(t)\}$. As $x(0) = -5$, we have

$$(s+9)X(s) + 5 = 2e^{-3s}.$$

As a result,

$$X(s) = -\frac{5}{s+9} + 2\frac{e^{-3s}}{s+9}. \tag{11.6.8}$$

Let us write the second term on the right-hand side as $2e^{-bs}\mathcal{L}\{f(t)\}$, where $b = 3$ and $\mathcal{L}\{f(t)\} = 1/(s+9)$. So, $f(t) = e^{-9t}$. According to the alternate form of the shift in t property,

$$e^{-3s}\mathcal{L}\{f(t)\} = \mathcal{L}\{f(t-3)u(t-3)\}$$

where $f(t-3) = e^{-9(t-3)}$. Thus, we have

$$\frac{e^{-3s}}{s+9} = e^{-3s}\mathcal{L}\{e^{-9t}\} = \mathcal{L}\{e^{-9(t-3)}u(t-3)\}.$$

Consequently, (11.6.8) can be written as

$$\mathcal{L}\{x(t)\} = \mathcal{L}\{-5e^{-9t} + 2e^{-9(t-3)}u(t-3)\}.$$

Therefore,

$$x(t) = -5e^{-9t} + 2e^{-9(t-3)}u(t-3) = \begin{cases} -5e^{-9t}, & \text{if } t < 3 \\ -5e^{-9t} + 2e^{-9(t-3)}, & \text{if } t > 3 \end{cases}$$

is the solution of the initial value problem. ♦

11.6.3 Derivatives of Laplace Transforms

Does taking the derivative of the Laplace transform of a function yield another Laplace transform? To answer that, let $F(s)$ denote the Laplace transform of a function $f(t)$. Formally differentiating $F(s)$ with respect to s, we obtain

$$\begin{aligned} F'(s) &= \frac{d}{ds}F(s) = \frac{d}{ds}\mathcal{L}\{f(t)\}(s) = \frac{d}{ds}\int_0^\infty e^{-st} f(t)\,dt \\ &= \int_0^\infty \frac{\partial}{\partial s}[e^{-st} f(t)]\,dt = \int_0^\infty e^{-st}(-t)f(t)\,dt = \mathcal{L}\{-tf(t)\}. \end{aligned} \tag{11.6.9}$$

Reversing the order of differentiation and integration by moving d/ds across the integral sign and changing it to $\partial/\partial s$ seems reasonable but that in itself does not make it true. Although we are not in a position to prove it because of the advanced mathematics that would be needed, this step is valid for the integrand $e^{-st} f(t)$ if the function $f(t)$ is piecewise continuous and of exponential order.[14] Consequently, if that is the case, we conclude from (11.6.9) that

$$\mathcal{L}\{-tf(t)\} = F'(s). \tag{11.6.10}$$

Are there similar formulas when higher powers of t are involved? To answer that, consider $\mathcal{L}\{(-t)^2 f(t)\}$. By (11.6.10),

$$\mathcal{L}\{(-t)^2 f(t)\} = \mathcal{L}\{(-t)[-tf(t)]\} = \frac{d}{ds}\mathcal{L}\{-tf(t)\} = \frac{d}{ds}[F'(s)] = F''(s).$$

Similarly, it can be shown that $\mathcal{L}\{(-t)^3 f(t)\} = F'''(s)$. The pattern that is developing suggests the formula

[14] Churchill [21, Sec. 12] states and proves such a result.

$$\mathcal{L}\{(-t)^n f(t)\} = F^{(n)}(s) \tag{11.6.11}$$

when n is a positive integer. We prove that this is the case with the following mathematical induction argument. Suppose (11.6.11) is true for an integer $n = k$ (as it is for $n = 1, 2, 3$). Then for $n = k + 1$, it follows from (11.6.10) and (11.6.11) that

$$\begin{aligned}\mathcal{L}\{(-t)^{k+1} f(t)\} &= \mathcal{L}\{(-t)[(-t)^k f(t)]\} \\ &= \frac{d}{ds}\mathcal{L}\{(-t)^k f(t)\} = \frac{d}{ds}F^{(k)}(s) = F^{(k+1)}(s).\end{aligned}$$

So if (11.6.11) is true for $n = k$, then it must also be true for $n = k + 1$. Therefore, since it has already been established that (11.6.11) is true for $n = 1, 2, 3$, it follows that it also must be true for the rest of the positive integers.

In short, because of the validity of differentiating inside the integral sign and the induction argument, we have the following property.

Derivatives of Laplace Transforms

Theorem 11.6.4 *Suppose a function $f(t)$ is piecewise continuous on $[0, \infty)$ and of exponential order γ. Let $F(s) = \mathcal{L}\{f(t)\}$. Then*

$$\mathcal{L}\{t^n f(t)\} = (-1)^n \frac{d^n}{ds^n} F(s) \tag{11.6.12}$$

for $s > \gamma$ and $n = 1, 2, 3, \ldots$.

Example 11.6.7 Find $\mathcal{L}\{t \sin bt\}$.

Solution Theorem 11.6.4 clearly applies here since $\sin bt$ is multiplied by t and we know that $\mathcal{L}\{\sin bt\} = b/(s^2 + b^2)$. In the notation of the theorem, $f(t) = \sin bt$ and $F(s) = b/(s^2 + b^2)$. Setting $n = 1$ in (11.6.12), we have

$$\mathcal{L}\{t \sin bt\} = -\frac{d}{ds}F(s) = -\frac{-2bs}{(s^2 + b^2)^2}.$$

Therefore,

$$\mathcal{L}\{t \sin bt\} = \frac{2bs}{(s^2 + b^2)^2}.$$

♦

Example 11.6.8 Find $\mathcal{L}\{t^2 e^{3t}\}$.

Solution Since $\mathcal{L}\{e^{3t}\} = 1/(s-3)$,

$$\mathcal{L}\{t^2 e^{3t}\} = (-1)^2 \frac{d^2}{ds^2}\left(\frac{1}{s-3}\right) = \frac{d}{ds}\left(\frac{-1}{(s-3)^2}\right) = \frac{2}{(s-3)^3}.$$

Another approach is to use the shift in s property; it is also requires less work. Since $\mathcal{L}\{t^2\} = 2/s^3$, it immediately follows from shifting s that

$$\mathcal{L}\{t^2 e^{3t}\} = \frac{2}{(s-3)^3}.$$

♦

11.6.4 Integrals of Laplace Transforms

According to Theorem 11.6.4, the Laplace transform of the product $-tf(t)$ can be obtained by differentiating $\mathcal{L}\{f(t)\}$. It turns out that the Laplace transform of the quotient $f(t)/t$ can be obtained by integrating $\mathcal{L}\{f(t)\}$, provided this quotient is well behaved in the sense stated in the next theorem.

Integral of a Laplace Transform

Theorem 11.6.5 *Suppose a function $f(t)$ is piecewise continuous on $[0, \infty)$ and of exponential order γ. Furthermore, suppose*

$$\lim_{t\to 0^+} \frac{f(t)}{t} \tag{11.6.13}$$

exists and is finite. Then for $s > \gamma$,

$$\mathcal{L}\left\{\frac{f(t)}{t}\right\}(s) = \int_s^\infty F(u)\,du \tag{11.6.14}$$

where $F(s) = \mathcal{L}\{f(t)\}(s)$.

Proof Let us begin with the right-hand side of (11.6.14). Integrating $F(u) = \mathcal{L}\{f\}(u)$ from $u = s$ to $u = T$, we have

$$\int_s^T F(u)\,du = \int_s^T \left(\int_0^\infty e^{-ut} f(t)\,dt\right) du.$$

Because of the hypotheses, we can reverse the order of integration.[15] In doing so, we obtain

$$\int_s^T F(u)\,du = \int_0^\infty f(t)\left(\int_s^T e^{-ut}\,du\right)dt. \tag{11.6.15}$$

Existence of the limit (11.6.13) and piecewise continuity of $f(t)$ on $[0,\infty)$ imply that $f(t)/t$ is piecewise continuous on $[0,\infty)$. Moreover, as $f(t)$ is of exponential order γ, so is $f(t)/t$. Thus, by Corollary 11.5.4, the Laplace transform of $f(t)/t$ exists for $s > \gamma$. Since

$$\int_s^T e^{-ut}\,du = -\frac{1}{t}e^{-ut}\Big|_{u=s}^{u=T} = \frac{1}{t}e^{-st} - \frac{1}{t}e^{-Tt},$$

we can write (11.6.15) as

$$\begin{aligned}\int_s^T F(u)\,du &= \int_0^\infty \frac{f(t)}{t}e^{-st}\,dt - \int_0^\infty \frac{f(t)}{t}e^{-Tt}\,dt\\ &= \mathcal{L}\left\{\frac{f(t)}{t}\right\}(s) - \mathcal{L}\left\{\frac{f(t)}{t}\right\}(T).\end{aligned}$$

Letting $T \to \infty$, we obtain (11.6.14) since it follows from Corollary 11.5.2 that $\mathcal{L}\{f(t)/t\}(T) \to 0$. ■

Example 11.6.9 Find the Laplace transform of $\dfrac{e^t - 1}{t}$.

Solution In Theorem 11.6.5, let $f(t) = e^t - 1$. Since

$$|e^t - 1| = e^t - 1 \le e^t$$

for $t \ge 0$, f is of exponential order 1. By L'Hôpital's Rule,

$$\lim_{t\to 0^+} \frac{f(t)}{t} = \lim_{t\to 0^+} \frac{e^t - 1}{t} = \lim_{t\to 0^+} e^t = 1.$$

Consequently, the hypotheses of Theorem 11.6.5 are satisfied.

[15] The reason for this is found in Churchill [21, Sec. 18]. However, it may be difficult for a reader at this level to understand the explanation given there since the concept of uniform convergence is used, which is a topic that is discussed in advanced calculus and real analysis textbooks, such as Bartle and Sherbert [6], Fulks [37], and Ross [62].

The Laplace transform of $f(t)$ is

$$F(s) = \mathcal{L}\{f(t)\} = \mathcal{L}\{e^t - 1\} = \frac{1}{s-1} - \frac{1}{s}.$$

Integrating $F(u)$ from $u = s$ to $u = T$, we have

$$\begin{aligned}\int_s^T F(u)\,du &= \int_s^T \left(\frac{1}{u-1} - \frac{1}{u}\right) du = \ln\left(\frac{u-1}{u}\right)\Bigg|_s^T \\ &= \ln\left(\frac{T-1}{T}\right) - \ln\left(\frac{s-1}{s}\right)\end{aligned}$$

for $s > 1$. Thus,

$$\int_s^\infty F(u)\,du = \lim_{T\to\infty} \int_s^T F(u)\,du = -\ln\left(\frac{s-1}{s}\right)$$

since

$$\lim_{T\to\infty} \ln\left(\frac{T-1}{T}\right) = \lim_{T\to\infty} \ln\left(1 - \frac{1}{T}\right) = \ln 1 = 0.$$

Therefore,

$$\mathcal{L}\left\{\frac{e^t - 1}{t}\right\} = \ln\left(\frac{s}{s-1}\right)$$

for $s > 1$. ♦

Example 11.6.10 For any nonzero constant b, show that

$$\mathcal{L}\left\{\frac{\sin bt}{t}\right\} = \cot^{-1}\left(\frac{s}{b}\right) \tag{11.6.16}$$

for $s > 0$.

Solution First let us check if $f(t) = \sin bt$ satisfies the conditions of Theorem 11.6.5. Of course, it is continuous. And $\sin bt = O(1)$; that is, it is of exponential order 0 since its range is $[-1, 1]$. Furthermore, due to the fact that $\sin\theta/\theta \to 1$ as $\theta \to 0$,[16] condition (11.6.13) is satisfied because

$$\lim_{t\to 0^+} \frac{f(t)}{t} = \lim_{t\to 0^+} \frac{\sin bt}{t} = b \cdot \lim_{t\to 0^+} \frac{\sin bt}{bt} = b \cdot 1 = b.$$

[16] Recall this limit is used with the difference quotient for $\sin t$ to show that its derivative is $\cos t$.

This limit also follows from L'Hôpital's Rule. Accordingly, all of the conditions of Theorem 11.6.5 are met. So, by (11.6.14),

$$\mathcal{L}\left\{\frac{\sin bt}{t}\right\} = \int_s^\infty \mathcal{L}\{\sin bt\}(u)\, du = \int_s^\infty \frac{b}{u^2+b^2}\, du$$
$$= \lim_{T\to\infty}\left[\tan^{-1}\left(\frac{T}{b}\right) - \tan^{-1}\left(\frac{s}{b}\right)\right].$$

Since $\tan^{-1} x \to \pi/2$ as $x \to \infty$,

$$\mathcal{L}\left\{\frac{\sin bt}{t}\right\} = \frac{\pi}{2} - \tan^{-1}\left(\frac{s}{b}\right).$$

Using the trigonometric identity (see Problem 39)

$$\tan^{-1} x + \cot^{-1} x = \pi/2,$$

we obtain (11.6.16). ♦

11.6.5 Laplace Transforms of Integrals

If a function $f(t)$ is piecewise continuous on $[0, \infty)$ and of exponential order γ, another well-known property of the Laplace transform can be derived by taking the transform of the derivative of the function $g(t) := \int_0^t f(\tau)\, d\tau$. Since f is piecewise continuous for $t \geq 0$, one of the versions of the Fundamental Theorem of Calculus states that g is continuous on $[0, \infty)$ and that

$$g'(t) = \frac{d}{dt}\int_0^t f(\tau)\, d\tau = f(t),$$

except at those values of t where f is discontinuous.[17] It follows that g' is piecewise continuous on $[0, \infty)$.

Is g of exponential order? If it is, then this along with its piecewise smoothness will guarantee the existence of its transform. Since f is of exponential order γ, there are constants $M_1 > 0$ and $t_1 \geq 0$ such that $|f(\tau)| \leq M_1 e^{\gamma\tau}$ for $\tau \geq t_1$. If the constant γ is not already positive, we can certainly replace it with a positive number. Since f is piecewise continuous, it is bounded on the finite interval $[0, t_1]$. In other words, there is a constant M_2 such that $|f(\tau)| \leq M_2$ for $0 \leq \tau \leq t_1$. Even more so then, $|f(\tau)| \leq M_2 e^{\gamma\tau}$ on $[0, t_1]$. Defining $M := \max\{M_1, M_2\}$, we have

[17] See Bartle and Sherbert [6, pp. 257-258], Ross [62, §34], or Gaughan [38, Ch. 5.6].

$|f(\tau)| \le Me^{\gamma\tau}$ on the entire interval $[0, \infty)$. Since $-|f(\tau)| \le f(\tau) \le |f(\tau)|$,

$$\left|\int_0^t f(\tau)\,d\tau\right| \le \int_0^t |f(\tau)|\,d\tau.$$

Thus,

$$|g(t)| \le \int_0^t |f(\tau)|\,d\tau \le M\int_0^t e^{\gamma\tau}\,d\tau = \frac{M}{\gamma}(e^{\gamma t} - 1) < \frac{M}{\gamma}e^{\gamma t}$$

for all $t \ge 0$. Hence, g is of exponential order γ. This together with the continuity of g on $[0, \infty)$ implies that the Laplace transform of g exists for $s > \gamma$ (see Corollary 11.5.4).

Earlier we ascertained that g' is piecewise continuous on $[0, \infty)$. This, the continuity of g, and $g = O(e^{\gamma t})$ guarantee that the transform of the first derivative formula (11.4.7) still applies since Theorem 11.4.3 can be extended to piecewise differentiable functions (cf. Churchill [21, p. 8]). Therefore,

$$\mathcal{L}\{g'(t)\} = s\mathcal{L}\{g(t)\} - g(0)$$

for $s > \gamma$. Since $g(0) = 0$,

$$\mathcal{L}\{f(t)\} = \mathcal{L}\{g'(t)\} = s\mathcal{L}\{g(t)\} = s\mathcal{L}\left\{\int_0^t f(\tau)\,d\tau\right\}.$$

As a result, we have the following property.

Laplace Transform of an Integral

Theorem 11.6.6 *If $f(t)$ is piecewise continuous on $[0, \infty)$ and of exponential order γ, then*

$$\mathcal{L}\left\{\int_0^t f(\tau)\,d\tau\right\} = \frac{1}{s}\mathcal{L}\{f(t)\} \qquad (11.6.17)$$

for $s > \gamma$.

Example 11.6.11 Take $f(t) \equiv 1$ on $[0, \infty)$ in (11.6.17). It certainly satisfies the hypotheses of Theorem 11.6.6. Hence,

$$\mathcal{L}\left\{\int_0^t 1\,d\tau\right\} = \frac{1}{s}\mathcal{L}\{1\}.$$

Since $\int_0^t d\tau = t$ and $\mathcal{L}\{1\} = 1/s$, we conclude that $\mathcal{L}\{t\} = 1/s^2$ for $s > 0$. ♦

Even though we had obtained this result in Sect. 11.2, we chose it as our first example in order to become comfortable with using Theorem 11.6.6. But the next example is new and not as simple.

Example 11.6.12 Find the Laplace transform of the sine integral function

$$\mathrm{Si}(t) = \int_0^t \frac{\sin \tau}{\tau}\, d\tau.$$

Solution The integrand $\tau^{-1} \sin \tau$ is piecewise continuous on $[0, \infty)$ as

$$\lim_{\tau \to 0^+} \frac{\sin \tau}{\tau} = 1.$$

And it is of exponential order $\gamma = 0$ because

$$\left| \frac{\sin \tau}{\tau} \right| \leq \frac{1}{\tau} \leq 1 \quad \text{for} \quad \tau \geq 1.$$

Consequently, the conditions of Theorem 11.6.6 are satisfied. It follows from (11.6.17) and Example 11.6.10 that

$$\mathcal{L}\left\{\mathrm{Si}(t)\right\} = \mathcal{L}\left\{\int_0^t \frac{\sin \tau}{\tau}\, d\tau\right\} = \frac{1}{s}\mathcal{L}\left\{\frac{\sin t}{t}\right\} = \frac{1}{s}\cot^{-1} s$$

for $s > 0$. ♦

11.6.6 Periodic Functions

Since periodic functions are used to model cyclic phenomena, it is important to know how to find their Laplace transforms. Recall from a trigonometry or precalculus course that a nonconstant function $f(t)$ is said to be ***periodic*** if a constant b exists such that

$$f(t + b) = f(t) \tag{11.6.18}$$

wherever $f(t)$ is defined. The smallest positive value of b for which (11.6.18) holds is called the ***period*** of f. For example, the sine function is periodic on the interval $(-\infty, \infty)$ with period 2π because $\sin(t + b) = \sin t$ for all values of t when $b = 2\pi$; moreover, 2π is the smallest positive number for which this is true. Compare this with the tangent function. Of course, $\tan t$ is undefined at the points $t = (n + 1/2)\pi$ for $n = 0, \pm 1, \pm 2, \pm 3, \ldots$. At all other points, $\tan(t + b) = \tan t$ when

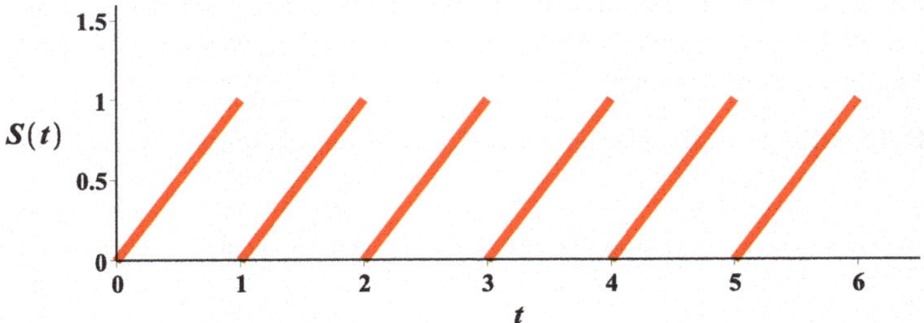

Fig. 11.2 A sawtooth wave

$b = \pi$ but for no other positive values of b less than π. Therefore, the period of $\tan t$ is π.

The graph of another periodic function defined for $t \geq 0$ is shown in Fig. 11.2. Its period is clearly 1. Its graph consists of an infinite succession of *sawtooth pulses* resembling the teeth of a saw. This waveform is called a *sawtooth wave*. Let us call this function S. Although it is not clear from the graph itself whether or not $S(t)$ is defined at the points $t = 0, 1, 2, 3, \ldots$, it is defined at all other positive values of t. Those values are easily ascertainable by knowing that its period is 1 and seeing from the graph that $S(t) = t$ on $(0, 1)$. As for those points $t = n$ where $n = 0, 1, 2, 3, \ldots$, suppose $S(n)$ is not defined. Then we can completely define the function S by simply stating that

$$S(t) = t, \quad 0 < t < 1$$

and that it is periodic on $[0, \infty)$ with period 1. Alternatively, we can express it in terms of the unit step functions (cf. Examples 7.2.4 and 7.2.5) as follows:

$$S(t) = tu(t) - u(t-1) - u(t-2) - u(t-3) - \cdots = tu(t) - \sum_{n=1}^{\infty} u(t-n)$$

for $t \geq 0$. However, if we make one minor change and let $S(0) = 0$, then

$$S(t) = t - \sum_{n=1}^{\infty} u(t-n).$$

The next theorem states that a piecewise continuous and periodic function on $[0, \infty)$ has a Laplace transform. Moreover, it tells us how to compute that transform.

Transform of a Periodic Function

Theorem 11.6.7 *If a function $f(t)$ is periodic on $[0, \infty)$ with period b and if it is piecewise continuous on $[0, b]$, then its Laplace transform exists for $s > 0$, where*

$$\mathcal{L}\{f(t)\} = \frac{1}{1 - e^{-bs}} \int_0^b e^{-st} f(t)\, dt. \tag{11.6.19}$$

Equivalently,

$$\mathcal{L}\{f(t)\} = \frac{\mathcal{L}\{f_b(t)\}}{1 - e^{-bs}} \tag{11.6.20}$$

for $s > 0$, where

$$f_b(t) := \begin{Bmatrix} f(t), & 0 < t < b \\ 0, & t > b \end{Bmatrix} = f(t)\big[u(t) - u(t-b)\big]. \tag{11.6.21}$$

Proof It follows from the hypotheses that $f(t)$ is bounded and piecewise continuous on $[0, \infty)$. Recall that a bounded function is of exponential order 0. Consequently, by Corollary 11.5.4, its Laplace transform exists for $s > 0$, where

$$\mathcal{L}\{f(t)\} = \int_0^b e^{-st} f(t)\, dt + \int_b^\infty e^{-st} f(t)\, dt.$$

With the change of variable $u = t - b$, the improper integral becomes

$$\int_b^\infty e^{-st} f(t)\, dt = \int_0^\infty e^{-s(u+b)} f(u+b)\, du = e^{-sb} \int_0^\infty e^{-su} f(u)\, du$$

since the period of f is b. Therefore,

$$\mathcal{L}\{f(t)\} = \int_0^b e^{-st} f(t)\, dt + e^{-sb} \mathcal{L}\{f(t)\}.$$

Formula (11.6.19) is the result when this is solved for $\mathcal{L}\{f(t)\}$.

Now observe that

$$\int_0^b e^{-st} f(t)\, dt = \int_0^b e^{-st} f_b(t)\, dt = \int_0^\infty e^{-st} f_b(t)\, dt = \mathcal{L}\{f_b(t)\}$$

since $f_b(t) = f(t)$ for $t \in (0, b)$ while $f_b(t) = 0$ for $t > b$. The result of replacing the integral in (11.6.19) with $\mathcal{L}\{f_b(t)\}$ is (11.6.20).

Expressed in terms of unit step functions,

$$f_b(t) = f(t)u(t) + \left\{\begin{matrix} 0, & t < b \\ -f(t), & t > b \end{matrix}\right\} = f(t)u(t) - f(t)u(t-b),$$

which is the right-hand side of (11.6.21). Note that the first term is $f(t)u(t)$ rather than $f(t)$ since $f_b(t)$ is undefined at $t = 0$. ■

Remarks As we work through the following examples, the following remarks should be kept in mind:

(i) If $f(t)$ is defined at $t = 0$ as well as at $t = nb$ for $n = 1, 2, 3, \ldots$, then we modify $f_b(t)$ as follows:

$$f_b(t) := \left\{\begin{matrix} f(t), & 0 \le t < b \\ 0, & t \ge b \end{matrix}\right\} = f(t)\big[1 - H(t-b)\big]. \tag{11.6.22}$$

The only difference between $u(t-b)$ and $H(t-b)$ is the jump discontinuity at $t = b$: at this point, $u(t-b)$ is undefined whereas $H(t-b) = 1$ (see (11.2.3) or (7.2.3)). However, changing the value of a function at a single point does not change the integral defining its Laplace transform. Consequently, (11.6.21) and (11.6.22) have the same Laplace transform.

(ii) If $f(t)$ is continuous on $[0, b]$, then it is continuous on the entire interval $[0, \infty)$, such as the full-rectified sine wave that will be considered in Example 11.6.15. The theorem still applies as every continuous function is trivially piecewise continuous since it has no jump or removable discontinuities (see Definition 11.5.3).

(iii) Unless indicated otherwise, assume, as in Chap. 7, that piecewise continuous functions are undefined at all locations of the jump discontinuities.

Example 11.6.13 Find the Laplace transform of the square-wave function whose graph is shown in Fig. 11.3.

Solution In order to be able to employ Theorem 11.6.7, we have to first represent the function algebraically. Calling it f, we see from its graph that it is periodic with period 2 and that

$$f(t) = \begin{cases} 1, & 0 < t < 1 \\ 0, & 1 < t < 2 \end{cases}$$

for $0 < t < 2$. Using the technique demonstrated in Sect. 7.2 for expressing a periodic function in terms of unit step functions, $f(t)$ can be represented by the

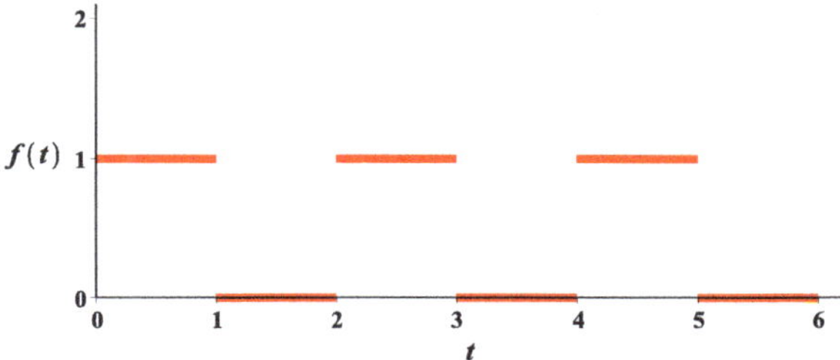

Fig. 11.3 A periodic square wave

infinite series

$$f(t) = u(t) - u(t-1) + u(t-2) - u(t-3) + \cdots = \sum_{n=0}^{\infty} (-1)^n u(t-n)$$

for $t \geq 0$. As to the computation of $\mathcal{L}\{f(t)\}$, we will demonstrate how to apply both of the formulas (11.6.19) and (11.6.20).

Method 1 Let us begin with (11.6.19). Since the period of $f(t)$ is $b = 2$, we have

$$\begin{aligned}
\mathcal{L}\{f(t)\} &= \frac{1}{1-e^{-2s}} \int_0^2 e^{-st} f(t)\, dt \\
&= \frac{1}{1-e^{-2s}} \left(\int_0^1 e^{-st} \cdot 1\, dt + \int_1^2 e^{-st} \cdot 0\, dt \right) \\
&= \frac{1}{1-e^{-2s}} \left(\frac{1}{s} - \frac{1}{s} e^{-s} \right) = \frac{1-e^{-s}}{s(1-e^{-2s})}.
\end{aligned}$$

Factoring the denominator and canceling the common factor, this simplifies to

$$\mathcal{L}\{f(t)\} = \frac{1-e^{-s}}{s(1-e^{-s})(1+e^{-s})} = \frac{1}{s(1+e^{-s})}.$$

Method 2 Formula (11.6.20) uses the function f_b defined by (11.6.21). Since $b = 2$,

$$f_2(t) = f(t)\big[u(t) - u(t-2)\big]. \tag{11.6.23}$$

Note, however, that the formula for $f(t)$ changes inside the fundamental interval $(0, 2)$ due to the jump discontinuity at $t = 1$. Accordingly, we rewrite this as

$$f_2(t) = f(t)\big[u(t) - u(t-1) + u(t-1) - u(t-2)\big]$$

$$= f(t)\big[u(t) - u(t-1)\big] + f(t)\big[u(t-1) - u(t-2)\big].$$

Since $f(t) \equiv 1$ on $(0, 1)$ and $f(t) \equiv 0$ on $(1, 2)$, this becomes

$$f_2(t) = 1 \cdot \big[u(t) - u(t-1)\big] + 0 \cdot \big[u(t-1) - u(t-2)\big].$$

which simplifies to

$$f_2(t) = u(t) - u(t-1). \tag{11.6.24}$$

Actually, there is a slight difference between (11.6.23) and (11.6.24), which occurs at the point $t = 2$. Formula (11.6.23) correctly indicates that $f_2(t)$ is undefined at $t = 2$, whereas (11.6.24) says it is equal to 0 there. Nevertheless, we can still use (11.6.24) to compute the Laplace transform of $f(t)$. This is due to the fact that two functions that differ at a finite number of points (or even at the points of some infinite sets, such as $t = 1, 2, 3, \ldots$) have the same Laplace transform, provided it exists for one of them (in which case it will exist for the other one too).[18]

And so using (11.6.24) to compute $\mathcal{L}\{f_2(t)\}$, we get

$$\mathcal{L}\{f_2(t)\} = \mathcal{L}\{u(t)\} - \mathcal{L}\{u(t-1)\} = \frac{1}{s} - \frac{e^{-s}}{s}.$$

Therefore,

$$\mathcal{L}\{f(t)\} = \frac{\mathcal{L}\{f_2(t)\}}{1 - e^{-2s}} = \frac{1 - e^{-s}}{s(1 - e^{-2s})} = \frac{1}{s(1 + e^{-s})},$$

as before. ♦

Example 11.6.14 Find the Laplace transform of the function $S(t)$ that is periodic on $[0, \infty)$ with period 1 and with $S(t) = t$ for $0 < t < 1$.

Solution 1 The graph of $S(t)$ is the sawtooth wave shown earlier in Fig. 11.2. Using (11.6.19) with $b = 1$,

$$\mathcal{L}\{S(t)\} = \frac{1}{1 - e^{-s}} \int_0^1 e^{-st} t \, dt.$$

Integration by parts yields

$$\int_0^1 e^{-st} t \, dt = \frac{1}{s^2}(1 - e^{-s}) - \frac{1}{s} e^{-s}. \tag{11.6.25}$$

---

[18] The reason for this is beyond the mathematical level of this book and is best left to a real analysis book.

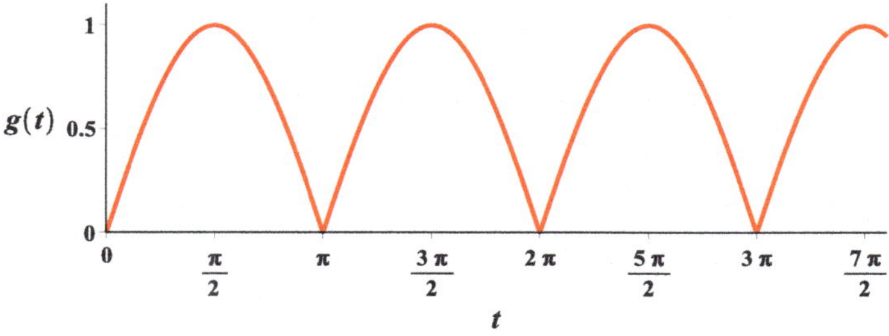

Fig. 11.4 A full-rectified sine wave

Thus,

$$\mathcal{L}\{S(t)\} = \frac{1}{1-e^{-s}}\left[\frac{1-e^{-s}}{s^2} - \frac{e^{-s}}{s}\right] = \frac{1}{s^2} - \frac{e^{-s}}{s(1-e^{-s})}.$$

Solution 2 Integrating by parts can be avoided by using formula (11.6.20) instead of (11.6.19). With $b = 1$ and $S(t) = t$ replacing $f(t)$ in (11.6.21), we have

$$S_1(t) = \begin{Bmatrix} t, & 0 < t < 1 \\ 0, & t > 1. \end{Bmatrix} = t\big[u(t) - u(t-1)\big].$$

So,

$$\mathcal{L}\{S_1(t)\} = \mathcal{L}\{tu(t)\} - \mathcal{L}\{tu(t-1)\}.$$

By the shift in t property,

$$\mathcal{L}\{S_1(t)\} = \frac{1}{s^2} - e^{-s}\mathcal{L}\{t+1\} = \frac{1}{s^2} - e^{-s}\left(\frac{1}{s^2} + \frac{1}{s}\right),$$

which is equal to (11.6.25). Substituting this expression for the numerator in (11.6.20), we get the same result for $\mathcal{L}\{S(t)\}$ that we obtained above in *Solution 1*.

♦

Example 11.6.15 The graph of a function g is the full-rectified sine wave shown in Fig. 11.4.[19] That is, $g(t) = \sin t$ for $0 \le t \le \pi$ and its period is π. Find $\mathcal{L}\{g(t)\}$.

[19] A ***full-rectified sine wave*** is obtained from a sine wave by inverting all of its negative (or positive) half-cycles. The full rectification shown in the figure is $|\sin t|$.

Solution Because g is defined at $t = 0$, we use (11.6.22) rather than (11.6.21):

$$g_\pi(t) = \begin{Bmatrix} \sin t, & 0 \le t < \pi \\ 0, & t \ge \pi \end{Bmatrix} = \sin t\big[1 - H(t-\pi)\big].$$

By the shift in t property,

$$\mathcal{L}\{g_\pi(t)\} = \mathcal{L}\{\sin t\} - \mathcal{L}\{\sin t \cdot H(t-\pi)\} = \frac{1}{s^2+1} - e^{-\pi s}\mathcal{L}\{\sin(t+\pi)\}.$$

Because of this and $\sin(t+\pi) = -\sin t$, it follows from (11.6.20) that

$$\mathcal{L}\{g(t)\} = \frac{1}{1-e^{-\pi s}}\left[\frac{1}{s^2+1} + e^{-\pi s}\frac{1}{s^2+1}\right] = \frac{1}{s^2+1}\left[\frac{1+e^{-\pi s}}{1-e^{-\pi s}}\right].$$

♦

If a periodic function with period b is defined piecewise over the fundamental interval $[0, b]$, such as the function $f(t)$ in Example 11.6.13, it is oftentimes less work to express $f_b(t)$ in terms of unit step functions and then compute $\mathcal{L}\{f_b(t)\}$ in (11.6.20) using the properties of the Laplace transform than to directly evaluate the integral in (11.6.19). A useful tool for this is the so-called box function, which was mentioned briefly in Sect. 7.2. A foreshadowing of this can be seen in how the function $f_2(t)$ was dealt with in Example 11.6.13 (see *Method* 2).

Box Function

Definition 11.6.8 Let $a, b \in \mathbb{R}$, where $a < b$. The ***box function*** $u_{a,b}(t)$ is defined by

$$u_{a,b}(t) := u(t-a) - u(t-b). \tag{11.6.26}$$

That is, $u_{a,b}(t) = 1$ for $a < t < b$ and $u_{a,b}(t) = 0$ for $t < a$ and $t > b$.

This function is also called the ***boxcar function*** or the ***window function*** by others. All of these are apt names for the function $u_{a,b}(t)$ in view of its graph: for example, with $a = 2$ and $b = 5$, a portion of the graph of

$$u_{2,5}(t) = u(t-2) - u(t-5) \tag{11.6.27}$$

is the "box" or "boxcar" shown in Fig. 11.5.

The effect of multiplying a function $f(t)$ by a box function $u_{a,b}(t)$ is to set all values outside the box ($t < a$ and $t > b$) to 0 but to retain all values of $f(t)$ within

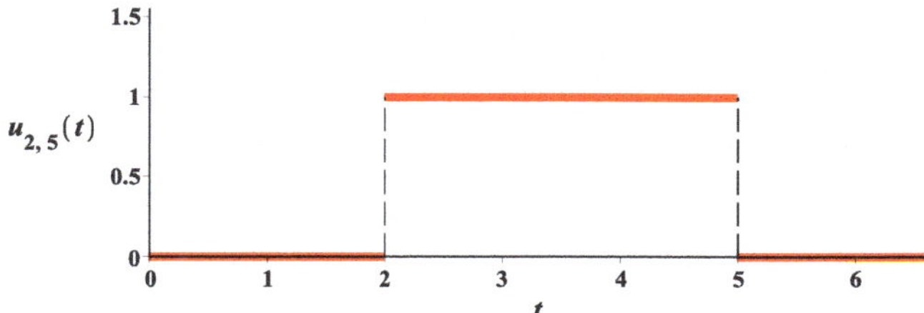

Fig. 11.5 Graph of the box function $u_{2,5}(t)$

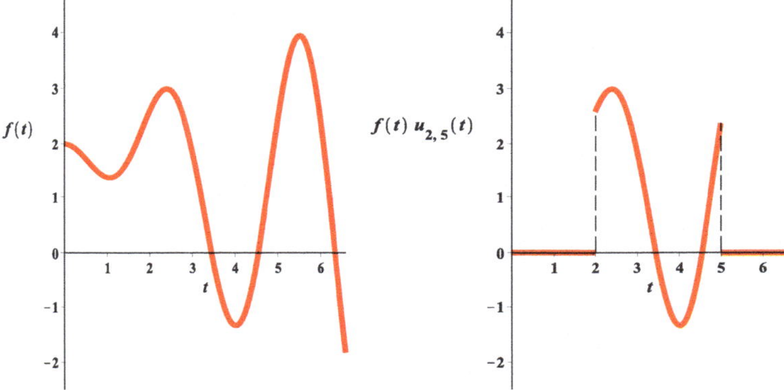

Fig. 11.6 Comparison of the graphs of $f(t)$ and $f(t)u_{2,5}(t)$

the box ($a < t < b$). For example, in Fig. 11.6, the graph on the right is the result of multiplying $f(t)$—whose graph is depicted on the left—by the box function $u_{2,5}(t)$.

Example 11.6.16 Find $\mathcal{L}\{h(t)\}$, where $h(t)$ is the triangular wave function whose graph is displayed in Fig. 11.7.

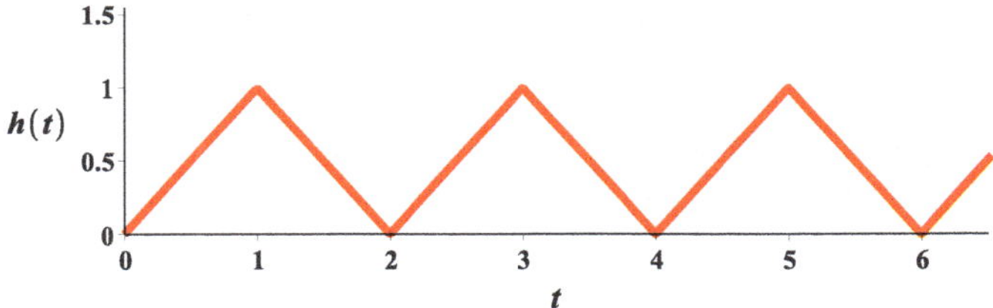

Fig. 11.7 A triangular wave

Solution Since this function is periodic with period 2, we can use Theorem 11.6.7. Setting $b = 2$, we have

$$h_2(t) = \begin{Bmatrix} h(t), & 0 < t < 2 \\ 0, & t > 2 \end{Bmatrix} = \begin{Bmatrix} t, & 0 < t < 1 \\ 2 - t, & 1 < t < 2 \\ 0, & t > 2 \end{Bmatrix}.$$

Expressed in terms of unit step functions, this is

$$\begin{aligned} h_2(t) &= tu(t) + \begin{Bmatrix} 0, & t < 1 \\ 2 - 2t, & t > 1 \end{Bmatrix} + \begin{Bmatrix} 0, & t < 2 \\ t - 2, & t > 2 \end{Bmatrix} \\ &= tu(t) - 2(t-1)u(t-1) + (t-2)u(t-2). \end{aligned}$$

Alternatively, employing box functions, $h_2(t) = tu_{0,1}(t) + (2 - t)u_{1,2}(t)$. That is,

$$\begin{aligned} h_2(t) &= t\big[u(t) - u(t-1)\big] + (2-t)\big[u(t-1) - u(t-2)\big] \\ &= tu(t) - 2(t-1)u(t-1) + (t-2)u(t-2), \end{aligned}$$

as before. Thus,

$$\begin{aligned} \mathcal{L}\{h_2(t)\} &= \mathcal{L}\{tu(t)\} - 2\mathcal{L}\{(t-1)u(t-1)\} + \mathcal{L}\{(t-2)u(t-2)\} \\ &= \frac{1}{s^2} - \frac{2}{s^2}e^{-s} + \frac{1}{s^2}e^{-2s} = \frac{1}{s^2}\left(1 - 2e^{-s} + e^{-2s}\right). \end{aligned}$$

Therefore, according to Theorem 11.6.7,

$$\mathcal{L}\{h(t)\} = \frac{1}{s^2(1 - e^{-2s})}\left(1 - 2e^{-s} + e^{-2s}\right).$$

Factoring the numerator and denominator, canceling the common factor $1 - e^{-s}$, and using the definition of the hyperbolic tangent function in (10.9.12), we end up with the result

$$\mathcal{L}\{h(t)\} = \frac{1}{s^2}\left(\frac{1 - e^{-s}}{1 + e^{-s}}\right) = \frac{1}{s^2}\left(\frac{e^{s/2} - e^{-s/2}}{e^{s/2} + e^{-s/2}}\right) = \frac{1}{s^2}\tanh\frac{s}{2}.$$

♦

11.6.7 Convolution

Consider the initial value problem

$$\frac{dx}{dt} - x = g(t), \quad x(0) = 0, \tag{11.6.28}$$

where the function g is piecewise continuous and of exponential order. Taking the Laplace transform of both sides of the differential equation in (11.6.28), using the initial condition, and solving for $\mathcal{L}\{x(t)\}$, we get

$$\mathcal{L}\{x(t)\} = \frac{1}{s-1}\,\mathcal{L}\{g(t)\}.$$

Since $1/(s-1) = \mathcal{L}\{e^t\}$, we can also write this as

$$\mathcal{L}\{x(t)\} = \mathcal{L}\{e^t\}\,\mathcal{L}\{g(t)\}. \tag{11.6.29}$$

Thus, the Laplace transform of the solution $x(t)$ of (11.6.28) is equal to the product of the Laplace transforms of e^t and $g(t)$. Unfortunately, at this point we are stymied as what to do next in order to find $x(t)$ even if $g(t)$ were known. So let us leave this for a moment and solve (11.6.28) using the integrating factor method or the variation of parameters formula given in Chap. 5. Using either of these, we find that

$$x(t) = \int_0^t e^{t-u} g(u)\,du. \tag{11.6.30}$$

Comparing (11.6.29) and (11.6.30), we conclude

$$\mathcal{L}\{e^t\}\,\mathcal{L}\{g(t)\} = \mathcal{L}\left\{\int_0^t e^{t-u} g(u)\,du\right\}. \tag{11.6.31}$$

The point to be made here is that the solutions of some initial value problems, such as (11.6.28), can be expressed in terms of integrals of the form

$$\int_0^t f(t-u)g(u)\,du.$$

In this example, $f(t) = e^t$. An integral of this type is known as a ***convolution integral***.

Convolution of Two Functions

Definition 11.6.9 Let $f(t)$ and $g(t)$ be piecewise continuous functions on $[0, \infty)$. The ***convolution*** of f and g is the integral function h defined by

$$h(t) := \int_0^t f(t-u)g(u)\,du. \tag{11.6.32}$$

Example 11.6.17 Compute the convolution of $f(t) = t$ and $g(t) = e^t$.

Solution Integrating by parts, we obtain

$$h(t) = \int_0^t (t-u)e^u\,du = \left[(t-u)e^u + e^u\right]_0^t = e^t - t - 1.$$

♦

Rather than using h or some other letter to denote the convolution integral, it is standard practice to use the symbol $f * g$. Thus,

$$(f * g)(t) := \int_0^t f(t-u)g(u)\,du.$$

With this notation, the convolution of t and e^t is written as

$$t * e^t = \int_0^t (t-u)e^u\,du = e^t - t - 1.$$

Example 11.6.18 Let $f(t) = 1$ and $g(t) = \cos t$. Compute $(f*g)(t)$ and $(g*f)(t)$.

Solution Since $f(t-u) = 1$,

$$(f * g)(t) = \int_0^t f(t-u)g(u)\,du = \int_0^t \cos u\,du = \sin t.$$

As for $(g * f)(t)$, we have

$$(g * f)(t) = \int_0^t g(t-u)f(u)\,du = \int_0^t \cos(t-u)\,du = -\int_t^0 \cos v\,dv = \sin t.$$

Observe that $(f * g)(t) = (g * f)(t)$. ♦

Convolution has the properties of commutativity, associativity, and distributivity as we explain in the following theorem.

Properties of Convolution

Theorem 11.6.10 *If $f(t)$, $g(t)$, and $h(t)$ are piecewise continuous functions on $[0, \infty)$, then the convolution operator has the following properties:*

(i) *Commutativity:* $f * g = g * f$
(ii) *Associativity:* $f * (g * h) = (f * g) * h$
(iii) *Distributivity:* $f * (g + h) = f * g + f * h$
(iv) $0 * f = f * 0 = 0.$

Proof To prove (i), simply change variables: let $v = t - u$. Then

$$\begin{aligned}(f * g)(t) &= \int_0^t f(t-u)g(u)\,du = -\int_t^0 f(v)g(t-v)\,dv \\ &= \int_0^t g(t-v)f(v)\,dv = (g * f)(t).\end{aligned}$$

The proof of (iv) is trivial:

$$(f * 0)(t) = \int_0^t f(t-u)\cdot 0\,du = \int_0^t 0\,du = 0,$$

which together with (i) proves (iv). The proofs of (ii) and (iii) are left as exercises. ■

Even though it may appear from Theorem 11.6.10 that the convolution operator $*$ has all of the properties of ordinary multiplication, it does not. For instance, $1*1 \neq 1$ since

$$1 * 1 = \int_0^t 1\cdot 1\,du = t.$$

Also, note for the function g in Example 11.6.18 that $1 * g \neq g$.

For the exponential function $f(t) = e^t$, we determined from (11.6.31) that

$$\mathcal{L}\{(f * g)(t)\} = \mathcal{L}\{f(t)\}\,\mathcal{L}\{g(t)\}.$$

In fact, as we prove next, this is true for any two piecewise continuous functions of exponential order.

Convolution Theorem

Theorem 11.6.11 *Let $f(t)$ and $g(t)$ be piecewise continuous functions on $[0, \infty)$ and of exponential order γ. Then*

$$\mathcal{L}\{(f * g)(t)\} = \mathcal{L}\{f(t)\}\,\mathcal{L}\{g(t)\} \tag{11.6.33}$$

for $s > \gamma$.

Proof According to Corollary 11.5.4, the Laplace transforms

$$\mathcal{L}\{f(t)\} = \int_0^\infty e^{-st} f(t)\,dt = F(s) \quad \text{and} \quad \mathcal{L}\{g(t)\} = \int_0^\infty e^{-st} g(t)\,dt = G(s)$$

exist for $s > \gamma$. Starting with the right-hand side of (11.6.33), we have

$$\begin{aligned}\mathcal{L}\{f(t)\}\,\mathcal{L}\{g(t)\} &= F(s)\int_0^\infty e^{-st} g(t)\,dt \\ &= \int_0^\infty e^{-st} F(s) g(t)\,dt = \int_0^\infty e^{-su} F(s) g(u)\,du,\end{aligned}$$

where we could move $F(s)$ inside the integral since it does not depend on the variable of integration. As a result, we have

$$\begin{aligned}\mathcal{L}\{f(t)\}\,\mathcal{L}\{g(t)\} &= \int_0^\infty e^{-su}\left(\int_0^\infty e^{-st} f(t)\,dt\right) g(u)\,du \\ &= \int_0^\infty \left(\int_0^\infty e^{-s(t+u)} f(t)\,dt\right) g(u)\,du.\end{aligned}$$

With the change of variable $v = t + u$, this becomes

$$\mathcal{L}\{f(t)\}\,\mathcal{L}\{g(t)\} = \int_0^\infty \left(\int_u^\infty e^{-sv} f(v-u)\,dv\right) g(u)\,du. \tag{11.6.34}$$

Now let us change the order of integration in (11.6.34). However, this is where the proof becomes quite technical and involves a result known as the *Weierstrass M -test for uniform convergence* of an improper integral, which is beyond the mathematical level of this book. Suffice it to say that it is legitimate to change the order of integration here (cf. [21, Sec. 13]). Thus,

$$\mathcal{L}\{f(t)\}\,\mathcal{L}\{g(t)\} = \int_0^\infty e^{-sv}\left(\int_0^v f(v-u) g(u)\,du\right) dv$$

$$= \int_0^\infty e^{-sv}(f * g)(v)\,dv = \int_0^\infty e^{-st}(f * g)(t)\,dt$$
$$= \mathcal{L}\{(f * g)(t)\}$$

for $s > \gamma$. ■

One important application of the convolution theorem is that it can be used to solve certain types of integral equations. An ***integral equation*** is an equation with the unknown function appearing under an integral sign. One such integral equation has the form

$$x(t) = f(t) + \int_0^t k(t-u)x(u)\,du, \tag{11.6.35}$$

where f and k represent known functions. The function k is called the ***kernel*** of the integral equation. This particular type of equation is referred to as a ***Volterra integral equation of convolution type***. The phrase "of convolution type" means that the kernel k depends only on the combination $t - u$. In the following example, we illustrate how to solve such an equation using the convolution theorem.

Example 11.6.19 Solve the Volterra integral equation

$$x(t) = t + \int_0^t (t-u)x(u)\,du. \tag{11.6.36}$$

Solution First take the Laplace transform of the equation and apply (11.6.33):

$$\mathcal{L}\{x(t)\} = \mathcal{L}\{t\} + \mathcal{L}\left\{\int_0^t (t-u)x(u)\,du\right\}$$
$$= \mathcal{L}\{t\} + \mathcal{L}\{t\}\,\mathcal{L}\{x(t)\} = \frac{1}{s^2} + \frac{1}{s^2}\mathcal{L}\{x(t)\}.$$

Now solve for $\mathcal{L}\{x(t)\}$:

$$\mathcal{L}\{x(t)\}\left(1 - \frac{1}{s^2}\right) = \frac{1}{s^2} \quad \Rightarrow \quad \mathcal{L}\{x(t)\} = \frac{1}{s^2-1}.$$

It follows from the partial fraction expansion

$$\frac{1}{s^2-1} = \frac{1}{2}\left(\frac{1}{s-1}\right) - \frac{1}{2}\left(\frac{1}{s+1}\right)$$

that

$$\frac{1}{s^2-1} = \frac{1}{2}\mathcal{L}\{e^t\} - \frac{1}{2}\mathcal{L}\{e^{-t}\}.$$

Therefore,

$$\mathcal{L}\{x(t)\} = \mathcal{L}\left\{\frac{1}{2}e^{t} - \frac{1}{2}e^{-t}\right\},$$

from which we conclude the solution is

$$x(t) = \frac{e^{t} - e^{-t}}{2} = \sinh t.$$

♦

There are also some types of integro-differential equations that can be solved with Laplace transforms. ***Integro-differential equations*** involves both derivatives and integrals of the unknown function. One type that we can solve with transforms is a ***Volterra integro-differential equation of convolution type*** of the form

$$x'(t) = f(t) + \int_0^t k(t-u)x(u)\,du. \tag{11.6.37}$$

Example 11.6.20 Find the solution of the Volterra integro-differential equation

$$x'(t) = t - \int_0^t e^{t-u}x(u)\,du$$

satisfying the initial condition $x(0) = 0$.

Solution Taking the Laplace transform of the equation, we have

$$\mathcal{L}\{x'(t)\} = \mathcal{L}\{t\} - \mathcal{L}\left\{\int_0^t e^{t-u}x(u)\,du\right\}.$$

Due to Theorem 11.4.3 and the convolution theorem, this becomes

$$s\mathcal{L}\{x(t)\} - x(0) = \frac{1}{s^2} - \mathcal{L}\{e^t\}\mathcal{L}\{x(t)\}.$$

Written in terms of $X(s) = \mathcal{L}\{x(t)\}$, this is

$$sX(s) = \frac{1}{s^2} - \frac{1}{s-1}X(s)$$

since $x(0) = 0$. Thus,

$$X(s) = \frac{s-1}{s^2(s^2 - s + 1)}.$$

A partial fraction expansion of $X(s)$ is

$$X(s) = -\frac{1}{s^2} + \frac{1}{s^2 - s + 1}.$$

In Example 11.6.3, we determined that

$$\frac{1}{s^2 - s + 1} = \mathcal{L}\left\{\frac{2}{\sqrt{3}} e^{t/2} \sin\left(\frac{\sqrt{3}}{2}t\right)\right\}.$$

Therefore, the solution is

$$x(t) = -t + \frac{2\sqrt{3}}{3} e^{t/2} \sin\left(\frac{\sqrt{3}}{2}t\right).$$

♦

11.7 Inverse Laplace Transforms

We have seen that in order to solve a differential or an integral equation using Laplace transforms, we have to first find the transform of the solution. Take for instance Example 11.4.10. There we considered the initial value problem

$$x' + 4x = 2 - 25\cos 3t, \quad x(0) = 2.$$

The objective at the time was to find its solution $x(t)$. The transform of $x(t)$, which is denoted by either $\mathcal{L}\{x(t)\}$ or $X(s)$, turned out to be

$$X(s) = \frac{2}{s+4} + \frac{2}{s(s+4)} - \frac{25s}{(s+4)(s^2+9)}.$$

Then, using a partial fraction expansion, we were able to find a continuous function whose Laplace transform is equal to $X(s)$. According to Lerch's theorem (Theorem 11.4.4), there is only one such continuous function. We showed in Example 11.4.10 that it is

$$x(t) = \frac{1}{2} + \frac{11}{2}e^{-4t} - 4\cos 3t - 3\sin 3t.$$

Because $\mathcal{L}\{x(t)\} = X(s)$, the function $x(t)$ is called an *inverse Laplace transform* of the function $X(s)$.

Since inverse Laplace transforms are a necessary component of the process for finding solutions of differential equations, let us introduce the notation for the inverse Laplace transform of a function and then look at a couple of examples to get acclimated to this new notation.

Inverse Laplace Transform

Definition 11.7.1 An ***inverse Laplace transform*** of a function $F(s)$, denoted by $\mathcal{L}^{-1}\{F(s)\}$, is a function $f(t)$ whose Laplace transform is $F(s)$ for $s > a$ for some $a \in \mathbb{R}$ (provided such a function f exists). That is,

$$\mathcal{L}^{-1}\{F(s)\} = f(t) \text{ for } t \geq 0 \quad \text{if} \quad \mathcal{L}\{f(t)\} = F(s) \text{ for } s > a.$$

Remarks Recall that the symbol $\mathbb{R}$ denotes the set of all real numbers. The function $f(t)$ may not be defined at all points in the interval $[0, \infty)$. Concerning this matter, see Definition 11.1.1 and the remark following it as well as the discussion about piecewise continuous functions in Sect. 11.5 and the examples in Sect. 11.6.

Example 11.7.1 Let $F(s) = \dfrac{1}{s-5}$. Find $\mathcal{L}^{-1}\{F(s)\}$.

Solution Since $\mathcal{L}\left\{e^{5t}\right\} = 1/(s-5)$ for $s > 5$,

$$\mathcal{L}^{-1}\left\{\frac{1}{s-5}\right\} = e^{5t} \quad \text{for} \quad t \geq 0.$$

◆

Example 11.7.2 Find $\mathcal{L}^{-1}\left\{\dfrac{6}{s^4}\right\}$.

Solution Recall that $\mathcal{L}\{t^n\} = n!/s^{n+1}$ for $s > 0$. In particular, $\mathcal{L}\{t^3\} = 3!/s^4$. Thus,

$$\mathcal{L}^{-1}\left\{\frac{6}{s^4}\right\} = \mathcal{L}^{-1}\left\{\frac{3!}{s^4}\right\} = t^3 \quad \text{for} \quad t \geq 0.$$

◆

Functions do not have unique inverse Laplace transforms. For example, the Laplace transform of the continuous function t^3 is $3!/s^4$. But so is the transform of the piecewise continuous function

$$h(t) := \begin{cases} t^3, & \text{if } t \neq 2, 7 \\ -1, & \text{if } t = 2 \\ 10, & \text{if } t = 7. \end{cases}$$

So is t^3 the inverse Laplace transform of $3!/s^4$? Or is $h(t)$? But hold on: there are infinitely many other possibilities as well. Which of them should we choose as the inverse transform? The answer to this question lies with the solutions of differential equations. Solutions of linear first-order initial value problems of the form

$$ax'(t) + bx(t) = g(t), \quad x(t_0) = x_0$$

and of linear second-order initial value problems of the form

$$ax''(t) + bx'(t) + cx(t) = g(t), \quad x(0) = x_0, \quad x'(0) = x_1$$

are continuous functions when the forcing function $g(t)$ is continuous. Consequently, using Laplace transform methods to solve such initial value problems necessitates choosing inverse transforms that are continuous functions. Since Lerch's theorem implies that a function $F(s)$ cannot have more than one inverse Laplace transform $f(t)$ that is continuous on the entire interval $[0, \infty)$, the following convention is adopted.

Convention: Inverse Laplace Transforms
Let $F(s)$ be a function with an inverse Laplace transform $f(t)$. In other words,

$$\mathcal{L}\{f(t)\} = F(s). \tag{11.7.1}$$

(i) If $f(t)$ is continuous on the entire interval $[0, \infty)$, then $f(t)$ is called *the inverse Laplace transform of* $F(s)$.
(ii) If no continuous function on $[0, \infty)$ satisfies (11.7.1), then *an inverse Laplace transform of* $F(s)$ is any piecewise continuous function $f(t)$ that satisfies (11.7.1).

Example 11.7.3 Find $\mathcal{L}^{-1}\left\{\dfrac{e^{-bs}}{s}\right\}$, where $b \geq 0$.

Solution From Example 11.2.3, we know that $\mathcal{L}\{u(t-b)\} = e^{-bs}/s$. Of course, the unit step function $u(t-b)$ is not continuous on the entire interval $[0, \infty)$ because of the discontinuity at $t = b$. If $b > 0$, a more general statement of Lerch's theorem than is given in Theorem 11.4.4 can be used to argue that there is no continuous function on $[0, \infty)$ whose Laplace transform is e^{-bs}/s. So, according to (ii) of the convention

$$\mathcal{L}^{-1}\left\{\frac{e^{-bs}}{s}\right\} = u(t-b) \qquad (b > 0). \tag{11.7.2}$$

However, if $b = 0$, then not only is $\mathcal{L}\{u(t)\} = 1/s$ but also $\mathcal{L}\{1\} = 1/s$. Of course, the constant function $f(t) = 1$ for all $t \geq 0$ is continuous on $[0, \infty)$, whereas $u(t)$ being undefined at $t = 0$ is not. Abiding by (i) of the convention, we have

$$\mathcal{L}^{-1}\left\{\frac{1}{s}\right\} = 1 \tag{11.7.3}$$

rather than $\mathcal{L}^{-1}\{1/s\} = u(t)$. ♦

The shift properties are important when it comes to finding the inverse Laplace transforms of many functions. To demonstrate this, let us restate the "shift in s" and "shift in t" properties in forms that are better suited for working with inverse transforms and then follow each of them with an example. The following version of the shift in s property is a restatement of Theorem 11.6.1.

Shift in s

Theorem 11.7.2 *If $\mathcal{L}^{-1}\{F(s)\} = f(t)$ for $s > a$, then, for any constant b,*

$$\mathcal{L}^{-1}\{F(s-b)\} = e^{bt} f(t) \tag{11.7.4}$$

for $s > a + b$.

Example 11.7.4 Find $\mathcal{L}^{-1}\left\{\dfrac{s-3}{(s-3)^2+25}\right\}$.

Solution If it were not for the shift in the argument of the Laplace transform, this would be merely a matter of determining an inverse Laplace transform of

$$F(s) = \frac{s}{s^2+5^2},$$

which is $\mathcal{L}^{-1}\{F(s)\} = \cos 5t$ for $s > 0$. But because this is $\mathcal{L}^{-1}\{F(s-3)\}$, take $b = 3$ and $f(t) = \cos 5t$ in (11.7.4). As a result, we get

$$\mathcal{L}^{-1}\left\{\frac{s-3}{(s-3)^2+25}\right\} = \mathcal{L}^{-1}\{F(s-3)\} = e^{3t}\cos 5t$$

for $s > 3$. ♦

From Theorem 11.6.3, we have the following version of the shift in t property.

Shift in t (Alternate Form)

Theorem 11.7.3 *If $\mathcal{L}^{-1}\{F(s)\} = f(t)$ for $s > a$, then, for any $b > 0$,*

$$\mathcal{L}^{-1}\{e^{-bs}F(s)\} = f(t-b)u(t-b) \tag{11.7.5}$$

for $s > a$.

Example 11.7.5 Find $\mathcal{L}^{-1}\{Y(s)\}$ for $Y(s) = \dfrac{e^{-5s}s}{s^2+4}$.

Solution Observe that $Y(s) = e^{-5s}F(s)$, where

$$F(s) = \frac{s}{s^2+4} = \mathcal{L}\{\cos 2t\}.$$

Taking $b = 5$ and $f(t) = \cos 2t$ in (11.7.5), we have

$$\mathcal{L}^{-1}\{Y(s)\} = f(t-5)u(t-5) = \cos(2(t-5)) \cdot u(t-5).$$

♦

It follows from Theorem 11.3.1 that the inverse Laplace transform inherits the following linearity property from the Laplace transform.

Linearity of the Inverse Laplace Transform

Theorem 11.7.4 *If $F_1(s)$ and $F_2(s)$ are functions whose inverse Laplace transforms exist for $s > b$, then for any constants c_1 and c_2, an inverse transform of the linear combination $c_1F_1(s) + c_2F_2(s)$ for $s > b$ is given by*

$$\mathcal{L}^{-1}\{c_1F_1(s) + c_2F_2(s)\} = c_1\mathcal{L}^{-1}\{F_1(s)\} + c_2\mathcal{L}^{-1}\{F_2(s)\}.$$

In other words, $\mathcal{L}^{-1}$ is a linear operator.

Example 11.7.6 Find $\mathcal{L}^{-1}\left\{\dfrac{5s}{s^2+9} - \dfrac{24}{s^4}\right\}$.

Solution By the above linearity property,

$$\mathcal{L}^{-1}\left\{\frac{5s}{s^2+9} - \frac{24}{s^4}\right\} = 5\mathcal{L}^{-1}\left\{\frac{s}{s^2+9}\right\} - \mathcal{L}^{-1}\left\{\frac{4\cdot 3!}{s^4}\right\}$$

$$= 5\mathcal{L}^{-1}\left\{\frac{s}{s^2+3^2}\right\} - 4\mathcal{L}^{-1}\left\{\frac{3!}{s^4}\right\} = 5\cos 3t - 4t^3.$$

♦

Example 11.7.7 Find $\mathcal{L}^{-1}\left\{\frac{2s}{s^2+5}\right\}$.

Solution First note that $\frac{s}{s^2+5} = \frac{s}{s^2+(\sqrt{5})^2} = \mathcal{L}\{\cos t\sqrt{5}\}$. Thus, by the linearity property,

$$\mathcal{L}^{-1}\left\{\frac{2s}{s^2+5}\right\} = 2\mathcal{L}^{-1}\left\{\frac{s}{s^2+(\sqrt{5})^2}\right\} = 2\cos(t\sqrt{5}).$$

♦

The following examples demonstrate that functions may have to be expressed in other forms in order to determine their inverse Laplace transforms. Some techniques that are useful in this regard are the same ones that aid in carrying out integrations of rational functions: the method of partial fractions and completion of the square of the denominator.

Example 11.7.8 Find an inverse Laplace transform of the function

$$F(s) = \frac{26s-13}{2s^2+7s-15}.$$

Solution When it comes to finding an inverse Laplace transform of a rational function, a rule of thumb is to start by factoring each reducible quadratic function in the denominator. After this is done, it becomes clear how to write the partial fraction expansion. For $F(s)$, we have

$$F(s) = \frac{26s-13}{(2s-3)(s+5)} = \frac{A}{2s-3} + \frac{B}{s+5}.$$

Clearing the fractions of their denominators, we obtain

$$A(s+5) + B(2s-3) = 26s - 13.$$

Set $s = -5$ to find B:

$$-13B = 26(-5) - 13 \quad \Rightarrow \quad B = (-2)(-5) + 1 = 11.$$

Set $s = 3/2$ to find A:

$$\frac{13}{2}A = 26\left(\frac{3}{2}\right) - 13 = 26 \quad \Rightarrow \quad A = \frac{(2)(26)}{13} = 4.$$

Thus,

$$\begin{aligned}\mathcal{L}^{-1}\{F(s)\} &= \mathcal{L}^{-1}\left\{\frac{4}{2s-3}\right\} + \mathcal{L}^{-1}\left\{\frac{11}{s+5}\right\} \\ &= \frac{4}{2}\mathcal{L}^{-1}\left\{\frac{1}{s-\frac{3}{2}}\right\} + 11\mathcal{L}^{-1}\left\{\frac{1}{s-(-5)}\right\} = 2e^{3t/2} + 11e^{-5t}.\end{aligned}$$

♦

Example 11.7.9 Find $\mathcal{L}^{-1}\{F(s)\}$ for

$$G(s) = \frac{1}{s^2+4s+8}.$$

Solution Unlike the previous example, partial fraction expansion will not work here because the quadratic function is irreducible. In a case such as this, the rule of thumb is to complete the square of the denominator as follows:

$$G(s) = \frac{1}{(s+2)^2+2^2}.$$

This does not quite have the form of any of the functions appearing in the right-hand column of Table 11.1. On the other hand, it is clearly related to the form $b/(s^2+b^2)$. All that is required then is a little mathematical legerdemain:

$$\frac{1}{(s+2)^2+2^2} = \frac{1}{2}\cdot\frac{2}{(s+2)^2+2^2}.$$

Due to the linearity of the inverse Laplace transform and the shift in s properties, the inverse Laplace transform of $G(s)$ is

$$\mathcal{L}^{-1}\{G(s)\} = \frac{1}{2}\mathcal{L}^{-1}\left\{\frac{2}{(s+2)^2+2^2}\right\} = \frac{1}{2}e^{-2t}\sin 2t.$$

♦

Example 11.7.10 Find an inverse transform of $H(s) = \dfrac{3s}{s^2+4s+13}$.

Solution As in the previous example, the quadratic function in the denominator is irreducible. Completing its square, we have

$$H(s) = \frac{3s}{(s+2)^2+3^2}.$$

$H(s)$ has the form $s/(s^2+b^2)$ except for the shift from s to $s+2$ in the denominator. This calls for another trick of the mathematical trade, which is to shift the s in the numerator as follows:

$$H(s) = \frac{3s}{(s+2)^2+3^2} = \frac{3(s+2)-3(2)}{(s+2)^2+3^2}$$

$$= 3\,\frac{s+2}{(s+2)^2+3^2} - 2\,\frac{3}{(s+2)^2+3^2}.$$

Therefore, by the linearity and shift in s properties,

$$\mathcal{L}^{-1}\{H(s)\} = 3\,\mathcal{L}^{-1}\left\{\frac{s+2}{(s+2)^2+3^2}\right\} - 2\,\mathcal{L}^{-1}\left\{\frac{3}{(s+2)^2+3^2}\right\}$$

$$= 3e^{-2t}\cos 3t - 2e^{-2t}\sin 3t.$$

◆

One of the lessons to take away from the next set of examples is the utility of the convolution theorem (Theorem 11.6.11) for finding inverse Laplace transforms.

Example 11.7.11 Find an inverse transform of

$$P(s) = \frac{1}{s^2\,(s^2+1)}.$$

Solution 1 By the convolution theorem,

$$P(s) = \frac{1}{s^2}\cdot\frac{1}{s^2+1} = \mathcal{L}\{t\}\,\mathcal{L}\{\sin t\} = \mathcal{L}\{t * \sin t\}.$$

So an inverse transform of $P(s)$ is

$$\mathcal{L}^{-1}\{P(s)\} = t * \sin t = \int_0^t (t-u)\sin u\,du.$$

Integrating by parts, we get

$$\mathcal{L}^{-1}\{P(s)\} = \big[-(t-u)\,\cos u - \sin u\big]_0^t = t - \sin t.$$

◆

Solution 2 A partial fraction expansion for $P(s)$ is

$$P(s) = \frac{A}{s} + \frac{B}{s^2} + \frac{C}{s^2+1},$$

where $A = 0$, $B = 1$, and $C = -1$. Thus, an inverse transform of $P(s)$ is

$$\mathcal{L}^{-1}\{P(s)\} = \mathcal{L}^{-1}\left\{\frac{1}{s^2}\right\} - \mathcal{L}^{-1}\left\{\frac{1}{s^2+1}\right\} = t - \sin t.$$

♦

Solution 3 Still another way of finding $\mathcal{L}^{-1}\{P(s)\}$ is to use the "Laplace transform of an integral property" (Theorem 11.6.6). It follows from (11.6.17) that

$$\mathcal{L}^{-1}\left\{\frac{F(s)}{s}\right\} = \int_0^t f(u)\,du,$$

where $F(s) = \mathcal{L}\{f(t)\}$. Let $F(s) = 1/(s^2+1)$. Then $f(t) = \sin t$. Thus,

$$\mathcal{L}^{-1}\left\{\frac{1}{s(s^2+1)}\right\} = \mathcal{L}^{-1}\left\{\frac{F(s)}{s}\right\} = \int_0^t \sin u\,du = 1 - \cos t.$$

Now apply this same property to the function $G(s) = 1/[s(s^2+1)]$:

$$\mathcal{L}^{-1}\left\{\frac{1}{s^2(s^2+1)}\right\} = \mathcal{L}^{-1}\left\{\frac{G(s)}{s}\right\} = \int_0^t (1 - \cos u)\,du.$$

Therefore,

$$\mathcal{L}^{-1}\{P(s)\} = \int_0^t (1 - \cos u)\,du = t - \sin t.$$

♦

We can see from comparing Solutions 1 and 2 in the preceding example that the convolution theorem is sometimes a viable alternative to the method of partial fractions for finding inverse Laplace transforms. The convolution theorem is also used in the following example.

Example 11.7.12 Find an inverse transform of

$$Q(s) = \frac{s}{(s^2+1)^2}.$$

Solution From

$$Q(s) = \frac{1}{s^2+1} \cdot \frac{s}{s^2+1} = \mathcal{L}\{\sin t\}\,\mathcal{L}\{\cos t\} = \mathcal{L}\{\sin t * \cos t\},$$

we have

$$\mathcal{L}^{-1}\{Q(s)\} = \sin t * \cos t = \int_0^t \sin(t-u)\cos u\, du.$$

Using the product identity

$$\sin\theta\cos\phi = \frac{1}{2}[\sin(\theta-\phi) + \sin(\theta+\phi)],$$

let us rewrite the integrand as

$$\sin(t-u)\cos u = \frac{1}{2}[\sin(t-2u) + \sin t],$$

which allows us to carry out the following integration:

$$\begin{aligned}\int_0^t \sin(t-u)\cos u\, du &= \frac{1}{2}\int_0^t [\sin(t-2u) + \sin t]\, du \\ &= \frac{1}{4}\cos(t-2u)\Big|_0^t + \frac{1}{2}t\sin t = \frac{1}{2}t\sin t.\end{aligned}$$

Therefore,

$$\mathcal{L}^{-1}\{Q(s)\} = \frac{1}{2}t\sin t.$$

♦

Most of the previous examples demonstrating how to find inverse Laplace transforms involved rational functions. Finding them for other types of functions may require strategies different from the ones that we have used up to now, such as the "derivatives of Laplace transforms" property.

Example 11.7.13 Find the inverse Laplace transform of

$$F(s) = \ln\left(\frac{s}{s-1}\right).$$

Solution We see from Theorem 11.6.4 with $n = 1$ that

$$\frac{d}{ds}F(s) = -\mathcal{L}\{tf(t)\},$$

where $f(t) = \mathcal{L}^{-1}\{F(s)\}$. Since

$$\frac{d}{ds}F(s) = \frac{d}{ds}\big[\ln s - \ln(s-1)\big] = \frac{1}{s} - \frac{1}{s-1},$$

we have

$$\mathcal{L}\{tf(t)\} = -F'(s) = \frac{1}{s-1} - \frac{1}{s}.$$

Thus, $tf(t) = e^t - 1$. As a result,

$$\mathcal{L}^{-1}\{F(s)\} = f(t) = \frac{e^t - 1}{t}.$$

Compare this to Example 11.6.9. ♦

11.7.1 Inverse Transforms and Periodicity

Theorem 11.6.7 deals with the Laplace transforms of piecewise-continuous, periodic functions. We ascertained that the transforms of such functions with period b can be computed using formula (11.6.20). In the next theorem, we change things around by starting off with a quotient function whose denominator looks exactly like the denominator of the function on the right-hand side of (11.6.20) and whose numerator is the Laplace transform of a function of the form (11.6.21).

Inverse Transforms and Periodic Functions

Theorem 11.7.5 *Let $F(s)$ be a function of the form*

$$F(s) = \frac{G(s)}{1 - e^{-bs}} \quad (s > 0) \tag{11.7.6}$$

where $b > 0$. If $G(s)$ has an inverse Laplace transform $g(t)$, where

$$g(t) = \begin{Bmatrix} \varphi(t), & 0 < t < b \\ 0, & t > b \end{Bmatrix} = \varphi(t)\big[u(t) - u(t-b)\big] \tag{11.7.7}$$

and $\varphi(t)$ is piecewise continuous on $[0, b]$, then

$$f(t) = \sum_{n=0}^{\infty} \varphi(t - nb)\,[u(t - nb) - u(t - (n+1)b)] \tag{11.7.8}$$

is an inverse Laplace transform of $F(s)$ for $s > 0$. Furthermore, $f(t)$ is periodic on $[0, \infty)$ with period b. Its value on the interval $(nb, (n+1)b)$ is

$$f(t) = \varphi(t - nb)$$

for $n = 0, 1, 2, 3, \ldots$.

Proof Since $\varphi(t)$ denotes a function that is defined on the interval $[0, b]$, its translation $\varphi(t - nb)$ is a function on $[nb, (n + 1)b]$. Note that the infinite series (11.7.8) does define a function on $[0, \infty)$ since for any given $t > 0$ only a finite number of the terms of this series are nonzero because $u(t - nb) = 0$ when $nb > t$. For a fixed value of n,

$$u(t - nb) - u(t - (n + 1)b) = 1$$

for $nb < t < (n + 1)b$, whereas for these same values of t

$$u(t - mb) - u(t - (m + 1)b) = 0$$

when $m \neq n$. Consequently, (11.7.8) simplifies to

$$f(t) = \varphi(t - nb)\,[u(t - nb) - u(t - (n + 1)b)] = \varphi(t - nb)$$

for $t \in (nb, (n + 1)b)$. Since the graph of $\varphi(t - nb)$ is the graph of $\varphi(t)$ shifted nb units to the right, the graph of $f(t)$ on the interval $(nb, (n + 1)b)$ is an exact copy of the graph of $f(t)$ on the interval $(0, b)$. As this is the case for every one of these intervals, the function $f(t)$ is periodic on $[0, \infty)$ with period b. This proves the last two statements of the theorem.

Since $f(t) = \varphi(t)$ on $(0, b)$ and $\varphi(t)$ is piecewise continuous on this interval, $f(t)$ is piecewise continuous on $[0, b]$. As a result, all of the conditions of Theorem 11.6.7 are met. Accordingly,

$$\mathcal{L}\{f(t)\} = \frac{\mathcal{L}\{f_b(t)\}}{1 - e^{-bs}} \tag{11.7.9}$$

for $s > 0$, where

$$f_b(t) = \left\{\begin{matrix} f(t), & 0 < t < b \\ 0, & t > b \end{matrix}\right\} = \left\{\begin{matrix} \varphi(t), & 0 < t < b \\ 0, & t > b \end{matrix}\right\}.$$

So if $G(s)$ is a function with an inverse Laplace transform $g(t)$, where $g(t)$ is given by (11.7.7), then $f_b(t) = g(t)$. As a result, we have $\mathcal{L}\{f_b(t)\} = \mathcal{L}\{g(t)\} = G(s)$. Therefore, $\mathcal{L}\{f(t)\} = F(s)$. ■

Example 11.7.14 Find an inverse Laplace transform of the function

$$F(s) = \frac{1 - e^{-s}}{s(1 + e^{-s})},$$

where $s > 0$.

Solution We begin by rewriting $F(s)$ so that it looks like (11.7.6):

$$F(s) = \frac{(1-e^{-s})^2}{s(1+e^{-s})(1-e^{-s})} = \frac{1-2e^{-s}+e^{-2s}}{s(1-e^{-2s})} = \frac{G(s)}{1-e^{-bs}},$$

where $b = 2$ and

$$G(s) = \frac{1-2e^{-s}+e^{-2s}}{s}.$$

An inverse transform of $G(s)$ is the function

$$g(t) = \mathcal{L}^{-1}\left\{\frac{1}{s}\right\} - 2\mathcal{L}^{-1}\left\{\frac{e^{-s}}{s}\right\} + \mathcal{L}^{-1}\left\{\frac{e^{-2s}}{s}\right\} = u(t) - 2u(t-1) + u(t-2).$$

What points to periodicity is that this can also be expressed as (11.7.7), where

$$\varphi(t) = \begin{cases} 1, & 0 < t < 1 \\ -1, & 1 < t < 2, \end{cases}$$

since

$$g(t) = \left\{\begin{array}{ll} 1, & 0 < t < 1 \\ -1, & 1 < t < 2 \\ 0, & t > 2 \end{array}\right\} = \left\{\begin{array}{ll} \varphi(t), & 0 < t < 2 \\ 0, & t > 2 \end{array}\right\} = \varphi(t)\big[u(t) - u(t-2)\big].$$

As a result, $\mathcal{L}^{-1}\{F(s)\}$ exists. It is given by (11.7.8) with $b = 2$. That is,

$$\mathcal{L}^{-1}\{F(s)\} = f(t) = \sum_{n=0}^{\infty} \varphi(t-2n)\big[u(t-2n) - u(t-2(n+1))\big],$$

where the value of $f(t)$ on $(2n, 2(n+1))$ is

$$f(t) = \varphi(t-2n) = \begin{cases} 1, & 2n < t < 2n+1 \\ -1, & 2n+1 < t < 2n+2. \end{cases}$$

Now let us rewrite the infinite series representing $f(t)$ by subtracting $u(t-(2n+1))$ and then adding it back again for $n = 1, 2, 3, \ldots$ and replacing $\varphi(t-2n)$ with the values 1 and -1 on the appropriate intervals as follows:

$$f(t) = \sum_{n=0}^{\infty} \varphi(t-2n)\big[u(t-2n) - u(t-(2n+1))$$

$$+ u(t-(2n+1)) - u(t-(2n+2))\big]$$

$$= \sum_{n=0}^{\infty} \left(\big[u(t-2n) - u(t-(2n+1))\big] - \big[u(t-(2n+1)) - u(t-(2n+2))\big]\right).$$

Writing out a few terms, we have

$$\begin{aligned} f(t) &= [u(t) - u(t-1)] - [u(t-1) - u(t-2)] + [u(t-2) - u(t-3)] \\ &\quad - [u(t-3) - u(t-4)] + [u(t-4) - u(t-5)] \\ &\quad - [u(t-5) - u(t-6)] + \cdots, \end{aligned}$$

which simplifies to

$$f(t) = u(t) - 2u(t-1) + 2u(t-2) - 2u(t-3) + 2u(t-4) - 2u(t-5) + \cdots.$$

Therefore,

$$\mathcal{L}^{-1}\{F(s)\} = f(t) = u(t) + 2\sum_{n=1}^{\infty}(-1)^n u(t-n).$$

The graph of $f(t)$ is a periodic rectangular wave with period 2, which resembles the graph of $r(t)$ in Fig. 7.9 in Chap. 7. It is in fact the graph of $r(t)$ shrunk vertically toward the t-axis by a factor of $1/2$. ♦

Example 11.7.15 Find $\mathcal{L}^{-1}\{F(s)\}$, where

$$F(s) = \frac{1}{s^2+1}\left[\frac{1+e^{-\pi s}}{1-e^{-\pi s}}\right] \quad (s > 0). \tag{11.7.10}$$

Solution This is (11.7.6) with $b = \pi$ and

$$G(s) = \frac{1+e^{-\pi s}}{s^2+1} = \frac{1}{s^2+1} + e^{-\pi s}\frac{1}{s^2+1}.$$

An inverse transform of $G(s)$ is

$$g(t) = \sin t + \sin(t-\pi)H(t-\pi),$$

which is equal to 0 at $t = 0$ and $t = \pi$. However, if in some application $g(t)$ is known to be undefined at $t = 0$ and $t = \pi$, then

$$g(t) = \sin(t)u(t) + \sin(t-\pi)u(t-\pi).$$

In this example, let us deal with the former. As $\sin(t-\pi) = -\sin t$,

$$g(t) = \sin t\,[1 - H(t-\pi)] = \begin{cases} \sin t, & 0 \le t < \pi \\ 0, & t \ge \pi. \end{cases}$$

Referring to (11.7.7), let $\varphi(t) = \sin t$. It then follows from (11.7.8) that an inverse transform of $F(s)$ is the function

$$f(t) = \sum_{n=0}^{\infty} \varphi(t-n\pi)\big[H(t-n\pi) - H(t-(n+1)\pi)\big].$$

The bracketed quantity is a box function (see Definition 11.6.8), where

$$H(t-n\pi) - H(t-(n+1)\pi) = \begin{cases} 1, & n\pi \le t < (n+1)\pi \\ 0, & t < n\pi \ \text{ or } \ t \ge (n+1)\pi. \end{cases}$$

Consequently,

$$f(t) = \varphi(t-n\pi) = \sin(t-n\pi) = \sin t \cos n\pi - \cos t \sin n\pi = (-1)^n \sin t$$

on the interval $[n\pi, (n+1)\pi)$, for $n = 1, 2, 3, \ldots$. Note that $\sin t \le 0$ on the intervals $[\pi, 2\pi)$, $[3\pi, 4\pi)$, $[5\pi, 6\pi), \ldots$. Thus, on these intervals, $f(t) = -\sin t \ge 0$. However, $f(t) = \sin t \ge 0$ on the intervals $[0, \pi)$, $[2\pi, 3\pi)$, $[4\pi, 5\pi), \ldots$. In other words, $f(t) = |\sin t|$. Therefore, we conclude that

$$\mathcal{L}^{-1}\{F(s)\} = f(t) = |\sin t|$$

for $t \ge 0$. ♦

Remark With this example we have come full circle: See Example 11.6.15, where we found that the Laplace transform of $f(t) = |\sin t|$ is given by (11.7.10).

11.7.2 Inverse Transforms and Infinite Series

A cursory reading of Theorem 11.7.5 and the accompanying examples could give the impression that inverse Laplace transforms of functions of the form (11.7.6) always exist and are periodic. However, as the next example shows, this is not the case.

Example 11.7.16 Find an inverse Laplace transform of

$$F(s) = \frac{s}{(s^2+1)(1-e^{-s})} \qquad (s > 0). \tag{11.7.11}$$

Solution By all outward appearances, it seems that Theorem 11.7.5 applies to $F(s)$ since it looks like (11.7.6). Suppose this to be the case, namely, that there is a periodic function $f(t)$ on $[0, \infty)$ with period b that is piecewise continuous on the interval $[0, b]$ and whose Laplace transform is $F(s)$. Then it would follow from Theorem 11.6.7 that

$$F(s) = \mathcal{L}\{f(t)\} = \frac{\mathcal{L}\{f_b(t)\}}{1 - e^{-bs}} \tag{11.7.12}$$

for $s > 0$, where

$$f_b(t) = \begin{cases} f(t), & 0 < t < b \\ 0, & t > b. \end{cases}$$

Moreover, we see from comparing (11.7.11) and (11.7.12) that the value of b would have to be 1 and that

$$\mathcal{L}\{f_1(t)\} = \frac{s}{s^2 + 1}.$$

This implies $f_1(t) = \cos t$, which in turn implies that $f(t) = \cos t$ since $f_1(t)$ and $f(t)$ agree on the interval $(0, 1)$. However, this cannot be because the period of $\cos t$ is 2π, not $b = 1$. Therefore, we conclude that should an inverse Laplace transform of (11.7.11) exist, it is not a periodic function.

Assuming that $F(s)$ has an inverse Laplace transform, albeit nonperiodic, let us call it $f(t)$ and forge ahead with the goal of finding it. We begin by rewriting (11.7.11) as

$$F(s) - e^{-s}F(s) = \frac{s}{s^2 + 1}. \tag{11.7.13}$$

Suppose that $F(s)$ is the result of modeling some technical application, where it is known that $f(t)$ is defined at $t = 0$ but not at $t = 1$. Then, taking the inverse Laplace transform of both sides of (11.7.13) and using the alternate form of the shift in t property (11.7.5), we have

$$f(t) - f(t-1)u(t-1) = \cos t.$$

In other words, $f(t)$ has to satisfy the *functional equation*[20]

$$f(t) = \cos t + f(t-1)u(t-1) \tag{11.7.14}$$

[20] Functional equations are not differential equations. Briefly, a functional equation is an equation that relates an unknown function at a point to its values elsewhere.

in order for it to be an inverse Laplace transform of $F(s)$. That is to say,

$$f(t) = \begin{cases} \cos t, & 0 \le t < 1 \\ \cos t + f(t-1), & t > 1. \end{cases} \tag{11.7.15}$$

So for $t \in [0, 1)$, $f(t) = \cos t$. Thus, for $t \in (1, 2)$,

$$f(t-1) = \cos(t-1)$$

since $t - 1 \in (0, 1)$. Then, it follows from (11.7.15) that

$$f(t) = \cos t + f(t-1) = \cos t + \cos(t-1)$$

on the interval $(1, 2)$. Therefore, for $t \in [0, 2)$,

$$f(t) = \cos t + \cos(t-1)u(t-1). \tag{11.7.16}$$

For $2 < t < 3$,

$$f(t-1) = \cos(t-1) + \cos(t-2)$$

as $t - 1 \in (1, 2)$. From this and (11.7.15), it follows that

$$f(t) = \cos(t) + f(t-1) = \cos t + \cos(t-1) + \cos(t-2)$$

on the interval $(2, 3)$. And so for $t \in [0, 3)$,

$$f(t) = \cos t + \cos(t-1)u(t-1) + \cos(t-2)u(t-2). \tag{11.7.17}$$

So far, we have shown that $f(t)$ is given by (11.7.16) on the interval $[0, 2)$ and by (11.7.17) on the interval $[0, 3)$. This suggests that

$$f(t) = \cos t + \sum_{k=1}^{n-1} \cos(t-k)u(t-k) \tag{11.7.18}$$

on $[0, n)$ for every integer $n \ge 1$. Note that for $n = 1$, the upper index of summation (stopping point) is less than the lower index (starting point). This particular summation is known as the ***empty sum*** since it has no summands. Its value is defined to be 0. As a result, we get $f(t) = \cos t$ from (11.7.18), as it should be on the interval $[0, 1)$.

Now let us prove that (11.7.18) holds on every interval $[0, n)$ with the following induction argument. For each integer $n \ge 1$, let $P(n)$ represent the statement: The inverse Laplace transform of $F(s)$ is given by (11.7.18) on the interval $[0, n)$. Suppose $P(n)$ is true for some $n \ge 1$. Keeping this supposition in mind, consider the interval $[0, n+1)$. If $t \in (n, n+1)$, then as $t-1 \in (n-1, n)$, it follows from (11.7.18) that

$$\begin{aligned} f(t-1) &= \cos(t-1) + \sum_{k=1}^{n-1} \cos(t-1-k)u(t-1-k) \\ &= \cos(t-1) + \sum_{k=1}^{n-1} \cos(t-(k+1))u(t-(k+1)) \\ &= \cos(t-1) + \sum_{k=2}^{n} \cos(t-k)u(t-k). \end{aligned}$$

Moreover, as $t > n$, $u(t-k) = 1$ for $k = 2, 3, \ldots, n$. And so

$$f(t-1) = \cos(t-1) + \sum_{k=2}^{n} \cos(t-k) = \sum_{k=1}^{n} \cos(t-k).$$

Consequently, for $t \in (n, n+1)$, we have

$$f(t) = \cos t + f(t-1) = \cos t + \sum_{k=1}^{n} \cos(t-k).$$

Therefore, for $t \in [0, n+1)$,

$$f(t) = \cos t + \sum_{k=1}^{n} \cos(t-k)u(t-k). \tag{11.7.19}$$

Thus we have established that if the statement $P(n)$ is true, then $P(n+1)$ must also be true. Since we already know that $P(1)$, $P(2)$, and $P(3)$ are true, it follows from the principle of mathematical induction that $P(n)$ is true for all integers $n \geq 1$. In other words, formula (11.7.18) is valid on $[0, n)$ for every integer $n \geq 1$.

Observe that the formula for $\mathcal{L}^{-1}\{F(s)\} = f(t)$ on the entire interval $[0, \infty)$ can be represented by the infinite series

$$f(t) = \cos t + \sum_{k=1}^{\infty} \cos(t-k)u(t-k). \tag{11.7.20}$$

The reason is that for a given finite interval $[0, n)$ only a finite number of the terms in (11.7.20) are actually nonzero because $u(t-k) = 0$ for $k \geq n$. As a result, (11.7.20) simplifies to (11.7.18) for $t \in [0, n)$.

In summary, we have shown that should an inverse Laplace transform of (11.7.11) exist, then the function $f(t)$ defined by (11.7.20) would be one, or for that matter, any function which only differs from it at a finite number of points on any given finite interval, such as the function

$$g(t) = \cos t + \sum_{k=1}^{\infty} \cos(t-k)H(t-k).$$

So this brings up the question: Is $f(t)$ really an inverse transform of the function $F(s)$? The answer is yes: It is left as an exercise to verify that $\mathcal{L}\{f(t)\} = F(s)$, assuming that the Laplace transform of the series in (11.7.20) can be computed by taking the Laplace transform of each individual term in the series and then summing the results. This is in fact a valid assumption: see Problem 101. ♦

The next theorem generalizes Theorem 11.7.5, which, among other things, will give us an alternative way to compute inverse Laplace transforms of functions of the form

$$F(s) = \frac{G(s)}{1 - ae^{-bs}}$$

that will be far more expeditious than what we have just experienced with Example 11.7.16.

Inverse Laplace Transforms Expressed as Series

Theorem 11.7.6 *Let $F(s)$ denote a function of the form*

$$F(s) = \frac{G(s)}{1 - ae^{-bs}}, \tag{11.7.21}$$

where a and b are constants with $a \neq 0$ and $b > 0$. If there exists a piecewise continuous function $g(t)$ on $(-\infty, \infty)$ of exponential order γ and if $G(s)$ is its Laplace transform, then the function

$$f(t) = g(t) + \sum_{n=1}^{\infty} a^n g(t-nb)u(t-nb) \tag{11.7.22}$$

with domain $[0, \infty)$ is an inverse Laplace transform of $F(s)$ for $s > \mu$, where μ is any constant satisfying the inequality $\mu > \max\{\gamma, b^{-1}\ln|a|\}$.

(i) *If $a = -1$ and if $g(t)$ is continuous and periodic with period b on $(-\infty, \infty)$, then the function $f(t)$ is periodic on $[0, \infty)$ with period $2b$.*

(continued)

(ii) *If $g(t)$ can be expressed in the form*

$$g(t) = \varphi(t)\big[u(t) - u(t-b)\big], \tag{11.7.23}$$

where $\varphi(t)$ is a piecewise continuous function on $(-\infty, \infty)$ of exponential order γ, then (11.7.22) *simplifies to $f(t) = \sum_{n=0}^{\infty} a^n g(t - nb)$. In other words, the Laplace transform of*

$$f(t) = \sum_{n=0}^{\infty} a^n \varphi(t - nb)\big[u(t - nb) - u(t - (n+1)b)\big] \tag{11.7.24}$$

is the function $F(s)$. Furthermore, $f(t)$ is periodic on $[0, \infty)$ when $|a| = 1$. If $a = 1$, the period is b; however, it is $2b$ if $a = -1$.

Remarks By hypothesis, the function $g(t)$ in the statement of the theorem is of exponential order γ. So, according to Definition 11.4.1, constants $K > 0$ and $t_1 \geq 0$ exist such that $|g(t)| \leq Ke^{\gamma t}$ for all $t \geq t_1$. Moreover, it also assumed that $g(t)$ is piecewise continuous on $(-\infty, \infty)$. Thus, it is bounded on the interval $[0, t_1]$. It follows from these two assumptions that a constant $M > 0$ exists such that

$$|g(t)| \leq Me^{\gamma t} \quad \text{for all} \quad t \geq 0.$$

Corollary 11.5.4 states if a function is piecewise continuous on $[0, \infty)$ and of exponential order γ, then its Laplace transform exists for all $s > \gamma$. So by positing that $g(t)$ is piecewise continuous and of exponential order γ, its Laplace transform exists for all $s > \gamma$.

Proof Note that the series on the right-hand side in (11.7.22) that we call $f(t)$ does in fact define a function on $[0, \infty)$ since for a given $t \geq 0$ only a finite number of the terms of this series are actually nonzero as $u(t - nb) = 0$ when $nb > t$. In order to prove the main conclusion of the theorem, namely, the "then" statement containing (11.7.22), we have to show that the Laplace transform of $f(t)$ is $F(s)$. First, let us establish that the function $f(t)$ is piecewise continuous on $[0, \infty)$. Obviously, due to the hypotheses, the very first term of the series, namely $g(t)$, is piecewise continuous on $[0, \infty)$. That is, it is continuous on every bounded interval $[0, T]$ except for at most a finite number of jump discontinuities. Consequently, all of the other terms of the series are also piecewise continuous on $[0, \infty)$ since

$$a^n g(t - nb)u(t - nb) = \begin{cases} 0, & t < nb \\ a^n g(t - nb), & t > nb \end{cases}$$

for a given $n \geq 1$. Moreover, since only a finite number of the terms defining $f(t)$ in (11.7.22) are actually nonzero on any given interval $[0, T]$, there is a positive integer N such that

$$f(t) = g(t) + \sum_{n=1}^{N} a^n g(t - nb)u(t - nb)$$

for $0 \leq t \leq T$. Consequently, $f(t)$ has at most a finite number of jump discontinuities on every bounded interval $[0, T]$. Thus, according to Definition 11.5.3, it is a piecewise continuous function on $[0, \infty)$.

Now let us show that $f(t)$ is of exponential order μ, where the constant μ satisfies the inequality after (11.7.22). Consider the partial sums

$$f_m(t) = g(t) + \sum_{n=1}^{m-1} a^n g(t - nb)u(t - nb)$$

for $m = 1, 2, 3, \ldots$, where the sum is 0 when $m = 1$ so that $f_1(t) = g(t)$. Taking the absolute value of both sides and applying the generalized triangle inequality, we have

$$|f_m(t)| \leq |g(t)| + \sum_{n=1}^{m-1} |a^n||g(t - nb)||u(t - nb)|.$$

Since $g(t)$ is piecewise continuous and of exponential order γ, a constant $M > 0$ exists such that $|g(t)| \leq Me^{\gamma t}$ for all $t \geq 0$ (see *Remarks*). Consequently,

$$|f_m(t)| \leq Me^{\gamma t} + \sum_{n=1}^{m-1} |a^n| Me^{\gamma(t-nb)} u(t - nb)$$

for all $t \geq 0$. This, together with $\mu > \gamma$ and

$$u(t - nb) = \begin{cases} 0, & t < nb \\ 1, & t > nb, \end{cases}$$

implies that

$$\begin{aligned} |f_m(t)| &\leq Me^{\mu t} + \sum_{n=1}^{m-1} |a^n| Me^{\mu(t-nb)} u(t - nb) \\ &\leq Me^{\mu t} + \sum_{n=1}^{m-1} |a^n| Me^{\mu(t-nb)} \end{aligned}$$

$$= Me^{\mu t}\left[1+\sum_{n=1}^{m-1}(|a|e^{-\mu b})^n\right] = Me^{\mu t}\left[1+\sum_{n=1}^{m-1}x^n\right]$$

for $t \geq 0$, where $x := |a|e^{-\mu b}$. The restrictions placed on the values of a, b, μ in the statement of the theorem imply that $0 < x < 1$. Moreover, as

$$1+\sum_{n=1}^{m-1}x^n = \frac{1-x^m}{1-x}$$

and since $|a|^m e^{-m\mu b} < 1$ for $m = 1, 2, 3, \ldots$, we have

$$|f_m(t)| < Me^{\mu t}\left[\frac{1-|a|^m e^{-m\mu b}}{1-|a|e^{-\mu b}}\right] < \frac{Me^{\mu t}}{1-|a|e^{-\mu b}} = Le^{\mu t},$$

where $L := M/(1-|a|e^{-\mu b})$. Since $f_m(t) \to f(t)$ as $m \to \infty$, we conclude that

$$|f(t)| \leq Le^{\mu t}$$

for $t \geq 0$. Thus $f(t)$ is of exponential order μ.

So far we have established that if a function $g(t)$ exists satisfying the hypotheses of the theorem, then the function $f(t)$ defined by (11.7.22) is piecewise continuous on $[0, \infty)$ and of exponential order μ. It follows from Corollary 11.5.4 that the Laplace transform of $f(t)$ exists for $s > \mu$. Likewise, because $g(t)$ is piecewise continuous and of exponential order γ, $\mathcal{L}\{g(t)\}$ also exists for $s > \mu$ as $\mu > \gamma$, which by hypothesis is the function $G(s)$.

Now we are all set to prove that the Laplace transform of $f(t)$ is the function $F(s)$ defined by (11.7.21). For $t > b$, it follows from (11.7.22) that

$$f(t-b) = g(t-b) + \sum_{n=1}^{\infty} a^n g(t-(n+1)b)u(t-(n+1)b)$$

$$= g(t-b) + \sum_{n=2}^{\infty} a^{n-1} g(t-nb)u(t-nb) = \sum_{n=1}^{\infty} a^{n-1} g(t-nb)u(t-nb).$$

Consequently,

$$f(t) - af(t-b)u(t-b) = f(t) - af(t-b)$$

$$= g(t) + \sum_{n=1}^{\infty} a^n g(t-nb)u(t-nb) - a\sum_{n=1}^{\infty} a^{n-1} g(t-nb)u(t-nb),$$

which simplifies to

$$f(t) - af(t-b)u(t-b) = g(t). \tag{11.7.25}$$

In fact, this functional equation is valid not only for $t > b$ but for all $t \geq 0$ because $f(t) = g(t)$ for $0 \leq t < b$.

Taking the Laplace transform of (11.7.25), we have

$$\mathcal{L}\{f(t)\} - a\mathcal{L}\{f(t-b)u(t-b)\} = \mathcal{L}\{g(t)\}. \tag{11.7.26}$$

Recall that $\mathcal{L}\{g(t)\} = G(s)$ for $s > \mu$. Letting $F(s)$ denote the Laplace transform of $f(t)$ and employing the alternate form of the shift in t property, this becomes

$$F(s) - ae^{-bs}F(s) = G(s).$$

Solving for $F(s)$, we obtain

$$F(s) = \frac{G(s)}{1 - ae^{-bs}}$$

for $s > \mu$. This is (11.7.21) with $G(s) = \mathcal{L}\{g(t)\}$. In other words, we have shown that the function $f(t)$ defined by (11.7.22) is truly an inverse Laplace transform of $F(s)$.

As for the proof of (i), let $a = -1$ and suppose the function $g(t)$ is continuous and periodic with period b on $(-\infty, \infty)$. Then it is of exponential order $\gamma = 0$ since it is bounded. And as $g(t - nb) = g(t)$, an inverse Laplace transform of $F(s)$ is

$$f(t) = g(t) + \sum_{n=1}^{\infty}(-1)^n g(t-nb)u(t-nb) = g(t)\left[1 + \sum_{n=1}^{\infty}(-1)^n u(t-nb)\right].$$

Thus,

$$f(t) = \begin{cases} g(t), & \text{if} \quad n = 0, 2, 4, \ldots \\ 0, & \text{if} \quad n = 1, 3, 5, \ldots \end{cases}$$

on the intervals $(nb, (n+1)b)$. In other words, $f(t)$ is a periodic function on $[0, \infty)$ with period $2b$, where its values on the fundamental interval $(0, 2b)$ are $f(t) = g(t)$ for $0 < t < b$ and $f(t) = 0$ for $b < t < 2b$.

Finally, consider (ii). The function $\varphi(t)$ in (11.7.23) is posited to be piecewise continuous on $(-\infty, \infty)$ and of exponential order γ. The function $g(t)$ inherits both of these attributes since

$$g(t) = \varphi(t)\big[u(t) - u(t-b)\big] = \begin{cases} \varphi(t), & 0 < t < b \\ 0, & t < 0 \text{ or } t > b, \end{cases}$$

which fulfills the conditions imposed on it in the main statement of the theorem. The terms of the sum in (11.7.22) are $a^n g(t-nb)u(t-nb)$, where

$$g(t-nb) = \begin{cases} \varphi(t-nb), & nb < t < (n+1)b \\ 0, & t < nb \ \text{or} \ t > (n+1)b. \end{cases}$$

Since $g(t-nb) = 0$ when $t < nb$, we see that

$$a^n g(t-nb)u(t-nb) = a^n g(t-nb).$$

As a result, (11.7.22) simplifies to

$$\begin{aligned} f(t) = g(t) + \sum_{n=1}^{\infty} a^n g(t-nb) &= \sum_{n=0}^{\infty} a^n g(t-nb) \\ &= \sum_{n=0}^{\infty} a^n \varphi(t-nb)\big[u(t-nb) - u(t-(n+1)b)\big], \end{aligned}$$

which is (11.7.24).

As for the periodicity of $f(t)$, suppose that $|a| = 1$.

- Suppose $a = 1$. Since the bracketed unit step functions in the terms of the series (11.7.24) are box functions, it follows for a given value of n that

$$f(t) = a^n \varphi(t-nb) = \varphi(t-nb)$$

for $t \in (nb, (n+1)b)$. Likewise, as $t+b \in ((n+1)b, (n+2)b)$ for these values of t, we have

$$f(t+b) = \varphi((t+b)-(n+1)b) = \varphi(t-nb) = f(t).$$

Thus, $f(t)$ is periodic on $[0, \infty)$ with period b.

- Suppose $a = -1$. Then

$$f(t) = a^n \varphi(t-nb) = (-1)^n \varphi(t-nb)$$

for $t \in (nb, (n+1)b)$. If n is odd, then $f(t) = -\varphi(t-nb)$ for $t \in (nb, (n+1)b)$. Since $t+b \in ((n+1)b, (n+2)b)$ and $n+1$ is even,

$$f(t+b) = (-1)^{n+1}\varphi((t+b)-(n+1)b) = \varphi(t-nb) = -f(t).$$

Likewise, supposing n is even leads to the same conclusion, namely, that $f(t)$ is antiperiodic on $(0, \infty)$ with antiperiod b. So, aside from the endpoints of the intervals $(nb, (n+1)b)$,

$$f(t+2b) = -f(t+b) = f(t)$$

for $t \in [0, \infty)$. In other words, $f(t)$ is periodic on $[0, \infty)$ with period $2b$.

This concludes the proof of Theorem 11.7.6. ■

In the following examples, Theorem 11.7.6 is applied to several functions of the form (11.7.21). Examples 11.7.17–11.7.19 illustrate the main statement of the theorem while Examples 11.7.20 and 11.7.21 illustrate the special cases (i) and (ii), respectively.

Example 11.7.17 Find an inverse Laplace transform of the function

$$F(s) = \frac{1}{s(1-e^{-ks})},$$

where $k > 0$.

Solution Comparing this particular function to the general form (11.7.21), we see that $a = 1$, $b = k$, and $G(s) = 1/s$. Recall $1/s = \mathcal{L}\{1\}$. Since a constant function is of exponential order $\gamma = 0$, it follows that this Laplace transform is valid for all $s > 0$ (see *Remarks* after Theorem 11.7.6). But we already knew that from computing $\mathcal{L}\{1\}$ in Example 11.2.2. Consequently, let $g(t) \equiv 1$ on $(-\infty, \infty)$. Because $\gamma = 0$ and $\ln|a|/b = \ln(1)/k = 0$, the inequality condition that appears after (11.7.22) is satisfied by any positive constant μ. Then, according to (11.7.22), the function

$$f(t) = g(t) + \sum_{n=1}^{\infty} a^n g(t-nk)u(t-nk) = 1 + \sum_{n=1}^{\infty} u(t-nk)$$

with domain $[0, \infty)$ is an inverse Laplace transform of $F(s)$ for all $s > \mu$. The graph of $f(t)$ resembles a flight of stairs where each stair has a rise of 1 and a tread depth of k, which for $k = 1$ is the graph shown in Fig. 7.21 in Chap. 7. If for some reason an inverse transform is preferred that is undefined at $t = 0$, then let

$$f(t) = u(t) + \sum_{n=1}^{\infty} u(t-nk) = \sum_{n=0}^{\infty} u(t-nk).$$

This latter version of the function and its graph can be found in the Table of Laplace Transforms in Abramowitz and Stegun [1, cf. 29.3.65].

Example 11.7.18 Find an inverse Laplace transform of

$$F(s) = \frac{10}{s^3(1+3e^{-2s})}.$$

Solution Referring to (11.7.21), we see that

$$a = -3, \ b = 2, \ \text{and } G(s) = \frac{10}{s^3}.$$

Let $g(t) = \mathcal{L}^{-1}\{G(s)\} = 5\mathcal{L}^{-1}\{2!/s^3\} = 5t^2$. This function satisfies the hypotheses of the theorem as it is continuous on $(-\infty, \infty)$ and of exponential order γ for any $\gamma > 0$ (see Example 11.4.1). So we can choose $\gamma = \frac{1}{2}\ln 3$. Since $|a| = 3$ and $b = 2$, the inequality condition after (11.7.22) is satisfied if $\mu > \gamma$. It follows from (11.7.22) that the function

$$\begin{aligned} f(t) &= g(t) + \sum_{n=1}^{\infty}(-3)^n g(t-2n)u(t-2n) = 5t^2 + \sum_{n=1}^{\infty} 5(-3)^n(t-2n)^2u(t-2n) \\ &= 5t^2 - 15(t-2)^2u(t-2) + 45(t-4)^2u(t-4) - 135(t-6)^2u(t-6) + \ldots \end{aligned}$$

for $t \geq 0$ is an inverse Laplace transform of $F(s)$ for all $s > \mu$. Its graph is shown in Fig. 11.8. The removable discontinuities (holes) of the function $f(t)$, which are located at $t = 2, 4, 6, 8, \ldots$, are encircled. ♦

Example 11.7.19 Find $\mathcal{L}^{-1}\{F_1(s)\}$, where

$$F_1(s) = \frac{s}{(s^2+1)(1-e^{-s})}.$$

Solution Once again referring to (11.7.21), we have $a = b = 1$ and

$$G(s) = \frac{s}{s^2+1}.$$

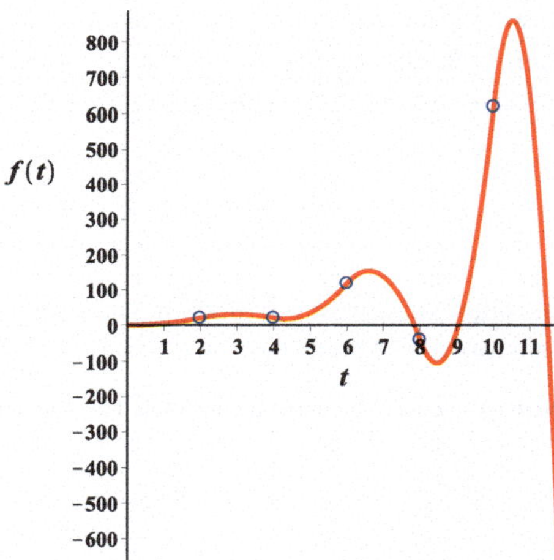

Fig. 11.8 Graph of $f(t)$ in Example 11.7.18

Let $g(t) = \mathcal{L}^{-1}\{G(s)\} = \cos t$, which is of exponential order $\gamma = 0$ since

$$|g(t)| = |\cos t| \le 1 \cdot e^{0 \cdot t}$$

for all $t \ge 0$. Since a and γ have the same values as in Example 11.7.17, we can assign μ any positive value. And so by (11.7.22) the function

$$f_1(t) = g(t) + \sum_{n=1}^{\infty} 1^n g(t-n)u(t-n) = \cos t + \sum_{n=1}^{\infty} \cos(t-n)u(t-n)$$

is an inverse transform of $F_1(s)$ for $s > \mu$. It is clear from this series that even though each of its terms include the 2π-periodic function $\cos t$, the inverse transform $f_1(t)$ itself is not periodic, which is visually corroborated by its graph shown on the left in Fig. 11.9. Recall this is the same result (see (11.7.20)) that was obtained earlier in Example 11.7.16 without the benefit of having Theorem 11.7.6 at the time. ♦

Example 11.7.20 Find $\mathcal{L}^{-1}\{F_2(s)\}$, where

$$F_2(s) = \frac{s}{(s^2+1)(1+e^{-2\pi s})}.$$

Solution Since $G(s)$ is the same as in the previous example, $g(t) = \cos t$ and $\gamma = 0$. And as $|a| = 1$, we may choose μ to be any positive number. Since $a = -1$ and the

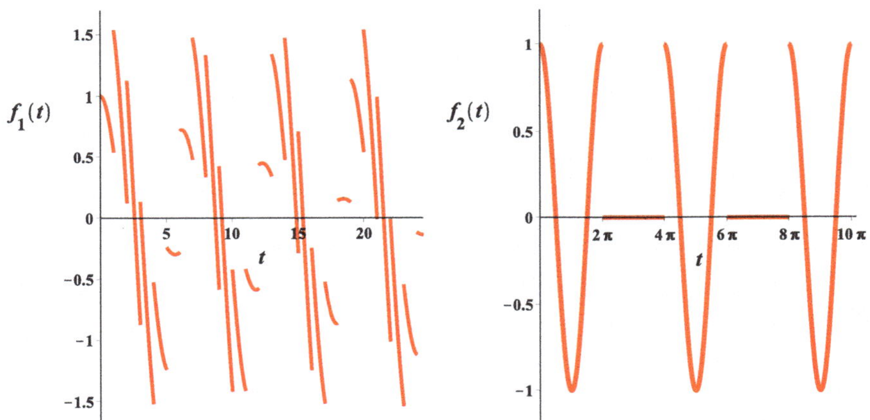

Fig. 11.9 Graphs of $f_1(t)$ and $f_2(t)$ in Examples 11.7.19 and 11.7.20

period of $g(t)$ is $b = 2\pi$, an inverse transform of $F_2(s)$ for $s > \mu$ is

$$f_2(t) = \cos t + \sum_{n=1}^{\infty} (-1)^n \cos(t - 2\pi n) u(t - 2\pi n)$$
$$= \cos t + \sum_{n=1}^{\infty} (-1)^n u(t - 2\pi n) \cos t,$$

which according to Theorem 11.7.6 (i) is periodic with period $2b = 4\pi$. Its graph is shown on the right in Fig. 11.9. Note that even though $F_1(s)$ and $F_2(s)$ are similar in appearance, the inverse transform of the latter is periodic, whereas that of the former is not. ♦

Example 11.7.21 Find an inverse Laplace transform of

$$F(s) = \frac{1 + s - e^s}{s^2(1 - e^s)}.$$

Solution At first it seems that Theorem 11.7.6 is not applicable since $b < 0$. But this can be rectified by multiplying the numerator and denominator of $F(s)$ by $-e^{-s}$, whereby $F(s)$ becomes

$$F(s) = \frac{1 - se^{-s} - e^{-s}}{s^2(1 - e^{-s})} = \frac{G(s)}{1 - e^{-s}}.$$

As a result, we can now use the theorem with $a = 1, b = 1$, and

$$G(s) = \frac{1 - se^{-s} - e^{-s}}{s^2} = \frac{1}{s^2} - \frac{e^{-s}}{s} - \frac{e^{-s}}{s^2}.$$

An inverse Laplace transform of $G(s)$ is

$$g(t) = tu(t) - u(t - 1) - (t - 1)u(t - 1) = t[u(t) - u(t - 1)],$$

which is (11.7.23) with $\varphi(t) = t$ and $b = 1$. And so an inverse Laplace transform of $F(s)$ is the function $f(t)$ defined by (11.7.24) with $a = b = 1$ and $\varphi(t - nb) = t - n$. That is,

$$f(t) = \sum_{n=0}^{\infty} (t - n) [u(t - n) - u(t - (n + 1))] = tu(t) - \sum_{n=1}^{\infty} u(t - n). \quad (11.7.27)$$

Moreover, as $a = 1$, $f(t)$ is periodic on $[0, \infty)$ with period $b = 1$. The nature of its graph becomes clear after realizing that the bracketed quantity is the box function

$u_{n,n+1}(t)$ (see Definition 11.6.8). That is,

$$u_{n,n+1}(t) = u(t-n) - u(t-(n+1)) = \begin{cases} 1, & n < t < n+1 \\ 0, & t < n \text{ or } t > n+1. \end{cases}$$

Consequently, $f(t) = t - n$ on the interval $(n, n+1)$. That is, $f(t) = t$ on $(0, 1)$, $f(t) = t - 1$ on $(1, 2)$, $f(t) = t - 2$ on $(2, 3)$, and so on. Its graph is the *sawtooth wave* that was shown earlier in Fig. 11.2. ♦

Remark Note that $f(t)$ in (11.7.27) is not defined at the points $t = 0, 1, 2, 3, \ldots$, which is due to the way the Heaviside function $u(t)$ is defined in (11.2.2). But if for some technical reason, $f(t)$ does in fact have values at some or all of these points, such as in physics and engineering applications, then it can be modified accordingly. For example, if $f(0) = 0$ but $f(t)$ is not defined at $t = 1, 2, 3, \ldots$, then we can alter (11.7.27) as follows:

$$f(t) = t[1 - u(t-1)] + \sum_{n=1}^{\infty}(t-n)[u(t-n) - u(t-(n+1))] = t - \sum_{n=1}^{\infty} u(t-n).$$

And that is because t and $tu(t)$ have the same Laplace transform. Or if we need to modify $f(t)$ so that $f(t) = 0$ at all of the points $t = 0, 1, 2, 3, \ldots$, then we can replace $u(t-n)$ with the modified Heaviside function $H(t-n)$, where (see (11.2.3))

$$H(t-n) = \begin{cases} 0, & t < n \\ 1, & t \geq n \end{cases}$$

since these two functions, differing only at a single point, have the same Laplace transform. As a result,

$$f(t) = \sum_{n=0}^{\infty}(t-n)[H(t-n) - H(t-(n+1))] = t - \sum_{n=1}^{\infty} H(t-n) = t - \lfloor t \rfloor$$

for $t \geq 0$. The symbol $\lfloor \cdot \rfloor$ denotes the ***floor function*** (or, ***greatest integer function***), which is defined as follows:

$$\lfloor t \rfloor := \max\{z, \text{an integer} : z \leq t\}. \tag{11.7.28}$$

In the next example, Theorem 11.7.6 will be used twice in order to find the solution of the initial value problem (11.7.29): first, to find the Laplace transform of the periodic forcing function $f(t)$ defined by (11.7.30) and then to obtain the solution by finding an inverse Laplace transform of a function of the form (11.7.21).

Example 11.7.22 Solve the initial value problem

$$x' + x = f(t), \quad x(0) = 0, \tag{11.7.29}$$

where

$$f(t) = \begin{cases} 1, & 0 \le t < 1 \\ -1, & 1 < t < 2 \end{cases} \tag{11.7.30}$$

and $f(t+2) = f(t)$ for $t > 0$.

Solution The graph of $f(t)$ resembles the graph of the rectangular wave function in Fig. 7.9 in Chap. 7 except its values alternate between 1 and -1. Since $f(t)$ has jump discontinuities at the points $t = 1, 2, 3 \ldots,$ what does it mean for this initial value problem to have a solution? The answer is that a solution is a function $x(t)$ that is differentiable on the open intervals $(n-1, n)$ for $n = 1, 2, 3, \ldots$ and which satisfies the differential equation on these intervals. Of course, $x(t)$ must also satisfy the prescribed initial condition that $x(0) = 0$.

Taking the Laplace transform of the differential equation, we get

$$sX(s) - x(0) + X(s) = F(s),$$

where $X(s) = \mathcal{L}\{x(t)\}$ and $F(s) = \mathcal{L}\{f(t)\}$. Thus, as $x(0) = 0$,

$$X(s) = \frac{F(s)}{s+1}. \tag{11.7.31}$$

In order to compute $F(s)$, let us write $f(t)$ as follows so that it looks like (11.7.22):

$$f(t) = 1 - 2u(t-1) + 2u(t-2) - 2u(t-3) + \cdots = 1 + 2\sum_{n=1}^{\infty}(-1)^n u(t-n).$$

Then rewriting this as

$$f(t) = -1 + 2 + 2\sum_{n=1}^{\infty}(-1)^n u(t-n)$$

and letting $g(t) \equiv 2$, $a = -1$, and $b = 1$, it assumes the form

$$f(t) = -1 + g(t) + \sum_{n=1}^{\infty} a^n g(t-nb)u(t-nb),$$

which, aside from the first term, is (11.7.22). It then follows from (11.7.21) that its Laplace transform is

$$\mathcal{L}\{f(t)\} = -\frac{1}{s} + \frac{G(s)}{1+e^{-s}},$$

where $G(s) = \mathcal{L}\{g(t)\} = 2\mathcal{L}\{1\} = 2/s$. Therefore,

$$F(s) = -\frac{1}{s} + \frac{2}{s(1+e^{-s})}. \tag{11.7.32}$$

Alternatively, another way of obtaining (11.7.32) is to use Theorem 11.6.7 as was done in computing the Laplace transform of the square-wave function in Example 11.6.13.

It follows from (11.7.31) and (11.7.32) that

$$\begin{aligned} X(s) &= \frac{1}{s+1}\left[-\frac{1}{s} + \frac{2}{s(1+e^{-s})}\right] = \frac{1}{s(s+1)}\left[-1 + \frac{2}{1+e^{-s}}\right] \\ &= \left[\frac{1}{s} - \frac{1}{s+1}\right]\left[-1 + \frac{2}{1+e^{-s}}\right], \end{aligned}$$

which can be rewritten as

$$X(s) = -\left(\frac{1}{s} - \frac{1}{s+1}\right) + 2\left(\frac{1}{s} - \frac{1}{s+1}\right)\frac{1}{1+e^{-s}}.$$

Since

$$\mathcal{L}^{-1}\left\{\frac{1}{s} - \frac{1}{s+1}\right\} = 1 - e^{-t},$$

we have

$$x(t) = \mathcal{L}^{-1}\{X(s)\} = -(1-e^{-t}) + 2\mathcal{L}^{-1}\left\{\frac{\frac{1}{s} - \frac{1}{s+1}}{1+e^{-s}}\right\}.$$

To find the inverse Laplace transform of the rightmost term, let us employ (11.7.21) and (11.7.22) once again:

$$\mathcal{L}^{-1}\left\{\frac{\frac{1}{s} - \frac{1}{s+1}}{1+e^{-s}}\right\} = 1 - e^{-t} + \sum_{n=1}^{\infty}(-1)^n\left[1 - e^{-(t-n)}\right]u(t-n).$$

Therefore,

$$x(t) = -(1 - e^{-t}) + 2(1 - e^{-t}) + 2\sum_{n=1}^{\infty}(-1)^n \left[1 - e^{-(t-n)}\right] u(t - n),$$

which simplifies to

$$x(t) = 1 - e^{-t} + 2\sum_{n=1}^{\infty}(-1)^n \left[1 - e^{-(t-n)}\right] u(t - n). \tag{11.7.33}$$

This is the solution of the initial value problem (11.7.29). Before we leave this example, let us examine the solution more closely. First of all, note that due to the way $u(t - n)$ is defined, $x(t)$ is undefined at $t = 1, 2, 3, \ldots$. If $0 \leq t < 1$, then as $u(t - n) = 0$ for $n \geq 1$,

$$x(t) = 1 - e^{-t}.$$

It follows from this that $x(0) = 0$, which shows that (11.7.33) does satisfy the initial condition. If $1 < t < 2$, then as $u(t - 1) = 1$ while $u(t - n) = 0$ for $n \geq 2$,

$$x(t) = 1 - e^{-t} - 2\left[1 - e^{-(t-1)}\right] = -1 - e^{-t} + 2e^{-(t-1)}.$$

In general, if $N < t < N + 1$ for some $N \geq 1$, then as $u(t - n) = 1$ for $n = 1, 2, \ldots, N$ while $u(t - n) = 0$ for $n \geq N + 1$, the solution (11.7.33) simplifies to

$$x(t) = 1 - e^{-t} + 2\sum_{n=1}^{N}(-1)^n \left[1 - e^{-(t-n)}\right]$$

on the interval $(N, N + 1)$. ♦

The final result in this section states that

$$\mathcal{L}^{-1}\left\{\sum_{n=0}^{\infty} a^n e^{-nbs} G(s)\right\}$$

can be found by interchanging the inverse Laplace transform operator $\mathcal{L}^{-1}$ and the summation operator Σ if certain conditions are met, the details of which are provided in the next theorem.

Inverse Laplace Transform of a Special Series

Theorem 11.7.7 *Suppose a function $G(s)$ has an inverse Laplace transform that is piecewise continuous on $(-\infty, \infty)$ and of exponential order γ. If a, b, μ are constants such that $a \neq 0$, $b > 0$, and $\mu > \max\{\gamma, b^{-1}\ln|a|\}$, then*

$$\mathcal{L}^{-1}\left\{\sum_{n=0}^{\infty} a^n e^{-nbs} G(s)\right\} = \sum_{n=0}^{\infty} a^n \mathcal{L}^{-1}\left\{e^{-nbs} G(s)\right\} \tag{11.7.34}$$

for $s > \mu$.

Proof It follows from the hypotheses and Corollary 11.5.4 that there is a function $g(t)$ that is piecewise continuous on $(-\infty, \infty)$ and of exponential order γ and whose Laplace transform is $G(s)$ for $s > \gamma$. The geometric series

$$\sum_{n=0}^{\infty} x^n = \frac{1}{1-x}, \qquad |x| < 1$$

with $x = ae^{-bs}$ is

$$\sum_{n=0}^{\infty} a^n e^{-nbs} = \frac{1}{1-ae^{-bs}}, \qquad |ae^{-bs}| < 1.$$

If $s > \mu$, then $s > b^{-1}\ln|a|$; so $|a|e^{-bs} < 1$. Thus,

$$\frac{G(s)}{1-ae^{-bs}} = \sum_{n=0}^{\infty} a^n e^{-nbs} G(s) \tag{11.7.35}$$

for $s > \mu$.

It follows from Theorem 11.7.6 that the inverse Laplace transform of the left-hand side of (11.7.35) exists for $s > \mu$. Thus,

$$\mathcal{L}^{-1}\left\{\frac{G(s)}{1-ae^{-bs}}\right\} = \mathcal{L}^{-1}\left\{\sum_{n=0}^{\infty} a^n e^{-nbs} G(s)\right\} \tag{11.7.36}$$

for $s > \mu$. Moreover from (11.7.22), we have

$$\mathcal{L}^{-1}\left\{\frac{G(s)}{1-ae^{-bs}}\right\} = g(t) + \sum_{n=1}^{\infty} a^n g(t-nb)u(t-nb). \tag{11.7.37}$$

By the alternate form of the shift in t property,

$$g(t - nb)u(t - nb) = \mathcal{L}^{-1}\left\{e^{-nbs}G(s)\right\}$$

for $n = 1, 2, 3, \ldots$. Consequently,

$$\begin{aligned} g(t) + \sum_{n=1}^{\infty} a^n g(t - nb)u(t - nb) &= \mathcal{L}^{-1}\{G(s)\} + \sum_{n=1}^{\infty} a^n \mathcal{L}^{-1}\left\{e^{-nbs}G(s)\right\} \\ &= \sum_{n=0}^{\infty} a^n \mathcal{L}^{-1}\left\{e^{-nbs}G(s)\right\}. \end{aligned} \tag{11.7.38}$$

Then by (11.7.37) and (11.7.38) we have

$$\mathcal{L}^{-1}\left\{\frac{G(s)}{1 - ae^{-bs}}\right\} = \sum_{n=0}^{\infty} a^n \mathcal{L}^{-1}\left\{e^{-nbs}G(s)\right\} \tag{11.7.39}$$

for $s > \mu$. Formula (11.7.34) then follows from (11.7.36) and (11.7.39). ■

Example 11.7.23 Find $\mathcal{L}^{-1}\{F(s)\}$, where

$$F(s) = \frac{2}{(s^2 + 4)(1 + e^{-\pi s})}. \tag{11.7.40}$$

Solution This resembles the function $F_2(s)$ in Example 11.7.20, where the purpose at the time was to demonstrate how to compute an inverse Laplace transform by way of Theorem 11.7.6 (i). But instead of using that theorem here, let us use Theorem 11.7.7 to illustrate how engineering and differential equations textbooks generally rely on the geometric series

$$\frac{1}{1 + x} = 1 - x + x^2 - x^3 + \ldots \quad (|x| < 1)$$

to find the inverse Laplace transform of a quotient whose denominator has a factor of the form $1 - ae^{-bs}$. With $x = e^{-\pi s}$, the series becomes

$$\frac{1}{1 + e^{-\pi s}} = 1 - e^{-\pi s} + e^{-2\pi s} - e^{-3\pi s} + \ldots = \sum_{n=0}^{\infty} (-1)^n e^{-n\pi s}.$$

Thus,

$$F(s) = \frac{G(s)}{1 + e^{-\pi s}} = \sum_{n=0}^{\infty} (-1)^n e^{-n\pi s} G(s),$$

where

$$G(s) = \frac{2}{s^2 + 4}.$$

The inverse Laplace transform of $G(s)$ is

$$g(t) = \mathcal{L}^{-1}\{G(s)\} = \mathcal{L}^{-1}\left\{\frac{2}{s^2 + 4}\right\} = \sin 2t,$$

which is continuous on $(-\infty, \infty)$ and of exponential order $\gamma = 0$. Then it follows from (11.7.34) with $a = -1$ and $b = \pi$ that

$$\mathcal{L}^{-1}\{F(s)\} = \mathcal{L}^{-1}\left\{\sum_{n=0}^{\infty}(-1)^n e^{-n\pi s}G(s)\right\} = \sum_{n=0}^{\infty}(-1)^n \mathcal{L}^{-1}\left\{e^{-n\pi s}G(s)\right\}$$

for $s > \mu$, where $\mu > \max\left\{0, \pi^{-1}\ln 1\right\} = 0$. By the alternate form of the shift in t property,

$$\begin{aligned}\mathcal{L}^{-1}\left\{e^{-n\pi s}G(s)\right\} &= g(t - n\pi)u(t - n\pi)\\ &= \sin(2(t - n\pi))u(t - n\pi) = u(t - n\pi)\sin 2t\end{aligned}$$

for $n = 1, 2, 3, \ldots$. Therefore,

$$\begin{aligned}\mathcal{L}^{-1}\{F(s)\} &= \mathcal{L}^{-1}\{G(s)\} + \sum_{n=1}^{\infty}(-1)^n \mathcal{L}^{-1}\left\{e^{-n\pi s}G(s)\right\}\\ &= \sin 2t + \sum_{n=1}^{\infty}(-1)^n u(t - n\pi)\sin 2t\end{aligned}$$

for $s > \mu$. ♦

11.8 Second-Order Linear Equations

We are able to solve first-order linear differential equations using the method of Laplace transforms because of Theorem 11.4.3 and the properties of the transform that are given in Sects. 11.6 and 11.7. The next theorem adds another property that will also enable us to solve second-order linear differential equations with Laplace transforms.

Laplace Transform of the Second Derivative

Theorem 11.8.1 *Suppose that the function $f(t)$ is twice differentiable on $[0, \infty)$ and both $f(t)$ and $f'(t)$ are of exponential order γ. If the second derivative $f''(t)$ is integrable on $[0, T]$ for every $T \geq 0$, then its Laplace transform exists for all $s > \gamma$ and is given by the formula*

$$\mathcal{L}\{f''(t)\} = s^2\mathcal{L}\{f(t)\} - sf(0) - f'(0). \tag{11.8.1}$$

Proof Since $f''(t)$ exists on $[0, \infty)$, both $f(t)$ and $f'(t)$ are continuous on $[0, \infty)$. Because they are also of exponential order γ, their Laplace transforms exist according to Corollary 11.5.4. Therefore, it follows from Theorem 11.4.3 that

$$\mathcal{L}\{f'(t)\} = s\mathcal{L}\{f(t)\} - f(0) \tag{11.8.2}$$

for all $s > \gamma$. Appealing again to Theorem 11.4.3, we see that the Laplace transform of $f''(t)$ exists by the integrability of $f''(t)$ on $[0, T]$ for every $T \geq 0$ and because $f'(t)$ is differentiable and of exponential order γ. Thus,

$$\mathcal{L}\{f''(t)\} = s\mathcal{L}\{f'(t)\} - f'(0)$$

for all $s > \gamma$. Then by (11.8.2), we have

$$\mathcal{L}\{f''(t)\} = s\,[s\mathcal{L}\{f(t)\} - f(0)] - f'(0),$$

which when multiplied out is (11.8.1). ∎

The following examples demonstrate how to use Theorem 11.8.1 and many of the properties presented in the previous sections to solve initial value problems involving second-order linear equations. At the beginning of each example, the main properties or techniques that will be used to solve the initial value problem are listed in boldface.

Example 11.8.1 (Linear Factors; Partial Fractions) Solve the initial value problem

$$x'' + 6x' + 5x = 36e^t, \quad x(0) = 0, \quad x'(0) = 10.$$

Solution Taking the Laplace transform of both sides of the differential equation and using the linearity property, we have

$$\mathcal{L}\{x''(t)\} + 6\mathcal{L}\{x'(t)\} + 5\mathcal{L}\{x(t)\} = 36\,\mathcal{L}\{e^t\},$$

which because of Theorems 11.8.1 and 11.4.3 can be written as

$$s^2X(s) - sx(0) - x'(0) + 6\big[sX(s) - x(0)\big] + 5X(s) = 36 \cdot \frac{1}{s-1},$$

where $X(s) = \mathcal{L}\{x(t)\}$. Because of the initial conditions, this simplifies to

$$(s^2 + 6s + 5)X(s) - s \cdot 0 - 10 = \frac{36}{s-1}.$$

Solving for $X(s)$, we obtain

$$X(s) = \frac{\dfrac{36}{s-1} + 10}{(s+1)(s+5)} = \frac{36 + 10(s-1)}{(s-1)(s+1)(s+5)} = \frac{10s + 26}{(s-1)(s+1)(s+5)}.$$

Expanding $X(s)$ in partial fractions, we have

$$\frac{10s + 26}{(s-1)(s+1)(s+5)} = \frac{A}{s-1} + \frac{B}{s+1} + \frac{C}{s+5}.$$

Clearing this equation of fractions, this becomes

$$A(s+1)(s+5) + B(s-1)(s+5) + C(s-1)(s+1) = 10s + 26.$$

Now we can easily find A by letting $s = 1$:

$$(2)(6)A = 10(1) + 26 \quad \Rightarrow \quad 12A = 36 \quad \Rightarrow \quad A = 3.$$

To find B, let $s = -1$:

$$(-2)(4)B = 10(-1) + 26 \quad \Rightarrow \quad B = -2.$$

Finally, let $s = -5$:

$$(-6)(-4)C = 10(-5) + 26 \quad \Rightarrow \quad C = -1.$$

Thus, the Laplace transform of $x(t)$ is

$$X(s) = \frac{3}{s-1} - \frac{2}{s+1} - \frac{1}{s+5}.$$

Finally, we obtain the solution $x(t)$ of the initial problem by taking the inverse Laplace transform of $X(s)$. And so

$$x(t) = \mathcal{L}^{-1}\{X(s)\} = 3e^t - 2e^{-t} - e^{-5t}.$$

♦

Example 11.8.2 (Irreducible Quadratic; Partial Fractions; Shift Property) Solve the initial value problem

$$x'' + 2x' + 10x = -25e^{3t}, \quad x(0) = 0, \quad x'(0) = 6.$$

Solution As in the previous example, we begin by transforming the equation:

$$\mathcal{L}\{x''(t)\} + 2\mathcal{L}\{x'(t)\} + 10\mathcal{L}\{x(t)\} = -25\mathcal{L}\{e^{3t}\}.$$

Let $X(s) = \mathcal{L}\{x(t)\}$. Using Table 11.1 and the formulas for the Laplace transforms of the first and second derivatives, this becomes

$$s^2X(s) - sx(0) - x'(0) + 2\big[sX(s) - x(0)\big] + 10X(s) = \frac{-25}{s-3}.$$

Solving for $X(s)$, we obtain

$$X(s) = \frac{6s - 43}{(s-3)\big[(s+1)^2 + 9\big]}.$$

The partial fraction expansion of $X(s)$ is

$$\frac{6s - 43}{(s-3)\big[(s+1)^2 + 9\big]} = \frac{A}{s-3} + \frac{B(s+1) + 3C}{(s+1)^2 + 3^2}.$$

Clearing fractions, we have

$$A\big[(s+1)^2 + 9\big] + \big[B(s+1) + 3C\big](s-3) = 6s - 43.$$

Setting s equal to 3, -1, and 0, we find $A = -1$, $B = 1$, and $C = 10/3$. Thus,

$$X(s) = -\frac{1}{s-3} + \frac{s+1}{(s+1)^2 + 3^2} + \frac{10}{3} \cdot \left[\frac{3}{(s+1)^2 + 3^2}\right].$$

Taking the inverse Laplace transform of $X(s)$ and using the shift in s property (see Theorem 11.7.2), we find that

$$x(t) = -e^{3t} + e^{-t}\cos 3t + \frac{10}{3}e^{-t}\sin 3t$$

is the solution. ♦

Example 11.8.3 (Dirac Delta Function; Irreducible Quadratic, Shift Properties) Solve the initial value problem

$$x'' + 6x' + 13x = 5\delta(t-2), \quad x(0) = 0, \quad x'(0) = 2.$$

Solution Taking the Laplace transform of the differential equation, we have

$$\mathcal{L}\{x''(t)\} + 6\mathcal{L}\{x'(t)\} + 13\mathcal{L}\{x(t)\} = 5\mathcal{L}\{\delta(t-2)\}.$$

Using the formulas for the transforms of the derivatives and Table 11.1, we obtain

$$s^2X(s) - sx(0) - x'(0) + 6\left[sX(s) - x(0)\right] + 13X(s) = 5e^{-2s},$$

where $X(s) = \mathcal{L}\{x(t)\}$. Thus,

$$(s^2 + 6s + 13)X(s) = 2 + 5e^{-2s}.$$

Solving for $X(s)$, we get

$$X(s) = \frac{2}{s^2 + 6s + 13} + 5e^{-2s} \cdot \frac{1}{s^2 + 6s + 13}.$$

Since $s^2 + 6s + 13$ is irreducible over the reals, complete the square obtaining

$$X(s) = \frac{2}{(s+3)^2 + 2^2} + \frac{5}{2}e^{-2s} \cdot \frac{2}{(s+3)^2 + 2^2}.$$

Now let us find the inverse Laplace transform of $X(s)$. Note that it involves the function

$$F(s) := \frac{2}{s^2 + 2^2}$$

in both terms, where $F(s) = \mathcal{L}\{\sin 2t\}$. By the shift in s property (Theorem 11.7.2),

$$\mathcal{L}^{-1}\left\{\frac{2}{(s+3)^2 + 2^2}\right\} = \mathcal{L}^{-1}\{F(s+3)\} = e^{-3t}\mathcal{L}^{-1}\{F(s)\} = e^{-3t}\sin 2t.$$

As for the second term, apply the shift in t property (Theorem 11.7.3), namely,

$$\mathcal{L}^{-1}\left\{e^{-bs}G(s)\right\} = g(t-b)u(t-b),$$

where $g(t) = \mathcal{L}^{-1}\{G(s)\}$. With $b = 2$ and $G(s) := F(s+3)$, we have

$$\mathcal{L}^{-1}\left\{e^{-2s}F(s+3)\right\} = \mathcal{L}^{-1}\left\{e^{-2s}G(s)\right\} = g(t-2)u(t-2),$$

where $g(t) = \mathcal{L}^{-1}\{F(s+3)\} = e^{-3t}\sin 2t$. As a result,

$$\mathcal{L}^{-1}\left\{e^{-2s} \cdot \frac{2}{(s+3)^2 + 2^2}\right\} = g(t-2)u(t-2) = e^{-3(t-2)}\sin(2(t-2))\,u(t-2).$$

Therefore,

$$x(t) = e^{-3t}\sin 2t + \frac{5}{2}e^{6-3t}\sin(2t-4)u(t-2).$$

In other words, the solution is

$$x(t) = \begin{cases} e^{-3t}\sin 2t, & \text{if } t < 2 \\ e^{-3t}\sin 2t + \frac{5}{2}e^{6-3t}\sin(2t-4), & \text{if } t > 2. \end{cases}$$

♦

Example 11.8.4 (Convolution Theorem) Find a formula for the solution of

$$x'' + 4x = g(t), \quad x(0) = 0, \quad x'(0) = 2,$$

where the function $g(t)$ is piecewise continuous on $[0, \infty)$ and of exponential order.

Solution The transformed equation is

$$s^2X(s) - sx(0) - x'(0) + 4X(s) = G(s),$$

where $X(s) = \mathcal{L}\{x(t)\}$ and $G(s) = \mathcal{L}\{g(t)\}$. Substituting the given initial values for $x(0)$ and $x'(0)$, we get

$$(s^2+4)X(s) - s\cdot 0 - 2 = G(s).$$

Thus,

$$X(s) = \frac{2}{s^2+4} + \frac{1}{s^2+4}\cdot G(s).$$

That is,

$$\mathcal{L}\{x(t)\} = \mathcal{L}\{\sin 2t\} + \frac{1}{2}\mathcal{L}\{\sin 2t\}\,\mathcal{L}\{g(t)\}.$$

By the convolution theorem,

$$\mathcal{L}\{x(t)\} = \mathcal{L}\{\sin 2t\} + \frac{1}{2}\mathcal{L}\{\sin(2t) * g(t)\}.$$

Consequently, the solution is

$$x(t) = \sin 2t + \frac{1}{2}\int_0^t \sin[2(t-u)]\,g(u)\,du.$$

♦

Example 11.8.5 (Periodic Forcing Function; Nonperiodic Solution) Solve the initial value problem

$$x'' + x = f(t), \quad x(0) = 0, \quad x'(0) = 0, \tag{11.8.3}$$

where

$$f(t) = \begin{cases} 1, & 0 \le t < 1 \\ 0, & 1 < t < 2 \end{cases} \tag{11.8.4}$$

and $f(t+2) = f(t)$ for $t > 0$. Other than f being defined and equal to 1 at the points $t = 0, 2, 4, \ldots$, its graph is the square wave with period 2 that is shown in Fig. 11.3.

Solution As usual, let us begin by applying the Laplace transform operator $\mathcal{L}$ to both sides of the differential equation:

$$\mathcal{L}\{x''(t)\} + \mathcal{L}\{x(t)\} = \mathcal{L}\{f(t)\}.$$

Transforming both terms on the left-hand side and letting $X(s) = \mathcal{L}\{x(t)\}$, we have

$$s^2X(s) - sx(0) - x'(0) + X(s) = \mathcal{L}\{f(t)\}.$$

As for the transform of the right-hand side, it can be easily found if we first express $f(t)$ as the series

$$f(t) = 1 - u(t-1) + u(t-2) - u(t-3) + \cdots = 1 + \sum_{n=1}^{\infty}(-1)^n u(t-n),$$

which is (11.7.22) with $a = -1$, $b = 1$, and $g(t) = 1$. It then follows from (11.7.21) that

$$\mathcal{L}\{f(t)\} = \frac{\mathcal{L}\{g(t)\}}{1+e^{-s}} = \frac{\mathcal{L}\{1\}}{1+e^{-s}} = \frac{1}{s(1+e^{-s})}.$$

This transform was also found in Example 11.6.13 by way of Theorem 11.6.7.

Picking up where we left off with the transformed equation, we now have

$$s^2X(s) - sx(0) - x'(0) + X(s) = \frac{1}{s(1+e^{-s})}.$$

Using the initial conditions and solving for $X(s)$, we obtain

$$X(s) = \frac{1}{s(1+e^{-s})(s^2+1)} = \frac{1}{s(s^2+1)} \cdot \frac{1}{1+e^{-s}}.$$

Since the partial fraction expansion of the first factor on the right-hand side of the equation is

$$\frac{1}{s(s^2+1)} = \frac{1}{s} - \frac{s}{s^2+1},$$

the inverse transform $X(s)$ can be rewritten as

$$X(s) = \left(\frac{1}{s} - \frac{s}{s^2+1}\right) \cdot \frac{1}{1+e^{-s}}. \tag{11.8.5}$$

So it is of the form

$$X(s) = \frac{G(s)}{1-ae^{-bs}},$$

where $a = -1$, $b = 1$, and

$$G(s) = \frac{1}{s} - \frac{s}{s^2+1}.$$

Consequently, an inverse transform of $X(s)$ can be obtained using Theorem 11.7.6. Since

$$g(t) = \mathcal{L}^{-1}\{G(s)\} = \mathcal{L}^{-1}\left\{\frac{1}{s}\right\} - \mathcal{L}^{-1}\left\{\frac{s}{s^2+1}\right\} = 1 - \cos t,$$

it follows from (11.7.22) that the solution of (11.8.3) is

$$\begin{aligned} x(t) &= g(t) + \sum_{n=1}^{\infty}(-1)^n g(t-n)u(t-n) \\ &= 1 - \cos t + \sum_{n=1}^{\infty}(-1)^n[1-\cos(t-n)]u(t-n). \end{aligned} \tag{11.8.6}$$

Observe that the infinite sum on the right-hand side of (11.8.6) is equal to 0 when $0 < t < 1$ since for these values of t, $u(t-n) = 0$ for $n = 1, 2, 3, \ldots$. So, on the interval $[0, 1)$, the solution $x(t)$ simplifies to $x(t) = 1 - \cos t$. It follows from this that $x(0) = 0$ and $x'(0) = 0$, which verifies that $x(t)$ does in fact satisfy the initial conditions. Verifying that $x(t)$ also satisfies the differential equation on every interval $(n-1, n)$ for $n = 1, 2, 3, \ldots$ is left to Problem 77.

This solution is less complicated than it first appears. Although the expression for $x(t)$ has infinitely many terms, only a finite number of them actually contribute nonzero values to the solution on any given finite interval $[0, T)$, such as the one mentioned above with $T = 1$. To drive this point home, let us take another value for T, say $T = 20$. Since $u(t-n) = 0$ for $t \in (0, 20)$ when $n \geq 20$, except at the points where it is undefined, the solution simplifies to

$$\begin{aligned} x(t) &= 1 - \cos t + \sum_{n=1}^{19} (-1)^n [1 - \cos(t-n)] u(t-n) \\ &= 1 - \cos t - [1 - \cos(t-1)] u(t-1) + [1 - \cos(t-2)] u(t-2) - \cdots \\ &\qquad + [1 - \cos(t-18)] u(t-18) - [1 - \cos(t-19)] u(t-19) \end{aligned}$$

for $0 \leq t < 20$. Its graph on this interval is shown in Fig. 11.10. The small circles on the graph enclose the removable discontinuities (holes) of $x(t)$, which are located at $t = 1, 2, 3, 4, \ldots$.

There is another way to determine the inverse transform of (11.8.5) and that is to imitate the procedure in Example 11.7.23, where we ended up using Theorem 11.7.7. With the geometric series

$$\frac{1}{1+y} = 1 - y + y^2 - y^3 + y^4 - y^5 + \cdots \qquad (|y| < 1), \tag{11.8.7}$$

we can express $X(s)$ in (11.8.5) as a series by substituting e^{-s} for y. Then

$$X(s) = \frac{G(s)}{1 + e^{-s}} = G(s)\big(1 - e^{-s} + e^{-2s} - e^{-3s} + \cdots\big) = \sum_{n=0}^{\infty} (-1)^n e^{-ns} G(s),$$

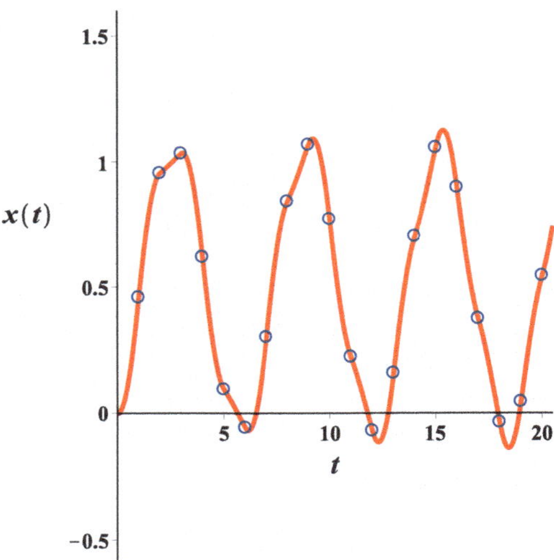

Fig. 11.10 Graph of the solution $x(t)$ of the initial value problem (11.8.3)

where as before

$$G(s) = \frac{1}{s} - \frac{s}{s^2+1}.$$

Since $g(t) = \mathcal{L}^{-1}\{G(s)\} = 1 - \cos t$ is continuous on $[0, \infty)$ and of exponential order $\gamma = 0$, we can use Theorem 11.7.7. Thus,

$$\mathcal{L}^{-1}\{X(s)\} = \mathcal{L}^{-1}\left\{\sum_{n=0}^{\infty}(-1)^n e^{-ns}G(s)\right\} = \sum_{n=0}^{\infty}(-1)^n\mathcal{L}^{-1}\left\{e^{-ns}G(s)\right\},$$

which is valid for $s > \max\left\{\gamma, \frac{1}{b}\ln|a|\right\} = 0$ since $\gamma = 0$, $a = -1$, and $b = 1$. By the alternate form of the shift in t property,

$$\mathcal{L}^{-1}\{e^{-ns}G(s)\} = g(t-n)u(t-n) = [1 - \cos(t-n)]u(t-n).$$

Therefore, the solution of (11.8.3) is

$$\begin{aligned} x(t) = \mathcal{L}^{-1}\{X(s)\} &= \mathcal{L}^{-1}\{G(s)\} + \sum_{n=1}^{\infty}(-1)^n\mathcal{L}^{-1}\left\{e^{-ns}G(s)\right\} \\ &= 1 - \cos t + \sum_{n=1}^{\infty}(-1)^n[1 - \cos(t-n)]u(t-n), \end{aligned}$$

which is (11.8.6). ♦

11.9 Equations with Variable Coefficients

Generally speaking, differential equations with variable coefficients cannot be solved with Laplace transforms. Nevertheless there are exceptions as the following examples will reveal. The Laplace transform of a second-order linear equation with coefficients that are constants or polynomials of the first degree (but no higher) is a first-order linear equation. The reason for this is due to the following formulas, which we derive using the formula for the first derivative of a Laplace transform (Theorem 11.6.4) and the formulas for the Laplace transforms of the first and second derivatives (Theorems 11.4.3 and 11.8.1). For $X(s) = \mathcal{L}\{x(t)\}$,

$$\mathcal{L}\{tx(t)\} = -\frac{d}{ds}\mathcal{L}\{x(t)\} = -X'(s), \tag{11.9.1}$$

$$\mathcal{L}\{tx'(t)\} = -\frac{d}{ds}\mathcal{L}\{x'(t)\} = -\frac{d}{ds}\big[sX(s) - x(0)\big] = -sX'(s) - X(s), \qquad (11.9.2)$$

and

$$\mathcal{L}\{tx''(t)\} = -\frac{d}{ds}\mathcal{L}\{x''(t)\} = -\frac{d}{ds}\big[s^2X(s) - sx(0) - x'(0)\big] \qquad (11.9.3)$$

$$= -s^2X'(s) - 2sX(s) + x(0).$$

Example 11.9.1 Show that the result of taking the Laplace transform of the second-order linear equation

$$tx'' + 2x' + tx = f(t)$$

is a first-order linear equation. Assume the transform of $f(t)$ exists.

Solution Using formulas (11.9.1) and (11.9.3), the Laplace transform of the equation is

$$-s^2X'(s) - 2sX(s) + x(0) + 2[sX(s) - x(0)] - X'(s) = F(s),$$

where $F(s) := \mathcal{L}\{f(t)\}$. Simplifying, we get

$$X'(s) = -\frac{F(s)}{s^2+1} - \frac{x(0)}{s^2+1}, \qquad (11.9.4)$$

a first-order linear equation. ♦

Example 11.9.2 Find a solution of

$$tx'' + 2x' + tx = 0 \qquad (11.9.5)$$

such that $x(0) = 1$.

Solution 1 In view of the previous example, the solution can be found by solving equation (11.9.4) with $x(0) = 1$ and $F(s) = \mathcal{L}\{f(t)\} = \mathcal{L}\{0\} = 0$, namely, the equation

$$X'(s) = -\frac{1}{s^2+1}. \qquad (11.9.6)$$

Let us rewrite this as

$$-\frac{d}{ds}X(s) = \frac{1}{s^2+1} = \mathcal{L}\{\sin t\}.$$

According to the formula for the first derivative of a transform,

$$-\frac{d}{ds}X(s) = -\frac{d}{ds}\mathcal{L}\{x(t)\} = \mathcal{L}\{tx(t)\}.$$

Thus,

$$\mathcal{L}\{tx(t)\} = \mathcal{L}\{\sin t\}.$$

Therefore,

$$tx(t) = \sin t.$$

At first, it appears we are done and that

$$x(t) = \frac{\sin t}{t}$$

is the solution. But $\sin t/t$ is undefined at $t = 0$, which seems to be a problem due to the initial condition. Fortunately, it has a removable discontinuity at $t = 0$ as

$$\lim_{t\to 0^+} \frac{\sin t}{t} = 1.$$

Therefore,

$$x(t) = \begin{cases} \dfrac{\sin t}{t}, & t > 0 \\ 1, & t = 0 \end{cases} \tag{11.9.7}$$

is a solution of Eq. (11.9.5) on $[0, \infty)$ with $x(0) = 1$.

Solution 2 An alternative is to first solve the transformed equation (11.9.6):

$$X(s) = -\int \frac{ds}{s^2 + 1} = -\tan^{-1} s + C.$$

Taking the limit of $X(s)$ as $s \to \infty$, we have

$$\lim_{s\to\infty} X(s) = -\frac{\pi}{2} + C.$$

Assuming that $X(s)$ is the Laplace transform of a function that is piecewise continuous on $[0, \infty)$ and of exponential order, we also know from Corollaries 11.5.2

and 11.5.4 that

$$\lim_{s\to\infty} X(s) = 0.$$

So, $C = \pi/2$. Therefore,

$$X(s) = \frac{\pi}{2} - \tan^{-1} s = \cot^{-1} s.$$

From (11.6.16), we have

$$\mathcal{L}^{-1}\{\cot^{-1} s\} = \frac{\sin t}{t}.$$

Once again we conclude that (11.9.7) solves (11.9.5) such that $x(0) = 1$. ♦

Example 11.9.3 Solve the initial value problem

$$x'' + tx' - 2x = 1, \quad x(0) = 0,\ x'(0) = 0. \tag{11.9.8}$$

Solution The transformed equation is

$$s^2X(s) - sx(0) - x'(0) - sX'(s) - X(s) - 2X(s) = \frac{1}{s}.$$

Taking into account the initial conditions, this becomes

$$X'(s) + \left(\frac{3}{s} - s\right) X(s) = -\frac{1}{s^2}. \tag{11.9.9}$$

An integrating factor for (11.9.9) is

$$\mu(s) = e^{\int (\frac{3}{s}-s)\, ds} = s^3 e^{-s^2/2}.$$

Multiplying (11.9.9) by $\mu(s)$, we have

$$\frac{d}{ds}\left[s^3 e^{-s^2/2} X(s)\right] = -se^{-s^2/2}.$$

Integration yields

$$s^3 e^{-s^2/2} X(s) = -\int se^{-s^2/2}\, ds + C = e^{-s^2/2} + C.$$

Thus,

$$X(s) = \frac{1}{s^3} + C\,\frac{e^{s^2/2}}{s^3}.$$

Assuming that $X(s)$ is the Laplace transform of a piecewise continuous function of exponential order,

$$\lim_{s\to\infty} X(s) = 0$$

according to Corollary 11.5.2. But this can only be the case if $C = 0$ since

$$\lim_{s\to\infty} \frac{e^{s^2/2}}{s^3} = \infty.$$

Therefore,

$$X(s) = \frac{1}{s^3}.$$

Consequently, the solution is

$$x(t) = \mathcal{L}^{-1}\{X(s)\} = \frac{1}{2}\,\mathcal{L}^{-1}\left\{\frac{2!}{s^3}\right\} = \frac{1}{2}t^2.$$

A quick check shows that this function does solve the differential equation and also satisfies the given initial conditions. ♦

Problems

You see, in every job that must be done, there is an element of fun. You find the fun and snap! The job's a game.

from Walt Disney's 1964 film *Mary Poppins* (Mary Poppins' advice to Jane and Michael)

Laplace Transforms from the Definition

Use Definition 11.1.1 to determine the Laplace transforms of the following functions.

1. $\cos bt$, where b is any real constant
2. $f(t) = \begin{cases} e^{2t}, & \text{if } t < 5 \\ 0, & \text{if } t > 5 \end{cases}$
3. $g(t) = \begin{cases} 0, & \text{if } t < 5 \\ e^{2t}, & \text{if } t > 5 \end{cases}$
4. $\sinh t$
5. $h(t) = \begin{cases} 1-t, & \text{if } t < 7 \\ 10, & \text{if } t > 7 \end{cases}$
6. $tu(t-2)$

Transforms Using Tables and Linearity

Use Table 11.1 and Theorem 11.3.1 to find the Laplace transforms of the following functions.

7. $\sin 2t + e^{6t}$
8. $3e^{-2t} + 2t^7 - 12$
9. $5u(t-2)$
10. $\delta(t-2)$
11. $\cosh(t/2)$, where $\cosh t := (e^t + e^{-t})/2$
12. $\frac{3}{4}\delta(t-5) - 2\cos\frac{4}{7}t$

Equivalent Properties

13. Show that the linearity property (11.3.1) is equivalent to the following two properties:

    (a) $\mathcal{L}\{f(t)+g(t)\} = \mathcal{L}\{f(t)\}+\mathcal{L}\{g(t)\}$

    (b) $\mathcal{L}\{cf(t)\} = c\mathcal{L}\{g(t)\}$,

where c is a constant. In other words, show that (11.3.1) implies (a) and (b) and that conversely (a) and (b) imply (11.3.1).

Using Tables and Properties of the Transform

In Problems 14 through 30, determine the Laplace transform of each of the following functions using Table 11.1 and the properties of the transform.

14. $e^{5t}t^3$
15. $e^{-2t}\sin\frac{t}{3}$
16. $10t^4 - \sin 2t$
17. $t\cos 2t$
18. $3t^4 - \sin 2t$
19. t^3e^{-t}
20. $1 + \frac{3}{2}e^{-5t} - \delta(t-3)$
21. $te^{2t}\sin 3t$
22. $t^2\sin t$
23. $2\cos\frac{t}{2} + \begin{cases} 0, & \text{if } t < 5 \\ 10, & \text{if } t > 5 \end{cases}$
24. $t\sin 5t + 4(t-1)^2$
25. $t^{10} + 3\delta(t-2)$
26. $5 + 8t\sin\frac{2}{3}t - \cosh 5t$
27. $h(t) = \begin{cases} 3+e^{5t}, & \text{if } t < 1 \\ e^{5t}, & \text{if } t > 1 \end{cases}$
28. $\frac{4}{13}t^2\sin 7t - 3t^4e^{-2t} + tu(t-4)$
29. $\cos(3t)u(t-\pi/6)$
30. $\sin(3t+2)$

Potpourri: Explanations and Proofs

31. Explain why $\cos bt$ is of exponential order 0.
32. Explain why $\sin bt = O(1)$ as $t \to \infty$.
33. Explain why e^t is of exponential order 1.
34. Show that $f(t) = O(e^{\gamma t})$ as $t \to \infty$ is equivalent to the statement: $e^{-\gamma t}|f(t)|$ is bounded for $t \geq t_1$ for some $t_1 \geq 0$.
35. Use the Laplace transform of the derivative property to determine $\mathcal{L}\{\cos bt\}$ from

$$\mathcal{L}\{\sin bt\} = \frac{b}{s^2+b^2}.$$

36. Use a half-angle identity to find $\mathcal{L}\{\sin^2 bt\}$. (If you are not familiar with the half-angle identities, look them up in a trigonometry book, find out how they are derived, and memorize them.) Compare your answer with the result obtained in one of the examples.

37. Let $f(t)$ be piecewise continuous on $[0, \infty)$ and of exponential order $\gamma \geq 0$. Generalize the transform of an integral property by determining the transform of $\int_a^\infty f(\tau)\,d\tau$, where $a \geq 0$.
38. Suppose $f(t)$ is continuous on $[0, \infty)$ and of exponential order γ. If $\mathcal{L}\{f(t)\} = 0$ for $s > \gamma$, show that $\mathcal{L}\{f(t)P(t)\} = 0$ for $s > \gamma$ for all polynomials $P(t)$.
39. Prove that

$$\tan^{-1} x + \cot^{-1} x = \pi/2.$$

Hint. Differentiate the function

$$f(x) := \tan^{-1} x + \cot^{-1} x.$$

What conclusions can be drawn about f from its derivative?
40. Complete the proof of Example 11.2.8 to show that the function $f(t) = t^{-1}$ has no inverse Laplace transform.

First-Order Linear Equations
In Problems 41 through 49, solve the initial value problems using the method of Laplace transforms. Then check your answers by using the integrating factor method to solve the first-order equations.

41. $x' - 5x = e^{2t},\ x(0) = 1$
42. $y' + 2y = 4e^{-3t},\ y(0) = -2$
43. $y' + \frac{1}{2}y = 0,\ y(0) = 4$
44. $x' + 3x = \cos 2t,\ x(0) = 1$
45. $x' - 2x = 10\cos t,\ x(0) = 3$
46. $x' - 10x = 29\cos 5t,\ x(0) = 1$
47. $x' + x = 8u(t - 10),\ x(0) = -5$
48. $x' - 3x = 6\delta(t - 5),\ x(0) = 2$
49. $x' - x = 4\delta(t - 5),\ x(0) = -8$

Inverse Laplace Transforms
Find an inverse Laplace transform of each of the following functions.

50. $\dfrac{3}{s^2 + 2}$
51. $\dfrac{s + 2}{s^2 + 4s - 5}$
52. $\dfrac{2s - 4}{s^2 - 4s + 18}$
53. $\dfrac{s}{s^2 + 4s + 20}$
54. $\dfrac{5s^2 + 3s + 39}{(s - 2)(s^2 + 9)}$
55. $\dfrac{8}{(s - 2)^5}$
56. $\dfrac{12s}{4s^2 + 8s + 5}$
57. $\dfrac{7}{s - 5} - \dfrac{2s}{s^2 + 16}$
58. $\dfrac{6}{s + 4} - \dfrac{5s}{s^2 + 7}$
59. $\dfrac{2s^2 - s + 5}{s^3 + s^2 - 6s}$
60. $\dfrac{1}{(s^2 + 1)^2}$ *Hint.* See Ex. 11.7.12.
61. $\dfrac{s^2}{(s^2 + 4)^2}$ *Hint.* See Ex. 11.7.12.

Second-Order Linear Equations
Use the Method of Laplace Transforms to solve the following initial value problems.

62. $\dfrac{d^2x}{dt^2} + 4\dfrac{dx}{dt} = 8,\ x(0) = 0,\ x'(0) = 0$
63. $\ddot{x} - 4\dot{x} + 5x = 6e^{3t},\ x(0) = 2,\ \dot{x}(0) = -5$
64. $x'' + 9x = 2e^t,\ x(0) = 1,\ x'(0) = 0$
65. $\dfrac{d^2x}{dt^2} + 5\dfrac{dx}{dt} - 14x = 0,\ x(0) = 5,\ x'(0) = 1$
66. $\ddot{x} + 2\dot{x} + 10x = \frac{1}{2}t^2 - 5e^{3t},\ x(0) = 0,\ \dot{x}(0) = 6$
67. $x'' + 4x' + 5x = 6\delta(t - 2),\ x(0) = 0,\ x'(0) = 1$
68. $\ddot{x} + 4x = 1 - 2u(t - 5),\ x(0) = -1,\ \dot{x}(0) = 0$

Transforms of Periodic Functions
The following functions are periodic on $[0, \infty)$. Find their Laplace transforms.

69. The function $f(t)$ with period b, where $f(t) = t$ on the interval $(0, b)$.
70. The rectangular wave function $r(t)$ with period 2, where

$$r(t) = \begin{cases} 2, & 0 < t < 1 \\ -2, & 1 < t < 2. \end{cases}$$

See Example 7.2.7 in Chap. 7.

71. The function $y(t)$ with period 2π, where

$$y(t) = \begin{cases} \sin t, & 0 \le t < \pi \\ 0, & \pi \le t < 2\pi. \end{cases}$$

See Fig. 7.20 in Chap. 7.

72. The function $g(t)$ with period 2, where

$$g(t) = \begin{cases} \cos 2\pi t, & 0 < t < 1 \\ -1, & 1 < t < 2. \end{cases}$$

Inverse Transforms Expressed as Series

Find an inverse Laplace transform of each of the following functions.

73. $F(s) = \dfrac{1}{s(1+e^{-s})}$

74. $F(s) = \dfrac{1-2e^{-s}+e^{-2s}}{s^2(1-e^{-2s})}$

75. $F(s) = \dfrac{s}{(s^2+1)(1-e^{-\pi s})}$

76. $F(s) = \dfrac{1}{s^2} - \dfrac{e^{-s}}{s(1-e^{-s})}$

Equations with Periodic Forcing Functions

77. In Example 11.8.5, verify that the function $x(t)$ given by (11.8.6) satisfies the differential equation in (11.8.3) on the intervals $(n, n-1)$ for $n = 1, 2, 3, \ldots$.

Use the method of Laplace transforms to determine the solutions of the following initial value problems, where the forcing function $f(t)$ is periodic on $[0, \infty)$. Hint. *See Examples 11.7.22 and* 11.8.5.

78. $x'' + x = f(t)$, $x(0) = 0, x'(0) = 0$, where

$$f(t) = \begin{cases} 1, & 0 \le t < 1 \\ -1, & 1 < t < 2 \end{cases}$$

and $f(t+2) = f(t)$ for $t > 0$.

79. Solve Problem 78 but using the initial conditions: $x(0) = 1, x'(0) = -3$. *Hint.* Find the solution of Problem 78 and then employ Theorem 10.7.1.

80. $x' + 2x = f(t)$, $x(0) = 0$ where

$$f(t) = \begin{cases} 2, & 0 \le t < 1 \\ 2-t, & 1 \le t \le 2 \end{cases}$$

and $f(t+2) = f(t)$ for $t > 0$.

81. $x'' + \frac{1}{2}x' + x = f(t)$ such that $x(0) = 0$ and $x'(0) = 0$, where $f(0) = 0$, $f(t) = t$ for $0 < t < \pi$, and $f(t+\pi) = f(t)$ for $t > 0$.

Equations with Variable Coefficients

Determine a solution $x(t)$ of each of the following second-order differential equations satisfying the accompanying condition(s) assuming that the Laplace transform of $x(t)$ and its derivatives exist.

82. $tx'' - 2x' + tx(t) = 0$, $x(0) = 0$ *Hint.* See Problem 60.

83. $tx'' - 2x' + tx(t) = 0$, $x(0) = 2$ *Hint.* See Problems 60 and 61.

84. $x'' + tx' - x(t) = 0, x(0) = 0$, $x'(0) = -2$

85. $2tx'' - (t+2)x' + 3x(t) = t - 1, x(0) = 0$

Volterra Equations

Solve each of the following Volterra integral and integro-differential equations of convolution type.

86. $x(t) = t - \int_0^t x(u)\,du$

87. $x(t) = 6t + \int_0^t \sin(t-u)\,x(u)\,du$

88. $x(t) = 5 - \int_0^t u x(t-u)\,du$

89. $x(t) = 4 + 4t + \frac{8}{3}\int_0^t (t-u)^3 x(u)\,du$

90. $x'(t) = 2\sin t + \int_0^t \cos(t-u)x(u)\,du$ with $x(0) = 8$

91. $x'(t) = 3t - \int_0^t (t-u)\,x(u)\,du$ with $x(0) = 0$

Inverse Transforms and Infinite Series

The forcing functions of the following differential and integro-differential equations are series. Find their solutions so that the accompanying initial conditions are satisfied.

92. $x'' + x = f(t)$, $x(0) = 0$, $x'(0) = 0$, where $f(t) = \cos t + \sum_{n=1}^{\infty} \cos(t-n)u(t-n)$.

93. $x'' + 2x' + x = f(t)$, $x(0) = 0, x'(0) = 0$, where

$$f(t) = (1-t)[1-u(t-2)]$$

$$+\sum_{n=1}^{\infty}\left(\frac{1}{2}\right)^n g(t-2n)$$

and $g(t) = (1-t)[u(t)-u(t-2)]$.

94. $x'(t) = f(t) + \int_0^t \cos(t-s)\, x(s)\, ds$ such that $x(0) = 0$, where

$$f(t) = \cos t + \sum_{n=1}^{\infty} \cos(t-n)u(t-n).$$

95. $x'(t) = f(t) + \int_0^t \cos(t-s)\, x(s)\, ds$ such that $x(0) = 0$, where

$$f(t) = \cos t + \sum_{n=1}^{\infty}(-1)^n \cos(t)u(t-2\pi n).$$

A Menagerie of Problems

96. Use Theorem 11.6.4 to find

$$\mathcal{L}^{-1}\left\{\ln\left(\frac{s+1}{s}\right)\right\}.$$

Hint. See Example 11.7.13.

97. Use Theorem 11.6.4 to find

$$\mathcal{L}^{-1}\left\{\ln\left(1+\frac{1}{s^2}\right)\right\}.$$

Hint. See Example 11.7.13.

98. Use Theorem 11.6.6 to find

$$\mathcal{L}^{-1}\left\{\frac{4}{s(s+5)}\right\}.$$

Check your answer by reworking this using a partial fraction expansion.

99. Use Theorem 11.6.6 and the result of Example 11.7.11 to compute

$$\mathcal{L}^{-1}\left\{\frac{1}{s^3(s^2+1)}\right\}.$$

100. Using the method of Laplace transforms, it was determined that the solution of the initial value problem in Example 11.7.22 is

$$x(t) = 1 - e^{-t}$$

on the interval $[0, 1)$ and that

$$x(t) = 1 - e^{-t} + 2\sum_{n=1}^{N}(-1)^n\left[1 - e^{-(t-n)}\right]$$

on the intervals $(N, N+1)$ for $N = 1, 2, \ldots$. Show that these same formulas can also be obtained with the integrating factor method that was presented in Chap. 5 for solving first-order linear equations.

101. Let $a, b \in \mathbb{R}$ with $a \neq 0$ and $b > 0$. If a function $g(t)$ is piecewise continuous on $[0, \infty)$ and of exponential order γ, show that

$$\mathcal{L}\left\{\sum_{n=1}^{\infty} a^n g(t-nb)u(t-nb)\right\} = \sum_{n=1}^{\infty} a^n \mathcal{L}\left\{g(t-nb)u(t-nb)\right\}.$$

A Delay Differential Equation

102. The equation

$$x'(t) + ax(t-1) = 0,$$

where a denotes a constant, is an example of a *linear delay differential equation*. Completing parts (a)–(d) below will yield a function $x(t)$ that is a solution of this equation and which satisfies the initial condition that $x(t) = x_0$ for $-1 \le t \le 0$, where x_0 is a given constant.

(a) Use Definition 11.1.1 and a change of variable to show

$$\mathcal{L}\{x(t-1)\} = e^{-s}\mathcal{L}\{x(t)\} + \frac{x_0}{s}(1-e^{-s}).$$

(b) Take the Laplace transform of both sides of the delay differential equation to show

$$X(s) = \frac{x_0}{s} - \frac{ax_0}{s(s+ae^{-s})},$$

where $X(s) = \mathcal{L}\{x(t)\}$.

(c) Show that

$$\mathcal{L}^{-1}\left\{\frac{1}{s(s+ae^{-s})}\right\} = \sum_{n=0}^{\infty}\frac{(-a)^n}{(n+1)!} \times (t-n)^{n+1}u(t-n).$$

Hint. Express the function as a series using (11.8.7) with $y = ae^{-s}/s$. Then apply Theorem 11.7.7.

(d) Use the results from (a)–(c) to determine an infinite series representation of $x(t)$ on the interval $[0,\infty)$.

(e) Use the result from (d) to show that the solution which is continuous on $[0,2)$ is

$$x(t) = x_0\big[1 - atH(t) + \frac{1}{2}a^2(t-1)^2H(t-1)\big]$$

and that

$$x(t) = x_0\big[1 - at + \frac{1}{2}a^2(t-1)^2\big]$$

for $1 \le t < 2$.

(f) Let $x(t)$ denote the solution that is continuous on the interval $[0, n+1)$. Show that

$$x(t) = x_0\sum_{j=0}^{n+1}\frac{(-a)^j}{j!}(t-j+1)^j$$

for $n \le t < n+1$.

(g) Use the formula given in (f) to find the solution of

$$x'(t) = -6x(t-1)$$

for $2 \le t \le 3$ satisfying the initial condition: $x(t) = 1$ on $[-1, 0]$.

Remark For those with access to *Maple*, the answer to (g) can be verified using a *Maple* worksheet from the Maplesoft Application Center. Its title and URL can be found in reference [10]. Another way to solve linear delay differential equations is also discussed in this worksheet: it is known as the *method of steps*. An introduction to this method and examples showing how to implement it are given in Sects. 11.1 and 11.2. To find the answer to (g), execute the entire worksheet and then go to Sect. 11.5 and replace $\varphi : t \to t$ with $\varphi : t \to 1$ and MethodSteps(1,1,3) with MethodSteps(0,1,3).

Computer Algebra System Problems

103. Use a CAS to compute the Laplace transform of the function $g(t)$ in Problem 72. Also, express $g(t)$ as an infinite series of unit step functions and graph it.

104. Use a CAS to graph the solution of the initial value problem given in Problem 81.

105. Use a CAS to graph the solution of the differential equation satisfying the initial conditions given in Problem 92.

Chapter 12
Planar Autonomous Homogeneous Linear Systems

The great importance of linearity lies in a combination of two circumstances. First, many tangible phenomena behave approximately linearly over restricted periods of time or restricted ranges of the variable, so that useful linear mathematical models can simulate their behavior. A pendulum swinging through a small angle is a nearly linear system. Second, linear equations can be handled by a wide variety of techniques that do not work with nonlinear equations.

Edward N. Lorenz[1]

Abstract This chapter introduces the *eigenvalue-eigenvector method* for finding solutions of autonomous homogeneous linear systems of differential equations, namely, systems of the form

$$\frac{dx}{dt} = ax + by, \quad \frac{dy}{dt} = cx + dy, \tag{1}$$

where a, b, c, d are constants. Prior knowledge of the matrix algebra that is used in this section to find solutions of (1) is not assumed in the text since many students enrolled in an introductory differential equations course have little to no knowledge of matrix algebra, especially if the course is taken immediately after the first two semesters of calculus. Accordingly, this textbook is written so that it can be read without previous exposure to matrices: any terminology and results from matrix algebra that are needed to understand this chapter have been woven into the text rather than being relegated to an appendix.

[1] See [54, p. 162]. Edward N. Lorenz (1917–2008) was an American research scientist and professor of meteorology at the Massachusetts Institute of Technology who wrote seminal papers on chaos theory during the 1960s and 1970s. He discovered the chaotic system whose equations are mentioned in Chap. 1 and which bear his name. The Lorenz system has a special limit set that is a collection of states known as the *Lorenz attractor*. The graph of the Lorenz attractor (see Appendix C) resembles a pair of outstretched butterfly wings; for that reason, it is also called the "butterfly" and has become an icon of chaos theory.

© The Author(s), under exclusive license to Springer Nature Switzerland AG 2026
L. C. Becker, *Ordinary Differential Equations: Concepts, Methods, and Models*,
https://doi.org/10.1007/978-3-032-15150-6_12

Sections 12.1 through 12.4 contain foundational information for understanding the rest of the chapter. Section 12.1 introduces terminology related to (1). An example in Sect. 12.2 illustrates that linear systems as simple as (1) can serve as mathematical models of certain types of physical systems, such as modeling the amount of sugar at any given time in each of several interconnected tanks composing a multi-tank system if pure water is pumped into the system while sugar water is being pumped out at the same rate. Section 12.3 defines vectors, matrices, and the operations involving them, such as matrix addition and multiplication, after which it is shown how to write (1) more compactly as the vector differential equation

$$\mathbf{x}' = A\mathbf{x}, \quad \text{where } \mathbf{x} = \begin{bmatrix} x \\ y \end{bmatrix} \text{ and } A = \begin{bmatrix} a & b \\ c & d \end{bmatrix}. \tag{2}$$

Section 12.4 introduces terms, such as *velocity vector field*, *direction field*, *phase point*, and *trajectory*, which aid in describing the state of an imaginary particle whose motion if governed by (2) for given values of a, b, c, and d.

Section 12.5 begins with an example of an uncoupled linear system and its solutions as a means to assist in devising a method to solve any system (1) when it is rewritten as (2). This leads to the definitions of *eigenvalues* and *eigenvectors* of a matrix followed by more definitions, such as *determinant*, *inverse*, and *characteristic equation* of a matrix. Topics presented in this section related to (2) and its solutions are the Superposition and Autonomous Principles, linear independence of functions, Wronskian of solutions, Abel's formula, just to name a few. The section culminates with theorems (and their proofs) regarding the solutions of (2) and an ample number of examples explaining how to find a general solution of a given equation taking into account whether the eigenvalues of the coefficient matrix A are real (distinct or repeated) or complex numbers.

The chapter concludes with Sect. 12.6, where linear systems (1) are used to model the temperatures of one- and two-room buildings.

12.1 Planar Linear Systems

A pair of first-order differential equations of the form

$$\begin{aligned} \frac{dx}{dt} &= ax + by \\ \frac{dy}{dt} &= cx + dy \end{aligned} \tag{12.1.1}$$

is known as a ***two-dimensional homogeneous linear system*** of first-order differential equations with constant coefficients a, b, c, and d. The coefficients, any of which may equal zero, are called the ***parameters*** of the system. The primary objective of this chapter is to come up with a method for finding all solutions of (12.1.1). A

solution is a pair of differential functions of the independent variable t, denoted by writing $x(t)$ and $y(t)$, that satisfy both equations in (12.1.1) on some interval of the t-axis when $x(t)$ and $y(t)$ are substituted for the dependent variables x and y, respectively. The term ***two-dimensional*** points out that the system consists of two equations with two dependent variables whose solutions can be depicted as curves in the xy-plane. This system is said to be ***autonomous***, which means that the right-hand sides of the two scalar equations constituting the system do not depend explicitly on t. From now on, whenever the term ***planar homogeneous linear system*** is used, it will refer to a system of the form (12.1.1). For the sake of brevity, we will sometimes shorten this to ***homogeneous linear system*** since all of the systems in this and the next chapter are two-dimensional.

A ***planar nonhomogeneous linear system*** of first-order differential equations with constant coefficients is a system of the form

$$\begin{aligned} \frac{dx}{dt} &= ax + by + f(t) \\ \frac{dy}{dt} &= cx + dy + g(t) \end{aligned} \tag{12.1.2}$$

where $f(t)$ and $g(t)$, as the notation indicates, are functions of t but not of x or y. Generally speaking, as above, we will drop the word "planar" and refer to (12.1.2) as a ***nonhomogeneous linear system***. Of course, the difference between (12.1.1) and (12.1.2) is the presence of f and g, one of which must be a nonzero function in order for (12.1.2) to be nonhomogeneous. If any one of these functions is nonconstant, then the system (12.1.2) is said to be ***nonautonomous***. Henceforth, whenever we speak of a ***linear system***, we will mean a system of first-order differential equations that look like either (12.1.1) or (12.1.2).

Linear systems of first-order differential equations are important in the applied sciences. Consider a mathematical model of some simple physical situation involving one independent variable, say t, and two dependent variables, say x and y, and for each value of t, the dependent variables x and y describe the state of the physical system. Suppose it can be argued that the rate of change of each dependent variable is equal to a *linear combination*[2] of x and y and that the rate of change of at least one of them, say x for instance, depends either on both x and y or at least on the other variable y. This interdependence of the dependent variables is expressed by saying that the two differential equations constituting the linear system are ***coupled***. An example of a situation modeled by a coupled linear system is described in the following section.

[2] A ***linear combination*** of x and y is an expression of the form $ax + by$, where a and b are parameters.

12.2 Flow of Sugar in Interconnected Tanks

Let us model the amount of sugar in each of the two interconnected tanks shown in Fig. 12.1. Suppose tank L, the tank on the left, initially contains 50 gallons of a sugar-water solution with a concentration of 0.20 pounds of sugar per gallon of sugar water, while tank R starts off with 100 gallons of pure water. Pure water is pumped into tank L from an external source at the rate of 1 gallon per minute while sugar water (initially pure water) is pumped completely out of the two-tank system from tank R at the same rate. Sugar water is also pumped from tank L into tank R at a rate of 3 gallons per minute and from tank R back into tank L at a rate of 2 gallons per minute. For the sake of having a simple yet viable model, let us assume that changes in concentrations will spread instantaneous and uniformly throughout each tank so that the solution in each tank is always a homogeneous mixture. Let us determine the system of differential equations whose solution would model the amount of sugar in each tank as a function of time.

Let $x(t)$ and $y(t)$ denote the respective amounts of sugar (in pounds) in tanks L and R at time t minutes. At time $t = 0$ minutes, the initial state of the system is

$$x(0) = (50 \text{ gal}) \left(0.20 \, \frac{\text{lb}}{\text{gal}}\right) = 10 \text{ lb} \quad \text{and} \quad y(0) = 0 \text{ lb}.$$

The combined inflow of both water and sugar water into tank L is 3 gallons per minute, while the rate of outflow of sugar water is also 3 gallons per minute. As a result, there will always be 50 gallons of sugar water in tank L. Consequently, the concentration of sugar in tank L at time t minutes is $x(t)/50$ pounds per gallon of sugar water. Hence, the rate at which sugar flows from tank L into tank R is

$$\left(3 \, \frac{\text{gal}}{\text{min}}\right) \left(\frac{x(t) \text{ lb}}{50 \text{ gal}}\right) = 0.06x(t) \, \frac{\text{lb}}{\text{min}}.$$

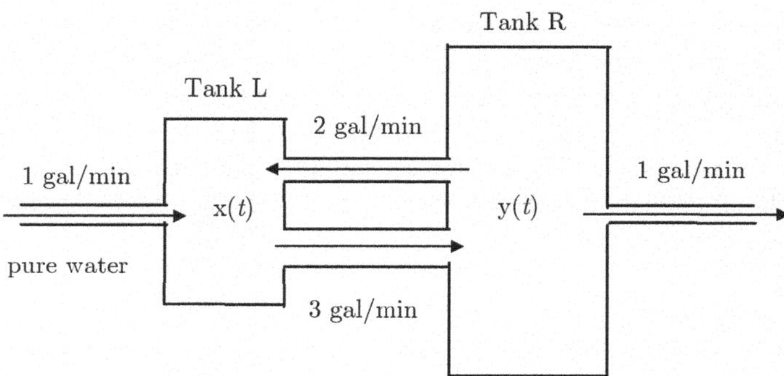

Fig. 12.1 Interconnected tanks

Likewise, since tank R always contains 100 gallons of sugar water with a sugar concentration of $y(t)/100$ pounds per gallon, sugar flows from tank R into tank L at the rate of

$$\left(2\ \frac{\text{gal}}{\text{min}}\right)\left(\frac{y(t)\ \text{lb}}{100\ \text{gal}}\right) = 0.02y(t)\ \frac{\text{lb}}{\text{min}}.$$

It follows from the conservation equation for one-compartment systems in Chap. 6 that the amount of sugar in tank L is modeled by the differential equation

$$\frac{dx}{dt} = -0.06x + 0.02y.$$

As for the sugar content of tank R, we just determined that sugar flows into it from tank L at a rate of $0.06x(t)$ pounds per minute, and it flows out of tank R into tank L at a rate of $0.02y(t)$ pounds per minute. But there is also a flow of sugar out of the two-tank system from tank R at the rate of

$$\left(1\ \frac{\text{gal}}{\text{min}}\right)\left(\frac{y(t)\ \text{lb}}{100\ \text{gal}}\right) = 0.01y(t)\ \frac{\text{lb}}{\text{min}}.$$

This results in a net outflow rate from tank R of $0.03y(t)$ pounds per minute. Consequently, the equation modeling the amount of sugar in tank R is

$$\frac{dy}{dt} = 0.06x - 0.03y.$$

In conclusion, the model for the sugar contents of both tanks is given by the homogeneous linear system

$$\begin{aligned} x' &= -0.06x + 0.02y \\ y' &= 0.06x - 0.03y. \end{aligned} \tag{12.2.1}$$

where $x(0) = 10$ and $y(0) = 0$. Observe that this looks like (12.1.1), which is the general form of planar homogeneous linear systems. In the following sections, we take up the matter of finding solutions of such systems.

12.3 Homogeneous Linear Systems as Vector Equations

One of the goals of this chapter is to devise a method that will enable us to find all solutions of the homogeneous linear system

$$x' = ax + by$$

$$y' = cx + dy, \tag{12.1.1}$$

where the prime symbol ($'$) denotes differentiation with respect to the independent variable t. By grouping like quantities together in ordered rectangular arrays, we can rewrite (12.1.1) in a compact form that not only saves space but is also well suited for the method of finding solutions that we will develop in subsequent sections. The coefficients of the dependent variables x and y are grouped together in an array in the order that they appear in the system. This results in a square array with two rows and two columns, which is enclosed (depending on the author) in either brackets or parentheses. That is to say, the array is written as follows:

$$\begin{bmatrix} a & b \\ c & d \end{bmatrix} \quad \text{or} \quad \begin{pmatrix} a & b \\ c & d \end{pmatrix}. \tag{12.3.1}$$

Such an ordered array of quantities is called a ***matrix***.[3] In this book, we use brackets rather than parentheses for matrices.

If we are dealing with two or more matrices (plural of matrix) in a problem, we can distinguish between them by naming them: usually by using uppercase letters, such as

$$A = \begin{bmatrix} a & b \\ c & d \end{bmatrix}, \quad B = \begin{bmatrix} 2 & 0 \\ 4 & -3 \\ 0 & 1 \end{bmatrix}.$$

The individual items in a matrix are called its ***entries*** (or ***elements***). For instance, the entries of A are a, b, c, and d. A matrix with m rows and n columns is called an ***m* x *n* matrix** (read "m by n matrix") or a ***matrix of order* (*m*, *n*)**.[4] For example, we can describe B by saying that it is a 3×2 matrix or a matrix of order $(3, 2)$. Since A has two rows and two columns, it is a 2×2 (read "two by two") matrix. Alternatively, it is said to be a ***matrix of order*** 2, which means that it is a ***square matrix*** (a matrix with the same number of columns as rows)—in this case, two rows and two columns. Since the entries of the matrix A are the coefficients of the homogeneous linear system (12.1.1), it is called the ***coefficient matrix*** of (12.1.1).

Since one of the goals of this section is to explain how to rewrite (12.1.1) in terms of matrices, let also put the two dependent variables x and y into a matrix with two rows and one column as follows:

$$\begin{bmatrix} x \\ y \end{bmatrix} \quad \text{or} \quad \begin{pmatrix} x \\ y \end{pmatrix}.$$

[3] Originally the word meant *womb*; a nontechnical use of *matrix* is to refer to that within which something else develops or is contained. As we will see, a matrix is more than just an array of numbers: from the matrix of coefficients of (12.1.1), we will be able to obtain information about the solutions of (12.1.1).

[4] The terms ***dimension*** or ***size*** are also used.

A column matrix of this form is known as a 2×1 ***column vector***.[5] A column vector is still a matrix. It is just a special type of matrix: one with a single column. The entries of a vector are called its ***components***: this column vector has the components x and y. Of course, we could also put x and y into a matrix with one row and two columns resulting in the 1×2 matrix

$$\begin{bmatrix} x & y \end{bmatrix} \quad \text{or} \quad \begin{pmatrix} x & y \end{pmatrix} .$$

This type of matrix is called a 1×2 ***row vector***. However, because of the way in which matrix multiplication will be defined shortly, convention dictates that the dependent variables are to be put into column vectors rather than into row vectors if they are involved in equations and calculations. Unless stated otherwise, the word *vector* in this book will refer to a *column vector*. In typewritten text, vectors are ordinarily labeled with boldface lowercase letters, such as

$$\mathbf{x} = \begin{bmatrix} x \\ y \end{bmatrix} . \tag{12.3.2}$$

Note that $\mathbf{x}$ (in boldface type) denotes a vector whereas x (in regular type) denotes the first component of $\mathbf{x}$. If both components are 0, then the vector is known as the ***zero vector*** and denoted by $\mathbf{0}$. That is,

$$\mathbf{0} := \begin{bmatrix} 0 \\ 0 \end{bmatrix} . \tag{12.3.3}$$

There are important operations involving matrices, such as addition and multiplication, that we will have need of later on. First, consider ***matrix addition***. It is defined only for matrices of the same order. The sum of two such matrices is defined to be the matrix of that order whose entries are obtained by adding corresponding entries of the two matrices. To illustrate this operation, consider the following examples of adding 2×1 vectors. The sum of

$$\mathbf{u} = \begin{bmatrix} 1 \\ -3 \end{bmatrix} \quad \text{and} \quad \mathbf{v} = \begin{bmatrix} -4 \\ 10 \end{bmatrix}$$

is

$$\mathbf{u} + \mathbf{v} = \begin{bmatrix} 1 \\ -3 \end{bmatrix} + \begin{bmatrix} -4 \\ 10 \end{bmatrix} = \begin{bmatrix} 1 + (-4) \\ -3 + 10 \end{bmatrix} = \begin{bmatrix} -3 \\ 7 \end{bmatrix} .$$

[5] Calling this a vector may come across as strange if you have previously encountered vectors in physics or engineering classes. We shall explain shortly why this particular kind of matrix is called a vector.

Likewise, the ***sum*** of $\mathbf{x}$ and $\mathbf{0}$ is

$$\mathbf{x}+\mathbf{0}=\begin{bmatrix}x\\y\end{bmatrix}+\begin{bmatrix}0\\0\end{bmatrix}=\begin{bmatrix}x+0\\y+0\end{bmatrix}=\begin{bmatrix}x\\y\end{bmatrix}=\mathbf{x}.$$

It is clear why the vector $\mathbf{0}$ is called the zero vector: it assumes the role of the *additive identity element* in vector addition.

In general, the sum of two $n \times 1$ vectors ($n \geq 2$) is an $n \times 1$ vector whose entries are obtained by adding corresponding entries of the two vectors. However, in this and the following chapter, we will only have to deal with 2×1 vectors and square matrices of order 2. So let us explain how to add two square matrices of order 2. Let

$$A=\begin{bmatrix}a & b\\c & d\end{bmatrix} \quad \text{and} \quad B=\begin{bmatrix}r & s\\t & u\end{bmatrix}. \tag{12.3.4}$$

The ***sum*** of A and B is defined to be the matrix obtained by adding the corresponding entries of A and B:

$$A+B=\begin{bmatrix}a & b\\c & d\end{bmatrix}+\begin{bmatrix}r & s\\t & u\end{bmatrix}=\begin{bmatrix}a+r & b+s\\c+t & d+u\end{bmatrix}. \tag{12.3.5}$$

Let O denote the 2×2 matrix whose four entries are 0, to wit:

$$O:=\begin{bmatrix}0 & 0\\0 & 0\end{bmatrix}$$

This is a ***zero matrix***. It follows from the definition of matrix addition that

$$A+O=O+A=A.$$

The operation of ***scalar multiplication*** generally involves a matrix of any order and a ***scalar*** (a real or complex number). But, as mentioned before, we only need to consider square matrices of order 2. The product of a scalar k and a matrix A, denoted by kA, is defined as follows:

$$kA:=k\begin{bmatrix}a & b\\c & d\end{bmatrix}=\begin{bmatrix}ka & kb\\kc & kd\end{bmatrix}. \tag{12.3.6}$$

Likewise, the product of a scalar k and the vector $\mathbf{x}$ in (12.3.2) is

$$k\mathbf{x}:=k\begin{bmatrix}x\\y\end{bmatrix}=\begin{bmatrix}kx\\ky\end{bmatrix}.$$

The vector $k\mathbf{x}$ is said to be a ***scalar multiple*** of the vector $\mathbf{x}$. Just for clarity, the term *scalar multiple* refers to the vector $k\mathbf{x}$, not to the scalar k. For instance, letting $k = 0$, $x = 1$, and $y = 2$, the previous equation becomes

$$k\mathbf{x} = 0\begin{bmatrix}1\\2\end{bmatrix} = \begin{bmatrix}0\\0\end{bmatrix} = \mathbf{0}.$$

The scalar multiple here is not the scalar 0 but rather the vector $\mathbf{0}$.

We can use the operations of matrix addition and scalar multiplication to prove important properties that will be required later on. The following is an example of such a property, which is a *left distributive law* for matrices.

Example 12.3.1 If A and B are matrices of the same order and k is a scalar, then

$$k(A + B) = kA + kB.$$

Proof We prove this only for matrices of order 2. Consider the matrices A and B given in (12.3.4). Then from (12.3.5) and (12.3.6) it follows that

$$kA + kB = k\begin{bmatrix}a & b\\c & d\end{bmatrix} + k\begin{bmatrix}r & s\\t & u\end{bmatrix} = \begin{bmatrix}ka & kb\\kc & kd\end{bmatrix} + \begin{bmatrix}kr & ks\\kt & ku\end{bmatrix}$$

$$= \begin{bmatrix}ka + kr & kb + ks\\kc + kt & kd + ku\end{bmatrix} = k\begin{bmatrix}a + r & b + s\\c + t & d + u\end{bmatrix} = k(A + B),$$

which completes the proof. ■

In fact, this left distributive law is true for any two matrices of the same order. The reason for limiting the proof to 2×2 matrices is that only matrices with two rows and one or two columns are dealt with in this and the next chapter. Moreover, no prior knowledge of matrix algebra is assumed on the part of the reader. So for the sake of simplicity and understanding, we will only provide proofs for 2×1 vectors and 2×2 matrices. Proofs for $m \times n$ matrices can be found in most linear (matrix) algebra and advanced differential equations textbooks.

There are other important properties of matrix addition and scalar multiplication that we will need later on. Their proofs rely on the properties of numbers and the operations (12.3.5) and (12.3.6) and are similar to the proof of the *left distributive law* in Example 12.3.1. A list of these properties follow.

Properties of Matrix Addition and Scalar Multiplication

Let r and s be scalars. Let A, B, and C be matrices of the same order.

1. $A + B = B + A$ (commutative law of addition)

(continued)

2. $(A+B)+C = A+(B+C)$ (associative law of addition)
3. $A+O = O+A = A$ (identity for addition)
4. $r(A+B) = rA+rB$ (a left distributive law)
5. $(r+s)A = rA+sA$ (a right distributive law)
6. $r(sA) = (rs)A = s(rA)$ (associative law of scalar multiplication)

Defining vectors as column and row matrices may be confusing to those who prefer to view vectors as geometric objects: directed line segments or arrows, which can be expressed in terms of two special vectors represented by the symbols **i** and **j**. So before we continue our discussion showing how linear systems can be rewritten compactly in terms of column vectors and matrices, let us first reconcile these two ways of viewing vectors. Let us begin with the arrow viewpoint. The vectors **i** and **j** are called the ***unit coordinate vectors*** since they are arrows of unit length (arrows of length one) that point in the positive x- and y-directions, respectively. They are also called the ***standard basis vectors*** in the xy-plane. All other vectors, or arrows, lying in the plane can be expressed in terms of them. For instance, consider the vector **v** with its tail at the origin and its head at the point $(1, 2)$. Expressed in terms of **i** and **j**, the vector **v** is

$$\mathbf{v} = \mathbf{i} + 2\mathbf{j}.$$

Another vector, say **u**, with its tail at the origin and its head at the point $(-2, 5)$ is written

$$\mathbf{u} = -2\mathbf{i} + 5\mathbf{j}.$$

But what if the tail of a vector is not located at the origin? How is it represented then? The answer to this is simply to move the tail of the vector, regardless of its original location, to the origin while keeping it pointing in the same direction. In this way, the length and direction of the moved vector is the same as that of the original vector. In essence, nothing has changed: the two vectors are really the same in the sense of pointing in the same direction and having the same length. Moving a vector in this manner is known as ***translating*** the vector. Every vector **x** in the plane can be represented in terms of the translated vector that points in the same direction and has the same length as **x** but with its tail at the origin. If the head of this translated vector is located at the point (x, y), then **x** is represented by the arrow

$$\mathbf{x} = x\mathbf{i} + y\mathbf{j}. \tag{12.3.7}$$

Note that it is the location of the head of the vector, after its tail has been moved to origin, that distinguishes it from all other vectors. Consequently, every vector in the plane can be represented by the coordinates (x, y) of the head of the equivalent translated vector. So an alternative notation for (12.3.7) is

$$\mathbf{x} = \begin{bmatrix} x \\ y \end{bmatrix}. \tag{12.3.8}$$

As mentioned before, this particular type of matrix is called a column vector. In short, (12.3.7) and (12.3.8) denote the exact same vector.

At times, in the interest of saving vertical space on a page, we will write the components of a vector horizontally separated by a comma and enclosed in *angle brackets*. In other words, by $\langle x, y \rangle$ we mean (12.3.8). For example,

$$\langle 2, 5 \rangle \quad \text{and} \quad \begin{bmatrix} 2 \\ 5 \end{bmatrix}$$

represent the same vector. As a general rule, we will use the column form of the vector when it is involved in computations and the horizontal form when it is embedded in the text. Keep in mind the differences between (x, y), $\langle x, y \rangle$, and $\begin{bmatrix} x & y \end{bmatrix}$. The latter is a row vector, $\langle x, y \rangle$ denotes a column vector, and (x, y) is a point in the plane.

The upshot of the foregoing discussion is that we can denote a vector $\mathbf{x}$ in one of two equivalent ways: either by $\langle x, y \rangle$, equivalently (12.3.8), or by $x\mathbf{i} + y\mathbf{j}$. What is convenient about the former notation is that we can represent and manipulate vectors in the efficient shorthand language of matrices without having to include the unit coordinate vectors $\mathbf{i}$ and $\mathbf{j}$.[6]

Since $\mathbf{i}$ is the vector of unit length that points in the positive x-direction, it is represented by the arrow with its tail at the origin and its head at the point $(1, 0)$. In vector notation, $\mathbf{i} = \langle 1, 0 \rangle$. That is,

$$\mathbf{i} = \begin{bmatrix} 1 \\ 0 \end{bmatrix}.$$

Similarly, since $\mathbf{j}$ denotes the vector of unit length that points in the positive y-direction, it is the vector $\langle 0, 1 \rangle$. That is,

$$\mathbf{j} = \begin{bmatrix} 0 \\ 1 \end{bmatrix}.$$

The two ways of designating vectors are equivalent. The following shows how to transition from vector notation to the $(\mathbf{i}, \mathbf{j})$-representation:

$$\begin{bmatrix} x \\ y \end{bmatrix} = \begin{bmatrix} x \\ 0 \end{bmatrix} + \begin{bmatrix} 0 \\ y \end{bmatrix} = x\begin{bmatrix} 1 \\ 0 \end{bmatrix} + y\begin{bmatrix} 0 \\ 1 \end{bmatrix} = x\mathbf{i} + y\mathbf{j}.$$

[6] Another advantage is the ease with which vectors can be generalized from 2- and 3-dimensional spaces to n-dimensional spaces, a topic that is probably best left to more advanced courses.

To transition from the $(\mathbf{i}, \mathbf{j})$-representation to vector notation, simply reverse the steps.

We have described the types of vectors and matrices that will aid us in our study of planar linear systems of first-order differential equations. Now let us describe how to write the system (12.1.1) using vector-matrix notation. Let $\mathbf{x}$ denote the vector $\langle x, y \rangle$. First, use the left-hand sides of the two equations constituting (12.1.1) to define the column vector

$$\begin{bmatrix} dx/dt \\ dy/dt \end{bmatrix} = \begin{bmatrix} x' \\ y' \end{bmatrix}. \tag{12.3.9}$$

Second, use the right-hand sides of the equations to define the vector

$$\begin{bmatrix} ax + by \\ cx + dy \end{bmatrix}. \tag{12.3.10}$$

Matrices and vectors of the same order are defined to be ***equal*** if and only if their corresponding components are equal. So,

$$\begin{bmatrix} x' \\ y' \end{bmatrix} = \begin{bmatrix} ax + by \\ cx + dy \end{bmatrix} \tag{12.3.11}$$

if and only if

$$x' = ax + by \quad \text{and} \quad y' = cx + dy.$$

Consequently, system (12.1.1) and the vector equation (12.3.11) are equivalent.

Matrix-Vector Product

Definition 12.3.1 Let $A = \begin{bmatrix} a & b \\ c & d \end{bmatrix}$ and $\mathbf{x} = \begin{bmatrix} x \\ y \end{bmatrix}$. The ***product*** $A\mathbf{x}$ is defined by

$$A\mathbf{x} := \begin{bmatrix} ax + by \\ cx + dy \end{bmatrix}. \tag{12.3.12}$$

Remark From (12.3.12), we see that

$$A\mathbf{x} = \begin{bmatrix} a & b \\ c & d \end{bmatrix} \begin{bmatrix} x \\ y \end{bmatrix} = x \begin{bmatrix} a \\ c \end{bmatrix} + y \begin{bmatrix} b \\ d \end{bmatrix}. \tag{12.3.13}$$

In other words, the product $A\mathbf{x}$ *is equal to a linear combination of the column vectors of* A, *where the scalar coefficient of the first* [resp. *second*] *column vector is the first* [resp. *second*] *component of* $\mathbf{x}$.

Since the left-hand side of (12.1.1) can be written as the vector whose components are the derivatives of the components of $\mathbf{x}$ with respect to t, it makes sense to define the derivative of $\mathbf{x}$ with respect to t as follows:

$$\mathbf{x}' = \frac{d}{dt}\mathbf{x} = \frac{d}{dt}\begin{bmatrix} x \\ y \end{bmatrix} := \begin{bmatrix} x' \\ y' \end{bmatrix} = \begin{bmatrix} dx/dt \\ dy/dt \end{bmatrix}.$$

Because of Definition 12.3.1, the vector (12.3.10) can be written as the product $A\mathbf{x}$. Putting all of these pieces together, we see that the system of differential equations (12.1.1) can be written in a more compact form: as the ***vector differential equation***

$$\mathbf{x}' = A\mathbf{x}. \tag{12.3.14}$$

In conclusion, the vector differential equation (12.3.14), where

$$\mathbf{x} = \begin{bmatrix} x \\ y \end{bmatrix} \quad \text{and} \quad A = \begin{bmatrix} a & b \\ c & d \end{bmatrix},$$

is just another way of writing the homogeneous linear system

$$\begin{aligned} x' &= ax + by \\ y' &= cx + dy. \end{aligned}$$

So, it really makes no difference as to whether Eq. (12.3.14) is called a "vector differential equation" or a "homogeneous linear system of differential equations."

12.3.1 Matrix Multiplication

We conclude this section with a short introduction to matrix multiplication since this operation, as we shall see in the next section, facilitates solving linear systems of first-order differential equations. We confine our discussion to 2×2 matrices and 2×1 column vectors since this chapter is only concerned with systems of two equations in two unknowns, However, most of the results in this and subsequent chapters can be extended to systems of n equations in n unknowns. These results can be found in more advanced textbooks dealing with the topics of differential equations and linear algebra.

As we shall see below, the definition of the product of two matrices is a consistent and natural extension of the definition of the product of a matrix with a vector that was given in Definition 12.3.1. For the purpose of stating this definition, it is advantageous to rename the entries of a matrix. Rather than using a, b, c, d to denote the entries of the 2×2 matrix A as before, let us use a_{ij} to denote the entry of A that is located in the ith row and jth column. In other words,

$$A = \begin{bmatrix} a_{11} & a_{12} \\ a_{21} & a_{22} \end{bmatrix}. \tag{12.3.15}$$

Likewise, let us use a double-subscripted b for the entries of the 2×2 matrix B. That is to say,

$$B = \begin{bmatrix} b_{11} & b_{12} \\ b_{21} & b_{22} \end{bmatrix}. \tag{12.3.16}$$

By defining the vectors

$$\mathbf{b_1} := \begin{bmatrix} b_{11} \\ b_{21} \end{bmatrix} \quad \text{and} \quad \mathbf{b_2} := \begin{bmatrix} b_{12} \\ b_{22} \end{bmatrix},$$

we can write (12.3.16) in the abbreviated form

$$B = \begin{bmatrix} \mathbf{b_1} & \mathbf{b_2} \end{bmatrix}, \tag{12.3.17}$$

provided the brackets of the vectors $\mathbf{b_1}$ and $\mathbf{b_2}$ are discarded when they are used as columns of a matrix. In other words, let

$$\begin{bmatrix} \mathbf{b_1} & \mathbf{b_2} \end{bmatrix} = \begin{bmatrix} \begin{bmatrix} b_{11} \\ b_{21} \end{bmatrix} & \begin{bmatrix} b_{12} \\ b_{22} \end{bmatrix} \end{bmatrix}$$

represent the matrix B given by (12.3.16).

For the sake of brevity, we will sometimes use $[a_{ij}]$ to refer to the 2×2 matrix on the right-hand side of (12.3.15). Likewise, $B = [b_{ij}]$ is an abbreviated version of (12.3.16).

Matrix Multiplication I

Definition 12.3.2 Let $A = [a_{ij}]$ and $B = [b_{ij}]$ be 2×2 matrices. The ***matrix product*** AB is the 2×2 matrix

$$AB := \begin{bmatrix} A\mathbf{b_1} & A\mathbf{b_2} \end{bmatrix}. \tag{12.3.18}$$

According to Definition 12.3.1, the product $A\mathbf{b_1}$ is

$$A\mathbf{b_1} = \begin{bmatrix} a_{11} & a_{12} \\ a_{21} & a_{22} \end{bmatrix} \begin{bmatrix} b_{11} \\ b_{21} \end{bmatrix} = \begin{bmatrix} a_{11}b_{11} + a_{12}b_{21} \\ a_{21}b_{11} + a_{22}b_{21} \end{bmatrix}.$$

Likewise,

$$A\mathbf{b_2} = \begin{bmatrix} a_{11} & a_{12} \\ a_{21} & a_{22} \end{bmatrix} \begin{bmatrix} b_{12} \\ b_{22} \end{bmatrix} = \begin{bmatrix} a_{11}b_{12} + a_{12}b_{22} \\ a_{21}b_{12} + a_{22}b_{22} \end{bmatrix}.$$

Consequently,

$$\begin{bmatrix} A\mathbf{b_1} & A\mathbf{b_2} \end{bmatrix} = \begin{bmatrix} a_{11}b_{11} + a_{12}b_{21} & a_{11}b_{12} + a_{12}b_{22} \\ a_{21}b_{11} + a_{22}b_{21} & a_{21}b_{12} + a_{22}b_{22} \end{bmatrix}.$$

As a result, the matrix product (12.3.18) is

$$AB = \begin{bmatrix} a_{11} & a_{12} \\ a_{21} & a_{22} \end{bmatrix} \begin{bmatrix} b_{11} & b_{12} \\ b_{21} & b_{22} \end{bmatrix} = \begin{bmatrix} a_{11}b_{11} + a_{12}b_{21} & a_{11}b_{12} + a_{12}b_{22} \\ a_{21}b_{11} + a_{22}b_{21} & a_{21}b_{12} + a_{22}b_{22} \end{bmatrix}.$$

The result of these computation show that another definition of matrix multiplication which is equivalent to Definition 12.3.2 is as follows:

Matrix Multiplication II

Definition 12.3.3 The ***matrix product*** AB is the matrix $C = [c_{ij}]$ of order 2, where

$$c_{ij} := \sum_{k=1}^{2} a_{ik}b_{kj} \tag{12.3.19}$$

for $i = 1, 2$ and $j = 1, 2$.

Example 12.3.2 Given the matrices

$$A = \begin{bmatrix} 2 & -1 \\ 3 & -2 \end{bmatrix} \quad \text{and} \quad B = \begin{bmatrix} 1 & 0 \\ 2 & 5 \end{bmatrix},$$

compute the matrix products: AB, BA, and A^2.

Solution Let C denote the matrix product AB. The entries in the first row of C are

$$c_{11} = \sum_{k=1}^{2} a_{1k}b_{k1} = a_{11}b_{11} + a_{12}b_{21} = (2)(1) + (-1)(2) = 0,$$

$$c_{12} = \sum_{k=1}^{2} a_{1k}b_{k2} = a_{11}b_{12} + a_{12}b_{22} = (2)(0) + (-1)(5) = -5.$$

In the same way, the entries in the second row are

$$c_{21} = \sum_{k=1}^{2} a_{2k}b_{k1} = a_{21}b_{11} + a_{22}b_{21} = (3)(1) + (-2)(2) = -1,$$

$$c_{22} = \sum_{k=1}^{2} a_{2k}b_{k2} = a_{21}b_{12} + a_{22}b_{22} = (3)(0) + (-2)(5) = -10.$$

Thus,

$$AB = \begin{bmatrix} 2 & -1 \\ 3 & -2 \end{bmatrix} \begin{bmatrix} 1 & 0 \\ 2 & 5 \end{bmatrix} = \begin{bmatrix} 0 & -5 \\ -1 & -10 \end{bmatrix}.$$

Actually, these computations can be carried out more quickly by hand without having to write out all of these summations: To compute the entry c_{ij}, multiply the entries in the ith row of A by the corresponding entries in the jth column of B and then add the results. The mnemonic "row times column" is a way to remember this. For example,

$$c_{21} = \begin{bmatrix} 3 & -2 \end{bmatrix} \begin{bmatrix} 1 \\ 2 \end{bmatrix} = [3(1) + (-2)(2)] = [-1].$$

Now drop the brackets to get $c_{21} = -1$.

Continuing, let D represent the matrix product BA. That is,

$$D = BA = \begin{bmatrix} 1 & 0 \\ 2 & 5 \end{bmatrix} \begin{bmatrix} 2 & -1 \\ 3 & -2 \end{bmatrix}.$$

Let us show how to compute d_{11}, which denotes the entry in the first row and first column of D. Using the "row times column" mnemonic for matrix multiplication, we have

$$d_{11} = \begin{bmatrix} 1 & 0 \end{bmatrix} \begin{bmatrix} 2 \\ 3 \end{bmatrix} = [(1)(2) + (0)(3)] = [2].$$

Thus, $d_{11} = 2$. Similar calculations yield

$$d_{12} = -1, \quad d_{21} = 19, \quad d_{22} = -12.$$

Therefore,

$$BA = \begin{bmatrix} 2 & -1 \\ 19 & -12 \end{bmatrix}.$$

We leave it to the reader to verify that

$$A^2 = AA = \begin{bmatrix} 2 & -1 \\ 3 & -2 \end{bmatrix} \begin{bmatrix} 2 & -1 \\ 3 & -2 \end{bmatrix} = \begin{bmatrix} 1 & 0 \\ 0 & 1 \end{bmatrix}.$$ ♦

Note that the matrix products AB and BA are not equal in the previous example. For some pairs of matrices, they are equal. When they are (that is, when $AB = BA$), the matrices are said to ***commute***. But, generally speaking, $AB \neq BA$. So it is important to remember: *Matrix multiplication is not in general commutative.*

Before we look at the next example, let us also mention that the uppercase letter I is used to denote square matrices that have 1's down the main diagonal but 0's elsewhere. Since only matrices of order 2 are needed in this chapter, I is reserved for the following 2×2 matrix:

$$I := \begin{bmatrix} 1 & 0 \\ 0 & 1 \end{bmatrix}. \tag{12.3.20}$$

It is left to the reader to verify that I commutes with every matrix A of order 2. That is,

$$IA = AI = A.$$

For this reason, I is called the 2×2 ***identity matrix***.

Some important properties of matrix multiplication will be needed later on, one of which we state now and prove in the following example.

Example 12.3.3 Let $A = [a_{ij}]$, $B = [b_{ij}]$, and $C = [c_{ij}]$ be 2×2 matrices. Prove the following *right distributive law*:

$$(A + B)C = AC + BC.$$

Proof The entry in the ith row and jth column of the matrix $(A + B)C$ is

$$\sum_{k=1}^{2} (a_{ik} + b_{ik})c_{kj}.$$

It follows from the distributive and commutative laws of real numbers that this sum is also equal to

$$\sum_{k=1}^{2}(a_{ik}c_{kj} + b_{ik}c_{kj}) = \sum_{k=1}^{2} a_{ik}c_{kj} + \sum_{k=1}^{2} b_{ik}c_{kj}.$$

Now observe that the sum on the right-hand side is the entry in the ith row and jth column of the matrix $AC + BC$. ■

The right distributive law and other important properties of matrix multiplication are listed below. Since their proofs are similar to the proof of the right distributive law, they are left to the reader.

Properties of Matrix Multiplication
Let k be a scalar. Let A, B, and C be 2×2 matrices.

1. $A(B + C) = AB + AC$ (a left distributive law)
2. $(A + B)C = AC + BC$ (a right distributive law)
3. $A(BC) = (AB)C$ (associative law of matrix multiplication)
4. $(kA)B = k(AB) = A(kB)$

These properties of matrix multiplication hold not only for 2×2 matrices but also for matrices of other orders providing each of the matrix products involved in a given law are defined. For example, the above left distributive law holds for any $m \times n$ matrix A and any $n \times p$ matrices B and C. For more information, consult any linear (matrix) algebra textbook.

12.4 Phase Planes: Vector and Direction Fields

Although we do not know yet how to find solutions of the homogeneous linear system

$$\frac{dx}{dt} = ax + by$$

$$\frac{dy}{dt} = cx + dy \tag{12.1.1}$$

for given values of the coefficients a, b, c, and d, we can still glean some information from this system about the overall behavior of its solutions. The first equation of the system gives the instantaneous rate at which x is changing with respect to the independent variable t. Likewise, the second equation gives the rate of change of y with respect to t. We can depict $\langle x', y' \rangle$ as a vector and show how it varies from point to point in the xy-plane.

In order to explain this further, let us imagine that the xy-plane is covered with an ultra-thin layer of water that flows in the plane but not above or below it. Furthermore, imagine the velocity of the water at a given point (x, y) is governed by the linear system (12.1.1), where the independent variable t represents time, and that the velocity of an imaginary particle located at (x, y) is the same as the velocity of the water at this point.

Let us assign to every point (x, y) in the plane a ***velocity vector*** $\langle x', y' \rangle$ given by a function $\mathbf{v}$ defined by the right-hand side of (12.1.1). That is,

$$\mathbf{v}(x, y) := \begin{bmatrix} x' \\ y' \end{bmatrix} = \begin{bmatrix} ax + by \\ cx + dy \end{bmatrix}.$$

With this function we can determine the velocity of a particle at a given point (x, y). Since the outputs of $\mathbf{v}(x, y)$ are vectors, this is an example of a ***vector function*** (or ***vector-valued function***). Let us refer to this particular type of vector function as the ***velocity vector field*** defined by the homogeneous linear system (12.1.1). A plot of the velocity vector field consists of velocity vectors $\mathbf{v}$ assigned to points on a grid of the xy-plane. If there are a sufficient number of these points, the plot aids in visualizing the ***flow*** of the velocity vector field; that is, the velocities of particles and the paths they trace out as time evolves. In summary, $\mathbf{v}(x, y)$ is the instantaneous velocity of an imaginary particle located at the point (x, y). Such a particle is called a ***phase point***. When the xy-plane is viewed as composed of phase points whose velocities and paths are governed by a system of differential equations, it is called a ***phase plane***.

We illustrate the foregoing with a linear system that has the following values assigned to the parameters in (12.1.1): $a = 2, b = 1, c = 1, d = -5$. This results in the system

$$\begin{aligned} x' &= 2x + y \\ y' &= x - 5y. \end{aligned} \tag{12.4.1}$$

Now let us choose a couple of phase points to calculate their instantaneous velocities. A phase point at $(1, 0)$ has the instantaneous velocity

$$\mathbf{v}(1, 0) = \begin{bmatrix} 2x + y \\ x - 5y \end{bmatrix} \Bigg|_{(x,y)=(1,0)} = \begin{bmatrix} 2(1) + 0 \\ 1 - 5(0) \end{bmatrix} = \begin{bmatrix} 2 \\ 1 \end{bmatrix}.$$

Or, expressed in terms of the $\mathbf{i}$ and $\mathbf{j}$ vectors, it is

$$\mathbf{v}(1, 0) = 2\mathbf{i} + \mathbf{j}.$$

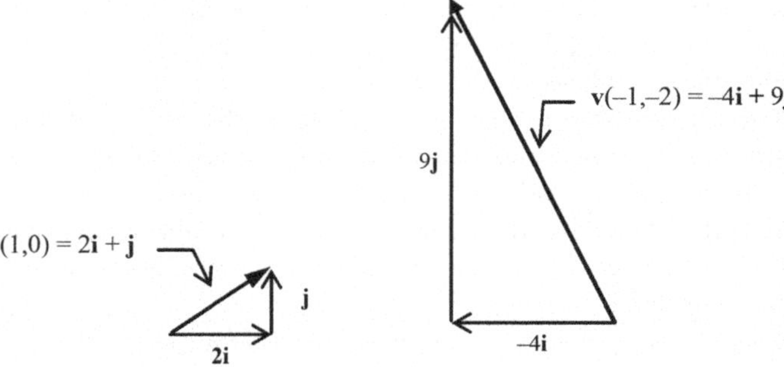

Fig. 12.2 Vectors $\mathbf{v}(1, 0)$ and $\mathbf{v}(-1, 2)$

To draw an arrow representing this vector, we first draw its component vectors, $2\mathbf{i}$ and $\mathbf{j}$, so that the head of one of them coincides with the tail of the other one. If we start with $2\mathbf{i}$, then the arrow shown on the left in Fig. 12.2 drawn from the tail of $2\mathbf{i}$ to the head of $\mathbf{j}$ represents $\mathbf{v}(1, 0)$. Similarly, the velocity of the phase point at $(-1, -2)$ is given by the vector

$$\mathbf{v}(-1, -2) = \begin{bmatrix} 2(-1) + (-2) \\ -1 - 5(-2) \end{bmatrix} = \begin{bmatrix} -4 \\ 9 \end{bmatrix} = -4\mathbf{i} + 9\mathbf{j},$$

which is represented by the arrow on the right in Fig. 12.2 labeled $\mathbf{v}(-1, -2)$.

The real benefit in having such vectors occurs when a sufficient number of them are drawn so as to make the velocity vector field defined by (12.4.1) easy to visualize. An example of this appears in the vector field on the left in Fig. 12.3, which consists of 25 velocity vectors.[7] It indicates how a phase point in the phase plane moves as it is carried by the flow of the vector field. Even so, there is still some uncertainty in exactly how a phase point moves from one location to another. However, if we were to include more vectors in the vector field it would become clearer. This is illustrated in Fig. 12.3 on the right which shows 225 vectors drawn over the same region of the phase plane. Because of these 200 additional vectors, we can more easily perceive how phase points move. Unfortunately there is now the disadvantage in attempting to portray the lengths of the velocity vectors relative to one another: the shorter vectors start looking more like points than arrows. For example, near the x-axis—particularly near the origin—it is nearly impossible to determine which directions the vectors are pointing.

As t increases (or decreases) in value, a phase point traces out a curve in the xy-plane. This curve is called a ***trajectory*** (or ***path*** or ***orbit***) of the system (12.4.1). That is, it consists of the set of points

$$\{(x(t), y(t)) : \ -\infty < t < \infty\}.$$

[7] These were drawn with *Maple*. Actually, the number of vectors that one can actually see is 24. The velocity vector at the origin is the zero vector: it is "invisible" because its length is zero.

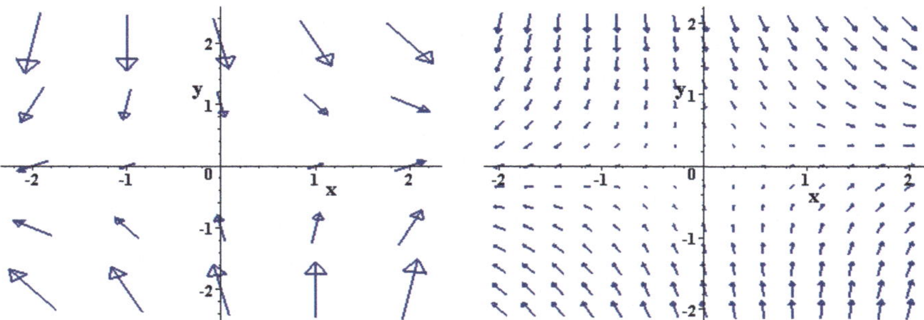

Fig. 12.3 A vector field with 25 vs. 225 vectors

A vector field, such as the one illustrated in Fig. 12.3, portrays the characteristics of the trajectories, one of which are the speeds of the phase points. The speed of a phase point at a given location on a trajectory is the length of the velocity vector there. Another characteristic is a trajectory's shape—whether it be a line, circle, hyperbola, or something more complicated—which is determined by changes in the direction of the phase point as it moves from one location to the next. At a given location in the phase plane, the phase point moves in the direction of the velocity vector as determined by the linear system of differential equations. Recall, from calculus and physics, that a velocity vector at a given point is tangent to the trajectory through that point.

If one's focus is on the curves that phase points trace out and the directions in which they move, then the actual lengths of the velocity vectors are irrelevant. And, as we showed in Fig. 12.3, attempting to show the relative lengths of the vectors may result in a plot that is difficult to interpret in certain regions of the phase plane. So it would be to our advantage to have plots of vector fields that clearly show the slopes and directions of the vectors but not their relative lengths. Instead of vector fields, we will usually use so-called direction fields. A ***direction field*** is simply a vector field that has been modified by replacing all nonzero vectors with arrows of the same length; however, the arrows still have the same orientations as the original vectors. In this way, we get a plot depicting the directions of vectors in the phase plane without the directions being masked by the lengths of the vectors. For example, the direction field corresponding to the vector field with 225 vectors in Fig. 12.3 is shown in Fig. 12.4. Since we can now see all the arrows, it is much easier to visualize the trajectories that the phase points trace out.

If phase points were placed at the points (0.5, 2) and (0.1, −2) in Fig. 12.4, they would trace out the trajectories that are shown on the left in Fig. 12.5. The direction in which a phase point moves is easy to ascertain from the vector field. Even so, it is helpful to a viewer if arrowheads are also added to trajectories to indicate the directions that phase points move as t increases, such as the two trajectories shown in the direction field on the right in Fig. 12.5. These are known as ***directed trajectories***.

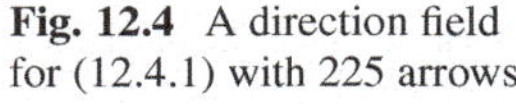

Fig. 12.4 A direction field for (12.4.1) with 225 arrows

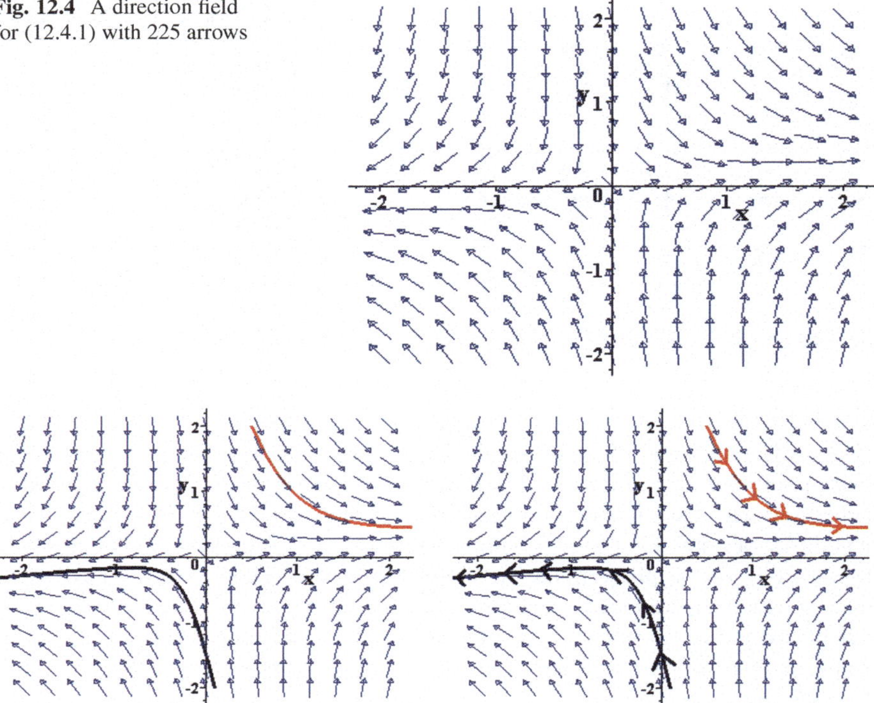

Fig. 12.5 Trajectories of (12.4.1)

12.5 Solutions of Homogeneous Linear Systems

12.5.1 Eigenvalues and Eigenvectors

Let us look at a couple of examples of homogeneous linear systems that will guide us to devise a method for finding solutions of these systems. First, consider

$$\begin{aligned} x' &= -2x \\ y' &= -2y, \end{aligned}$$

which written as a vector differential equation is

$$\mathbf{x}' = A\mathbf{x}, \quad \text{where } A = \begin{bmatrix} -2 & 0 \\ 0 & -2 \end{bmatrix}. \tag{12.5.1}$$

That is,

$$\begin{bmatrix} x' \\ y' \end{bmatrix} = \begin{bmatrix} -2 & 0 \\ 0 & -2 \end{bmatrix} \begin{bmatrix} x \\ y \end{bmatrix}.$$

This system is quite easy to solve because the dependent variables x and y do not appear together in either equation. A system, such as this, in which each of its equations contains one, and only one, dependent variable is said to be ***uncoupled***. In other words, this system consists of two first-order linear equations that have absolutely nothing to do with each other. Their general solutions are

$$x = k_1 e^{-2t} \quad \text{and} \quad y = k_2 e^{-2t},$$

where k_1 and k_2 are arbitrary constants. This means that all solutions of (12.5.1) are of the form

$$\mathbf{x}(t) = \begin{bmatrix} x(t) \\ y(t) \end{bmatrix} = \begin{bmatrix} k_1 e^{-2t} \\ k_2 e^{-2t} \end{bmatrix} = e^{-2t} \begin{bmatrix} k_1 \\ k_2 \end{bmatrix} \tag{12.5.2}$$

or

$$\mathbf{x}(t) = e^{-2t}\mathbf{v},$$

where $\mathbf{v} := \langle k_1, k_2 \rangle$. This resembles the general solution

$$x(t) = e^{-2t} k_1$$

of the scalar equation $x' = -2x$. However, k_1 is a scalar constant whereas $\mathbf{v}$ is a vector constant.

Now consider the following system:

$$\begin{aligned} x' &= -2x + y \\ y' &= x - 2y. \end{aligned} \tag{12.5.3}$$

So we need to figure out a way to solve the corresponding vector differential equation

$$\begin{bmatrix} x' \\ y' \end{bmatrix} = \begin{bmatrix} -2 & 1 \\ 1 & -2 \end{bmatrix} \begin{bmatrix} x \\ y \end{bmatrix}. \tag{12.5.4}$$

As we alluded to earlier in the paragraph following (12.3.14), this is the same thing as solving the homogeneous linear system (12.5.3). This system, however, is more complicated to solve than the previous one because it consists of *coupled* equations. A ***coupled system*** of two differential equations has at least one equation containing both dependent variables. This means that somehow we will have to involve both equations to solve for each dependent variable. This differs from the uncoupled system (12.5.1) in that we were able to solve for each dependent variable by just solving the equation containing it without involving the other.

The solutions (12.5.2) of the uncoupled system (12.5.1) suggest that (12.5.4) may also have solutions of the form

$$\mathbf{x}(t) = e^{\lambda t}\mathbf{v} \tag{12.5.5}$$

where λ denotes a scalar constant and $\mathbf{v}$ a vector constant. However, it is unlikely that the value of λ is still -2 because the 0's along the secondary diagonal of A are now 1's; so the value of λ depends on them as well as on the -2's along the main diagonal. What we are saying is that it is still reasonable to expect that we may be able to find solutions of (12.5.4) of the form

$$\mathbf{x}(t) = e^{\lambda t}\mathbf{v} = e^{\lambda t}\begin{bmatrix} v_1 \\ v_2 \end{bmatrix} = \begin{bmatrix} e^{\lambda t}v_1 \\ e^{\lambda t}v_2 \end{bmatrix},$$

where v_1 and v_2 denote the components of $\mathbf{v}$, but with a different value for λ. Written in terms of the components, we surmise that the substitution of the scalar functions

$$x = e^{\lambda t}v_1 \quad \text{and} \quad y = e^{\lambda t}v_2$$

for the dependent variables x and y will ultimately lead to solutions. Substituting these into (12.5.4), we get

$$\begin{aligned} \lambda e^{\lambda t}v_1 &= -2e^{\lambda t}v_1 + e^{\lambda t}v_2 \\ \lambda e^{\lambda t}v_2 &= e^{\lambda t}v_1 - 2e^{\lambda t}v_2, \end{aligned}$$

which simplifies to

$$\begin{aligned} -2v_1 + v_2 &= \lambda v_1 \\ v_1 - 2v_2 &= \lambda v_2 \end{aligned}$$

after dividing both sides of each equation by $e^{\lambda t}$. In matrix-vector notation, this is

$$\begin{bmatrix} -2 & 1 \\ 1 & -2 \end{bmatrix}\begin{bmatrix} v_1 \\ v_2 \end{bmatrix} = \lambda\begin{bmatrix} v_1 \\ v_2 \end{bmatrix};$$

or

$$A\mathbf{v} = \lambda\mathbf{v},$$

where

$$A = \begin{bmatrix} -2 & 1 \\ 1 & -2 \end{bmatrix} \quad \text{and} \quad \mathbf{v} = \begin{bmatrix} v_1 \\ v_2 \end{bmatrix}.$$

Since it will become important later on, let us point out that by using the identity matrix I defined by (12.3.20), there is another way to write $A\mathbf{v} = \lambda\mathbf{v}$. Since

$$I\mathbf{v} = \begin{bmatrix} 1 & 0 \\ 0 & 1 \end{bmatrix} \begin{bmatrix} v_1 \\ v_2 \end{bmatrix} = \begin{bmatrix} v_1 \\ v_2 \end{bmatrix} = \mathbf{v},$$

we can insert I on the right-hand side of $A\mathbf{v} = \lambda\mathbf{v}$ as follows: $A\mathbf{v} = \lambda I\mathbf{v}$. As a result, we have

$$(A - \lambda I)\mathbf{v} = \mathbf{0}.$$

Note that it is important here to have I after λ in this equation because λI is a matrix, whereas λ is a scalar. In other words, $A - \lambda I$ makes sense but $A - \lambda$ does not.

The previous examples indicate that solutions of the vector differential equation

$$\mathbf{x}' = A\mathbf{x} \tag{12.5.6}$$

are vector functions of the form (12.5.5). Accordingly, let us substitute $e^{\lambda t}\mathbf{v}$ into (12.5.6) to see if this is truly the case. It is a solution if and only if

$$\frac{d}{dt}\left(e^{\lambda t}\mathbf{v}\right) = A\left(e^{\lambda t}\mathbf{v}\right).$$

It is left as an exercise to verify that

$$\frac{d}{dt}\left(e^{\lambda t}\mathbf{v}\right) = \lambda e^{\lambda t}\mathbf{v}$$

and that

$$A\left(e^{\lambda t}\mathbf{v}\right) = e^{\lambda t}A\mathbf{v}.$$

Therefore, $e^{\lambda t}\mathbf{v}$ is a solution if and only if

$$e^{\lambda t}A\mathbf{v} = \lambda e^{\lambda t}\mathbf{v}.$$

We conclude that $\mathbf{x}(t) = e^{\lambda t}\mathbf{v}$ is a solution of (12.5.6) if and only if λ and $\mathbf{v}$ satisfy the vector algebraic equation

$$A\mathbf{v} = \lambda\mathbf{v},$$

which brings us to the next definition.

Eigenvalues and Eigenvectors

Definition 12.5.1 A scalar λ is called an ***eigenvalue*** of a matrix A if there is a nonzero vector $\mathbf{v}$ such that

$$A\mathbf{v} = \lambda\mathbf{v}. \tag{12.5.7}$$

The vector $\mathbf{v}$ is called an ***eigenvector*** of A corresponding to λ.

Eigenvalue is the English translation of the German word *eigenwert*. The famous German mathematician David Hilbert was the first to use the words *eigenwert* and *eigenvektor* in a research paper on linear integral equations that was published in 1904. In English, "eigen" is "own" and "wert" is "value." Another translation of "eigen" is "peculiar" in the sense of being a characteristic of a person or thing. As we will see, the eigenvalues of a matrix are special numbers that characterize the matrix. Depending on the academic discipline, an eigenvalue is also called a *characteristic root*, *characteristic value*, *proper value*, or *latent root*.

Observe that the zero vector $\mathbf{v} = \mathbf{0}$ is always a solution of (12.5.7). Even so, by definition, *the zero vector is not an eigenvector*. Otherwise, as

$$A\mathbf{0} = \lambda\mathbf{0}$$

for every value of $\lambda \in \mathbb{R}$, we would have to regard every scalar as an eigenvalue. Moreover, it is not even important in this regard since it merely gives the solution

$$\mathbf{x}(t) = e^{\lambda t}\mathbf{v} = e^{\lambda t}\mathbf{0} = \mathbf{0}.$$

This solution is called the ***zero solution*** (or ***trivial solution***) of the vector differential equation (12.5.6).

Another important observation about the eigenvectors of a given matrix A is that those corresponding to an eigenvalue λ form something akin to a closed community. By this we mean that if the vectors $\mathbf{u}$ and $\mathbf{v}$ are members of this community, then so is the linear combination $c_1\mathbf{u} + c_2\mathbf{v}$ for every pair of scalars c_1 and c_2. This is because

$$A(c_1\mathbf{u} + c_2\mathbf{v}) = c_1A\mathbf{u} + c_2A\mathbf{v} = c_1\lambda\mathbf{u} + c_2\lambda\mathbf{v} = \lambda(c_1\mathbf{u} + c_2\mathbf{v}).$$

In other words, if $\mathbf{u}$ and $\mathbf{v}$ are eigenvectors corresponding to λ, then so is $c_1\mathbf{u} + c_2\mathbf{v}$. So this community of vectors is shut off from vectors not in the community because it is impossible to leave or enter the community by taking linear combinations of vectors. The mathematical term for such a community (set) of vectors is ***closed set***.

We have just learned that if **u** and **v** are eigenvectors of a matrix A corresponding to an eigenvalue λ, then any linear combination of **u** and **v** is also an eigenvector of A corresponding to λ. Well almost. There is an exception: if we set $c_1 = c_2 = 0$, we end up with **0**, the zero vector, which is not considered an eigenvector as we already stated above. So we have to append **0** to the set to truly make it a closed set. This closed set is a subset of all vectors of the form $\langle x, y\rangle$.

The term ***2-dimensional Euclidean space*** (or simply ***Euclidean 2-space***) is used to designate the entire collection of vectors having two components, namely the set

$$\left\{\begin{bmatrix} x \\ y \end{bmatrix} : x, y \in \mathbb{R}\right\}, \tag{12.5.8}$$

where $\mathbb{R}$ denotes the set of all real numbers. The set (12.5.8) is denoted by $\mathbb{R}^2$. Using the angle brackets notation,

$$\mathbb{R}^2 = \{\langle x, y\rangle : x, y \in \mathbb{R}\}. \tag{12.5.9}$$

By identifying the vector $\langle x, y\rangle$ with the point (x, y) in the Cartesian plane, we can also view $\mathbb{R}^2$ as the entire set of ordered pairs (x, y) of real numbers. For example, the ordered pair (3, 4) not only represents a point in the Cartesian plane but can also be viewed as defining the vector with its tail at the origin (0, 0) and its head at the point (3, 4). This is the geometrical representation of the vector $\langle 3, 4\rangle = 3\mathbf{i} + 4\mathbf{j}$.[8] Thus every point in the Cartesian plane corresponds to a vector—and vice versa. Because of this one-to-one correspondence between vectors and points, we can also define $\mathbb{R}^2$ by

$$\mathbb{R}^2 = \{(x, y) : x, y \in \mathbb{R}\}. \tag{12.5.10}$$

In short, we can view $\mathbb{R}^2$ as the set of all vectors with two components, namely, (12.5.8) (or (12.5.9))—or we can view it as the set of all points in the Cartesian plane, namely, (12.5.10). Thus, *Euclidean 2-space* is really the same as the *Cartesian plane*.

From now on, the symbol $\mathbb{R}^2$ refers not only to the entire set of vectors

$$\begin{bmatrix} x \\ y \end{bmatrix} = x\mathbf{i} + y\mathbf{j} \qquad (x, y \in \mathbb{R})$$

but also to the operations of addition and scalar multiplication of vectors that were defined earlier in Sect. 12.3. Since a vector $\langle x, y\rangle$ can be depicted in the

[8] Keep in mind that vectors are regarded as ***equivalent*** (or ***equal***) if they have the same length and the same direction even though they may occupy different positions in the Cartesian plane. For instance, the vector with its tail at the point (0, -1) and its head at the point (3, 3) is equal to the vector with its tail at (0, 0) and its head at (3, 4).

Cartesian plane (2-dimensional Euclidean space), it is said to be a *two-dimensional vector*; and since the Cartesian plane can be considered as a depiction of all two-dimensional vectors, it is said to be a two-dimensional ***vector space***.[9] The vector space $\mathbb{R}^2$ is clearly closed since linear combinations of two-dimensional vectors are two-dimensional vectors themselves and not something else. Since we are now employing the word *space*, it makes sense to say that the closed subset of all eigenvectors corresponding to a given eigenvalue λ together with the zero vector is a subspace of the vector space $\mathbb{R}^2$. This particular subspace is known as the ***eigenspace*** of A corresponding to λ.

Eigenspace

Definition 12.5.2 If λ is an eigenvalue of a matrix A, then the set of vectors

$$E(\lambda) := \left\{\mathbf{v} \in \mathbb{R}^2 \colon A\mathbf{v} = \lambda\mathbf{v}\right\}$$

is known as the ***eigenspace*** of A corresponding to λ. In other words, $E(\lambda)$ consists of all the eigenvectors of A corresponding to λ together with the zero vector.

For a given matrix A, let us explain how to compute its eigenvalues. Employing a trick involving the identity matrix, we can rewrite (12.5.7) as

$$(A - \lambda I)\mathbf{v} = \mathbf{0} \tag{12.5.11}$$

because

$$A\mathbf{v} = \lambda\mathbf{v} \;\Leftrightarrow\; A\mathbf{v} - \lambda\mathbf{v} = \mathbf{0} \;\Leftrightarrow\; A\mathbf{v} - \lambda I\mathbf{v} = \mathbf{0} \;\Leftrightarrow\; (A - \lambda I)\mathbf{v} = \mathbf{0}.$$

The trick is to insert the identity matrix I between λ and $\mathbf{v}$ before factoring since the difference $A - \lambda$ is not defined. From this we see that (12.5.7) and (12.5.11) are equivalent equations. However, it is (12.5.11) that we need to find the eigenvalues and eigenvectors of A and that is because of a well-known result of matrix algebra. This result involves the so-called determinant of a matrix. Before we can state the result, we will first have to define what is meant by the determinant of a matrix.

[9] $\mathbb{R}^2$ is only one kind of vector space. A set of elements—be they the classical vectors $\mathbb{R}^2$ and $\mathbb{R}^3$ of physics or something altogether different, such as a set of functions in some other kind of mathematical setting—together with the operations of addition and scalar multiplication that are defined so as to meet a certain set of prescribed criteria is called a ***vector space***. We leave the precise definition of a vector space to a more advanced course, such as a linear (or matrix) algebra course.

Determinant of a 2×2 **Matrix**

Definition 12.5.3 The ***determinant*** of $A = \begin{bmatrix} a & b \\ c & d \end{bmatrix}$ is the number $ad - bc$. It is denoted by either $\det A$ or $|A|$. That is,

$$\det A = \det \begin{bmatrix} a & b \\ c & d \end{bmatrix} = ad - bc$$

or

$$|A| = \begin{vmatrix} a & b \\ c & d \end{vmatrix} = ad - bc.$$

Example 12.5.1 Find the determinant of the matrix

$$A = \begin{bmatrix} -2 & 3 \\ -4 & 5 \end{bmatrix}.$$

Solution

$$\det A = \begin{vmatrix} -2 & 3 \\ -4 & 5 \end{vmatrix} = (-2)(5) - (3)(-4) = 2.$$ ♦

An important property of the determinant of the product of two square matrices of the same order is stated in the next theorem. However, since a general proof would require more knowledge of matrix algebra than can be covered here, we prove it only for square matrices of order 2.

Multiplicative Property

Theorem 12.5.4 *For two square matrices A and B of the same order,*

$$\det(AB) = \det A \cdot \det B. \tag{12.5.12}$$

Proof The determinant of AB, where

$$A = \begin{bmatrix} a & b \\ c & d \end{bmatrix} \quad \text{and} \quad B = \begin{bmatrix} e & f \\ g & h \end{bmatrix},$$

is

$$\det(AB) = \begin{vmatrix} ae+bg & af+bh \\ ce+dg & cf+dh \end{vmatrix} = (ae+bg)(cf+dh) - (af+bh)(ce+dg)$$
$$= adeh - adfg - bceh + bcfg.$$

The right-hand side of (12.5.12) is

$$\det A \cdot \det B = \begin{vmatrix} a & b \\ c & d \end{vmatrix} \begin{vmatrix} e & f \\ g & h \end{vmatrix} = (ad - bc)(eh - fg)$$
$$= adeh - adfg - bceh + bcfg.$$

Comparing the two results, we see that $\det(AB) = \det A \cdot \det B$. ■

The next theorem is the result that we have been working toward for the express purpose of computing eigenvalues. Even though it is stated for square matrices of any order, the proof given here is restricted to matrices of order 2.

Solutions of $A\mathbf{v} = \mathbf{0}$

Theorem 12.5.5 *The equation*

$$A\mathbf{v} = \mathbf{0} \tag{12.5.13}$$

has the unique solution $\mathbf{v} = \mathbf{0}$ *if and only if* $\det A \neq 0$. *It has infinitely many solutions if and only if* $\det A = 0$.

Proof Let

$$A = \begin{bmatrix} a & b \\ c & d \end{bmatrix}$$

be a given 2×2 matrix. Let v_1 and v_2 denote the components of a vector $\mathbf{v}$. Writing out (12.5.13), we have

$$\begin{bmatrix} a & b \\ c & d \end{bmatrix} \begin{bmatrix} v_1 \\ v_2 \end{bmatrix} = \begin{bmatrix} 0 \\ 0 \end{bmatrix},$$

which after matrix multiplication becomes

$$\begin{bmatrix} av_1 + bv_2 \\ cv_1 + dv_2 \end{bmatrix} = \begin{bmatrix} 0 \\ 0 \end{bmatrix}.$$

Thus we are looking for the solutions of the linear system of equations

$$\begin{aligned} av_1 + bv_2 &= 0 \\ cv_1 + dv_2 &= 0. \end{aligned} \tag{12.5.14}$$

Clearly the pair $v_1 = 0$ and $v_2 = 0$ is always a solution of (12.5.14). In other words, the zero vector, namely $\mathbf{0} = \langle 0, 0\rangle$, is always a solution of (12.5.13).

Suppose $\det A \neq 0$. Multiplying the first equation of (12.5.14) by d and the second one by $-b$ and adding the resulting equations, we obtain $\det A \cdot v_1 = 0$. Similarly, $\det A \cdot v_2 = 0$. It follows that $v_1 = v_2 = 0$. In short, this shows that if $\det A \neq 0$, then $\mathbf{v} = \mathbf{0}$ is the only solution of (12.5.13). The contrapositive of this is: If (12.5.13) has a nonzero solution, then $\det A = 0$.

Now suppose $\det A = 0$. If both a and b are equal to zero, then (12.5.14) clearly has infinitely many solutions. Likewise, (12.5.14) has infinitely many solutions if both c and d are equal to zero. That leaves the case: at least one coefficient in the first equation and at least one coefficient in the second equation are nonzero. For instance, suppose both b and d are nonzero. Then as $\det A = 0$, we have

$$\frac{a}{b} = \frac{c}{d}.$$

So, for every $s \in \mathbb{R}$,

$$v_1 = s \quad \text{and} \quad v_2 = -\frac{a}{b}s$$

is a solution of (12.5.14). From this we conclude that there are infinitely many solutions of (12.5.14) if $\det A = 0$. Moreover, the contrapositive of this implies that if $\mathbf{v} = \mathbf{0}$ is the only solution of (12.5.14), then $\det A \neq 0$. ■

A square matrix A is said to be ***nonsingular*** if $\det A \neq 0$. If $\det A = 0$, then A is said to be a ***singular*** matrix. Theorem 12.5.5 expressed in these terms states that $A\mathbf{v} = \mathbf{0}$ has infinitely many solutions when A is singular but only the zero solution when A is nonsingular.

Example 12.5.2 Use Theorem 12.5.5 to find the eigenspaces of the matrix

$$A = \begin{bmatrix} -2 & 1 \\ 1 & -2 \end{bmatrix},$$

which is the coefficient matrix of the vector differential equation (12.5.4). Also, find the solutions of (12.5.4) corresponding to these eigenspaces.

Solution In view of Definition 12.5.2, we begin with $A\mathbf{v} = \lambda\mathbf{v}$. Or, as this is equivalent to $(A - \lambda I)\mathbf{v} = \mathbf{0}$, we have

$$\begin{bmatrix} -2-\lambda & 1 \\ 1 & -2-\lambda \end{bmatrix} \begin{bmatrix} v_1 \\ v_2 \end{bmatrix} = \begin{bmatrix} 0 \\ 0 \end{bmatrix}. \tag{12.5.15}$$

According to Theorem 12.5.5, this matrix equation has nonzero solutions if and only if

$$\det(A - \lambda I) = \begin{vmatrix} -2-\lambda & 1 \\ 1 & -2-\lambda \end{vmatrix} = 0.$$

Hence,

$$(-2-\lambda)(-2-\lambda) - 1 = 0 \quad \Rightarrow \quad (\lambda + 2)^2 - 1 = 0. \tag{12.5.16}$$

Consequently,

$$\lambda + 2 = \pm 1.$$

Thus, the eigenvalues of A are $\lambda_1 = -1$ and $\lambda_2 = -3$.

Now let us compute the eigenvectors corresponding to these eigenvalues. Substituting $\lambda_1 = -1$ for λ in (12.5.15), we obtain

$$\begin{bmatrix} -1 & 1 \\ 1 & -1 \end{bmatrix} \begin{bmatrix} v_1 \\ v_2 \end{bmatrix} = \begin{bmatrix} 0 \\ 0 \end{bmatrix},$$

from which we have

$$-v_1 + v_2 = 0 \quad \text{and} \quad v_1 - v_2 = 0.$$

Since these two equations are redundant, the only condition placed on v_1 and v_2 is that they be equal. As a result, we may regard either one of them as a ***free variable***. Choosing v_1 to be the free variable, we have $v_1 = c_1$, where c_1 denotes a real number. This gives the solution $\langle v_1, v_2 \rangle = \langle c_1, c_1 \rangle$. However, we have to exclude the value $c_1 = 0$ since the zero vector is not regarded as an eigenvector. Therefore, the eigenvectors corresponding to the eigenvalue $\lambda_1 = -1$ consist of all vectors of the form

$$\mathbf{v} = \begin{bmatrix} v_1 \\ v_2 \end{bmatrix} = \begin{bmatrix} c_1 \\ c_1 \end{bmatrix} = c_1 \begin{bmatrix} 1 \\ 1 \end{bmatrix}, \tag{12.5.17}$$

where $c_1 \in \mathbb{R}$ is nonzero.

The eigenspace of A corresponding to $\lambda_1 = -1$ is obtained by appending the zero vector to (12.5.17). In other words, include $c_1 = 0$. Denoting this eigenspace by $E(\lambda_1)$ or $E(-1)$, it is the set of vectors

$$E(-1) = \left\{ c_1 \begin{bmatrix} 1 \\ 1 \end{bmatrix} : c_1 \in \mathbb{R} \right\}.$$

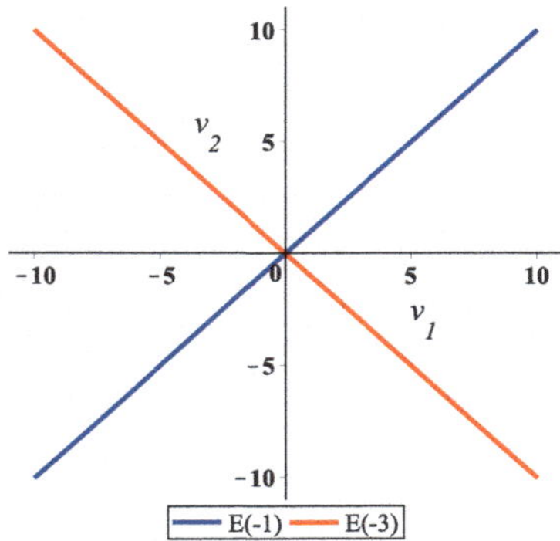

Fig. 12.6 Eigenspaces $E(-1)$ and $E(-3)$

Since $v_1 = c_1$ and $v_2 = c_2$, its graph is the line $v_2 = v_1$ in Fig. 12.6. From (12.5.5), it follows that the solutions of (12.5.4) corresponding to $E(-1)$ are

$$\mathbf{x}(t) = e^{\lambda_1 t}\mathbf{v} = c_1 \begin{bmatrix} 1 \\ 1 \end{bmatrix} e^{-t}. \tag{12.5.18}$$

Thus, the components of $\mathbf{x}(t)$ are $x(t) = c_1 e^{-t}$ and $y(t) = c_1 e^{-t}$.

The other eigenspace of A is found by substituting $\lambda_2 = -3$ for λ in (12.5.15):

$$\begin{bmatrix} 1 & 1 \\ 1 & 1 \end{bmatrix} \begin{bmatrix} v_1 \\ v_2 \end{bmatrix} = \begin{bmatrix} 0 \\ 0 \end{bmatrix}.$$

Thus,

$$v_1 + v_2 = 0 \quad \Rightarrow \quad v_2 = -v_1.$$

Let $v_1 = c_2$. Then $v_2 = -c_2$. And so $\langle v_1, v_2 \rangle = \langle c_2, -c_2 \rangle$. Hence, the eigenspace $E(\lambda_2)$ corresponding to the eigenvalue $\lambda_2 = -3$ consists of the set of vectors

$$E(-3) = \left\{ c_2 \begin{bmatrix} 1 \\ -1 \end{bmatrix} : c_2 \in \mathbb{R} \right\}.$$

Its graph is the line $v_2 = -v_1$ in Fig. 12.6. The corresponding solutions are

$$\mathbf{x}(t) = e^{\lambda_2 t}\mathbf{v} = c_2 \begin{bmatrix} 1 \\ -1 \end{bmatrix} e^{-3t}. \tag{12.5.19}$$

So $x(t) = c_2 e^{-3t}$ and $y(t) = -c_2 e^{-3t}$. ♦

Theorem 12.5.5 tells us that the matrix equation

$$A\mathbf{v} = \mathbf{b} \tag{12.5.20}$$

has the unique solution $\mathbf{v} = \mathbf{0}$ when the matrix A is nonsingular and $\mathbf{b} = \mathbf{0}$. So does it still have a unique solution when $\mathbf{b}$ is a nonzero vector? And if it does, what is it? To answer this, let us start by supposing that a square matrix X exists such that $XA = I$, where I is the identity matrix. Then we could easily obtain the solution of (12.5.20) by merely left multiplying it by the matrix X as follows:

$$A\mathbf{v} = \mathbf{b} \quad \Rightarrow \quad XA\mathbf{v} = X\mathbf{b} \quad \Rightarrow \quad I\mathbf{v} = X\mathbf{b} \quad \Rightarrow \quad \mathbf{v} = X\mathbf{b}.$$

For this to be the case, A must be nonsingular, which we prove next.

Lemma 12.5.6 *If A and X are matrices of the same order and $XA = I$, then* $\det A \neq 0$.

Proof Since $XA = I$, it follows from Theorem 12.5.4 that

$$\det X \det A = \det I.$$

And since $\det I = 1$, $\det X \det A = 1$. Thus, $\det A \neq 0$. ■

Question In view of Lemma 12.5.6, there is the matter of its converse. If $\det A \neq 0$, is there a square matrix X such that $XA = I$?

Answer In order to answer this, let

$$A = \begin{bmatrix} a & b \\ c & d \end{bmatrix}$$

be a given matrix with $\det A \neq 0$. Now let us ascertain if the matrix equation $XA = I$ has a solution. In other words, for given entries a, b, c, d of A, are there values for the entries w, x, y, z of X such that

$$\begin{bmatrix} w & x \\ y & z \end{bmatrix} \begin{bmatrix} a & b \\ c & d \end{bmatrix} = \begin{bmatrix} 1 & 0 \\ 0 & 1 \end{bmatrix}?$$

Multiplying the matrices on the left-hand side, we have

$$\begin{bmatrix} aw + cx & bw + dx \\ ay + cz & by + dz \end{bmatrix} = \begin{bmatrix} 1 & 0 \\ 0 & 1 \end{bmatrix}.$$

Equating corresponding entries, we obtain two systems of linear equations, namely,

$$\begin{aligned} aw + cx &= 1 \\ bw + dx &= 0 \end{aligned} \quad \text{and} \quad \begin{aligned} ay + cz &= 0 \\ by + dz &= 1 \end{aligned} \tag{12.5.21}$$

for which we seek solutions. Multiplying the top equation of the system on the left by d and the bottom one by $-c$ and adding the results, we get

$$(ad - bc)w = d \quad \text{or} \quad \det A \cdot w = d.$$

Since $\det A \neq 0$,

$$w = \frac{d}{\det A}.$$

And so,

$$x = -\frac{b}{d}w = -\frac{b}{\det A}.$$

In other words, the system on the left has the unique solution

$$w = \frac{d}{\det A}, \qquad x = -\frac{b}{\det A}.$$

Similarly, the system on the right of (12.5.21) has the unique solution

$$y = -\frac{c}{\det A}, \qquad z = \frac{a}{\det A}.$$

We conclude that $XA = I$ has the unique solution $X = B$, where

$$B := \begin{bmatrix} \dfrac{d}{\det A} & -\dfrac{b}{\det A} \\ -\dfrac{c}{\det A} & \dfrac{a}{\det A} \end{bmatrix} = \frac{1}{\det A}\begin{bmatrix} d & -b \\ -c & a \end{bmatrix}.$$

Although $BA = I$, we cannot jump to the conclusion that $AB = I$ since matrix multiplication is not in general commutative. However, for this particular matrix B it is true that $AB = I$, which we see from

$$AB = \begin{bmatrix} a & b \\ c & d \end{bmatrix} \cdot \frac{1}{\det A}\begin{bmatrix} d & -b \\ -c & a \end{bmatrix} = \frac{1}{\det A}\begin{bmatrix} ad - bc & 0 \\ 0 & -bc + ad \end{bmatrix}$$

$$= \frac{1}{\det A}\begin{bmatrix} \det A & 0 \\ 0 & \det A \end{bmatrix} = \begin{bmatrix} 1 & 0 \\ 0 & 1 \end{bmatrix}.$$

Thus, we have $AB = I$ as well as $BA = I$.

We determined that $XA = I$ has the unique solution $X = B$. Also, we just verified that $AB = I$. Let us rule out the existence of another matrix C with $AC = I$. If that were the case, then we would have $AC = I = BA$. By the associative law of matrix multiplication,

$$B(AC) = (BA)C.$$

Since $B(AC) = BI = B$ and $(BA)C = IC = C$, it follows that $C = B$. ♦

Lemma 12.5.6 and the results obtained in answering the question following the proof of the lemma bring us to the next definition and theorem.

Invertible Matrix

Definition 12.5.7 A square matrix A is said to be ***invertible*** if there exists a matrix B of the same order as A such that $AB = BA = I$. The matrix B is called the ***inverse*** of A and is denoted by A^{-1}.

Remark The foregoing analysis shows that the inverse of an invertible 2×2 matrix is unique. That is, if a matrix A is invertible, then there is one, and only one, matrix A^{-1} such that $AA^{-1} = A^{-1}A = I$. This result, as do many of the results in this section, also applies to $n \times n$ matrices.

Inverse of a 2×2 Matrix

Theorem 12.5.8 *Let*

$$A = \begin{bmatrix} a & b \\ c & d \end{bmatrix} \tag{12.5.22}$$

be a given matrix. If $\det A \neq 0$*, then* A *is invertible. Moreover,*

$$A^{-1} = \frac{1}{\det A} \begin{bmatrix} d & -b \\ -c & a \end{bmatrix}. \tag{12.5.23}$$

Example 12.5.3 Is the matrix

$$A = \begin{bmatrix} 4 & 5 \\ -1 & -2 \end{bmatrix}$$

invertible? If so, find its inverse.

Solution The determinant of A is

$$\det A = \begin{vmatrix} 4 & 5 \\ -1 & -2 \end{vmatrix} = 4(-2) - 5(-1) = -3.$$

Since $\det A \neq 0$, Theorem 12.5.8 tells us that it is invertible. Furthermore, its inverse is

$$A^{-1} = \frac{1}{-3}\begin{bmatrix} -2 & -5 \\ 1 & 4 \end{bmatrix} = \begin{bmatrix} \frac{2}{3} & \frac{5}{3} \\ -\frac{1}{3} & -\frac{4}{3} \end{bmatrix}.$$ ♦

Recall that a matrix is said to be *nonsingular* if its determinant is nonzero. It then follows from Lemma 12.5.6 and Theorem 12.5.8 that nonsingular matrices are invertible, and conversely. Thus we have the following statement.

Corollary 12.5.9 *A square matrix is invertible if and only if it is nonsingular.*

Before we continue with our discussion of eigenvalues and eigenvectors, here is another important result about invertible matrices that will be useful later on.

Inverse of a Product

Theorem 12.5.10 *Let A and B be invertible matrices. Then AB is invertible and*

$$(AB)^{-1} = B^{-1}A^{-1}. \tag{12.5.24}$$

Proof Since A and B are invertible, matrices A^{-1} and B^{-1} exist such that

$$AA^{-1} = A^{-1}A = I \quad \text{and} \quad BB^{-1} = B^{-1}B = I.$$

Applying the associative law of matrix multiplication a couple of times, we have

$$(AB)(B^{-1}A^{-1}) = [(AB)B^{-1}]A^{-1} = [A(BB^{-1})]A^{-1} = (AI)A^{-1} = AA^{-1} = I.$$

Similar reasoning yields $(B^{-1}A^{-1})(AB) = I$. Consequently,

$$(AB)(B^{-1}A^{-1}) = (B^{-1}A^{-1})(AB) = I.$$

This last result shows AB is invertible and $B^{-1}A^{-1}$ is its inverse. ■

In Example 12.5.2 we found the eigenvalues of the coefficient matrix A in (12.5.4) by solving the equation $\det(A - \lambda I) = 0$. We see from (12.5.16) that $\det(A - \lambda I)$ is a polynomial in the variable λ, namely,

$$\det(A - \lambda I) = (\lambda + 2)^2 - 1 = \lambda^2 + 4\lambda + 3.$$

Such polynomials are known as *characteristic polynomials*. Note that the characteristic polynomial of a 2×2 coefficient matrix is a quadratic polynomial (i.e., a polynomial of degree 2).

Characteristic Polynomial

Definition 12.5.11 The ***characteristic polynomial*** of a matrix A is

$$p(\lambda) = \det(A - \lambda I). \tag{12.5.25}$$

Recall that the roots of the quadratic equation

$$(\lambda + 2)^2 - 1 = 0$$

are the eigenvalues that we used to find solutions of the vector differential equation (12.5.4), namely, the solutions (12.5.18) and (12.5.19). In the context of calculating eigenvalues, an equation such as this one is known as a *characteristic equation*. For 2×2 matrices, the characteristic equation is a quadratic equation. Since eigenvalues are part of the recipe for finding the solutions of vector differential equations, let us record how to calculate them with the next theorem.

Characteristic Equation and Eigenvalues of a Matrix

Theorem 12.5.12 *The eigenvalues of a matrix A are the roots of the equation*

$$\det(A - \lambda I) = 0, \tag{12.5.26}$$

which is called the ***characteristic equation*** *of A.*

Example 12.5.4 Find the characteristic polynomial and the eigenvalues of the matrix

$$A = \begin{bmatrix} 4 & 5 \\ -1 & -2 \end{bmatrix}.$$

Solution Since

$$A - \lambda I = \begin{bmatrix} 4 & 5 \\ -1 & -2 \end{bmatrix} - \lambda \begin{bmatrix} 1 & 0 \\ 0 & 1 \end{bmatrix} = \begin{bmatrix} 4-\lambda & 5 \\ -1 & -2-\lambda \end{bmatrix},$$

the characteristic polynomial is

$$\det(A - \lambda I) = (4-\lambda)(-2-\lambda) + 5 = (\lambda - 4)(\lambda + 2) + 5 = \lambda^2 - 2\lambda - 3.$$

The eigenvalues are obtained by solving the characteristic equation $\det(A - \lambda I) = 0$, that is,

$$\lambda^2 - 2\lambda - 3 = 0 \quad \Rightarrow \quad (\lambda + 1)(\lambda - 3) = 0.$$

Therefore, the eigenvalues are $\lambda_1 = -1$ and $\lambda_2 = 3$. ♦

12.5.2 Superposition and Autonomous Principles

By setting $c_1 = 1$ and $c_2 = 1$ in (12.5.18) and (12.5.19), we obtain the following two particular solutions of the vector differential equation (12.5.4):

$$\mathbf{x}^1(t) = \begin{bmatrix} 1 \\ 1 \end{bmatrix} e^{-t} = \begin{bmatrix} e^{-t} \\ e^{-t} \end{bmatrix} \quad \text{and} \quad \mathbf{x}^2(t) = \begin{bmatrix} 1 \\ -1 \end{bmatrix} e^{-3t} = \begin{bmatrix} e^{-3t} \\ -e^{-3t} \end{bmatrix}. \tag{12.5.27}$$

It is left as an exercise to show that any ***linear combination*** of $\mathbf{x}^1(t)$ and $\mathbf{x}^2(t)$ is also a solution of (12.5.4). In other words,

$$c_1\mathbf{x}^1(t) + c_2\mathbf{x}^2(t) = c_1 \begin{bmatrix} 1 \\ 1 \end{bmatrix} e^{-t} + c_2 \begin{bmatrix} 1 \\ -1 \end{bmatrix} e^{-3t} \tag{12.5.28}$$

is a solution for all values of c_1 and c_2. In fact, this is true of all homogeneous linear vector differential equations: every linear combination of solutions of a given homogeneous linear vector differential equation is also a solution of the equation. This result is known as the ***Superposition Principle***.

Superposition Principle

Theorem 12.5.13 *If* $\mathbf{x}^1(t)$ *and* $\mathbf{x}^2(t)$ *are solutions of* $\mathbf{x}' = A\mathbf{x}$ *on a common interval* J*, then so is* $c_1\mathbf{x}^1(t) + c_2\mathbf{x}^2(t)$ *for all* $c_1, c_2 \in \mathbb{R}$.

Proof By hypothesis, $\mathbf{x}^1(t)$ and $\mathbf{x}^2(t)$ are solutions on an interval J. That is,

$$\frac{d}{dt}\mathbf{x}^1(t) = A\mathbf{x}^1(t) \quad \text{and} \quad \frac{d}{dt}\mathbf{x}^2(t) = A\mathbf{x}^2(t)$$

for $t \in J$. Consequently,

$$\begin{aligned}\frac{d}{dt}\left[c_1\mathbf{x}^1(t) + c_2\mathbf{x}^2(t)\right] &= c_1\frac{d}{dt}\mathbf{x}^1(t) + c_2\frac{d}{dt}\mathbf{x}^2(t)\\ &= c_1A\mathbf{x}^1(t) + c_2A\mathbf{x}^2(t) = A\left[c_1\mathbf{x}^1(t) + c_2\mathbf{x}^2(t)\right]\end{aligned}$$

for $t \in J$. ■

The superposition principle says that if we have a solution of $\mathbf{x}' = A\mathbf{x}$, then we can generate more solutions by merely multiplying this solution by scalar constants. Furthermore, if we have a second solution, then we may be able to generate even more by taking various linear combinations of the two solutions. So that brings up the question of the existence of solutions in the first place. Does the homogeneous linear system $\mathbf{x}' = A\mathbf{x}$ always have solutions? Of course, the zero vector function ($\mathbf{x}(t) \equiv \mathbf{0}$) is always a solution. So the real question is: Does $\mathbf{x}' = A\mathbf{x}$ always have nontrivial solutions? The answer, as we see in the next theorem, is yes.

Existence and Uniqueness of Solutions

Theorem 12.5.14 *Let A be a constant matrix. For $t_0 \in \mathbb{R}$ and a constant vector $\mathbf{x}_0$, the vector differential equation*

$$\mathbf{x}' = A\mathbf{x}$$

has a solution with the value $\mathbf{x}_0$ at $t = t_0$. Moreover, there is only one such solution and it satisfies the differential equation for $-\infty < t < \infty$.

Remarks This theorem can be proved with Picard's method of successive approximations, much in the same way that Theorem 9.3.2 was proved in Chap. 9. However, we leave the details of the proof to more advanced books on differential equations. For instance, Burton [18, pp. 25–27] uses Picard's method to prove that the theorem not only applies to $\mathbf{x}' = A\mathbf{x}$ but also to $\mathbf{x}' = A(t)\mathbf{x}$ when $A(t)$ is a matrix-valued function that is continuous and bounded on $\mathbb{R}$.

The uniqueness part implies that two solutions are identical if they are equal at some t. That is, if $\mathbf{x}(t)$ and $\overline{\mathbf{x}}(t)$ are solutions and if $\mathbf{x}(t_0) = \overline{\mathbf{x}}(t_0)$ at some $t = t_0$, then

$$\mathbf{x}(t) \equiv \overline{\mathbf{x}}(t)$$

on $(-\infty, \infty)$. A special case of this (which we will refer to shortly) is that if $\mathbf{x}(t)$ is a solution which is equal to $\mathbf{0}$ at some $t = t_0$, then $\mathbf{x}(t) = \mathbf{0}$ for all $-\infty < t < \infty$. That is because $\mathbf{x}(t)$ and the zero solution $\overline{\mathbf{x}}(t) \equiv \mathbf{0}$ both satisfy the differential equation and are equal to each other at $t = t_0$. It also follows that if $\mathbf{x}(t)$ is a nontrivial solution, then $\mathbf{x}(t) \neq \mathbf{0}$ for all $-\infty < t < \infty$.

Neither of the right-hand sides of the pair of equations making up the homogeneous linear system (12.1.1) depends explicitly on the independent variable t. Such a system is said to be ***autonomous***. Now compare this system to the nonhomogeneous linear system (12.1.2) in Sect. 12.1. It is said to be ***nonautonomous*** because the right-hand sides of the equations composing this system do depend explicitly on t. In fact, a system is considered nonautonomous if the right-hand side of at least one of the equations depends explicitly on the independent variable.

Autonomous equations have the property that any "time translation" of a solution is also a solution. What precisely is meant by this is stated more precisely in the next theorem known as the *Autonomous Principle*.

Autonomous Principle

Theorem 12.5.15 *Let A be a constant matrix, $t_0 \in \mathbb{R}$, and $\mathbf{x}_0$ a constant vector. If $\boldsymbol{\varphi}(t)$ is the unique solution of*

$$\frac{d\mathbf{x}}{dt} = A\mathbf{x}, \quad \mathbf{x}(0) = \mathbf{x}_0, \tag{12.5.29}$$

then $\boldsymbol{\varphi}(t - t_0)$ is the unique solution of

$$\frac{d\mathbf{x}}{dt} = A\mathbf{x}, \quad \mathbf{x}(t_0) = \mathbf{x}_0. \tag{12.5.30}$$

Proof Let $u := t - t_0$. By the chain rule, the derivative of $\boldsymbol{\varphi}(t - t_0)$ with respect to t is

$$\frac{d}{dt}\boldsymbol{\varphi}(t - t_0) = \frac{d}{dt}\boldsymbol{\varphi}(u) = \frac{d}{du}\boldsymbol{\varphi}(u) \cdot \frac{du}{dt} = A\boldsymbol{\varphi}(u) \cdot \frac{d}{dt}(t - t_0) = A\boldsymbol{\varphi}(t - t_0).$$

Thus, $\boldsymbol{\varphi}(t - t_0)$ is a solution of the differential equation $\mathbf{x}' = A\mathbf{x}$. Furthermore, it is the solution of the initial value problem (12.5.30) since its value at $t = t_0$ is

$$\boldsymbol{\varphi}(t_0 - t_0) = \boldsymbol{\varphi}(0) = \mathbf{x}_0.$$

According to Theorem 12.5.14, it is the only solution. ■

Example 12.5.5 Find the solution of the homogeneous linear system

$$
\begin{aligned}
x' &= x \\
y' &= 4x - 3y
\end{aligned}
\tag{12.5.31}
$$

satisfying the initial conditions: $x(0) = 1$, $y(0) = 0$. What is the solution of the system if the initial conditions are: $x(5) = 1$, $y(5) = 0$?

Solution Since the system is partially uncoupled (the first equation does not involve the dependent variable y), there is no need to deal with eigenvalues and eigenvectors. Clearly, $\varphi(t) = e^t$ is the solution of the initial value problem

$$x' = x, \quad x(0) = 1.$$

Substituting $\varphi(t)$ for x in the second equation, we obtain the initial value problem

$$y' + 3y = 4e^t, \quad y(0) = 0.$$

It is left as an exercise to verify that the solution of this is $\psi(t) = e^t - e^{-3t}$. Therefore, the solution of system (12.5.31) with $x(0) = 1$ and $y(0) = 0$ is

$$x = \varphi(t) = e^t, \quad y = \psi(t) = e^t - e^{-3t}.$$

Then it follows from the Autonomous Principle that the solution of (12.5.31) such that $x(5) = 1$ and $y(5) = 0$ is

$$x = \varphi(t-5) = e^{t-5}, \quad y = \psi(t-5) = e^{t-5} - e^{-3(t-5)}. \qquad \blacklozenge$$

12.5.3 Linear Independence and General Solutions

Suppose $\mathbf{x}^1(t)$ and $\mathbf{x}^2(t)$ denote two solutions of $\mathbf{x}' = A\mathbf{x}$ that are not proportional to each other; that is, neither is a scalar multiple of the other. In other words, suppose it is not the case that $\mathbf{x}^1(t) = k_1\mathbf{x}^2(t)$ or that $\mathbf{x}^2(t) = k_2\mathbf{x}^1(t)$ for some $k_1, k_2 \in \mathbb{R}$. Now let c_1 and c_2 be a pair of scalars, both of which are nonzero. By the Superposition Principle, the linear combination $c_1\mathbf{x}^1(t) + c_2\mathbf{x}^2(t)$ is also solution. Obviously this solution is neither proportional to $\mathbf{x}^1(t)$ nor to $\mathbf{x}^2(t)$. From this it becomes clear that more solutions can be generated with two solutions that are not proportional to each other than with two that are. The terminology that is normally used to state that two solutions are not proportional to each other is to say that they are ***linearly independent***. On the other hand, if one of the solutions is proportional to the other, then they are said to be ***linearly dependent***. Since more solutions can be generated with the former than with the latter, we need to focus our attention on linearly independent solutions rather than on linearly dependent ones. Naturally the question arises as to whether there are solutions that cannot be expressed as linear combinations of two linearly independent solutions. Before that can be answered, we need to know more about linear independence vis-à-vis linear dependence. First let us clearly differentiate between vector functions that are linearly dependent and those that are linearly independent, irrespective of whether or not they are solutions.

Linearly Dependent Functions

Definition 12.5.16 Two vector functions $\mathbf{u}(t)$ and $\mathbf{v}(t)$ that are defined on an interval J are said to be ***linearly dependent*** on J if one of them is a scalar multiple of the other. An equivalent statement is that they are ***linearly dependent*** on J if there are constants $c_1, c_2 \in \mathbb{R}$, not both zero, such that $c_1\mathbf{u}(t) + c_2\mathbf{v}(t) = \mathbf{0}$ for all $t \in J$.

The terms *linear independence* and *linear dependence* are antonyms in the sense that two functions are either linearly dependent or linearly independent; so if we negate the statements in Definition 12.5.16, we obtain the following statements defining linear independence.

Linearly Independent Functions

Definition 12.5.17 Two vector functions $\mathbf{u}(t)$ and $\mathbf{v}(t)$ are ***linearly independent*** on an interval J if neither is a scalar multiple of the other. Equivalently, they are ***linearly independent*** on J if

$$c_1\mathbf{u}(t) + c_2\mathbf{v}(t) = \mathbf{0}$$

holds for all $t \in J$ only if $c_1 = 0$ and $c_2 = 0$.

Example 12.5.6 Consider the vector functions

$$\mathbf{x}^1(t) = \begin{bmatrix} e^{-t} \\ e^{-t} \end{bmatrix}, \quad \mathbf{x}^2(t) = \begin{bmatrix} e^{-3t} \\ -e^{-3t} \end{bmatrix}, \quad \mathbf{x}^3(t) = \begin{bmatrix} 2e^{-3t} \\ -2e^{-3t} \end{bmatrix} \tag{12.5.32}$$

on an arbitrary interval J. Clearly $\mathbf{x}^1(t)$ and $\mathbf{x}^2(t)$ are not scalar multiples of each other; so they are linearly independent functions on J. Likewise, $\mathbf{x}^1(t)$ and $\mathbf{x}^3(t)$ are linearly independent on J. Since $\mathbf{x}^3(t) = 2\mathbf{x}^2(t)$ for all values of $t \in J$, $\mathbf{x}^2(t)$ and $\mathbf{x}^3(t)$ are linearly dependent on J. ♦

Example 12.5.7 Explain why any vector function that is defined on an interval J and a vector function that is identically zero on J are linearly dependent on this interval.

Explanation If a vector function $\mathbf{u}(t)$ is defined on J and $\mathbf{v}(t)$ is a vector function such that $\mathbf{v}(t) \equiv \mathbf{0}$ on J, then

$$0 \cdot \mathbf{u}(t) + 1 \cdot \mathbf{v}(t) = \mathbf{0} + 1 \cdot \mathbf{0} = \mathbf{0}$$

for $t \in J$. Thus, $\mathbf{u}(t)$ and $\mathbf{v}(t)$ satisfy Definition 12.5.16 with $c_1 = 0$ and $c_2 = 1$. ♦

When we evaluate vector functions, such as those in Example 12.5.6 at a given value of t, we obtain constant vectors. The notions of linear dependence and linear independence also carry over to constant vectors.

Linear Dependent Vectors

Definition 12.5.18 Two vectors $\mathbf{u}$ and $\mathbf{v}$ are said to be ***linearly dependent*** if one of them is a scalar multiple of the other. Equivalently, $\mathbf{u}$ and $\mathbf{v}$ are ***linearly dependent*** if there exist scalars c_1 and c_2, not both zero, such that $c_1\mathbf{u} + c_2\mathbf{v} = \mathbf{0}$.

Vectors that are not linearly dependent are said to be *linearly independent*. Spelled out in detail, we have the following definition.

Linear Independent Vectors

Definition 12.5.19 Two vectors $\mathbf{u}$ and $\mathbf{v}$ are ***linearly independent*** if neither is a scalar multiple of the other. Equivalently, $\mathbf{u}$ and $\mathbf{v}$ are ***linearly independent*** if $c_1\mathbf{u} + c_2\mathbf{v} \neq \mathbf{0}$, unless both scalars are equal to zero. That is, $c_1\mathbf{u} + c_2\mathbf{v} = \mathbf{0}$ implies $c_1 = c_2 = 0$.

From a geometric point of view, two nonzero linearly dependent vectors are parallel. Specifically, if $\mathbf{v} = c\mathbf{u}$, then $\mathbf{v}$ is the vector whose length is $|c|$ times the length of $\mathbf{u}$. The vectors point in the same direction if $c > 0$ and in opposite directions if $c < 0$ (see Problem 17). Two linear independent vectors point in different directions.

Example 12.5.8 Consider the vectors: $\mathbf{u} = \langle 1, 1 \rangle$, $\mathbf{v} = \langle 1, -1 \rangle$, and $\mathbf{w} = \langle -2, 2 \rangle$. Since $\mathbf{u}$ and $\mathbf{v}$ are not scalar multiples of each other, they are linearly independent. Likewise $\mathbf{u}$ and $\mathbf{w}$ are linearly independent. Since $\mathbf{w} = -2\mathbf{v}$, the vectors $\mathbf{v}$ and $\mathbf{w}$ are linearly dependent. ♦

It is relatively simple to ascertain whether two vectors are linearly dependent or independent by merely appealing to Definitions 12.5.18 and 12.5.19. However, there is also a useful test which is at times less tedious to use, such as illustrated below in Example 12.5.9. It involves the determinant of the 2×2 matrix $\begin{bmatrix}\mathbf{u} & \mathbf{v}\end{bmatrix}$, where the entries in the first and second columns are the components of $\mathbf{u}$ and $\mathbf{v}$, respectively. For example, for $\mathbf{u} = \langle 1, 1\rangle$ and $\mathbf{v} = \langle 2, -1\rangle$,

$$\begin{bmatrix}\mathbf{u} & \mathbf{v}\end{bmatrix} = \begin{bmatrix}1 & 2\\ 1 & -1\end{bmatrix}. \tag{12.5.33}$$

Determinant Test for Linearly Independent Vectors

Theorem 12.5.20 *Let* $\mathbf{u}, \mathbf{v} \in \mathbb{R}^2$. *Let* $\begin{bmatrix}\mathbf{u} & \mathbf{v}\end{bmatrix}$ *denote the* 2×2 *matrix that has as its first and second columns the components of* $\mathbf{u}$ *and* $\mathbf{v}$, *respectively.*

(i) $\mathbf{u}$ *and* $\mathbf{v}$ *are linearly independent if and only if* $\det\begin{bmatrix}\mathbf{u} & \mathbf{v}\end{bmatrix} \neq 0$.
(ii) $\mathbf{u}$ *and* $\mathbf{v}$ *are linearly dependent if and only if* $\det\begin{bmatrix}\mathbf{u} & \mathbf{v}\end{bmatrix} = 0$.

Proof Let $M := \begin{bmatrix}\mathbf{u} & \mathbf{v}\end{bmatrix}$ with $\mathbf{u} = \langle u_1, u_2\rangle$ and $\mathbf{v} = \langle v_1, v_2\rangle$. Taking a linear combination of $\mathbf{u}$ and $\mathbf{v}$ and setting it equal to the zero vector, we have $c_1\mathbf{u}+c_2\mathbf{v} = \mathbf{0}$, which can be written as the matrix-vector product (see (12.3.13))

$$\begin{bmatrix}\mathbf{u} & \mathbf{v}\end{bmatrix}\begin{bmatrix}c_1\\ c_2\end{bmatrix} = \begin{bmatrix}0\\ 0\end{bmatrix}.$$

Expressed in terms of the matrix M and the vector $\mathbf{c} := \langle c_1, c_2\rangle$, this is $M\mathbf{c} = \mathbf{0}$. By Theorem 12.5.5, the zero vector $\mathbf{c} = \mathbf{0}$ is the unique solution of $M\mathbf{c} = \mathbf{0}$ if and only if $\det M \neq 0$. That is, the trivial solution ($c_1 = 0$, $c_2 = 0$) is the unique solution of $c_1\mathbf{u}+c_2\mathbf{v} = \mathbf{0}$ if and only if $\det M \neq 0$. Therefore, $\mathbf{u}$ and $\mathbf{v}$ are linearly independent vectors if and only if $\det M \neq 0$. This concludes the proof since (i) and (ii) are logically equivalent statements. ■

Example 12.5.9 Applying the test to $\mathbf{u} = \langle 1, 1\rangle$, $\mathbf{v} = \langle 1, -1\rangle$, and $\mathbf{w} = \langle -2, 2\rangle$, we find that

$$\det\begin{bmatrix}\mathbf{u} & \mathbf{v}\end{bmatrix} = \begin{vmatrix}1 & 1\\ 1 & -1\end{vmatrix} = -2, \quad \det\begin{bmatrix}\mathbf{u} & \mathbf{w}\end{bmatrix} = \begin{vmatrix}1 & -2\\ 1 & 2\end{vmatrix} = 4, \text{ and}$$

$$\det\begin{bmatrix}\mathbf{v} & \mathbf{w}\end{bmatrix} = \begin{vmatrix}1 & -2\\ -1 & 2\end{vmatrix} = 0.$$

Thus, the pairs $\{\mathbf{u}, \mathbf{v}\}$ and $\{\mathbf{u}, \mathbf{w}\}$ are linearly independent, whereas $\mathbf{v}$ and $\mathbf{w}$ are linearly dependent. This confirms the results obtained earlier in Example 12.5.8. ♦

Example 12.5.10 The pair of vector functions

$$\mathbf{u}(t) = \langle t^2 - 1, t + 1\rangle \quad \text{and} \quad \mathbf{v}(t) = \langle 2t - 2, t\rangle$$

are clearly linearly independent on the interval $(-\infty, \infty)$. Are there values of t for which $\mathbf{u}(t)$ and $\mathbf{v}(t)$ are linearly dependent vectors?

Solution For a fixed value of t, the Determinant Test says that $\mathbf{u}(t)$ and $\mathbf{v}(t)$ are linearly dependent vectors if

$$\det\left[\mathbf{u}(t)\ \ \mathbf{v}(t)\right] = \begin{vmatrix} t^2 - 1 & 2t - 2 \\ t + 1 & t \end{vmatrix} = 0.$$

This gives the equation

$$t(t^2 - 1) - 2(t - 1)(t + 1) = 0 \quad \text{or} \quad (t - 1)(t + 1)(t - 2) = 0.$$

Therefore, $\mathbf{u}(t)$ and $\mathbf{v}(t)$ are linearly dependent vectors when $t = -1$, 1, and 2.
Letting $t = -1$, we get the vectors

$$\mathbf{u}(0) = \begin{bmatrix} 0 \\ 0 \end{bmatrix} \quad \text{and} \quad \mathbf{v}(0) = \begin{bmatrix} -4 \\ -1 \end{bmatrix}.$$

Similarly, $t = 1$ yields the vectors $\langle 0, 2\rangle$ and $\langle 0, 1\rangle$. And $t = 2$ yields $\langle 3, 3\rangle$ and $\langle 2, 2\rangle$. This confirms that each of these pairs are linearly dependent vectors. ♦

It is clear from Definition 12.5.16 that two vector functions that are linearly dependent on an interval J will turn into linearly dependent vectors when they are evaluated at a point belonging to J. Contrast this with Example 12.5.10 which presents a pair of vector functions that are linearly independent on $\mathbb{R}$ but which become linearly dependent vectors when they are evaluated at the three values $t = \pm 1$ and $t = 2$. Problem 18 at the end of this chapter presents another pair of vector functions $\mathbf{u}(t)$ and $\mathbf{v}(t)$ that are linearly independent on $\mathbb{R}$, but the vectors $\mathbf{u}(t_0)$ and $\mathbf{v}(t_0)$ are linearly dependent for every $t_0 \in \mathbb{R}$. However, as the proof of the next lemma shows, if $\mathbf{u}(t_0)$ and $\mathbf{v}(t_0)$ are linearly independent vectors, then the vector functions $\mathbf{u}(t)$ and $\mathbf{v}(t)$ are also linearly independent.

Linearly Independent Functions and Vectors

Lemma 12.5.21 *Let $\mathbf{u}(t)$ and $\mathbf{v}(t)$ be vector functions that are defined on an interval J. If for some $t_0 \in J$, the vectors $\mathbf{u}(t_0)$ and $\mathbf{v}(t_0)$ are linearly independent, then the vector functions $\mathbf{u}(t)$ and $\mathbf{v}(t)$ are linearly independent on J.*

Proof Suppose, contrary to the conclusion of the lemma, the vector functions $\mathbf{u}(t)$ and $\mathbf{v}(t)$ are linearly dependent on J. That is, there are constants $c_1, c_2 \in \mathbb{R}$, not both zero, such that $c_1\mathbf{u}(t) + c_2\mathbf{v}(t) = \mathbf{0}$ for all $t \in J$. For $t = t_0$, this equation becomes

$$c_1\mathbf{u}(t_0) + c_2\mathbf{v}(t_0) = \mathbf{0}.$$

But this says $\mathbf{u}(t_0)$ and $\mathbf{v}(t_0)$ are linearly dependent vectors, which contradicts the hypothesis of the lemma. ■

Example 12.5.11 Consider the vector functions

$$\mathbf{u}(t) = \begin{bmatrix} e^{-t} \\ e^{-t} \end{bmatrix} \quad \text{and} \quad \mathbf{v}(t) = \begin{bmatrix} e^{-3t} \\ -e^{-3t} \end{bmatrix}.$$

on $\mathbb{R}$. Evaluating them at $t = 0$, we have

$$\mathbf{u}(0) = \begin{bmatrix} 1 \\ 1 \end{bmatrix} \quad \text{and} \quad \mathbf{v}(0) = \begin{bmatrix} 1 \\ -1 \end{bmatrix}.$$

These vectors are clearly linearly independent. Thus, it follows from Lemma 12.5.21 that $\mathbf{u}(t)$ and $\mathbf{v}(t)$ are linearly independent on $\mathbb{R}$. ♦

Theorem 12.5.14 states that a solution of $\mathbf{x}' = A\mathbf{x}$ satisfies this equation for all $t \in \mathbb{R}$. So whenever we state that two solutions of this equation are linearly independent, we mean they are linearly independent on all of $\mathbb{R}$, even if "on $\mathbb{R}$" is not explicitly stated. Because of Lemma 12.5.21, we can determine if a pair of solutions are linearly independent by merely evaluating them at some convenient value of t and then checking whether the resulting vectors are linearly independent. The next theorem states that the converse of Lemma 12.5.21 is true provided both vector functions are solutions of the same differential equation $\mathbf{x}' = A\mathbf{x}$.

Vector Test for Linearly Independent Solutions

Theorem 12.5.22 *Let $t_0 \in \mathbb{R}$. If $\mathbf{x}^1(t)$ and $\mathbf{x}^2(t)$ are solutions of $\mathbf{x}' = A\mathbf{x}$, then they are linearly independent if and only if the vectors $\mathbf{x}^1(t_0)$ and $\mathbf{x}^2(t_0)$ are linearly independent.*

Proof The proof of the "if" part of the theorem has already been taken care of with Lemma 12.5.21. To prove the "only if" part, suppose the solutions $\mathbf{x}^1(t)$ and $\mathbf{x}^2(t)$ are linearly independent on $\mathbb{R}$. Moreover, suppose—contrary to the statement of the theorem—that the vectors $\mathbf{x}^1(t_0)$ and $\mathbf{x}^2(t_0)$ are linearly dependent. Then there are

$c_1, c_2 \in \mathbb{R}$, not both zero, such that

$$c_1\mathbf{x}^1(t_0) + c_2\mathbf{x}^2(t_0) = \mathbf{0}.$$

For these particular constants, define the vector function $\mathbf{x}(t)$ by

$$\mathbf{x}(t) := c_1\mathbf{x}^1(t) + c_2\mathbf{x}^2(t).$$

It is also a solution by the Superposition Principle. Letting $t = t_0$, we have

$$\mathbf{x}(t_0) = c_1\mathbf{x}^1(t_0) + c_2\mathbf{x}^2(t_0) = \mathbf{0}.$$

From the uniqueness part of Theorem 12.5.14, it follows that $\mathbf{x}(t) \equiv \mathbf{0}$, that is,

$$c_1\mathbf{x}^1(t) + c_2\mathbf{x}^2(t) = \mathbf{0} \quad \text{for all } -\infty < t < \infty.$$

But this says that $\mathbf{x}^1(t)$ and $\mathbf{x}^2(t)$ are linearly dependent solutions as c_1 and c_2 are not both zero. However, this contradicts the supposition that they are linearly independent solutions. Therefore, we are forced to conclude that the vectors $\mathbf{x}^1(t_0)$ and $\mathbf{x}^2(t_0)$ are linearly independent, which concludes the proof of the theorem. ■

It is clear from perusing Theorems 12.5.20 and 12.5.22 that they can be combined to give us the following simple test to determine whether or not two solutions are linearly independent.

Determinant Test for Linearly Independent Solutions

Theorem 12.5.23 *Let $t_0 \in \mathbb{R}$. Two solutions $\mathbf{x}^1(t)$ and $\mathbf{x}^2(t)$ of $\mathbf{x}' = A\mathbf{x}$ are linearly independent if and only if*

$$\det\left[\mathbf{x}^1(t_0) \;\; \mathbf{x}^2(t_0)\right] \neq 0.$$

Example 12.5.12 Recall that the vector functions

$$\mathbf{x}^1(t) = \langle e^{-t}, e^{-t}\rangle \quad \text{and} \quad \mathbf{x}^2(t) = \langle e^{-3t}, -e^{-3t}\rangle$$

are solutions of equation (12.5.4) (see (12.5.27)). Choosing $t_0 = 0$, we have

$$\det\left[\mathbf{x}^1(0) \;\; \mathbf{x}^2(0)\right] = \begin{vmatrix} 1 & 1 \\ 1 & -1 \end{vmatrix} = -2.$$

Therefore, $\mathbf{x}^1(t)$ and $\mathbf{x}^2(t)$ are linearly independent on $\mathbb{R}$. ♦

We can see from Theorem 12.5.23 and the previous example the importance of the function $\det\left[\mathbf{x}^1(t) \;\; \mathbf{x}^2(t)\right]$. In fact, we are reminded of the Wronskian function in Chap. 10 that can be used to determine if two differentiable scalar-valued functions are linearly independent (see Theorem 10.9.2), which is an integral part of the method of variation of parameters. So it should not come as no surprise that this particular determinant function is also called the Wronskian.

Wronskian

Definition 12.5.24 The ***Wronskian*** of two vector functions $\mathbf{u}(t)$ and $\mathbf{v}(t)$ that are defined on an interval J is the real-valued function $W[\mathbf{u}, \mathbf{v}](t)$ that is defined on J by

$$W[\mathbf{u}, \mathbf{v}](t) := \det\left[\mathbf{u}(t)\ \ \mathbf{v}(t)\right]. \tag{12.5.34}$$

In Theorem 12.5.23, t_0 denotes an arbitrary real number; consequently, two solutions of $\mathbf{x}' = A\mathbf{x}$ are linearly independent on $\mathbb{R}$ if and only if their Wronskian is never zero. Furthermore, if the Wronskian of two solutions is ever zero at some value of t, then the two solutions have to be linearly dependent. As a result, we have the next theorem.

Wronskian and Linear Independence

Theorem 12.5.25 *Let $\mathbf{x}^1(t)$ and $\mathbf{x}^2(t)$ be solutions of $\mathbf{x}' = A\mathbf{x}$.*

(i) *$\mathbf{x}^1(t)$ and $\mathbf{x}^2(t)$ are linearly independent if and only if $W[\mathbf{x}^1, \mathbf{x}^2](t) \neq 0$ for all $t \in \mathbb{R}$.*
(ii) *$\mathbf{x}^1(t)$ and $\mathbf{x}^2(t)$ are linearly dependent if and only if $W[\mathbf{x}^1, \mathbf{x}^2](t) = 0$ for all $t \in \mathbb{R}$.*

From now on, for the sake of brevity, we will denote the Wronskian of two vector functions $\mathbf{u}(t)$ and $\mathbf{v}(t)$ by $W(t)$ rather than by $W[\mathbf{u}, \mathbf{v}](t)$, unless it is important to identify the two functions in order to avoid ambiguity. There is a well-known formula for the Wronskian, namely, (12.5.35) below. Its proof is left to Problem 22.

Abel's Formula

Theorem 12.5.26 *Let $W(t)$ denote the Wronskian of any two solutions of the system $\mathbf{x}' = A\mathbf{x}$. Then, for any $t_0 \in \mathbb{R}$,*

$$W(t) = W(t_0)e^{(\operatorname{tr} A)(t-t_0)} \tag{12.5.35}$$

for all $t \in \mathbb{R}$, where the **trace of** A [*abbr.* tr A] *denotes the sum of the entries along the main diagonal of A.*

Remark Abel's formula implies that the Wronskian of two solutions of $\mathbf{x}' = A\mathbf{x}$ is either identically zero or never zero on $\mathbb{R}$ (as does Theorem 12.5.25).

Example 12.5.13 Find the Wronskian of any two solutions of the system

$$\begin{aligned} x' &= 5x - 2y \\ y' &= 2x + y. \end{aligned} \tag{12.5.36}$$

Solution The trace of the coefficient matrix

$$A = \begin{bmatrix} 5 & -2 \\ 2 & 1 \end{bmatrix}$$

is $\operatorname{tr} A = 5 + 1 = 6$. If, in Abel's formula, we choose $t_0 = 0$, then the Wronskian of any two solutions is

$$W(t) = W(0)e^{(\operatorname{tr} A)t} = W(0)e^{6t}.$$

The exponential factor is the same for all pairs of solutions; however, the value of $W(0)$ depends on the particular pair. For example, it turns out (see Example 12.5.23) that

$$\mathbf{x}^1(t) = \begin{bmatrix} e^{3t} \\ e^{3t} \end{bmatrix} \quad \text{and} \quad \mathbf{x}^2(t) = \begin{bmatrix} (2t+1)e^{3t} \\ 2te^{3t} \end{bmatrix}.$$

are linearly independent solutions of (12.5.36). Computing their Wronskian, we have

$$W[\mathbf{x}^1, \mathbf{x}^2](t) = \begin{bmatrix} e^{3t} & (2t+1)e^{3t} \\ e^{3t} & 2te^{3t} \end{bmatrix} = 2te^{6t} - (2t+1)e^{6t} = -e^{6t}.$$

Thus, for this pair of solutions, $W(0) = W[\mathbf{x}^1, \mathbf{x}^2](0) = \text{-}1$.

Now consider a different pair of solutions, such as $\mathbf{x}^1(t)$ and $5\mathbf{x}^1(t)$. Computing the Wronskian of these two solutions, we have

$$W[\mathbf{x}^1, 5\mathbf{x}^1](t) = \begin{bmatrix} e^{3t} & -5e^{3t} \\ e^{3t} & -5e^{3t} \end{bmatrix} = 0.$$

Viewing this as $W[\mathbf{x}^1, 5\mathbf{x}^1](t) = 0{\cdot}e^{6t}$, we see that $W(0) = 0$. Note that this calculation was completely unnecessary since $\mathbf{x}^1(t)$ and $5\mathbf{x}^1(t)$ are linearly dependent. Consequently, $W[\mathbf{x}^1, 5\mathbf{x}^1](t) \equiv 0$ on $\mathbb{R}$ by Theorem 12.5.25. ♦

Example 12.5.14 For the linear system in Example 12.5.13, compute the Wronskian of the solutions $\mathbf{x}^1(t)$ and $\mathbf{x}^2(t)$ satisfying the initial conditions $\mathbf{x}^1(-1) = \langle 1, 1 \rangle$ and $\mathbf{x}^2(-1) = \langle 0, 3 \rangle$.

Solution Since the initial conditions are given at $t_0 = -1$,

$$W(t_0) = W[\mathbf{x}^1, \mathbf{x}^2](-1) = \det\left[\mathbf{x}^1(-1)\ \ \mathbf{x}^2(-1)\right] = \begin{vmatrix} 1 & 0 \\ 1 & 3 \end{vmatrix} = 3.$$

By Abel's formula,

$$W(t) = W(t_0)e^{(\operatorname{tr} A)(t-t_0)} = 3e^{6(t+1)}.$$

♦

The question now arises if all solutions of a given homogeneous linear system can be expressed in terms of two linearly independent solutions, as is the case with second-order linear homogeneous differential equations. The next theorem provides a definitive answer.

General Solution of a Homogeneous Linear System

Theorem 12.5.27 *Let $\mathbf{x}^1(t)$ and $\mathbf{x}^2(t)$ be linearly independent solutions of*

$$\mathbf{x}' = A\mathbf{x} \tag{12.5.37}$$

where A is a 2×2 constant matrix. Then every solution $\mathbf{x}(t)$ of (12.5.37) *can be expressed as a linear combination of $\mathbf{x}^1(t)$ and $\mathbf{x}^2(t)$. In other words, there are constants c_1 and c_2 such that*

$$\mathbf{x}(t) = c_1\mathbf{x}^1(t) + c_2\mathbf{x}^2(t) \tag{12.5.38}$$

for all $-\infty < t < \infty$.

Proof In addition to $\mathbf{x}^1(t)$ and $\mathbf{x}^2(t)$, let $\mathbf{x}(t)$ be another solution of the homogeneous linear system (12.5.37). That is, $\mathbf{x}(t)$ is a vector function that satisfies this system for all $t \in \mathbb{R}$. Let the vector $\langle k, l\rangle$ denote its value at $t = 0$. Moreover, let $\mathbf{u}$ and $\mathbf{v}$ be vectors that are the values of $\mathbf{x}^1(t)$ and $\mathbf{x}^2(t)$ at $t = 0$, respectively. Since, by hypothesis, $\mathbf{x}^1(t)$ and $\mathbf{x}^2(t)$ are solutions, it follows from the Superposition Principle that all vector functions of the form $c_1\mathbf{x}^1(t) + c_2\mathbf{x}^2(t)$, where $c_1, c_2 \in \mathbb{R}$, are solutions of (12.5.37). Let us see if at least one of these functions has the same value as $\mathbf{x}(t)$ at $t = 0$. Since $c_1\mathbf{x}^1(t) + c_2\mathbf{x}^2(t)$ and $\mathbf{x}(t)$ are equal to $c_1\mathbf{u} + c_2\mathbf{v}$ and $\langle k, l\rangle$ at $t = 0$, respectively, this would mean that there would have to be values for c_1 and c_2 satisfying the vector equation

$$c_1\begin{bmatrix} u_1 \\ u_2 \end{bmatrix} + c_2\begin{bmatrix} v_1 \\ v_2 \end{bmatrix} = \begin{bmatrix} k \\ l \end{bmatrix},$$

which is equivalent to the linear system of equations

$$c_1u_1 + c_2v_1 = k$$
$$c_1u_2 + c_2v_2 = l.$$

Let us find values for c_1 and c_2, if any, using the elimination method for solving two linear algebraic equations in two unknowns. To eliminate the unknown c_2, we multiply the first equation by v_2, the second one by $-v_1$, and then add them. The result is

$$(u_1v_2 - u_2v_1)c_1 = kv_2 - lv_1.$$

Similarly, we can eliminate the unknown c_1 to obtain the equation

$$(u_1v_2 - u_2v_1)c_2 = lu_1 - ku_2.$$

Since **u** and **v** are linearly independent vectors (see Theorem 12.5.20),

$$\det\begin{bmatrix}\mathbf{u} & \mathbf{v}\end{bmatrix} = u_1v_2 - u_2v_1 \neq 0. \tag{12.5.39}$$

Therefore, with

$$c_1 = \frac{kv_2 - lv_1}{u_1v_2 - u_2v_1} \quad \text{and} \quad c_2 = \frac{lu_1 - ku_2}{u_1v_2 - u_2v_1},$$

the linear combination $c_1\mathbf{x}^1(t) + c_2\mathbf{x}^2(t)$ is a solution that is equal to the solution $\mathbf{x}(t)$ at $t = 0$.[10] But since they are equal at $t = 0$, they are in fact equal for all values of $t \in \mathbb{R}$ (see Theorem 12.5.14). ■

What Theorem 12.5.27 tells us is that the set of all linear combinations (12.5.38) is the complete set of solutions of (12.5.37). A particular solution belonging to this set is obtained by assigning values to the constants c_1 and c_2. Since there are no other solutions other than the ones belonging to this set, (12.5.38) is said to be a ***general solution*** of (12.5.37). In the language of linear (or matrix) algebra, any two linearly independent solutions $\mathbf{x}^1(t)$ and $\mathbf{x}^2(t)$ of (12.5.37) are said to ***span*** the space of all solutions of (12.5.37).

Example 12.5.15 Recall from (12.5.27) that

$$\mathbf{x}^1(t) = \begin{bmatrix}1\\1\end{bmatrix}e^{-t} \quad \text{and} \quad \mathbf{x}^2(t) = \begin{bmatrix}1\\-1\end{bmatrix}e^{-3t}$$

[10] Because of (12.5.39), the tedium of using the elimination method could have been avoided with Cramer's rule. See Problem 85 in Chap. 10.

are solutions of (12.5.4), namely, the vector differential equation

$$\mathbf{x}' = \begin{bmatrix} -2 & 1 \\ 1 & -2 \end{bmatrix} \mathbf{x}.$$

According to Example 12.5.6 (cf. Examples 12.5.11 and 12.5.12), they are linearly independent. As a result, it follows from Theorem 12.5.27 that

$$\mathbf{x}(t) = c_1 \begin{bmatrix} 1 \\ 1 \end{bmatrix} e^{-t} + c_2 \begin{bmatrix} 1 \\ -1 \end{bmatrix} e^{-3t} \tag{12.5.40}$$

or

$$\mathbf{x}(t) = \begin{bmatrix} x(t) \\ y(t) \end{bmatrix} = \begin{bmatrix} c_1 e^{-t} + c_2 e^{-3t} \\ c_1 e^{-t} - c_2 e^{-3t} \end{bmatrix}$$

is a general solution. In other words, this is the form of all solutions of (12.5.4). A particular solution is obtained when a pair of values is assigned to c_1 and c_2. For example, with $c_1 = 2$ and $c_2 = -5$, we have the particular solution

$$\mathbf{x}(t) = \begin{bmatrix} x(t) \\ y(t) \end{bmatrix} = 2 \begin{bmatrix} 1 \\ 1 \end{bmatrix} e^{-t} - 5 \begin{bmatrix} 1 \\ -1 \end{bmatrix} e^{-3t} = \begin{bmatrix} 2e^{-t} - 5e^{-3t} \\ 2e^{-t} + 5e^{-3t} \end{bmatrix}. \tag{12.5.41}$$

This is a particular vector function that satisfies (12.5.4) for all $-\infty < t < \infty$. Or we can say that the pair of functions

$$x(t) = 2e^{-t} - 5e^{-3t} \quad \text{and} \quad y(t) = 2e^{-t} + 5e^{-3t}$$

is a particular solution of the homogeneous linear system

$$\begin{aligned} x' &= -2x + y \\ y' &= x - 2y. \end{aligned}$$

◆

12.5.4 *Trajectories and Phase Portraits*

A way to get a sense of the behavior of the solutions of the vector differential equation (12.5.37) (that is, $\mathbf{x}' = A\mathbf{x}$) is to view its ***phase portrait***. This is a plot of a direction field for the system together with the graphs of a sufficient number of its solutions in the phase plane so that all of the various types of graphs are displayed. While phase portraits of homogeneous linear systems will be fully addressed later on in Chap. 13, Fig. 12.7 is a preview of one: it shows a direction field for equation (12.5.4) in the preceding example and the graphs of a representative sample of its solutions.

Let $\mathbf{x}(t) = \langle x(t), y(t)\rangle$ be a particular solution of (12.5.4), such as (12.5.41). The corresponding set of points

$$\{(x(t), y(t)) \in \mathbb{R}^2 : -\infty < t < \infty\}$$

is called the ***trajectory*** of this solution. The graph of this set is a curve in the phase plane. It is helpful to view the independent variable t as "time" (even though it may not really represent time) and a trajectory as the path that a phase point takes as it moves in accordance with $\mathbf{x}' = A\mathbf{x}$. With this interpretation, it makes sense to think of

$$\frac{d}{dt}\mathbf{x}(t) = \langle x'(t), y'(t)\rangle$$

as the ***velocity*** of the phase point along a given trajectory.

The eigenspaces of a coefficient matrix A are indispensable when it comes to graphing trajectories of $\mathbf{x}' = A\mathbf{x}$. Even though there are infinitely many trajectories, the goal is to graph a sufficient number of them so as to effectively communicate what all of them would look like. Consider again Example 12.5.15. The eigenspace $E(\lambda_1)$ corresponding to the eigenvalue $\lambda_1 = -1$ is the subspace ***spanned*** by the vector $\langle 1, 1\rangle$, that is, the set of all vectors $c_1\langle 1, 1\rangle$ generated when c_1 assumes all possible real values. The graph of the set of points $c_1(1, 1)$ is the line in Fig. 12.7 through the origin $(0, 0)$ and the point $(1, 1)$. This line is the geometric representation of the eigenspace $E(-1)$. If a point is selected on this line, aside from the origin, then its position is $c_1(1, 1)$ for some nonzero value of c_1. If we think of t as time and consider a phase point at $c_1(1, 1)$ when $t = 0$, then it will move along the line as t changes and its position is given by the vector

$$\mathbf{x}(t) = c_1\langle 1, 1\rangle e^{-t} = c_1\begin{bmatrix}1\\1\end{bmatrix}e^{-t}.$$

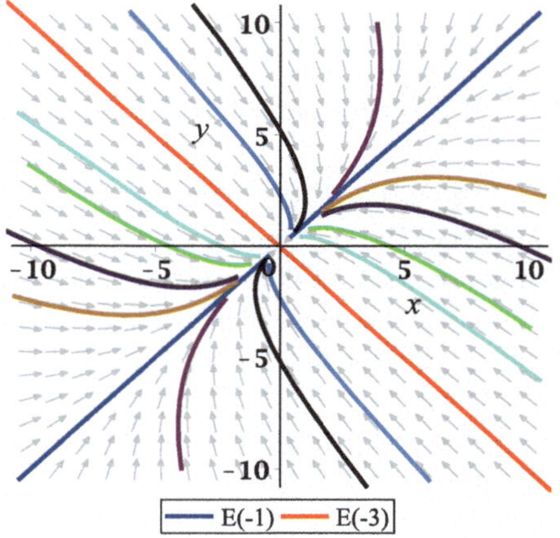

Fig. 12.7 Phase portrait of the linear system (12.5.4)

Since $c_1\langle 1, 1\rangle e^{-t} \to \langle 0, 0\rangle$ as $t \to \infty$, this phase point moves toward the origin with increasing t. The direction field in Fig. 12.7 clearly substantiates this. On the other hand, the phase point moves away from the origin along the line as $t \to -\infty$.

Now suppose a phase point is placed at the origin at $t = 0$. Then, by setting $t = 0$ and $\mathbf{x}(0) = \langle 0, 0\rangle$ in (12.5.40), we get $c_1 = c_2 = 0$. Consequently, $\mathbf{x}(t) = \langle 0, 0\rangle$ for all $-\infty < t < \infty$, which means the phase point remains stuck at the origin as t changes. This special solution is known as the ***equilibrium solution*** while the origin is said to be an ***equilibrium point***.

The eigenspace corresponding to the eigenvalue $\lambda_2 = -3$, denoted by $E(-3)$, is the subspace spanned by the vector $\langle 1, -1\rangle$, namely, the set of vectors $c_2\langle 1, -1\rangle$ generated when c_2 takes on all possible real values. The graph of all the points $c_2(1, -1)$ is the line in Fig. 12.7 that passes through the origin and the point $(1, -1)$. The phase point whose position at time t is $\mathbf{x}(t) = c_2\langle 1, -1\rangle e^{-3t}$ approaches the origin along this line as t increases. All of the other trajectories, namely, those besides the equilibrium point $(0, 0)$ and the other four lying on the lines $E(-3)$ and $E(-1)$, are the graphs of the solutions (12.5.40) when both c_1 and c_2 are nonzero. Note that every one of these trajectories approaches the origin with increasing t.

In summary, Fig. 12.7 is a ***phase portrait*** of the system (12.5.4), which portrays all of the qualitatively different trajectories of this system.

12.5.5 Classification of Eigenvalues

Earlier we determined that the coefficient matrix of (12.5.4) has two distinct real eigenvalues, namely, -1 and -3. Eigenvalues, however, are not always distinct. For instance, consider the eigenvalues of

$$A = \begin{bmatrix} -2 & 0 \\ 0 & -2 \end{bmatrix},$$

the coefficient matrix of (12.5.1). Recall that the eigenvalues of a matrix A are the roots of the characteristic equation $\det(A - \lambda I) = 0$. Thus, the characteristic equation for this particular matrix is the quadratic equation

$$\begin{vmatrix} -2-\lambda & 0 \\ 0 & -2-\lambda \end{vmatrix} = 0 \quad \Rightarrow \quad (-2-\lambda)^2 = 0 \quad \Rightarrow \quad (\lambda + 2)^2 = 0.$$

If this were an algebra course, we would say that the preceding equation has only one solution, namely, $\lambda = -2$. However, for our purposes it is better to take a different stance. The factored form of the characteristic equation is

$$(\lambda + 2)(\lambda + 2) = 0.$$

Setting each factor equal to zero and solving for λ, we obtain $\lambda = -2$ and $\lambda = -2$. Viewed in this way, the roots of the characteristic equation are -2 and -2, albeit they are the same number. Someone might insist that there is really just one root; even so, it will be more useful to adopt the view that a quadratic equation always has exactly two roots. So, for this example, we say that -2 is a ***repeated root***. The terms ***double root*** and ***root of multiplicity*** 2 are also used.

Therefore, every 2×2 matrix is viewed as having exactly two eigenvalues because eigenvalues are the roots of quadratic equations. Sometimes these roots are distinct real numbers; sometimes they are repeated real numbers as above; and sometimes they are complex numbers (numbers that are not real numbers). Let us delve into this further to see how these different cases can arise.

Recall that the eigenvalues of the matrix

$$A = \begin{bmatrix} a & b \\ c & d \end{bmatrix}$$

are the solutions of its characteristic equation

$$\det(A - \lambda I) = \begin{vmatrix} a - \lambda & b \\ c & d - \lambda \end{vmatrix} = 0,$$

which simplifies to

$$\lambda^2 - (a + d)\lambda + (ad - bc) = 0.$$

Notice that $a + d$ is the trace of the matrix A and $ad - bc$ its determinant. So all that we have to do in order to obtain the characteristic equation of A is to calculate these two quantities and to commit Eq. (12.5.42) below to memory.

Characteristic Equation via Determinant and Trace

Theorem 12.5.28 *The characteristic equation of a matrix A of order* 2 *is*

$$\lambda^2 - (\operatorname{tr} A)\,\lambda + \det A = 0. \tag{12.5.42}$$

Example 12.5.16 Find the characteristic equation of

$$A = \begin{bmatrix} -2 & 1 \\ 1 & -2 \end{bmatrix}.$$

Solution Since

$$\operatorname{tr} A = -2 - 2 = -4 \quad \text{and} \quad \det A = (-2)(-2) - (1)(1) = 3,$$

the characteristic equation is

$$\lambda^2 + 4\lambda + 3 = 0. \qquad \blacklozenge$$

Since the eigenvalues of a 2×2 matrix A are the solutions of the quadratic equation (12.5.42), they are given by the quadratic formula

$$\lambda = \frac{-(-\operatorname{tr} A) \pm \sqrt{(-\operatorname{tr} A)^2 - 4 \det A}}{2}.$$

That is, the eigenvalues are

$$\lambda_1 = \frac{1}{2}\left(\operatorname{tr} A + \sqrt{\Delta}\right) \quad \text{and} \quad \lambda_2 = \frac{1}{2}\left(\operatorname{tr} A - \sqrt{\Delta}\right), \tag{12.5.43}$$

where Δ denotes the *discriminant* of the quadratic formula, namely,

$$\Delta := (\operatorname{tr} A)^2 - 4 \det A. \tag{12.5.44}$$

Whether the eigenvalues are real or complex numbers depends on the sign of Δ:

1. If $\Delta > 0$, the eigenvalues λ_1 and λ_2 are real and distinct.
2. If $\Delta = 0$, the eigenvalues are real and equal with

$$\lambda_1 = \lambda_2 = \frac{1}{2}\operatorname{tr} A.$$

3. If $\Delta < 0$, the eigenvalues are the complex numbers

$$\lambda_1 = \alpha + i\omega \quad \text{and} \quad \lambda_2 = \alpha - i\omega,$$

where $i := \sqrt{-1}$ and

$$\alpha := \frac{1}{2}\operatorname{tr} A \quad \text{and} \quad \omega := \frac{1}{2}\sqrt{4 \det A - (\operatorname{tr} A)^2}.$$

The real number α is called the ***real part*** of λ_1, which is also the real part of λ_2. The real number ω is called the ***imaginary part*** of λ_1. However, the imaginary part of λ_2 is not ω, but rather $-\omega$. The complex eigenvalue $\alpha - i\omega$ is said to be the ***complex conjugate*** of $\alpha + i\omega$, and conversely. The pair $\alpha \pm i\omega$ is referred to as a ***complex conjugate pair***. See "Complex Eigenvalues and General Solution" in the next subsection for a more detailed discussion of complex eigenvalues.

From the solutions (12.5.43) of the characteristic equation, it is readily seen that the sum of the eigenvalues of a matrix is equal to its trace. Moreover, their product is equal to the determinant of the matrix.

Sum and Product of Eigenvalues

Theorem 12.5.29 *If* λ_1 *and* λ_2 *are the eigenvalues of a matrix* A *of order* 2, *then*

$$\operatorname{tr} A = \lambda_1 + \lambda_2 \quad \textit{and} \quad \det A = \lambda_1 \lambda_2. \tag{12.5.45}$$

Proof It follows from (12.5.43) that

$$\lambda_1 + \lambda_2 = \frac{1}{2}\left(\operatorname{tr} A + \sqrt{\Delta}\right) + \frac{1}{2}\left(\operatorname{tr} A - \sqrt{\Delta}\right) = \operatorname{tr} A.$$

Furthermore,

$$\begin{aligned}\lambda_1\lambda_2 &= \frac{1}{4}\left(\operatorname{tr} A + \sqrt{\Delta}\right)\left(\operatorname{tr} A - \sqrt{\Delta}\right) = \frac{1}{4}\left((\operatorname{tr} A)^2 - \Delta\right) \\ &= \frac{1}{4}\left((\operatorname{tr} A)^2 - \left[(\operatorname{tr} A)^2 - 4\det A\right]\right) = \det A.\end{aligned}$$

This completes the proof. ■

Sometimes all that we really need to know is whether the eigenvalues of a matrix are real or complex numbers—and whether they are positive, negative, or zero if they are real numbers or whether their real parts are positive, negative, or zero if they are a complex conjugate pair. The next result points out that this information is ascertainable from simply knowing the signs of the determinant and trace of the coefficient matrix.

Classification of Eigenvalues

Corollary 12.5.30 *Let* λ_1 *and* λ_2 *be the eigenvalues of a matrix* A *of order* 2.

(i) *If* $\det A > 0$ *and* $\operatorname{tr} A < 0$, *the eigenvalues are either negative real numbers or a complex conjugate pair with negative real parts. In other words, either* $\lambda_1 < 0$ *and* $\lambda_2 < 0$ *or* $\lambda_1 = \alpha + i\omega$ *and* $\lambda_2 = \alpha - i\omega$ *with* $\alpha < 0$.
(ii) *If* $\det A > 0$ *and* $\operatorname{tr} A > 0$, *the eigenvalues are either positive real numbers or a complex conjugate pair with positive real parts.*
(iii) *If* $\det A > 0$ *and* $\operatorname{tr} A = 0$, *the eigenvalues are a complex conjugate pair with zero real parts. That is,* $\lambda_1 = i\omega$ *and* $\lambda_2 = -i\omega$.

(continued)

(iv) *If* $\det A < 0$, *the eigenvalues are real numbers with opposite signs.*
(v) *If* $\det A = 0$ *and* $\operatorname{tr} A = 0$, *both eigenvalues are zero.*
(vi) *If* $\det A = 0$ *and* $\operatorname{tr} A \neq 0$, *one eigenvalue is zero while the other is equal to* $\operatorname{tr} A$.

Proof Consider (i): suppose $\det A > 0$ and $\operatorname{tr} A < 0$. There are three cases to consider. If $\Delta > 0$, then λ_1 and λ_2 are real and distinct. Consequently, $\lambda_1\lambda_2 = \det A > 0$ and $\lambda_1 + \lambda_2 = \operatorname{tr} A < 0$ imply λ_1 and λ_2 are distinct negative numbers. If $\Delta = 0$, then $\lambda_1 = \lambda_2 = \frac{1}{2}\operatorname{tr} A < 0$. If $\Delta < 0$, then λ_1 and λ_2 are complex conjugate numbers with their real parts equal to the negative number $\frac{1}{2}\operatorname{tr} A$. This completes the proof of (i). Statement (ii) mirrors (i). The same is true of its proof.

Now consider (iii), that is, suppose $\det A > 0$ and $\operatorname{tr} A = 0$. Then the discriminant Δ is negative since

$$\Delta = (\operatorname{tr} A)^2 - 4\det A = -4\det A < 0.$$

Consequently, the eigenvalues of A are a complex conjugate pair, say $\alpha \pm i\omega$. It then follows from (12.5.45) and $\operatorname{tr} A = 0$ that

$$\operatorname{tr} A = (\alpha + i\omega) + (\alpha - i\omega) = 2\alpha = 0.$$

So, as $\alpha = 0$, the eigenvalues are the pure imaginary numbers $\pm i\omega$.

The proofs of (iv), (v), and (vi) are left as exercises. ■

Example 12.5.17 Let

$$A = \begin{bmatrix} -1 & -2 \\ -3 & 4 \end{bmatrix} \quad \text{and} \quad B = \begin{bmatrix} 1 & -2 \\ 3 & 4 \end{bmatrix}.$$

Since $\det A = -10$, the eigenvalues of A are real numbers with opposite signs. As for the matrix B, $\det B = 10$ and $\operatorname{tr} B = 5$. Thus, as both quantities are positive, the eigenvalues of B are either positive real numbers or a complex conjugate pair with positive real parts. In fact, they are the latter since the discriminant is negative:

$$\Delta = (\operatorname{tr} B)^2 - 4\det B = 5^2 - 4(10) < 0.$$

In fact, the eigenvalues of A are -2 and 5 and those of B are $\frac{5}{2} \pm \frac{\sqrt{15}}{2}$. ♦

Now that we know both eigenvalues of a matrix A of order 2 can be either (i) distinct real numbers or (ii) repeated real numbers or (iii) a pair of complex conjugate numbers, let us investigate each of these cases in detail and determine how they affect the solutions of the vector differential equation $x' = Ax$.

12.5.6 Eigenvalue-Eigenvector Method

In the last subsection, it was pointed out that the eigenvalues λ_1 and λ_2 of a matrix A are distinct numbers when the discriminant of the characteristic equation of A is nonzero. According to Theorem 12.5.27, a general solution of $\mathbf{x}' = A\mathbf{x}$ can be expressed in terms of two solutions—but not just any two—but two that are linearly independent. Finding two solutions is no problem—at least not when the eigenvalues are real and distinct—because as we have discovered, all we have to do is to find two eigenvectors $\mathbf{u}$ and $\mathbf{v}$ of A corresponding to λ_1 and λ_2 respectively; for then

$$\mathbf{x}^1(t) = e^{\lambda_1 t}\mathbf{u} \quad \text{and} \quad \mathbf{x}^2(t) = e^{\lambda_2 t}\mathbf{v} \tag{12.5.46}$$

are two solutions. So the question is as to the linearly independence of $\mathbf{x}^1(t)$ and $\mathbf{x}^2(t)$. By Theorem 12.5.22, they are provided $\mathbf{u} = \mathbf{x}^1(0)$ and $\mathbf{v} = \mathbf{x}^2(0)$ are linearly independent vectors. Indeed they are, as the next theorem asserts, if as we are now supposing $\lambda_1 \neq \lambda_2$.

Distinct Eigenvalues and Linear Independence

Theorem 12.5.31 *Suppose the eigenvalues λ_1 and λ_2 of a 2×2 matrix A are not equal. If $\mathbf{u}$ and $\mathbf{v}$ are eigenvectors corresponding to λ_1 and λ_2, respectively, then they are linearly independent.*

Proof First, set a linear combination of the vectors $\mathbf{u}$ and $\mathbf{v}$ equal to the zero vector:

$$c_1\mathbf{u} + c_2\mathbf{v} = \mathbf{0}. \tag{12.5.47}$$

Next, multiply both sides of (12.5.47) by A:

$$A(c_1\mathbf{u} + c_2\mathbf{v}) = A \cdot \mathbf{0} \quad \Rightarrow \quad c_1 A\mathbf{u} + c_2 A\mathbf{v} = \mathbf{0}.$$

Since

$$A\mathbf{u} = \lambda_1\mathbf{u} \quad \text{and} \quad A\mathbf{v} = \lambda_2\mathbf{v},$$

this becomes

$$c_1\lambda_1\mathbf{u} + c_2\lambda_2\mathbf{v} = \mathbf{0}. \tag{12.5.48}$$

Now multiply (12.5.47) by $-\lambda_2$ and add the result to (12.5.48):

$$c_1(\lambda_1 - \lambda_2)\mathbf{u} = \mathbf{0}.$$

Since $\lambda_1 - \lambda_2 \neq 0$ and $\mathbf{u} \neq \mathbf{0}$, it follows that $c_1 = 0$. This implies $c_2 = 0$ as $\mathbf{v} \neq \mathbf{0}$. Therefore, $\mathbf{u}$ and $\mathbf{v}$ are linearly independent vectors. ■

12.5.6.1 Distinct Real Eigenvalues and General Solution

Suppose the discriminant of the characteristic equation of a matrix A of order 2 is a positive number. Then both of its eigenvalues λ_1 and λ_2 are distinct real numbers. By Theorem 12.5.31, any two eigenvectors $\mathbf{u}$ and $\mathbf{v}$ of A corresponding to λ_1 and λ_2, respectively, are linearly independent. Thus, $\mathbf{x}^1(t) = e^{\lambda_1 t}\mathbf{u}$ and $\mathbf{x}^2(t) = e^{\lambda_2 t}\mathbf{v}$ are not only solutions of $\mathbf{x}' = A\mathbf{x}$ but also are linearly independent by Theorem 12.5.22. Therefore, by Theorem 12.5.27,

$$\mathbf{x}(t) = c_1 e^{\lambda_1 t}\mathbf{u} + c_2 e^{\lambda_2 t}\mathbf{v}$$

is a general solution of $\mathbf{x}' = A\mathbf{x}$. As a result, we have the following theorem.

Distinct Real Eigenvalues and General Solution

Theorem 12.5.32 *If the eigenvalues λ_1 and λ_2 of a 2×2 matrix A are distinct real numbers, then a general solution of*

$$\mathbf{x}' = A\mathbf{x}$$

is

$$\mathbf{x}(t) = c_1 e^{\lambda_1 t}\mathbf{u} + c_2 e^{\lambda_2 t}\mathbf{v},$$

where $\mathbf{u}$, $\mathbf{v}$ are eigenvectors of A corresponding to λ_1, λ_2, respectively, and c_1, c_2 are arbitrary constants.

Example 12.5.18 Find a general solution of

$$\mathbf{x}' = \begin{bmatrix} -2 & 1 \\ 1 & -2 \end{bmatrix}\mathbf{x}. \tag{12.5.4}$$

Solution In Example 12.5.2, we found:

1. $\lambda_1 = -1$ and $\lambda_2 = -3$ are the eigenvalues of $A = \begin{bmatrix} -2 & 1 \\ 1 & -2 \end{bmatrix}$.
2. $\mathbf{u} = \langle 1, 1\rangle$ is an eigenvector corresponding to $\lambda_1 = -1$.
3. $\mathbf{v} = \langle 1, -1\rangle$ is an eigenvector corresponding to $\lambda_2 = -3$.

Hence, by Theorem 12.5.32, a general solution is

$$\mathbf{x}(t) = c_1 e^{-t} \begin{bmatrix} 1 \\ 1 \end{bmatrix} + c_2 e^{-3t} \begin{bmatrix} 1 \\ -1 \end{bmatrix}.$$ ♦

Example 12.5.19 Find a general solution of the homogeneous linear system

$$\begin{aligned} x' &= -2x + 2y \\ y' &= 2x + y. \end{aligned}$$

Solution The corresponding vector differential equation is $\mathbf{x}' = A\mathbf{x}$, where $\mathbf{x} = \langle x, y \rangle$ and

$$A = \begin{bmatrix} -2 & 2 \\ 2 & 1 \end{bmatrix}.$$

The eigenvalues of A are the solutions of the characteristic equation

$$\det(A - \lambda I) = \begin{vmatrix} -2-\lambda & 2 \\ 2 & 1-\lambda \end{vmatrix} = (\lambda + 2)(\lambda - 1) - 4 = \lambda^2 + \lambda - 6 = 0.$$

Factoring, we get

$$(\lambda - 2)(\lambda + 3) = 0.$$

Thus, the eigenvalues are $\lambda_1 = 2$ and $\lambda_2 = -3$.

Now let us find eigenvectors corresponding to these eigenvalues. An eigenvector $\mathbf{u}$ corresponding to λ_1 is a nonzero vector that satisfies the equation $A\mathbf{u} = \lambda_1\mathbf{u}$, which can be rewritten as

$$(A - \lambda_1 I)\mathbf{u} = \begin{bmatrix} -2-\lambda_1 & 2 \\ 2 & 1-\lambda_1 \end{bmatrix} \begin{bmatrix} u_1 \\ u_2 \end{bmatrix} = \begin{bmatrix} 0 \\ 0 \end{bmatrix}.$$

With $\lambda_1 = 2$, this becomes

$$\begin{bmatrix} -4 & 2 \\ 2 & -1 \end{bmatrix} \begin{bmatrix} u_1 \\ u_2 \end{bmatrix} = \begin{bmatrix} 0 \\ 0 \end{bmatrix}.$$

As a result, we have the redundant equations (see Problem 19)

$$\begin{aligned} -4u_1 + 2u_2 &= 0 \\ 2u_1 - u_2 &= 0. \end{aligned}$$

Viewing u_1 as the free variable, we solve for u_2 obtaining $u_2 = 2u_1$. Letting $u_1 = 1$, we obtain the eigenvector $\mathbf{u} = \langle 1, 2 \rangle$. Had we assigned some other nonzero value to u_1, we would have merely obtained another eigenvector corresponding to $\lambda_1 = 2$ but one that would be a multiple of $\mathbf{u}$.

An eigenvector $\mathbf{v}$ corresponding to $\lambda_2 = -3$ satisfies the vector equation

$$(A - \lambda_2 I)\mathbf{v} = \begin{bmatrix} -2-\lambda_2 & 2 \\ 2 & 1-\lambda_2 \end{bmatrix} \begin{bmatrix} v_1 \\ v_2 \end{bmatrix} = \begin{bmatrix} 0 \\ 0 \end{bmatrix}$$

or

$$\begin{bmatrix} 1 & 2 \\ 2 & 4 \end{bmatrix} \begin{bmatrix} v_1 \\ v_2 \end{bmatrix} = \begin{bmatrix} 0 \\ 0 \end{bmatrix}.$$

Hence, $v_1 = -2v_2$. Setting $v_2 = 1$ yields the eigenvector $\mathbf{v} = \langle -2, 1 \rangle$.

It follows from Theorem 12.5.32 that a general solution is

$$\mathbf{x}(t) = c_1 e^{2t} \begin{bmatrix} 1 \\ 2 \end{bmatrix} + c_2 e^{-3t} \begin{bmatrix} -2 \\ 1 \end{bmatrix}.$$

Therefore,

$$x(t) = c_1 e^{2t} - 2c_2 e^{-3t}$$
$$y(t) = 2c_1 e^{2t} + c_2 e^{-3t}.$$

♦

12.5.6.2 Repeated Real Eigenvalues and General Solution

Now let us investigate how to find the general solution of $\mathbf{x}' = A\mathbf{x}$ when the eigenvalues of the matrix A are repeated real numbers, which happens when the discriminant of the characteristic equation of A is equal to zero. To get a feel for this case, we will consider two examples of linear systems that present two different situations.

Example 12.5.20 Once again consider the vector differential equation (12.5.1):

$$\mathbf{x}' = A\mathbf{x} = \begin{bmatrix} -2 & 0 \\ 0 & -2 \end{bmatrix} \mathbf{x}.$$

Recall that it was pointed out in the introduction to the previous subsection that this coefficient matrix has the repeated eigenvalue $\lambda = -2$. Substituting this value for λ in the equation $(A - \lambda I)\mathbf{u} = \mathbf{0}$ in order to find corresponding eigenvectors, we have

$$(A + 2I)\mathbf{u} = \mathbf{0} \quad \text{or} \quad \begin{bmatrix} 0 & 0 \\ 0 & 0 \end{bmatrix} \begin{bmatrix} u_1 \\ u_2 \end{bmatrix} = \begin{bmatrix} 0 \\ 0 \end{bmatrix}.$$

Clearly any vector $\mathbf{u}$ satisfies this equation; in other words, every nonzero vector is an eigenvector. What we need in order to write down a general solution are two linearly independent eigenvectors. Any two will do, such as $\langle 1, 0\rangle$ and $\langle 0, 1\rangle$. Accordingly,

$$\mathbf{x}^1(t) = e^{-2t}\begin{bmatrix}1\\0\end{bmatrix} \quad \text{and} \quad \mathbf{x}^2(t) = e^{-2t}\begin{bmatrix}0\\1\end{bmatrix}$$

are two linearly independent solutions. By Theorem 12.5.27,

$$\mathbf{x}(t) = c_1e^{-2t}\begin{bmatrix}1\\0\end{bmatrix} + c_2e^{-2t}\begin{bmatrix}0\\1\end{bmatrix} = \begin{bmatrix}c_1e^{-2t}\\c_2e^{-2t}\end{bmatrix}$$

is a general solution.

This is what we had already obtained with much less work at the beginning of Sect. 12.5 (see (12.5.2)), where we recognized that (12.5.1) represents a system of two uncoupled scalar differential equations. But the point of this example is that a matrix with a repeated eigenvalue may have two linearly independent eigenvectors associated with that eigenvalue. But this is usually not the case. ♦

Example 12.5.21 Consider the linear system

$$\begin{aligned} x' &= 2x + y \\ y' &= 2y. \end{aligned} \tag{12.5.49}$$

The equation $\det(A - \lambda I) = 0$ for the coefficient matrix $A = \begin{bmatrix}2 & 1\\0 & 2\end{bmatrix}$ is

$$\begin{vmatrix}2-\lambda & 1\\0 & 2-\lambda\end{vmatrix} = 0 \quad \Rightarrow \quad (\lambda - 2)^2 = 0.$$

Thus, $\lambda = 2$ is a repeated eigenvalue of A. Substituting $\lambda = 2$ into $(A - \lambda I)\mathbf{u} = 0$, we get

$$\begin{bmatrix}0 & 1\\0 & 0\end{bmatrix}\begin{bmatrix}u_1\\u_2\end{bmatrix} = \begin{bmatrix}0\\0\end{bmatrix}$$

or simply

$$\begin{aligned} 0u_1 + 1u_2 &= 0 \\ 0u_1 + 0u_2 &= 0. \end{aligned}$$

Since $u_2 = 0$ and any value of u_1 satisfies both equations, any vector of the form $\langle u_1, 0\rangle$ with $u_1 \neq 0$ is an eigenvector of A. If we choose $u_1 = 1$, then a solution of

(12.5.49) is

$$\mathbf{x}^1(t) = e^{2t}\begin{bmatrix}1\\0\end{bmatrix}.$$

Since the vectors $\langle u_1, 0\rangle$ are the only eigenvectors corresponding to $\lambda = 2$, they are in fact all of the eigenvectors of A since this matrix does not have another eigenvalue (so no associated eigenvectors). What sets this linear system apart from previously considered linear systems is that any two eigenvectors of A are linearly dependent. In other words, no two eigenvectors of this system are linearly independent. This presents a problem for finding two linearly independent solutions of linear systems like this one.

So the task is this: find a second solution $\mathbf{x}^2(t)$ so that it and the solution $\mathbf{x}^1(t)$ are linearly independent. Even though we do not know yet how to accomplish this with eigenvalues and eigenvectors, there is another way to do this for this particular system since it is partially uncoupled. A general solution of the last equation in (12.5.49) is $y(t) = c_2e^{2t}$. Substitution of this function for y in the first equation of (12.5.49) results in the linear first-order scalar equation

$$\frac{dx}{dt} - 2x = c_2e^{2t}.$$

Multiplying by the integrating factor e^{-2t} and integrating, we obtain

$$\frac{d}{dt}(e^{-2t}x) = c_2 \quad \Rightarrow \quad e^{-2t}x = c_2t + c_1 \quad \Rightarrow \quad x(t) = c_1e^{2t} + c_2te^{2t}.$$

Letting $x(t)$ and $y(t)$ denote the components of a vector function $\mathbf{x}(t)$, we have

$$\mathbf{x}(t) = \begin{bmatrix}x(t)\\y(t)\end{bmatrix} = \begin{bmatrix}c_1e^{2t} + c_2te^{2t}\\c_2e^{2t}\end{bmatrix} = c_1e^{2t}\begin{bmatrix}1\\0\end{bmatrix} + c_2e^{2t}\begin{bmatrix}t\\1\end{bmatrix}. \qquad (12.5.50)$$

Setting $c_1 = 0$ and $c_2 = 1$ gives a second solution $\mathbf{x}^2(t)$:

$$\mathbf{x}^2(t) = e^{2t}\begin{bmatrix}t\\1\end{bmatrix}.$$

Clearly, $\mathbf{x}^1(t) = e^{2t}\langle 1, 0\rangle$ and $\mathbf{x}^2(t) = e^{2t}\langle t, 1\rangle$ are linearly independent. Therefore, $\mathbf{x}(t)$ is a general solution of the linear system. ♦

This last example clearly shows that there are linear systems $\mathbf{x}' = A\mathbf{x}$ where no two eigenvectors of the coefficient matrix A are linearly independent. Unfortunately, at this point, the eigenvalue-eigenvector method we have considered so far does not address this situation. However, the general solution that we obtained in the example will help us figure out a way of coming up with a second linearly independent solution that relies only on the eigenvalues and eigenvectors of the coefficient matrix of the system without having to resort to an ad hoc method like in the example. With that said, consider the general linear system

$$\mathbf{x}' = \begin{bmatrix} a & b \\ c & d \end{bmatrix} \mathbf{x}$$

supposing that A has a repeated eigenvalue λ and that no two of its eigenvectors are linearly independent. If $\mathbf{u}$ denotes one of the eigenvectors, then $A\mathbf{u} = \lambda\mathbf{u}$ and

$$\mathbf{x}^1(t) = e^{\lambda t}\mathbf{u}$$

is a solution. Now the goal is to find a second linearly independent solution $\mathbf{x}^2(t)$. This is where the general solution (12.5.50) of the previous example provides some guidance. First, let us rewrite the vector $\langle t, 1\rangle$ in (12.5.50) as $t\langle 1, 0\rangle + \langle 0, 1\rangle$. Then (12.5.50) becomes

$$\mathbf{x}(t) = \begin{bmatrix} x(t) \\ y(t) \end{bmatrix} = c_1 e^{2t} \begin{bmatrix} 1 \\ 0 \end{bmatrix} + c_2 e^{2t} \left(t \begin{bmatrix} 1 \\ 0 \end{bmatrix} + \begin{bmatrix} 0 \\ 1 \end{bmatrix} \right).$$

Second, let $\mathbf{u} = \langle 1, 0\rangle$ and $\mathbf{v} = \langle 0, 1\rangle$. Then

$$\mathbf{x}(t) = c_1 e^{2t}\mathbf{u} + c_2 e^{2t}(t\mathbf{u} + \mathbf{v}).$$

Recall that $\mathbf{u}$ is an eigenvector—but $\mathbf{v}$ is not!

The general solution (12.5.50) of system (12.5.49) rewritten as above suggests that if λ is a repeated eigenvalue of a coefficient matrix A and no two eigenvectors of A are linearly independent that there is a second linearly independent solution of $\mathbf{x}' = A\mathbf{x}$ of the form

$$\mathbf{x}^2(t) = e^{\lambda t}(t\mathbf{u} + \mathbf{v}),$$

where $\mathbf{u}$ denotes any eigenvector of A and $\mathbf{v}$ represents some other vector. Let us determine what condition $\mathbf{v}$ must satisfy for this to be the case. For $\mathbf{x}^2(t)$ to be a solution, it must satisfy $\mathbf{x}' = A\mathbf{x}$. That is,

$$\frac{d}{dt}\mathbf{x}^2(t) = A\mathbf{x}^2(t) \quad \Rightarrow \quad \frac{d}{dt}\big[e^{\lambda t}(t\mathbf{u} + \mathbf{v})\big] = A\big[e^{\lambda t}(t\mathbf{u} + \mathbf{v})\big]$$

for all $-\infty < t < \infty$. Carrying out the differentiation and expanding the right-hand side, we get

$$e^{\lambda t}\mathbf{u} + \lambda e^{\lambda t}t\mathbf{u} + \lambda e^{\lambda t}\mathbf{v} = e^{\lambda t}tA\mathbf{u} + e^{\lambda t}A\mathbf{v}.$$

Since $A\mathbf{u} = \lambda\mathbf{u}$, this simplifies to

$$e^{\lambda t}(\mathbf{u} + \lambda\mathbf{v}) = e^{\lambda t}A\mathbf{v} \quad \Rightarrow \quad A\mathbf{v} = \mathbf{u} + \lambda\mathbf{v}.$$

From this we conclude that $\mathbf{v}$ must be a solution of the equation

$$(A - \lambda I)\mathbf{v} = \mathbf{u}, \tag{12.5.51}$$

where $\mathbf{u}$ is an eigenvector of A.

Finally, there is the matter of showing that $\mathbf{u}$ and $\mathbf{v}$ are linearly independent vectors. If they were not, then $\mathbf{v} = k\mathbf{u}$ for some $k \in \mathbb{R}$ from which it would follow that

$$(A - \lambda I)\mathbf{v} = (A - \lambda I)(k\mathbf{u}) = k(A - \lambda I)\mathbf{u} = k\,\mathbf{0} = \mathbf{0}$$

since $\mathbf{u}$ is an eigenvector of A. But this contradicts (12.5.51) since $\mathbf{u} \neq \mathbf{0}$. Therefore, $\mathbf{u}$ and $\mathbf{v}$ are linearly independent. This implies that the solutions $\mathbf{x}^1(t)$ and $\mathbf{x}^2(t)$ are linearly independent since $\mathbf{x}^1(0) = \mathbf{u}$ and $\mathbf{x}^2(0) = \mathbf{v}$ (cf. Theorem 12.5.22). The result of the foregoing analysis is summarized in (ii) of the next theorem.

Repeated Real Eigenvalues and General Solution

Theorem 12.5.33 *Let λ be a repeated eigenvalue of a 2×2 matrix A.*

(i) *If there are two linearly independent eigenvectors* $\mathbf{u}$ *and* $\mathbf{v}$ *corresponding to the eigenvalue λ, then a general solution of*

$$\mathbf{x}' = A\mathbf{x} \tag{12.5.52}$$

is

$$\mathbf{x}(t) = c_1 e^{\lambda t}\mathbf{u} + c_2 e^{\lambda t}\mathbf{v}, \tag{12.5.53}$$

where c_1, c_2 are arbitrary constants.

(ii) *If no two eigenvectors corresponding to λ are linearly independent, then a general solution of* (12.5.52) *is*

$$\mathbf{x}(t) = c_1 e^{\lambda t}\mathbf{u} + c_2 e^{\lambda t}(t\mathbf{u} + \mathbf{v}), \tag{12.5.54}$$

where $\mathbf{u}$ *is an eigenvector, $c_1, c_2 \in \mathbb{R}$, and* $\mathbf{v}$ *is any vector satisfying the matrix equation*

$$(A - \lambda I)\mathbf{v} = \mathbf{u}. \tag{12.5.55}$$

Remarks

1. Example 12.5.20 illustrates case (i).
2. Multiplying (12.5.55) on the left by the matrix $A - \lambda I$, we get

$$(A - \lambda I)^2 \mathbf{v} = (A - \lambda I)\mathbf{u} = \mathbf{0}$$

since $\mathbf{u}$ is an eigenvector of A. From (12.5.51), we see that $(A - \lambda I)\mathbf{v} \neq \mathbf{0}$ as $\mathbf{u} \neq \mathbf{0}$. Thus,

$$(A - \lambda I)^2 \mathbf{v} = \mathbf{0} \quad \text{but} \quad (A - \lambda I)\mathbf{v} \neq \mathbf{0}. \tag{12.5.56}$$

A vector $\mathbf{v}$ (necessarily nonzero) satisfying (12.5.56), where λ is an eigenvalue of a matrix A, is called a ***generalized eigenvector*** corresponding to λ.

Example 12.5.22 Let us revisit (12.5.49), namely, the linear system

$$x' = 2x + y$$
$$y' = 2y,$$

in order to illustrate how to use Theorem 12.5.33 to find its general solution.

Solution Recall $\mathbf{u} = \langle 1, 0 \rangle$ is an eigenvector corresponding to the repeated eigenvalue $\lambda = 2$ and $\mathbf{x}^1(t) = e^{2t}\langle 1, 0 \rangle$ is a solution of (12.5.49). To find a general solution, we first find a vector $\mathbf{v}$ satisfying (12.5.55):

$$\begin{bmatrix} 0 & 1 \\ 0 & 0 \end{bmatrix} \begin{bmatrix} v_1 \\ v_2 \end{bmatrix} = \begin{bmatrix} 1 \\ 0 \end{bmatrix}.$$

Thus, $v_2 = 1$. Since v_1 is not restricted by some condition, we can assign it any value. Letting $v_1 = 0$, $\mathbf{v} = \langle 0, 1 \rangle$. By Theorem 12.5.33 (ii), a general solution is

$$\begin{aligned} \mathbf{x}(t) &= c_1 e^{\lambda t}\mathbf{u} + c_2 e^{\lambda t}(t\mathbf{u} + \mathbf{v}) \\ &= c_1 e^{2t} \begin{bmatrix} 1 \\ 0 \end{bmatrix} + c_2 e^{2t} \left(t \begin{bmatrix} 1 \\ 0 \end{bmatrix} + \begin{bmatrix} 0 \\ 1 \end{bmatrix} \right) = \begin{bmatrix} c_1 e^{2t} + c_2 t e^{2t} \\ c_2 e^{2t} \end{bmatrix}, \end{aligned}$$

which is what we obtained earlier in Example 12.5.21. ♦

Example 12.5.23 Find a general solution of the linear system

$$x' = 5x - 2y$$
$$y' = 2x + y.$$

Solution The coefficient matrix is

$$A = \begin{bmatrix} 5 & -2 \\ 2 & 1 \end{bmatrix}.$$

Since $\operatorname{tr} A = 6$ and $\det A = 9$, the characteristic equation (see Theorem 12.5.28) is

$$\lambda^2 - 6\lambda + 9 = 0 \quad \Rightarrow \quad (\lambda - 3)^2 = 0.$$

Thus, $\lambda = 3$ is a repeated eigenvalue. The eigenvectors corresponding to this eigenvalue are all of the nonzero vectors such that $A\mathbf{u} = 3\mathbf{u}$, which can be rewritten as

$$(A - 3I)\mathbf{u} = \mathbf{0},$$

which is

$$\begin{bmatrix} 2 & -2 \\ 2 & -2 \end{bmatrix} \begin{bmatrix} u_1 \\ u_2 \end{bmatrix} = \begin{bmatrix} 0 \\ 0 \end{bmatrix}.$$

Thus,

$$2u_1 - 2u_2 = 0, \quad u_2 = u_1.$$

As a result, all eigenvectors of A consists of the set of vectors $k\langle 1, 1\rangle$, where k is any real number. In particular, $\mathbf{u} = \langle 1, 1\rangle$ is an eigenvector. Therefore, one solution of the system is

$$\mathbf{x}^1(t) = e^{3t}\mathbf{u} = e^{3t} \begin{bmatrix} 1 \\ 1 \end{bmatrix}.$$

Since no two eigenvectors of A are linearly independent, let us look for a second solution $\mathbf{x}^2(t)$ of the form

$$\mathbf{x}^2(t) = e^{3t}(t\mathbf{u} + \mathbf{v}),$$

where $\mathbf{v}$ is a solution of $(A - 3I)\mathbf{v} = \mathbf{u}$, namely,

$$\begin{bmatrix} 2 & -2 \\ 2 & -2 \end{bmatrix} \begin{bmatrix} v_1 \\ v_2 \end{bmatrix} = \begin{bmatrix} 1 \\ 1 \end{bmatrix}.$$

Thus,

$$2v_1 - 2v_2 = 1 \quad \Rightarrow \quad v_2 = v_1 - \frac{1}{2}.$$

Let $v_1 = 1/2$. Then $v_2 = 0$. So, $\mathbf{v} = \langle 1/2, 0\rangle$. Therefore, a second solution is

$$\mathbf{x}^2(t) = e^{3t}(t\mathbf{u} + \mathbf{v}) = e^{3t}\left(t\begin{bmatrix}1\\1\end{bmatrix} + \begin{bmatrix}\frac{1}{2}\\0\end{bmatrix}\right) = e^{3t}\begin{bmatrix}t+\frac{1}{2}\\t\end{bmatrix}.$$

Or, any scalar multiple of this solution could serve as $\mathbf{x}^2(t)$, such as

$$2e^{3t}\begin{bmatrix}t+\frac{1}{2}\\t\end{bmatrix} = e^{3t}\begin{bmatrix}2t+1\\2t\end{bmatrix}.$$

According to Theorem 12.5.33 (ii), $\mathbf{x}^1(t)$ and $\mathbf{x}^2(t)$ are linearly independent solutions. Therefore, a general solution of the linear system is

$$\mathbf{x}(t) = c_1e^{3t}\begin{bmatrix}1\\1\end{bmatrix} + c_2e^{3t}\begin{bmatrix}2t+1\\2t\end{bmatrix}.$$

Its components are

$$x(t) = (c_1 + c_2)e^{3t} + 2c_2te^{3t}, \quad y(t) = c_1e^{3t} + 2c_2te^{3t}.$$ ♦

12.5.6.3 Complex Eigenvalues and General Solution

Now that we know what to do when the discriminant of the characteristic equation (12.5.42) is positive or equal to zero, let us take up the case when it is negative, that is, suppose $(\operatorname{tr} A)^2 - 4\det A < 0$. Then the eigenvalues are

$$\frac{\operatorname{tr} A \pm \sqrt{(\operatorname{tr} A)^2 - 4\det A}}{2} = \frac{\operatorname{tr} A \pm \sqrt{-1}\sqrt{4\det A - (\operatorname{tr} A)^2}}{2}.$$

Note that the number

$$\sqrt{4\det A - (\operatorname{tr} A)^2}$$

is real. On the other hand, $\sqrt{-1}$ is not because its square is negative. It is known as an ***imaginary number*** (or ***complex number***) and is denoted by the letter i.[11] And so

$$i := \sqrt{-1} \quad \text{and} \quad i^2 = -1.$$

[11] Except in some disciplines, such as electrical engineering, where j is generally used because i is reserved for representing electrical current.

Letting

$$\alpha := \frac{1}{2}\operatorname{tr} A \quad \text{and} \quad \omega := \frac{1}{2}\sqrt{4\det A - (\operatorname{tr} A)^2},$$

we can write the two eigenvalues as $\alpha + i\omega$ and $\alpha - i\omega$, thereby making them less wieldy to deal with.

Numbers expressed in the form $a + ib$, where b is a nonzero real number, are called ***complex numbers***. Since $\omega > 0$, the complex number i appears in both eigenvalues which makes both of them complex numbers. It follows that it is impossible for a 2×2 matrix to have one real eigenvalue and one complex eigenvalue. In other words, both eigenvalues must be complex numbers or both must be real numbers. Since complex eigenvalues always occur as an $\alpha \pm i\omega$ pair, they are said to be ***complex conjugates*** of each other. The word *conjugate* used in this sense means *joined together in a pair*. Ordinarily, a bar is placed over a complex number to denote its complex conjugate; with this notation, the two eigenvalues are

$$\lambda = \alpha + i\omega \quad \text{and} \quad \overline{\lambda} = \alpha - i\omega.$$

The real number α is called the ***real part*** of the complex number λ. Typically, the notation $\operatorname{Re}\lambda$ is used to denote the real part of λ: thus, $\operatorname{Re}\lambda = \alpha$. The real number ω next to i is called the ***imaginary part*** of λ and we write $\operatorname{Im}\lambda = \omega$. Note that the imaginary part does not include the i. Likewise, $\operatorname{Re}\overline{\lambda} = \alpha$ and $\operatorname{Im}\overline{\lambda} = -\omega$.

Complex constants are handled in the same way as are real constants when it comes to algebraic manipulations and applications of the rules of differentiation. So it seems reasonable to expect that Theorem 12.5.31 also applies to distinct complex eigenvalues. If that is the case, then two eigenvectors, say $\mathbf{v}$ and $\overline{\mathbf{v}}$, corresponding to the eigenvalues λ and $\overline{\lambda}$, respectively, are linearly independent. Let us proceed formally by supposing this and that all of the previous results and theorems also pertain to complex eigenvalues and eigenvectors and complex-valued solutions. Then, it would follow from Theorems 12.5.22, 12.5.27, and 12.5.32 that a general solution of the linear system $\mathbf{x}' = A\mathbf{x}$ expressed in terms of vector functions with complex components is

$$\mathbf{x}(t) = c_1 e^{\lambda t}\mathbf{v} + c_2 e^{\overline{\lambda} t}\overline{\mathbf{v}} = c_1 e^{(\alpha + i\omega)t}\mathbf{v} + c_2 e^{(\alpha - i\omega)t}\overline{\mathbf{v}}. \tag{12.5.57}$$

But does this solution make sense? For instance, we know the meaning of $\exp(kt)$ when k is a real constant, but what does it mean when k is a complex constant? We will address this by first constructing a relatively simple linear sustem with complex eigenvalues and complex-valued solutions: the homogeneous linear system (12.5.58) below to see if it has any real-valued solutions. And if so, can they be obtained using complex numbers and complex-valued functions?

Consider the linear system

$$x' = y, \quad y' = -x. \tag{12.5.58}$$

First let us determine the eigenvalues λ and $\overline{\lambda}$ together with corresponding eigenvectors $\mathbf{v}$ and $\overline{\mathbf{v}}$ to ascertain if we can make some sense of (12.5.57) for this system. From

$$\det(A - \lambda I) = \det\left(\begin{bmatrix} 0 & 1 \\ -1 & 0 \end{bmatrix} - \lambda \begin{bmatrix} 1 & 0 \\ 0 & 1 \end{bmatrix}\right) = \begin{vmatrix} -\lambda & 1 \\ -1 & -\lambda \end{vmatrix} = 0,$$

we obtain the characteristic equation

$$\lambda^2 + 1 = 0.$$

Thus, the eigenvalues are $\lambda = i$ and $\overline{\lambda} = -i$. An eigenvector $\mathbf{v} = \langle v_1, v_2 \rangle$ corresponding to $\lambda = i$ is a solution of $(A - \lambda I)\mathbf{v}) = 0$:

$$\begin{bmatrix} -i & 1 \\ -1 & -i \end{bmatrix} \begin{bmatrix} v_1 \\ v_2 \end{bmatrix} = \begin{bmatrix} 0 \\ 0 \end{bmatrix}.$$

Hence, $v_2 = iv_1$. Setting $v_1 = 1$ results in the eigenvector $\mathbf{v} = \langle 1, i \rangle$. Similarly, an eigenvector $\overline{\mathbf{v}} = \langle \overline{v}_1, \overline{v}_2 \rangle$ corresponding to $\overline{\lambda} = -i$ is a solution of

$$\begin{bmatrix} i & 1 \\ -1 & i \end{bmatrix} \begin{bmatrix} \overline{v}_1 \\ \overline{v}_2 \end{bmatrix} = \begin{bmatrix} 0 \\ 0 \end{bmatrix}.$$

So $\overline{v}_2 = -i\overline{v}_1$. Setting $\overline{v}_1 = 1$ yields the eigenvector $\overline{\mathbf{v}} = \langle 1, -i \rangle$. As a result, it follows from (12.5.57) that

$$\mathbf{x}(t) = c_1 e^{\lambda t}\mathbf{v} + c_2 e^{\overline{\lambda} t}\overline{\mathbf{v}} = c_1 e^{it} \begin{bmatrix} 1 \\ i \end{bmatrix} + c_2 e^{-it} \begin{bmatrix} 1 \\ -i \end{bmatrix}. \tag{12.5.59}$$

So it appears that

$$\begin{aligned} x(t) &= c_1 e^{it} + c_2 e^{-it} \\ y(t) &= c_1 i e^{it} - c_2 i e^{-it}. \end{aligned} \tag{12.5.60}$$

is a complex-valued general solution, provided that there is a way of defining e^{it} and e^{-it} that makes sense and is compatible with previous results. ♦

With the hope of sorting this out, let us look for real-valued solutions of (12.5.58). Note that by differentiating the first equation in (12.5.58), we can eliminate y:

$$x'' = y' = -x \quad \Rightarrow \quad x'' = -x. \tag{12.5.61}$$

To find a solution of (12.5.61), all we have to do is to find a function whose second derivative is the negative of itself? Certainly $\sin t$ is such a function, as are all multiples of $\sin t$, such as $3 \sin t$ and $\pi \sin t$. Are there solutions that are not multiples of $\sin t$? Of course, $\cos t$ and all multiples of it. The point is that we now have two linearly independent real-valued solutions of (12.5.61), namely, $\sin t$ and $\cos t$.

Into the mix, let us add the initial conditions $x(0) = 1$, $x'(0) = 0$ to (12.5.58); then look for the solution of the initial value problem

$$\begin{aligned} x' &= y; \quad x(0) = 1, \\ y' &= -x; \quad x'(0) = 0. \end{aligned} \tag{12.5.62}$$

With the elimination of y, this is equivalent to the initial value problem

$$x'' + x = 0; \quad x(0) = 1, \ x'(0) = 0. \tag{12.5.63}$$

Now we can get the solution of (12.5.62) by finding the solution of (12.5.63). Knowing already that $\sin t$ and $\cos t$ are solutions of the second-order differential equation makes this an easy task. The solution $x(t) = \cos t$ is the one we need here since it satisfies both initial conditions. Then, of course,

$$y(t) = x'(t) = -\sin t.$$

But we can also can get complex-valued solutions of (12.5.62) by setting $t = 0$ in (12.5.60):

$$\begin{aligned} x(0) &= c_1 e^{i0} + c_2 e^{-i0} = c_1 + c_2 = 1 \\ y(0) &= c_1 i e^{i0} - c_2 i e^{-i0} = c_1 i - c_2 i = 0 \end{aligned}$$

as $e^{i0} = e^0 = 1$. Clearly, $c_1 = c_2 = 1/2$. As a result,

$$\begin{aligned} x(t) &= \frac{1}{2} e^{it} + \frac{1}{2} e^{-it} \\ y(t) &= \frac{1}{2} i e^{it} - \frac{1}{2} i e^{-it} = \frac{i(e^{it} - e^{-it})}{2} \end{aligned}$$

or

$$x(t) = \frac{e^{it} + e^{-it}}{2}, \quad y(t) = -\frac{e^{it} - e^{-it}}{2i}$$

is a solution of (12.5.62).

So apparently we have two different solutions of the initial value problem (12.5.62), namely, these complex-valued solutions and $x(t) = \cos t$ and $y(t) = -\sin t$. However, by the existence and uniqueness theorem (Theorem 12.5.14), the solution of this initial value problem is unique. Since that is the case, we see that e^{it} has to be defined in such a way so that

$$\cos t = \frac{e^{it} + e^{-it}}{2} \quad \text{and} \quad \sin t = \frac{e^{it} - e^{-it}}{2i}. \tag{12.5.64}$$

Adding $\cos t$ and $i \sin t$, we have

$$\cos t + i \sin t = \frac{e^{it} + e^{-it}}{2} + \frac{e^{it} - e^{-it}}{2} = e^{it}.$$

This formula, which is known as ***Euler's formula***, defines e^{it}.

Euler's Formula

Definition 12.5.34 For $t \in \mathbb{R}$,

$$e^{it} := \cos t + i \sin t. \tag{12.5.65}$$

The formula

$$e^{-it} = \cos t - i \sin t. \tag{12.5.66}$$

is obtained by subtracting $i \sin t$ from $\cos t$ using formulas (12.5.64)—or replacing t with $-t$ in Euler's formula. Note that e^{-it} is the complex conjugate of e^{it}.

Furthermore, by replacing the real variable t with $\pm \omega t$ in both (12.5.65) and (12.5.66), where ω denotes any real constant, we have

$$e^{\pm i\omega t} = \cos \omega t \pm i \sin \omega t. \tag{12.5.67}$$

Defining $e^{(\alpha \pm i\omega)t}$ to be $e^{\alpha t} \cdot e^{\pm i\omega t}$, we have the following generalized version.

Euler's Formula: Generalized Version

Definition 12.5.35 Let α and ω be real constants.

$$e^{(\alpha \pm i\omega)t} := e^{\alpha t} \left(\cos \omega t \pm i \sin \omega t\right). \tag{12.5.68}$$

for all $t \in \mathbb{R}$.

The differentiation formulas

$$\frac{d}{dt}e^{it} = ie^{it} \quad \text{and} \quad \frac{d}{dt}e^{-it} = -ie^{-it} \tag{12.5.69}$$

are consequences of Euler's formula and (12.5.66). For instance,

$$\frac{d}{dt}e^{it} = \frac{d}{dt}(\cos t + i\sin t) = -\sin t + i\cos t = i(\cos t + i\sin t = ie^{it}.$$

Moreover, it follows from these differentiation formulas and (12.5.68) that

$$\frac{d}{dt}e^{(\alpha \pm i\omega)t} = (\alpha \pm i\omega)e^{(\alpha \pm i\omega)t}. \tag{12.5.70}$$

Therefore, for both real and complex constants k,

$$\frac{d}{dt}e^{kt} = ke^{kt}.$$

Other important properties of the complex exponential function are the following: Let $\theta, \phi \in \mathbb{R}$ and $n = 0, \pm 1, \pm 2, \ldots$. Then

1. $e^{i\phi}e^{\pm i\theta} = e^{i(\phi \pm \theta)}$ (Problem 25).
2. $\left(e^{i\theta}\right)^n = e^{in\theta}$ (Problem 26).
3. $(\cos\theta + i\sin\theta)^n = \cos n\theta + i\sin n\theta$ (***De Moivre's formula***: see Problem 27).

We can learn more about complex eigenvalues and eigenvectors by examining the solutions of the linear system (12.5.58) and observing the following. First, of course, the eigenvalues and eigenvectors are complex.[12] Second, the eigenvector

$$\overline{\mathbf{v}} = \begin{bmatrix} 1 \\ -i \end{bmatrix} = \begin{bmatrix} 1 \\ 0 \end{bmatrix} - i\begin{bmatrix} 0 \\ 1 \end{bmatrix}$$

corresponding to the eigenvalue $\overline{\lambda} = -i$ is the complex conjugate of the eigenvector

$$\mathbf{v} = \begin{bmatrix} 1 \\ i \end{bmatrix} = \begin{bmatrix} 1 \\ 0 \end{bmatrix} + i\begin{bmatrix} 0 \\ 1 \end{bmatrix}$$

corresponding to the eigenvalue $\lambda = i$. This is the reason for placing the complex conjugate bar over $\mathbf{v}$ in the first place. Third, the two vector solutions

$$\mathbf{x}^1(t) = e^{it}\begin{bmatrix} 1 \\ i \end{bmatrix} = \begin{bmatrix} \cos t + i\sin t \\ -\sin t + i\cos t \end{bmatrix}$$

[12] An eigenvector is ***complex*** if at least one of its components is a complex number.

and

$$\mathbf{x}^2(t) = e^{-it}\begin{bmatrix} 1 \\ -i \end{bmatrix} = \begin{bmatrix} \cos t - i\sin t \\ -\sin t - i\cos t \end{bmatrix}$$

are complex conjugates of each other. Fourth, even though $\mathbf{x}^1(t)$ and $\mathbf{x}^2(t)$ are complex-valued solutions, we can obtain real-valued solutions by taking the following linear combinations of them:[13]

$$\frac{1}{2}\mathbf{x}^1(t) + \frac{1}{2}\mathbf{x}^2(t) = \frac{1}{2}\begin{bmatrix} \cos t + i\sin t \\ -\sin t + i\cos t \end{bmatrix} + \frac{1}{2}\begin{bmatrix} \cos t - i\sin t \\ -\sin t - i\cos t \end{bmatrix} = \begin{bmatrix} \cos t \\ -\sin t \end{bmatrix}$$

and

$$\frac{1}{2i}\mathbf{x}^1(t) - \frac{1}{2i}\mathbf{x}^2(t) = \frac{1}{2i}\begin{bmatrix} \cos t + i\sin t \\ -\sin t + i\cos t \end{bmatrix} - \frac{1}{2i}\begin{bmatrix} \cos t - i\sin t \\ -\sin t - i\cos t \end{bmatrix} = \begin{bmatrix} \sin t \\ \cos t \end{bmatrix}.$$

And note this: These two real-valued vector solutions are the real and imaginary parts of the complex-valued solution $\mathbf{x}^1(t)$. Furthermore, they are linearly independent since they are obviously not real scalar multiples of one another. Therefore, the real-valued vector function

$$\mathbf{x}(t) = c_1\begin{bmatrix} \cos t \\ -\sin t \end{bmatrix} + c_2\begin{bmatrix} \sin t \\ \cos t \end{bmatrix}.$$

is also a general solution of

$$\mathbf{x}' = \begin{bmatrix} 0 & 1 \\ -1 & 0 \end{bmatrix}\mathbf{x},$$

which is the vector form of (12.5.58).

We have to wonder if the foregoing observations and remarks for (12.5.58) will carry over to any linear system with complex eigenvalues. Let us delve into this further by taking a look at a linear system $\mathbf{x}' = A\mathbf{x}$, where A is a 2×2 matrix with real entries whose eigenvalues are $\lambda = \alpha + i\omega$ and $\overline{\lambda} = \alpha - i\omega$. We see from $A\mathbf{v} = \lambda\mathbf{v}$ that an eigenvector $\mathbf{v}$ corresponding to a complex eigenvalue λ must be complex: otherwise, if $\mathbf{v}$ were real, then $A\mathbf{v}$ would be real but $\lambda\mathbf{v}$ would be complex. Accordingly, we conclude:

1. Eigenvectors corresponding to complex eigenvalues are complex vectors.

Taking the complex conjugate of both sides of $A\mathbf{v} = \lambda\mathbf{v}$, we have

$$\overline{A\mathbf{v}} = \overline{\lambda\mathbf{v}} \quad \Rightarrow \quad A\overline{\mathbf{v}} = \overline{\lambda}\overline{\mathbf{v}}$$

[13] Linear combinations of solutions are also solutions by the Superposition Principle.

since the conjugate of a product is the product of the conjugates (see Problem 30). Since all of the entries of the matrix A are real, $\overline{A} = A$ and so $A\overline{\mathbf{v}} = \overline{\lambda}\overline{\mathbf{v}}$. This proves:

2. If $\mathbf{v}$ is an eigenvector corresponding to a complex eigenvalue $\lambda = \alpha + i\omega$ of a matrix A with real entries, then its complex conjugate $\overline{\mathbf{v}}$ is an eigenvector corresponding to the eigenvalue $\overline{\lambda} = \alpha - i\omega$ of A.

The corresponding solutions are

$$\mathbf{x}^1(t) = e^{\lambda t}\mathbf{v} = e^{(\alpha+i\omega)t}\mathbf{v} \quad \text{and} \quad \mathbf{x}^2(t) = e^{\overline{\lambda}t}\overline{\mathbf{v}} = e^{(\alpha-i\omega)t}\overline{\mathbf{v}}.$$

Since $\mathbf{x}^1(0) = \mathbf{v}$ and $\mathbf{x}^2(0) = \overline{\mathbf{v}}$ are linearly independent vectors (Theorem 12.5.31), it follows from Theorem 12.5.22 that $\mathbf{x}^1(t)$ and $\mathbf{x}^2(t)$ are linearly independent solutions. Let the vectors $\mathbf{v}^1$ and $\mathbf{v}^2$ denote the real and imaginary parts of the vector $\mathbf{v}$. That is, $\mathbf{v} = \mathbf{v}^1 + i\mathbf{v}^2$. Replacing $\mathbf{v}$ in $\mathbf{x}^1(t)$ with $\mathbf{v}^1 + i\mathbf{v}^2$, multiplying out, and separating the real terms from the imaginary terms, we obtain

$$\begin{aligned}\mathbf{x}^1(t) &= e^{(\alpha+i\omega)t}\mathbf{v} = e^{\alpha t}(\cos\omega t + i\sin\omega t)(\mathbf{v}^1 + i\mathbf{v}^2)\\ &= e^{\alpha t}(\mathbf{v}^1\cos\omega t - \mathbf{v}^2\sin\omega t) + ie^{\alpha t}(\mathbf{v}^1\sin\omega t + \mathbf{v}^2\cos\omega t).\end{aligned}$$

Consequently, we see that the real and imaginary parts of $\mathbf{x}^1(t)$ are

$$\operatorname{Re}\mathbf{x}^1(t) = e^{\alpha t}(\mathbf{v}^1\cos\omega t - \mathbf{v}^2\sin\omega t) \tag{12.5.71}$$

and

$$\operatorname{Im}\mathbf{x}^1(t) = e^{\alpha t}(\mathbf{v}^1\sin\omega t + \mathbf{v}^2\cos\omega t). \tag{12.5.72}$$

We conclude:

3. A complex-valued solution $\mathbf{x}^1(t) = e^{\lambda t}\mathbf{v} = e^{(\alpha+i\omega)t}\mathbf{v}$ can be separated into real and imaginary parts as follows: Let $\mathbf{v}^1$ and $\mathbf{v}^2$ denote the real and imaginary parts of $\mathbf{v}$. Then

$$\mathbf{x}^1(t) = \operatorname{Re}\mathbf{x}^1(t) + i\operatorname{Im}\mathbf{x}^1(t), \tag{12.5.73}$$

where $\operatorname{Re}\mathbf{x}^1(t)$ and $\operatorname{Im}\mathbf{x}^1(t)$ are given by (12.5.71) and (12.5.72), respectively.

Since $\mathbf{x}^2(t)$ is the complex conjugate of $\mathbf{x}^1(t)$,

$$\mathbf{x}^2(t) = \operatorname{Re}\mathbf{x}^1(t) - i\operatorname{Im}\mathbf{x}^1(t). \tag{12.5.74}$$

From (12.5.73) and (12.5.74), it follows that

$$\operatorname{Re}\mathbf{x}^1(t) = \frac{1}{2}\left(\mathbf{x}^1(t) + \mathbf{x}^2(t)\right); \quad \operatorname{Im}\mathbf{x}^1(t) = \frac{1}{2i}\left(\mathbf{x}^1(t) - \mathbf{x}^2(t)\right). \tag{12.5.75}$$

Since this shows that Re $\mathbf{x}^1(t)$ and Im $\mathbf{x}^1(t)$ are linear combinations of the solutions $\mathbf{x}^1(t)$ and $\mathbf{x}^2(t)$, they, too, are solutions by the Superposition Principle—moreover, they are real-valued solutions. Furthermore, as $\mathbf{x}^1(t)$ and $\mathbf{x}^2(t)$ are linearly independent, so are they. The proof of this is left as an exercise. Thus, we conclude:

4. The functions Re $\mathbf{x}^1(t)$ and Im $\mathbf{x}^1(t)$ given by (12.5.75) are linearly independent real-valued solutions of the linear system $\mathbf{x}' = A\mathbf{x}$. Moreover, it follows from Theorem 12.5.27 that a general solution of this system is

$$\mathbf{x}(t) = c_1 \operatorname{Re} \mathbf{x}^1(t) + c_2 \operatorname{Im} \mathbf{x}^1(t) = c_1 \operatorname{Re} \left(e^{\lambda t}\mathbf{v}\right) + c_2 \operatorname{Im} \left(e^{\lambda t}\mathbf{v}\right).$$

The foregoing discussion and results are summarized in the following theorem.

Complex Eigenvalues and General Solution

Theorem 12.5.36 *Let $\lambda = \alpha + i\omega$ be a complex eigenvalue of a matrix A of order 2 with real entries. Let $\mathbf{v}$ be an eigenvector corresponding to λ.*

(i) *A real-valued general solution of $\mathbf{x}' = A\mathbf{x}$ is a linear combination of the real and imaginary parts of the complex-valued solution $e^{\lambda t}\mathbf{v}$. That is,*

$$\mathbf{x}(t) = c_1 \operatorname{Re} \left(e^{\lambda t}\mathbf{v}\right) + c_2 \operatorname{Im} \left(e^{\lambda t}\mathbf{v}\right),$$

where c_1, c_2 are arbitrary real constants.

(ii) *The real and imaginary parts of $e^{\lambda t}\mathbf{v}$ are given by* (12.5.71) *and* (12.5.72)*, respectively.*

Example 12.5.24 Find a general solution of the system

$$\begin{aligned} x' &= x + 2y \\ y' &= -2x + y. \end{aligned}$$

Solution Since the coefficient matrix is

$$A = \begin{bmatrix} 1 & 2 \\ -2 & 1 \end{bmatrix},$$

the characteristic equation is

$$\det(A - \lambda I) = \begin{vmatrix} 1-\lambda & 2 \\ -2 & 1-\lambda \end{vmatrix} = 0 \quad \Rightarrow \quad (\lambda - 1)^2 + 4 = 0.$$

Therefore, the eigenvalues are

$$\lambda = 1 + 2i \quad \text{and} \quad \overline{\lambda} = 1 - 2i.$$

To get two real-valued linearly independent solutions, all we have to do, according to Theorem 12.5.36, is to first find a complex-valued solution corresponding to one of the eigenvalues, after which we separate it into real and imaginary parts. A complex-valued solution $\mathbf{x}(t) = e^{\lambda t}\mathbf{v}$ corresponding to the eigenvalue $\lambda = 1 + 2i$ is

$$\mathbf{x}(t) = e^{(1+2i)t}\mathbf{v},$$

where $\mathbf{v}$ is a solution of

$$\big[A - (1 + 2i)I\big]\mathbf{v} = \mathbf{0} \quad \text{or} \quad \begin{bmatrix} -2i & 2 \\ -2 & -2i \end{bmatrix}\begin{bmatrix} v_1 \\ v_2 \end{bmatrix} = \begin{bmatrix} 0 \\ 0 \end{bmatrix}.$$

Hence, $v_2 = iv_1$. Selecting v_1 as the free variable and setting it equal to 1, we get the eigenvector $\mathbf{v} = \langle 1, i\rangle$. Therefore,

$$\begin{aligned} \mathbf{x}(t) &= e^{(1+2i)t}\begin{bmatrix} 1 \\ i \end{bmatrix} = e^{t}(\cos 2t + i\sin 2t)\begin{bmatrix} 1 \\ i \end{bmatrix} \\ &= e^{t}\begin{bmatrix} \cos 2t + i\sin 2t \\ -\sin 2t + i\cos 2t \end{bmatrix} = e^{t}\begin{bmatrix} \cos 2t \\ -\sin 2t \end{bmatrix} + ie^{t}\begin{bmatrix} \sin 2t \\ \cos 2t \end{bmatrix}. \end{aligned}$$

By Theorem 12.5.36, a real-valued general solution is

$$\mathbf{x}(t) = c_1 e^{t}\begin{bmatrix} \cos 2t \\ -\sin 2t \end{bmatrix} + c_2 e^{t}\begin{bmatrix} \sin 2t \\ \cos 2t \end{bmatrix}.$$

Thus,

$$x(t) = e^{t}(c_1\cos 2t + c_2\sin 2t), \quad y(t) = e^{t}(-c_1\sin 2t + c_2\cos 2t).$$ ♦

Example 12.5.25 Find a general solution of

$$\ddot{x} + \omega^2 x = 0 \quad (\omega > 0). \tag{12.5.76}$$

This is the equation of motion of a simple harmonic oscillator whose general solution has already been determined in Chap. 10 using the method of characteristic roots. Nevertheless, let us again find a general solution in order to illustrate the methods of this chapter.

Solution First convert (12.5.76) to an equivalent system by defining $y = \dot{x}$. Then $\dot{y} = \ddot{x} = -\omega^2 x$. Thus, an equivalent system is

$$\begin{aligned} \dot{x} &= y \\ \dot{y} &= -\omega^2 x. \end{aligned} \tag{12.5.77}$$

In vector form, this is

$$\dot{\mathbf{x}} = A\mathbf{x}$$

where

$$\mathbf{x} = \langle x, y \rangle \quad \text{and} \quad A = \begin{bmatrix} 0 & 1 \\ -\omega^2 & 0 \end{bmatrix}.$$

Since $\operatorname{tr} A = 0$ and $\det A = \omega^2$, the characteristic equation is

$$\lambda^2 + \omega^2 = 0.$$

Consequently, the eigenvalues are

$$\lambda = i\omega \quad \text{and} \quad \overline{\lambda} = -i\omega.$$

A complex-valued solution corresponding to $\lambda = i\omega$ is $e^{i\omega t}\mathbf{v}$, where $\mathbf{v} = \langle v_1, v_2 \rangle$ is a nonzero vector satisfying $A\mathbf{v} = (i\omega)\mathbf{v}$, that is,

$$\begin{bmatrix} -i\omega & 1 \\ -\omega^2 & -i\omega \end{bmatrix} \begin{bmatrix} v_1 \\ v_2 \end{bmatrix} = \begin{bmatrix} 0 \\ 0 \end{bmatrix}.$$

From this, we see that $v_2 = i\omega v_1$. Setting $v_1 = 1$, we get $\mathbf{v} = \langle 1, i\omega \rangle$. Thus,

$$e^{i\omega t}\mathbf{v} = e^{i\omega t} \begin{bmatrix} 1 \\ i\omega \end{bmatrix} = (\cos \omega t + i \sin \omega t) \begin{bmatrix} 1 \\ i\omega \end{bmatrix} = \begin{bmatrix} \cos \omega t \\ -\omega \sin \omega t \end{bmatrix} + i \begin{bmatrix} \sin \omega t \\ \omega \cos \omega t \end{bmatrix}.$$

It follows from Theorem 12.5.36 that a real-valued solution of (12.5.77) is

$$\mathbf{x}(t) = c_1 \operatorname{Re}\left(e^{i\omega t}\mathbf{v}\right) + c_2 \operatorname{Im}\left(e^{i\omega t}\mathbf{v}\right) = c_1 \begin{bmatrix} \cos \omega t \\ -\omega \sin \omega t \end{bmatrix} + c_2 \begin{bmatrix} \sin \omega t \\ \omega \cos \omega t \end{bmatrix}.$$

We conclude that a general solution of (12.5.76) is

$$x(t) = c_1 \cos \omega t + c_2 \sin \omega t.$$

♦

12.6 Modeling Room Temperatures

12.6.1 One-Room Building

The goal of this section is to derive a temperature model for a two-room building using a system of two linear differential equations. Then we will illustrate how to employ the methods of this and the next chapter to find solutions of this temperature model in order to predict the temperatures of both rooms at any time provided the temperatures are known at some earlier time. Unless it is already clear from the start how to construct a mathematical model of some real-world phenomenon, one could attempt to first model something similar but simpler: here that would be a one-room building. If a temperature model for the one-room building can accurately predict the inside temperature, then that would presumably give us some idea how to tackle a two-room building. For that reason, let us begin with a one-room building where $T(t)$ denotes the temperature of the room at time t. In order to avoid having to contend with too many details right from the start, let us suppose for now that the building's HVAC (Heating, Ventilation, and Air Conditioning) system is switched off and that all of the appliances, computers, lights, and the like are also turned off and that no one is in the building. Since there are no sources of heat in the room, its temperature will be solely determined by the ambient (outside) temperature, which is subject to the whim of nature. For the sake of simplicity and to illustrate the results of the previous sections in this chapter, let us suppose that nature is mathematically kind to us in that the outdoor temperature is and will remain a constant T_a degrees.

With the HVAC system off, the temperature of the room will fall if it exceeds the outdoor temperature since heat from the room will be lost to the outdoors, and it will continue to fall until it is the same as the outdoor temperature. On the other hand, if the temperature outdoors is higher than the temperature of the room, the room's temperature will rise until it attains the higher outdoor temperature since the outdoors is now supplying the room with heat. ***Heat*** is thermal energy, the form of energy associated with temperature, that moves in the direction of decreasing temperature, such as the thermal energy transferred from a hot frying pan to a child's wayward finger (Ouch!) or the energy lost by a warm room when it is cold outdoors. In the SI system of units, the unit of heat is the *joule* (newton-meter). However, in technical applications, such as heating and cooling systems for buildings, many industries still use and prefer the English system of units. In this system, the unit of heat is the ***British thermal unit*** (***BTU***), which is the amount of heat that is needed to raise the temperature of 1 pound of water from 63 to 64 °F. This is approximately 252 calories[14] or 1055 joules.

[14] The ***15 °C calorie*** (also called the "small calorie" or "gram-calorie") was originally defined as the amount of heat that will raise the temperature of 1 gram of water from 14.5 to 15.5 °C. This was the definition until it was realized that heat was a form of energy. After that it was decided that the SI unit for heat should be the same as for other forms of energy. The current definition of the 15 °C calorie is that it is exactly 4.1855 joules. Incidentally, the "calorie" for measuring the

Regardless of what units of heat are chosen, let Q represent the total quantity of heat gained or lost by a body or system of bodies, such as the heat gained by an ice-cold can of soda taken from the refrigerator or the heat lost by an unheated room, which is composed of many bodies: the walls, floor, ceiling, the objects in it, and the enclosed air. We follow the convention that Q has a positive numerical value when a body or system gains heat and a negative value when a body or system loses heat. From now on, we will use the term "system" to mean a collection of bodies or just one body in order to avoid the clumsy phrase "body or system of bodies." If Q can be regarded as a function of the time t, its time derivative (that is, $\dot{Q} = dQ/dt$) represents the rate at which heat is gained or lost by a system and is called the ***heat transfer rate***. If $\dot{Q}$ is positive, its value is the rate at which heat is gained by a system; if $\dot{Q}$ is negative, its absolute value is the rate at which the system loses heat.

When a system absorbs a certain quantity of heat, which we denote by ΔQ, its temperature will rise by a certain amount, say ΔT. The ***heat capacity*** of a system is defined as the ratio of ΔQ to ΔT.[15] Using C to denote this quantity, we have

$$C := \frac{\Delta Q}{\Delta T}. \tag{12.6.1}$$

In other words, the heat capacity is the quantity of heat that is needed to raise a system's temperature by 1 degree. In the English system of units, the unit of heat capacity is 1 BTU per degree Fahrenheit (BTU/°F). In reality, the value of (12.6.1) depends not only on ΔT but also on the initial temperature.[16] Despite this, for temperatures that are not extreme and for moderate temperature changes, experiments indicate that the heat capacities of certain kinds of systems are more or less constant. In other words, for these systems, the difference quotient $\Delta Q/\Delta T$ is nearly constant. Consequently,

$$\frac{dQ}{dT} = \lim_{\Delta T \to 0} \frac{\Delta Q}{\Delta T} \approx \frac{\Delta Q}{\Delta T} = C.$$

In short, for moderate temperatures and temperature changes, the derivative dQ/dT is essentially constant[17] and gives the heat capacity of a body; that is,

---

energy content of food is actually the so-called "large calorie", which is 1 kilocalorie (1000 small calories).

[15] See any general physics textbook, such as Halliday and Resnick [43, Ch. 18].

[16] There are also other conditions affecting the heat capacity of a system, such as pressure. The assertions concerning heat capacity and the units for heat (calorie, BTU, etc.) are for systems subjected to a constant 1 atmosphere of pressure as heat is being added to the system.

[17] In truth, the derivative dQ/dT is the actual definition of the heat capacity of an object. As a rule, it is not constant. The difference quotient $\Delta Q/\Delta T$ should really be called the ***average heat capacity*** over the temperature range ΔT; as mentioned before, its value depends on the location of ΔT on the temperature scale. See Wikipedia (*https://en.wikipedia.org/wiki/Heat_capacity*) or a thermodynamics textbook, such as Battino and Wood [7, pp. 78–79.].

$$C = \frac{dQ}{dT}. \tag{12.6.2}$$

As a result, we can regard the heat capacity of a room as a constant. Because of this and the chain rule of differentiation, this constant, which we now denote by C_R, allows us to relate the heat transfer rate $\dot{Q}$ to the rate of change of the temperature T of the room as follows:

$$\frac{dQ}{dt} = \frac{dQ}{dT} \cdot \frac{dT}{dt} = C_R \frac{dT}{dt}. \tag{12.6.3}$$

Now we are all set to use the foregoing information to construct a mathematical model of the time-varying temperature $T(t)$ inside a one-room building. Suppose the outside temperature is a constant T_a degrees and that at some point in time, which we designate as $t = 0$, the room's temperature T_0 is higher than the outside temperature T_a. Suppose further that the HVAC system in the one-room building is turned off. So the room receives no heat. With this additional information, we can model the room temperature by knowing the rate at which it changes. To this end, it seems reasonable to hypothesize that the rate at which the room loses heat to the outdoors is proportional to the difference in the two temperatures. Expressed in the language of differential equations, this hypothesis is

$$\frac{dQ}{dt} = -K_a(T - T_a), \tag{12.6.4}$$

where the constant of proportionality K_a is taken to be positive. The negative sign is affixed to the right-hand side of (12.6.4) because the heat transfer rate on the left-hand side is a negative quantity since heat is being lost to the outdoors, whereas both $T - T_a$ and K_a on the right-hand side are positive. Let us refer to the constant K_a as the ***heat transfer coefficient***.[18] It is a constant in that it does not depend on t, T_a, or T. It is a parameter in that it does depend on the physical properties of the room being modeled: the number of doors and windows in the room and how well they are sealed, the type of insulation used in the walls and ceiling, and so forth. It follows from (12.6.3) and (12.6.4) that

$$C_R \frac{dT}{dt} = -K_a(T - T_a).$$

Thus, we have

$$\frac{dT}{dt} = -\frac{K_a}{C_R}(T - T_a).$$

[18] Strictly speaking, the coefficient K_a is not really the heat transfer coefficient as it is defined in a branch of applied thermodynamics called *heat transfer*. However, it is a multiple of the bona fide heat transfer coefficient. So we have no misgivings in applying that term to K_a. See White [77, p. 20, p. 511].

So, due to (12.6.3), hypothesis (12.6.4) is essentially Newton's law of cooling,[19] which, in the context of this example, says that the rate of change of the room's temperature is proportional to the difference between the room's temperature and the outside temperature. Defining the constant λ_a as the ratio K_a/C_R, the model becomes

$$\frac{dT}{dt} = -\lambda_a(T - T_a). \tag{12.6.5}$$

Equation (12.6.5) is a linear differential equation that can be solved by separating variables or employing an integrating factor. Be that as it may, we can also solve the equation quickly with this simple trick: since T_a is a constant, we can rewrite it as

$$\frac{d}{dt}(T - T_a) = -\lambda_a(T - T_a). \tag{12.6.6}$$

This resembles the radioactive decay model

$$\frac{dx}{dt} = -\lambda x, \tag{12.6.7}$$

which has the solution

$$x(t) = x_0 e^{-\lambda t}, \tag{12.6.8}$$

where x_0 is the initial value of $x(t)$. With $x(t) := T(t) - T_a$ and λ replaced by λ_a, (12.6.8) becomes

$$T(t) - T_a = (T_0 - T_a)\, e^{-\lambda_a t}$$

since

$$x_0 = x(0) = T(0) - T_a = T_0 - T_a,$$

where T_0 is the initial room temperature. Therefore, the solution of (12.6.6) is

$$T(t) = T_a + (T_0 - T_a)e^{-\lambda_a t}. \tag{12.6.9}$$

Since $\lambda_a > 0$, it follows from this solution that $T(t) \to T_a$ as $t \to \infty$. So this part of the model is in accord with reality: with the HVAC system off, the room will eventually cool down to the same temperature as it is outside. Moreover, the model predicts that the room temperature decreases exponentially to the outside temperature. A term connected to exponential functions, such as (12.6.8) and

[19] Newton's law of cooling is discussed in detail in Chap. 1.

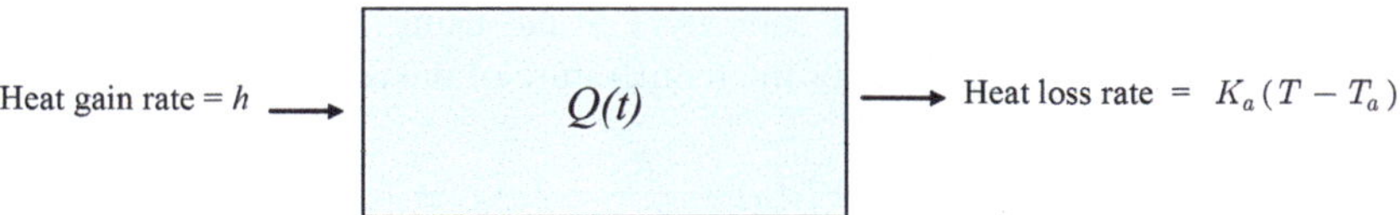

Fig. 12.8 Heat gain and loss rates

(12.6.9), that is used in engineering and physics is the so-called time constant. The ***time constant*** for a quantity undergoing exponential decay is the time it takes for the quantity to decay by a factor of $1/e$, that is, to approximately 36.8% of its initial value. For the quantity $x(t)$ in Eq. (12.6.7), the time constant is $1/\lambda$ since

$$x\left(\frac{1}{\lambda}\right) = x_0 e^{-\lambda(1/\lambda)} = \frac{1}{e}x_0 \approx 0.368x_0.$$

It follows that the *time constant for heat transfer* between the room and the outdoors is

$$\frac{1}{\lambda_a} = \frac{C_R}{K_a}. \qquad (12.6.10)$$

This is the time that it takes for the temperature difference $T - T_a$ to decay from the initial temperature difference $T_0 - T_a$ to $(T_0 - T_a)/e$.

Now that we are done analyzing a scenario with the HVAC system turned off, let us consider the same scenario again but this time with the furnace part of the HVAC system turned on. In other words, imagine that it is very cold outside and that someone turns on the furnace at time $t = 0$ because it is a chilly T_0 degrees in the room. Suppose the furnace delivers heat to the room at a constant rate of h BTUs per hour.[20] So during a time interval $[0, t]$, not only does the room lose heat to the outside but it also gains heat from the furnace. Figure 12.8 depicts the rate at which heat is delivered to the room and the hypothesized rate (given by (12.6.4)) at which the room loses heat, where $Q(t)$ denotes the amount of heat in the room at time t.

It follows that the rate at which the heat in the room is changing is

$$\frac{dQ}{dt} = h - K_a(T - T_a). \qquad (12.6.11)$$

Because of (12.6.3), we can rewrite this as

$$C_R\frac{dT}{dt} = h - K_a(T - T_a). \qquad (12.6.12)$$

[20] According to a number of websites, a home furnace in the Mid-South region of the United States needs to generate between 40 and 50 BTUs of heat per square foot every hour. So a home that is around 2000 square feet requires a furnace rated between 80,000 to 100,000 BTUs per hour..

Dividing both sides by the heat capacity C_R and using (12.6.10), we obtain the following differential equation for the temperature of the room:

$$\frac{dT}{dt} = \frac{h}{C_R} - \frac{K_a}{C_R}(T - T_a) = \frac{h}{C_R} - \lambda_a(T - T_a). \tag{12.6.13}$$

It is left as an exercise (see Problem 48) to verify that the solution of (12.6.13) is

$$T(t) = T_a + (T_0 - T_a)e^{-\lambda_a t} + \frac{h}{\lambda_a C_R}(1 - e^{-\lambda_a t}). \tag{12.6.14}$$

So the room's temperature will top out at

$$\lim_{t\to\infty} T(t) = T_a + \frac{h}{\lambda_a C_R}. \tag{12.6.15}$$

Example 12.6.1 A furnace heats a one-room cabin. The heat capacity of the room is 5000 BTU/°F and the time constant for heat transfer to the outdoors is 4 hours. If the outside temperature is 32 °F and the furnace supplies heat to the room at a rate of 57,500 BTU/hr without shutting off, what temperature will the room reach?

Solution Since the time constant is 4 hours, $\lambda_a = 1/(4\,\text{hr})$. According to (12.6.15), the temperature of the room will eventually exceed the outside temperature by $h/(\lambda_a C_R)$ degrees. Thus the room will reach the temperature

$$T_a + \frac{h}{\lambda_a C_R} = 32\,°\text{F} + \frac{57500\ \text{BTU/hr}}{(0.25\ \text{hr}^{-1})(5000\ \text{BTU/°F})} = 78\,°\text{F}.$$ ♦

12.6.2 *Two-Room Building*

Now we are ready to tackle modeling the temperatures inside two-room buildings, such as a small house in the dead of winter. The two rooms in the house could very well consist of a heated living space and an unheated, well-insulated attic bedroom. In fact, the living space may actually consist of several rooms but their proximity to one another and to a centrally located furnace that heats them uniformly may justify treating them as one room. Consider the house depicted in Fig. 12.9 where the following items affect the temperatures in the living area and the attic bedroom:

- The living space is heated by a furnace, which supplies heat at a rate of 60,000 BTUs per hour.
- The heat capacity of the living space is 1500 BTU/°F.
- The time constant for heat loss from the living space through the walls, windows, and doors to the outdoors is 2.5 hours.
- The time constant for heat loss from the living space through the ceiling to the attic bedroom is 5 hours.

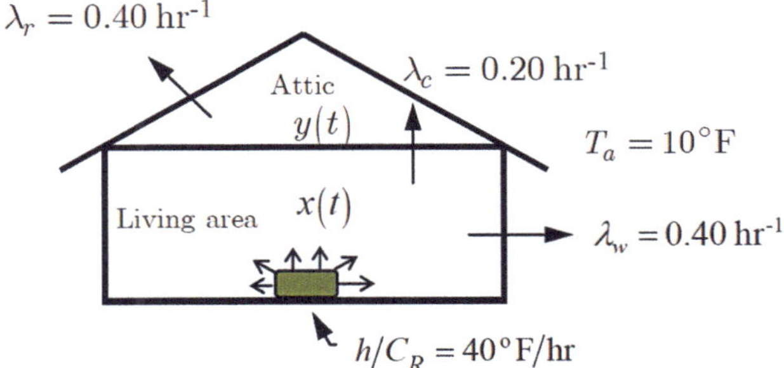

Fig. 12.9 Heated house with attic bedroom

- The time constant for heat loss from the attic bedroom through the roof to the outdoors is 2.5 hours.
- Heat loss through the floor of the house is negligible.
- Other than the furnace, all other sources of heat within the building (such as from major appliances running, lights turned on, the presence of one or two people, etc.) are negligible.

Suppose one morning the outside temperature is 10 °F and it remains constant throughout the day. Let $x(t)$ and $y(t)$ denote the temperatures at time t in the living space and the attic bedroom, respectively. Let us construct a differential equations model of the temperatures $x(t)$ and $y(t)$. We see from (12.6.13) that were it not for heat losses described above, the furnace would raise the temperature of the living space at the rate of

$$\frac{h}{C_R} = 60{,}000\ \frac{\text{BTU}}{\text{hr}} \cdot \frac{°\text{F}}{1500\ \text{BTU}} = 40\ °\text{F/hr}. \tag{12.6.16}$$

However, the living space does lose heat to the outdoors. According to equation (12.6.5), the heat loss through the walls, windows, and doors counteracts the temperature rise in the living space from the furnace by contributing to a temperature drop there at the rate of $\lambda_w(x - 10)$ °F/hr. Since the time constant for the heat loss is $1/\lambda_w = 2.5$ hours, $\lambda_w = 0.40\ \text{hr}^{-1}$. Thus,

$$\lambda_w(x - 10) = 0.40(x - 10).$$

The living space also loses heat to the attic bedroom through the ceiling. This also contributes to a temperature drop in the living space at the rate of

$$\lambda_c(x - y) = 0.20(x - y)$$

since $1/\lambda_c = 5$ hours is the time constant for heat loss through the ceiling. Of course, the heat lost by the living space through the ceiling is heat gained by the attic bedroom, which causes the temperature to rise there. However, this heat gain is offset by the loss of heat to the outdoors, which contributes to a temperature drop in the attic bedroom at the rate of

$$\lambda_r(y-10) = 0.40(y-10),$$

since $\lambda_r = \lambda_w$ for this particular two-room house.

It follows from the foregoing analysis that the temperature $x(t)$ in the living space is modeled by the differential equation

$$\frac{dx}{dt} = 40 - 0.40(x-10) - 0.20(x-y). \tag{12.6.17}$$

Likewise, the temperature $y(t)$ in the attic bedroom is modeled by the differential equation

$$\frac{dy}{dt} = 0.20(x-y) - 0.40(y-10). \tag{12.6.18}$$

Simplifying both differential equations, we obtain

$$\begin{aligned}\frac{dx}{dt} &= -0.60x + 0.20y + 44\\ \frac{dy}{dt} &= 0.20x - 0.60y + 4.\end{aligned} \tag{12.6.19}$$

This is a linear system of differential equations of the form

$$\begin{aligned}\frac{dx}{dt} &= ax + by + f(t)\\ \frac{dy}{dt} &= cx + dy + g(t),\end{aligned} \tag{12.6.20}$$

where $a = -0.60$, $b = c = 0.20$, and $d = -0.60$. Moreover, this is a nonhomogeneous linear system because of the nonzero functions $f(t) \equiv 44$ and $g(t) \equiv 4$. We will learn how to solve a system like this in the next chapter.

12.6.2.1 Broken Furnace

Let us alter the previous scenario by supposing the outdoor temperature is a constant 0 °F and that the furnace is inoperable. This alteration is readily handled by merely replacing $h/C_R = 40$ and $T_a = 10$ in Eqs. (12.6.17) and (12.6.18) with $h/C_R = 0$ and $T_a = 0$. So for this situation the room temperatures are modeled by the homogeneous linear system

$$\begin{aligned}\frac{dx}{dt} &= -0.60x + 0.20y\\ \frac{dy}{dt} &= 0.20x - 0.60y.\end{aligned} \tag{12.6.21}$$

Example 12.6.2 Outdoors it is a frigid 0 °F. In a small house, the respective temperatures in the living space and the attic bedroom are 78 and 30 °F. Suppose the furnace in the living space suddenly breaks down. If the linear homogeneous system (12.6.21) is an appropriate model of the temperatures of the living space and the attic bedroom, predict what these temperatures will be 15 minutes after the furnace malfunction if the outdoor temperature does not change.

Solution Written as a vector differential equation, the homogeneous linear system (12.6.21) is $\mathbf{x}' = A\mathbf{x}$, where $\mathbf{x} = \langle x, y\rangle$ and

$$A = \begin{bmatrix} -0.60 & 0.20 \\ 0.20 & -0.60 \end{bmatrix} = \begin{bmatrix} -\frac{3}{5} & \frac{1}{5} \\ \frac{1}{5} & -\frac{3}{5} \end{bmatrix}.$$

It left to the reader to verify that $\lambda_1 = -4/5$ and $\lambda_2 = -2/5$ are the eigenvalues of A with corresponding eigenvectors $\mathbf{v}^1 = \langle 1, -1\rangle$ and $\mathbf{v}^2 = \langle 1, 1\rangle$. It follows from Theorem 12.5.32 that a general solution of (12.6.21) is

$$\mathbf{x}(t) = c_1 e^{-\frac{4}{5}t}\begin{bmatrix} 1 \\ -1 \end{bmatrix} + c_2 e^{-\frac{2}{5}t}\begin{bmatrix} 1 \\ 1 \end{bmatrix}.$$

Let $t = 0$ designate the point in time when the furnace breaks down. Then, as $\mathbf{x}(0) = \langle 78, 30\rangle$,

$$c_1\begin{bmatrix} 1 \\ -1 \end{bmatrix} + c_2\begin{bmatrix} 1 \\ 1 \end{bmatrix} = \begin{bmatrix} 78 \\ 30 \end{bmatrix}.$$

This has the solution $c_1 = 24$ and $c_2 = 54$. Consequently,

$$\begin{bmatrix} x(t) \\ y(t) \end{bmatrix} = 24e^{-\frac{4}{5}t}\begin{bmatrix} 1 \\ -1 \end{bmatrix} + 54e^{-\frac{2}{5}t}\begin{bmatrix} 1 \\ 1 \end{bmatrix}.$$

In other words, the model predicts the temperatures $x(t)$ and $y(t)$ for $t \geq 0$ are

$$\begin{aligned} x(t) &= 54e^{-\frac{2}{5}t} + 24e^{-\frac{4}{5}t}\\ y(t) &= 54e^{-\frac{2}{5}t} - 24e^{-\frac{4}{5}t}. \end{aligned}$$

So, after 15 minutes, the temperature of the living space is

$$x(1/4) = 54e^{-\frac{1}{10}} + 24e^{-\frac{1}{5}} = 68.5°\text{F}$$

and the temperature of the attic bedroom is

$$y(1/4) = 54e^{-\frac{1}{10}} - 24e^{-\frac{1}{5}} = 29.2°\text{F}.$$

♦

12.6.2.2 Thermostats and Changing Temperatures

The system of differential equations (12.6.19) models the temperatures of the living area and attic bedroom in the house depicted in Fig. 12.9 assuming that the outdoor temperature is a constant 10 °F and the furnace delivers heat to the living space at the constant rate given by (12.6.16). Now suppose there is a thermostat that turns the furnace off when the temperature $x(t)$ in the living area exceeds a preset temperature, say 68 °F, and turns it back on again when $x(t)$ dips below 68 °F. This changes the rate of delivery of heat: from the constant rate $h = 60{,}000$ BTUs per hour to

$$h(t) = \begin{cases} 60{,}000, & \text{if } x(t) < 68 \\ 0, & \text{if } x(t) \geq 68 \end{cases}$$

BTUs per hour. Consequently, we have to modify the model by changing (12.6.16) to

$$\frac{h(t)}{C_R} = \begin{cases} 40, & \text{if } x(t) < 68 \\ 0, & \text{if } x(t) \geq 68 \end{cases} \tag{12.6.22}$$

and $f(t)$ in (12.6.20) from $f(t) \equiv 44$ to

$$f(t) = 4 + \frac{h(t)}{C_R} = \begin{cases} 44, & \text{if } x(t) < 68 \\ 4, & \text{if } x(t) \geq 68. \end{cases} \tag{12.6.23}$$

So, because of this thermostatic control[21] of the inside temperatures, we have to modify system (12.6.19) modeling the temperatures of the living space and attic bedroom as follows:

[21] A more realistic description of thermostatic control and how to incorporate it into the temperature model that we have been discussing here can be found in [25].

$$\frac{dx}{dt} = -0.60x + 0.20y + f(t)$$
$$\frac{dy}{dt} = 0.20x - 0.60y + 4,$$

where $f(t)$ is given by (12.6.23).

Finally, we briefly mention how simple variations in the outdoor temperature could be integrated into the temperature model. The "10's" inside two of the parenthesized terms in Eqs. (12.6.17) and (12.6.18) are there because we considered a scenario in which the outdoor temperature remains a constant $T_a = 10\,°\text{F}$. But, of course, the outdoor temperature changes with the passage of time. For instance, suppose during some twenty-four hour period that the sinusoidal function

$$T_a(t) = 10 + 20\sin(\pi t/24) \quad (0 \le t \le 24)$$

models the outdoor temperature. We can incorporate this time-varying temperature in the model by replacing both of the "10's" in the equations with the function $T_a(t)$. Furthermore, we can also include the thermostatic control of the temperature in the living space by replacing the constant term "40" in (12.6.17) with the piecewise-defined function $h(t)/C_R$ given by (12.6.22). The modified model is

$$\frac{dx}{dt} = \frac{h(t)}{C_R} - 0.40\big(x - T_a(t)\big) - 0.20(x - y)$$
$$\frac{dy}{dt} = 0.20(x - y) - 0.40\big(y - T_a(t)\big),$$

which simplifies to

$$\frac{dx}{dt} = -0.60x + 0.20y + f(t)$$
$$\frac{dy}{dt} = 0.20x - 0.60y + g(t),$$

where the nonhomogeneous terms are

$$f(t) = 4 + 8\sin\left(\frac{\pi t}{24}\right) + \frac{h(t)}{C_R} \quad \text{and} \quad g(t) = 4 + 8\sin\left(\frac{\pi t}{24}\right).$$

Problems

Results! Why, man, I have gotten a lot of results! I know several thousand things that won't work.

Thomas Alva Edison (Source: [32, p. 616])

Mathematical Model of the Flow of Sugar in Two Interconnected Tanks

1. Two tanks containing sugar water are interconnected by pipes. Initially, tank L contains 50 gallons of a sugar-water solution with a concentration of 0.10 pounds of sugar per gallon of sugar water. Tank R starts off with 100 gallons of sugar water with a concentration of 0.25 pounds of sugar per gallon.

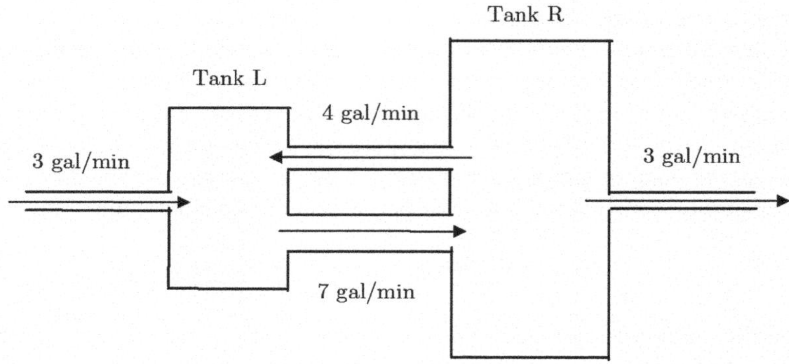

Pure water is pumped into tank L from an external source at the rate of 3 gallons per minute, while sugar water is pumped completely out of the two-tank system from tank R at the same rate. Sugar water is also pumped from tank L into tank R at a rate of 7 gallons per minute and from tank R back into tank L at a rate of 4 gallons per minute. Assume that changes in concentrations will spread instantaneous and uniformly throughout each tank so that the solution in each tank is always a homogeneous mixture. Determine the system of differential equations and the initial conditions whose solution would model the amount of sugar in each tank as a function of time.

Matrices

2. Let $A = \begin{bmatrix} a & b \\ c & d \end{bmatrix}$, $B = \begin{bmatrix} r & s \\ t & u \end{bmatrix}$, $\mathbf{u} = \begin{bmatrix} u_1 \\ u_2 \end{bmatrix}$, and $\mathbf{v} = \begin{bmatrix} v_1 \\ v_2 \end{bmatrix}$.

   (a) Prove matrix addition is commutative:

      i. $\mathbf{u} + \mathbf{v} = \mathbf{v} + \mathbf{u}$.

      ii. $A + B = B + A$.

   (b) Prove: $A(k\mathbf{u}) = kA\mathbf{u}$, where $k \in \mathbb{R}$.

   (c) Prove the distributive law: $A(\mathbf{u} + \mathbf{v}) = A\mathbf{u} + A\mathbf{v}$.

3. Show that the matrices

$$\begin{bmatrix} a & b \\ -b & a \end{bmatrix} \quad \text{and} \quad \begin{bmatrix} c & d \\ -d & c \end{bmatrix}$$

commute for all values of a, b, c, and d.

4. Let $A = \begin{bmatrix} 1 & 1 \\ 2 & 2 \end{bmatrix}$ and $B = \begin{bmatrix} -3 & 4 \\ 3 & -4 \end{bmatrix}$. Compute AB and BA.
5. Find the determinant of each of the matrices below. If a matrix is invertible, find its inverse.
(a) $\begin{bmatrix} 5 & -8 \\ -4 & -2 \end{bmatrix}$ (b) $\begin{bmatrix} 0 & 3/8 \\ -2/3 & 8 \end{bmatrix}$
(c) $\begin{bmatrix} -4/3 & -1/2 \\ 16 & 6 \end{bmatrix}$ (d) $\begin{bmatrix} 1/2 & -8 \\ -6 & 3/4 \end{bmatrix}$.
6. The characteristic equation of a 2×2 matrix A is $\lambda^2 - (\operatorname{tr} A)\,\lambda + \det A = 0$. Verify that A satisfies its own characteristic equation in the sense that
$$A^2 - (\operatorname{tr} A)\,A + (\det A)I = O,$$
where I is the 2×2 identity matrix. This result is the ***Cayley-Hamilton theorem*** for 2×2 matrices.
7. Let A and B be 2×2 matrices.
   (a) If $A\mathbf{x} = \mathbf{0}$ for all $\mathbf{x} \in \mathbb{R}^2$, prove that $A = O$.
   (b) If $A\mathbf{x} = B\mathbf{x}$ for all $\mathbf{x} \in \mathbb{R}^2$, prove that $A = B$.

Eigenvalues and Eigenvectors

8. For a matrix A of order 2, a real number λ, and a vector $\mathbf{v}$ with two components, prove:
   a. $A(e^{\lambda t}\mathbf{v}) = e^{\lambda t}A\mathbf{v}$
   b. $\dfrac{d}{dt}(e^{\lambda t}\mathbf{v}) = \lambda e^{\lambda t}\mathbf{v}$.
9. Find the eigenvalues and corresponding eigenspaces of $A = \begin{bmatrix} 3 & 5 \\ -2 & -4 \end{bmatrix}$.
10. In quantum mechanics, the matrix
$$\mathbf{L}_z := \begin{bmatrix} \frac{1}{2}\hbar & 0 \\ 0 & -\frac{1}{2}\hbar \end{bmatrix}$$
is known as a ***Pauli spin matrix***. The symbol $\hbar$ represents the constant $h/2\pi$, where h is *Planck's constant*. The two eigenvalues of $\mathbf{L}_z$ are the only possible values of the z-components of the spin angular momentum of an electron (cf. [52, pp. 186–189]). Find both eigenvalues of $\mathbf{L}_z$ and an eigenvector corresponding to each of them.
11. Let $M = \begin{bmatrix} -2 & 3 \\ -4 & 5 \end{bmatrix}$. Compute the trace and determinant of M and then use these values to obtain the characteristic equation of M. Check the result by computing the characteristic equation of M directly from $\det(M - \lambda I) = 0$.
12. Complete the proof of Corollary 12.5.30 by proving parts (iv)–(vi).
13. Determine the eigenvalues of the matrices A and B in Example 12.5.17.
14. Let $A = \begin{bmatrix} -2 & 6 \\ 3 & 1 \end{bmatrix}$.
   (a) Find the eigenvalues λ_1 and λ_2 of the matrix A.
   (b) Find eigenvectors corresponding to λ_1 and λ_2.
   (c) Find a solution $\mathbf{x}^1(t)$ of $\mathbf{x}' = A\mathbf{x}$ corresponding to λ_1.
   (d) Find a solution $\mathbf{x}^2(t)$ of $\mathbf{x}' = A\mathbf{x}$ corresponding to λ_2.
   (e) Use the Superposition Principle to find infinitely many solutions of the homogeneous linear system
$$x' = -2x + 6y$$
$$y' = 3x + y.$$

Linear Dependence and Independence

15. Prove statements (a) and (b) are equivalent. That is, prove (a) implies (b) and conversely.
   (a) Two vectors $\mathbf{u}$ and $\mathbf{v}$ are linearly dependent if and only if one of the vectors is a scalar multiple of the other.
   (b) Two vectors $\mathbf{u}$ and $\mathbf{v}$ are linearly dependent if and only if there exist scalars c_1 and c_2, not both zero, such that $c_1\mathbf{u} + c_2\mathbf{v} = \mathbf{0}$.
16. Prove: Two nonzero vectors $\mathbf{u}$ and $\mathbf{v}$ are linearly dependent if scalars $c_1 \neq 0$ and $c_2 \neq 0$ exist such that $c_1\mathbf{u} + c_2\mathbf{v} = \mathbf{0}$.

17. Prove that two nonzero vectors are linearly dependent if and only if they are parallel.
18. Are the vector functions

$$\mathbf{u(t)} = \begin{bmatrix} t \\ |t| \end{bmatrix} \quad \text{and} \quad \mathbf{v(t)} = \begin{bmatrix} |t| \\ t \end{bmatrix}$$

linearly dependent or independent on the interval $(-\infty, 0)$? What about on the intervals $(0, \infty)$ and $(-\infty, \infty)$?

19. Let $A = \begin{bmatrix} a & b \\ c & d \end{bmatrix}$. Prove that the row vectors $\begin{bmatrix} a & b \end{bmatrix}$ and $\begin{bmatrix} c & d \end{bmatrix}$ are linearly dependent if and only if $\det A = 0$.
20. Prove that two vectors $\mathbf{u}$ and $\mathbf{v}$ ***span*** the plane $\mathbb{R}^2$ if $\mathbf{u}$ and $\mathbf{v}$ are linearly independent. In other words, prove that each vector $\mathbf{w} \in \mathbb{R}^2$ can be written as a linear combination of $\mathbf{u}$ and $\mathbf{v}$.
21. Does a planar homogeneous linear system always have two linearly independent solutions? Explain.

Abel's Formula and Wronskians

22. Prove Abel's formula by completing the following parts.

    a. If $W(t)$ is the Wronskian of two solutions of $\mathbf{x}' = A\mathbf{x}$, show that $W'(t) = \operatorname{tr}(A)\, W(t)$.

    b. Solve the differential equation in part (a) to obtain (12.5.35).

23. Find the Wronskian $W(t)$ of the pair of solutions of the linear system

$$x' = x + 3y$$
$$y' = 4x - 3y$$

for which $W(0) = 8$.

24. Find the Wronskian of the solutions $\mathbf{x}^1(t)$ and $\mathbf{x}^2(t)$ of the linear system

$$x' = 2x + 3y$$
$$y' = -3x + 7y$$

satisfying the initial conditions:

$$\mathbf{x}^1(1) = \langle 1, 2 \rangle; \quad \mathbf{x}^2(1) = \langle 0, 3 \rangle.$$

Complex Exponential Functions

25. Use Euler's formula together with the trigonometric addition and subtraction identities to prove that

$$e^{i\phi} e^{\pm i\theta} = e^{i(\phi \pm \theta)},$$

where $\phi, \theta \in \mathbb{R}$.

26. Use mathematical induction to prove the formula

$$\left(e^{i\theta}\right)^n = e^{in\theta}$$

holds for $n = 0, \pm 1, \pm 2, \ldots$ with the convention that

$$\left(e^{i\theta}\right)^n = 1$$

when $n = 0$. *Hint*. See Problem 25.

27. The well-known formula

$$(\cos\theta + i \sin\theta)^n = \cos n\theta + i \sin n\theta,$$

where $\theta \in \mathbb{R}$ and n is an integer, is known as ***de Moivre's formula***. Prove it using Problem 26 and Euler's formula.

28. Use de Moivre's formula (Problem 27) to derive the following identities:

    a. $\cos 3\theta = \cos^3\theta - 3\cos\theta \sin^2\theta$

    b. $\sin 3\theta = 3\cos^2\theta \sin\theta - \sin^3\theta$.

Complex Eigenvalues and Solutions

29. Use $e^{(\alpha+i\omega)t} = e^{\alpha t}(\cos\omega t + i \sin\omega t)$ to show that

$$\frac{d}{dt} e^{(\alpha+i\omega)t} = (\alpha + i\omega) e^{(\alpha+i\omega)t}.$$

30. Let λ and γ be complex scalars, $\mathbf{v}$ a complex vector, and A a matrix. Prove that the conjugate of a product is equal to the product of the conjugates:

$$\overline{\lambda\gamma} = \overline{\lambda}\,\overline{\gamma}, \quad \overline{A\mathbf{v}} = \overline{A}\,\overline{\mathbf{v}}, \quad \overline{\lambda\mathbf{v}} = \overline{\lambda}\,\overline{\mathbf{v}}.$$

31. Let $\mathbf{f}(t)$ and $\mathbf{g}(t)$ be real-valued vector functions. Prove that if $\mathbf{x} = \mathbf{f} + i\mathbf{g}$ is a complex-valued solution of $\mathbf{x}' = A\mathbf{x}$, then $\mathbf{f}$ and $\mathbf{g}$ are real-valued solutions of this differential equation.

Solutions of Homogeneous Linear Systems

In Problems 32 through 45, find a general solution of the homogeneous linear system.

32. $x' = 5x + 3y$
$y' = -3y$
33. $x' = x + 3y$
$y' = 4x - 3y$
34. $x' = 3x - 4y$
$y' = -2x + 3y$
35. $x' = 2x - 3y$
$y' = 3x - 4y$
36. $x' = 2x - 5y$
$y' = x - 2y$
37. $x' = -3x + 2y$
$y' = -x - y$
38. $x' = 4x + 5y$
$y' = -x + 6y$
39. $x' = x - y$
$y' = x + 3y$
40. $x' = -2x + 6y$
$y' = 3x + y$
41. $x' = 2x - 5y$
$y' = 4x - 2y$
42. $x' = 3x - y$
$y' = x + y$
43. $x' = x + 5y$
$y' = -2x - y$
44. $x' = y$
$y' = 10x - 3y$
45. $x' = 4x + 3y$
$y' = -4x - 4y$

Application: Flow of Sugar in Tanks

46. Solve the linear system (12.2.1) so that $x(0) = 10$ and $y(0) = 0$, where $x(t)$ and $y(t)$ denote the amount of sugar (in pounds) at time t (in minutes) in tanks L and R (see Fig. 12.1), respectively. Compute the amount of sugar in each tank at $t = 100$ minutes.
47. Solve the linear system modeling the situation described in Problem 1. Compute the amount of sugar in tank L and in tank R at $t = 30$ minutes.

Application: Room Temperatures

48. Show that the solution of (12.6.13) is given by (12.6.14).
49. Solve the homogeneous linear system (12.6.21) satisfying the conditions that $x(0) = 60\,°\text{F}$ and $y(0) = 40\,°\text{F}$. What are the temperatures in the living space and attic bedroom after 1 hour?
50. Matthew wants to work on his car in his stand-alone garage. Inside and outside the garage, it is a chilly 50°F. So he brings in a portable heater that produces 5200 BTUs per hour. The heat capacity of the garage is 1000 BTUs per °F and its time constant is 2 hours. With the portable heater turned on, determine the temperature in the garage as a function of time. What temperature will the garage ultimately reach if the heater is left on?
51. Rather than using the *time constant* for a building, we could use the *half-life*, defining it to be the time that it takes for the difference between the inside and outside temperatures to be reduced by half. Determine the relationship between these two quantities.

Computer Algebra System Problems

52. Use a CAS to solve Problem 46.
53. Use a CAS to solve Problem 49.

Chapter 13
Matrix Exponential Functions and Series

I have photographed many people: artists, writers, and scientists, among others. In speaking about their work, mathematicians use the words "elegance", "truth", and "beauty" more than everyone else combined.

Mariana Cook[1]

Abstract This chapter investigates aspects of two-dimensional linear systems that are not considered in the previous chapter. Some of the topics discussed in Chap. 12 raise questions that are addressed in this chapter. For instance, $e^{at}c$ is a solution of the differential equation $x' = ax$ for every real number c. So, is $e^{At}\mathbf{v}$ a solution of the vector differential equation

$$\mathbf{x}' = A\mathbf{x} \tag{1}$$

for every 2×1 constant vector $\mathbf{v}$, where A is a constant matrix of order 2? But hold on, A is a matrix; so what would e^{At} even mean? Supposing it could be defined in a way that is natural and compatible with the definition of its scalar counterpart e^{at}, there is then the question as to its differentiability. If it does have a derivative, is it Ae^{At}? Is there a *variation of parameter formula* for

$$\mathbf{x}' = A\mathbf{x} + \mathbf{b}(t) \tag{2}$$

that generalizes the one for nonhomogeneous first-order linear scalar equations?

Answers to these and related questions begin in Sects. 13.1 and 13.2 with new ideas and definitions, most notably the *fundamental matrix solutions* known as *principal matrix solutions*. Differentiation rules for matrix functions with differentiable entries are stated and proved, after which these rules are used to prove properties of the *principal matrix solution* denoted by $Z(t)$, which is defined to be the matrix function whose columns are linearly independent for all values of t and which

[1] Mariana Cook (born 1955) is an American photographer. See [22, Preface].

© The Author(s), under exclusive license to Springer Nature Switzerland AG 2026

L. C. Becker, *Ordinary Differential Equations: Concepts, Methods, and Models*,
https://doi.org/10.1007/978-3-032-15150-6_13

satisfies the following matrix differential equation and initial condition:

$$\frac{d}{dt}X(t) = AX(t), \quad X(0) = I, \tag{3}$$

where I is the identity matrix. Rather than defining e^{At} by a power series, as is commonly done, it is defined to be the principal matrix solution $Z(t)$, that is, $e^{At} := Z(t)$. Consequently, the properties of e^{At} are the same as those of $Z(t)$.

In Sect. 13.3, it is shown using a *similarity transformation* that every 2×2 matrix A can be changed into a matrix J called a *Jordan* (*canonical*) *matrix*, of which there are only four types. Several examples using one of the main results in this chapter (Theorem 13.3.4) show how to use Jordan matrices to find solutions of (1). In Sect. 13.4, these matrices also aid in finding the power series expansion of e^{At}. In Sect. 13.5, *phase portraits* of (1) for several different matrices A and of the corresponding *canonical linear systems* $\mathbf{y}' = J\mathbf{y}$ are obtained by finding formulas for the trajectories of these systems near the equilibrium point $(0, 0)$ and then graphing them.

The chapter concludes with the derivation of the *variation of parameters formula*

$$\mathbf{x}(t) = e^{A(t-t_0)}\mathbf{x}_0 + \int_{t_0}^{t} e^{A(t-s)}\mathbf{b}(s)\, ds. \tag{4}$$

Several examples and an application illustrate how to use this formula to obtain a solution of a nonhomogeneous equation (2) so that it satisfies a given initial condition.

13.1 Fundamental and Principal Matrix Solutions

Let $\mathbf{x}^1(t)$ and $\mathbf{x}^2(t)$ be any two solutions of a given vector differential equation

$$\mathbf{x}' = A\mathbf{x} \tag{13.1.1}$$

where A is a real constant 2×2 matrix (all entries are real numbers). Let $X(t)$ be the ***matrix function*** (i.e., matrix-valued function) with $\mathbf{x}^1(t)$ and $\mathbf{x}^2(t)$ as its columns:

$$X(t) := \begin{bmatrix} \mathbf{x}^1(t) & \mathbf{x}^2(t) \end{bmatrix}. \tag{13.1.2}$$

It follows from Theorem 12.5.14 that $\mathbf{x}^1(t)$ and $\mathbf{x}^2(t)$ are differentiable at every point $t \in \mathbb{R}$. It then makes sense to define $X'(t)$ to be the matrix function that results from differentiating both column vectors of $X(t)$:

$$X'(t) := \begin{bmatrix} \frac{d}{dt}\mathbf{x}^1(t) & \frac{d}{dt}\mathbf{x}^2(t) \end{bmatrix}.$$

Since $\mathbf{x}^1(t)$ and $\mathbf{x}^2(t)$ are solutions, it follows from Definition 12.3.2 that

$$X'(t) = \left[A\mathbf{x}^1(t) \;\; A\mathbf{x}^2(t)\right] = A\left[\mathbf{x}^1(t) \;\; \mathbf{x}^2(t)\right] = AX(t).$$

Consequently, $X(t)$ is said to be a ***matrix solution*** of (13.1.1) because it satisfies the matrix differential equation

$$X' = AX \tag{13.1.3}$$

for all $t \in \mathbb{R}$. A matrix solution $X(t)$ is said to be a ***fundamental matrix solution*** if its columns $\mathbf{x}^1(t)$ and $\mathbf{x}^2(t)$ are linearly independent. As we will soon discover, fundamental matrix solutions are very important and useful for a number of reasons.

Example 13.1.1 In Examples 12.5.2 and 12.5.15, we determined that

$$\mathbf{x}^1(t) = \begin{bmatrix} 1 \\ 1 \end{bmatrix} e^{-t} \quad \text{and} \quad \mathbf{x}^2(t) = \begin{bmatrix} 1 \\ -1 \end{bmatrix} e^{-3t}$$

are solutions of (12.5.4), namely, of the vector differential equation

$$\mathbf{x}' = \begin{bmatrix} -2 & 1 \\ 1 & -2 \end{bmatrix} \mathbf{x}.$$

Consequently,

$$X(t) = \left[\mathbf{x}^1(t) \;\; \mathbf{x}^2(t)\right] = \begin{bmatrix} e^{-t} & e^{-3t} \\ e^{-t} & -e^{-3t} \end{bmatrix} \tag{13.1.4}$$

is a matrix solution of (12.5.4). To fully understand this, let us verify that this truly is the case:

$$\begin{aligned} AX(t) &= \begin{bmatrix} -2 & 1 \\ 1 & -2 \end{bmatrix} \begin{bmatrix} e^{-t} & e^{-3t} \\ e^{-t} & -e^{-3t} \end{bmatrix} = \begin{bmatrix} -2e^{-t} + e^{-t} & -2e^{-3t} - e^{-3t} \\ e^{-t} - 2e^{-t} & e^{-3t} + 2e^{-3t} \end{bmatrix} \\ &= \begin{bmatrix} -e^{-t} & -3e^{-3t} \\ -e^{-t} & 3e^{-3t} \end{bmatrix} = \begin{bmatrix} \frac{d}{dt}e^{-t} & \frac{d}{dt}e^{-3t} \\ \frac{d}{dt}e^{-t} & \frac{d}{dt}(-e^{-3t}) \end{bmatrix} \\ &= \frac{d}{dt}\begin{bmatrix} e^{-t} & e^{-3t} \\ e^{-t} & -e^{-3t} \end{bmatrix} = X'(t). \end{aligned}$$

This verifies that the matrix function $X(t)$ does satisfy (13.1.3).

The solutions $\mathbf{x}^1(t)$ and $\mathbf{x}^2(t)$ are linearly independent on $(-\infty, \infty)$ since neither is a scalar multiple of the other (cf. Definition 12.5.17). In this regard, see also

Example 12.5.12. Therefore, (13.1.4) is not only a matrix solution but a fundamental matrix solution as well.

The matrix function

$$\begin{bmatrix} -5e^{-3t} & e^{-3t} \\ 5e^{-3t} & -e^{-3t} \end{bmatrix}$$

is also a matrix solution, but it is not a fundamental matrix solution since its columns are linearly dependent. ♦

Whether a matrix solution is a fundamental matrix solution or not can be easily discerned from its determinant.

A Fundamental Matrix Solution Test

Theorem 13.1.1 *Let $X(t)$ be a matrix solution of the vector differential equation $\mathbf{x}' = A\mathbf{x}$. If $\det X(t_0) \neq 0$ for some $t_0 \in \mathbb{R}$, then $X(t)$ is a fundamental matrix solution of the equation. Furthermore, $\det X(t) \neq 0$ for all $t \in \mathbb{R}$.*

Proof By hypothesis, the columns $\mathbf{x}^1(t)$ and $\mathbf{x}^2(t)$ of $X(t)$ are solutions of $\mathbf{x}' = A\mathbf{x}$ and $\det X(t_0) \neq 0$ for some $t_0 \in \mathbb{R}$. It follows from this and Theorem 12.5.23 that $\mathbf{x}^1(t)$ and $\mathbf{x}^2(t)$ are not only solutions but are linearly independent solutions as well. Therefore, $X(t)$ is a fundamental matrix solution of $\mathbf{x}' = A\mathbf{x}$. By Theorem 12.5.25, $\det X(t) \neq 0$ for all $t \in \mathbb{R}$. ■

Theorem 12.5.27 tells us that if $\mathbf{x}^1(t)$ and $\mathbf{x}^2(t)$ are linearly independent solutions of the vector differential equation (13.1.1), then a general solution of this equation is given by (12.5.38) for $-\infty < t < \infty$, that is, by

$$\mathbf{x}(t) = c_1\mathbf{x}^1(t) + c_2\mathbf{x}^2(t).$$

This can be expressed (cf. Definition 12.3.1 and (12.3.13)) as the matrix-vector product

$$\mathbf{x}(t) = \begin{bmatrix} \mathbf{x}^1(t) & \mathbf{x}^2(t) \end{bmatrix} \begin{bmatrix} c_1 \\ c_2 \end{bmatrix} = X(t)\mathbf{c}$$

where $\mathbf{c} := \langle c_1, c_2 \rangle$. Thus we have the following result.

General Solution

Theorem 13.1.2 *If $X(t)$ is a fundamental matrix solution of $\mathbf{x}' = A\mathbf{x}$, then a general solution of this equation is*

$$\mathbf{x}(t) = X(t)\mathbf{c} \tag{13.1.5}$$

for all $t \in \mathbb{R}$, where $\mathbf{c} = \langle c_1, c_2 \rangle$ is a vector of arbitrary constants.

Example 13.1.2 Find a fundamental matrix solution of the linear system

$$\begin{aligned} x' &= x + 2y \\ y' &= -2x + y. \end{aligned} \tag{13.1.6}$$

Express the general solution of this system in terms of this matrix.

Solution The system expressed as a vector differential equation is

$$\mathbf{x}' = \begin{bmatrix} 1 & 2 \\ -2 & 1 \end{bmatrix} \mathbf{x}.$$

In Example 12.5.24 we determined that

$$\mathbf{x}^1(t) = e^t \begin{bmatrix} \cos 2t \\ -\sin 2t \end{bmatrix} \quad \text{and} \quad \mathbf{x}^2(t) = e^t \begin{bmatrix} \sin 2t \\ \cos 2t \end{bmatrix}$$

are linearly independent solutions. As a result, a fundamental matrix solution is

$$X(t) = \begin{bmatrix} e^t \cos 2t & e^t \sin 2t \\ -e^t \sin 2t & e^t \cos 2t \end{bmatrix}.$$

So, in terms of $X(t)$, a general solution of (13.1.6) is

$$\mathbf{x}(t) = X(t)\mathbf{c} = \begin{bmatrix} e^t \cos 2t & e^t \sin 2t \\ -e^t \sin 2t & e^t \cos 2t \end{bmatrix} \begin{bmatrix} c_1 \\ c_2 \end{bmatrix}$$

for $-\infty < t < \infty$. That is,

$$x(t) = c_1 e^t \cos 2t + c_2 e^t \sin 2t, \quad y(t) = -c_1 e^t \sin 2t + c_2 e^t \cos 2t$$

are solutions of the system (13.1.6) for all $c_1, c_2 \in \mathbb{R}$. ♦

Invertibility of Fundamental Matrix Solutions

Theorem 13.1.3 *Every fundamental matrix solution of* $\mathbf{x}' = A\mathbf{x}$ *is invertible.*

Proof Let $X(t)$ be a fundamental matrix solution of $\mathbf{x}' = A\mathbf{x}$. Then, by definition, the columns $\mathbf{x}^1(t)$ and $\mathbf{x}^2(t)$ of $X(t)$ are linearly independent. By Theorem 12.5.25, the matrix $X(t)$ is nonsingular for all $t \in \mathbb{R}$. And so, by Corollary 12.5.9, $X(t)$ is also invertible for all $t \in \mathbb{R}$. ■

Fundamental matrix solutions of a given vector differential equation are not unique; however, any two of them are related through some invertible constant matrix.

Fundamental Matrix Solutions of the Same Equation

Theorem 13.1.4 *If* $X(t)$ *and* $Y(t)$ *are fundamental matrix solutions of the same vector differential equation* $\mathbf{x}' = A\mathbf{x}$*, then there is an invertible constant matrix* C *such that*

$$Y(t) = X(t)C \tag{13.1.7}$$

for all $t \in \mathbb{R}$.

Proof Since $Y(t)$ is a matrix of order 2, $Y(t) = \begin{bmatrix}\mathbf{y}^1(t) & \mathbf{y}^2(t)\end{bmatrix}$, where the vector function $\mathbf{y}^i(t)$ designates the ith column for $i = 1, 2$. Since $\mathbf{y}^i(t)$ is a solution of $\mathbf{x}' = A\mathbf{x}$, it follows from Theorem 13.1.2 that

$$\mathbf{y}^i(t) = X(t)\mathbf{c}^i$$

for some constant vector $\mathbf{c}^i$. Thus, $Y(t) = \begin{bmatrix}X(t)\mathbf{c}^1 & X(t)\mathbf{c}^2\end{bmatrix}$, which can be written as (see Definition 12.3.2)

$$Y(t) = X(t)\begin{bmatrix}\mathbf{c}^1 & \mathbf{c}^2\end{bmatrix} = X(t)C,$$

where $C := \begin{bmatrix}\mathbf{c}^1 & \mathbf{c}^2\end{bmatrix}$. By Theorem 12.5.4,

$$\det Y(t) = \det X(t) \cdot \det C.$$

Since $\mathbf{y}^1(t)$ and $\mathbf{y}^2(t)$ are linearly independent solutions of $\mathbf{x}' = A\mathbf{x}$, Theorem 12.5.25 tells us that

$$\det Y(t) = \det\left[\mathbf{y}^1(t)\ \ \mathbf{y}^2(t)\right] \neq 0$$

for all $t \in \mathbb{R}$. So $\det C \neq 0$. By Corollary 12.5.9, the matrix C is invertible. ■

Many important results concerning the solutions of $\mathbf{x}' = A\mathbf{x}$ can be stated in terms of a special fundamental matrix solution. It generalizes the notion of the *principal (scalar) solution* of a first-order scalar linear equation $x' = a(t)x$, which is defined in Chap. 5 (Definition 5.5.1). Since we are now dealing with matrices, we shall refer to it as the *principal matrix solution* of $\mathbf{x}' = A\mathbf{x}$.

Principal Matrix Solution

Definition 13.1.5 The ***principal matrix solution of*** $\mathbf{x}' = A\mathbf{x}$ ***at the point*** t_0, denoted by $Z(t, t_0)$, refers to the fundamental matrix solution of this differential equation whose value at $t = t_0$ is the identity matrix I. In other words,

$$\frac{d}{dt}Z(t, t_0) = AZ(t, t_0) \quad \text{and} \quad Z(t_0, t_0) = I.$$

For the sake of brevity, let $Z(t) := Z(t, 0)$, namely, the principal matrix solution of $\mathbf{x}' = A\mathbf{x}$ at $t = 0$.

Remarks

1. The principal matrix solution $Z(t, t_0)$ is unique (cf. Theorem 12.5.14) in that it is the only fundamental matrix solution that is equal to the identity matrix I at $t = t_0$.
2. Some authors do not use the term "principal matrix solution" for $Z(t, t_0)$ but refer to it as the fundamental matrix whose value at t_0 is I. Nevertheless, we find it convenient to have a name for this special matrix as do many prominent authors and researchers, such as Burton [18], Hale [41], and Hale and Koçak [42]. Moreover, "principal matrix solution" is commonly abbreviated as PMS.
3. Other authors, particularly those involved in engineering and physics, refer to $Z(t, t_0)$ as the ***state transition matrix*** because, as we will see later in Theorem 13.1.10, it relates the state $\mathbf{x}(t)$ of a physical system modeled by (13.1.1) at a time t to its state $\mathbf{x}(t_0)$ at another time t_0 through the equation $\mathbf{x}(t) = Z(t, t_0)\mathbf{x}(t_0)$.
4. The definition of the principal matrix solution $Z(t, t_0)$ for the linear autonomous differential equation $\mathbf{x}' = A\mathbf{x}$ carries over to the linear nonautonomous differential equation $\mathbf{x}' = A(t)\mathbf{x}$, where $A(t)$ is a matrix of continuous functions (cf.

[18, p. 29]). In fact, the definition can also be extended to the linear integro-differential equation

$$\mathbf{x}'(t) = A(t)\mathbf{x}(t) + \int_{t_0}^{t} B(t, u)\mathbf{x}(u)\,du,$$

where $A(t)$ and $B(t, u)$ are matrices of continuous functions. See [9] and [19] for more information.

PMS via a Fundamental Matrix Solution

Theorem 13.1.6 *If $X(t)$ is a fundamental matrix solution of $\mathbf{x}' = A\mathbf{x}$, then*

$$Z(t, t_0) = X(t)X^{-1}(t_0). \tag{13.1.8}$$

Proof We see from Theorem 13.1.4 that the fundamental matrix solutions $X(t)$ and $Z(t, t_0)$ are related via the equation $Z(t, t_0) = X(t)C$, where C is some invertible matrix. With $t = t_0$, we have

$$X(t_0)C = Z(t_0, t_0) = I \quad \Rightarrow \quad C = X^{-1}(t_0),$$

which proves the theorem. ■

The following result is an immediate consequence of this theorem.

Corollary 13.1.7 *If $X(t)$ and $Y(t)$ are fundamental matrix solutions of a vector differential equation $\mathbf{x}' = A\mathbf{x}$, then*

$$X(t)X^{-1}(s) = Y(t)Y^{-1}(s) \tag{13.1.9}$$

for all $s, t \in \mathbb{R}$.

Example 13.1.3 Find the principal matrix solution $Z(t)$ of

$$\begin{aligned} x' &= x - 2y \\ y' &= x + 4y. \end{aligned} \tag{13.1.10}$$

Also, find $Z(t, t_0)$ for a given $t_0 \in \mathbb{R}$.

Solution The equivalent vector differential equation is $\mathbf{x}' = A\mathbf{x}$, where

$$A = \begin{bmatrix} 1 & -2 \\ 1 & 4 \end{bmatrix}. \qquad (13.1.11)$$

Since $\operatorname{tr} A = 5$ and $\det A = 6$, we see from (12.5.42) that the characteristic equation is

$$\lambda^2 - 5\lambda + 6 = 0.$$

So the eigenvalues are the distinct positive real numbers: $\lambda_1 = 2$ and $\lambda_2 = 3$.

An eigenvector corresponding to $\lambda_1 = 2$ is a nonzero vector $\mathbf{u}$ such that $A\mathbf{u} = 2\mathbf{u}$:

$$(A - 2I)\mathbf{u} = \begin{bmatrix} -1 & -2 \\ 1 & 2 \end{bmatrix} \begin{bmatrix} u_1 \\ u_2 \end{bmatrix} = \begin{bmatrix} 0 \\ 0 \end{bmatrix}.$$

A solution is $u_1 = 2$ and $u_2 = -1$. Consequently, a solution of (13.1.10) is

$$\mathbf{x}^1(t) = e^{2t} \begin{bmatrix} 2 \\ -1 \end{bmatrix} = \begin{bmatrix} 2e^{2t} \\ -e^{2t} \end{bmatrix}.$$

An eigenvector corresponding to $\lambda_2 = 3$ is a nonzero vector $\mathbf{v}$ such that $A\mathbf{v} = 3\mathbf{v}$:

$$(A - 3I)\mathbf{v} = \begin{bmatrix} -2 & -2 \\ 1 & 1 \end{bmatrix} \begin{bmatrix} v_1 \\ v_2 \end{bmatrix} = \begin{bmatrix} 0 \\ 0 \end{bmatrix}.$$

Clearly, $v_1 = 1$ and $v_2 = -1$ is a solution. Thus, another solution of (13.1.10) is

$$\mathbf{x}^2(t) = e^{3t} \begin{bmatrix} 1 \\ -1 \end{bmatrix} = \begin{bmatrix} e^{3t} \\ -e^{3t} \end{bmatrix}.$$

Since $\mathbf{x}^1(t)$ and $\mathbf{x}^2(t)$ are linearly independent solutions, a fundamental matrix solution of (13.1.10) is

$$X(t) = \begin{bmatrix} 2e^{2t} & e^{3t} \\ -e^{2t} & -e^{3t} \end{bmatrix}.$$

Once we have the inverse of $X(t)$, we can find $Z(t)$ by way of Theorem 13.1.6. Referring to (12.5.23),

$$X^{-1}(t) = \frac{1}{\det X(t)} \begin{bmatrix} -e^{3t} & -e^{3t} \\ e^{2t} & 2e^{2t} \end{bmatrix} = -e^{-5t} \begin{bmatrix} -e^{3t} & -e^{3t} \\ e^{2t} & 2e^{2t} \end{bmatrix} = \begin{bmatrix} e^{-2t} & e^{-2t} \\ -e^{-3t} & -2e^{-3t} \end{bmatrix}.$$

Therefore,

$$Z(t) = Z(t,0) = X(t)X^{-1}(0) = \begin{bmatrix} 2e^{2t} & e^{3t} \\ -e^{2t} & -e^{3t} \end{bmatrix} \begin{bmatrix} 1 & 1 \\ -1 & -2 \end{bmatrix},$$

which multiplies out to

$$Z(t) = \begin{bmatrix} 2e^{2t} - e^{3t} & 2e^{2t} - 2e^{3t} \\ e^{3t} - e^{2t} & 2e^{3t} - e^{2t} \end{bmatrix}.$$

Likewise, as $Z(t, t_0) = X(t)X^{-1}(t_0)$, $Z(t, t_0)$ can be found by computing $X^{-1}(t_0)$. However, since we already have $Z(t)$, we can bypass these tedious calculations with the Autonomous Principle (Theorem 12.5.15). Since the columns $\mathbf{z}^1(t)$ and $\mathbf{z}^2(t)$ of $Z(t)$ are solutions of (13.1.10) with $\mathbf{z}^1(0) = \langle 1, 0\rangle$ and $\mathbf{z}^2(0) = \langle 0, 1\rangle$, their translations $\mathbf{z}^1(t - t_0)$ and $\mathbf{z}^2(t - t_0)$ are also solutions of (13.1.10) but with the respective values $\langle 1, 0\rangle$ and $\langle 0, 1\rangle$ at $t = t_0$. In other words,

$$Z(t, t_0) = \left[\mathbf{z}^1(t - t_0) \;\; \mathbf{z}^2(t - t_0)\right].$$

Therefore, $Z(t, t_0) = Z(t - t_0)$. As a result, we have

$$Z(t, t_0) = \begin{bmatrix} 2e^{2(t-t_0)} - e^{3(t-t_0)} & 2e^{2(t-t_0)} - 2e^{3(t-t_0)} \\ e^{3(t-t_0)} - e^{2(t-t_0)} & 2e^{3(t-t_0)} - e^{2(t-t_0)} \end{bmatrix}, \tag{13.1.12}$$

which concludes this example. ♦

Principal matrix solutions enjoy properties that facilitate finding the solutions of linear systems. To prove these properties, we will rely on the next theorem.

Differentiable Matrix Functions

Theorem 13.1.8 *Let $A(t)$ and $B(t)$ be square matrix functions of the same order whose entries are differentiable scalar functions on $\mathbb{R}$. Then:*

(i) $\dfrac{d}{dt}\big(A(t) + B(t)\big) = A'(t) + B'(t).$

(ii) $\dfrac{d}{dt}\big(A(t)B(t)\big) = A(t)B'(t) + A'(t)B(t).$

(iii) *If A is a constant matrix, then* $\dfrac{d}{dt}\big(AB(t)\big) = AB'(t).$

(iv) *If B is a constant matrix, then* $\dfrac{d}{dt}\big(A(t)B\big) = A'(t)B.$

(v) *If $A(t)$ is invertible on $\mathbb{R}$, then* $\dfrac{d}{dt}A^{-1}(t) = -A^{-1}(t)\left(\dfrac{d}{dt}A(t)\right)A^{-1}(t).$

Remark Since matrix multiplication is not in general commutative, it is important to retain the order of the matrices in (ii)–(v).

Proof Let $A(t) = \left[a_{ij}(t)\right]$, where $a_{ij}(t)$ denotes the entry in the ith row and jth column of $A(t)$. Likewise, let $B(t) = \left[b_{ij}(t)\right]$. Since the derivatives of $A(t)$ and $B(t)$ are obtained by differentiating each of their entries, it follows that

$$\begin{aligned}\frac{d}{dt}\big(A(t)+B(t)\big) &= \frac{d}{dt}\big(\left[a_{ij}(t)\right]+\left[b_{ij}(t)\right]\big) = \frac{d}{dt}\left[a_{ij}(t)+b_{ij}(t)\right]\\ &= \left[a'_{ij}(t)+b'_{ij}(t)\right] = \left[a'_{ij}(t)\right]+\left[b'_{ij}(t)\right] = \frac{d}{dt}A(t)+\frac{d}{dt}B(t).\end{aligned}$$

To prove (ii), use the definition of matrix multiplication and the product rule for differentiable scalar functions. The details are left to the reader in Problem 1. Clearly, (ii) implies (iii) and (iv).

As for (v), we have $A(t)A^{-1}(t) = I$ for all $t \in \mathbb{R}$. Thus,

$$\frac{d}{dt}\big(A(t)A^{-1}(t)\big) = \frac{d}{dt}I.$$

By (ii),

$$A(t)\frac{d}{dt}A^{-1}(t) + \left(\frac{d}{dt}A(t)\right)A^{-1}(t) = O$$

or

$$A(t)\frac{d}{dt}A^{-1}(t) = -\left(\frac{d}{dt}A(t)\right)A^{-1}(t).$$

Multiplying both sides of this equation on the left by $A^{-1}(t)$, we obtain (v). ■

Properties of Principal Matrix Solutions

Theorem 13.1.9 *For a given constant square matrix A, let $Z(t,s)$ denote the principal matrix solution of $\mathbf{x}' = A\mathbf{x}$ at $t = s$. In particular, let $Z(t) = Z(t,0)$. Let B be a constant square matrix of the same order as A that commutes with A. Then, for all $t, s, \tau \in \mathbb{R}$,*

(i) $\dfrac{\partial}{\partial t}Z(t,s) = AZ(t,s)$ (ii) $Z(t,s)Z(s,\tau) = Z(t,\tau)$

(iii) $Z(t+s) = Z(t)Z(s)$ (iv) $Z(t,s) = Z(t-s)$

(v) $Z^{-1}(t,s) = Z(s,t)$ (vi) $Z^{-1}(t) = Z(-t)$

(vii) $BZ(t,s) = Z(t,s)B$ (viii) $\dfrac{\partial}{\partial s}Z(t,s) = -AZ(t,s)$.

Remark Although proofs are given for square matrices of order 2, nearly all of the statements in the theorems and corollaries in this chapter are actually valid for square matrices of any order. That is the reason why the term "square matrix" is sometimes used instead of "2×2 matrix" or "square matrix of order 2."

Proof Property (i) is merely a matter of notation. Since $Z(t, s)$ is a fundamental matrix solution of equation $\mathbf{x}' = A\mathbf{x}$ for a given value of s, $Z'(t, s) = AZ(t, s)$. Now (i) is just another way of expressing this, where the partial derivative symbol "$\partial/\partial t$" indicates that the value of s remains fixed as t changes in value.

As for (ii), let $X(t)$ be any fundamental matrix solution of $\mathbf{x}' = A\mathbf{x}$. Then, by Theorem 13.1.6, we have

$$Z(t, s)Z(s, \tau) = X(t)X^{-1}(s)X(s)X^{-1}(\tau) = X(t)X^{-1}(\tau) = Z(t, \tau).$$

To prove (iii), let s have an arbitrarily fixed value. Since both $Z(t)$ and $Z(t + s)$ are fundamental matrix solutions of $\mathbf{x}' = A\mathbf{x}$, an invertible constant matrix C exists so that (see Theorem 13.1.4)

$$Z(t + s) = Z(t)C$$

for all $t \in \mathbb{R}$. Setting $t = 0$, this becomes $Z(s) = Z(0)C = C$ since $Z(0) = I$. Statement (iii) is the result when $Z(s)$ is substituted for C in the previous equation.

Now consider (iv). Because $Z(t)$ represents the fundamental matrix solution of the equation $\mathbf{x}' = A\mathbf{x}$ whose value is I at $t = 0$, it follows from the Autonomous Principle that $Z(t - s)$ is the solution whose value at $t = s$ is I. In other words, it is the PMS of $\mathbf{x}' = A\mathbf{x}$ at $t = s$. Therefore, $Z(t - s) = Z(t, s)$.

To prove (v), set $\tau = t$ in (ii). Then,

$$Z(t, s)Z(s, t) = Z(t, t) = I.$$

Thus, $Z(s, t) = Z^{-1}(t, s)$.

The proof of (vi) is left to Problem 2.

As for the proof (vii), let $X(t) := BZ(t, s) - Z(t, s)B$ for an arbitrarily fixed value of s. Differentiating $X(t)$ with respect to t and using the hypothesis that $BA = AB$, we have

$$\begin{aligned} X'(t) &= \frac{d}{dt}\left[BZ(t, s) - Z(t, s)B\right] = BZ'(t, s) - Z'(t, s)B \\ &= BAZ(t, s) - AZ(t, s)B = A\left[BZ(t, s) - Z(t, s)B\right] = AX(t). \end{aligned}$$

This shows $X(t)$ is a matrix solution of $\mathbf{x}' = A\mathbf{x}$. Since

$$X(s) = BZ(s, s) - Z(s, s)B = BI - IB = B - B = O,$$

where O is the zero matrix, we conclude that $X(t) \equiv O$ because of uniqueness of solutions (cf. Theorem 12.5.14). So $BZ(t,s) - Z(t,s)B \equiv O$, which simplifies to (vii).

Finally, in order to prove (viii), let us begin with (v) written as $Z^{-1}(s,t) = Z(t,s)$. Taking the partial derivative of both of its sides with respect to s and then referring to item (v) from Theorem 13.1.8, we have

$$\frac{\partial}{\partial s} Z(t,s) = \frac{\partial}{\partial s} Z^{-1}(s,t) = -Z^{-1}(s,t)\left(\frac{\partial}{\partial s} Z(s,t)\right) Z^{-1}(s,t),$$

which because of (i) is

$$\frac{\partial}{\partial s} Z(t,s) = -Z^{-1}(s,t)\big(AZ(s,t)\big)Z^{-1}(s,t).$$

By (vii), this simplifies to

$$\frac{\partial}{\partial s} Z(t,s) = -Z^{-1}(s,t)Z(s,t)AZ^{-1}(s,t) = -AZ^{-1}(s,t)$$

since A commutes with itself. Therefore, we end up with

$$\frac{\partial}{\partial s} Z(t,s) = -AZ(t,s)$$

because of (v). ∎

Example 13.1.4 Use Example 13.1.3 to illustrate property (vi) of Theorem 13.1.9, namely that $Z^{-1}(t) = Z(-t)$.

Solution We determined that the principal matrix solution of $\mathbf{x}' = A\mathbf{x}$ at $t = 0$, where A is given by (13.1.11), is

$$Z(t) = \begin{bmatrix} 2e^{2t} - e^{3t} & 2e^{2t} - 2e^{3t} \\ e^{3t} - e^{2t} & 2e^{3t} - e^{2t} \end{bmatrix}.$$

Its inverse is

$$Z^{-1}(t) = \frac{1}{\det Z(t)} \begin{bmatrix} 2e^{3t} - e^{2t} & -2e^{2t} + 2e^{3t} \\ -e^{3t} + e^{2t} & 2e^{2t} - e^{3t} \end{bmatrix},$$

where

$$\det Z(t) = (2e^{2t} - e^{3t})(2e^{3t} - e^{2t}) - (2e^{2t} - 2e^{3t})(e^{3t} - e^{2t}) = e^{5t}.$$

Thus,

$$Z^{-1}(t) = e^{-5t}\begin{bmatrix} 2e^{3t} - e^{2t} & -2e^{2t} + 2e^{3t} \\ -e^{3t} + e^{2t} & 2e^{2t} - e^{3t} \end{bmatrix} = \begin{bmatrix} 2e^{-2t} - e^{-3t} & -2e^{-3t} + 2e^{-2t} \\ -e^{-2t} + e^{-3t} & 2e^{-3t} - e^{-2t} \end{bmatrix}$$

$$= \begin{bmatrix} 2e^{2(-t)} - e^{3(-t)} & 2e^{2(-t)} - 2e^{3(-t)} \\ e^{3(-t)} - e^{2(-t)} & 2e^{3(-t)} - e^{2(-t)} \end{bmatrix} = Z(-t).$$ ♦

Example 13.1.5 Use Example 13.1.3 again to illustrate property (viii) of Theorem 13.1.9, namely that

$$\frac{\partial}{\partial s}Z(t, s) = -AZ(t, s).$$

Solution The matrix A in this example is given by (13.1.11) and its PMS at $t = t_0$ by (13.1.12). Using these two matrices to calculate the the right-hand side of (viii), we obtain

$$-AZ(t, s) = -\begin{bmatrix} 1 & -2 \\ 1 & 4 \end{bmatrix}\begin{bmatrix} 2e^{2(t-s)} - e^{3(t-s)} & 2e^{2(t-s)} - 2e^{3(t-s)} \\ e^{3(t-s)} - e^{2(t-s)} & 2e^{3(t-s)} - e^{2(t-s)} \end{bmatrix}$$

$$= \begin{bmatrix} -4e^{2(t-s)} + 3e^{3(t-s)} & -4e^{2(t-s)} + 6e^{3(t-s)} \\ -3e^{3(t-s)} + 2e^{2(t-s)} & -6e^{3(t-s)} + 2e^{2(t-s)} \end{bmatrix}.$$

The left-hand side of (viii) is

$$\frac{\partial}{\partial s}Z(t, s) = \frac{\partial}{\partial s}\begin{bmatrix} 2e^{2(t-s)} - e^{3(t-s)} & 2e^{2(t-s)} - 2e^{3(t-s)} \\ e^{3(t-s)} - e^{2(t-s)} & 2e^{3(t-s)} - e^{2(t-s)} \end{bmatrix}.$$

It is easy to check that the matrix which results from carrying out this differentiation is the same as that obtained for $-AZ(t, s)$. ♦

Now consider the problem of finding the solution of a given differential equation $\mathbf{x}' = A\mathbf{x}$ satisfying the initial condition $\mathbf{x}(t_0) = \mathbf{x}_0$. According to Theorem 13.1.2, a general solution of the equation is

$$\mathbf{x}(t) = X(t)\mathbf{c},$$

where $X(t)$ is a fundamental matrix solution of the equation and $\mathbf{c}$ is a vector of arbitrary constants. So, with $t = t_0$,

$$\mathbf{x}(t_0) = X(t_0)\mathbf{c} \quad \text{or} \quad X(t_0)\mathbf{c} = \mathbf{x}_0.$$

It follows from Theorem 13.1.3 that $X(t)$ is invertible for all $t \in \mathbb{R}$. In particular, $X(t_0)$ is invertible. To solve for $\mathbf{c}$, let us multiply both sides of $X(t_0)\mathbf{c} = \mathbf{x}_0$ on the left by $X^{-1}(t_0)$, thereby obtaining

$$\mathbf{c} = X^{-1}(t_0)\mathbf{x}_0.$$

Consequently, the solution satisfying the initial condition is

$$\mathbf{x}(t) = X(t)X^{-1}(t_0)\mathbf{x}_0,$$

which, because of Theorem 13.1.6, can also be written as

$$\mathbf{x}(t) = Z(t, t_0)\mathbf{x}_0,$$

where $Z(t, t_0)$ is the PMS at $t = t_0$. As a result, we have the following theorem.

Solutions of Initial Value Problems via PMSs

Theorem 13.1.10 *The (unique) solution of the initial value problem*

$$\mathbf{x}' = A\mathbf{x}, \quad \mathbf{x}(t_0) = \mathbf{x}_0 \tag{13.1.13}$$

for $t \in \mathbb{R}$ is

$$\mathbf{x}(t) = Z(t, t_0)\mathbf{x}_0, \tag{13.1.14}$$

where $Z(t, t_0)$ is the principal matrix solution of $\mathbf{x}' = A\mathbf{x}$ at t_0.

Example 13.1.6 Find the solution of

$$\begin{aligned} x' &= x - 2y \\ y' &= x + 4y. \end{aligned} \tag{13.1.10}$$

satisfying the initial conditions: $x(1) = -1$, $y(1) = 2$.

Solution Setting $t_0 = 1$ and $\mathbf{x}_0 = \langle -1, 2\rangle$ in (13.1.14) and using the PMS given by (13.1.12), we have $\mathbf{x}(t) = Z(t, 1)\mathbf{x}_0$ or

$$\mathbf{x}(t) = \begin{bmatrix} 2e^{2(t-1)} - e^{3(t-1)} & 2e^{2(t-1)} - 2e^{3(t-1)} \\ e^{3(t-1)} - e^{2(t-1)} & 2e^{3(t-1)} - e^{2(t-1)} \end{bmatrix} \begin{bmatrix} -1 \\ 2 \end{bmatrix} = \begin{bmatrix} 2e^{2(t-1)} - 3e^{3(t-1)} \\ -e^{2(t-1)} + 3e^{3(t-1)} \end{bmatrix}.$$

Therefore,

$$x(t) = 2e^{2(t-1)} - 3e^{3(t-1)} \quad \text{and} \quad y(t) = -e^{2(t-1)} + 3e^{3(t-1)}.$$

♦

13.2 Matrix Exponential Functions

Something quite familiar stands out when Theorem 13.1.9 is perused and it is this: the properties of $Z(t) = Z(t, 0)$ look like the properties of the natural exponential function. This can be seen by replacing the matrices A and B with the scalars a and b, respectively, and by replacing the matrix function $Z(t)$ with e^{at}. Then it becomes obvious that the properties listed in the theorem for $Z(t)$ become the properties of the natural exponential function e^{at} although the condition regarding commutativity for property (vii) is no longer needed. The point to be made is this: The principal matrix solution of $\mathbf{x}' = A\mathbf{x}$ at $t = 0$, which is defined to be the matrix solution of

$$X' = AX, \quad X(0) = I$$

and denoted by $Z(t)$, is the matrix analogue of the principal (scalar) solution of the scalar differential equation $x' = ax$, namely, the solution of

$$x' = ax, \quad x(0) = 1,$$

which is $z(t) = e^{at}$ (cf. Definition 5.5.1). This suggests that the concept of the exponential function e^{at} can be extended to matrix functions. Accordingly, we define next what is meant by the *matrix exponential function*, which is denoted by e^{At}.

Matrix Exponential Function

Definition 13.2.1 Let A be a square matrix. The ***matrix exponential function*** e^{At} is defined as the principal matrix solution of $\mathbf{x}' = A\mathbf{x}$ at $t = 0$. Simply put,

$$e^{At} := Z(t). \tag{13.2.1}$$

Remark It follows from Theorem 13.1.2 that a general solution of the linear system $\mathbf{x}' = A\mathbf{x}$ is $\mathbf{x}(t) = e^{At}\mathbf{c}$ for all $t \in \mathbb{R}$, where $\mathbf{c} = \langle c_1, c_2 \rangle$ is a vector of arbitrary constants.

Example 13.2.1 Compute the matrix exponential function e^{At} for

$$A = \begin{bmatrix} \lambda_1 & 0 \\ 0 & \lambda_2 \end{bmatrix}, \tag{13.2.2}$$

where λ_1 and λ_2 denote any pair of real numbers.

Solution Consider $\mathbf{x}' = A\mathbf{x}$, where A is the matrix given by (13.2.2) and $\mathbf{x} = \langle x, y \rangle$. This is the matrix representation of the uncoupled system of differential equations

$$x' = \lambda_1 x, \quad y' = \lambda_2 y.$$

Since a general solution of this system is $x(t) = c_1 e^{\lambda_1 t}$ and $y(t) = c_2 e^{\lambda_2 t}$, it follows that $\mathbf{x}^1(t) = \langle e^{\lambda_1 t}, 0 \rangle$ and $\mathbf{x}^2(t) = \langle 0, e^{\lambda_2 t} \rangle$ are solutions of $\mathbf{x}' = A\mathbf{x}$. Moreover, they are linearly independent on $(-\infty, \infty)$. So, as $\begin{bmatrix} \mathbf{x}^1(0) & \mathbf{x}^2(0) \end{bmatrix} = I$, the matrix $Z(t)$ defined by $\begin{bmatrix} \mathbf{x}^1(t) & \mathbf{x}^2(t) \end{bmatrix}$ is the PMS of $\mathbf{x}' = A\mathbf{x}$ at $t = 0$. Accordingly, by (13.2.1),

$$e^{At} = Z(t) = \begin{bmatrix} e^{\lambda_1 t} & 0 \\ 0 & e^{\lambda_2 t} \end{bmatrix} \tag{13.2.3}$$

is the matrix exponential function associated with the matrix A. ♦

Example 13.2.2 Compute e^{Bt} for the matrix

$$B = \begin{bmatrix} 0 & 1 \\ 0 & 0 \end{bmatrix}.$$

Solution The linear system corresponding to $\mathbf{x}' = B\mathbf{x}$ is

$$x' = y, \quad y' = 0,$$

which has the general solution $x(t) = k_1 t + k_2$, $y(t) = k_1$. From this it follows that $\mathbf{x}^1(t) = \langle 1, 0 \rangle$ and $\mathbf{x}^2(t) = \langle t, 1 \rangle$ are linearly independent solutions of $\mathbf{x}' = B\mathbf{x}$. So, as $\begin{bmatrix} \mathbf{x}^1(0) & \mathbf{x}^2(0) \end{bmatrix} = I$, the matrix $Z(t) = \begin{bmatrix} \mathbf{x}^1(t) & \mathbf{x}^2(t) \end{bmatrix}$ is the PMS of $\mathbf{x}' = B\mathbf{x}$ at $t = 0$. Therefore,

$$e^{Bt} = \begin{bmatrix} 1 & t \\ 0 & 1 \end{bmatrix}.$$ ♦

Transcribing the properties of $Z(t)$ listed in Theorem 13.1.9 into matrix exponential notation, we obtain the following list of properties of e^{At}.

Properties of e^{At}

Theorem 13.2.1 *Let A and B be constant square matrices of the same order. Then, for all $r, s, t \in \mathbb{R}$,*

(i) $e^{A0} = e^{O} = I$ (ii) $e^{At} e^{As} = e^{A(t+s)}$

(continued)

(iii) $Be^{At} = e^{At}B$ *if* $AB = BA$ (iv) $Ae^{At} = e^{At}A$

(v) $\left(e^{At}\right)^{-1} = e^{-At}$ (vi) $e^{rIt} = e^{rt}I$

(vii) $\frac{d}{dt}e^{At} = Ae^{At}$ (viii) $\frac{d}{dt}e^{-At} = -Ae^{-At}$.

Proof It follows from Definition 13.2.1 that $e^{A0} = Z(0) = I$. Properties (ii)–(v) and (vii) follow from Theorem 13.1.9. For example, the property $Z(t)Z(s) = Z(t+s)$ becomes $e^{At}e^{As} = e^{A(t+s)}$ in matrix exponential notation. If $AB = BA$, then with $s = 0$ in Theorem 13.1.9 (vii) we get $BZ(t) = Z(t)B$. This yields (iii) and (iv). The proofs of (v), (vi), and (vii) are left to Problem 6. As for (viii), it follows from Theorem 13.1.8 (v) and from the above properties (v), (vii), and (iv) that

$$\frac{d}{dt}e^{-At} = -e^{-At}\left(\frac{d}{dt}e^{At}\right)e^{-At} = -e^{-At}\left(Ae^{At}\right)e^{-At} = -e^{-At}A = -Ae^{-At}.$$

Compare this to Problem 4. ■

Even though the matrix exponential function e^{At} has many of the same properties as that of the analogous scalar exponential function, we have to be careful when dealing with it. For instance, generally speaking, $Be^{At} \neq e^{At}B$ and $e^{At}e^{Bt} \neq e^{(A+B)t}$. An example of the latter is given later on in Example 13.4.3. However, these quantities are equal if the matrices A and B commute, which we see in Theorem 13.2.1 (iii) and in the next theorem.

Product of Matrix Exponential Functions

Theorem 13.2.2 *The matrix equation*

$$e^{At}e^{Bt} = e^{(A+B)t} \tag{13.2.4}$$

holds for $-\infty < t < \infty$ *if and only if* $AB = BA$.

Proof Recall from Definition 13.2.1 that e^{At} is defined to be $Z(t)$, namely, the PMS of $\mathbf{x}' = A\mathbf{x}$ at $t = 0$. And so $e^{(A+B)t}$ is the PMS of

$$\mathbf{x}' = (A+B)\mathbf{x} \tag{13.2.5}$$

at $t = 0$.

Now let us differentiate the left-hand side of (13.2.4). It follows from the product rule for matrix functions (Theorem 13.1.8 (ii)) and from Theorem 13.2.1 (vii) that

$$\frac{d}{dt}\left(e^{At}e^{Bt}\right) = e^{At}\frac{d}{dt}e^{Bt} + \frac{d}{dt}\left(e^{At}\right)e^{Bt} = e^{At}Be^{Bt} + Ae^{At}e^{Bt}.$$

If A and B commute, then it follows from Theorem 13.2.1 (iii) and from the right distributive law and the commutative law of addition for matrices that

$$\frac{d}{dt}\left(e^{At}e^{Bt}\right) = Be^{At}e^{Bt} + Ae^{At}e^{Bt} = (B+A)e^{At}e^{Bt} = (A+B)e^{At}e^{Bt}.$$

This shows $e^{At}e^{Bt}$ is a matrix solution of (13.2.5). In fact, by Theorem 13.1.1, it is a fundamental matrix solution of (13.2.5) since its determinant is nonzero at a point: at $t = 0$ its value is

$$\det\left(e^{At}e^{Bt}\right) = \det\left(e^{A0}e^{B0}\right) = \det I = 1.$$

Moreover, it is the PMS of (13.2.5) at $t = 0$ since its value there is I.

Since we determined $e^{At}e^{Bt}$ as well as $e^{(A+B)t}$ is the PMS of (13.2.5) at $t = 0$, we conclude that they are equal for $-\infty < t < \infty$ (see the first remark after Definition 13.1.5).

The proof of the converse is left to Problem 8. ■

Example 13.2.3 Compute e^{Jt} for the matrix

$$J = \begin{bmatrix} \lambda & 1 \\ 0 & \lambda \end{bmatrix}, \tag{13.2.6}$$

where $\lambda \in \mathbb{R}$.

Solution We could take the eigenvalue-eigenvector approach as in Example 13.1.3 to compute e^{Jt} (see Problem 3). However, the point of this example is to illustrate Theorem 13.2.2. First, observe that the sum of the matrices

$$A = \begin{bmatrix} \lambda & 0 \\ 0 & \lambda \end{bmatrix} \quad \text{and} \quad B = \begin{bmatrix} 0 & 1 \\ 0 & 0 \end{bmatrix}$$

is the matrix J:

$$A + B = \begin{bmatrix} \lambda & 0 \\ 0 & \lambda \end{bmatrix} + \begin{bmatrix} 0 & 1 \\ 0 & 0 \end{bmatrix} = \begin{bmatrix} \lambda & 1 \\ 0 & \lambda \end{bmatrix} = J.$$

Second, the matrix exponential functions e^{At} and e^{Bt} have already been computed in Examples 13.2.1 and 13.2.2, respectively. As a result, we have

$$e^{At} = \begin{bmatrix} e^{\lambda t} & 0 \\ 0 & e^{\lambda t} \end{bmatrix}, \quad e^{Bt} = \begin{bmatrix} 1 & t \\ 0 & 1 \end{bmatrix}.$$

Third, A and B commute:

$$AB = \begin{bmatrix} \lambda & 0 \\ 0 & \lambda \end{bmatrix} \begin{bmatrix} 0 & 1 \\ 0 & 0 \end{bmatrix} = \begin{bmatrix} 0 & \lambda \\ 0 & 0 \end{bmatrix} = \begin{bmatrix} 0 & 1 \\ 0 & 0 \end{bmatrix} \begin{bmatrix} \lambda & 0 \\ 0 & \lambda \end{bmatrix} = BA.$$

The upshot is that we can now use Theorem 13.2.2 to compute e^{Jt}:

$$e^{Jt} = e^{(A+B)t} = e^{At}e^{Bt} = \begin{bmatrix} e^{\lambda t} & 0 \\ 0 & e^{\lambda t} \end{bmatrix} \begin{bmatrix} 1 & t \\ 0 & 1 \end{bmatrix} = \begin{bmatrix} e^{\lambda t} & te^{\lambda t} \\ 0 & e^{\lambda t} \end{bmatrix}.$$

Therefore, the matrix exponential function e^{Jt} with J given by (13.2.6) is

$$e^{Jt} = e^{\lambda t} \begin{bmatrix} 1 & t \\ 0 & 1 \end{bmatrix} \tag{13.2.7}$$

for all $t \in \mathbb{R}$. ♦

Example 13.2.4 Let $\alpha, \beta \in \mathbb{R}$, where $\beta \neq 0$, and

$$M = \begin{bmatrix} \alpha & \beta \\ -\beta & \alpha \end{bmatrix}. \tag{13.2.8}$$

Compute e^{Mt}.

Solution The matrix M is the sum of the matrices

$$A = \begin{bmatrix} \alpha & 0 \\ 0 & \alpha \end{bmatrix} \quad \text{and} \quad B = \begin{bmatrix} 0 & \beta \\ -\beta & 0 \end{bmatrix}.$$

It is left as an exercise to show $AB = BA$. From Example 13.2.1, we have

$$e^{At} = \begin{bmatrix} e^{\alpha t} & 0 \\ 0 & e^{\alpha t} \end{bmatrix} = e^{\alpha t} \begin{bmatrix} 1 & 0 \\ 0 & 1 \end{bmatrix} = e^{\alpha t} I.$$

Now let us compute e^{Bt} by obtaining two linearly independent solutions of the vector differential equation

$$\mathbf{x}' = B\mathbf{x}.$$

which is equivalent to the system of equations

$$x' = \beta y, \quad y' = -\beta x.$$

Differentiating the first of these, we have

$$x'' = \beta y' = \beta(-\beta x)$$

or

$$x'' + \beta^2 x = 0,$$

which is the equation of the simple harmonic oscillator. Having encountered this equation many times before, we already know that

$$x(t) = c_1 \cos \beta t + c_2 \sin \beta t$$

is a general solution. Thus, as $x' = \beta y$,

$$y(t) = \frac{1}{\beta}(-c_1 \beta \sin \beta t + c_2 \beta \cos \beta t) = -c_1 \sin \beta t + c_2 \cos \beta t.$$

Corresponding to the pairs of values $(c_1, c_2) = (1, 0)$ and $(c_1, c_2) = (0, 1)$ are the particular solutions

$$\langle x(t), y(t) \rangle = \langle \cos \beta t, -\sin \beta t \rangle \quad \text{and} \quad \langle x(t), y(t) \rangle = \langle \sin \beta t, \cos \beta t \rangle.$$

As a result,

$$e^{Bt} = \begin{bmatrix} \cos \beta t & \sin \beta t \\ -\sin \beta t & \cos \beta t \end{bmatrix}$$

since its columns are linearly independent solutions of $\mathbf{x}' = B\mathbf{x}$ and its value at $t = 0$ is the identity matrix I.

Therefore, it follows from Theorem 13.2.2 that

$$e^{Mt} = e^{At} e^{Bt} = e^{\alpha t} I \begin{bmatrix} \cos \beta t & \sin \beta t \\ -\sin \beta t & \cos \beta t \end{bmatrix} = e^{\alpha t} \begin{bmatrix} \cos \beta t & \sin \beta t \\ -\sin \beta t & \cos \beta t \end{bmatrix} \tag{13.2.9}$$

for all $t \in \mathbb{R}$. ♦

Before we move on to the next section, let us point out the following. Recall from Definition 13.2.1 that $e^{At} := Z(t)$. It follows from this and Theorem 13.1.9 (iv) that

$$Z(t, t_0) = Z(t - t_0) = e^{A(t-t_0)}. \tag{13.2.10}$$

Because of this and Theorem 13.1.10, the solution $\mathbf{x}(t) = Z(t, t_0)\mathbf{x}_0$ of

$$\mathbf{x}' = A\mathbf{x}, \quad \mathbf{x}(t_0) = \mathbf{x}_0$$

can also be expressed in terms of a matrix exponential function as follows:

$$\mathbf{x}(t) = e^{A(t-t_0)}\mathbf{x}_0. \tag{13.2.11}$$

13.3 Jordan Canonical Matrices

We continue with our investigation of the linear system

$$\mathbf{x}' = A\mathbf{x} \tag{13.3.1}$$

where

$$\mathbf{x} = \begin{bmatrix} x_1 \\ x_2 \end{bmatrix} \quad \text{and} \quad A = \begin{bmatrix} a_{11} & a_{12} \\ a_{21} & a_{22} \end{bmatrix}.$$

Throughout this chapter, as in the previous one, let us assume that the coefficient matrix A is a ***real matrix*** (that is, a matrix with real entries).

Of course, there are infinitely many coefficient matrices A. Consequently, there are infinitely many phase portraits, one for each A. But surprisingly, each of these phase portraits can be classified as belonging to just a few types, where those of a given type resemble each other and correspond to linear systems whose solutions exhibit the same qualitative behavior. As we shall see, the reason for this boils down to the fact that every 2×2 coefficient matrix A can be transformed into a matrix belonging to one of four basic (canonical) forms. These are known as the *Jordan canonical forms of the matrix A*. A complete explanation of this together with illustrative examples takes takes up most of this section. Phase portraits of (13.3.1) are discussed in Sect. 13.5.

We begin with a change in vector variables: from $\mathbf{x} = \langle x_1, x_2 \rangle$ to $\mathbf{y} = \langle y_1, y_2 \rangle$ so that these variables are related by the change-of-variable formula

$$\mathbf{x} = P\mathbf{y} \tag{13.3.2}$$

for some suitably chosen constant square matrix P of order 2. Our goal is to find P so that (13.3.1) transforms into a vector differential equation

$$\mathbf{y}' = J\mathbf{y} \tag{13.3.3}$$

with the matrix J being as simple as possible. It turns out that for a given coefficient matrix A and with just the right P, the matrix J has one of only four possible forms, to wit:

$$\begin{bmatrix} \lambda_1 & 0 \\ 0 & \lambda_2 \end{bmatrix}, \quad \begin{bmatrix} \lambda & 0 \\ 0 & \lambda \end{bmatrix}, \quad \begin{bmatrix} \lambda & 1 \\ 0 & \lambda \end{bmatrix}, \quad \begin{bmatrix} \alpha & \beta \\ -\beta & \alpha \end{bmatrix}, \tag{13.3.4}$$

where $\lambda_1, \lambda_2, \lambda, \alpha, \beta$ are real numbers, $\lambda_1 \neq \lambda_2$, and $\beta > 0$. Moreover, as we shall see, the reason why there are only four of these forms comes down to the eigenvalues of A. Since A is a 2×2 matrix, it can have (i) two distinct real eigenvalues, (ii) a real eigenvalue of multiplicity two, or (iii) a complex conjugate pair of eigenvalues (see Sect. 12.5.5).

We also require the matrix P to be invertible so that we can switch back and forth between $\mathbf{x}$ and $\mathbf{y}$ using (13.3.2) and the inverse transformation

$$\mathbf{y} = P^{-1}\mathbf{x}. \tag{13.3.5}$$

This will also allow us to compare phase portraits in the x_1x_2-plane with their counterparts in the y_1y_2-plane.

Now let us see how equation (13.3.1) is changed by the transformation (13.3.2). Differentiating (13.3.5) with respect to t and then employing (13.3.1) and (13.3.2), we obtain

$$\mathbf{y}' = P^{-1}\mathbf{x}' = P^{-1}A\mathbf{x} = P^{-1}A(P\mathbf{y}) = (P^{-1}AP)\mathbf{y}. \tag{13.3.6}$$

In short, the transformation (13.3.2) changes $\mathbf{x}' = A\mathbf{x}$ to $\mathbf{y}' = J\mathbf{y}$, where

$$J := P^{-1}AP.$$

So if P is an invertible matrix and if there is a vector function $\mathbf{y}(t)$ that is a solution of equation (13.3.6), then $\mathbf{x}(t) = P\mathbf{y}(t)$ is a solution of (13.3.1) since

$$\mathbf{x}'(t) = (P\mathbf{y}(t))' = P\mathbf{y}'(t) = P(P^{-1}AP)\mathbf{y}(t) = A(P\mathbf{y}(t)) = A\mathbf{x}(t).$$

Conversely, if $\mathbf{x}(t)$ is a solution of (13.3.1), then it follows from (13.3.5) and (13.3.2) that $\mathbf{y}(t) = P^{-1}\mathbf{x}(t)$ is a solution of $\mathbf{y}' = (P^{-1}AP)\mathbf{y} = J\mathbf{y}$. Thus, we have proved the first two items of the following theorem.

Transformation of Solutions

Theorem 13.3.1 *Let A and P be square matrices of order 2, where P is invertible. Let $J := P^{-1}AP$.*

(i) *If $\mathbf{y}(t)$ is a solution of $\mathbf{y}' = J\mathbf{y}$, then $P\mathbf{y}(t)$ is a solution of $\mathbf{x}' = A\mathbf{x}$.*
(ii) *If $\mathbf{x}(t)$ is a solution of $\mathbf{x}' = A\mathbf{x}$, then $P^{-1}\mathbf{x}(t)$ is a solution of $\mathbf{y}' = J\mathbf{y}$.*

(continued)

(iii) *A and J have the same eigenvalues.*
(iv) $J^k = P^{-1}A^kP$ (*k, a positive integer*)
(v) $e^{At} = Pe^{Jt}P^{-1}$
(vi) $e^{Jt} = P^{-1}e^{At}P$

Remark Instead of memorizing $J = P^{-1}AP$ and possibly misremembering it as $J = PAP^{-1}$, memorize $PJ = AP$ and that P is invertible. A silly mnemonic for the former is "***PJ****s are* ***A****bsolutely* ***P****erfect* on a cold wintry night."

Proof Consider (iii). Since the eigenvalues of J are the roots of $\det(J - \lambda I) = 0$, all we have to do is show that the characteristic polynomials of J and A are equal. This is where Theorem 12.5.4 comes in handy:

$$\begin{aligned}\det(J - \lambda I) &= \det(P^{-1}AP - \lambda P^{-1}IP) = \det(P^{-1}(A - \lambda I)P) \\ &= \det P^{-1} \cdot \det(A - \lambda I) \cdot \det P.\end{aligned}$$

Since $\det P^{-1} \cdot \det P = \det(P^{-1}P) = 1$, we have $\det(J - \lambda I) = \det(A - \lambda I)$, which completes the proof of (iii).

Item (iv) is trivially true for $k = 1$. It is also true for $k = 2$ since

$$J^2 = (P^{-1}AP)(P^{-1}AP) = P^{-1}A(PP^{-1})AP = P^{-1}A^2P.$$

If (iv) is true for $k = n$, then $J^n = P^{-1}A^nP$, which implies that

$$J^{n+1} = JJ^n = (P^{-1}AP)(P^{-1}AP)^n = (P^{-1}AP)(P^{-1}A^nP) = P^{-1}A^{n+1}P.$$

So, by mathematical induction, (iv) is true.

As for the proof of (v), let $\mathbf{x_0}$ denote any vector in $\mathbb{R}^2$. By (13.2.11), the unique solution of the equation $\mathbf{x}' = A\mathbf{x}$ such that $\mathbf{x}(0) = \mathbf{x}_0$ is

$$\mathbf{x}(t) = e^{At}\mathbf{x_0}. \tag{13.3.7}$$

It follows from item (ii), which was proved earlier, that the function $\mathbf{y}(t) = P^{-1}\mathbf{x}(t)$ is a solution of $\mathbf{y}' = J\mathbf{y}$, where $J = P^{-1}AP$. Furthermore, referring to (13.2.11) again, the unique solution of the equation $\mathbf{y}' = J\mathbf{y}$ such that $\mathbf{y}(0) = P^{-1}\mathbf{x}(0) = P^{-1}\mathbf{x_0}$ is

$$\mathbf{y}(t) = e^{Jt}\mathbf{y}(0) = e^{Jt}P^{-1}\mathbf{x_0}.$$

Or, as $\mathbf{y}(t) = P^{-1}\mathbf{x}(t)$, this becomes $P^{-1}\mathbf{x}(t) = e^{Jt}P^{-1}\mathbf{x_0}$, from which we obtain

$$\mathbf{x}(t) = Pe^{Jt}P^{-1}\mathbf{x_0}. \tag{13.3.8}$$

Thus, from (13.3.7) and (13.3.8), we see that

$$e^{At}\mathbf{x_0} = Pe^{Jt}P^{-1}\mathbf{x_0}.$$

Since this is true for all $\mathbf{x}_0 \in \mathbb{R}^2$, we have statement (v). (cf. Problem 7 in Chap. 12).

Finally, statement (v) implies (vi). ■

The relationship $J = P^{-1}AP$ between the matrices A and J that we encounter in Theorem 13.3.1 is an important theme in linear algebra and differential equations, so much so that there is a name for it, which is given in the following definition.

Similar Matrices

Definition 13.3.2 Let C and D be 2×2 matrices. The matrix C is said to be ***similar*** to D if there is an invertible 2×2 matrix P such that $C = P^{-1}DP$. The change-of-variable formulas $\mathbf{x} = P\mathbf{y}$ and $\mathbf{y} = P^{-1}\mathbf{x}$ are known as ***similarity transformations*** and P and P^{-1} are called ***(similarity) transformation matrices***.

Remark If C is similar to D, then D is also similar to C (cf. Problem 9). So instead of saying "C is similar to D" or that "D is also similar to C", we can simply say that C and D are *similar matrices*. With this terminology, we can describe the relationship between A and J in Theorem 13.3.1 by simply stating that they are similar matrices.

Example 13.3.1 Let $J = P^{-1}AP$, where

$$A = \begin{bmatrix} 3 & 5 \\ -2 & -4 \end{bmatrix} \quad \text{and} \quad P = \begin{bmatrix} 1 & 5 \\ -1 & -2 \end{bmatrix}.$$

Compute the matrices J, e^{Jt}, and e^{At}.

Solution The matrix P is truly invertible since it is nonsingular ($\det P \neq 0$). By Theorem 12.5.8,

$$P^{-1} = \frac{1}{3}\begin{bmatrix} -2 & -5 \\ 1 & 1 \end{bmatrix}.$$

Therefore,

$$J = P^{-1}AP = \frac{1}{3}\begin{bmatrix} -2 & -5 \\ 1 & 1 \end{bmatrix}\begin{bmatrix} 3 & 5 \\ -2 & -4 \end{bmatrix}\begin{bmatrix} 1 & 5 \\ -1 & -2 \end{bmatrix}$$

$$= \frac{1}{3}\begin{bmatrix} 4 & 10 \\ 1 & 1 \end{bmatrix}\begin{bmatrix} 1 & 5 \\ -1 & -2 \end{bmatrix} = \frac{1}{3}\begin{bmatrix} -6 & 0 \\ 0 & 3 \end{bmatrix} = \begin{bmatrix} -2 & 0 \\ 0 & 1 \end{bmatrix}.$$

It follows from Example 13.2.1 that

$$e^{Jt} = \begin{bmatrix} e^{-2t} & 0 \\ 0 & e^{t} \end{bmatrix}.$$

We can use item (v) in Theorem 13.3.1 to compute e^{At}. Since

$$Pe^{Jt}P^{-1} = \frac{1}{3}\begin{bmatrix} 1 & 5 \\ -1 & -2 \end{bmatrix}\begin{bmatrix} e^{-2t} & 0 \\ 0 & e^{t} \end{bmatrix}\begin{bmatrix} -2 & -5 \\ 1 & 1 \end{bmatrix}$$

$$= \frac{1}{3}\begin{bmatrix} e^{-2t} & 5e^{t} \\ -e^{-2t} & -2e^{t} \end{bmatrix}\begin{bmatrix} -2 & -5 \\ 1 & 1 \end{bmatrix},$$

we have

$$e^{At} = Pe^{Jt}P^{-1} = \frac{1}{3}\begin{bmatrix} -2e^{-2t} + 5e^{t} & -5e^{-2t} + 5e^{t} \\ 2e^{-2t} - 2e^{t} & 5e^{-2t} - 2e^{t} \end{bmatrix}.$$

Note the right-hand side is equal to the identity matrix at $t = 0$, as it should be. ♦

It turns out that there is a real invertible 2×2 matrix P for each 2×2 real matrix A such that $P^{-1}AP$ looks like one of the four matrices listed earlier in this section (see (13.3.4)). These matrices are called the ***Jordan canonical forms of*** 2×2 ***matrices***. In other words, for every 2×2 matrix A, a matrix P exists such that $P^{-1}AP = J$, where J is one of the four matrices listed in the next theorem. The matrix J is known as a ***Jordan (canonical) matrix*** and said to be a ***Jordan (canonical) form of*** A.

Jordan Canonical Forms of Matrices of Order 2

Theorem 13.3.3 *For each real* 2×2 *matrix A, there is a real invertible* 2×2 *matrix P such that the matrix J defined by*

$$J = P^{-1}AP \tag{13.3.9}$$

(continued)

has the form of one of the following four Jordan canonical matrices:

$$\text{(i)} \begin{bmatrix} \lambda_1 & 0 \\ 0 & \lambda_2 \end{bmatrix} \quad \text{(ii)} \begin{bmatrix} \lambda & 0 \\ 0 & \lambda \end{bmatrix} \quad \text{(iii)} \begin{bmatrix} \lambda & 1 \\ 0 & \lambda \end{bmatrix} \quad \text{(iv)} \begin{bmatrix} \alpha & \beta \\ -\beta & \alpha \end{bmatrix}, \tag{13.3.10}$$

where λ_1, λ_2, λ, α, β *are real numbers,* $\lambda_1 \neq \lambda_2$, *and* $\beta \neq 0$.

Proof Recall that a given real 2×2 matrix A always has two eigenvalues, which are either (a) real and distinct, (b) real and repeated, or (c) a complex conjugate pair. First let us take up the the case of distinct real eigenvalues.

(a) *Distinct real eigenvalues*: Suppose the eigenvalues of a matrix A are distinct real numbers. Then calling them λ_1 and λ_2, let us define the matrix J by

$$J = \begin{bmatrix} \lambda_1 & 0 \\ 0 & \lambda_2 \end{bmatrix}.$$

Let $\mathbf{v}^1$ and $\mathbf{v}^2$ be eigenvectors of A corresponding to λ_1 and λ_2, respectively. So,

$$A\mathbf{v}^1 = \lambda_1 \mathbf{v}^1 \quad \text{and} \quad A\mathbf{v}^2 = \lambda_2 \mathbf{v}^2.$$

By Theorem 12.5.31, these two eigenvectors are linearly independent. Thus the matrix $\begin{bmatrix} \mathbf{v}^1 & \mathbf{v}^2 \end{bmatrix}$ is invertible (cf. Theorem 12.5.20 and Corollary 12.5.9), which suggests this matrix might serve as the matrix P in (13.3.9). So the question at the moment is: Is $P^{-1}AP$ with $P = \begin{bmatrix} \mathbf{v}^1 & \mathbf{v}^2 \end{bmatrix}$ equal to the matrix J?

To answer this, let us compare the matrix products AP and PJ. Computing AP, we get

$$AP = A \begin{bmatrix} \mathbf{v}^1 & \mathbf{v}^2 \end{bmatrix} = \begin{bmatrix} A\mathbf{v}^1 & A\mathbf{v}^2 \end{bmatrix} = \begin{bmatrix} \lambda_1 \mathbf{v}^1 & \lambda_2 \mathbf{v}^2 \end{bmatrix}.$$

Before computing PJ, let $\mathbf{u}^1 := \langle \lambda_1, 0 \rangle$ and $\mathbf{u}^2 := \langle 0, \lambda_2 \rangle$. Then the matrix J can be written as $\begin{bmatrix} \mathbf{u}^1 & \mathbf{u}^2 \end{bmatrix}$. Since

$$P\mathbf{u}^1 = \begin{bmatrix} \mathbf{v}^1 & \mathbf{v}^2 \end{bmatrix} \begin{bmatrix} \lambda_1 \\ 0 \end{bmatrix} = \lambda_1 \mathbf{v}^1 \quad \text{and} \quad P\mathbf{u}^2 = \begin{bmatrix} \mathbf{v}^1 & \mathbf{v}^2 \end{bmatrix} \begin{bmatrix} 0 \\ \lambda_2 \end{bmatrix} = \lambda_2 \mathbf{v}^2,$$

we have

$$PJ = P \begin{bmatrix} \lambda_1 & 0 \\ 0 & \lambda_2 \end{bmatrix} = \begin{bmatrix} P\mathbf{u}^1 & P\mathbf{u}^2 \end{bmatrix} = \begin{bmatrix} \lambda_1 \mathbf{v}^1 & \lambda_2 \mathbf{v}^2 \end{bmatrix}.$$

Consequently, $AP = PJ$. Since P is invertible,

$$P^{-1}AP = J = \begin{bmatrix} \lambda_1 & 0 \\ 0 & \lambda_2 \end{bmatrix},$$

which is the Jordan canonical matrix labeled (i) in (13.3.10).

(b) *Repeated real eigenvalues*: If A has a repeated real eigenvalue λ, then there are two possibilities to consider: either there are two linearly independent eigenvectors of A corresponding to λ or no two eigenvectors of A are linearly independent.

Suppose there are two linearly independent eigenvectors $\mathbf{v}^1$ and $\mathbf{v}^2$ corresponding to λ (as in Example 12.5.20). Let $P = \begin{bmatrix} \mathbf{v}^1 & \mathbf{v}^2 \end{bmatrix}$ and define J by

$$J = \begin{bmatrix} \lambda & 0 \\ 0 & \lambda \end{bmatrix}.$$

Aside from λ replacing both λ_1 and λ_2, this looks exactly like case (a), which means that the argument used there also applies here; so

$$P^{-1}AP = J = \begin{bmatrix} \lambda & 0 \\ 0 & \lambda \end{bmatrix},$$

which is the Jordan canonical matrix (ii). Now observe this: Since $J = \lambda I$,

$$A = PJP^{-1} = \lambda PIP^{-1} = \lambda PP^{-1} = \lambda I = J.$$

From this we conclude: A matrix with a repeated real eigenvalue and two linearly independent eigenvectors is only similar to itself. The only 2×2 matrices with a real repeated eigenvalue and two linearly independent eigenvectors are Jordan canonical matrices of the form (ii).

Now consider the possibility that no two eigenvectors of A are linearly independent (as in Example 12.5.21). Let $\mathbf{v}^1$ be any eigenvector of A corresponding to λ. It follows from Theorem 12.5.33 (ii) and its proof that there is a vector $\mathbf{v}^2$ such that it and $\mathbf{v}^1$ are linearly independent and that

$$(A - \lambda I)\mathbf{v}^2 = \mathbf{v}^1.$$

Let $P = \begin{bmatrix} \mathbf{v}^1 & \mathbf{v}^2 \end{bmatrix}$. So, as $A\mathbf{v}^2 = \mathbf{v}^1 + \lambda \mathbf{v}^2$, we have

$$AP = \begin{bmatrix} A\mathbf{v}^1 & A\mathbf{v}^2 \end{bmatrix} = \begin{bmatrix} \lambda \mathbf{v}^1 & \mathbf{v}^1 + \lambda \mathbf{v}^2 \end{bmatrix}.$$

For this case, let $J = \begin{bmatrix}\mathbf{w^1} & \mathbf{w^2}\end{bmatrix}$ with $\mathbf{w}^1 = \langle \lambda, 0\rangle$ and $\mathbf{w}^2 = \langle 1, \lambda\rangle$. Then the product PJ is

$$PJ = P\begin{bmatrix}\mathbf{w^1} & \mathbf{w^2}\end{bmatrix} = \begin{bmatrix}P\mathbf{w^1} & P\mathbf{w^2}\end{bmatrix},$$

where

$$P\mathbf{w^1} = \begin{bmatrix}\mathbf{v^1} & \mathbf{v^2}\end{bmatrix}\begin{bmatrix}\lambda \\ 0\end{bmatrix} = \lambda\mathbf{v^1} \quad \text{and} \quad P\mathbf{w^2} = \begin{bmatrix}\mathbf{v^1} & \mathbf{v^2}\end{bmatrix}\begin{bmatrix}1 \\ \lambda\end{bmatrix} = \mathbf{v^1} + \lambda\mathbf{v^2}.$$

Thus, $PJ = \begin{bmatrix}\lambda\mathbf{v^1} & \mathbf{v^1} + \lambda\mathbf{v^2}\end{bmatrix}$. Consequently, $AP = PJ$. Finally, as P is invertible,

$$P^{-1}AP = J = \begin{bmatrix}\lambda & 1 \\ 0 & \lambda\end{bmatrix},$$

which is Jordan matrix (iii).

(c) *Complex eigenvalues*: Now suppose $\alpha \pm i\beta$, where $\beta \neq 0$, are the eigenvalues of a matrix A (recall that complex eigenvalues always occur in conjugate pairs). Let $\mathbf{v}$ denote a complex eigenvector corresponding to the eigenvalue $\alpha + i\beta$. In other words,

$$A\mathbf{v} = (\alpha + i\beta)\mathbf{v} \tag{13.3.11}$$

In the last subsection of Sect. 12.5, which deals with complex eigenvalues, we ascertained that eigenvectors corresponding to complex eigenvalues are always complex vectors. Thus,

$$\mathbf{v} = \mathbf{v^1} + i\mathbf{v^2},$$

where $\mathbf{v^1}$ and $\mathbf{v^2}$ are the real and imaginary parts of $\mathbf{v}$, respectively, with $\mathbf{v^2} \neq \mathbf{0}$. Moreover, $\mathbf{v^1} \neq \mathbf{0}$; for otherwise (13.3.11) or

$$A(\mathbf{v^1} + i\mathbf{v^2}) = (\alpha + i\beta)(\mathbf{v^1} + i\mathbf{v^2}) \tag{13.3.12}$$

would simplify to

$$A\mathbf{v^2} = (\alpha + i\beta)\mathbf{v^2}, \tag{13.3.13}$$

which is not possible since $A\mathbf{v^2}$ is a real vector, whereas $(\alpha + i\beta)\mathbf{v^2}$ is a complex vector.

It is also important to realize that $\mathbf{v^1}$ and $\mathbf{v^2}$ are linearly independent vectors. If they were not, then a nonzero real constant k would exist such that $\mathbf{v^1} = k\mathbf{v^2}$. But then (13.3.12) would be

$$A(k\mathbf{v^2} + i\mathbf{v^2}) = (\alpha + i\beta)(k\mathbf{v^2} + i\mathbf{v^2})$$

which simplifies to the nonsensical equation (13.3.13). Since $\mathbf{v}^1$ and $\mathbf{v}^2$ are linearly independent, the matrix $\begin{bmatrix}\mathbf{v}^1 & \mathbf{v}^2\end{bmatrix}$ is invertible.

Taking the real and imaginary parts of (13.3.12), we obtain the pair of equations

$$A\mathbf{v}^1 = \alpha\mathbf{v}^1 - \beta\mathbf{v}^2 = \begin{bmatrix}\mathbf{v}^1 & \mathbf{v}^2\end{bmatrix}\begin{bmatrix}\alpha \\ -\beta\end{bmatrix}$$

and

$$A\mathbf{v}^2 = \beta\mathbf{v}^1 + \alpha\mathbf{v}^2 = \begin{bmatrix}\mathbf{v}^1 & \mathbf{v}^2\end{bmatrix}\begin{bmatrix}\beta \\ \alpha\end{bmatrix}.$$

Thus, letting $P = \begin{bmatrix}\mathbf{v}^1 & \mathbf{v}^2\end{bmatrix}$, we have

$$AP = A\begin{bmatrix}\mathbf{v}^1 & \mathbf{v}^2\end{bmatrix} = \begin{bmatrix}A\mathbf{v}^1 & A\mathbf{v}^2\end{bmatrix} = \begin{bmatrix}\alpha\mathbf{v}^1 - \beta\mathbf{v}^2 & \beta\mathbf{v}^1 + \alpha\mathbf{v}^2\end{bmatrix} = \begin{bmatrix}\mathbf{v}^1 & \mathbf{v}^2\end{bmatrix}\begin{bmatrix}\alpha & \beta \\ -\beta & \alpha\end{bmatrix}.$$

With J as the name of the second matrix on the right-hand side, we have $AP = PJ$. Furthermore, because P is invertible, we can also write this as $P^{-1}AP = J$. This completes the proof of the theorem. ■

Theorem 13.3.3 states that every 2×2 matrix A is similar to one of the four Jordan canonical matrices listed in that theorem. Furthermore, its proof shows that the eigenvectors of A can be used to obtain a similarity transformation matrix P so that $P^{-1}AP = J$. The next theorem is a consequence of this result, Theorem 13.3.1 (v), and the results of the computations in Examples 13.2.1, 13.2.3, and 13.2.4.

The Matrix Pairs (A, e^{At}) and (J, e^{Jt})

Theorem 13.3.4 *Let A be a 2×2 matrix with real entries. In each of the following cases, J is a Jordan canonical form of A and P is a 2×2 transformation matrix.*

(i) *If λ_1 and λ_2 are distinct real eigenvalues of the matrix A, let $\mathbf{v}^1$ and $\mathbf{v}^2$ be any two linearly independent eigenvectors of A corresponding to λ_1 and λ_2, respectively. Let $P = \begin{bmatrix}\mathbf{v}^1 & \mathbf{v}^2\end{bmatrix}$. The matrices A and e^{At} are related to the matrices*

$$J = \begin{bmatrix}\lambda_1 & 0 \\ 0 & \lambda_2\end{bmatrix} \quad \text{and} \quad e^{Jt} = \begin{bmatrix}e^{\lambda_1 t} & 0 \\ 0 & e^{\lambda_2 t}\end{bmatrix}$$

through the equations $A = PJP^{-1}$ and $e^{At} = Pe^{Jt}P^{-1}$.

(continued)

(ii) *If λ is a repeated real eigenvalue of A and if there are two linearly independent eigenvectors of A corresponding to λ, then*

$$J = \begin{bmatrix} \lambda & 0 \\ 0 & \lambda \end{bmatrix} \quad \text{and} \quad e^{Jt} = \begin{bmatrix} e^{\lambda t} & 0 \\ 0 & e^{\lambda t} \end{bmatrix}.$$

Moreover, $A = J$ and $e^{Jt} = e^{At}$.

(iii) *If λ is a repeated real eigenvalue of A and if no two eigenvectors of A are linearly independent, let $P = \begin{bmatrix} \mathbf{v}^1 & \mathbf{v}^2 \end{bmatrix}$, where $\mathbf{v}^1$ is an eigenvector of A and $\mathbf{v}^2$ is a generalized eigenvector of A satisfying the equation*

$$(A - \lambda I)\mathbf{v}^2 = \mathbf{v}^1.$$

The matrices A and e^{At} are related to

$$J = \begin{bmatrix} \lambda & 1 \\ 0 & \lambda \end{bmatrix} \quad \text{and} \quad e^{Jt} = e^{\lambda t} \begin{bmatrix} 1 & t \\ 0 & 1 \end{bmatrix}$$

through the equations $A = PJP^{-1}$ and $e^{At} = Pe^{Jt}P^{-1}$.

(iv) *If $\lambda = \alpha + i\beta$, where $\beta \neq 0$, is a complex eigenvalue of A, let $\mathbf{v}$ be an eigenvector of A corresponding to λ. This eigenvector is complex and can be expressed in the form $\mathbf{v} = \mathbf{v}^1 + i\mathbf{v}^2$, where $\mathbf{v}^1$ and $\mathbf{v}^2$ are linearly independent real vectors. Let $P = \begin{bmatrix} \mathbf{v}^1 & \mathbf{v}^2 \end{bmatrix}$. The matrices A and e^{At} are related to*

$$J = \begin{bmatrix} \alpha & \beta \\ -\beta & \alpha \end{bmatrix} \quad \text{and} \quad e^{Jt} = e^{\alpha t} \begin{bmatrix} \cos\beta t & \sin\beta t \\ -\sin\beta t & \cos\beta t \end{bmatrix}$$

through the equations $A = PJP^{-1}$ and $e^{At} = Pe^{Jt}P^{-1}$.

Remarks The Jordan canonical matrices in parts (i) and (iv) of this theorem are not unique. In the case of (i), suppose we alter J by interchanging the eigenvalues along its main diagonal. Then it becomes

$$J = \begin{bmatrix} \lambda_2 & 0 \\ 0 & \lambda_1 \end{bmatrix}.$$

This is also a Jordan canonical form of A. However, in order to still have $PJ = AP$, the columns of the transformation matrix P in (i) must also be interchanged. Then the new P becomes $P = \begin{bmatrix} \mathbf{v}^2 & \mathbf{v}^1 \end{bmatrix}$. The reason for this can be seen from an inspection of the proof of Theorem 13.3.3 (i). For these altered matrices, it is still the case that

$e^{At} = Pe^{Jt}P^{-1}$, where

$$e^{Jt} = \begin{bmatrix} e^{\lambda_2 t} & 0 \\ 0 & e^{\lambda_1 t} \end{bmatrix}.$$

As for part (iv) of the theorem, suppose we use the complex conjugate eigenvalue $\alpha - i\beta$ instead of $\alpha + i\beta$. Just as we proceeded with the proof of part (c) in Theorem 13.3.3, let $\mathbf{v}$ be a complex eigenvector corresponding to the complex eigenvalue $\alpha - i\beta$. In other words, let $\mathbf{v}$ be any vector satisfying the equation

$$A\mathbf{v} = (\alpha - i\beta)\mathbf{v}.$$

Following the steps of the proof for (c) beginning with (13.3.11) and making the necessary changes, we find that the matrix

$$J = \begin{bmatrix} \alpha & -\beta \\ \beta & \alpha \end{bmatrix}$$

is another Jordan form of A and that the corresponding matrix exponential function is

$$e^{Jt} = e^{\alpha t} \begin{bmatrix} \cos \beta t & -\sin \beta t \\ \sin \beta t & \cos \beta t \end{bmatrix}.$$

Both of these forms of the Jordan canonical matrix for complex eigenvalues are used; for example, see Arrowsmith and Place [2, p. 53] and Hale and Koçak [42, p. 231].

Example 13.3.2 Find a matrix J that is a Jordan canonical form of

$$A = \begin{bmatrix} 1 & 6 \\ 2 & 2 \end{bmatrix}.$$

Then find the matrix exponential functions e^{Jt} and e^{At}.

Solution Since $\operatorname{tr} A = 3$ and $\det A = -10$, the characteristic polynomial of A is

$$\det(A - \lambda I) = \lambda^2 - (\operatorname{tr} A)\,\lambda + \det A = \lambda^2 - 3\lambda - 10 = (\lambda - 5)(\lambda + 2).$$

So $\lambda_1 = 5$ and $\lambda_2 = -2$ are the eigenvalues of A. It then follows from part (i) of Theorem 13.3.4 that

$$J = \begin{bmatrix} 5 & 0 \\ 0 & -2 \end{bmatrix} \quad \text{and} \quad e^{Jt} = \begin{bmatrix} e^{5t} & 0 \\ 0 & e^{-2t} \end{bmatrix}.$$

At this point we need to find a transformation matrix $P = \begin{bmatrix} \mathbf{v}^1 & \mathbf{v}^2 \end{bmatrix}$, where $\mathbf{v}^1$ and $\mathbf{v}^1$ are linearly independent eigenvectors of A corresponding to $\lambda_1 = 5$ and

$\lambda_2 = -2$, respectively, so that we will be able to compute $e^{At} = Pe^{Jt}P^{-1}$. An eigenvector $\mathbf{v^1}$ corresponding to $\lambda_1 = 5$ is a solution of

$$(A - 5I)\mathbf{v} = \mathbf{0} \quad \text{or} \quad \begin{bmatrix} -4 & 6 \\ 2 & -3 \end{bmatrix} \begin{bmatrix} v_1 \\ v_2 \end{bmatrix} = \begin{bmatrix} 0 \\ 0 \end{bmatrix}.$$

By inspection, a solution is $\mathbf{v^1} = \langle 3, 2 \rangle$. An eigenvector $\mathbf{v^2}$ corresponding to $\lambda_2 = -2$ is a solution of

$$(A + 2I)\mathbf{v} = \mathbf{0} \quad \text{or} \quad \begin{bmatrix} 3 & 6 \\ 2 & 4 \end{bmatrix} \begin{bmatrix} v_1 \\ v_2 \end{bmatrix} = \begin{bmatrix} 0 \\ 0 \end{bmatrix}.$$

A solution of this matrix equation is $\mathbf{v^1} = \langle 2, -1 \rangle$. Therefore,

$$P = \begin{bmatrix} \mathbf{v^1} & \mathbf{v^2} \end{bmatrix} = \begin{bmatrix} 3 & 2 \\ 2 & -1 \end{bmatrix}.$$

Its inverse is

$$P^{-1} = \frac{1}{\det P} \begin{bmatrix} -1 & -2 \\ -2 & 3 \end{bmatrix} = \frac{1}{7} \begin{bmatrix} 1 & 2 \\ 2 & -3 \end{bmatrix}.$$

As a result,

$$e^{At} = Pe^{Jt}P^{-1} = \frac{1}{7} \begin{bmatrix} 3 & 2 \\ 2 & -1 \end{bmatrix} \begin{bmatrix} e^{5t} & 0 \\ 0 & e^{-2t} \end{bmatrix} \begin{bmatrix} 1 & 2 \\ 2 & -3 \end{bmatrix},$$

which multiplied out is

$$e^{At} = \frac{1}{7} \begin{bmatrix} 3e^{5t} + 4e^{-2t} & 6e^{5t} - 6e^{-2t} \\ 2e^{5t} - 2e^{-2t} & 4e^{5t} + 3e^{-2t} \end{bmatrix}.$$ ♦

Example 13.3.3 Let $A = \begin{bmatrix} 3 & 2 \\ 0 & 3 \end{bmatrix}$.

(a) Find J, the Jordan canonical form of A.
(b) Find a transformation matrix P so that $P^{-1}AP = J$.
(c) Find e^{Jt} and e^{At}.
(d) Find a general solution of the differential equation $\mathbf{x}' = A\mathbf{x}$.

Solution (a) The characteristic equation of A is

$$\det(A - \lambda I) = \begin{vmatrix} 3 - \lambda & 2 \\ 0 & 3 - \lambda \end{vmatrix} = (\lambda - 3)^2 = 0.$$

Thus, $\lambda = 3$ is a repeated eigenvalue of A. It follows from Theorem 13.3.4 that one of the matrices

$$\begin{bmatrix} 3 & 0 \\ 0 & 3 \end{bmatrix}, \begin{bmatrix} 3 & 1 \\ 0 & 3 \end{bmatrix}$$

is the Jordan canonical form of A. It cannot be the first matrix; otherwise, it would follow from part (ii) of the theorem that $A = J$, which is not the case. Thus, the Jordan canonical form of A is

$$J = \begin{bmatrix} 3 & 1 \\ 0 & 3 \end{bmatrix}.$$

(b) Eigenvectors of A corresponding to $\lambda = 3$ are solutions of

$$(A - 3I)\mathbf{v} = \begin{bmatrix} 0 & 2 \\ 0 & 0 \end{bmatrix} \begin{bmatrix} v_1 \\ v_2 \end{bmatrix} = \begin{bmatrix} 0 \\ 0 \end{bmatrix},$$

such as $\mathbf{v^1} = \langle 1, 0 \rangle$. In fact, all eigenvectors corresponding to $\lambda = 3$ are of the form $\langle k, 0 \rangle$, where $k \in \mathbb{R}$; so no two eigenvectors of A are linearly independent. By Theorem 13.3.4 (iii), a transformation matrix is $P = \left[\mathbf{v^1}\ \mathbf{v^2}\right]$, where $\mathbf{v^2}$ is a generalized eigenvector corresponding to $\lambda = 3$, that is, a vector satisfying the matrix equation

$$(A - 3I)\mathbf{v^2} = \mathbf{v^1} \quad \text{or} \quad \begin{bmatrix} 0 & 2 \\ 0 & 0 \end{bmatrix} \begin{bmatrix} v_1 \\ v_2 \end{bmatrix} = \begin{bmatrix} 1 \\ 0 \end{bmatrix}.$$

Since $v_1 = 0$ and $v_2 = 1/2$ is a solution, a generalized eigenvector is $\mathbf{v^2} = \langle 0, 1/2 \rangle$. Therefore,

$$P = \left[\mathbf{v^1}\ \mathbf{v^2}\right] = \begin{bmatrix} 1 & 0 \\ 0 & 1/2 \end{bmatrix} \quad \text{and} \quad P^{-1} = \begin{bmatrix} 1 & 0 \\ 0 & 2 \end{bmatrix}.$$

It is left as an exercise to verify that $P^{-1}AP = J$.

(c) It also follows from Theorem 13.3.4 (iii) that

$$e^{Jt} = e^{3t} \begin{bmatrix} 1 & t \\ 0 & 1 \end{bmatrix} = \begin{bmatrix} e^{3t} & te^{3t} \\ 0 & e^{3t} \end{bmatrix}.$$

And that

$$e^{At} = Pe^{Jt}P^{-1} = \begin{bmatrix} 1 & 0 \\ 0 & 1/2 \end{bmatrix} \begin{bmatrix} e^{3t} & te^{3t} \\ 0 & e^{3t} \end{bmatrix} \begin{bmatrix} 1 & 0 \\ 0 & 2 \end{bmatrix},$$

which multiplies out to

$$e^{At} = \begin{bmatrix} e^{3t} & 2te^{3t} \\ 0 & e^{3t} \end{bmatrix}.$$

(d) Since the matrix function e^{At} is a principal matrix solution of $\mathbf{x}' = A\mathbf{x}$, it follows from Theorem 13.1.2 and Definition 13.2.1 that a general solution of this equation is

$$\mathbf{x}(t) = e^{At}\mathbf{c}, \tag{13.3.14}$$

where $\mathbf{c} = \langle c_1, c_2 \rangle$ is a vector of arbitrary constants. That is, a general solution is

$$\mathbf{x}(t) = \begin{bmatrix} x_1(t) \\ x_2(t) \end{bmatrix} = \begin{bmatrix} e^{3t} & 2te^{3t} \\ 0 & e^{3t} \end{bmatrix} \begin{bmatrix} c_1 \\ c_2 \end{bmatrix} = \begin{bmatrix} c_1e^{3t} + 2c_2te^{3t} \\ c_2e^{3t} \end{bmatrix}$$

for all $t \in \mathbb{R}$. ♦

Example 13.3.4 Find a Jordan canonical matrix J corresponding to the matrix

$$A = \begin{bmatrix} 1/2 & -15/2 \\ 3/2 & 7/2 \end{bmatrix}$$

and a matrix P so that $P^{-1}AP = J$. Compute e^{Jt} and e^{At}.

Solution Since $\operatorname{tr} A = 4$ and $\det A = 13$, the characteristic equation of A is

$$\lambda^2 - (\operatorname{tr} A)\,\lambda + \det A = \lambda^2 - 4\lambda + 13 = 0.$$

Its solutions are $\lambda = 2 + 3i$ and $\overline{\lambda} = 2 - 3i$. In the notation of Theorem 13.3.4 (iv), $\alpha = 2$ and $\beta = 3$. Therefore, a Jordan canonical form of A is

$$J = \begin{bmatrix} \alpha & \beta \\ -\beta & \alpha \end{bmatrix} = \begin{bmatrix} 2 & 3 \\ -3 & 2 \end{bmatrix}$$

and

$$e^{Jt} = e^{\alpha t} \begin{bmatrix} \cos\beta t & \sin\beta t \\ -\sin\beta t & \cos\beta t \end{bmatrix} = e^{2t} \begin{bmatrix} \cos 3t & \sin 3t \\ -\sin 3t & \cos 3t \end{bmatrix}.$$

Let $\mathbf{v}$ be a complex eigenvector corresponding to the eigenvalue $\lambda = 2 + 3i$, that is, any solution of $(A - (2 + 3i)I)\mathbf{v} = \mathbf{0}$:

$$\begin{bmatrix} -3/2 - 3i & -15/2 \\ 3/2 & 3/2 - 3i \end{bmatrix} \begin{bmatrix} v_1 \\ v_2 \end{bmatrix} = \begin{bmatrix} 0 \\ 0 \end{bmatrix}.$$

Let $v_2 = 1$. Then $\frac{3}{2}v_1 + (\frac{3}{2} - 3i) = 0$, which has the solution $v_1 = -1 + 2i$. Thus,

$$\mathbf{v} = \begin{bmatrix} -1+2i \\ 1 \end{bmatrix}.$$

Separating this vector into its real and imaginary parts, we get

$$\mathbf{v} = \begin{bmatrix} -1 \\ 1 \end{bmatrix} + i \begin{bmatrix} 2 \\ 0 \end{bmatrix} = \mathbf{v^1} + i\mathbf{v^2},$$

where $\mathbf{v^1} = \langle -1, 1 \rangle$ and $\mathbf{v^2} = \langle 2, 0 \rangle$. Thus,

$$P = \begin{bmatrix} \mathbf{v^1} & \mathbf{v^2} \end{bmatrix} = \begin{bmatrix} -1 & 2 \\ 1 & 0 \end{bmatrix} \quad \text{and} \quad P^{-1} = \frac{1}{\det P} \begin{bmatrix} 0 & -2 \\ -1 & -1 \end{bmatrix} = \begin{bmatrix} 0 & 1 \\ \frac{1}{2} & \frac{1}{2} \end{bmatrix}.$$

Now that we have P, P^{-1}, and e^{Jt}, we can compute e^{At} as follows:

$$e^{At} = Pe^{Jt}P^{-1} = \begin{bmatrix} -1 & 2 \\ 1 & 0 \end{bmatrix} \begin{bmatrix} e^{2t}\cos 3t & e^{2t}\sin 3t \\ -e^{2t}\sin 3t & e^{2t}\cos 3t \end{bmatrix} \begin{bmatrix} 0 & 1 \\ 1/2 & 1/2 \end{bmatrix}.$$

Carrying out the multiplication, we obtain

$$e^{At} = \begin{bmatrix} e^{2t}\cos 3t - \frac{1}{2}e^{2t}\sin 3t & -\frac{5}{2}e^{2t}\sin 3t \\ \frac{1}{2}e^{2t}\sin 3t & e^{2t}\cos 3t + \frac{1}{2}e^{2t}\sin 3t \end{bmatrix}. \quad \blacklozenge$$

13.4 Matrix Exponential Series

Anyone conversant with calculus knows that the power series

$$1 + at + \frac{a^2}{2!}t^2 + \frac{a^3}{3!}t^3 + \cdots + \frac{a^n}{n!}t^n + \cdots \qquad (a \in \mathbb{R})$$

converges to e^{at} for all $t \in \mathbb{R}$. This is what is meant by

$$e^{at} = \sum_{n=0}^{\infty} \frac{a^n}{n!}t^n \qquad (-\infty < t < \infty). \tag{13.4.1}$$

In this section, we show the same is true of matrices: the power series expansion (13.4.1) still makes sense if a, which denotes any real number, is replaced by a matrix A with real entries. We will prove this first for each of the Jordan canonical matrices displayed in (13.3.10), after which we will prove it for any matrix A. But first we need to define what is meant by the convergence of a series of constant matrices.

Convergence of a Matrix Series

Definition 13.4.1 Let $B(n)$ and S be 2×2 matrices with entries $b_{ij}(n)$ and s_{ij}, respectively, for $n = 0, 1, 2, \ldots$. That is,

$$B(n) = \begin{bmatrix} b_{11}(n) & b_{12}(n) \\ b_{21}(n) & b_{22}(n) \end{bmatrix} \quad \text{and} \quad S = \begin{bmatrix} s_{11} & s_{12} \\ s_{21} & s_{22} \end{bmatrix}.$$

The matrix series $\sum_{n=0}^{\infty} B(n)$ is said to ***converge*** to S if and only if

$$\sum_{n=0}^{\infty} b_{ij}(n) = s_{ij}$$

for $i = 1, 2$ and $j = 1, 2$.

Power Series Expansion of e^{Jt}

Lemma 13.4.2 *If J is a Jordan canonical matrix, then*

$$e^{Jt} = I + tJ + \frac{t^2}{2!}J^2 + \frac{t^3}{3!}J^3 + \cdots = \sum_{n=0}^{\infty} \frac{t^n}{n!}J^n \tag{13.4.2}$$

for $-\infty < t < \infty$, where $J^0 := I$.

Proof Let J denote any one of the four possible Jordan (canonical) matrices of order 2 that are displayed in (13.3.10). We will prove that (13.4.2) is the power series expansion of e^{Jt}. We begin with the Jordan matrix labeled (i) in (13.3.10). For this particular matrix, it is left to the reader (see Problem 30)) to show that

$$J^n = \begin{bmatrix} \lambda_1^n & 0 \\ 0 & \lambda_2^n \end{bmatrix} \qquad (n = 1, 2, 3, \ldots). \tag{13.4.3}$$

From (13.2.3), we have

$$e^{Jt} = \begin{bmatrix} e^{\lambda_1 t} & 0 \\ 0 & e^{\lambda_2 t} \end{bmatrix}.$$

Using the Maclaurin series for $e^{\lambda_1 t}$ and $e^{\lambda_2 t}$, we can rewrite this as

$$e^{Jt} = \begin{bmatrix} 1+\lambda_1 t+\frac{1}{2!}\lambda_1^2t^2+\frac{1}{3!}\lambda_1^3t^3+\cdots & 0 \\ 0 & 1+\lambda_2 t+\frac{1}{2!}\lambda_2^2t^2+\frac{1}{3!}\lambda_2^3t^3+\cdots \end{bmatrix}$$

$$= \begin{bmatrix} 1 & 0 \\ 0 & 1 \end{bmatrix} + t\begin{bmatrix} \lambda_1 & 0 \\ 0 & \lambda_2 \end{bmatrix} + \frac{t^2}{2!}\begin{bmatrix} \lambda_1^2 & 0 \\ 0 & \lambda_2^2 \end{bmatrix} + \frac{t^3}{3!}\begin{bmatrix} \lambda_1^3 & 0 \\ 0 & \lambda_2^3 \end{bmatrix} + \cdots .$$

Or, because of (13.4.3), this is the same as

$$e^{Jt} = I + tJ + \frac{t^2}{2!}J^2 + \frac{t^3}{3!}J^3 + \cdots ,$$

which concludes the proof for the Jordan matrix (i). This proof takes care of the Jordan matrix (ii) as well: simply replace every occurrence of λ_1 and λ_2 with λ.

Now suppose J is the Jordan matrix (13.3.10) (iii). It is left to the reader (see Problem 30) to show that

$$J^n = \begin{bmatrix} \lambda^n & n\lambda^{n-1} \\ 0 & \lambda^n \end{bmatrix} \qquad (n = 1, 2, 3, \ldots). \tag{13.4.4}$$

From (13.2.7), we have

$$e^{Jt} = e^{\lambda t}\begin{bmatrix} 1 & t \\ 0 & 1 \end{bmatrix} = \begin{bmatrix} e^{\lambda t} & te^{\lambda t} \\ 0 & e^{\lambda t} \end{bmatrix}.$$

As a result,

$$e^{Jt} = \begin{bmatrix} 1+\lambda t+\frac{1}{2!}\lambda^2t^2+\frac{1}{3!}\lambda^3t^3+\cdots & t+\lambda t^2+\frac{1}{2!}\lambda^2t^3+\frac{1}{3!}\lambda^3t^4+\cdots \\ 0 & 1+\lambda t+\frac{1}{2!}\lambda^2t^2+\frac{1}{3!}\lambda^3t^3+\cdots \end{bmatrix}$$

$$= \begin{bmatrix} 1 & 0 \\ 0 & 1 \end{bmatrix} + t\begin{bmatrix} \lambda & 1 \\ 0 & \lambda \end{bmatrix} + \frac{t^2}{2!}\begin{bmatrix} \lambda^2 & 2\lambda \\ 0 & \lambda^2 \end{bmatrix} + \frac{t^3}{3!}\begin{bmatrix} \lambda^3 & 3\lambda^2 \\ 0 & \lambda^3 \end{bmatrix} + \cdots .$$

It follows from this and (13.4.4) that the power series expansion of e^{Jt} for the Jordan matrix J in (13.3.10) (iii) is (13.4.2).

The final Jordan matrix to consider is (13.3.10) (iv), that is, the matrix

$$J = \begin{bmatrix} \alpha & \beta \\ -\beta & \alpha \end{bmatrix}.$$

where $\alpha, \beta \in \mathbb{R}$. Computing powers of J, as done earlier in (13.4.3) and (13.4.4), is laborious for this particular matrix. Fortunately, we can avoid the drudgery of doing so by using a proof involving complex numbers. The reason for this is that any 2×2 matrix of this form can be regarded as a complex number. This is explained in Appendix B. There we argue that matrix calculations involving such a matrix can be carried out by replacing it and the attendant matrix algebraic operations with the complex number $\alpha + i\beta$ and the corresponding complex algebraic operations, after which the result (a complex number) that is obtained is converted back to the corresponding matrix. Examples of converting back and forth between matrices and the corresponding complex numbers in order to simplify calculations are found in Appendix B. A left-right arrow is used to indicate the correspondence between J and the complex number $\alpha - i\beta$ as follows:

$$\begin{bmatrix} \alpha & \beta \\ -\beta & \alpha \end{bmatrix} \leftrightarrow \alpha - i\beta. \tag{13.4.5}$$

With this brief explanation, we offer the following proof.

Since we already know from (13.2.9) that

$$e^{Jt} = e^{\alpha t} \begin{bmatrix} \cos\beta t & \sin\beta t \\ -\sin\beta t & \cos\beta t \end{bmatrix}, \tag{13.4.6}$$

we have

$$e^{Jt} = \begin{bmatrix} e^{\alpha t}\cos\beta t & e^{\alpha t}\sin\beta t \\ -e^{\alpha t}\sin\beta t & e^{\alpha t}\cos\beta t \end{bmatrix} \leftrightarrow e^{\alpha t}\cos\beta t - i e^{\alpha t}\sin\beta t.$$

Consequently,

$$e^{Jt} \leftrightarrow e^{(\alpha - i\beta)t} \tag{13.4.7}$$

since according to Euler's formula

$$e^{\alpha t}\cos\beta t - i e^{\alpha t}\sin\beta t = e^{\alpha t}(\cos\beta t - i\sin\beta t) = e^{\alpha t} \cdot e^{-i\beta t} = e^{(\alpha - i\beta)t}.$$

It is shown in complex analysis textbooks that the power series expansion

$$e^z = 1 + z + \frac{z^2}{2!} + \frac{z^3}{3!} + \cdots + \frac{z^n}{n!} + \cdots$$

holds for all complex numbers z. Thus,

$$e^{(\alpha - i\beta)t} = 1 + (\alpha - i\beta)t + \frac{1}{2!}(\alpha - i\beta)^2 t^2 + \cdots + \frac{1}{n!}(\alpha - i\beta)^n t^n + \cdots.$$

It then follows from (13.4.7) and the correspondence $\alpha - i\beta \leftrightarrow J$ that

$$e^{Jt} = 1 + tJ + \frac{t^2}{2!}J^2 + \cdots + \frac{t^n}{n!}J^n + \cdots,$$

thereby concluding the proof for the Jordan matrix (13.3.10) (iv). ■

We have just finished proving that the power series expansion of e^{At} is

$$e^{At} = \sum_{n=0}^{\infty} \frac{t^n}{n!} A^n$$

if A is one of the four Jordan matrices displayed in (13.3.10). As it turns out, this power series expansion is valid not only for Jordan matrices of order 2 but for any real constant matrix A of order n. We will only prove the $n = 2$ case; however, before we can proceed with a proof, we will need the next lemma.

Properties of Matrix Series

Lemma 13.4.3 *Let $K = [k_{ij}]$, $B(n) = [b_{ij}(n)]$, and $C(n) = [c_{ij}(n)]$ be 2×2 matrices, where $i = 1, 2$, $j = 1, 2$, and $n = 0, 1, 2, \ldots$. If the series of matrices $\sum_{n=0}^{\infty} B(n)$ and $\sum_{n=0}^{\infty} C(n)$ converge to matrices S and T, respectively, then*

(i) $$\sum_{n=0}^{\infty} [B(n) \pm C(n)] = S \pm T = \sum_{n=0}^{\infty} B(n) \pm \sum_{n=0}^{\infty} C(n).$$

(ii) $$\sum_{n=0}^{\infty} KB(n) = KS = K\sum_{n=0}^{\infty} B(n).$$

(iii) $$\sum_{n=0}^{\infty} B(n)K = SK = \left(\sum_{n=0}^{\infty} B(n)\right)K.$$

Proof The proof of (i) is left as an exercise. Suppose $\sum_{n=0}^{\infty} B(n)$ converges to a matrix S, where $B(n) = [b_{ij}(n)]$ and $S = [s_{ij}]$. Then by Definition 13.4.1,

$$\sum_{n=0}^{\infty} B(n) = \begin{bmatrix} b_{11}(0) \; b_{12}(0) \\ b_{21}(0) \; b_{22}(0) \end{bmatrix} + \begin{bmatrix} b_{11}(1) \; b_{12}(1) \\ b_{21}(1) \; b_{22}(1) \end{bmatrix} + \cdots + \begin{bmatrix} b_{11}(n) \; b_{12}(n) \\ b_{21}(n) \; b_{22}(n) \end{bmatrix} + \ldots$$

$$= \begin{bmatrix} \sum_{n=0}^{\infty} b_{11}(n) & \sum_{n=0}^{\infty} b_{12}(n) \\ \sum_{n=0}^{\infty} b_{21}(n) & \sum_{n=0}^{\infty} b_{22}(n) \end{bmatrix} = \begin{bmatrix} s_{11} & s_{12} \\ s_{21} & s_{22} \end{bmatrix} = S.$$

The left-hand side of (ii) is

$$\sum_{n=0}^{\infty} KB(n) = \sum_{n=0}^{\infty} \begin{bmatrix} k_{11} & k_{12} \\ k_{21} & k_{22} \end{bmatrix} \begin{bmatrix} b_{11}(n) & b_{12}(n) \\ b_{21}(n) & b_{22}(n) \end{bmatrix}$$

$$= \sum_{n=0}^{\infty} \begin{bmatrix} k_{11}b_{11}(n) + k_{12}b_{21}(n) & k_{11}b_{12}(n) + k_{12}b_{22}(n) \\ k_{21}b_{11}(n) + k_{22}b_{21}(n) & k_{21}b_{12}(n) + k_{22}b_{22}(n) \end{bmatrix}$$

$$= \begin{bmatrix} \sum_{n=0}^{\infty} \left(k_{11}b_{11}(n) + k_{12}b_{21}(n)\right) & \sum_{n=0}^{\infty} \left(k_{11}b_{12}(n) + k_{12}b_{22}(n)\right) \\ \sum_{n=0}^{\infty} \left(k_{21}b_{11}(n) + k_{22}b_{21}(n)\right) & \sum_{n=0}^{\infty} \left(k_{21}b_{12}(n) + k_{22}b_{22}(n)\right) \end{bmatrix}.$$

Using the properties of convergent scalar series (cf. [68, Ch. 11] or any similar calculus textbook), we have

$$\sum_{n=0}^{\infty} \left(k_{11}b_{11}(n) + k_{12}b_{21}(n)\right) = k_{11} \sum_{n=0}^{\infty} b_{11}(n) + k_{12} \sum_{n=0}^{\infty} b_{21}(n).$$

Applying this to the other three entries is the previous matrix, we obtain

$$\sum_{n=0}^{\infty} KB(n) =$$

$$\begin{bmatrix} k_{11} \sum_{n=0}^{\infty} b_{11}(n) + k_{12} \sum_{n=0}^{\infty} b_{21}(n) & k_{11} \sum_{n=0}^{\infty} b_{12}(n) + k_{12} \sum_{n=0}^{\infty} b_{22}(n) \\ k_{21} \sum_{n=0}^{\infty} b_{11}(n) + k_{22} \sum_{n=0}^{\infty} b_{21}(n) & k_{21} \sum_{n=0}^{\infty} b_{12}(n) + k_{22} \sum_{n=0}^{\infty} b_{22}(n) \end{bmatrix}$$

or

$$\sum_{n=0}^{\infty} KB(n) = \begin{bmatrix} k_{11} & k_{12} \\ k_{21} & k_{22} \end{bmatrix} \begin{bmatrix} \sum_{n=0}^{\infty} b_{11}(n) & \sum_{n=0}^{\infty} b_{12}(n) \\ \sum_{n=0}^{\infty} b_{21}(n) & \sum_{n=0}^{\infty} b_{22}(n) \end{bmatrix}$$

$$= \begin{bmatrix} k_{11} & k_{12} \\ k_{21} & k_{22} \end{bmatrix} \begin{bmatrix} s_{11} & s_{12} \\ s_{21} & s_{22} \end{bmatrix} = KS = K \sum_{n=0}^{\infty} B(n).$$

The proof of item (iii) is very much like that of (ii) and is left as an exercise. ■

Power Series Expansion of e^{At}

Theorem 13.4.4 *If A is a real constant $n \times n$ matrix, then*

$$e^{At} = I + tA + \frac{t^2}{2!}A^2 + \cdots = \sum_{n=0}^{\infty} \frac{t^n}{n!} A^n \tag{13.4.8}$$

for $-\infty < t < \infty$, where $A^0 = I$.

Proof Let A be a given real constant 2×2 matrix. Then, by Theorem 13.3.3, there is a Jordan canonical matrix J and a real invertible matrix P exists such that $P^{-1}AP = J$. Now we have already established with Lemma 13.4.2 that

$$e^{Jt} = \sum_{n=0}^{\infty} \frac{t^n}{n!} J^n.$$

It follows from (ii) of Lemma 13.4.3 that

$$Pe^{Jt} = P\sum_{n=0}^{\infty} \frac{t^n}{n!} J^n = \sum_{n=0}^{\infty} \frac{t^n}{n!} PJ^n$$

and from (iii) that

$$Pe^{Jt}P^{-1} = \left(\sum_{n=0}^{\infty} \frac{t^n}{n!} PJ^n\right)P^{-1} = \sum_{n=0}^{\infty} \frac{t^n}{n!} PJ^nP^{-1}.$$

This is (13.4.8) as $Pe^{Jt}P^{-1} = e^{At}$ and $PJ^nP^{-1} = A^n$ (see items (iv) and (v) in Theorem 13.3.1). ■

Example 13.4.1 Compute the matrix e^{At} for

$$A = \begin{bmatrix} 0 & -1 \\ 1 & 0 \end{bmatrix}$$

by first using (13.4.8) to find the power series expansion of e^{At} and then replacing the four entries of this matrix by the functions to which the series in these entries converge. Check the result by computing e^{At} using Theorem 13.3.4.

Solution Computing A^2 and A^3, we obtain

$$A^2 = \begin{bmatrix} 0 & -1 \\ 1 & 0 \end{bmatrix} \begin{bmatrix} 0 & -1 \\ 1 & 0 \end{bmatrix} = \begin{bmatrix} -1 & 0 \\ 0 & -1 \end{bmatrix}$$

$$A^3 = A \cdot A^2 = \begin{bmatrix} 0 & -1 \\ 1 & 0 \end{bmatrix} \begin{bmatrix} -1 & 0 \\ 0 & -1 \end{bmatrix} = \begin{bmatrix} 0 & 1 \\ -1 & 0 \end{bmatrix}.$$

It is left as an exercise to verify that

$$A^4 = \begin{bmatrix} 1 & 0 \\ 0 & 1 \end{bmatrix}, \quad A^5 = \begin{bmatrix} 0 & -1 \\ 1 & 0 \end{bmatrix} = A, \quad A^6 = A^2, \quad A^7 = A^3 \;\ldots .$$

So from (13.4.8), we have

$$e^{At} = \begin{bmatrix} 1 & 0 \\ 0 & 1 \end{bmatrix} + t \begin{bmatrix} 0 & -1 \\ 1 & 0 \end{bmatrix} + \frac{t^2}{2!} \begin{bmatrix} -1 & 0 \\ 0 & -1 \end{bmatrix} + \frac{t^3}{3!} \begin{bmatrix} 0 & 1 \\ -1 & 0 \end{bmatrix} + \frac{t^4}{4!} \begin{bmatrix} 1 & 0 \\ 0 & 1 \end{bmatrix}$$
$$+ \frac{t^5}{5!} \begin{bmatrix} 0 & -1 \\ 1 & 0 \end{bmatrix} + \frac{t^6}{6!} \begin{bmatrix} -1 & 0 \\ 0 & -1 \end{bmatrix} + \frac{t^7}{7!} \begin{bmatrix} 0 & 1 \\ -1 & 0 \end{bmatrix} + \cdots$$

or

$$e^{At} = \begin{bmatrix} 1 - \frac{1}{2!}t^2 + \frac{1}{4!}t^4 - \frac{1}{6!}t^6 + \cdots & -t + \frac{1}{3!}t^3 - \frac{1}{5!}t^5 + \frac{1}{7!}t^7 - \cdots \\ t - \frac{1}{3!}t^3 + \frac{1}{5!}t^5 - \frac{1}{7!}t^7 + \cdots & 1 - \frac{1}{2!}t^2 + \frac{1}{4!}t^4 - \frac{1}{6!}t^6 + \cdots \end{bmatrix}.$$

It follows from the Maclaurin series for $\cos t$ and $\sin t$ (cf. (9.2.13) and (9.2.14)) that

$$e^{At} = \begin{bmatrix} \cos t & -\sin t \\ \sin t & \cos t \end{bmatrix}. \tag{13.4.9}$$

Another way to compute e^{At} is via the principal matrix solution $Z(t)$ of the vector differential equation $\mathbf{x}' = A\mathbf{x}$ since $e^{At} := Z(t)$ (Definition 13.2.1). This computation is left to the reader (see Problem 5).

A third alternative is to use Theorem 13.3.4. Since the characteristic equation of A is

$$\det(A - \lambda I) = \begin{vmatrix} -\lambda & -1 \\ 1 & -\lambda \end{vmatrix} = \lambda^2 + 1 = 0,$$

the eigenvalues of A are $\pm i$. The corresponding Jordan matrix is given by item (iv) in Theorem 13.3.4 with $\alpha = 0$ and $\beta = 1$. That is,

$$J = \begin{bmatrix} 0 & 1 \\ -1 & 0 \end{bmatrix}.$$

Note that $A = -J$. Thus, $e^{At} = e^{-Jt}$. It follows from property (v) in Theorem 13.2.1 that $e^{-Jt} = (e^{Jt})^{-1}$. Referring again to Theorem 13.3.4 (iv), we see that

$$e^{Jt} = \begin{bmatrix} \cos t & \sin t \\ -\sin t & \cos t \end{bmatrix}.$$

Therefore,

$$e^{At} = (e^{Jt})^{-1} = \begin{bmatrix} \cos t & \sin t \\ -\sin t & \cos t \end{bmatrix}^{-1} = \begin{bmatrix} \cos t & -\sin t \\ \sin t & \cos t \end{bmatrix}. \qquad \blacklozenge$$

Generally speaking, relying on Theorem 13.4.4 to compute e^{At} is impractical when it comes to ascertaining the sum of the power series in (13.4.8)—unless there are recognizable patterns and the entries of e^{At} turn out to be series whose sums are well-known from calculus as in Example 13.4.1. It is usually easier to either (a) find the Jordan canonical form of A via Theorem 13.3.4 and then use the information given there to compute the appropriate transformation matrix P and obtain e^{At} from $Pe^{Jt}P^{-1}$ or (b) compute the principal matrix solution $Z(t)$ of $\mathbf{x}' = A\mathbf{x}$ using Theorem 13.1.6 since $e^{At} = Z(t)$. Both (a) and (b) have been illustrated with examples in the previous two sections. As a review and for the sake of comparison, let us use both methods to compute e^{At} for the matrix A in the following example.

Example 13.4.2 Compute e^{At} for

$$A = \begin{bmatrix} 1 & 2 \\ -1 & -1 \end{bmatrix}.$$

Solution 1 (via the Jordan form of A). Since $\operatorname{tr} A = 0$ and $\det A = 1$, the characteristic equation of A is

$$\lambda^2 + 1 = 0.$$

So $\lambda = i$ and $\overline{\lambda} = -i$ are the eigenvalues of A. Accordingly, by Theorem 13.3.4, the Jordan canonical form of A is

$$J = \begin{bmatrix} 0 & 1 \\ -1 & 0 \end{bmatrix}.$$

Moreover,

$$e^{Jt} = \begin{bmatrix} \cos t & \sin t \\ -\sin t & \cos t \end{bmatrix}.$$

If $\mathbf{v^1}+i\mathbf{v^2}$ is an eigenvector corresponding to $\lambda = i$, where $\mathbf{v^1}$ and $\mathbf{v^2}$ are real vectors, then $P = \begin{bmatrix}\mathbf{v^1} & \mathbf{v^2}\end{bmatrix}$ is a transformation matrix relating A and J by $J = P^{-1}AP$. With e^{Jt} and P, we can compute e^{At} since $e^{At} = Pe^{Jt}P^{-1}$. But first we have to find P, which we do next.

Eigenvectors of A corresponding to $\lambda = i$ are solutions of $(A - iI)\mathbf{v} = \mathbf{0}$:

$$\begin{bmatrix} 1-i & 2 \\ -1 & -1-i \end{bmatrix}\begin{bmatrix} v_1 \\ v_2 \end{bmatrix} = \begin{bmatrix} 0 \\ 0 \end{bmatrix}.$$

It is left as an exercise to verify that $\mathbf{v} = \langle 1+i, -1\rangle$ is a solution (eigenvector). Since

$$\mathbf{v} = \begin{bmatrix} 1+i \\ -1 \end{bmatrix} = \begin{bmatrix} 1 \\ -1 \end{bmatrix} + i\begin{bmatrix} 1 \\ 0 \end{bmatrix},$$

we have $\mathbf{v} = \mathbf{v^1} + i\mathbf{v^2}$, where $\mathbf{v^1} = \langle 1, -1\rangle$ and $\mathbf{v^2} = \langle 1, 0\rangle$. As a result, we have

$$P = \begin{bmatrix}\mathbf{v^1} & \mathbf{v^2}\end{bmatrix} = \begin{bmatrix} 1 & 1 \\ -1 & 0 \end{bmatrix} \quad \text{and} \quad P^{-1} = \frac{1}{\det P}\begin{bmatrix} 0 & -1 \\ 1 & 1 \end{bmatrix} = \begin{bmatrix} 0 & -1 \\ 1 & 1 \end{bmatrix}.$$

Therefore,

$$\begin{aligned} Pe^{Jt}P^{-1} &= \begin{bmatrix} 1 & 1 \\ -1 & 0 \end{bmatrix}\begin{bmatrix} \cos t & \sin t \\ -\sin t & \cos t \end{bmatrix}\begin{bmatrix} 0 & -1 \\ 1 & 1 \end{bmatrix} \\ &= \begin{bmatrix} \cos t - \sin t & \sin t + \cos t \\ -\cos t & -\sin t \end{bmatrix}\begin{bmatrix} 0 & -1 \\ 1 & 1 \end{bmatrix}. \end{aligned}$$

Carrying out this final multiplication, we conclude that

$$e^{At} = \begin{bmatrix} \cos t + \sin t & 2\sin t \\ -\sin t & \cos t - \sin t \end{bmatrix}.$$

Solution 2 (via the PMS) By definition, $e^{At} = Z(t)$, where $Z(t)$ is the PMS of the linear system $\mathbf{x}' = A\mathbf{x}$ at $t = 0$. Since $Z(t) = X(t)X^{-1}(0)$, where $X(t)$ is any fundamental matrix solution of $\mathbf{x}' = A\mathbf{x}$ (see Theorem 13.1.6), let us find one of these fundamental matrix solutions. Normally we begin by computing the eigenvalues of A and the corresponding eigenvectors; however, we already know from the previous "Solution 1" computations that $\mathbf{v} = \langle 1+i, -1\rangle$ is a complex eigenvector corresponding to the eigenvalue $\lambda = i$. Recall from Theorem 12.5.36 that the real and imaginary parts of a complex-valued solution $e^{\lambda t}\mathbf{v}$ of $\mathbf{x}' = A\mathbf{x}$ are linearly independent real-valued solutions of the system.

Since $\mathbf{v} = \langle 1+i, -1\rangle$ is an eigenvector corresponding to the eigenvalue $\lambda = i$, a complex-valued solution is

$$e^{\lambda t}\mathbf{v} = e^{it}\begin{bmatrix} 1+i \\ -1 \end{bmatrix} = (\cos t + i\sin t)\begin{bmatrix} 1+i \\ -1 \end{bmatrix},$$

which we rewrite as

$$e^{\lambda t}\mathbf{v} = \begin{bmatrix} \cos t - \sin t \\ -\cos t \end{bmatrix} + i\begin{bmatrix} \cos t + \sin t \\ -\sin t \end{bmatrix}.$$

From this we obtain the linearly independent real-valued solutions

$$\operatorname{Re}\left(e^{\lambda t}\mathbf{v}\right) = \begin{bmatrix} \cos t - \sin t \\ -\cos t \end{bmatrix} \quad \text{and} \quad \operatorname{Im}\left(e^{\lambda t}\mathbf{v}\right) = \begin{bmatrix} \cos t + \sin t \\ -\sin t \end{bmatrix}.$$

As a result, a fundamental matrix solution of $\mathbf{x}' = A\mathbf{x}$ is

$$X(t) = \begin{bmatrix} \cos t - \sin t & \cos t + \sin t \\ -\cos t & -\sin t \end{bmatrix}.$$

Thus $Z(t) = X(t)X^{-1}(0)$, where

$$X(0) = \begin{bmatrix} 1 & 1 \\ -1 & 0 \end{bmatrix} \quad \text{and} \quad X^{-1}(0) = \begin{bmatrix} 0 & -1 \\ 1 & 1 \end{bmatrix}.$$

Therefore, as $e^{At} = Z(t)$,

$$e^{At} = \begin{bmatrix} \cos t - \sin t & \cos t + \sin t \\ -\cos t & -\sin t \end{bmatrix}\begin{bmatrix} 0 & -1 \\ 1 & 1 \end{bmatrix} = \begin{bmatrix} \cos t + \sin t & 2\sin t \\ -\sin t & \cos t - \sin t \end{bmatrix},$$

which is what we obtained earlier (see *Solution 1*). ♦

If a matrix A is ***nilpotent*** (that is, if $A^m = O$ for some positive integer m), then using (13.4.8) may be the best way to proceed as the next example illustrates.

Example 13.4.3 Compute e^{At} and e^{Bt}, where

$$A = \begin{bmatrix} 0 & 0 \\ 1 & 0 \end{bmatrix} \quad \text{and} \quad B = \begin{bmatrix} 0 & -1 \\ 0 & 0 \end{bmatrix}.$$

Show that $e^{At}e^{Bt} \neq e^{(A+B)t}$. Does this contradict Theorem 13.2.2?

Solution The matrix A is nilpotent since

$$A^2 = \begin{bmatrix} 0 & 0 \\ 1 & 0 \end{bmatrix} \begin{bmatrix} 0 & 0 \\ 1 & 0 \end{bmatrix} = \begin{bmatrix} 0 & 0 \\ 0 & 0 \end{bmatrix}.$$

Obviously then,

$$A^n = \begin{bmatrix} 0 & 0 \\ 0 & 0 \end{bmatrix} \quad \text{for} \quad n = 2, 3, 4, \ldots .$$

Likewise, $B^2 = O$; so $B^n = O$ for $n \geq 2$. By Theorem 13.4.4,

$$e^{At} = I + tA = \begin{bmatrix} 1 & 0 \\ 0 & 1 \end{bmatrix} + t \begin{bmatrix} 0 & 0 \\ 1 & 0 \end{bmatrix} = \begin{bmatrix} 1 & 0 \\ t & 1 \end{bmatrix}$$

and

$$e^{Bt} = I + tB = \begin{bmatrix} 1 & 0 \\ 0 & 1 \end{bmatrix} + t \begin{bmatrix} 0 & -1 \\ 0 & 0 \end{bmatrix} = \begin{bmatrix} 1 & -t \\ 0 & 1 \end{bmatrix}.$$

Thus,

$$e^{At}e^{Bt} = \begin{bmatrix} 1 & 0 \\ t & 1 \end{bmatrix} \begin{bmatrix} 1 & -t \\ 0 & 1 \end{bmatrix} = \begin{bmatrix} 1 & -t \\ t & 1 - t^2 \end{bmatrix}.$$

Now what about $e^{(A+B)t}$? Note that the sum $A + B = \begin{bmatrix} 0 & -1 \\ 1 & 0 \end{bmatrix}$ happens to be the subject of Example 13.4.1, where we determined that

$$\exp\left(\begin{bmatrix} 0 & -1 \\ 1 & 0 \end{bmatrix} t \right) = \begin{bmatrix} \cos t & -\sin t \\ \sin t & \cos t \end{bmatrix}.$$

And so for these two matrices, $e^{At}e^{Bt} \neq e^{(A+B)t}$. Note that this does not contradict Theorem 13.2.2 because the matrices A and B in this example do not commute. ♦

13.5 Phase Portraits of Homogeneous Linear Systems

In Sect. 12.5 we graphed some trajectories of (12.5.4), namely the system

$$\begin{aligned} x_1' &= -2x_1 + x_2 \\ x_2' &= x_1 - 2x_2, \end{aligned}$$

on the same phase plane in Fig. 12.7. Obviously it is impossible to graph all of the trajectories of this system, but we did graph a sufficient number of them so that anyone viewing Fig. 12.7 would know what all of the trajectories basically look like and their most prominent features. Such a portrayal of the trajectories of a given system is known as a ***phase portrait*** of the system. In this section, we will analyze phase portraits of

$$\mathbf{y}' = J\mathbf{y}, \tag{13.5.1}$$

where $\mathbf{y} = \langle y_1, y_2 \rangle$ and J denotes a matrix that has the form of one of the four Jordan canonical matrices that are displayed in Theorem 13.3.3. There is good reason to do so and that is due to Theorems 13.3.1 and 13.3.4, which tell us that every linear system $\mathbf{x}' = A\mathbf{x}$ can be transformed into a corresponding ***canonical linear system*** $\mathbf{y}' = J\mathbf{y}$ by means of a similarity transformation $y = P^{-1}x$. Moreover, we intuit that the trajectories of $\mathbf{x}' = A\mathbf{x}$ in the x_1x_2-phase plane will look essentially like those of $\mathbf{y}' = J\mathbf{y}$ in the y_1y_2-phase plane.

Starting with the Jordan canonical matrix given below in item **[1]**, we will investigate what the phase portraits of the canonical system $\mathbf{y}' = J\mathbf{y}$ look like for all four Jordan matrices. However, instead of beginning with the first Jordan matrix that is displayed in (13.3.10), let us first take a look at the simplest phase portraits, which as we might suspect are those corresponding to the second Jordan matrix in (13.3.10).

[1] Let J be a Jordan canonical matrix that looks like (ii) in Theorem 13.3.3, where $\lambda \neq 0$. It follows from Theorem 13.1.10, Definition 13.2.1, and Theorem 13.3.4 that the solution of

$$\mathbf{y}' = J\mathbf{y} = \begin{bmatrix} \lambda & 0 \\ 0 & \lambda \end{bmatrix} \mathbf{y} \tag{13.5.2}$$

such that $\mathbf{y}(0) = \langle a, b \rangle$, where $(a, b) \in \mathbb{R}^2$, is

$$\mathbf{y}(t) = e^{Jt}\mathbf{y}(0) = \begin{bmatrix} e^{\lambda t} & 0 \\ 0 & e^{\lambda t} \end{bmatrix} \begin{bmatrix} a \\ b \end{bmatrix} = \begin{bmatrix} ae^{\lambda t} \\ be^{\lambda t} \end{bmatrix}. \tag{13.5.3}$$

That is,

$$y_1(t) = ae^{\lambda t}, \quad y_2(t) = be^{\lambda t} \tag{13.5.4}$$

for all $t \in \mathbb{R}$. Of course, we could also have obtained this solution by merely solving the pair of uncoupled equations

$$y_1' = \lambda y_1, \quad y_2' = \lambda y_2.$$

Consequently, the trajectory of the solution $\mathbf{y}(t)$ through a given point (a, b) is the set of points

$$\left\{(y_1(t), y_2(t)) : -\infty < t < \infty\right\} = \left\{(ae^{\lambda t}, be^{\lambda t}) : -\infty < t < \infty\right\}. \tag{13.5.5}$$

To find out what a typical phase portrait of the linear system (13.5.2) looks like, let us first eliminate the independent variable t. If $a \neq 0$, then

$$y_2(t) = be^{\lambda t} = b\left(\frac{y_1(t)}{a}\right) = \frac{b}{a}y_1(t).$$

Thus the trajectory through the point (a, b) lies on the line $y_2 = (b/a)y_1$ if $a \neq 0$.

If $a = 0$, then the trajectory through a point $(0, b)$ is confined to the y_2-axis since the solution is $y_1(t) = 0,\ y_2(t) = be^{\lambda t}$. A special situation occurs when a and b are both 0, for then the solution is the constant solution

$$y_1(t) = 0,\ y_2(t) = 0 \ \text{ for all } -\infty < t < \infty. \tag{13.5.6}$$

A constant solution, such as (13.5.6), is said to be an ***equilibrium solution*** (or ***steady state solution***) of the system. The trajectory of the solution (13.5.6) is

$$\left\{(y_1(t), y_2(t)) : -\infty < t < \infty\right\} = \left\{(0, 0)\right\}.$$

The point $(0, 0)$ is said to be an ***equilibrium point*** (or ***critical point***) of the system (13.5.2). In fact, it is the only equilibrium point of (13.5.2) (see Problem 28).

Although there are trajectories on every line $y_2 = (b/a)y_1$, the line itself is not a trajectory but rather is composed of three trajectories. To see this, suppose $\lambda < 0$. Now choose an arbitrary point (a, b) in Quadrant I ($a > 0, b > 0$). Then it follows from (13.5.5) that the trajectory through this point is the (open) semi-infinite line

$$y_2 = (b/a)y_1 \quad (0 < y_1 < \infty)$$

since $(ae^{\lambda t}, be^{\lambda t}) \to (0, 0)$ as $t \to \infty$ while $ae^{\lambda t} \to \infty$ and $be^{\lambda t} \to \infty$ as $t \to -\infty$. For instance, the semi-infinite line through the point $(3, 3)$ in Fig. 13.1 is such a trajectory. On the other hand, the trajectory through a point (a, b) in Quadrant III ($a < 0, b < 0$) is the semi-infinite line

$$y_2 = (b/a)y_1 \quad (-\infty < y_1 < 0).$$

An example is the trajectory through the point $(-3, -3)$ in Fig. 13.1. Also, we have already explained that the set $\left\{(0, 0)\right\}$ is a trajectory, despite it consisting of just a single point. So we see that every line $y_2 = (b/a)y_1$ consists of three trajectories. The same is true of the vertical line $y_1 = 0$.

The phase portrait in Fig. 13.1 displays nine trajectories, eight of which are semi-infinite lines radiating from the equilibrium point $(0, 0)$. The other trajectory,

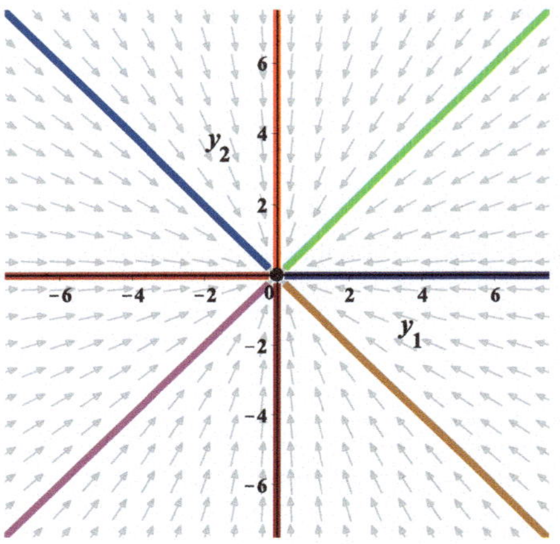

Fig. 13.1 Phase portrait of (13.5.2) near the stable star node (0, 0)

depicted by the solid circle in the figure, is the set consisting of a single point, namely, (0, 0). Aside from it, all trajectories of (13.5.2) are the infinitely many semi-infinite lines radiating from (0, 0). The equilibrium point (0, 0) is called a ***stable star node***, where the word "stable" indicates that all of the phase points on these trajectories move toward it as t increases. But had we supposed $\lambda > 0$, then the phase points on these trajectories would move away from (0, 0) as t increases. In that case, the equilibrium point (0, 0) is called an ***unstable star node***.

Finally, observe that every nonzero vector $\mathbf{v} \in \mathbb{R}^2$ is an eigenvector of J as

$$J\mathbf{v} = \begin{bmatrix} \lambda & 0 \\ 0 & \lambda \end{bmatrix} \mathbf{v} = \lambda \begin{bmatrix} 1 & 0 \\ 0 & 1 \end{bmatrix} \mathbf{v} = \lambda \mathbf{v}.$$

So the eigenspace of J corresponding to λ is

$$E(\lambda) = \left\{ \mathbf{v} \in \mathbb{R}^2 \colon J\mathbf{v} = \lambda \mathbf{v} \right\} = \mathbb{R}^2.$$

[2] We determined that when J is the Jordan canonical matrix (ii) in Theorem 13.3.3 and $\lambda \neq 0$ that all of the nontrivial trajectories of the linear system $\mathbf{y}' = J\mathbf{y}$ lie on the half lines radiating from (0, 0) as depicted in Fig. 13.1. The equilibrium point (0, 0), being a trajectory in its own right, can be regarded as a trivial trajectory.

Let us now turn to Jordan canonical matrices of the form (i) in the theorem. For these matrices, $\lambda_1 \neq \lambda_2$. Let us also assume that $\det J \neq 0$, in which case both λ_1 and λ_2 are nonzero.[2] It follows from Definition 13.2.1, Theorem 13.1.10, and Theorem 13.3.4 that the solution of the canonical system

$$\mathbf{y}' = J\mathbf{y} = \begin{bmatrix} \lambda_1 & 0 \\ 0 & \lambda_2 \end{bmatrix} \mathbf{y} \tag{13.5.7}$$

[2] The case $\det J = 0$ is the subject of Problems 35 and 36.

satisfying the initial condition $\mathbf{y}(0) = \langle a, b\rangle$ is

$$\mathbf{y}(t) = e^{Jt}\mathbf{y}(0) = \begin{bmatrix} e^{\lambda_1 t} & 0 \\ 0 & e^{\lambda_2 t} \end{bmatrix} \begin{bmatrix} a \\ b \end{bmatrix} = \begin{bmatrix} ae^{\lambda_1 t} \\ be^{\lambda_2 t} \end{bmatrix}. \tag{13.5.8}$$

Thus, for all $t \in \mathbb{R}$,

$$y_1(t) = ae^{\lambda_1 t}, \quad y_2(t) = be^{\lambda_2 t}. \tag{13.5.9}$$

With (13.5.8) expressed in the form

$$\mathbf{y}(t) = ae^{\lambda_1 t}\begin{bmatrix} 1 \\ 0 \end{bmatrix} + be^{\lambda_2 t}\begin{bmatrix} 0 \\ 1 \end{bmatrix},$$

we readily see that $\langle 1, 0\rangle$ and $\langle 0, 1\rangle$ are eigenvectors of J corresponding to the eigenvalues λ_1 and λ_2, respectively (see Problem 32). The eigenspaces of J corresponding to these eigenvalues are

$$E(\lambda_1) = \left\{c\begin{bmatrix} 1 \\ 0 \end{bmatrix} : c \in \mathbb{R}\right\} \quad \text{and} \quad E(\lambda_2) = \left\{c\begin{bmatrix} 0 \\ 1 \end{bmatrix} : c \in \mathbb{R}\right\}. \tag{13.5.10}$$

In other words, the eigenspace $E(\lambda_1)$ is the set of all the vectors lying on the y_1-axis while the eigenspace $E(\lambda_2)$ is the set of all the vectors lying on the y_2-axis.

Now let us determine all of the possible trajectories corresponding to the solutions of the canonical system (13.5.7). To this end, imagine a phase point dropped onto a point (a, b) at some instant and its subsequent motion. If this point is $(0, 0)$, then it follows from (13.5.9) that $y_1(t) = 0$ and $y_2(t) = 0$ for all $t \in (-\infty, \infty)$. Thus $(0, 0)$ is an equilibrium point. Furthermore, it is the only equilibrium point of (13.5.7) since $\det J \neq 0$ (see Problem 28). If $a > 0$ and $b = 0$, then $y_1(t) = ae^{\lambda_1 t} > 0$ and $y_2(t) = 0$. So the trajectory of the solution $\mathbf{y}(t)$ through the point $(a, 0)$ is

$$\{(y_1(t), y_2(t)) \in \mathbb{R}^2 : -\infty < t < \infty\} = \{(ae^{\lambda_1 t}, 0) \in \mathbb{R}^2 : -\infty < t < \infty\}.$$

Since the range of the function $ae^{\lambda_1 t}$ is $(0, \infty)$, the trajectory completely covers the positive y_1-axis. But if $a < 0$ and $b = 0$, then the trajectory containing the point $(a, 0)$ completely covers the negative y_1-axis. Consequently, the y_1-axis is composed of three trajectories: two semi-infinite line trajectories separated by the equilibrium point $(0, 0)$. Likewise, a pair of semi-infinite line trajectories and the equilibrium point $(0, 0)$ completely cover the y_2-axis.

Now imagine at some moment that a phase point finds itself at a point (a, b), where $a \neq 0$. Eliminating the independent variable t in (13.5.9), we have

$$y_2(t) = be^{\lambda_2 t} = b\big(e^{\lambda_1 t}\big)^{\lambda_2/\lambda_1} = b\left(\frac{y_1(t)}{a}\right)^{\lambda_2/\lambda_1}.$$

So the trajectory through the point (a, b) lies on the curve

$$y_2 = b\left(\frac{y_1}{a}\right)^{\lambda_2/\lambda_1}. \tag{13.5.11}$$

As we will see in the examples below, the signs of λ_1 and λ_2 determine whether this curve is composed of trajectories or is itself a trajectory. If $b = 0$, the curve is simply the line $y_2 = 0$, which as we already know consists of three trajectories.

Because of the uniqueness of solutions of a linear system (cf. Theorem 12.5.14), different trajectories never intersect. That being the case, a trajectory in one of the quadrants cannot cross the coordinate axes because every point on an axis belongs to a trajectory confined to that axis. Apart from the three trajectories lying on the y_2-axis, all of the other trajectories of the canonical linear system (13.5.7) lie on the curves given by (13.5.11), with $y_2 = 0$ being one of them.

Using formula (13.5.11) and information gleaned from the preceding analysis, we can graph a sufficient number of trajectories of the canonical system (13.5.7) to obtain examples of phase portraits when (i) λ_1 and λ_2 are both negative, (ii) both are positive, and (iii) λ_1 and λ_2 have opposite signs.

Example 13.5.1 (Canonical System: $\lambda_2 < \lambda_1 < 0$) Let $\lambda_1 = -1$ and $\lambda_2 = -2$, in which case (13.5.7) becomes

$$\mathbf{y}' = J\mathbf{y} = \begin{bmatrix} -1 & 0 \\ 0 & -2 \end{bmatrix}\mathbf{y}. \tag{13.5.12}$$

According to (13.5.10), the eigenspace $E(-1)$ consists of all the vectors lying on the y_1-axis while the eigenspace $E(-2)$ are those vectors lying on the y_2-axis.

It follows from (13.5.11) that all trajectories, aside from those confined to the y_1- and y_2-axes, lie on the parabolas

$$y_2 = \frac{b}{a^2}y_1^2. \tag{13.5.13}$$

Imagine placing a phase point (imaginary particle) at a location (a, b) in the phase plane, where $ab \neq 0$, and watching it move as t changes. Since both a and b have nonzero values, this phase point is on the parabola (13.5.13). If we stipulate that $t = 0$ marks the instant that it is located at (a, b), then its location at other values of t is given by (13.5.9), that is,

$$(y_1(t), y_2(t)) = \left(ae^{-t}, be^{-2t}\right). \tag{13.5.14}$$

With this it is easy to determine the direction that the phase point moves along the parabola (13.5.13). As t increases, it moves toward the equilibrium point $(0, 0)$ since

$$y_1(t) = ae^{-t} \to 0 \quad \text{and} \quad y_2(t) = be^{-2t} \to 0 \quad \text{as} \quad t \to \infty,$$

regardless of the values of a and b. The direction field shown in the phase portrait in Fig. 13.2 attests to this. As t decreases, the phase point moves away from $(0, 0)$.

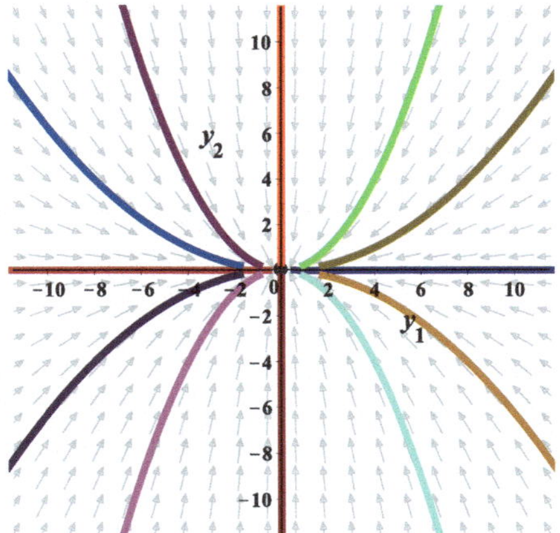

Fig. 13.2 Phase portrait of (13.5.12) near the stable node (0, 0)

For instance, consider (13.5.14) again. Suppose the phase point is initially located at a point (a, b) in the fourth quadrant. Then, as $a > 0$ and $b < 0$,

$$y_1(t) = ae^{-t} \to \infty \quad \text{and} \quad y_2(t) = be^{-2t} \to -\infty \quad \text{as} \quad t \to -\infty.$$

Analyses of the other quadrants yield the same conclusion, namely, that a phase point always moves away from the equilibrium point (0, 0) as $t \to -\infty$.

A number of trajectories lying on left and right halves of the family of parabolas given by equation (13.5.13) are shown in the phase portrait in Fig. 13.2. The equilibrium point (0, 0) in this example is called a ***stable node***. ♦

In the next example, we plot a phase portrait of a noncanonical linear system whose corresponding Jordan canonical form is the linear system (13.5.12). We can exploit the relationship between these two systems by using the information already obtained in the previous example. From the first two items in Theorem 13.3.1, we see that a similarity transformation maps solutions to solutions, and consequently, trajectories to trajectories. Since we have already ascertained the trajectories of (13.5.12), our strategy for obtaining the trajectories of the noncanonical linear system (13.5.15) (below) will be to find a similarity transformation $\mathbf{x} = P\mathbf{y}$ that will map the trajectories of (13.5.12) to those of (13.5.15). This similarity transformation will also map the eigenspaces of (13.5.12) to the eigenspaces of (13.5.15) (see Problem 34).

Example 13.5.2 (Noncanonical System: $\lambda_2 < \lambda_1 < 0$) Plot a phase portrait of the linear system

$$\begin{aligned} x_1' &= -x_1 - x_2 \\ x_2' &= -2x_2. \end{aligned} \tag{13.5.15}$$

Solution The coefficient matrix of this system is

$$A = \begin{bmatrix} -1 & -1 \\ 0 & -2 \end{bmatrix}. \tag{13.5.16}$$

Since $\operatorname{tr} A = -3$ and $\det A = 2$, the characteristic equation of A is

$$\lambda^2 + 3\lambda + 2 = 0,$$

which has the roots -1 and -2. It is left as an exercise to verify that $\mathbf{v^1} = \langle 1, 0 \rangle$ and $\mathbf{v^2} = \langle 1, 1 \rangle$ are eigenvectors of A corresponding to the eigenvalues $\lambda_1 = -1$ and $\lambda_2 = -2$, respectively. Now define the matrix P by

$$P := \begin{bmatrix} \mathbf{v^1} & \mathbf{v^2} \end{bmatrix} = \begin{bmatrix} 1 & 1 \\ 0 & 1 \end{bmatrix}.$$

It follows from Theorem 13.3.4 that the Jordan canonical matrix

$$J = \begin{bmatrix} \lambda_1 & 0 \\ 0 & \lambda_2 \end{bmatrix} = \begin{bmatrix} -1 & 0 \\ 0 & -2 \end{bmatrix}$$

is similar to the matrix A in that $J = P^{-1}AP$.

Since $\mathbf{v^1}$ and $\mathbf{v^2}$ are linearly independent eigenvectors, there are two eigenspaces of A, one of which is the eigenspace $\mathcal{E}(-1) = \{c\mathbf{v^1} \colon c \in \mathbb{R}\}$ corresponding to $\lambda_1 = -1$. The other is the eigenspace $\mathcal{E}(-2) = \{c\mathbf{v^2} \colon c \in \mathbb{R}\}$ corresponding to $\lambda_2 = -2$. That is,

$$\mathcal{E}(-1) = \left\{ c\begin{bmatrix} 1 \\ 0 \end{bmatrix} : c \in \mathbb{R} \right\} \quad \text{and} \quad \mathcal{E}(-2) = \left\{ c\begin{bmatrix} 1 \\ 1 \end{bmatrix} : c \in \mathbb{R} \right\}. \tag{13.5.17}$$

Since we already ascertained in the previous example what the trajectories of the corresponding canonical system (13.5.12) are and what they look like, let us avail ourselves of that knowledge here to obtain the trajectories of (13.5.15) by using the similarity transformation $\mathbf{x} = P\mathbf{y}$ to map the trajectories of (13.5.12) to those of (13.5.15). But before we do so, let us confirm that this transformation does map eigenspaces to eigenspaces (cf. Problem 34).

First note that the transformation maps a vector $\mathbf{y} = \langle y_1, y_2 \rangle$ to the vector

$$\mathbf{x} = P\mathbf{y} = \begin{bmatrix} 1 & 1 \\ 0 & 1 \end{bmatrix} \begin{bmatrix} y_1 \\ y_2 \end{bmatrix} = \begin{bmatrix} y_1 + y_2 \\ y_2 \end{bmatrix}. \tag{13.5.18}$$

In other words, the first component of any vector $\langle y_1, y_2 \rangle$ is changed by the addition of its second component; however, the second component is unchanged. This transformation causes a ***horizontal shear*** of the y_1y_2-phase plane (cf. [63,

Fig. 2.1, p. 94]). Recall from the previous example that the eigenspaces of J are

$$E(-1) = \left\{ \begin{bmatrix} c \\ 0 \end{bmatrix} : c \in \mathbb{R} \right\} \quad \text{and} \quad E(-2) = \left\{ \begin{bmatrix} 0 \\ c \end{bmatrix} : c \in \mathbb{R} \right\}.$$

Let us apply the transformation to a vector $\mathbf{y} = \langle c, 0 \rangle$, which belongs to $E(-1)$. It follows from (13.5.18) that

$$\mathbf{x} = P\mathbf{y} = \begin{bmatrix} c + 0 \\ 0 \end{bmatrix} = \begin{bmatrix} c \\ 0 \end{bmatrix}.$$

That is, the transformation $\mathbf{x} = P\mathbf{y}$ maps $E(-1)$ onto itself. Thus, $\mathcal{E}(-1) = E(-1)$. Applying the transformation to $\mathbf{y} = \langle 0, c \rangle$, a vector belonging to $E(-2)$, we get

$$\mathbf{x} = P\mathbf{y} = \begin{bmatrix} 0 + c \\ c \end{bmatrix} = \begin{bmatrix} c \\ c \end{bmatrix}.$$

Thus the transformation maps the eigenspace $E(-2)$ to the set of vectors

$$\left\{ \begin{bmatrix} c \\ c \end{bmatrix} : c \in \mathbb{R} \right\} = \left\{ c \begin{bmatrix} 1 \\ 1 \end{bmatrix} : c \in \mathbb{R} \right\},$$

which is the eigenspace $\mathcal{E}(-2)$ that we had already obtained earlier in (13.5.17). In other words, it rotates $E(-2)$, the eigenspace of J corresponding to $\lambda_2 = -2$, through an angle of 45° clockwise about the origin: the result of which is $\mathcal{E}(-2)$, the eigenspace of A corresponding to $\lambda_2 = -2$. To sum up, the vectors belonging to the eigenspace $\mathcal{E}(-1)$ are on the x_1-axis while those belonging to $\mathcal{E}(-2)$ are on the line $x_2 = x_1$. See Fig. 13.3.

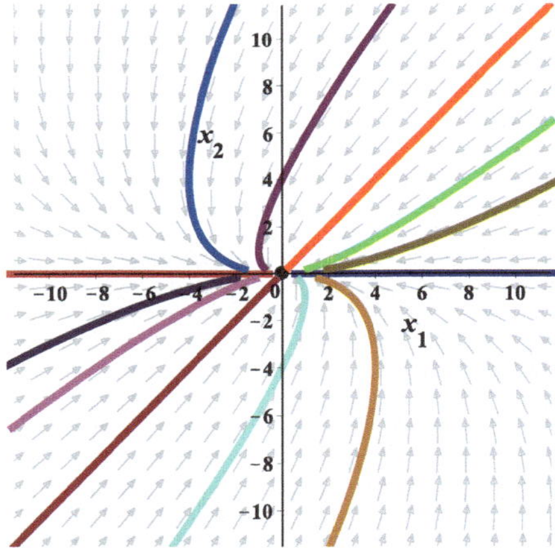

Fig. 13.3 Phase portrait of (13.5.15) near the stable node (0, 0)

Now that we know that the transformation $\mathbf{x} = P\mathbf{y}$ does not affect the trajectories on the y_1-axis but does rotate those on the y_2-axis, let us find out how it affects the other trajectories of (13.5.12). Applying it to the solution of $\mathbf{y}' = J\mathbf{y}$ whose value at $t = 0$ is $\mathbf{y}(0) = \langle a, b\rangle$, which from (13.5.8) is

$$\mathbf{y}(t) = \begin{bmatrix} ae^{-t} \\ be^{-2t} \end{bmatrix}, \tag{13.5.19}$$

we get

$$\mathbf{x}(t) = P\mathbf{y}(t) = \begin{bmatrix} 1 & 1 \\ 0 & 1 \end{bmatrix}\begin{bmatrix} ae^{-t} \\ be^{-2t} \end{bmatrix} = a\begin{bmatrix} 1 \\ 0 \end{bmatrix}e^{-t} + b\begin{bmatrix} 1 \\ 1 \end{bmatrix}e^{-2t}. \tag{13.5.20}$$

This is the solution of $\mathbf{x}' = A\mathbf{x}$ whose value at $t = 0$ is $\mathbf{x}(0) = \langle a + b, b\rangle$. For example, let $a = -4$ and $b = 1$ in (13.5.19) and (13.5.20). Then (13.5.19) is the solution of the canonical equation $\mathbf{y}' = J\mathbf{y}$ such that $\mathbf{y}(0) = \langle -4, 1\rangle$. Its graph, shown in Fig. 13.2, is the leftmost trajectory in Quadrant II passing through the point $(a, b) = (-4, 1)$ when $t = 0$. As we see above, the transformation $\mathbf{x} = P\mathbf{y}$ maps (13.5.19) to

$$\mathbf{x} = -4\begin{bmatrix} 1 \\ 0 \end{bmatrix}e^{-t} + \begin{bmatrix} 1 \\ 1 \end{bmatrix}e^{-2t} = \begin{bmatrix} -4e^{-t} + e^{-2t} \\ e^{-2t} \end{bmatrix}, \tag{13.5.21}$$

which is the solution of $\mathbf{x}' = A\mathbf{x}$ such that $\mathbf{x}(0) = \langle -3, 1\rangle$. Its graph, shown in Fig. 13.3, is the leftmost trajectory in the upper half-plane passing through the point $(-3, 1)$ when $t = 0$.

An alternative to using the parametric equations (13.5.20) to graph trajectories of (13.5.15) is to first eliminate the independent variable t from the pair of equations

$$x_1 = ae^{-t} + be^{-2t}, \quad x_2 = be^{-2t}$$

to get the equation

$$x_1 = x_2 + a\sqrt{\frac{x_2}{b}}. \tag{13.5.22}$$

With this, the graphs of trajectories, such as those shown in Fig. 13.3, can be obtained by assigning values to a and b and then graphing the resulting equation. For example, with $a = -4$ and $b = 1$, (13.5.22) becomes

$$x_1 = x_2 - 4\sqrt{x_2}. \tag{13.5.23}$$

Its graph is the curve in the upper half-plane in Fig. 13.4, which is the leftmost trajectory in the upper half-plane in Fig. 13.3. Switching the signs of a and b to

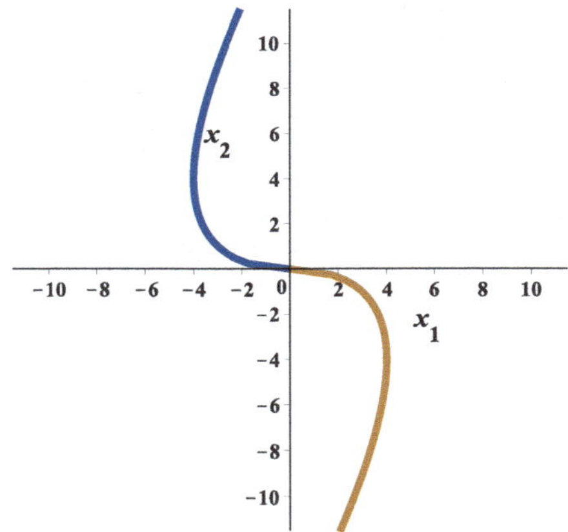

Fig. 13.4 Graphs of (13.5.23) and (13.5.24)

$a = 4$ and $b = -1$, equation (13.5.22) becomes

$$x_1 = x_2 + 4\sqrt{-x_2}. \tag{13.5.24}$$

Its graph is the curve in the lower half-plane in Fig. 13.4, which is the rightmost trajectory in the lower half-plane in Fig. 13.3.

Observe that the phase portrait of $\mathbf{x}' = A\mathbf{x}$ shown in Fig. 13.3 resembles the phase portrait of its corresponding canonical equation $\mathbf{y}' = J\mathbf{y}$ in Fig. 13.2. The trajectories of $\mathbf{y}' = J\mathbf{y}$ have been rotated and deformed by the similarity transformation to become those of $\mathbf{x}' = A\mathbf{x}$, but the phase portraits still have the same qualitative features. Moreover, every phase point in the x_1x_2-plane, other than one located at the equilibrium point $(0, 0)$, moves toward $(0, 0)$ as t increases, just as it is the case with the phase points in the y_1y_2-plane. As a result, the equilibrium point $(0, 0)$ of $\mathbf{x}' = A\mathbf{x}$ is said to be a ***stable node***. ♦

Example 13.5.3 (Canonical System: $\lambda_1 > \lambda_2 > 0$) Let $\lambda_1 = 2$ and $\lambda_2 = 1$, in which case (13.5.7) becomes

$$\mathbf{y}' = J\mathbf{y} = \begin{bmatrix} 2 & 0 \\ 0 & 1 \end{bmatrix} \mathbf{y}. \tag{13.5.25}$$

According to (13.5.10), the eigenspaces of J corresponding to these eigenvalues are

$$E(\lambda_1) = E(2) = \left\{ c \begin{bmatrix} 1 \\ 0 \end{bmatrix} : c \in \mathbb{R} \right\} \quad \text{and} \quad E(\lambda_2) = E(1) = \left\{ c \begin{bmatrix} 0 \\ 1 \end{bmatrix} : c \in \mathbb{R} \right\}.$$

Choosing 8 distinct points (a, b) not belonging to the eigenspaces and graphing the solutions (see (13.5.9))

$$y_1(t) = ae^{2t}, \quad y_2(t) = be^t$$

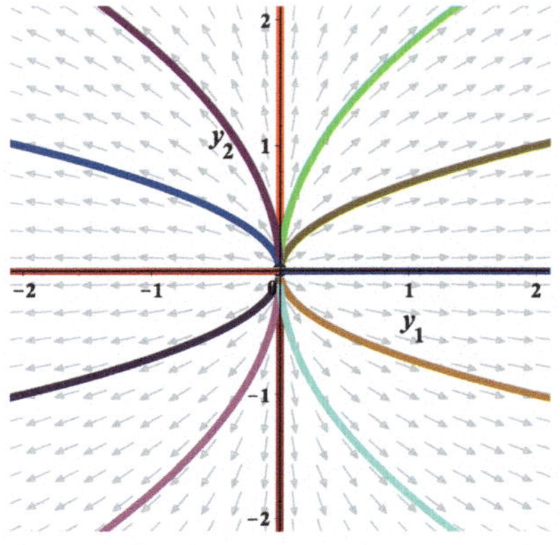

Fig. 13.5 Phase portrait of (13.5.25) near the unstable node (0, 0)

using the computer algebra system *Maple*, we obtain the phase portrait shown in Fig. 13.5. The equilibrium point (0, 0) is an ***unstable node*** since

$$|y_1(t)| = |ae^{\lambda_1 t}| = |a|e^{2t} \to \infty, \quad |y_2(t)| = |be^{\lambda_2 t}| = |b|e^{t} \to \infty$$

as $t \to \infty$.

Finally, we only have to refer to (13.5.11) to see that the equation of the trajectory through a point (a, b) that is not located on the y_2-axis is

$$y_2 = b\sqrt{\frac{y_1}{a}}. \qquad \blacklozenge$$

Example 13.5.4 (Canonical System: $\lambda_2 < 0 < \lambda_1$) Let $\lambda_1 = 1$ and $\lambda_2 = -2$. Then (13.5.7) becomes

$$\mathbf{y}' = J\mathbf{y} = \begin{bmatrix} 1 & 0 \\ 0 & -2 \end{bmatrix} \mathbf{y}. \tag{13.5.26}$$

As in the previous (canonical) examples, the eigenspaces $E(1)$ and $E(-2)$ consist of all vectors on the y_1- and y_2-axes, respectively. Choosing 8 distinct points (a, b) not belonging to $E(1)$ and $E(2)$ and graphing the solutions

$$y_1(t) = ae^t, \quad y_2(t) = be^{-2t}, \tag{13.5.27}$$

we obtain the phase portrait shown in Fig. 13.6. It follows from (13.5.11) that trajectories which are not confined to the y_1- and y_2-axes lie on the curves defined by the equation

$$y_2 = b\left(\frac{y_1}{a}\right)^{-2} = \frac{a^2 b}{y_1^2}. \tag{13.5.28}$$

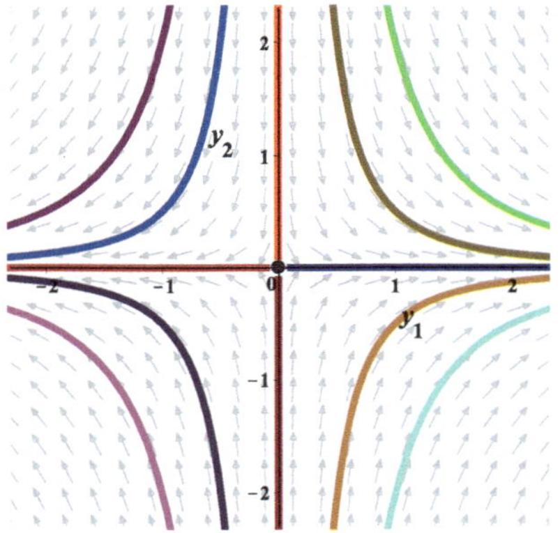

Fig. 13.6 Phase portrait of (13.5.26) near the saddle point (0, 0)

If both a and b are nonzero, then

$$|y_1(t)| = |a|e^t \to \infty \text{ and } y_2(t) = be^{-2t} \to 0$$

as $t \to \infty$ and

$$y_1(t) = ae^t \to 0 \text{ and } |y_2(t)| = |b|e^{-2t} \to \infty$$

as $t \to -\infty$. It follows from (13.5.27) and the limiting behavior of these solutions that a trajectory through a point (a, b) in a given quadrant is confined to this quadrant for all time t and that every phase point on the trajectory approaches the y_1-axis in forward time and the y_2-axis in backward time. If $a \neq 0$ and $b = 0$, then $|y_1(t)| = |a|e^t \to \infty$ and $y_2(t) \equiv 0$. Thus, a trajectory through a point $(a, 0)$ on the y_1-axis is confined to this axis and a phase point on this trajectory moves away from the equilibrium point $(0, 0)$ in forward time but towards it in backward time. If $a = 0$ and $b \neq 0$, then $y_1(t) \equiv 0$ and $y_2(t) \to 0$ as $t \to \infty$. So a trajectory through a point $(0, b)$ on the y_2-axis is confined to this axis and a phase point on this trajectory moves toward the equilibrium point $(0, 0)$ in forward time but away from it in backward time. Since all phase points (except for those on the y_2-axis) veer away from the equilibrium point $(0, 0)$ in forward time, the equilibrium point $(0, 0)$ is unstable. Whenever trajectories behave near an equilibrium point as they do in this example, the equilibrium point is said to be a ***saddle point***. Thus the equilibrium point $(0, 0)$ of the linear system (13.5.26) is a saddle point. ♦

[3] Let us construct a phase portrait of the canonical system

$$\mathbf{y}' = J\mathbf{y} = \begin{bmatrix} \lambda & 1 \\ 0 & \lambda \end{bmatrix} \mathbf{y}, \tag{13.5.29}$$

where J is the canonical matrix (iii) in Theorem 13.3.3 and λ represents a given real number. Since λ is a repeated eigenvalue of J and the only eigenvectors corresponding to λ are the nonzero scalar multiples of the vector $\langle 1, 0\rangle$, the matrix J has just one eigenspace, namely, the set of vectors

$$E(\lambda) = \left\{ c \begin{bmatrix} 1 \\ 0 \end{bmatrix} : c \in \mathbb{R} \right\}.$$

One way to find the solution $\mathbf{y}(t)$ of (13.5.29) so that $\mathbf{y}(0) = \langle a, b\rangle$, where a and b are a pair of given real numbers, is to refer to Theorem 13.1.10, Definition 13.2.1, and Theorem 13.3.4 (iii), from which we conclude that

$$\mathbf{y}(t) = e^{Jt}\mathbf{y}(0) = e^{\lambda t} \begin{bmatrix} 1 & t \\ 0 & 1 \end{bmatrix} \begin{bmatrix} a \\ b \end{bmatrix} = \begin{bmatrix} a + bt \\ b \end{bmatrix} e^{\lambda t}. \tag{13.5.30}$$

That is,

$$y_1(t) = (a + bt)e^{\lambda t}, \quad y_2(t) = be^{\lambda t} \tag{13.5.31}$$

for all $t \in \mathbb{R}$. Another way of obtaining this solution is by solving the pair of equations

$$y_1' = \lambda y_1 + y_2, \quad y_2' = \lambda y_2 \tag{13.5.32}$$

via two integrations and then finding the values of the constants of integration so that $y_1(0) = a$ and $y_2(0) = b$.

Suppose $\lambda < 0$ (the cases $\lambda = 0$ and $\lambda > 0$ are left as exercises). If a and b are both equal to 0, then it follows from (13.5.31) that $y_1(t) \equiv 0$ and $y_2(t) \equiv 0$. If $a \neq 0$ and $b = 0$, then the solution is

$$y_1(t) = ae^{\lambda t} \text{ and } y_2(t) \equiv 0.$$

Since $\lambda < 0$, $y_1(t) \to 0$ as $t \to \infty$. It follows that there are three trajectories on the y_1-axis. One of them is the trivial trajectory $\{(0, 0)\}$, which separates the other two: the semi-infinite line trajectory that covers the positive y_1-axis and the semi-infinite line trajectory that covers the negative y_1-axis. All phase points on the semi-infinite line trajectories approach the equilibrium point $(0, 0)$ as $t \to \infty$.

If $b \neq 0$, then

$$y_1(t) = (a + bt)e^{\lambda t} \to 0 \text{ and } y_2(t) = be^{\lambda t} \to 0$$

and

$$\frac{dy_2}{dy_1} = \frac{y_2'(t)}{y_1'(t)} = \frac{b\lambda}{\lambda(a + bt) + b} \to 0$$

as $t \to \infty$. In other words, every phase point in the y_1y_2-plane that is not on the y_1-axis moves toward the origin and in so doing describes a trajectory that flattens out near the origin. It follows from the second differential equation in (13.5.32) that the y_2-coordinates of phase points moving in the upper half of the plane decrease with increasing t, whereas the y_2-coordinates of those moving in the lower half of the plane increase. Furthermore, it follows from the first equation in (13.5.32) that the y_1-coordinate of every phase point moving above the line $y_2 = -\lambda y_1$ increases as t increases, whereas this same coordinate decreases for a phase point moving below this line. From this analysis of the motion of phase points in various regions of the $y_1 y_2$-plane, we deduce the following:

1. Every phase point that is not confined to the y_1-axis will reverse its horizontal direction of motion the moment it crosses the line $y_2 = -\lambda y_1$. Consequently, the value of the y_1-coordinate of a phase point on a given trajectory attains its maximum value on this line if the trajectory lies in the upper half plane. If the trajectory lies in the lower half plane, then the y_1-coordinate of a phase point attains its minimum value on this line.
2. Every trajectory, aside from the three on the y_1-axis, crosses the line $y_2 = -\lambda y_1$ exactly once and the part of the trajectory that is above the line $y_2 = -\lambda y_1$ is concave down while the part below this line is concave up.

Figure 13.7 shows the graphs of a number of solutions of (13.5.29) with $\lambda = -1$, which helps clarify the foregoing analysis and statements.

Example 13.5.5 (Canonical System: $\lambda < 0$) With $\lambda = -1$, the canonical system (13.5.29) is

$$\mathbf{y}' = J\mathbf{y} = \begin{bmatrix} -1 & 1 \\ 0 & -1 \end{bmatrix} \mathbf{y}. \tag{13.5.33}$$

From (13.5.31), we see that its solutions are

$$y_1(t) = (a + bt)e^{-t}, \quad y_2(t) = be^{-t},$$

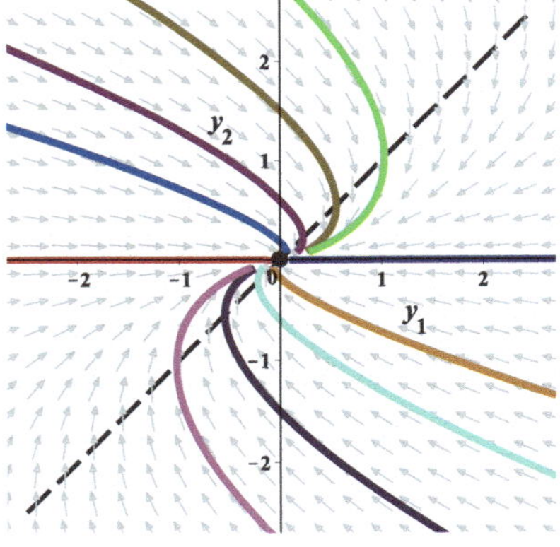

Fig. 13.7 Phase portrait of (13.5.33) near the stable degenerate node (0, 0)

where $a, b \in \mathbb{R}$. Graphs of 11 of these solutions, including the point $(0, 0)$ (graph of the trivial solution $y_1(t) \equiv 0$, $y_2(t) \equiv 0$) are shown in Fig. 13.7. The dashed line is the graph of $y_2 = y_1$, on which the y_1-reversals in direction take place. An equilibrium point with trajectories behaving near it as they do in this example near $(0, 0)$ is called a ***degenerate node*** or an ***improper node***. It is a stable degenerate mode if $\lambda < 0$ and an unstable degenerate mode if $\lambda > 0$. ♦

[4] The final Jordan matrix J to consider is matrix (iv) in Theorem 13.3.3. It is the canonical form of a matrix with complex eigenvalues $\alpha \pm i\beta$, where $\alpha, \beta \in \mathbb{R}$ with $\beta \neq 0$. Let us construct phase portraits of the corresponding canonical system

$$\mathbf{y}' = J\mathbf{y} = \begin{bmatrix} \alpha & \beta \\ -\beta & \alpha \end{bmatrix} \mathbf{y}, \tag{13.5.34}$$

which is the matrix representation of the linear system

$$y_1' = \alpha y_1 + \beta y_2, \quad y_2' = -\beta y_1 + \alpha y_2. \tag{13.5.35}$$

It follows from Theorem 13.3.4 or Theorem 12.5.36 (cf. Example 12.5.24) that

$$\begin{aligned} y_1(t) &= e^{\alpha t}(c_1 \cos \beta t + c_2 \sin \beta t) \\ y_2(t) &= e^{\alpha t}(-c_1 \sin \beta t + c_2 \cos \beta t) \end{aligned} \tag{13.5.36}$$

is a general solution of this system for $-\infty < t < \infty$.

With (13.5.36) we can find the solution of (13.5.34) whose trajectory passes through a given point (y_1, y_2). However, as we shall see, the graph of a solution and its behavior as t changes becomes clearer when the graph is expressed in polar coordinates (r, θ). That being the case, let r denote the distance from the origin to the point (y_1, y_2):

$$r = \sqrt{y_1^2 + y_2^2}.$$

Let θ denote the angle measured from the positive y_1-axis to the ray (half-line) that starts at the origin and passes through the point (y_1, y_2). This angle can be determined from

$$\cos\theta = \frac{y_1}{r} \quad \text{and} \quad \sin\theta = \frac{y_2}{r}. \tag{13.5.37}$$

Given a point (a, b) in the y_1y_2-plane, let $\mathbf{y}(t) = \langle y_1(t), y_2(t) \rangle$ denote the solution of the linear system such that $\mathbf{y}(0) = \langle a, b \rangle$. Since

$$r(t) = \sqrt{y_1^2(t) + y_2^2(t)},$$

let r_0 denote the value of $r(t)$ at $t = 0$. And so

$$r_0 = r(0) = \sqrt{y_1^2(0) + y_2^2(0)} = \sqrt{a^2 + b^2}.$$

With θ_0 denoting the value of $\theta(t)$ when $t = 0$, we have

$$\cos\theta_0 = \frac{y_1(0)}{r(0)} = \frac{a}{r_0} \quad \text{and} \quad \sin\theta_0 = \frac{y_2(0)}{r(0)} = \frac{b}{r_0}.$$

It follows from (13.5.36) that the solution of (13.5.35) satisfying the initial conditions $y_1(0) = a$ and $y_2(0) = b$ is

$$y_1(t) = r_0 e^{\alpha t}\cos(\beta t - \theta_0), \quad y_2(t) = -r_0 e^{\alpha t}\sin(\beta t - \theta_0) \tag{13.5.38}$$

for $-\infty < t < \infty$. The details showing this are left as an exercise (see Problem 37). It is also left to the reader to show that

$$r(t) = r_0 e^{\alpha t} \quad \text{and} \quad \theta(t) = \theta_0 - \beta t. \tag{13.5.39}$$

Actually there is a less tedious way of deriving (13.5.39), which does not even involve (13.5.38). To derive the first function in (13.5.39), simply differentiate the equation

$$r^2 = y_1^2 + y_2^2$$

implicitly with respect to t. Then, from (13.5.35), we have

$$\begin{aligned}\frac{d}{dt}r^2 &= \frac{d}{dt}(y_1^2 + y_2^2) = 2y_1y_1' + 2y_2y_2' = 2y_1(\alpha y_1 + \beta y_2) + 2y_2(-\beta y_1 + \alpha y_2)\\ &= 2\alpha(y_1^2 + y_2^2) = 2\alpha r^2.\end{aligned}$$

Since

$$\frac{d}{dt}r^2 = 2r\frac{dr}{dt},$$

this simplifies to

$$\frac{dr}{dt} = \alpha r. \tag{13.5.40}$$

The solution of (13.5.40) with $r(0) = r_0$ is

$$r(t) = r_0 e^{\alpha t}. \tag{13.5.41}$$

To derive the second function in (13.5.39), differentiate both sides of

$$\sin\theta = \frac{y_2}{r}$$

implicitly with respect to t:

$$\cos\theta \frac{d\theta}{dt} = \frac{ry_2' - y_2 r'}{r^2}.$$

Then, because of (13.5.35), (13.5.37), and (13.5.40), this simplifies to

$$\frac{d\theta}{dt} = \frac{r(-\beta y_1 + \alpha y_2) - \alpha r y_2}{r^2 \cos\theta} = \frac{-\beta y_1}{r\cos\theta} = -\beta.$$

The solution of this differential equation with $\theta(0) = \theta_0$ is the second function in (13.5.39), namely,

$$\theta(t) = -(\beta t - \theta_0). \tag{13.5.42}$$

Observe that we can express r directly in terms of θ by solving (13.5.42) for t and substituting the result into (13.5.41). Thus, as $t = -(\theta - \theta_0)/\beta$,

$$r(\theta) = r_0 e^{-\frac{\alpha}{\beta}(\theta - \theta_0)} \tag{13.5.43}$$

for $-\infty < \theta < \infty$. If $\alpha \neq 0$, this equation can be rewritten in the form $r = \rho e^{k\theta}$, or as $\theta = (1/k)\ln(r/\rho)$, where $\rho > 0$ and $k \neq 0$ are constants. Its graph is known as a ***logarithmic spiral*** (cf. [79]).

Example 13.5.6 (Canonical System: $\alpha < 0$, $\beta > 0$) Graph the solution of the system

$$\mathbf{y}' = J\mathbf{y} = \begin{bmatrix} -1 & 2 \\ -2 & -1 \end{bmatrix} \mathbf{y} \tag{13.5.44}$$

satisfying the initial condition $\mathbf{y}(0) = \langle -10, 0 \rangle$. Graph a phase portrait of this system.

Solution. Comparing the coefficient matrices in (13.5.44) and (13.5.34), we see that $\alpha = -1$ and $\beta = 2$. Since $y_1(0) = -10$ and $y_2(0) = 0$,

$$r_0 = \sqrt{y_1^2(0) + y_2^2(0)} = 10 \quad \text{and} \quad \theta_0 = \pi.$$

With these values, the polar equation (13.5.43) becomes

$$r = 10e^{\frac{1}{2}(\theta - \pi)}, \tag{13.5.45}$$

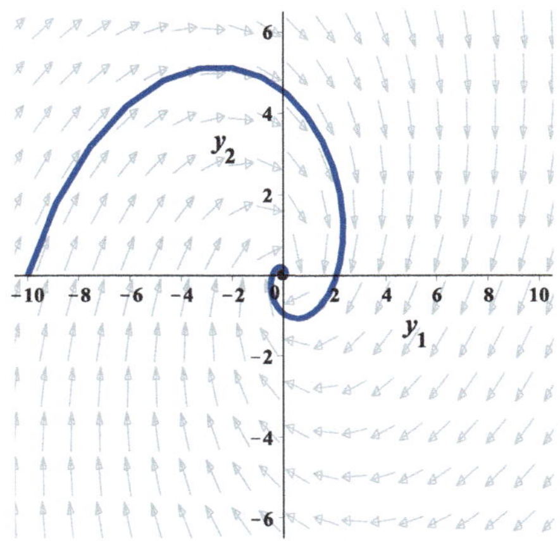

Fig. 13.8 Graph of (13.5.45) for $-\infty < \theta \leq \pi$

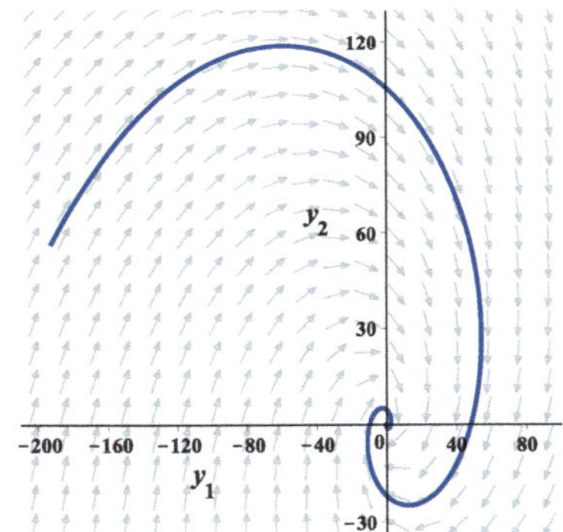

Fig. 13.9 Graph of (13.5.45) for $-3 \leq t < \infty$

where $-\infty < \theta < \infty$. Its graph for $-\infty < \theta \leq \pi$ is the logarithmic spiral shown in Fig. 13.8.

Substituting the values of α, β, r_0, and θ_0 into (13.5.38), we obtain the following solution of (13.5.44):

$$y_1(t) = 10e^{-t}\cos(2t - \pi), \quad y_2(t) = -10e^{-t}\sin(2t - \pi). \tag{13.5.46}$$

This simplifies to

$$y_1(t) = -10e^{-t}\cos 2t, \quad y_2(t) = 10e^{-t}\sin 2t. \tag{13.5.47}$$

The logarithmic spiral shown in Fig. 13.9 is the graph of (13.5.47) for $-3 \leq t < \infty$. As $\theta(t) = -(2t - \pi)$, it is also the graph of (13.5.45) for $-\infty < \theta \leq \pi + 6$.

For initial conditions other than $(a, b) = (-10, 0)$, we have from (13.5.38) that

$$y_1(t) = r_0 e^{-t}\cos(2t - \theta_0), \quad y_2(t) = -r_0 e^{-t}\sin(2t - \theta_0),$$

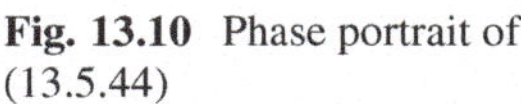

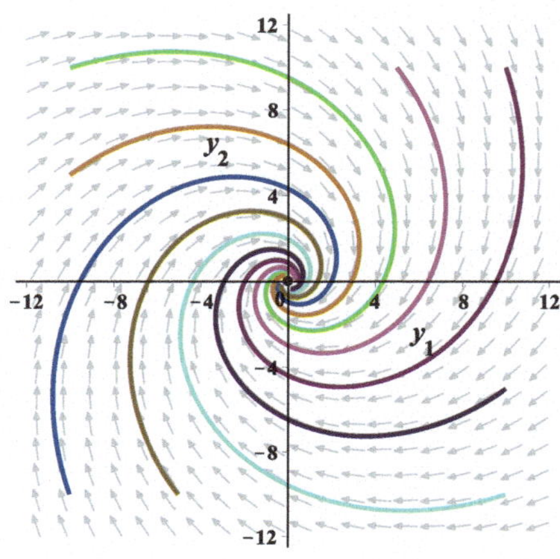

Fig. 13.10 Phase portrait of (13.5.44)

from which we see that $y_1(t) \to 0$ and $y_2(t) \to 0$ as $t \to \infty$. Because of this and $r(\theta) = r_0 e^{\frac{1}{2}(\theta-\theta_0)}$, we conclude that all trajectories of (13.5.44) are logarithmic spirals that approach the equilibrium point (0, 0) as $t \to \infty$. A phase portrait of (13.5.44) near (0, 0) is shown in Fig. 13.10. This type of equilibrium point is called a **stable focus**. It is also known as a **spiral sink** or a **stable spiral point**. ♦

In general, it follows from (13.5.39) that the equilibrium point (0, 0) of the canonical system (13.5.34) is a stable focus if $\alpha < 0$, as in Example 13.5.6, but an unstable focus if $\alpha > 0$. However, if $\alpha = 0$, then the trajectories of (13.5.34) neither approach (0, 0) nor move away from it since the equation of the trajectory through a point (a, b) is $r = r_0$, where $r_0 = \sqrt{a^2 + b^2}$, or

$$y_1^2 + y_2^2 = a^2 + b^2$$

in Cartesian coordinates. Consequently, a phase portrait of

$$\mathbf{y}' = \begin{bmatrix} 0 & \beta \\ -\beta & 0 \end{bmatrix} \mathbf{y} \tag{13.5.48}$$

consists of concentric circles in the $y_1 y_2$-plane with the equilibrium point (0, 0) as their common center. Figure 13.11 shows a phase portrait of (13.5.48) when $\beta = 2$ (see the next example). An equilibrium point that is enclosed by closed orbits, such as circles enclosing (0, 0), is called a ***center***. A center is said to be ***neutrally stable*** because a phase point on any trajectory remains within a certain distance from the center but does not get arbitrarily close to it.

Example 13.5.7 (Noncanonical System: $\alpha = 0$, $\beta > 0$) Plot a phase portrait of the linear system

$$x_1' = x_2$$

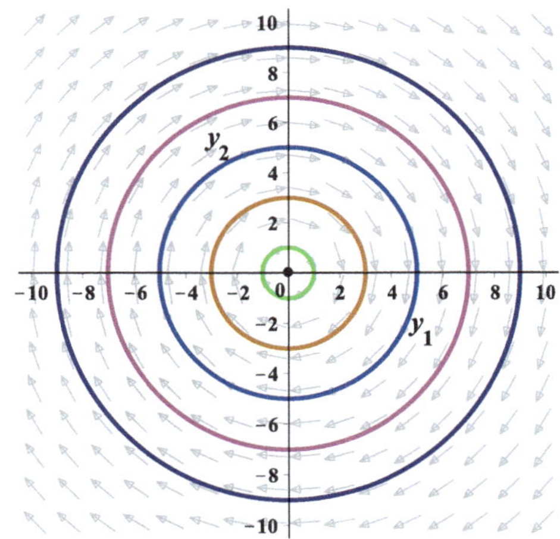

Fig. 13.11 Phase portrait of (13.5.48) with $\beta = 2$

$$x_2' = -4x_1. \tag{13.5.49}$$

Solution The coefficient matrix of this system is

$$A = \begin{bmatrix} 0 & 1 \\ -4 & 0 \end{bmatrix}. \tag{13.5.50}$$

Since the characteristic equation is

$$\lambda^2 + 4 = 0,$$

the eigenvalues of A are $\pm 2i$. Therefore, it follows from Theorem 13.3.4 (iv) with $\alpha = 0$ and $\beta = 2$ that

$$J = \begin{bmatrix} 0 & 2 \\ -2 & 0 \end{bmatrix} \tag{13.5.51}$$

is the Jordan canonical form of A and

$$e^{Jt} = \begin{bmatrix} \cos 2t & \sin 2t \\ -\sin 2t & \cos 2t \end{bmatrix}. \tag{13.5.52}$$

Consequently, the trajectory of the solution $\mathbf{y}(t) = \langle y_1(t), y_2(t)\rangle$ of the canonical system $\mathbf{y}' = J\mathbf{y}$ that passes through a point $(a, 0)$ in the y_1y_2-phase plane when $t = 0$ is

$$\mathbf{y}(t) = e^{Jt}\mathbf{y}(0) = \begin{bmatrix} \cos 2t & \sin 2t \\ -\sin 2t & \cos 2t \end{bmatrix} \begin{bmatrix} a \\ 0 \end{bmatrix} = \begin{bmatrix} a\cos 2t \\ -a\sin 2t \end{bmatrix}. \tag{13.5.53}$$

Since $y_1(t) = a\cos 2t$ and $y_2(t) = -a\sin 2t$,

$$y_1^2(t) + y_2^2(t) = a^2\cos^2 2t + a^2\sin^2 2t = a^2.$$

Thus, the trajectory through a point $(a, 0)$ on the y_1-axis is a circle of radius a with its center at $(0, 0)$. By assigning various values to a and graphing the corresponding trajectories, we can generate a phase portrait of the canonical system $\mathbf{y}' = J\mathbf{y}$, such as in Fig. 13.11 which shows five circular trajectories through the points $(a, 0)$ for $a = 1, 3, 5, 7, 9$.

Recall from Theorems 13.3.1 and 13.3.4 that solutions of the noncanonical system (13.5.49) can be obtained from the solutions $\mathbf{y}(t)$ of the canonical system $\mathbf{y}' = J\mathbf{y}$ by means of a similarity transformation $\mathbf{x}(t) = P\mathbf{y}(t)$, where $P = \begin{bmatrix}\mathbf{v}^1 & \mathbf{v}^2\end{bmatrix}$ is a matrix whose columns $\mathbf{v}^1$ and $\mathbf{v}^2$ are the real and imaginary parts of a complex eigenvector $\mathbf{v}$ of the matrix A corresponding to the eigenvalue $2i$. Such an eigenvector is any solution of the matrix equation

$$\begin{bmatrix}-2i & 1\\ -4 & -2i\end{bmatrix}\begin{bmatrix}v_1\\ v_2\end{bmatrix} = \begin{bmatrix}0\\ 0\end{bmatrix}.$$

Since $v_1 = 1$ and $v_2 = 2i$ is a solution, $\mathbf{v} = \langle 1, 2i\rangle$ is an eigenvector. Thus,

$$\mathbf{v} = \begin{bmatrix}1\\ 2i\end{bmatrix} = \begin{bmatrix}1\\ 0\end{bmatrix} + i\begin{bmatrix}0\\ 2\end{bmatrix}.$$

In other words, $\mathbf{v} = \mathbf{v}^1 + i\mathbf{v}^2$, where $\mathbf{v}^1 = \langle 1, 0\rangle$ and $\mathbf{v}^2 = \langle 0, 2\rangle$. So,

$$P = \begin{bmatrix}\mathbf{v}^1 & \mathbf{v}^2\end{bmatrix} = \begin{bmatrix}1 & 0\\ 0 & 2\end{bmatrix}.$$

Therefore, a solution of (13.5.49) is

$$\mathbf{x}(t) = P\mathbf{y}(t) = \begin{bmatrix}1 & 0\\ 0 & 2\end{bmatrix}\begin{bmatrix}a\cos 2t\\ -a\sin 2t\end{bmatrix} = \begin{bmatrix}a\cos 2t\\ -2a\sin 2t\end{bmatrix}.$$

Since $x_1(t) = a\cos 2t$ and $x_2(t) = -2a\sin 2t$, we have

$$\frac{x_1^2(t)}{a^2} + \frac{x_2^2(t)}{4a^2} = 1,$$

which is the equation of an ellipse. So the similarity transformation $\mathbf{x} = P\mathbf{y}$ maps the circle $y_1^2 + y_2^2 = a^2$ in the y_1y_2-phase plane to the ellipse

$$\frac{x_1^2}{a^2} + \frac{x_2^2}{(2a)^2} = 1 \tag{13.5.54}$$

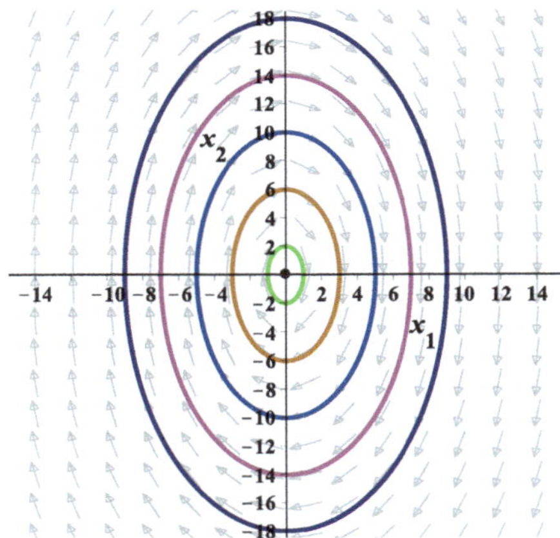

Fig. 13.12 Phase portrait of (13.5.49)

in the x_1x_2-phase plane. Since the vertices of this ellipse are the points $(0, \pm 2a)$, the major axis is on the x_2-axis while $(\pm a, 0)$ are the endpoints of the minor axis on the x_1-axis. That this is an ellipse is hardly surprising since

$$P\mathbf{y} = \begin{bmatrix} 1 & 0 \\ 0 & 2 \end{bmatrix} \begin{bmatrix} y_1 \\ y_2 \end{bmatrix} = \begin{bmatrix} y_1 \\ 2y_2 \end{bmatrix}$$

stretches the circular trajectories vertically away from the y_1-axis by a factor of 2. Figure 13.12 shows five concentric elliptical trajectories through the points $(a, 0)$ for $a = 1, 3, 5, 7, 9$.

Alternative Solution Another way to derive equation (13.5.54) is to convert system (13.5.49) into an equivalent second-order differential equation as follows: Take the derivative of the first equation in the system with respect to t obtaining $x_1'' = x_2'$; then replace x_2' with $-4x_1$, thereby obtaining

$$x_1'' + 4x_1 = 0, \quad \text{where} \quad x_1' = x_2. \tag{13.5.55}$$

Now recall from Chap. 10 that a physical system with a time-dependent attribute that can be modeled by a second-order homogeneous linear equation like this one is known as a *simple harmonic oscillator* (cf. (10.1.3)). Since a general solution of (13.5.55) is

$$x_1(t) = c_1 \cos 2t + c_2 \sin 2t$$

and as

$$x_2(t) = x_1'(t) = -2c_1 \sin 2t + 2c_2 \cos 2t,$$

it follows that the solution of (13.5.49) with $x_1(0) = a$ and $x_2(0) = 0$ is

$$x_1(t) = a\cos 2t, \quad x_2(t) = -2a\sin 2t.$$

By squaring these functions, it becomes apparent that $4x_1^2 + x_2^2 = 4a^2$, which is another form of equation (13.5.54). ♦

Figure 13.12 is a phase portrait of the linear system (13.5.49). But it can also be regarded as a phase portrait of the linear second-order differential equation (13.5.55) if the vertical axis of the phase plane is called the x_1'-axis instead of the x_2-axis. In classical mechanics literature (e.g., see [57, p. 144 footnote]), the vertical axis of the phase plane showing the graphs of trajectories of $x'' + \omega^2 x = 0$ is sometimes chosen to be x'/ω instead of x', in which case the trajectories in this phase plane are concentric circles rather than concentric ellipses. The reason for this is that this choice for the vertical axis is tantamount to the change of variables: $y_1 = x$ and $y_2 = x'/\omega$. With $\omega = 2$, the elliptical trajectories are shrunk vertically toward the y_1-axis by a factor of $1/2$ resulting in trajectories that are circles in the y_1y_2-phase plane.

The phase portraits in Figs. 13.11 and 13.12 consists of concentric circles and ellipses. The eigenvalues of both of the coefficient matrices associated with these phase portraits are the pure imaginary numbers $\pm 2i$. In general, when eigenvalues are purely imaginary, all trajectories, other than the equilibrium point (0, 0), will be either circles or ellipses. Details are provided in the following theorem.

Elliptical and Circular Trajectories

Theorem 13.5.1 *The eigenvalues of a matrix*

$$A = \begin{bmatrix} a_{11} & a_{12} \\ a_{21} & a_{22} \end{bmatrix} \tag{13.5.56}$$

are pure imaginary numbers if and only if $\det A > 0$ *and* $\operatorname{tr} A = 0$. *If a given matrix A has pure imaginary eigenvalues, then a phase portrait of the system* $\mathbf{x}' = A\mathbf{x}$ *consists of concentric ellipses with the equilibrium point* $(0, 0)$ *as their common center, unless* $a_{11} = a_{22} = 0$ *and* $a_{21} = -a_{12}$ *in which case it consists of concentric circles. The equation of each trajectory of the system is*

$$a_{21}x_1^2 - 2a_{11}x_1x_2 - a_{12}x_2^2 = K \tag{13.5.57}$$

for some value of K.

Proof For a proof of the first statement, see Corollary 12.5.30. As for the second one, suppose for the rest of this proof that $\det A > 0$ and $\operatorname{tr} A = 0$ so that the eigenvalues of A are purely imaginary. Then, as

$$\operatorname{tr} A = a_{11} + a_{22} = 0,$$

the characteristic equation $\lambda^2 - (\operatorname{tr} A)\,\lambda + \det A = 0$ simplifies to $\lambda^2 = -\det A$, where

$$\det A = a_{11}a_{22} - a_{12}a_{21} = -a_{11}^2 - a_{12}a_{21} > 0.$$

This implies that a_{12} and a_{21} are nonzero with opposite signs. Defining $\beta := \sqrt{\det A}$, the eigenvalues of A are

$$\lambda = i\beta \text{ and } \overline{\lambda} = -i\beta.$$

Since $a_{22} = -a_{11}$,

$$A = \begin{bmatrix} a_{11} & a_{12} \\ a_{21} & -a_{11} \end{bmatrix}.$$

In order to ascertain what the trajectories of $\mathbf{x}' = A\mathbf{x}$ look like, let us start with the Jordan canonical form of A (see Theorem 13.3.4 (iv)), namely, the matrix

$$J = \begin{bmatrix} 0 & \beta \\ -\beta & 0 \end{bmatrix}.$$

Referring to the paragraph containing the canonical system (13.5.48), the equation of the solution $\mathbf{y}(t) = \langle y_1(t), y_2(t)\rangle$ of $\mathbf{y}' = J\mathbf{y}$ through a given point (a, b) in the y_1y_2-plane satisfies the equation

$$y_1^2 + y_2^2 = a^2 + b^2. \tag{13.5.58}$$

It follows from Theorems 13.3.1 and 13.3.4 that if $\mathbf{y}(t)$ is a solution of $\mathbf{y}' = J\mathbf{y}$, then $P\mathbf{y}(t)$ is a solution of $\mathbf{x}' = A\mathbf{x}$ if $P = \begin{bmatrix}\mathbf{v}^1 & \mathbf{v}^2\end{bmatrix}$, where $\mathbf{v}^1$ and $\mathbf{v}^2$ are linearly independent real vectors and $\mathbf{v}^1 + i\mathbf{v}^2$ is a complex eigenvector of A corresponding to the eigenvalue $\lambda = i\beta$. A complex eigenvector $\mathbf{v} = \mathbf{v}^1 + i\mathbf{v}^2$ is a solution of $(A - i\beta I)\mathbf{v} = \mathbf{0}$, that is, of the matrix equation

$$\begin{bmatrix} a_{11} - i\beta & a_{12} \\ a_{21} & -a_{11} - i\beta \end{bmatrix}\begin{bmatrix} v_1 \\ v_2 \end{bmatrix} = \begin{bmatrix} 0 \\ 0 \end{bmatrix}.$$

It is left as an exercise to verify that $\mathbf{v} = \langle a_{11} + i\beta, a_{21}\rangle$ is a solution. So

$$\mathbf{v} = \begin{bmatrix} a_{11} + i\beta \\ a_{21} \end{bmatrix} = \begin{bmatrix} a_{11} \\ a_{21} \end{bmatrix} + i\begin{bmatrix} \beta \\ 0 \end{bmatrix} = \mathbf{v}^1 + i\mathbf{v}^2,$$

where $\mathbf{v^1} = \langle a_{11}, a_{21}\rangle$ and $\mathbf{v^2} = \langle \beta, 0\rangle$. Thus,

$$P = \begin{bmatrix}\mathbf{v^1} & \mathbf{v^2}\end{bmatrix} = \begin{bmatrix} a_{11} & \beta \\ a_{21} & 0 \end{bmatrix}.$$

Accordingly, if $\mathbf{y}(t) = \langle y_1(t), y_2(t)\rangle$ is a solution of $\mathbf{y}' = J\mathbf{y}$, then

$$\mathbf{x}(t) = P\mathbf{y}(t) = \begin{bmatrix} a_{11} & \beta \\ a_{21} & 0 \end{bmatrix}\begin{bmatrix} y_1(t) \\ y_2(t) \end{bmatrix} = \begin{bmatrix} a_{11}y_1(t) + \beta y_2(t) \\ a_{21}y_1(t) \end{bmatrix}$$

is a solution of $\mathbf{x}' = A\mathbf{x}$. With this we find that the components of the two solutions $\mathbf{x}(t)$ and $\mathbf{y}(t)$ are related by the equations

$$y_1(t) = \frac{1}{a_{21}}x_2(t) \quad \text{and} \quad y_2(t) = \frac{1}{\beta}x_1(t) - \frac{a_{11}}{\beta a_{21}}x_2(t).$$

Expressed in terms of x_1 and x_2, equation (13.5.58) is

$$\left(\frac{x_2}{a_{21}}\right)^2 + \left(\frac{x_1}{\beta} - \frac{a_{11}}{\beta a_{21}}x_2\right)^2 = a^2 + b^2$$

or

$$\frac{x_1^2}{\beta^2} - \frac{2a_{11}}{\beta^2 a_{21}}x_1x_2 + \left(\frac{1}{a_{21}^2} + \frac{a_{11}^2}{\beta^2 a_{21}^2}\right)x_2^2 = a^2 + b^2.$$

Multiplying both sides of the latter equation by $\beta^2 a_{21}^2$ and then replacing $\beta^2 + a_{11}^2$ with $-a_{12}a_{21}$, we have

$$a_{21}^2x_1^2 - 2a_{11}a_{21}x_1x_2 - a_{12}a_{21}x_2^2 = \beta^2 a_{21}^2(a^2 + b^2).$$

Recall that both a_{12} and a_{21} represent nonzero numbers. Dividing both sides of the previous equation by a_{21}, we have

$$a_{21}x_1^2 - 2a_{11}x_1x_2 - a_{12}x_2^2 = k \tag{13.5.59}$$

where $k := \beta^2 a_{21}(a^2 + b^2)$.[3] In sum, the similarity transformation $\mathbf{x} = P\mathbf{y}$ maps the circular trajectory (13.5.58) to the curve that this equation defines. Conversely, this curve can be mapped back to the circular trajectory with the transformation $\mathbf{y} = P^{-1}\mathbf{x}$. Exactly what this curve looks like is addressed in the remainder of this proof.

---

[3] Another derivation of (13.5.59) involving an exact differential equation is the subject of Problem 48.

First let us suppose that $a_{11} = 0$. Then equation (13.5.59) simplifies to

$$a_{21}x_1^2 - a_{12}x_2^2 = k.$$

Multiplying both sides of this equation by -a_{12}, we get

$$-a_{12}a_{21}x_1^2 + a_{12}^2x_2^2 = -a_{12}k.$$

Since $-a_{12}a_{21} = \det A = \beta^2$ and

$$-a_{12}k = -a_{12}a_{21}\beta^2(a^2 + b^2) = \beta^4(a^2 + b^2),$$

equation (13.5.59) becomes

$$\beta^2x_1^2 + a_{12}^2x_2^2 = \beta^4(a^2 + b^2).$$

This is the equation of an ellipse, which is more recognizable when it is written in the form

$$\frac{x_1^2}{c^2} + \frac{x_2^2}{d^2} = 1, \tag{13.5.60}$$

where

$$c^2 = \beta^2(a^2 + b^2) \quad \text{and} \quad d^2 = \frac{\beta^4(a^2 + b^2)}{a_{12}^2}.$$

This ellipse is centered at (0, 0) with its major and minor axes on the x_1 and x_2 axes.

Suppose, in addition to $a_{11} = 0$, that $a_{21} = -a_{12}$. Then

$$a_{12}^2 = -a_{12}a_{21} = \beta^2 \quad \text{and} \quad d^2 = \frac{\beta^4(a^2 + b^2)}{\beta^2} = c^2.$$

Therefore, (13.5.60) simplifies to

$$x_1^2 + x_2^2 = c^2,$$

which is the equation of a circle with center (0, 0) and radius $c = \beta\sqrt{a^2 + b^2}$.

There is still the case $a_{11} \neq 0$ to consider. But before doing so, note that equation (13.5.59) can be written in the form

$$Ax_1^2 + Bx_1x_2 + Cx_2^2 = 1. \tag{13.5.61}$$

where the uppercase letters represent constants. If $B^2 - 4AC < 0$ and $B \neq 0$, then the graph of equation (13.5.61) is a rotated ellipse centered at the origin. The presence of the x_1x_2 term ($B \neq 0$) indicates that the major and minor axes of the ellipse are not on the coordinate axes. This statement can be proved by starting out with the standard form of the equation of an ellipse centered at the origin with its major and minor axes on the coordinate axes and then rotating the coordinate system through an angle θ counterclockwise about the origin (see the problem at the end of Appendix B). The details are left to the interested reader.

Now let us return to equation (13.5.59) and suppose that $a_{11} \neq 0$. Also, suppose $a^2+b^2 \neq 0$ so that $k \neq 0$. Dividing both sides of the equation by k, we get (13.5.61) with $A = a_{21}/k$, $B = -2a_{11}/k$, and $C = -a_{12}/k$. Note $B \neq 0$ since $a_{11} \neq 0$. This together with

$$B^2 - 4AC = -\frac{4}{k^2}\det A < 0$$

implies that the graph of (13.5.59) is a rotated ellipse centered at the origin. ■

Example 13.5.8 Find the solution of the linear system

$$\begin{aligned} x_1' &= 2x_1 - 4x_2 \\ x_2' &= 2x_1 - 2x_2 \end{aligned} \tag{13.5.62}$$

satisfying the initial conditions $x_1(0) = 3$ and $x_2(0) = 1$. Find the equation of the corresponding trajectory. Also, find the equations of the trajectories through the points $(0, 0)$ and $(-4, -6)$.

Solution The coefficient matrix of this linear system is

$$A = \begin{bmatrix} 2 & -4 \\ 2 & -2 \end{bmatrix}.$$

Since $\det A = 4$ and $\operatorname{tr} A = 0$, it follows from Theorem 13.5.1 that the eigenvalues of A are purely imaginary. They are in fact $\lambda = 2i$ and $\overline{\lambda} = -2i$ since the characteristic equation of A is

$$\lambda^2 + 4 = 0.$$

Therefore, by Theorem 13.1.10, the solution of this system satisfying the initial conditions $x_1(0) = 3$ and $x_2(0) = 1$ is

$$\mathbf{x}(t) = e^{At}\mathbf{x}_0,$$

where $\mathbf{x}_0 = \langle 3, 1\rangle$. So now we need to find e^{At}.

In light of Theorem 13.3.4 (iv), we can compute e^{At} by finding a Jordan canonical form of A. To do that, let us find a solution of the equation $(A - 2iI)\mathbf{v} = \mathbf{0}$ so as to have an eigenvector corresponding to the eigenvalue $\lambda = 2i$. It is left as an exercise to verify that $\mathbf{v} = \langle 1 + i, 1 \rangle$ is a solution of

$$\begin{bmatrix} 2 - 2i & -4 \\ 2 & -2 - 2i \end{bmatrix} \begin{bmatrix} v_1 \\ v_2 \end{bmatrix} = \begin{bmatrix} 0 \\ 0 \end{bmatrix}.$$

Separating the real and imaginary parts of $\mathbf{v}$, we have

$$\mathbf{v} = \begin{bmatrix} 1 \\ 1 \end{bmatrix} + i \begin{bmatrix} 1 \\ 0 \end{bmatrix} = \mathbf{v^1} + i\mathbf{v^2},$$

where $\mathbf{v^1} = \langle 1, 1 \rangle$ and $\mathbf{v^1} = \langle 1, 0 \rangle$. Therefore,

$$P = \begin{bmatrix} \mathbf{v^1} & \mathbf{v^2} \end{bmatrix} = \begin{bmatrix} 1 & 1 \\ 1 & 0 \end{bmatrix}.$$

As $\det P = -1$,

$$P^{-1} = \frac{1}{\det P} \begin{bmatrix} 0 & -1 \\ -1 & 1 \end{bmatrix} = \begin{bmatrix} 0 & 1 \\ 1 & -1 \end{bmatrix}.$$

Referring to Theorem 13.3.4 (iv) again, we see that a Jordan canonical form of A is

$$J = \begin{bmatrix} 0 & 2 \\ -2 & 0 \end{bmatrix}.$$

Thus,

$$e^{Jt} = \begin{bmatrix} \cos 2t & \sin 2t \\ -\sin 2t & \cos 2t \end{bmatrix}.$$

Therefore,

$$\begin{aligned} e^{At} = Pe^{Jt}P^{-1} &= \begin{bmatrix} 1 & 1 \\ 1 & 0 \end{bmatrix} \begin{bmatrix} \cos 2t & \sin 2t \\ -\sin 2t & \cos 2t \end{bmatrix} \begin{bmatrix} 0 & 1 \\ 1 & -1 \end{bmatrix} \\ &= \begin{bmatrix} \cos 2t + \sin 2t & -2\sin 2t \\ \sin 2t & \cos 2t - \sin 2t \end{bmatrix}. \end{aligned}$$

And so

$$\mathbf{x}(t) = e^{At}\mathbf{x}_0 = \begin{bmatrix} \cos 2t + \sin 2t & -2\sin 2t \\ \sin 2t & \cos 2t - \sin 2t \end{bmatrix} \begin{bmatrix} 3 \\ 1 \end{bmatrix} = \begin{bmatrix} 3\cos 2t + \sin 2t \\ \cos 2t + 2\sin 2t \end{bmatrix}.$$

Therefore, the solution of (13.5.62) such that $x_1(0) = 3$ and $x_2(0) = 1$ is

$$x_1(t) = 3\cos 2t + \sin 2t \quad \text{and} \quad x_2(t) = \cos 2t + 2\sin 2t.$$

The equation of the trajectory of this solution is

$$x_1^2 - 2x_1x_2 + 2x_2^2 = 5, \tag{13.5.63}$$

which can be found by eliminating the independent variable t. The details are left to the reader.

Rather than finding trajectories by eliminating the independent variable, the work is less tedious if Theorem 13.5.1 is used, which is applicable here since the eigenvalues of the matrix A are purely imaginary. It follows from this theorem that the equation of every trajectory of (13.5.62) is given by (13.5.57), where $a_{21} = 2$, $a_{11} = 2$, and $a_{12} = -4$. That is, the equation is

$$2x_1^2 - 4x_1x_2 + 4x_2^2 = K. \tag{13.5.64}$$

For the trajectory through the point (3, 1), the value of K is

$$K = 2 \cdot 3^2 - 4(3)(1) + 4 \cdot 1^2 = 10.$$

Consequently, the equation of the trajectory through (3, 1) is

$$2x_1^2 - 4x_1x_2 + 4x_2^2 = 10, \tag{13.5.65}$$

which simplifies to (13.5.63). Its graph is the smaller of the two elliptical trajectories depicted in Fig. 13.13.

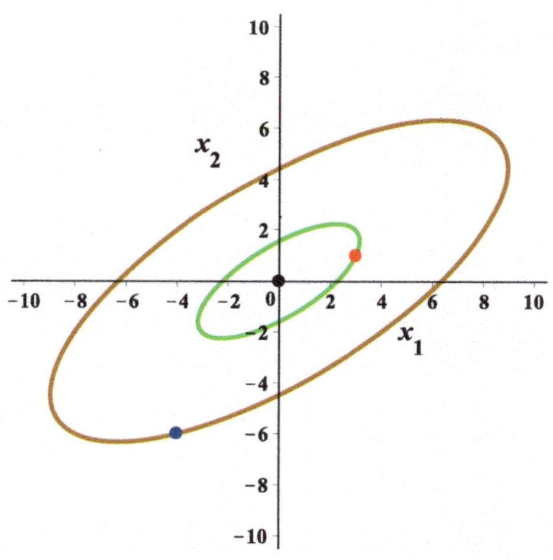

Fig. 13.13 Three trajectories of (13.5.62)

Setting $x_1 = 0$ and $x_2 = 0$ in (13.5.64), we get $K = 0$. So the equation now becomes

$$2x_1^2 - 4x_1x_2 + 4x_2^2 = 0.$$

Since $x_1 = 0$, $x_2 = 0$ is its only solution, the trajectory containing $(0, 0)$ consists only of that point, which must be the case since it is the equilibrium point of the system. This point trajectory is depicted by the small black dot at the origin in Fig. 13.13. The larger elliptical trajectory shown in the figure is the trajectory through the point $(-4, -6)$, which is the graph of

$$x_1^2 - 2x_1x_2 + 2x_2^2 = 40. \qquad \blacklozenge$$

13.6 Nonautonomous Linear Systems

Now that we have several methods at our disposal for obtaining the solutions of the autonomous vector differential equation $\mathbf{x}' = A\mathbf{x}$, where A is a given constant matrix, let us explain how these solutions can be employed to find solutions of an associated nonautonomous vector differential equation

$$\mathbf{x}' = A\mathbf{x} + \mathbf{b}(t), \tag{13.6.1}$$

where $\mathbf{b}(t)$ is a given vector function of t. Keep in mind that this function does not involve the dependent variable $\mathbf{x}$. Adding the term $\mathbf{b}(t)$ to the right-hand side of the equation $\mathbf{x}' = A\mathbf{x}$ changes it from a homogeneous equation to a nonhomogeneous equation, as well as from an autonomous equation to a nonautonomous equation.

In order to fully understand the ensuing discussion, let us start off by reviewing how to find the solutions of the scalar counterpart of (13.6.1), namely, the nonhomogeneous linear equation $x' = ax + b(t)$, where a is a given constant and $b(t)$ a given scalar function of t. Recall from Definition 5.5.1 that the *principal solution* of the linear homogeneous equation

$$\frac{dx}{dt} = ax \quad (a,\ \text{a constant}) \tag{13.6.2}$$

at $t = t_0$ is

$$z(t, t_0) = e^{a(t-t_0)} \tag{13.6.3}$$

for $-\infty < t < \infty$. Also, recall from Theorem 5.5.2 that for a given $x_0 \in \mathbb{R}$, the solution of the initial value problem

$$\frac{dx}{dt} = ax + b(t), \quad x(t_0) = x_0 \tag{13.6.4}$$

can be expressed in terms of $z(t, t_0)$ via the *variation of parameters formula*

$$x(t) = z(t, t_0)x_0 + \int_{t_0}^{t} z(t, s)b(s)\, ds = e^{a(t-t_0)}x_0 + \int_{t_0}^{t} e^{a(t-s)}b(s)\, ds \qquad (13.6.5)$$

for $-\infty < t < \infty$.

As one might suspect, there is also a counterpart to formula (13.6.5) for the nonautonomous vector differential equation (13.6.1). As we will see, it too has an integral term like the one in (13.6.5); but, of course, the integral involves vector functions rather than scalar functions. So let us first define what is meant by the integral of a vector function $\mathbf{f}(t) = \langle f_1(t), f_2(t)\rangle$. If both components of $\mathbf{f}(t)$ are integrable on an interval $[a, b]$, then the integral of $\mathbf{f}(t)$ is defined to be

$$\int_a^b \mathbf{f}(t)\, dt := \Big\langle \int_a^b f_1(t)\, dt, \int_a^b f_2(t)\, dt \Big\rangle.$$

It is left as an exercise to verify that integration of vector functions is a linear operation. That is, if $\mathbf{f}_1(t)$ and $\mathbf{f}_2(t)$ are integrable vector functions, then so is the linear combination $c_1\mathbf{f}_1(t) + c_2\mathbf{f}_2(t)$ and

$$\int_a^b \big[c_1\mathbf{f}_1(t)\, dt + c_2\mathbf{f}_2(t)\big]\, dt = c_1 \int_a^b \mathbf{f}_1(t)\, dt + c_2 \int_a^b \mathbf{f}_2(t)\, dt$$

for arbitrary $c_1, c_2 \in \mathbb{R}$.

The vector version of the variation of parameters formula and its proof that is given in the next theorem is an adaptation of Theorem 5.5.2 for scalar linear equations. The following three results will be needed in various parts of the proof.

Matrix and Vector Calculus Results

Lemma 13.6.1 *Let t_0 be a fixed point in $\mathbb{R}$. Let $M(t)$ be a 2×2 matrix function and $\mathbf{v}(t)$ a 2×1 vector function.*

(i) *If $M(t)$ and $\mathbf{v}(t)$ are differentiable on $\mathbb{R}$, then*

$$\frac{d}{dt}\,[M(t)\mathbf{v}(t)] = M(t)\mathbf{v}'(t) + M'(t)\mathbf{v}(t).$$

(ii) *If $M(t) \equiv B$, where B is a constant matrix, and $\mathbf{v}(t)$ is continuous on $\mathbb{R}$, then*

$$\int_{t_0}^{t} B\mathbf{v}(s)\, ds = B \int_{t_0}^{t} \mathbf{v}(s)\, ds.$$

(continued)

(iii) *If* $\mathbf{v}(t)$ *is continuous on* $\mathbb{R}$, *then*

$$\frac{d}{dt}\int_{t_0}^{t} \mathbf{v}(s)\,ds = \mathbf{v}(t).$$

Proof The proofs of statements (i) and (iii) are left to Problems 1, 49, and 50. As for (ii), with b_{ij} denoting the entries of the constant matrix B, the left-hand side of the formula is

$$\int_{t_0}^{t} B\mathbf{v}(s)\,ds = \int_{t_0}^{t} \begin{bmatrix} b_{11} & b_{12} \\ b_{21} & b_{22} \end{bmatrix} \begin{bmatrix} v_1(s) \\ v_2(s) \end{bmatrix} ds = \int_{t_0}^{t} \begin{bmatrix} b_{11}v_1(s) + b_{12}v_2(s) \\ b_{21}v_1(s) + b_{22}v_2(s) \end{bmatrix} ds$$

$$= \begin{bmatrix} \int_{t_0}^{t} \left(\sum_{j=1}^{2} b_{1j}v_j(s) \right) ds \\ \int_{t_0}^{t} \left(\sum_{j=1}^{2} b_{2j}v_j(s) \right) ds \end{bmatrix} = \begin{bmatrix} \sum_{j=1}^{2} \int_{t_0}^{t} b_{1j}v_j(s)\,ds \\ \sum_{j=1}^{2} \int_{t_0}^{t} b_{2j}v_j(s)\,ds \end{bmatrix}.$$

This is equal to the right-hand side of the formula as

$$B\int_{t_0}^{t} \mathbf{v}(s)\,ds = \begin{bmatrix} b_{11} & b_{12} \\ b_{21} & b_{22} \end{bmatrix} \begin{bmatrix} \int_{t_0}^{t} v_1(s)\,ds \\ \int_{t_0}^{t} v_2(s)\,ds \end{bmatrix} = \begin{bmatrix} b_{11}\int_{t_0}^{t} v_1(s)\,ds + b_{12}\int_{t_0}^{t} v_2(s)\,ds \\ b_{21}\int_{t_0}^{t} v_1(s)\,ds + b_{22}\int_{t_0}^{t} v_2(s)\,ds \end{bmatrix}$$

$$= \begin{bmatrix} \sum_{j=1}^{2} b_{1j}\int_{t_0}^{t} v_j(s)\,ds \\ \sum_{j=1}^{2} b_{2j}\int_{t_0}^{t} v_j(s)\,ds \end{bmatrix},$$

which proves statement (ii). ∎

Variation of Parameters Formula

Theorem 13.6.2 *Let A be a constant* 2×2 *matrix and* $\mathbf{b}(t)$ *a continuous* 2×1 *vector function on* $(-\infty, \infty)$. *Let* $t_0 \in \mathbb{R}$ *and* $\mathbf{x}_0 \in \mathbb{R}^2$. *The solution of the initial value problem*

$$\mathbf{x}' = A\mathbf{x} + \mathbf{b}(t), \quad \mathbf{x}(t_0) = \mathbf{x}_0 \tag{13.6.6}$$

is unique and given by the variation of parameters formula

$$\mathbf{x}(t) = e^{A(t-t_0)}\mathbf{x}_0 + \int_{t_0}^{t} e^{A(t-s)}\mathbf{b}(s)\,ds \tag{13.6.7}$$

(continued)

for all $t \in \mathbb{R}$. Another form of this formula is

$$\mathbf{x}(t) = Z(t, t_0)\mathbf{x}_0 + \int_{t_0}^{t} Z(t, s)\mathbf{b}(s)\, ds, \tag{13.6.8}$$

where $Z(t, s)$ is the principal matrix solution of the associated homogeneous equation $\mathbf{x}' = A\mathbf{x}$ at $t = s$.

Remarks The variation of parameters formula (as with most results in this chapter and the previous one) is actually valid for matrices A of order n and $n \times 1$ vector functions $\mathbf{b}(t)$, where n can be any natural number; however, in keeping with the mathematical level of this textbook, we take $n = 2$. We will only have to prove (13.6.7) because (13.6.8) is just another form of this formula since $e^{A(t-s)} = Z(t, s)$ (see Problem 7).

Proof The proof of Theorem 5.5.2 and in particular the solution of (13.6.4) rewritten in terms of the principal solution $z(t, t_0)$ (see (5.5.13)) suggests that there may be a solution of (13.6.6) of the form

$$\mathbf{x}(t) = Z(t)\mathbf{v}(t) = e^{At}\mathbf{v}(t), \tag{13.6.9}$$

where $\mathbf{v}(t)$ is a differentiable vector function of t. To ascertain if this truly is the case, let us substitute (13.6.9) and its derivative into the vector differential equation (13.6.1) to see if such a function $\mathbf{v}(t)$ exists and what it would have to be in order for (13.6.9) to be a solution. Making these substitutions, we have

$$\frac{d}{dt}\left[e^{At}\mathbf{v}(t)\right] = Ae^{At}\mathbf{v}(t) + \mathbf{b}(t).$$

Carrying out the differentiation (see Lemma 13.6.1 (i) and Theorem 13.2.1 (vii)), we obtain

$$e^{At}\mathbf{v}'(t) + Ae^{At}\mathbf{v}(t) = Ae^{At}\mathbf{v}(t) + \mathbf{b}(t).$$

Thus, $\mathbf{x}(t) = e^{At}\mathbf{v}(t)$ is a solution of (13.6.6) if a differentiable function $\mathbf{v}(t)$ exists satisfying the following differential equation and initial condition:

$$e^{At}\mathbf{v}'(t) = \mathbf{b}(t), \quad e^{At_0}\mathbf{v}(t_0) = \mathbf{x}_0.$$

Since e^{At} is invertible for all $t \in \mathbb{R}$ (see Theorem 13.2.1 (v)), we can rewrite this as

$$\mathbf{v}'(t) = e^{-At}\mathbf{b}(t), \quad \mathbf{v}(t_0) = e^{-At_0}\mathbf{x}_0. \tag{13.6.10}$$

The integration

$$\int_{t_0}^{t} \mathbf{v}'(s)\,ds = \int_{t_0}^{t} e^{-As}\mathbf{b}(s)\,ds$$

yields

$$\mathbf{v}(t) = e^{-At_0}\mathbf{x}_0 + \int_{t_0}^{t} e^{-As}\mathbf{b}(s)\,ds,$$

which by (13.6.9) is

$$e^{-At}\mathbf{x}(t) = e^{-At_0}\mathbf{x}_0 + \int_{t_0}^{t} e^{-As}\mathbf{b}(s)\,ds.$$

Multiplying both sides of the equation on the left by e^{At} (see Theorem 13.2.1 (ii)), we obtain the solution

$$\mathbf{x}(t) = e^{A(t-t_0)}\mathbf{x}_0 + e^{At}\int_{t_0}^{t} e^{-As}\mathbf{b}(s)\,ds. \tag{13.6.11}$$

It follows from Lemma 13.6.1 (ii) that this can also be written as (13.6.7). Furthermore, as stated earlier in *Remarks*, (13.6.7) and (13.6.8) are merely two different ways of expressing the same formula.

Before proving that (13.6.11) is the only solution of (13.6.6), let us verify that it is a solution. First, it does satisfy the initial condition since (see Theorem 13.2.1 (i))

$$\mathbf{x}(t_0) = e^{A0}\mathbf{x}_0 + e^{At_0}\int_{t_0}^{t_0} e^{-As}\mathbf{b}(s)\,ds = I\mathbf{x}_0 + e^{At_0}\begin{bmatrix}0\\0\end{bmatrix} = \mathbf{x}_0.$$

Second, let us verify that it solves the differential equation. Differentiating (13.6.11) and using the appropriate results listed in Theorem 13.2.1 and Lemma 13.6.1, we have

$$\begin{aligned}
\mathbf{x}'(t) &= \frac{d}{dt}\left[e^{A(t-t_0)}\mathbf{x}_0 + e^{At}\int_{t_0}^{t} e^{-As}\mathbf{b}(s)\,ds\right]\\
&= \frac{d}{dt}e^{A(t-t_0)}\mathbf{x}_0 + \frac{d}{dt}\left[e^{At}\int_{t_0}^{t} e^{-As}\mathbf{b}(s)\,ds\right]\\
&= Ae^{A(t-t_0)}\mathbf{x}_0 + e^{At}\frac{d}{dt}\int_{t_0}^{t} e^{-As}\mathbf{b}(s)\,ds + Ae^{At}\int_{t_0}^{t} e^{-As}\mathbf{b}(s)\,ds\\
&= A\left[e^{A(t-t_0)}\mathbf{x}_0 + e^{At}\int_{t_0}^{t} e^{-As}\mathbf{b}(s)\,ds\right] + e^{At}\frac{d}{dt}\int_{t_0}^{t} e^{-As}\mathbf{b}(s)\,ds
\end{aligned}$$

$$= A\mathbf{x}(t) + e^{At}e^{-At}\mathbf{b}(t) = A\mathbf{x}(t) + \mathbf{b}(t).$$

This confirms that the function given by (13.6.11), which is equivalent to (13.6.7) and (13.6.8)), solves the differential equation in (13.6.6).

Now that we have established that (13.6.7) is a solution of the initial value problem (13.6.6), the only thing left to do is to show that it is the only solution. To that end, let $\boldsymbol{\psi}(t) = \mathbf{x}(t) - \tilde{\mathbf{x}}(t)$, where $\mathbf{x}(t)$ and $\tilde{\mathbf{x}}(t)$ are both solutions of (13.6.6). Differentiating $\boldsymbol{\psi}(t)$, we get

$$\begin{aligned}\boldsymbol{\psi}'(t) &= \mathbf{x}'(t) - \tilde{\mathbf{x}}'(t) = A\mathbf{x}(t) + \mathbf{b}(t) - A\tilde{\mathbf{x}}'(t) - \mathbf{b}(t)\\ &= A\left[\mathbf{x}(t) - \tilde{\mathbf{x}}'(t)\right] = A\boldsymbol{\psi}(t),\end{aligned}$$

where $\boldsymbol{\psi}(t_0) = \mathbf{x}(t_0) - \tilde{\mathbf{x}}(t_0) = \mathbf{0}$ since $\tilde{\mathbf{x}}(t_0) = \mathbf{x}_0 = \mathbf{x}(t_0)$. By Theorem 13.1.10, the unique solution of

$$\boldsymbol{\psi}'(t) = A\boldsymbol{\psi}(t), \quad \boldsymbol{\psi}(t_0) = \mathbf{0}$$

is $\boldsymbol{\psi}(t) \equiv \mathbf{0}$. Therefore, $\tilde{\mathbf{x}}(t) \equiv \mathbf{x}(t)$. That is, the only solution of the initial value problem (13.6.6) is (13.6.7). ■

Example 13.6.1 Solve the initial value problem

$$\begin{aligned} x_1' &= 4x_1 - 5x_2 + 2t, \quad & x_1(0) &= 1\\ x_2' &= 2x_1 - 2x_2, & x_2(0) &= 0. \end{aligned} \tag{13.6.12}$$

Solution In order to compute the solution of this initial value problem by means of the variation of parameters formula, we must first compute the matrix exponential function e^{At} corresponding to the coefficient matrix

$$A = \begin{bmatrix} 4 & -5 \\ 2 & -2 \end{bmatrix}.$$

Since e^{At} is defined to be the principal matrix solution $Z(t)$ of $\mathbf{x}' = A\mathbf{x}$, we have the option of either computing $Z(t)$ using Theorem 13.1.6 or computing the Jordan canonical matrix J corresponding to A and then obtaining e^{At} with Theorem 13.3.4. Let us opt for the latter in order to gain more experience dealing with Jordan matrices.

The characteristic equation of A is

$$\det(A - \lambda I) = \begin{vmatrix} 4-\lambda & -5 \\ 2 & -2-\lambda \end{vmatrix} = \lambda^2 - 2\lambda + 2 = 0,$$

which has the solutions

$$\lambda = 1 + i \quad \text{and} \quad \overline{\lambda} = 1 - i.$$

It then follows from Theorem 13.3.4 that the Jordan canonical form of A is the matrix

$$J = \begin{bmatrix} 1 & 1 \\ -1 & 1 \end{bmatrix}$$

and that

$$e^{Jt} = e^t \begin{bmatrix} \cos t & \sin t \\ -\sin t & \cos t \end{bmatrix}.$$

Moreover, the theorem states that if $\mathbf{v^1} + i\mathbf{v^2}$ is an eigenvector corresponding to the eigenvalue $\lambda = 1+i$, where $\mathbf{v^1}$ and $\mathbf{v^2}$ are real vectors, then the matrix $P = \begin{bmatrix} \mathbf{v^1} & \mathbf{v^2} \end{bmatrix}$ relates A and J by $A = PJP^{-1}$.

Since the eigenvectors of A corresponding to $\lambda = 1 + i$ are vectors $\mathbf{v} \neq \mathbf{0}$ such that $(A - (1 + i)I)\mathbf{v} = \mathbf{0}$, let us find a nonzero solution of

$$\begin{bmatrix} 3 - i & -5 \\ 2 & -3 - i \end{bmatrix} \begin{bmatrix} v_1 \\ v_2 \end{bmatrix} = \begin{bmatrix} 0 \\ 0 \end{bmatrix},$$

which is the matrix form of the linear system

$$\begin{aligned} (3 - i)v_1 - 5v_2 &= 0 \\ 2v_1 + (-3 - i)v_2 &= 0. \end{aligned}$$

Setting $v_2 = 1$, we find that $v_1 = \frac{3}{2} + \frac{1}{2}i$ satisfies both equations of the system. As a result, an eigenvector corresponding to $\lambda = 1 + i$ is

$$\mathbf{v} = \begin{bmatrix} v_1 \\ v_2 \end{bmatrix} = \begin{bmatrix} \frac{3}{2} + \frac{1}{2}i \\ 1 \end{bmatrix}.$$

Separating the real and imaginary parts of $\mathbf{v}$, we get

$$\mathbf{v} = \begin{bmatrix} \frac{3}{2} \\ 1 \end{bmatrix} + i \begin{bmatrix} \frac{1}{2} \\ 0 \end{bmatrix} = \mathbf{v^1} + i\mathbf{v^2}$$

where

$$\mathbf{v^1} = \begin{bmatrix} \frac{3}{2} \\ 1 \end{bmatrix} \quad \text{and} \quad \mathbf{v^2} = \begin{bmatrix} \frac{1}{2} \\ 0 \end{bmatrix}.$$

Thus,

$$P = \begin{bmatrix} \mathbf{v}^1 & \mathbf{v}^2 \end{bmatrix} = \begin{bmatrix} \frac{3}{2} & \frac{1}{2} \\ 1 & 0 \end{bmatrix} \quad \text{and} \quad P^{-1} = \frac{1}{\det P} \begin{bmatrix} 0 & -\frac{1}{2} \\ -1 & \frac{3}{2} \end{bmatrix} = \begin{bmatrix} 0 & 1 \\ 2 & -3 \end{bmatrix}.$$

By Theorem 13.3.4,

$$e^{At} = Pe^{Jt}P^{-1} = \begin{bmatrix} \frac{3}{2} & \frac{1}{2} \\ 1 & 0 \end{bmatrix} \begin{bmatrix} e^t \cos t & e^t \sin t \\ -e^t \sin t & e^t \cos t \end{bmatrix} \begin{bmatrix} 0 & 1 \\ 2 & -3 \end{bmatrix},$$

which yields

$$e^{At} = \begin{bmatrix} e^t \cos t + 3e^t \sin t & -5e^t \sin t \\ 2e^t \sin t & e^t \cos t - 3e^t \sin t \end{bmatrix}.$$

Now let us employ the variation of parameters formula (13.6.7) to find the solution of the initial value problem (13.6.12). Since $t_0 = 0$, $\mathbf{x}_0 = \langle x_1(0), x_2(0) \rangle = \langle 1, 0 \rangle$, and $\mathbf{b}(t) = \langle 2t, 0 \rangle$, the solution is

$$\mathbf{x}(t) = e^{At}\mathbf{x}_0 + \int_0^t e^{A(t-s)}\mathbf{b}(s)\, ds = e^{At} \begin{bmatrix} 1 \\ 0 \end{bmatrix} + \int_0^t e^{A(t-s)} \begin{bmatrix} 2s \\ 0 \end{bmatrix} ds,$$

where

$$e^{At}\mathbf{x}_0 = e^{At} \begin{bmatrix} 1 \\ 0 \end{bmatrix} = \begin{bmatrix} e^t \cos t + 3e^t \sin t \\ 2e^t \sin t \end{bmatrix}$$

and

$$e^{A(t-s)}\mathbf{b}(s) = e^{A(t-s)} \begin{bmatrix} 2s \\ 0 \end{bmatrix} = 2s \begin{bmatrix} e^{t-s} \cos (t-s) + 3e^{t-s} \sin (t-s) \\ 2e^{t-s} \sin (t-s) \end{bmatrix}$$

$$= 2e^t \begin{bmatrix} se^{-s} \cos (t-s) + 3se^{-s} \sin (t-s) \\ 2se^{-s} \sin (t-s) \end{bmatrix}.$$

The integrals of both components of $e^{A(t-s)}\mathbf{b}(s)$ from $s = 0$ to $s = t$ are

$$2e^t \int_0^t \left[se^{-s} \cos (t-s) + 3se^{-s} \sin (t-s) \right] ds = e^t \sin t - 3e^t \cos t + 2t + 3$$

and

$$2e^t \int_0^t 2se^{-s} \sin (t-s)\, ds = -2e^t \cos t + 2t + 2.$$

As a result,

$$\mathbf{x}(t) = \begin{bmatrix} e^t \cos t + 3e^t \sin t + e^t \sin t - 3e^t \cos t + 2t + 3 \\ 2e^t \sin t - 2e^t \cos t + 2t + 2 \end{bmatrix}$$

$$= \begin{bmatrix} 4e^t \sin t - 2e^t \cos t + 2t + 3 \\ 2e^t \sin t - 2e^t \cos t + 2t + 2 \end{bmatrix}.$$

Therefore, the solution of (13.6.12) is

$$x_1(t) = 4e^t \sin t - 2e^t \cos t + 2t + 3, \quad x_2(t) = 2e^t \sin t - 2e^t \cos t + 2t + 2. \ \blacklozenge$$

Example 13.6.2 Solve the initial value problem

$$\begin{aligned} x_1' &= 3x_1 + 2x_2 - 6t, & x_1(0) &= -1/3 \\ x_2' &= -3x_1 - 4x_2 - 2e^{2t}, & x_2(0) &= 1/10. \end{aligned} \tag{13.6.13}$$

Solution Written as a vector differential equation, the system is

$$\mathbf{x}' = A\mathbf{x} + \mathbf{b}(t),$$

where

$$A = \begin{bmatrix} 3 & 2 \\ -3 & -4 \end{bmatrix} \quad \text{and} \quad \mathbf{b}(t) = \begin{bmatrix} -6t \\ -2e^{2t} \end{bmatrix}.$$

From the characteristic polynomial

$$\det(A - \lambda I) = \begin{vmatrix} 3 - \lambda & 2 \\ -3 & -4 - \lambda \end{vmatrix} = \lambda^2 + \lambda - 6,$$

we obtain the characteristic equation

$$(\lambda + 3)(\lambda - 2) = 0,$$

from which it follows that the eigenvalues of A are $\lambda_1 = -3$ and $\lambda_2 = 2$.

We could begin as in the previous example by first finding the Jordan canonical matrix J corresponding to A in order to compute $Z(t) = e^{At}$. However, let us use Theorem 13.1.6 instead for the purpose of reviewing this theorem and Definition 13.2.1 and to have an alternative to using Jordan and transformation matrices. Setting $t_0 = 0$ in (13.1.8), we have

$$Z(t) = X(t)X^{-1}(0),$$

where $X(t)$ is any fundamental matrix solution of the corresponding homogeneous equation $\mathbf{x}' = A\mathbf{x}$.

An eigenvector corresponding to $\lambda_1 = -3$ is any nonzero vector $\mathbf{u} = \langle u_1, u_2 \rangle$ such that $A\mathbf{u} = \lambda_1 \mathbf{u}$, which we can rewrite as

$$(A + 3I)\mathbf{u} = \begin{bmatrix} 6 & 2 \\ -3 & -1 \end{bmatrix} \begin{bmatrix} u_1 \\ u_2 \end{bmatrix} = \begin{bmatrix} 0 \\ 0 \end{bmatrix}.$$

Thus, $u_2 = -3u_1$. Choosing $u_1 = 1$, we obtain the eigenvector $\mathbf{u} = \langle 1, -3 \rangle$. It follows that

$$\mathbf{x}^1(t) = e^{-3t} \begin{bmatrix} 1 \\ -3 \end{bmatrix} = \begin{bmatrix} e^{-3t} \\ -3e^{-3t} \end{bmatrix}$$

is a solution of $\mathbf{x}' = A\mathbf{x}$.

As for the eigenvalue $\lambda_2 = 2$, a corresponding eigenvector is any nonzero vector $\mathbf{v} = \langle v_1, v_2 \rangle$ that satisfies $(A - 2I)\mathbf{v} = \mathbf{0}$, that is, the matrix equation

$$\begin{bmatrix} 1 & 2 \\ -3 & -6 \end{bmatrix} \begin{bmatrix} v_1 \\ v_2 \end{bmatrix} = \begin{bmatrix} 0 \\ 0 \end{bmatrix}.$$

Consequently, $v_1 = -2v_2$. Choosing $v_2 = -1$, we obtain the eigenvector $\mathbf{v} = \langle 2, -1 \rangle$. And so another solution of $\mathbf{x}' = A\mathbf{x}$ is

$$\mathbf{x}^2(t) = e^{2t} \begin{bmatrix} 2 \\ -1 \end{bmatrix} = \begin{bmatrix} 2e^{2t} \\ -e^{2t} \end{bmatrix}.$$

Since the solutions $\mathbf{x}^1(t)$ and $\mathbf{x}^2(t)$ are linearly independent, a fundamental matrix solution of $\mathbf{x}' = A\mathbf{x}$ is

$$X(t) = \left[\mathbf{x^1(t)} \;\; \mathbf{x^2(t)}\right] = \begin{bmatrix} e^{-3t} & 2e^{2t} \\ -3e^{-3t} & -e^{2t} \end{bmatrix}.$$

From this and

$$X(0) = \begin{bmatrix} 1 & 2 \\ -3 & -1 \end{bmatrix}, \quad X^{-1}(0) = \frac{1}{5} \begin{bmatrix} -1 & -2 \\ 3 & 1 \end{bmatrix},$$

it follows that

$$Z(t) = X(t)X^{-1}(0) = \frac{1}{5} \begin{bmatrix} e^{-3t} & 2e^{2t} \\ -3e^{-3t} & -e^{2t} \end{bmatrix} \begin{bmatrix} -1 & -2 \\ 3 & 1 \end{bmatrix}.$$

Thus,

$$Z(t) = \begin{bmatrix} -\frac{1}{5}e^{-3t} + \frac{6}{5}e^{2t} & -\frac{2}{5}e^{-3t} + \frac{2}{5}e^{2t} \\ \frac{3}{5}e^{-3t} - \frac{3}{5}e^{2t} & \frac{6}{5}e^{-3t} - \frac{1}{5}e^{2t} \end{bmatrix}.$$

The variation of parameters formula (13.6.8) with $t_0 = 0$, $\mathbf{x}(0) = \langle -1/3, 1/10 \rangle$, and $\mathbf{b}(t) = \langle -6t, -2e^{2t} \rangle$ yields the following solution of (13.6.13):

$$\mathbf{x}(t) = Z(t, t_0)\mathbf{x}_0 + \int_{t_0}^{t} Z(t, s)\mathbf{b}(s)\, ds = Z(t) \begin{bmatrix} -\frac{1}{3} \\ \frac{1}{10} \end{bmatrix} + \int_0^t Z(t-s) \begin{bmatrix} -6s \\ -2e^{2s} \end{bmatrix} ds$$

since $Z(t, t_0) = Z(t, 0) = Z(t)$ and $Z(t, s) = Z(t - s)$ (see Theorem 13.1.9). Multiplying out the first term and the integrand in the second term, we get

$$Z(t) \begin{bmatrix} -\frac{1}{3} \\ \frac{1}{10} \end{bmatrix} = \begin{bmatrix} -\frac{1}{5}e^{-3t} + \frac{6}{5}e^{2t} & -\frac{2}{5}e^{-3t} + \frac{2}{5}e^{2t} \\ \frac{3}{5}e^{-3t} - \frac{3}{5}e^{2t} & \frac{6}{5}e^{-3t} - \frac{1}{5}e^{2t} \end{bmatrix} \begin{bmatrix} -\frac{1}{3} \\ \frac{1}{10} \end{bmatrix} = \begin{bmatrix} \frac{2}{75}e^{-3t} - \frac{9}{25}e^{2t} \\ -\frac{2}{25}e^{-3t} + \frac{9}{50}e^{2t} \end{bmatrix}$$

and

$$Z(t-s) \begin{bmatrix} -6s \\ -2e^{2s} \end{bmatrix} = \begin{bmatrix} -\frac{1}{5}e^{-3(t-s)} + \frac{6}{5}e^{2(t-s)} & -\frac{2}{5}e^{-3(t-s)} + \frac{2}{5}e^{2(t-s)} \\ \frac{3}{5}e^{-3(t-s)} - \frac{3}{5}e^{2(t-s)} & \frac{6}{5}e^{-3(t-s)} - \frac{1}{5}e^{2(t-s)} \end{bmatrix} \begin{bmatrix} -6s \\ -2e^{2s} \end{bmatrix}$$

$$= \begin{bmatrix} \frac{6}{5}se^{-3(t-s)} - \frac{36}{5}se^{2(t-s)} + \frac{4}{5}e^{-3t+5s} - \frac{4}{5}e^{2t} \\ -\frac{18}{5}se^{-3(t-s)} + \frac{18}{5}se^{2(t-s)} - \frac{12}{5}e^{-3t+5s} + \frac{2}{5}e^{2t} \end{bmatrix}.$$

Integrating the components of the integrand from $s = 0$ to $s = t$, we obtain

$$\int_0^t \left(\tfrac{6}{5}se^{-3(t-s)} - \tfrac{36}{5}se^{2(t-s)} + \tfrac{4}{5}e^{-3t+5s} - \tfrac{4}{5}e^{2t}\right) ds$$
$$= -\tfrac{2}{75}e^{-3t} - \tfrac{41}{25}e^{2t} - \tfrac{4}{5}te^{2t} + 4t + \tfrac{5}{3}$$

and

$$\int_0^t \left(-\tfrac{18}{5}se^{-3(t-s)} + \tfrac{18}{5}se^{2(t-s)} - \tfrac{12}{5}e^{-3t+5s} + \tfrac{2}{5}e^{2t}\right) ds$$
$$= \tfrac{2}{25}e^{-3t} + \tfrac{21}{50}e^{2t} + \tfrac{2}{5}te^{2t} - 3t - \tfrac{1}{2}.$$

Thus,

$$\mathbf{x}(t) = \begin{bmatrix} \frac{2}{75}e^{-3t} - \frac{9}{25}e^{2t} - \frac{2}{75}e^{-3t} - \frac{41}{25}e^{2t} - \frac{4}{5}te^{2t} + 4t + \frac{5}{3} \\ -\frac{2}{25}e^{-3t} + \frac{9}{50}e^{2t} + \frac{2}{25}e^{-3t} + \frac{21}{50}e^{2t} + \frac{2}{5}te^{2t} - 3t - \frac{1}{2} \end{bmatrix}$$

$$= \begin{bmatrix} -2e^{2t} - \frac{4}{5}te^{2t} + 4t + \frac{5}{3} \\ \frac{3}{5}e^{2t} + \frac{2}{5}te^{2t} - 3t - \frac{1}{2} \end{bmatrix}.$$

Therefore, the solution of the initial value problem (13.6.13) is

$$x_1(t) = -2e^{2t} - \frac{4}{5}te^{2t} + 4t + \frac{5}{3}, \quad x_2(t) = \frac{3}{5}e^{2t} + \frac{2}{5}te^{2t} - 3t - \frac{1}{2}. \quad \blacklozenge$$

The previous two examples reveal that the calculations involved employing the variation of parameters formula to solve the initial value problem (13.6.6) is tedious and time-consuming. However it is less so if the coefficient matrix A happens to be a diagonal matrix. But even if it is not, the number of calculations can be reduced a bit if the Jordan canonical form of A is a diagonal matrix or an upper triangular matrix as in (i), (ii), and (iii) in Theorem 13.3.3. When this is the case, we can use the similarity transformation $\mathbf{x} = P\mathbf{y}$, where P is a transformation matrix as spelled out in detail in Theorem 13.3.4. Then the nonhomogeneous equation $\mathbf{x}' = A\mathbf{x}+\mathbf{b}(t)$ in (13.6.6) transforms to

$$P\mathbf{y}' = AP\mathbf{y} + \mathbf{b}(t),$$

where $\mathbf{y}(t_0) = P^{-1}\mathbf{x}(t_0)$. Thus, $\mathbf{y}_0 = P^{-1}\mathbf{x}_0$. Multiplying both sides of this equation on the left by P^{-1}, we have

$$\mathbf{y}' = P^{-1}AP\mathbf{y} + P^{-1}\mathbf{b}(t).$$

Let J denote the matrix $P^{-1}AP$ since it is a Jordan canonical form of A. As a result, we end up with the equation

$$\mathbf{y}' = J\mathbf{y} + P^{-1}\mathbf{b}(t). \tag{13.6.14}$$

At this juncture, we could employ the variation of parameters formula

$$\mathbf{y}(t) = e^{J(t-t_0)}\mathbf{y}_0 + \int_{t_0}^{t} e^{J(t-s)}P^{-1}\mathbf{b}(s)\,ds, \tag{13.6.15}$$

from which we could then obtain the solution $\mathbf{x}(t) = P\mathbf{y}(t)$ of (13.6.6). Even though the first term of this transformed formula is simpler than the first term of the original variation of parameters formula (13.6.7), carrying out the integrations in the second term are still laborious. However, we can avoid this by solving the differential equation (13.6.14) directly, which we illustrate next by reworking Example 13.6.2.

Example 13.6.3 Solve the initial value problem (13.6.13) in Example 13.6.2 again but without using either version of the variation of parameters formula in Theorem 13.6.2. Instead use an appropriate similarity transformation $\mathbf{x} = P\mathbf{y}$ to transform the vector differential equation in that example into equation (13.6.14).

Then find the solution $\mathbf{y}(t)$ of this equation satisfying the initial condition $\mathbf{y}(0) = P^{-1}\mathbf{x}(0)$ by any means aside from using the transformed version of the variation of parameters formula, namely, (13.6.15). Finally, find the solution of (13.6.13) by computing $\mathbf{x}(t) = P\mathbf{y}(t)$.

Solution We have already determined that the eigenvalues of the coefficient matrix

$$A = \begin{bmatrix} 3 & 2 \\ -3 & -4 \end{bmatrix}$$

of the linear system in (13.6.13) are $\lambda_1 = -3$ and $\lambda_2 = 2$ and that $\mathbf{u} = \langle 1, -3\rangle$ and $\mathbf{v} = \langle 2, -1\rangle$ are eigenvectors corresponding to λ_1 and λ_2, respectively. It follows from Theorem 13.3.4 that the Jordan canonical form of A is the diagonal matrix

$$J = \begin{bmatrix} \lambda_1 & 0 \\ 0 & \lambda_2 \end{bmatrix} = \begin{bmatrix} -3 & 0 \\ 0 & 2 \end{bmatrix}.$$

Furthermore, $J = P^{-1}AP$, where

$$P = \begin{bmatrix} \mathbf{u} & \mathbf{v} \end{bmatrix} = \begin{bmatrix} 1 & 2 \\ -3 & -1 \end{bmatrix} \quad \text{and} \quad P^{-1} = \frac{1}{5}\begin{bmatrix} -1 & -2 \\ 3 & 1 \end{bmatrix} = \begin{bmatrix} -\frac{1}{5} & -\frac{2}{5} \\ \frac{3}{5} & \frac{1}{5} \end{bmatrix}.$$

With $\mathbf{y} = \langle y_1, y_2\rangle$ and $\mathbf{b}(t) = \langle -6t, -2e^{2t}\rangle$, the transformed differential equation (13.6.14) is

$$\begin{bmatrix} y_1' \\ y_2' \end{bmatrix} = \begin{bmatrix} -3 & 0 \\ 0 & 2 \end{bmatrix}\begin{bmatrix} y_1 \\ y_2 \end{bmatrix} + \begin{bmatrix} -\frac{1}{5} & -\frac{2}{5} \\ \frac{3}{5} & \frac{1}{5} \end{bmatrix}\begin{bmatrix} -6t \\ -2e^{2t} \end{bmatrix}.$$

As $\mathbf{x} = P\mathbf{y}$, the initial condition in (13.6.13) becomes

$$\mathbf{y}(0) = P^{-1}\mathbf{x}(0) = \begin{bmatrix} -\frac{1}{5} & -\frac{2}{5} \\ \frac{3}{5} & \frac{1}{5} \end{bmatrix}\begin{bmatrix} -\frac{1}{3} \\ \frac{1}{10} \end{bmatrix} = \begin{bmatrix} \frac{2}{75} \\ -\frac{9}{50} \end{bmatrix}.$$

Thus, we have the uncoupled linear system

$$\begin{aligned} y_1' + 3y_1 &= \frac{6}{5}t + \frac{4}{5}e^{2t} \\ y_2' - 2y_2 &= -\frac{18}{5}t - \frac{2}{5}e^{2t} \end{aligned} \tag{13.6.16}$$

subject to the initial condition

$$y_1(0) = \frac{2}{75}, \quad y_2(0) = -\frac{9}{50}. \tag{13.6.17}$$

Although there are different options from which to choose to solve the scalar linear equations in (13.6.16), let us use the integrating factor method. Multiplying the first equation in (13.6.16) by the integrating factor e^{3t}, we get

$$\frac{d}{dt}\left(e^{3t}y_1\right) = \frac{6}{5}te^{3t} + \frac{4}{5}e^{5t}.$$

The result of integrating this is

$$e^{3t}y_1 = \frac{6}{5}\int te^{3t}\,dt + \frac{4}{5}\int e^{5t}\,dt = \frac{2}{5}te^{3t} - \frac{2}{15}e^{3t} + \frac{4}{25}e^{5t} + C.$$

The value of C is determined by the initial condition: setting $t = 0$ and $y_1 = 2/75$, we get

$$\frac{2}{75} = -\frac{2}{15} + \frac{4}{25} + C$$

from which it follows that $C = 0$. And so

$$y_1(t) = \frac{4}{25}e^{2t} + \frac{2}{5}t - \frac{2}{15}. \tag{13.6.18}$$

Multiplying the second equation of the system (13.6.16) by the integrating factor e^{-2t}, we obtain

$$\frac{d}{dt}\left(e^{-2t}y_2\right) = -\frac{18}{5}te^{-2t} - \frac{2}{5}.$$

Integration of this equation yields

$$e^{-2t}y_2 = -\frac{18}{5}\int te^{-2t}\,dt - \frac{2}{5}\int dt = \frac{9}{5}te^{-2t} + \frac{9}{10}e^{-2t} - \frac{2}{5}t + K.$$

Setting $t = 0$ and $y_2 = -9/50$, we have

$$-\frac{9}{50} = \frac{9}{10} + K.$$

Thus $K = -\frac{27}{25}$. Consequently,

$$y_2(t) = -\frac{2}{5}te^{2t} - \frac{27}{25}e^{2t} + \frac{9}{5}t + \frac{9}{10}. \tag{13.6.19}$$

Now we can easily obtain the solution of (13.6.13) by returning to the original variable:

$$\mathbf{x}(t) = P\mathbf{y}(t) = \begin{bmatrix} 1 & 2 \\ -3 & -1 \end{bmatrix} \begin{bmatrix} \frac{4}{25}e^{2t} + \frac{2}{5}t - \frac{2}{15} \\ -\frac{2}{5}te^{2t} - \frac{27}{25}e^{2t} + \frac{9}{5}t + \frac{9}{10} \end{bmatrix}$$

$$= \begin{bmatrix} -2e^{2t} - \frac{4}{5}te^{2t} + 4t + \frac{5}{3} \\ \frac{3}{5}e^{2t} + \frac{2}{5}te^{2t} - 3t - \frac{1}{2} \end{bmatrix}.$$

This is the solution that we found in Example 13.6.2 but with fewer calculations. ♦

As a final example, we will solve a nonhomogeneous linear system that models the temperatures inside a heated two-room cabin.

Example 13.6.4 Mike and Alicia check into a cabin vacation rental that is located in the Smoky Mountains near Maggie Valley, North Carolina. Their cabin consists of a room on the ground floor and an upstairs bedroom. It is winter and the temperature outside is a constant 10 °F. Upon entering the cabin, the couple finds that it is as cold inside the cabin as outside because the furnace had been turned off. So Mike immediately turns it on. Determine the temperatures 5 hours later in the living area and the upstairs bedroom if the system of differential equations (cf. (12.6.19) in Sect. 12.6)

$$\begin{aligned} x_1' &= -0.60x_1 + 0.20x_2 + 44 \\ x_2' &= 0.20x_1 - 0.60x_2 + 4 \end{aligned} \tag{13.6.20}$$

models the temperatures $x_1(t)$ and $x_2(t)$ of the downstairs room and the upstairs bedroom, respectively, when the furnace is running.

Solution This system is equivalent to the vector differential equation

$$\mathbf{x}' = A\mathbf{x} + \mathbf{b}(t), \tag{13.6.21}$$

where $\mathbf{x} = \langle x_1, x_2 \rangle$, $\mathbf{b}(t) = \langle 44, 4 \rangle$, and

$$A = \begin{bmatrix} -0.60 & 0.20 \\ 0.20 & -0.60 \end{bmatrix}. \tag{13.6.22}$$

We already know from Example 12.6.2 that the eigenvalues of A are $\lambda_1 = -4/5$ and $\lambda_2 = -2/5$. Consequently, the Jordan canonical form of A is

$$J = \begin{bmatrix} -\frac{4}{5} & 0 \\ 0 & -\frac{2}{5} \end{bmatrix}.$$

We also determined in the example that eigenvectors of A corresponding to λ_1 and λ_2 are $\mathbf{v}^1 = \langle 1, -1\rangle$ and $\mathbf{v}^2 = \langle 1, 1\rangle$, respectively. According to Theorem 13.3.4, $J = P^{-1}AP$, where

$$P = \begin{bmatrix}\mathbf{v}^1 & \mathbf{v}^2\end{bmatrix} = \begin{bmatrix} 1 & 1 \\ -1 & 1 \end{bmatrix} \quad \text{and} \quad P^{-1} = \frac{1}{2}\begin{bmatrix} 1 & -1 \\ 1 & 1 \end{bmatrix}.$$

Now let us change variables from $\mathbf{x} = \langle x_1, x_2\rangle$ to $\mathbf{y} = \langle y_1, y_2\rangle$ using the transformation $\mathbf{x} = P\mathbf{y}$. Then it follows from (13.6.14) that $\mathbf{x}' = A\mathbf{x} + \mathbf{b}(t)$ transforms to

$$\begin{bmatrix} y_1' \\ y_2' \end{bmatrix} = \begin{bmatrix} -\frac{4}{5} & 0 \\ 0 & -\frac{2}{5} \end{bmatrix}\begin{bmatrix} y_1 \\ y_2 \end{bmatrix} + \frac{1}{2}\begin{bmatrix} 1 & -1 \\ 1 & 1 \end{bmatrix}\begin{bmatrix} 44 \\ 4 \end{bmatrix} = \begin{bmatrix} -\frac{4}{5} & 0 \\ 0 & -\frac{2}{5} \end{bmatrix}\begin{bmatrix} y_1 \\ y_2 \end{bmatrix} + \begin{bmatrix} 20 \\ 24 \end{bmatrix}.$$

This is equivalent to the uncoupled linear system

$$\begin{aligned} y_1' &= -\frac{4}{5}y_1 + 20 \\ y_2' &= -\frac{2}{5}y_2 + 24. \end{aligned} \tag{13.6.23}$$

Let $t = 0$ indicate the moment Mike turns on the furnace. Since the temperatures corresponding to this time are $x_1(0) = 10$ and $x_2(0) = 10$, the initial condition for (13.6.21) is $\mathbf{x}(0) = \mathbf{x_0}$, where

$$\mathbf{x_0} = \langle x_1(0), x_2(0)\rangle = \langle 10, 10\rangle.$$

Since $\mathbf{y} = P^{-1}\mathbf{x}$,

$$\mathbf{y_0} = P^{-1}\mathbf{x_0} = P^{-1}\langle 10, 10\rangle = \langle 0, 10\rangle.$$

Thus, $y_1(0) = 0$ and $y_2(0) = 10$.

Employing any of the methods discussed in this book for solving the two first-order linear equations constituting the system in (13.6.23) so that $y_1(0) = 0$ and $y_2(0) = 10$, we find that

$$y_1(t) = 25 - 25e^{-\frac{4}{5}t}, \quad y_2(t) = 60 - 50e^{-\frac{2}{5}t}.$$

Consequently, the solution of (13.6.20) such that $x_1(0) = 10$ and $x_2(0) = 10$ is

$$\mathbf{x}(t) = P\mathbf{y}(t) = \begin{bmatrix} 1 & 1 \\ -1 & 1 \end{bmatrix}\begin{bmatrix} y_1(t) \\ y_2(t) \end{bmatrix} = \begin{bmatrix} y_1(t) + y_2(t) \\ -y_1(t) + y_2(t) \end{bmatrix}.$$

As a result,

$$x_1(t) = y_1(t) + y_2(t) = 85 - 25e^{-\frac{4}{5}t} - 50e^{-\frac{2}{5}t}$$

$$x_2(t) = -y_1(t) + y_2(t) = 35 + 25e^{-\frac{4}{5}t} - 50e^{-\frac{2}{5}t}. \qquad (13.6.24)$$

So, this model predicts that after 5 hours the respective temperatures in the downstairs room and the upstairs bedroom will be

$$x_1(5) = 85 - 25e^{-4} - 50e^{-2} = 77.8\,°\text{F}$$

and

$$x_2(5) = 35 + 25e^{-4} - 50e^{-2} = 28.7\,°\text{F}.$$ ♦

Remark The solution of the vector differential equation (13.6.21) satisfying the initial condition $\mathbf{x}(0) = \langle 10, 10\rangle$ can also be obtained with the variation of parameters formula (13.6.7). Letting $t_0 = 0$, $\mathbf{x_0} = \langle 10, 10\rangle$, and $\mathbf{b}(t) = \langle 44, 4\rangle$, this formula becomes

$$\mathbf{x}(t) = e^{At}\begin{bmatrix}10\\10\end{bmatrix} + \int_0^t e^{A(t-s)}\begin{bmatrix}44\\4\end{bmatrix}ds, \qquad (13.6.25)$$

where A is the coefficient matrix (13.6.22). It is left as an exercise (see Problem 55) to show that this also yields the solution (13.6.24).

Problems

In the manner of Winnie the Pooh, we can imagine dropping a 'Pooh point' onto the plane and watching it move according to the flow. The path followed by such a phase point can be represented by an oriented curve on the plane corresponding to a trajectory ...

D. K. Arrowsmith & C. M. Place, *Dynamical Systems* [2, p. 27]

Principal Matrix Solutions

1. Let $A(t)$ and $B(t)$ be 2×2 matrix functions. Let $\mathbf{v}(t)$ be a 2×1 vector function. Suppose the entries in all three functions are differentiable on an interval I. Prove for $t \in I$ that

   (a) $\frac{d}{dt}[A(t)B(t)] = A(t)B'(t) + A'(t)B(t)$;

   (b) $\frac{d}{dt}[B(t)\mathbf{v}(t)] = B(t)\mathbf{v}'(t) + B'(t)\mathbf{v}(t)$;

   (c) $\frac{d}{dt}[CB(t)] = CB'(t)$, where C is a constant 2×2 matrix.

2. Prove property (vi) in Theorem 13.1.9.
3. Use the eigenvalue-eigenvector method to compute the PMS of $\mathbf{x}' = J\mathbf{x}$ at $t = t_0$, where

$$J = \begin{bmatrix} \lambda & 1 \\ 0 & \lambda \end{bmatrix} \quad (\lambda \in \mathbb{R}).$$

4. Prove: $\frac{d}{dt}Z(-t) = -AZ(-t)$.

Jordan Canonical Matrices and Matrix Exponential Functions

5. Let

$$A = \begin{bmatrix} 0 & -1 \\ 1 & 0 \end{bmatrix}.$$

   Use Definition 13.2.1 to compute e^{At}.
6. In Theorem 13.2.1, prove properties (v), (vi), and (vii).
7. Justify that $Z(t, s) = e^{A(t-s)}$.
8. Prove the converse of Theorem 13.2.2. *Hint*. Using the appropriate properties in Theorems 13.1.8 and 13.2.1, take the second derivative of both sides of (13.2.4). Then set $t = 0$.
9. Let A and B be matrices of order 2. Matrix A is said to be ***similar*** to matrix B if an invertible matrix P of order 2 exists such that $A = P^{-1}BP$. Show that *similarity* is a relation on the set of matrices of order 2 that is *reflexive*, *symmetric*, and *transitive*. That is, for matrices A, B, and C, show the following:

   (a) *Reflexive*: A is similar to itself.

   (b) *Symmetric*: If A is similar to B, then B is similar to A.

   (c) *Transitive*: If A is similar to B and B is similar to C, then A is similar to C.

Remark A relation that is reflexive, symmetric, and transitive is known as an ***equivalence relation***.

In Problems 10 through 15, find a matrix J that is a Jordan canonical form of A. Also, find the matrix exponential function e^{Jt}.

10. $A = \begin{bmatrix} 3 & 5 \\ -2 & -4 \end{bmatrix}$
11. $A = \begin{bmatrix} 1 & 1 \\ 1 & 1 \end{bmatrix}$
12. $A = \begin{bmatrix} 0 & 1 \\ -1 & 0 \end{bmatrix}$
13. $A = \begin{bmatrix} 2 & 4 \\ 1 & 5 \end{bmatrix}$
14. $A = \begin{bmatrix} 4 & 3 \\ 0 & 4 \end{bmatrix}$
15. $A = \begin{bmatrix} 2 & 0 \\ 1 & 0 \end{bmatrix}$

In Problems 16–21, compute e^{Jt}, where J is the Jordan canonical form of A. Then compute e^{At}. Use either Theorems 13.3.4 or 13.1.6 (cf. Definition 13.2.1). Note. *Both theorems are used in Example 13.4.2.*

16. $A = \begin{bmatrix} 2 & 3 \\ 3 & 2 \end{bmatrix}$
17. $A = \begin{bmatrix} -7 & -1 \\ 1 & -9 \end{bmatrix}$
18. $A = \begin{bmatrix} 4 & 0 \\ -5 & 4 \end{bmatrix}$
19. $A = \begin{bmatrix} -1 & 3 \\ 2 & 4 \end{bmatrix}$
20. $A = \begin{bmatrix} 2 & 1 \\ -1 & 2 \end{bmatrix}$
21. $A = \begin{bmatrix} 3 & 2 \\ -3 & -4 \end{bmatrix}$
22. If J is the Jordan canonical form of a real 2×2 matrix A, prove that $\operatorname{tr} A = \operatorname{tr} J$ and $\det A = \det J$.

23. Is the similarity transformation matrix P in Theorem 13.3.4 unique? Explain your answer.

24. Let J denote the Jordan canonical form of a real 2×2 matrix A. Prove the following statements.

 (a) If λ_1 and λ_2 are distinct real eigenvalues of A, then
 $$(J - \lambda_1 I)(J - \lambda_2 I) = O.$$

 (b) If λ is a repeated real eigenvalue of A, then
 $$(J - \lambda I)^2 = O.$$

 (c) If $\lambda = \alpha + i\beta$ and $\overline{\lambda} = \alpha - i\beta$, where $\beta > 0$, are eigenvalues of A, then
 $$(J - \lambda I)(J - \overline{\lambda} I) = O.$$

 (d) Every Jordan canonical matrix satisfies the equation
 $$J^2 - (\operatorname{tr} J)\, J + (\det J) I = O.$$
 Hint. Use (a), (b), (c) and Theorems 12.5.28 and 12.5.29.

 (e) Every real 2×2 matrix A satisfies its own characteristic equation. That is to say,
 $$A^2 - (\operatorname{tr} A)\, A + (\det A) I = O.$$
 Hint. Use (d), Problem 22, and Theorem 13.3.1 (iv).

Remark A result in linear algebra called the ***Cayley-Hamilton theorem*** states: *Every matrix of order n satisfies its own characteristic equation.* For a matrix A of order 2 with the characteristic equation
$$\lambda^2 - (\operatorname{tr} A)\, \lambda + \det A = 0,$$
this means that A satisfies the matrix equation in part (e). For this equation to make sense, the identity matrix I needs to be included in the third term of this equation.

Matrix Exponential Series

25. Let $A = \begin{bmatrix} -1 & 1/2 \\ 8 & 2 \end{bmatrix}$.

 (a) Use (13.4.8) to find the power series expansion of the matrix exponential function e^{At} as in Example 13.4.1. Find at least the first three nonzero terms in each of the entries of e^{At}.

 (b) Compute e^{At} using either the Jordan canonical form of A or the PMS as in Example 13.4.2. From this we obtain the sum of the power series for each of the entries in the matrix series representation of e^{At} in part (a).

 (c) Derive a formula for A^n where $n = 1, 2, 3, 4, \ldots$.
 Hint. See Theorem 13.3.1 (iv) and (13.4.3).

26. Work Problem 25 with $A = \begin{bmatrix} 3 & 2 \\ 0 & 3 \end{bmatrix}$.
 Hint. For part (c), see (13.4.4).

27. Compute A^4 for
 $$A = \begin{bmatrix} \cos\theta & \sin\theta \\ -\sin\theta & \cos\theta \end{bmatrix}.$$
 Use the result and de Moivre's formula to derive the trig identity:
 $$\cos 4\theta = \cos^4\theta - 6\cos^2\theta \sin^2\theta + \sin^4\theta.$$
 Derive a similar identity for $\sin 4\theta$.
 Hint. See (13.4.5) and the examples in Appendix B.

Miscellaneous Problems and Proofs

28. Prove: If $\det A \neq 0$, then a linear system $\mathbf{x}' = A\mathbf{x}$ has one and only one equilibrium point, namely, $(0, 0)$.

29. What are the equilibrium points of a linear system $\mathbf{x}' = A\mathbf{x}$ if A is a square zero matrix (a square matrix all of whose entries are 0)?

30. Use mathematical induction to prove (13.4.3) and (13.4.4).

31. Work the problem at the conclusion of Appendix B.

32. Let λ_1 and λ_2 be two real numbers (distinct or equal). Let $\mathbf{v}^1$ and $\mathbf{v}^2$ be linearly independent vectors. Prove that if the vector function

$$\mathbf{x}(t) = c_1 e^{\lambda_1 t}\mathbf{v}^1 + c_2 e^{\lambda_2 t}\mathbf{v}^2$$

is a general solution of the equation $\mathbf{x}' = A\mathbf{x}$, then λ_1 and λ_2 are eigenvalues of A with eigenvectors $\mathbf{v}^1$ and $\mathbf{v}^2$, respectively.

33. Let λ be a repeated eigenvalue of a 2×2 matrix A. If there are two linearly independent eigenvectors of A corresponding to λ, prove that every vector $\mathbf{v} \in \mathbb{R}^2$ is an eigenvector of A. What is the eigenspace of A corresponding to λ? *Hint*. See Problem 20 in Chap. 12.

Canonical Systems and Phase Portraits

34. Let A be a real 2×2 matrix. Let J be a Jordan canonical form of A and P be a real invertible matrix such that $P^{-1}AP = J$. Prove that the transformation $\mathbf{x} = P\mathbf{y}$ maps an eigenspace of J to an eigenspace of A.

35. Determine the equilibrium points of the canonical linear system (13.5.7), where $\lambda_1 \neq \lambda_2$ and $\det J = 0$. In particular, sketch a phase portrait of this system if $\lambda_1 < 0$ and $\lambda_2 = 0$.

36. Work Problem 35 with $\lambda_1 = 0$ and $\lambda_2 < 0$.

37. Show that the solution of (13.5.35) such that $y_1(0) = a$ and $y_2(0)) = b$ is given by (13.5.38). Use this to derive (13.5.39).

38. Let the ordered pair (a, b) denote a point in the y_1y_2-plane. For each of the Jordan canonical matrices listed below, find the solution of the linear system $\mathbf{y}' = J\mathbf{y}$ such that $y_1(0) = a$ and $y_2(0) = b$. Construct a phase portrait of the system by assigning a sufficient number of different values to a and b and graphing the corresponding solutions. Either graph the solutions by hand using a calculator as needed or use a computer algebra system.

(a) $J = \begin{bmatrix} 2 & 0 \\ 0 & 2 \end{bmatrix}$

(b) $J = \begin{bmatrix} 1 & 1 \\ 0 & 1 \end{bmatrix}$

(c) $J = \begin{bmatrix} -2 & 0 \\ 0 & 3 \end{bmatrix}$

(d) $J = \begin{bmatrix} 2 & -3 \\ 3 & 2 \end{bmatrix}$

(e) $J = \begin{bmatrix} 4 & 1 \\ 0 & 4 \end{bmatrix}$

(f) $J = \begin{bmatrix} 3 & -2 \\ 2 & 3 \end{bmatrix}$

Solutions of Linear Systems

In Problems 39 through 44, complete the following parts for each matrix A: (a) Find a matrix J that is a Jordan canonical form of A and then find the matrix exponential function e^{Jt}. (b) Find a similarity transformation matrix P so that $P^{-1}AP = J$. (c) Compute the matrix exponential function e^{At}. (d) Use e^{At} to find the solution of the differential equation $\mathbf{x}' = A\mathbf{x}$ satisfying the initial condition $\mathbf{x}(0) = \langle 0, 4\rangle$.

39. $A = \begin{bmatrix} 1 & -2 \\ 0 & 3 \end{bmatrix}$

40. $A = \begin{bmatrix} -2 & 5 \\ -6 & 9 \end{bmatrix}$

41. $A = \begin{bmatrix} 2 & -1 \\ 2 & 4 \end{bmatrix}$

42. $A = \begin{bmatrix} 2 & -5 \\ 4 & -2 \end{bmatrix}$

43. $A = \begin{bmatrix} 2 & 0 \\ 1 & 2 \end{bmatrix}$

44. $A = \begin{bmatrix} 2 & 1 \\ -4 & -2 \end{bmatrix}$

Elliptical and Circular Trajectories

45. Solve the initial value problem

$$x_1' = 2x_1 - 5x_2$$
$$x_2' = 4x_1 - 2x_2$$
$$x_1(0) = 2,\ x_2(0) = 4.$$

Find the equation of the trajectory through $(2, 4)$ and graph it.

46. Solve the initial value problem

$$x_1' = 2x_1 + 5x_2$$
$$x_2' = -x_1 - 2x_2$$
$$x_1(0) = -2, \; x_2(0) = 1.$$

Find the equation of the trajectory through $(-2, 1)$ and graph it.

47. Explain why the graph of the second-degree equation (13.5.61) cannot be a circle if the x_1x_2-term is present.

48.

(a) Use the Chain Rule

$$\frac{dx_2}{dt} = \frac{dx_2}{dx_1} \cdot \frac{dx_1}{dt}$$

to convert the linear system

$$\begin{bmatrix} x_1' \\ x_2' \end{bmatrix} = \begin{bmatrix} a_{11} & a_{12} \\ a_{21} & a_{22} \end{bmatrix} \begin{bmatrix} x_1 \\ x_2 \end{bmatrix}$$

into the differential equation

$$M(x_1, x_2) + N(x_1, x_2)\frac{dx_2}{dx_1} = 0.$$

Find $M(x_1, x_2)$ and $N(x_1, x_2)$.

(b) Suppose the eigenvalues of the coefficient matrix of the linear system are pure imaginary numbers. Show that the differential equation in part (a) is exact. *Hint.* See the first statement in Theorem 13.5.1.

(c) Show that the solutions of the differential equation in part (a) are defined implicitly by (13.5.59) when the eigenvalues of the coefficient matrix are purely imaginary.

Nonautonomous Linear Systems

49. Prove item (i) in Lemma 13.6.1.

50. Prove item (iii) in Lemma 13.6.1

51. Solve the initial value problem

$$x_1' = 2x_1 - 10\sin t$$
$$x_2' = 2x_2 + 9te^{-t}$$
$$x_1(0) = -1, \; x_2(0) = 2.$$

52. Solve the initial value problem

$$x_1' = x_1 + x_2 + e^{-2t}$$
$$x_2' = 4x_1 - 2x_2 - e^{t}$$
$$x_1(0) = 5, \; x_2(0) = 0.$$

53. Solve the initial value problem

$$x_1' = 3x_1 - x_2 + 2$$
$$x_2' = 9x_1 - 3x_2 + t^{-1}$$
$$x_1(1) = 2, \; x_2(1) = 1.$$

54. Solve the initial value problem

$$x_1' = 4x_1 + 5x_2 + 4e^{t}\cos t$$
$$x_2' = -2x_1 - 2x_2$$
$$x_1(0) = 0, \; x_2(0) = 0.$$

55. The solution (13.6.24) of the initial value problem

$$x_1' = -0.60x_1 + 0.20x_2 + 44$$
$$x_2' = 0.20x_1 - 0.60x_2 + 4$$
$$x_1(0) = 10, \; x_2(0) = 10.$$

was obtained in Example 13.6.4 by uncoupling the system of differential equations with a transformation. Now use the variation of parameters formula to obtain this solution. See the remark at the end of the example.

56. Determine the limits

$$\lim_{t\to\infty} x_1(t) \;\text{ and }\; \lim_{t\to\infty} x_2(t)$$

if $x_1(t)$ and $x_2(t)$ are solutions of the following initial value problem:

$$x_1' = -\frac{1}{2}x_1 + \frac{1}{4}x_2 + 13$$
$$x_2' = \frac{1}{4}x_1 - \frac{3}{4}x_2 + 50$$
$$x_1(0) = 80, \; x_2(0) = 100.$$

Appendix A
Derivatives: Rules and Formulas

A.1 Derivative Rules

Let c be a constant. Let f and g be functions that are differentiable on an interval I. For $x \in I$, the following rules hold:

Constant Multiple Rule

$$\frac{d}{dx}[cf(x)] = cf'(x)$$

The derivative of a constant times a function is equal to the constant times the derivative of the function.

Sum and Difference Rules

$$\frac{d}{dx}[f(x) \pm g(x)] = f'(x) \pm g'(x)$$

The derivative of a sum (difference) is equal to the sum (difference) of the derivatives.

Product Rule

$$\frac{d}{dx}[f(x)g(x)] = f(x)g'(x) + g(x)f'(x)$$

The derivative of a product is equal to the first function times the derivative of the second function plus the second function times the derivative of the first function.

© The Author(s), under exclusive license to Springer Nature Switzerland AG 2026
L. C. Becker, *Ordinary Differential Equations: Concepts, Methods, and Models*,
https://doi.org/10.1007/978-3-032-15150-6

Quotient Rule

$$\frac{d}{dx}\left[\frac{f(x)}{g(x)}\right] = \frac{g(x)f'(x) - f(x)g'(x)}{[g(x)]^2} \quad \text{(provided } g(x) \neq 0\text{)}$$

Mnemonic: *Low d'high minus high d'low over the square of what's below.*

Chain Rule

Let f and g be defined on intervals I and J, respectively, with $f(I) \subseteq J$. If f is differentiable at $x \in I$ and g is differentiable at $f(x) \in J$, then the composite function $g \circ f$ is differentiable at x and

$$(g \circ f)'(x) = g'(f(x)) \cdot f'(x).$$

The derivative of a composite function is equal to the derivative of the outer function evaluated at the inner function times the derivative of the inner function.

Alternatively, let $y = g(u)$ and $u = f(x)$. Then

$$\frac{dy}{dx} = \frac{dy}{du} \cdot \frac{du}{dx},$$

where dy/du is evaluated at $u = f(x)$.

Power Rule (General Form)

If n is a real number and f a differentiable function, then

$$\frac{d}{dx}[f(x)]^n = n[f(x)]^{n-1} \cdot f'(x).$$

Alternatively, let $u = f(x)$. Then

$$\frac{d}{dx}u^n = nu^{n-1}\frac{du}{dx}.$$

A.2 Derivative Formulas

In Table A.1, k and n denote real numbers; b denotes a positive real number.

Caveat Unfortunately, the two formulas in the last row of the table may be slightly different in another book, table of integrals, or the result of using computer algebra

Table A.1 Basic derivatives

$f(x)$	$f'(x)$	$f(x)$	$f'(x)$
k	0	x	1
$\sqrt{x}$	$\frac{1}{2\sqrt{x}}$	x^n	nx^{n-1}
e^x	e^x	b^x	$b^x \ln b$
$\ln\|x\|$	$\frac{1}{x}$	$\log_b \|x\|$	$\frac{1}{x \ln b}$
$\sin x$	$\cos x$	$\cos x$	$-\sin x$
$\tan x$	$\sec^2 x$	$\cot x$	$-\csc^2 x$
$\sec x$	$\sec x \tan x$	$\csc x$	$-\csc x \cot x$
$\sin^{-1} x$	$\frac{1}{\sqrt{1-x^2}}$	$\cos^{-1} x$	$-\frac{1}{\sqrt{1-x^2}}$
$\tan^{-1} x$	$\frac{1}{1+x^2}$	$\cot^{-1} x$	$-\frac{1}{1+x^2}$
$\sec^{-1} x$	$\frac{1}{\|x\|\sqrt{x^2-1}}$	$\csc^{-1} x$	$-\frac{1}{\|x\|\sqrt{x^2-1}}$

system software. The reason for this is that there is no universal agreement on what the ranges of the inverse secant and inverse cosecant functions should be. The formulas given here are the result of defining the inverse secant function by

$$y = \sec^{-1} x \ \left(|x| \geq 1\right) \quad \text{if and only if} \quad \sec y = x \text{ and } y \in \left[0, \tfrac{\pi}{2}\right) \cup \left(\tfrac{\pi}{2}, \pi\right]$$

and the inverse cosecant function by

$$y = \csc^{-1} x \ \left(|x| \geq 1\right) \quad \text{if and only if} \quad \csc y = x \text{ and } y \in \left[-\tfrac{\pi}{2}, 0\right) \cup \left(0, \tfrac{\pi}{2}\right].$$

For details, see [71, pp. 519–528]. However, as mentioned above, other choices for the ranges of these two inverse trig functions may result in a slightly different formula. For instance, if the range of the function $y = \sec^{-1} x$ is chosen to be $\left[0, \frac{\pi}{2}\right) \cup \left[\pi, \frac{3\pi}{2}\right)$ and the range of $y = \csc^{-1} x$ is chosen to be $\left(0, \frac{\pi}{2}\right] \cup \left(\pi, \frac{3\pi}{2}\right]$ as in [68, p. 66, pp. 214–216], then the differentiation formulas are

$$\frac{d}{dx} \sec^{-1} x = \frac{1}{x\sqrt{x^2-1}} \quad \text{and} \quad \frac{d}{dx} \csc^{-1} x = -\frac{1}{x\sqrt{x^2-1}}.$$

Because of the Chain Rule, the differentiation formulas for the functions listed in Table A.1 can be generalized. If u is a differentiable function of x, then

$$\frac{d}{dx} u^n = n u^{n-1} \frac{du}{dx} \qquad \frac{d}{dx} \sqrt{u} = \frac{1}{2\sqrt{u}} \frac{du}{dx}$$

$$\frac{d}{dx} e^u = e^u \frac{du}{dx} \qquad \frac{d}{dx} \ln|u| = \frac{1}{u} \frac{du}{dx}$$

$$\frac{d}{dx}\sin u = \cos u \frac{du}{dx} \qquad \frac{d}{dx}\cos u = -\sin u \frac{du}{dx}$$

$$\frac{d}{dx}\tan u = \sec^2 u \frac{du}{dx} \qquad \frac{d}{dx}\sec u = \sec u \tan u \frac{du}{dx}$$

$$\frac{d}{dx}\cos^{-1} u = -\frac{1}{\sqrt{1-u^2}}\frac{du}{dx} \qquad \frac{d}{dx}\sec^{-1} u = \frac{1}{|u|\sqrt{u^2-1}}\frac{du}{dx}$$

and so on.

Appendix B
Complex Numbers as Matrices

B.1 Complex Numbers

A ***complex number*** is a number that is expressed in the form $a + ib$, where a and b are real numbers and the symbol i designates a non-real number that is defined to be a solution of the equation $x^2 = -1$. Thus, $i^2 = -1$. The real numbers a and b are known as the ***real part*** and the ***imaginary part*** of $a + ib$, respectively. The symbol $\mathbb{C}$ is commonly used to represent the set of all complex numbers, namely, the set

$$\mathbb{C} := \{a + ib : a, b \in \mathbb{R}\}, \tag{B.1}$$

where $\mathbb{R}$ denotes the set of all real numbers. Generally speaking, when referring to an arbitrary complex number, we will place the imaginary part of a complex number after the symbol i, as in $a + ib$; however, if the value of b is known, then we will usually place b before i, as in $2 + 3i$. This is not a steadfast rule; irrespective of the placement of b, the complex numbers $a + bi$ and $a + ib$ are one and the same. When b is negative and its value is known, it is customary to write $a - |b|i$ rather than $a + (-|b|)i$. For example, with $a = 2$ and $b = -3$, we write $2 - 3i$ rather than $2 + (-3)i$, unless there is some reason to do otherwise. The complex number $0 + ib$ is usually shortened to ib (or to bi). And when $b = 1$, then $0 + 1i$ is shortened to i; so $i \in \mathbb{C}$. As we shall soon see, the algebraic operations on $\mathbb{C}$ of addition, subtraction, multiplication, and division are defined in such a way so as to be consistent with the corresponding operations on $\mathbb{R}$. In order to define these operations, let $a + ib$ and $c + id$ denote any two complex numbers in the following discussion.

The operations of addition and subtraction on $\mathbb{C}$ are defined by

$$(a + ib) + (c + id) = (a + c) + i(b + d) \tag{B.2}$$

$$(a + ib) - (c + id) = (a - c) + i(b - d). \tag{B.3}$$

© The Author(s), under exclusive license to Springer Nature Switzerland AG 2026

L. C. Becker, *Ordinary Differential Equations: Concepts, Methods, and Models*,
https://doi.org/10.1007/978-3-032-15150-6

The complex number $0 + i0$ is distinctive in that

$$(a + ib) + (0 + i0) = a + ib$$

for all $a+ib \in \mathbb{C}$. In other words, it leaves unchanged any complex number to which it is added. This is analogous to the distinctive role that *zero* plays in $\mathbb{R}$, namely, that

$$a + 0 = a$$

for all $a \in \mathbb{R}$. We express the distinctiveness of these numbers by saying that 0 is the ***additive identity*** for $\mathbb{R}$ and $0 + i0$ is the ***additive identity*** or ***zero*** for $\mathbb{C}$. A complex number $a + ib$ is ***nonzero*** if $a^2 + b^2 \neq 0$, which is the case if $a \neq 0$ or $b \neq 0$.

Multiplication of two complex numbers $a + ib$ and $c + id$ is defined by

$$(a + ib)(c + id) = (ac - bd) + i(ad + bc). \tag{B.4}$$

Since real numbers obey the commutative laws of addition and multiplication, so do the complex numbers. It is left to the reader to verify this:

$$(a+ib)+(c+id) = (c+id)+(a+ib) \quad \text{and} \quad (a+ib)(c+id) = (c+id)(a+ib).$$

According to (B.4), the product of the two complex numbers $a + ib$ and $1 + i0$ is

$$(a + ib)(1 + i0) = a + ib.$$

This shows that the complex number $1+i0$ is the ***multiplicative identity*** for complex numbers, just as the real number 1 is the ***multiplicative identity*** for real numbers. Fortunately, there is really no need to memorize the product on the right-hand side of (B.4): it can be obtained by simply multiplying the two binomials on the left-hand side (as is normally done when the product involves only real numbers) and replacing i^2 with -1.

Finally, consider the operation of division. To divide a complex number $a + ib$ by a nonzero complex number $c + id$, first multiply the numerator and denominator by the complex conjugate $c - id$ to eliminate i in the denominator:

$$\frac{a + ib}{c + id} = \frac{(a + ib) \cdot (c - id)}{(c + id) \cdot (c - id)} = \frac{(ac + bd) + i(bc - ad)}{c^2 + d^2}.$$

Then express the resulting complex number in the form $x + iy$:

$$\frac{a + ib}{c + id} = \frac{ac + bd}{c^2 + d^2} + i\frac{bc - ad}{c^2 + d^2}. \tag{B.5}$$

In other words, to compute the quotient of two complex numbers, multiply both of these numbers by the complex conjugate of the ***divisor*** (the complex number in the denominator) and then separate the real and imaginary parts of the result. It is easy to show that the product of the right-hand side of (B.5) and the divisor $c + id$ is the dividend $a + ib$. This verifies that the right-hand side is truly the ***quotient*** of $a + ib$ by $c + id$.

If the imaginary parts of two complex numbers are zero, then the sum and difference in (B.2) and (B.3) simplify to

$$(a + i0) + (c + i0) = (a + c) + i0 \tag{B.6}$$

$$(a + i0) - (c + i0) = (a - c) + i0 \tag{B.7}$$

while the product and quotient (B.4) and (B.5) simplify to

$$(a + i0) \cdot (c + i0) = ac + i0 \tag{B.8}$$

$$\frac{a + i0}{c + i0} = \frac{ac}{c^2} + i0 = \frac{a}{c} + i0. \tag{B.9}$$

We can readily see from (B.6)–(B.9) that the algebraic operations on the set

$$\mathbb{C}_0 := \{(a + i0) : a \in \mathbb{R}\},$$

a subset of $\mathbb{C}$, mirror the algebraic operations on $\mathbb{R}$. This suggests that we pair every real number a with the corresponding complex number $a + i0$. Let us denote these pairings with a left-right arrow as follows:

$$a + i0 \;\leftrightarrow\; a \tag{B.10}$$

In so doing, we have set up a ***one-to-one correspondence*** between the real numbers $\mathbb{R}$ and the complex numbers in $\mathbb{C}_0$: there is precisely one real number corresponding to each complex number in $\mathbb{C}_0$; and conversely, there is precisely one complex number in $\mathbb{C}_0$ corresponding to each real number. Because of this and the identical algebraic structures of $\mathbb{R}$ and $\mathbb{C}_0$, we say that the system $\mathbb{C}_0$ is ***isomorphic*** to the system of real numbers.

Inasmuch as every real number can be expressed in the notation of $\mathbb{C}_0$ because of this isomorphism, it makes sense to regard $\mathbb{C}_0$ as the set of real numbers embedded in $\mathbb{C}$. Switching from $\mathbb{R}$ to $\mathbb{C}_0$ to express real numbers in $\mathbb{C}$, and vice versa, is somewhat akin to translating back and forth between two human languages. For example, consider English-speaking students learning German. Many students make out vocabulary lists consisting of pairings of German and English words. However, this is a rough analogy because such vocabulary lists with words put into one-to-one correspondence have the unintended consequence of miscommunication or unintelligibility. Take, for instance, a vocabulary list with the following pairings:

ich ↔ I; mir ↔ (to) me; bin ↔ am; ist ↔ is; kalt ↔ cold.

Now suppose an English-speaking exchange student is waiting for a bus in Berlin and it is a cold winter day. That student might say "Ich bin kalt" to a Berliner (resident of Berlin) who is also waiting for the bus. According to the above vocabulary list, this translates to "I am cold." However, in this situation, this translation is incorrect. The correct way to say "I am cold" in German is "Mir ist kalt." The Berliner would probably have understood what the student had meant; however, "Ich bin kalt" means "I am cold as ice" in the sense of being unfriendly in temperament. Even *Google Translate* often fails to render correct translations. In contrast, (B.10) always translates $\mathbb{R}$ to $\mathbb{C}_0$ perfectly, and vice versa.

Human languages often incorporate words from other languages, despite the objections of purists. For example, the German words "kindergarten" and "sauerkraut" are now part of the English language, while the English word "T-shirt" and the slang word "cool" (having various meanings depending on the social group) are now German words. Similarly, because of the one-to-one correspondence (B.10) and the identical algebraic structures of $\mathbb{R}$ and $\mathbb{C}_0$, the real numbers can be incorporated into the system of complex numbers by regarding $a \in \mathbb{R}$ and $a + i0 \in \mathbb{C}_0$ as the same number. With this viewpoint, real numbers are also complex numbers, just as all adopted German words, like kindergarten, are also English words. Likewise, the subset of complex numbers $\mathbb{C}_0$ are also real numbers. Consequently, the sets $\mathbb{C}_0$ and $\mathbb{R}$ are not only isomorphic but can be regarded as being the same numbers. For instance, if the result of some calculations in $\mathbb{C}$ is $5 + i0$, then we can also say the result is 5. In conclusion, since a and $a + i0$ represent the same number in two different "number languages," it is standard practice to write

$$a + i0 = a. \tag{B.11}$$

B.2 Matrix Representation

There are other ways of defining complex numbers: one of them being by way of the set of all ordered pairs of numbers with algebraic operations defined on this set that are consistent with the algebraic operations on $\mathbb{R}$ (cf. [16, Ch. 1]). But the one that turns out to be particularly useful for completing the proof of Lemma 13.4.2 in Chap. 13 involves the set of 2×2 matrices of the form

$$\mathbb{C}_M := \left\{ \begin{bmatrix} a & -b \\ b & a \end{bmatrix} : a, b \in \mathbb{R} \right\}.$$

By using the symbols $0'$, $1'$, i', a' to denote the following special matrices

$$0' := \begin{bmatrix} 0 & 0 \\ 0 & 0 \end{bmatrix}, \quad 1' := \begin{bmatrix} 1 & 0 \\ 0 & 1 \end{bmatrix}, \quad i' := \begin{bmatrix} 0 & -1 \\ 1 & 0 \end{bmatrix}, \quad a' := a1' = \begin{bmatrix} a & 0 \\ 0 & a \end{bmatrix} \tag{B.12}$$

that are themselves members of $\mathbb{C}_M$, we can express the rest of the matrices in $\mathbb{C}_M$ in terms of these four matrices as follows:

$$\begin{bmatrix} a & -b \\ b & a \end{bmatrix} = \begin{bmatrix} a & 0 \\ 0 & a \end{bmatrix} + \begin{bmatrix} 0 & -b \\ b & 0 \end{bmatrix} = \begin{bmatrix} a & 0 \\ 0 & a \end{bmatrix} + \begin{bmatrix} 0 & -1 \\ 1 & 0 \end{bmatrix}\begin{bmatrix} b & 0 \\ 0 & b \end{bmatrix}.$$

Thus,

$$\begin{bmatrix} a & -b \\ b & a \end{bmatrix} = a' + i'b'. \tag{B.13}$$

Addition, subtraction, and multiplication of matrices in $\mathbb{C}_M$ are defined in the usual way. Let us begin with the sum and difference of the special matrices

$$a' = \begin{bmatrix} a & 0 \\ 0 & a \end{bmatrix} \quad \text{and} \quad c' = \begin{bmatrix} c & 0 \\ 0 & c \end{bmatrix}.$$

The sum $a' + c'$ is defined as

$$a' + c' := \begin{bmatrix} a & 0 \\ 0 & a \end{bmatrix} + \begin{bmatrix} c & 0 \\ 0 & c \end{bmatrix} = \begin{bmatrix} a+c & 0 \\ 0 & a+c \end{bmatrix} = (a+c)',$$

while the difference $a' - c'$ is

$$a' - c' := \begin{bmatrix} a & 0 \\ 0 & a \end{bmatrix} - \begin{bmatrix} c & 0 \\ 0 & c \end{bmatrix} = \begin{bmatrix} a-c & 0 \\ 0 & a-c \end{bmatrix} = (a-c)'.$$

Observe that the prime symbol distributes over sums and differences; that is,

$$(a \pm c)' = a' \pm c'.$$

As a result, we see from (B.13) that the sum of the matrices

$$a' + i'b' = \begin{bmatrix} a & -b \\ b & a \end{bmatrix} \quad \text{and} \quad c' + i'd' = \begin{bmatrix} c & -d \\ d & c \end{bmatrix}$$

expressed in terms of this prime notation is

$$\begin{bmatrix} a & -b \\ b & a \end{bmatrix} + \begin{bmatrix} c & -d \\ d & c \end{bmatrix} = \begin{bmatrix} a+c & -(b+d) \\ b+d & a+c \end{bmatrix} = (a+c)' + i'(b+d)'.$$

But that is not all: we can also distribute the prime symbol over the sum on the right-hand side obtaining

$$(a' + i'b') + (c' + i'd') = (a' + c') + i'(b' + d'). \tag{B.14}$$

Comparing this to (B.2), we see this is exactly like addition in $\mathbb{C}$. As for subtraction, it is left as an exercise to show that

$$(a' + i'b') - (c' + i'd') = (a' - c') + i'(b' - d'), \tag{B.15}$$

which, aside from the prime symbol, is (B.3).

Next we see that the prime symbol also distributes over products, which follows from

$$a'c' = \begin{bmatrix} a & 0 \\ 0 & a \end{bmatrix} \begin{bmatrix} c & 0 \\ 0 & c \end{bmatrix} = \begin{bmatrix} ac & 0 \\ 0 & ac \end{bmatrix} = (ac)'.$$

Accordingly,

$$\begin{aligned} (a' + i'b')(c' + i'd') &= \begin{bmatrix} a & -b \\ b & a \end{bmatrix} \begin{bmatrix} c & -d \\ d & c \end{bmatrix} = \begin{bmatrix} ac - bd & -(ad + bc) \\ ad + bc & ac - bd \end{bmatrix} \\ &= (ac - bd)' + i'(ad + bc)'. \end{aligned}$$

Since the prime symbol distributes over sums, differences, and products, we end up with

$$(a' + i'b')(c' + i'd') = (a'c' - b'd') + i'(a'd' + b'c'). \tag{B.16}$$

Observe from (B.4) that this is exactly how we would multiply $a' + i'b'$ and $c' + i'd'$ had they been complex numbers instead of matrices from $\mathbb{C}_M$.

It is left as an exercise to show that the matrices $0'$ and $1'$ are, respectively, the additive and multiplicative identities for $\mathbb{C}_M$. That is to say, $0'$ plays the same role in $\mathbb{C}_M$ that $0 = 0 + 0i$ does in $\mathbb{C}$. Likewise, $1' \in \mathbb{C}_M$ is the analog of $1 = 1 + 0i \in \mathbb{C}$. It follows from the calculation

$$(i')^2 = \begin{bmatrix} 0 & -1 \\ 1 & 0 \end{bmatrix} \begin{bmatrix} 0 & -1 \\ 1 & 0 \end{bmatrix} = \begin{bmatrix} -1 & 0 \\ 0 & -1 \end{bmatrix} = -\begin{bmatrix} 1 & 0 \\ 0 & 1 \end{bmatrix}$$

that $(i')^2 = -1'$. In other words, the role of i' in $\mathbb{C}_M$ is the same as the role of i in $\mathbb{C}$.

Finally, let us ascertain how division should be defined for $\mathbb{C}_M$ so that it is consonant with the definition of division for $\mathbb{C}$. Let $a' + i'b'$ be a matrix in $\mathbb{C}_M$. If it is nonzero, then it has an inverse since (cf. Theorem 12.5.8)

$$\det(a' + i'b') = \det \begin{bmatrix} a & -b \\ b & a \end{bmatrix} = a^2 + b^2 \neq 0.$$

This suggests that we define the division of two matrices in $\mathbb{C}_M$ as follows: To divide a matrix $a' + i'b'$ by a nonzero matrix $c' + i'd'$, multiply $a' + i'b'$ by the inverse of the matrix $c' + i'd'$. Carrying out this multiplication, we obtain

$$\begin{aligned}\frac{a' + i'b'}{c' + i'd'} &= (a' + i'b')(c' + i'd')^{-1} = \begin{bmatrix} a & -b \\ b & a \end{bmatrix} \begin{bmatrix} c & -d \\ d & c \end{bmatrix}^{-1} \\ &= \frac{1}{c^2 + d^2} \begin{bmatrix} a & -b \\ b & a \end{bmatrix} \begin{bmatrix} c & d \\ -d & c \end{bmatrix} = \frac{1}{c^2 + d^2} \begin{bmatrix} ac + bd & -(bc - ad) \\ bc - ad & ac + bd \end{bmatrix} \\ &= \begin{bmatrix} \frac{ac+bd}{c^2+d^2} & -\frac{bc-ad}{c^2+d^2} \\ \frac{bc-ad}{c^2+d^2} & \frac{ac+bd}{c^2+d^2} \end{bmatrix} = \left(\frac{ac + bd}{c^2 + d^2}\right)' + i' \left(\frac{bc - ad}{c^2 + d^2}\right)'.\end{aligned}$$

When $b = d = 0$, this simplifies to

$$\frac{a'}{c'} = \left(\frac{a}{c}\right)'.$$

Since the prime symbol distributes over division as well as over sums, differences, and products, we have

$$\frac{a' + i'b'}{c' + i'd'} = \frac{a'c' + b'd'}{(c')^2 + (d')^2} + i' \frac{b'c' - a'd'}{(c')^2 + (d')^2}. \tag{B.17}$$

Observe that without the prime symbol this looks exactly like (B.5) .

Comparing the algebraic operations (B.14)–(B.17) for the set of matrices $\mathbb{C}_M$ with the algebraic operations (B.2)–(B.5) for the set of complex numbers $\mathbb{C}$, we conclude that the algebraic structures of these two sets are identical. As a result, for every pair (a, b) of real numbers, the complex number $a + ib$ corresponds to the matrix

$$a' + i'b' = \begin{bmatrix} a & -b \\ b & a \end{bmatrix},$$

and vice versa. As with $\mathbb{R}$ and $\mathbb{C}_0$, we indicate this correspondence with a left-right arrow:

$$\begin{bmatrix} a & -b \\ b & a \end{bmatrix} \leftrightarrow a + ib. \tag{B.18}$$

From this, it follows that

$$\begin{bmatrix} a & b \\ -b & a \end{bmatrix} = \begin{bmatrix} a & -(-b) \\ -b & a \end{bmatrix} \leftrightarrow a + (-b)i = a - ib. \tag{B.19}$$

In sum, we have set up a one-to-one correspondence between the set of complex numbers $\mathbb{C}$ and the set of matrices $\mathbb{C}_M$. This and the identical algebraic structures of $\mathbb{C}_M$ and $\mathbb{C}$ means that the system of matrices $\mathbb{C}_M$ is isomorphic to the system of complex numbers $\mathbb{C}$. The practical implication of this isomorphism is that a problem in $\mathbb{C}_M$ may be converted to an equivalent problem in $\mathbb{C}$ by way of (B.18), where the computations are then carried out, after which the result is converted back to $\mathbb{C}_M$. Likewise, a problem in $\mathbb{C}$ may be converted to the equivalent problem in $\mathbb{C}_M$. The following examples illustrate these back-and-forth conversions.

Example B.2.1 Let $A = \begin{bmatrix} a & -b \\ b & a \end{bmatrix}$ and $B = \begin{bmatrix} c & d \\ -d & c \end{bmatrix}$. With $a = 1, b = 2, c = 3$, and $d = 4$, compute the products AB and BA directly and compare the results. Also, use the correspondences (B.18) and (B.19) to compute these products. Explain why $AB = BA$ for all values of a, b, c, and d.

Solution The products AB and BA are

$$AB = \begin{bmatrix} 1 & -2 \\ 2 & 1 \end{bmatrix} \begin{bmatrix} 3 & 4 \\ -4 & 3 \end{bmatrix} = \begin{bmatrix} 11 & -2 \\ 2 & 11 \end{bmatrix}$$

and

$$BA = \begin{bmatrix} 3 & 4 \\ -4 & 3 \end{bmatrix} \begin{bmatrix} 1 & -2 \\ 2 & 1 \end{bmatrix} = \begin{bmatrix} 11 & -2 \\ 2 & 11 \end{bmatrix}.$$

Thus, for these particular matrices, $AB = BA$.

Since $\mathbb{C}_M$ and $\mathbb{C}$ are isomorphic, these products can also be computed using (B.18) and (B.19). Consider AB. Since $A \leftrightarrow 1 + 2i$ and $B \leftrightarrow 3 - 4i$, we have

$$AB \leftrightarrow (1 + 2i)(3 - 4i) = 11 + 2i.$$

Thus,

$$AB = \begin{bmatrix} 11 & -2 \\ 2 & 11 \end{bmatrix}.$$

Since multiplication of complex numbers is commutative, we have

$$AB \leftrightarrow (1 + 2i)(3 - 4i) = (3 - 4i)(1 + 2i) \leftrightarrow BA,$$

which implies that $AB = BA$.

Generally speaking, matrix multiplication is not commutative on the set of all matrices of order 2. But it is on the subset $\mathbb{C}_M$ because of

$$AB \leftrightarrow (a + ib)(c - id) = (c - id)(a + ib) \leftrightarrow BA$$

and the one-to-one correspondence between $\mathbb{C}_M$ and $\mathbb{C}$. ♦

The next two examples are much better illustrations of how the isomorphism between $\mathbb{C}_M$ and $\mathbb{C}$ can be exploited to simplify the calculations of certain types of problems.

Example B.2.2 Compute A^{21} for the matrix $A = \begin{bmatrix} 1 & -\sqrt{3} \\ \sqrt{3} & 1 \end{bmatrix}$.

Solution We see from (B.18) that $1 + i\sqrt{3}$ is the corresponding complex number. Thus,

$$\begin{bmatrix} 1 & -\sqrt{3} \\ \sqrt{3} & 1 \end{bmatrix}^{21} \leftrightarrow (1 + i\sqrt{3})^{21}.$$

By Euler's formula,

$$1 + i\sqrt{3} = 2\left(\cos\frac{\pi}{3} + i\sin\frac{\pi}{3}\right) = 2e^{i\frac{\pi}{3}}.$$

Since $(re^{i\theta})^n = r^n e^{in\theta}$ for $n = 0, \pm 1, \pm 2, \ldots$, we have[1]

$$(2e^{i\frac{\pi}{3}})^{21} = 2^{21}e^{i7\pi} = 2^{21}(\cos 7\pi + i\sin 7\pi) = -2^{21} + 0i.$$

The matrix in $\mathbb{C}_M$ corresponding to this complex number is

$$\begin{bmatrix} -2^{21} & 0 \\ 0 & -2^{21} \end{bmatrix}.$$

Therefore,

$$A^{21} = \begin{bmatrix} -2^{21} & 0 \\ 0 & -2^{21} \end{bmatrix}.$$

♦

The next example relies on results in the sections in Chap. 13 dealing with Jordan canonical matrices and matrix exponential series.

Example B.2.3 Compute the coefficient of t^4 in the power series expansion of e^{Jt} for

$$J = \begin{bmatrix} a & b \\ -b & a \end{bmatrix},$$

where a and b are real numbers. Compute the first five nonzero terms in the Maclaurin series expansion of $e^{at}\cos bt$.

---

[1] See any complex variables textbook, such as [16].

Solution From Theorem 13.4.4, we have

$$e^{Jt} = \sum_{n=0}^{\infty} \frac{t^n}{n!} J^n = I + tJ + \frac{t^2}{2!} J^2 + \frac{t^3}{3!} J^3 + \frac{t^4}{4!} J^4 + \dots$$

for $-\infty < t < \infty$. Since $J \leftrightarrow a - bi$,

$$J^4 \leftrightarrow (a - ib)^4,$$

where

$$\begin{aligned}(a - ib)^4 &= a^4 + 4a^3(-ib) + 6a^2(-ib)^2 + 4a(-ib)^3 + (-ib)^4 \\ &= (a^4 - 6a^2b^2 + b^4) - i(4a^3b - 4ab^3).\end{aligned}$$

It follows from (B.19) that the coefficient of t^4 is

$$\frac{1}{4!} J^4 = \frac{1}{4!} \begin{bmatrix} a^4 - 6a^2b^2 + b^4 & 4a^3b - 4ab^3 \\ -(4a^3b - 4ab^3) & a^4 - 6a^2b^2 + b^4 \end{bmatrix}.$$

As for the second computation, we see from Theorem 13.3.4 that

$$e^{Jt} = \begin{bmatrix} e^{at}\cos bt & e^{at}\sin bt \\ -e^{at}\sin bt & e^{at}\cos bt \end{bmatrix}.$$

Furthermore,

$$\begin{aligned} e^{Jt} &= \begin{bmatrix} 1 & 0 \\ 0 & 1 \end{bmatrix} + \begin{bmatrix} a & b \\ -b & a \end{bmatrix} t + \frac{1}{2!} \begin{bmatrix} a^2 - b^2 & 2ab \\ -2ab & a^2 - b^2 \end{bmatrix} t^2 \\ &+ \frac{1}{3!} \begin{bmatrix} a^3 - 3ab^2 & 3a^2b - b^3 \\ -(3a^2b - b^3) & a^3 - 3ab^2 \end{bmatrix} t^3 \\ &+ \frac{1}{4!} \begin{bmatrix} a^4 - 6a^2b^2 + b^4 & 4a^3b - 4ab^3 \\ -(4a^3b - 4ab^3) & a^4 - 6a^2b^2 + b^4 \end{bmatrix} t^4 + \dots . \end{aligned}$$

Comparing the entries of e^{Jt} with its power series expansion, we obtain

$$\begin{aligned} e^{at}\cos bt &= 1 + at + \frac{1}{2}(a^2 - b^2)t^2 + \frac{1}{6}(a^3 - 3ab^2)t^3 \\ &+ \frac{1}{24}(a^4 - 6a^2b^2 + b^4)t^4 + \dots . \end{aligned}$$

♦

Problem

(a) Let $\langle x', y' \rangle$ be the vector that is obtained by rotating a vector $\langle x, y \rangle$ about the origin through an angle θ in the counterclockwise direction. Show with trigonometry that x' and y' are related to x and y by

$$\begin{bmatrix} x' \\ y' \end{bmatrix} = \begin{bmatrix} \cos\theta & -\sin\theta \\ \sin\theta & \cos\theta \end{bmatrix} \begin{bmatrix} x \\ y \end{bmatrix}.$$

In other words, multiplying a vector $\langle x, y \rangle$ by the ***rotation matrix***

$$R_\theta := \begin{bmatrix} \cos\theta & -\sin\theta \\ \sin\theta & \cos\theta \end{bmatrix}$$

rotates it about the origin counterclockwise through the angle θ.

(b) Let z be a given complex number. Use the matrix R_θ and the isomorphism of the matrices $\mathbb{C}_M$ and the complex numbers $\mathbb{C}$ to determine the complex number w so that the result of multiplying z by w is a rotation of z about the origin counterclockwise through the angle θ. Check your answer by referring to a complex analysis textbook, such as [16], or by typing some appropriate keywords into Google.

Appendix C
The Lorenz Equations

... the struggle between order and chaos is constant in material reality; the cycle gives the victory to one, and then to the other. Neither ever emerges finally supreme, and reality remains this unstable equilibrium of conflicting forces. Man survives by adjusting himself, by submitting himself to the equilibrium. John L. McKenzie, S.J.[1]

C.1 Introduction

In 1978, Edward N. Lorenz, a meteorologist with the Massachusetts Institute of Technology, presented a paper entitled "On the Prevalence of Aperiodicity in Simple Systems" to a group of mathematicians attending a seminar on global analysis in Calgary, Canada. The paper (cf. [55]) published a year later, gives an account of his efforts in using a digital computer to study the behavior of the solutions of the following system of three ordinary differential equations that have come to be known as the ***Lorenz system*** or the ***Lorenz equations***:

$$
\begin{aligned}
\frac{dx}{dt} &= \sigma(y - x) \\
\frac{dy}{dt} &= rx - y - xz \\
\frac{dz}{dt} &= xy - bz.
\end{aligned}
\tag{C.1}
$$

[1] See [59, Ch. V]. Father McKenzie (1928–2008), a Jesuit, taught at a Catholic seminary and was considered one of the foremost biblical scholars in America.

© The Author(s), under exclusive license to Springer Nature Switzerland AG 2026

L. C. Becker, *Ordinary Differential Equations: Concepts, Methods, and Models*,
https://doi.org/10.1007/978-3-032-15150-6

Contrary to the commonly accepted notion at the time that trajectories of all solutions of systems of ordinary differential equations eventually settle down by approaching certain well-understood limiting sets, Lorenz discovered solutions of (C.1) exhibiting long-term erratic behavior. He referred to these as ***aperiodic solutions***. Their trajectories are atypical in that they do not approach the expected limiting sets, such as equilibrium points or the orbits of periodic solutions, as $t \to \infty$. Instead, they are attracted to a limiting set of points resembling a pair of butterfly wings to the casual observer; in reality, these wings are, in Lorenz's words, "an infinite complex of surfaces." Even though this limiting set of points is bounded, it has infinite surface area, yet zero volume. It is called the ***Lorenz (strange) attractor*** in his honor. Once on the attractor, trajectories will remain there, wandering aimlessly and unpredictably. Moreover, neighboring trajectories will diverge rapidly from one another, even if they began virtually in the same place. This phenomenon is known as ***sensitive dependence on initial conditions***. You may have heard of the ***Butterfly Effect***; this is a nontechnical name for the same phenomenon.[2]

Systems of ordinary differential equations exhibiting aperiodicity and sensitive dependence on initial conditions are called ***chaotic systems***. This was not the first chaotic system that Lorenz had discovered. The first one, discussed in a 1963 paper[3] entitled "Deterministic Nonperiodic Flow," (cf. [56]) was derived from a model of fluid convection. This convective model is a system of twelve equations and had been used to model the atmosphere and to imitate the unpredictability of the weather. However, the convective system is complex and its twelve-dimensional trajectories are impossible for someone to visualize. So Lorenz endeavored to reduce the convective system to a more tractable system with as few equations as possible but with solutions sensitive to initial conditions and exhibiting aperiodic behavior, just like the convective system, albeit no longer modeling the three-dimensional structure of the atmosphere. The result of his endeavor is the three equations that constitute the Lorenz system (C.1). Today, ownership of these celebrated equations is claimed not only by meteorologists and mathematicians but also by physicists, chemists, biologists, engineers, and economists.

C.2 The Lorenz Equations

The original derivation of the Lorenz equations in the 1963 paper requires knowledge of fluid dynamics and areas of mathematics, such as partial differential

[2] See the chapter entitled "The Butterfly Effect" in *Chaos: Making a New Science* by James Gleick [39] for an informal discussion of Lorenz's equations, the Butterfly Effect, aperiodicity, strange attractors—and a delightful account of Lorenz's investigations.

[3] This paper [56] is one of the most cited papers in the literature on ***chaos***, the term used to describe the subject dealing with chaotic systems, irrespective of the scientific field to which the system belongs.

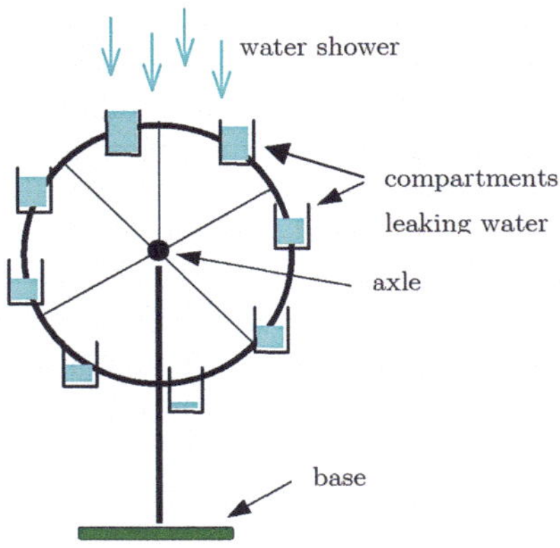

Fig. C.1 Lorenz's water wheel

equations, that are beyond the mathematical level of this book. However, in the 1979 paper, Lorenz was able to derive the equations by modeling the motion of a water wheel that was constructed by Willem Malkus, a colleague of Lorenz and a professor of applied mathematics, in his basement laboratory to convince skeptics that the Lorenz equations do indeed model actual physical phenomena. This derivation does not involve partial differential equations, only ordinary differential equations, and only uses a conservation of mass argument and the physics of the rotational motion of a body about a fixed axis. In what follows, we first describe this laboratory water wheel, hereafter called Lorenz's water wheel, and the observations that Lorenz had made regarding its motion. We then go through Lorenz's 1979 derivation,[4] although with added details and explanations and a few modifications.[5]

The simplest Lorenz water wheel[6] resembles a scaled-down noria, a wheel with buckets fastened to its rim, used in certain parts of the world such as Spain and the Orient to raise water from a stream and transfer it to irrigation ditches. But the Lorenz water wheel is defective in that its buckets, or rather compartments, leak. All of its compartments are identical and symmetrically positioned around the rim. The wheel is free to rotate in a vertical plane about its axle. A pump produces a steady, uniformly distributed shower of water falling over the wheel. See Fig. C.1.

Whether or not the wheel rotates depends on the flow rate of the falling water. Unless water is delivered fast enough so that the leaky compartments fill to the point of sufficiently unbalancing the wheel, it will not rotate. But with a sufficiently fast flow rate, the compartments near the top of the wheel will eventually contain more water than their counterparts near the bottom making the wheel top-heavy enough to overcome the frictional forces in the axle bearings, thereby setting the wheel

[4] This derivation is also based on Colin Sparrow's presentation in Appendix B of his monograph [67] on Lorenz equations.

[5] The reader may wish to consult an introductory college level physics textbook, such as [43], in order to learn more about rotational dynamics.

[6] Strogatz [70, Ch. 9] describes a more sophisticated design used at MIT.

into rotation. By virtue of its symmetry, rotation in either direction, clockwise or counterclockwise, is possible. Which way it starts depends on the initial conditions. At a certain flow rate, the rotation of the water wheel is ***chaotic*** in the sense that irregular reversals in the direction of rotation continually take place. As the wheel rotates in one direction, it eventually slows down and then reverses its direction of rotation. It will then rotate in the other direction for a while, but again it will slow down and reverse its direction. These reversals of direction of rotation occur erratically. The wheel reverses direction when it has slowed down to the point that it does not have enough rotational inertia to carry the compartments holding the most water over the top of the wheel. Let us now look at the derivation of the equations of motion.

To determine the rotational state of the wheel at a given instant, we need to set up a fixed frame of reference. And we need to consider the total mass of the wheel since it is an integral part of the equations governing the motion of the wheel. For simplicity, we ignore the mass of the rim, the parts of the wheel connecting the rim to the axle, and the compartments, except for the water therein. In effect we are envisioning a wheel composed of lightweight materials whose total mass is negligible in comparison to the mass of the water. With this simplification, the wheel's entire mass consists of water only, which is non-uniformly distributed along a massless rim. Even though there is no water at the places on the rim between compartments, we assume that it is distributed continuously around the rim. Let us label the geometric center of the rim with the letter O. And although the rim has extent, we model it with a circle of radius R. With this geometrizing, the real water wheel is modeled by a circle around which the water's mass is continuously and non-uniformly distributed. See Fig. C.2.

As we derive the equations of motion, the word *wheel* will be used as a synonym for this circular model. Let us also model the wheel's axle by the line through O perpendicular to the vertical plane containing the wheel. Let this line serve as the z-axis of a reference frame fixed in space with point O as its origin. Let the vertical line in the vertical plane through O be the x-axis, where the positive direction is upward. Let the horizontal line in the vertical plane through O be the y-axis, where

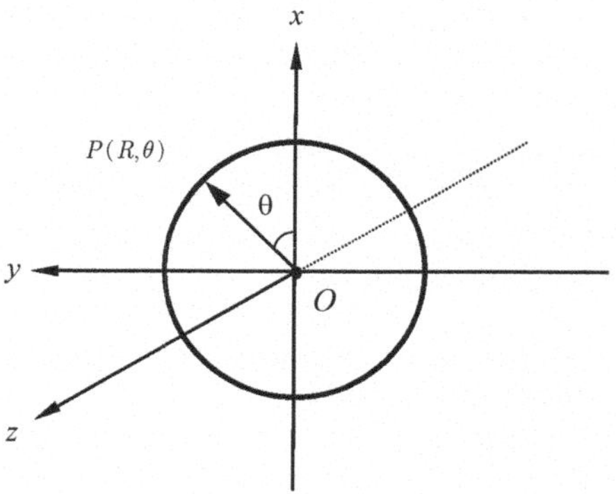

Fig. C.2 Circular water wheel

the positive direction is to the left. We use the ***right-hand rule*** to determine the direction of the positive z-axis: the direction that the extended right thumb points as the fingers of the right hand curl around the z-axis and push the x-axis toward the y-axis through the smaller angle (i.e., 90°) between them. Following this rule makes the reference frame a right-handed Cartesian coordinate system.

We indicate the positions of fixed points in the plane of the wheel by giving their polar coordinates with respect to the center of the wheel, where the ray $\theta = 0$ lies on the positive x-axis. If a line segment from O to a point P on the rim (circle) makes an angle θ with the positive x-axis, then its polar coordinates are (R, θ); let us label this point $P(R, \theta)$. Keep in mind that $P(R, \theta)$ is a point fixed in space which coincides, at a given moment, with some point on the wheel. As the wheel rotates, different points on the wheel move into the position designated by $P(R, \theta)$, but $P(R, \theta)$ itself does not move. Let us observe the convention that the positive (negative) direction of rotation is counterclockwise (clockwise).[7] And so θ increases with counterclockwise motion and decreases with clockwise motion.

Let us assume that there is some continuously differentiable function modeling how the water's mass is distributed around the wheel. In other words, it is a density function giving the mass of water per unit angle around the wheel. Naturally this function would depend on both the time t and angle θ. We let $\rho(t, \theta)$ denote this function and call it the ***mass density function***.

Because of the shower of water falling over the wheel and the leakage of water from the compartments, the wheel is both gaining and losing mass. Lorenz posited that a fixed point $P(R, \theta)$ loses mass at a rate proportional to the mass density function $\rho(t, \theta)$. Following Lorenz, let $\beta > 0$ be the constant of proportionality such that

- $\beta\rho(t, \theta)$ is the rate of loss of mass at the point $P(R, \theta)$.

Thus water leaks from an arc of radius R fixed in space between the rays $\theta = \theta_1$ and $\theta = \theta_2$ at the rate

$$\int_{\theta_1}^{\theta_2} \beta\rho(t, \theta)\, d\theta.$$

This implies that the rate with which water leaks from a compartment is proportional to the mass of water in it.[8] Even though the shower of water, before its contact with the wheel, is uniformly distributed, the rates with which compartments gain water at any given instant vary with their spatial positions at that instant. For a given compartment, this rate is determined by the degree to which the compartments

[7] The parenthetical words mean to read the sentence again with *negative* replacing *positive* and *clockwise* replacing *counterclockwise*.

[8] While this is plausible, in reality not all compartments with holes in their bottoms leak at this rate. According to Strogatz [70, Ch. 9], Malkus attached thin tubes to the holes to make the compartments leak in this linear fashion.

above it obstruct the shower and the proportion of water it receives directly from the shower versus the proportion of leakage it receives from the obstructing compartments. Lorenz assumed that the rate with which mass is added to a fixed point depends only on its position (R, θ).[9] Again following Lorenz, let us assume the existence of a continuously differentiable function A, where

- $A(\theta)$ is the rate of accretion of mass at the point $P(R, \theta)$.

As we described earlier, the motion of the wheel is chaotic alternating from rotations in one direction for a while to being momentarily at rest to rotations in the opposite direction. These reversals in direction of rotation occur erratically. In other words, the wheel's *angular velocity*, denoted by $\omega(t)$, changes at irregular times. Obviously, to determine the rotational state of the wheel at any given instant, we will need to find the differential equations governing the evolution of the angular velocity in time. Finding these equations is really our goal since a certain linear change of variables, as it turns out, will transform them into the Lorenz equations. Generally, the angular velocity is a vector quantity. However, since the water wheel can only rotate about the z-axis, the angular velocity can only point in one of two directions. So the plus and minus signs suffice for representing its vector character: A positive (negative) $\omega(t)$ means that at time t the angular velocity points in the direction of the positive (negative) z-axis and the rotation of the wheel is counterclockwise (clockwise).

Just as the linear momentum of a body can be changed by external forces, the angular momentum of a body, free to rotate about a fixed axis, can be changed by external torques. Let us define what is meant by torque. Suppose a force $\mathbf{F}$ acts at a point P. Let $\mathbf{r}$ be the position vector from a point O to P. The ***torque*** exerted by $\mathbf{F}$ about O is the vector

- which points in the direction of the extended right thumb as the fingers of the right hand curl around the line through O perpendicular to the plane formed by $\mathbf{r}$ and $\mathbf{F}$ so as to push $\mathbf{r}$ into $\mathbf{F}$ through the smaller angle between them

and

- with a magnitude $rF \sin\varphi$, where r, F are the magnitudes of $\mathbf{r}$, $\mathbf{F}$, respectively, and φ is the smaller angle between $\mathbf{r}$ and $\mathbf{F}$.[10]

For the reader familiar with vector cross products, the torque just described is $\mathbf{r} \times \mathbf{F}$.

Angular momentum is related to angular velocity much in the same way that linear momentum is related to velocity (more on this in the next paragraph). The vector sum of all the external torques, or the net torque, acting on a body changes the body's angular momentum in accordance with ***Newton's second law of rotational motion***, which states that the net torque is equal to the time rate of change of the

[9] Actually Lorenz was more specific than this; he assumed that the rate is proportional to the point's height.

[10] That is, the angle between them when their tails coincide.

angular momentum. In other words,

$$\boldsymbol{\tau} = \frac{d\mathbf{L}}{dt}, \tag{C.2}$$

where $\mathbf{L}$ is the angular momentum of the body and $\boldsymbol{\tau}$ is the net torque acting on the body. This is the analog of the following more recognizable form of Newton's second law of motion:

$$\mathbf{F} = \dot{\mathbf{p}}.$$

The angular momentum $\mathbf{L}$ and the net torque $\boldsymbol{\tau}$ are the rotational analogs of the linear momentum $\mathbf{p}$ and the net force $\mathbf{F}$, respectively. Once $\mathbf{L}$ and $\boldsymbol{\tau}$ have been determined for the rotating water wheel, (C.2) will yield one of the equations governing its motion.

Angular momentum is a vector quantity. However, in dealing with the water wheel we can dispense with vector notation since the wheel rotates about a fixed axis, namely, the z-axis. According to rotational mechanics, the ***angular momentum*** of a body rotating about a fixed axis with angular velocity ω is $L = I\omega$,[11] where I, a positive scalar quantity, is the *moment of inertia* of the body.[12] Thus, for such a body, its angular momentum points in only one of two directions, namely, in the direction of its angular velocity, which is along the axis of rotation—the sign of L gives the direction. In the case of the water wheel, a positive (negative) angular velocity means a positive (negative) angular momentum; consequently, both vectors point in the direction of the positive (negative) z-axis and the body rotates counterclockwise (clockwise). $L = I\omega$ is the rotational analog of $p = mv$, the linear momentum of a body of mass m in translational motion with velocity v. The momentum of inertia I is a quantitative measure of a body's rotational inertia, or resistance to a change in its rotational motion, just as the mass m of a body is the quantitative measure of its translational inertia, or resistance to a change in its translational motion.

To find the moment of inertia of the water wheel, we first need to know how to find it for any system of particles rotating about a fixed axis. By definition, the ***moment of inertia*** for such a system is the sum

$$I = \sum_{i=1}^{n} m_i r_i^2, \tag{C.3}$$

where m_i is the mass of a particle which lies at a perpendicular distance of r_i from the axis of rotation. Thus, to find the wheel's moment of inertia, we must first partition, or divide, the wheel into a lot of little pieces that can be considered

---

[11] For an explanation of this formula, see the sections dealing with rotational dynamics in any general physics book, such as *Fundamentals of Physics* [43] by Halliday and Resnick.

[12] Moment of inertia is discussed in the next paragraph.

as particles, all of which are rotating about the z-axis. Then we use the mass density function $\rho(t, \theta)$ to approximate the mass of each piece. Obviously each piece must be small enough for this to be a good approximation. Mathematically, this is accomplished by partitioning the interval $-\pi \le \theta \le \pi$ into n subintervals, each of the same length. This has the effect of dividing the wheel into n arcs, each subtending an angle of $2\pi/n$. The first arc lies between the rays

$$\theta_0 = -\pi \quad \text{and} \quad \theta_1 = -\pi + \frac{2\pi}{n},$$

the second one between

$$\theta_1 = -\pi + \frac{2\pi}{n} \quad \text{and} \quad \theta_2 = -\pi + 2\left(\frac{2\pi}{n}\right),$$

the third one between

$$\theta_2 = -\pi + 2\left(\frac{2\pi}{n}\right) \quad \text{and} \quad \theta_3 = -\pi + 3\left(\frac{2\pi}{n}\right),$$

and so on. In general, the ith arc, for $i = 1$ to $i = n$, lies between the rays

$$\theta_{i-1} = -\pi + (i-1)\frac{2\pi}{n} \quad \text{and} \quad \theta_i = -\pi + i\frac{2\pi}{n}.$$

The last arc, or nth one, lies between the rays

$$\theta_{n-1} = -\pi + (n-1)\frac{2\pi}{n} \quad \text{and} \quad \theta_n = -\pi + n\frac{2\pi}{n} = \pi.$$

If n is large enough, the mass of the ith arc, denoted by Δm_i, is approximately $\rho(t, \theta_i)\Delta\theta$, where

$$\Delta\theta = \theta_i - \theta_{i-1} = \frac{2\pi}{n}.$$

Naturally, the larger n is the smaller will be the arcs (pieces) and the better will be the approximation. Then by (C.3) the moment of inertia I of the wheel about the z-axis is approximately

$$I \approx \sum_{i=1}^{n} \Delta m_i R^2 \approx R^2 \sum_{i=1}^{n} \rho(t, \theta_i)\Delta\theta, \tag{C.4}$$

since all of the mass arcs are at a constant distance of $r_i = R$ from the z-axis. And since the approximation becomes better the larger n is, we let n increase without bound with the result that the Riemann sum in (C.4) approaches the integral

$$m(t) := \int_{-\pi}^{\pi} \rho(t,\theta)\, d\theta. \tag{C.5}$$

The quantity $m(t)$ gives the total mass of the wheel at time t. Therefore, at time t, the moment of inertia $I(t)$ about the z-axis is

$$I(t) = R^2 m(t). \tag{C.6}$$

Consequently, the wheel's angular momentum about the z-axis at time t is

$$L(t) = I(t)\omega(t) = R^2 m(t)\omega(t). \tag{C.7}$$

Now we need to find the net torque acting on the water wheel so that we can apply Newton's second law for rotational motion to obtain an equation of motion that models the angular acceleration of the wheel. First we consider all of the torques resulting from the nonuniform distribution of the mass around the wheel. Recall that Δm_i is the mass of the ith arc in the partition of the wheel into n arcs. The weight of this arc is $W_i = -\Delta m_i g$, where g is the acceleration due to gravity. Its weight is the downward force exerted on the arc by the earth; so it points in the direction of the negative x-axis, hence the negative sign. Since

$$\Delta m_i(t) \approx \rho(t,\theta_i)\Delta\theta,$$

the weight of the ith arc is

$$W_i(t) = -\Delta m_i g \approx -\rho(t,\theta_i)\Delta\theta\, g. \tag{C.8}$$

W_i contributes to the clockwise or counterclockwise rotation of the wheel about the z-axis. The measure of its contribution is given by the torque that it exerts on the wheel about the z-axis.

Note that when a force acts at a point $P(R,\theta_i)$ and points in the direction of the negative x-axis, the angle φ is either $\pi - \theta_i$ or $\pi + \theta_i$. Since $r = R$ and

$$F_i = |W_i(t)| \approx \rho(t,\theta_i)\Delta\theta\, g,$$

the torque $\Delta T_i(t)$ that the weight $W_i(t)$ exerts on the wheel at time t is

$$\Delta T_i(t) = \begin{cases} rF_i \sin\varphi \approx R\big(\rho(t,\theta_i)\Delta\theta\, g\big)\sin(\pi - \theta_i), & \text{if} \quad 0 \le \theta_i \le \pi \\ -rF_i \sin\varphi \approx -R\big(\rho(t,\theta_i)\Delta\theta\, g\big)\sin(\pi + \theta_i), & \text{if} \ -\pi \le \theta_i < 0. \end{cases}$$

As $\sin(\pi \mp \theta_i) = \pm\sin\theta_i$, this simplifies to

$$\Delta T_i(t) \approx R\rho(t,\theta_i)\Delta\theta\, g \sin\theta_i$$

for any value of θ_i. For angles $0 < \theta_i < \pi$, the torque is positive as $\sin\theta_i > 0$, which means it contributes to counterclockwise rotation of the wheel. Otherwise, the torque is either zero or negative, the latter contributing to clockwise rotation. The torque resulting from the weights of all n arcs is given by the Riemann sum:

$$\sum_{i=1}^{n} \Delta T_i = Rg \sum_{i=1}^{n} \rho(t, \theta_i) \sin\theta_i \Delta\theta. \tag{C.9}$$

This is an approximation of the actual torque, which is the limit of (C.9) as $n \to \infty$:

$$Rg \int_{-\pi}^{\pi} \rho(t, \theta) \sin\theta \, d\theta = Rgu(t),$$

where $u(t)$ denotes the integral on the left-hand side.

To sum up, the actual torque resulting from the nonuniform distribution of the weight of the water around the wheel is a constant multiple of the time-dependent integral $u(t)$:

$$\text{Torque due to gravity} = Rgu(t), \tag{C.10}$$

where

$$u(t) = \int_{-\pi}^{\pi} \rho(t, \theta) \sin\theta \, d\theta. \tag{C.11}$$

There is yet another torque to consider: the torque exerted on the wheel by the friction in the axle bearings. Let us assume that it is proportional to the wheel's angular momentum:

$$\text{Torque due to friction} = -kL(t),$$

where k is a positive constant. The negative sign means that this torque opposes the wheel's motion. Finally, we obtain the net torque $\tau(t)$ by adding the two torques:

$$\tau(t) = \text{Torque due to gravity} + \text{Torque due to friction} = Rgu(t) - kL(t). \tag{C.12}$$

Since the angular momentum and net torque vectors have only z-components, it follows from (C.2), (C.7), and (C.12) that the equation of motion of the wheel is

$$\frac{d}{dt}(R^2 m\omega) = Rgu - kR^2 m\omega. \tag{C.13}$$

Let us call this the ***torque equation***. Note that it contains the dependent variable $\omega(t)$, the wheel's angular velocity. Recall that finding ordinary differential equations governing the evolution of $\omega(t)$ is our basic goal: (C.13) is such an equation.

Unfortunately, it also contains two other dependent variables: $m(t)$ and $u(t)$. Consequently, we need to come up with two other ordinary differential equations in order to have a system of three equations with three dependent variables, with $\omega(t)$ as one of those variables. With this in mind, let us differentiate $m(t)$ with the goal of finding another ordinary differential equation. From (C.5) we obtain

$$\frac{dm}{dt} = \frac{d}{dt}\int_{-\pi}^{\pi} \rho(t,\theta)\, d\theta = \int_{-\pi}^{\pi} \frac{\partial}{\partial t}\rho(t,\theta))\, d\theta. \tag{C.14}$$

The interchange in the order of differentiation and integration is made possible by a rule from advanced calculus called ***Leibniz's rule***[13] for differentiating integrals. For there to be any hope of this turning into an ordinary differential equation, we need some kind of equation involving the partial derivative $\partial\rho/\partial t$ that will enable us to carry out the integration. So let us determine the time rate of change of the mass density function $\rho(t,\theta)$ at a fixed point $P(R,\theta)$.

Consider an arc fixed in space covering a portion of the wheel, specifically, an arc of radius R from θ to $\theta + \Delta\theta$. The amount of water mass contained in the arc changes from t to $t + \Delta t$ as the result of the following three processes taking place during this time.

1. The arc gains mass from the source. If $\Delta\theta$ is small enough, $A(\theta)$, the accretive rate function, is nearly constant over the arc. Hence, the rate at which the arc gains mass is approximately $A(\theta)\Delta\theta$. Therefore the arc gains approximately $[A(\theta)\Delta\theta]\Delta t$ of water mass.
2. The arc loses mass from leakage. Recall that $P(r,\theta)$ loses mass at a rate proportional to $\rho(t,\theta)$. Again if $\Delta\theta$ is small enough, the rate at which the arc leaks mass is approximately $\beta\rho(t,\theta)\Delta\theta$. Thus, the arc loses approximately $[\beta\rho(t,\theta)\Delta\theta]\Delta t$ of mass.
3. The arc gains mass at one end but loses it at the other end as a result of the wheel rotating. Suppose the wheel is rotating counterclockwise ($\omega > 0$). For a small enough Δt, the following masses are moved into and out of the arc by the rotating wheel:

   (a) At the point $P(r,\theta)$, the wheel moves approximately $\rho(t,\theta)\omega\Delta t$ of mass into the arc.
   (b) At $P(r,\theta+\Delta\theta)$, the wheel moves approximately $\rho(t,\theta+\Delta\theta)\omega\Delta t$ of mass out of the arc.

   Thus the net amount gained by the arc is $\rho(t,\theta)\omega\Delta t - \rho(t,\theta+\Delta\theta)\omega\Delta t$. Of course, this is really a loss if this quantity is negative. The reader can verify that this same expression holds when the rotation is clockwise ($\omega < 0$).

[13] See an advanced calculus book such as Fulks [37, p. 487].

The mass in the arc at time $t + \Delta t$ is approximately $\rho(t + \Delta t, \theta)\Delta\theta$ while at time t it is $\rho(t, \theta)\Delta\theta$. Since mass must be conserved, we end up with the equation

$$\rho(t + \Delta t, \theta)\Delta\theta \approx \rho(t, \theta)\Delta\theta + [A(\theta)\Delta\theta]\Delta t - \beta[\rho(t, \theta)\Delta\theta]\Delta t + \rho(t, \theta)\omega\Delta t - \rho(t, \theta + \Delta\theta)\omega\Delta t.$$

Dividing this by $\Delta\theta\,\Delta t$, we get

$$\frac{\rho(t + \Delta t, \theta) - \rho(t, \theta)}{\Delta t} \approx A(\theta) - \beta\rho(t, \theta) - \omega(t)\left[\frac{\rho(t, \theta + \Delta\theta) - \rho(t, \theta)}{\Delta\theta}\right].$$

As $\Delta t \to 0$, this becomes

$$\frac{\partial}{\partial t}\rho(t, \theta) \approx A(\theta) - \beta\rho(t, \theta) - \omega(t)\left[\frac{\rho(t, \theta + \Delta\theta) - \rho(t, \theta)}{\Delta\theta}\right].$$

Then, by letting $\Delta\theta \to 0$, we obtain a partial differential equation involving $\partial\rho/\partial t$ as we had hoped for, namely,

$$\frac{\partial\rho}{\partial t} = A - \beta\rho - \omega\frac{\partial\rho}{\partial\theta}. \tag{C.15}$$

Since this equation expresses conservation of mass, we will call it the ***equation of continuity*** in deference to the conservation arguments for deriving similar equations in fluid dynamics.

Whether or not (C.15) will make it possible for us to derive two more ordinary differential equations to complement the torque equation and to give us the system that we are seeking still remains to be seen. So let us forge on. Substituting the right-hand side of (C.15) for $\partial\rho/\partial t$ in (C.14), we obtain

$$\begin{aligned}\frac{dm}{dt} &= \int_{-\pi}^{\pi}\left(A - \beta\rho - \omega\frac{\partial\rho}{\partial\theta}\right)d\theta \\ &= \int_{-\pi}^{\pi} A(\theta)\,d\theta - \beta\int_{-\pi}^{\pi}\rho(t, \theta)\,d\theta - \omega(t)\big[\rho(t, \pi) - \rho(t, -\pi)\big].\end{aligned}$$

Since $\rho(t, \pi) = \rho(t, -\pi)$, this simplifies to

$$\frac{dm}{dt} = \alpha - \beta m, \quad \text{where } \alpha := \int_{-\pi}^{\pi} A(\theta)\,d\theta. \tag{C.16}$$

Since this contains only the dependent m and is a linear equation, we can easily solve for m by using the integrating factor $e^{\beta t}$. Suppose there is no water in the compartments at the start of the experiment. In that case, $m(0) = 0$ and the solution of (C.16) satisfying this initial condition is

$$m(t) = \frac{\alpha}{\beta}(1 - e^{-\beta t}). \tag{C.17}$$

The effect of having (C.17) is that the torque equation (C.13) can now be regarded as an ordinary differential equation with two dependent variables instead of three. With the product rule and some algebraic manipulations, we can rewrite it as

$$\frac{d\omega}{dt} + \left(\frac{\alpha}{m} - \beta + k\right)\omega = \frac{g}{mR}u, \tag{C.18}$$

where m is a known function of t given by (C.17).

In (C.18) we have an ordinary differential equation with the two dependent variables ω and u. Just as we found a differential equation by differentiating m, let us see if we can obtain another one by differentiating u. Applying Leibniz's rule to (C.11), we obtain

$$\begin{aligned}\frac{du}{dt} &= \int_{-\pi}^{\pi} \frac{\partial}{\partial t}\rho(t,\theta)\sin\theta\, d\theta = \int_{-\pi}^{\pi}\left(A - \beta\rho - \omega\frac{\partial\rho}{\partial\theta}\right)\sin\theta\, d\theta \\ &= \int_{-\pi}^{\pi} A(\theta)\sin\theta\, d\theta - \beta u - \omega\int_{-\pi}^{\pi}\frac{\partial\rho}{\partial\theta}\sin\theta\, d\theta. \end{aligned} \tag{C.19}$$

The first integral on the right-hand side is a parameter which depends on the rate at which the water is pumped onto the wheel; let us name it α_s. The second integral can be integrated by parts as follows:

$$\begin{aligned}\int_{-\pi}^{\pi}\frac{\partial\rho}{\partial\theta}\sin\theta\, d\theta &= \left[\rho(t,\theta)\sin\theta\right]_{-\pi}^{\pi} \\ &\quad - \int_{-\pi}^{\pi}\rho(t,\theta)\cos\theta\, d\theta = -\int_{-\pi}^{\pi}\rho(t,\theta)\cos\theta\, d\theta.\end{aligned}$$

Since no more can be done as far as expressing (C.19) in terms of ω or u, the two dependent variables that we already have, let us introduce a new dependent variable, call it v, to complement u:

$$v(t) := \int_{-\pi}^{\pi}\rho(t,\theta)\cos\theta\, d\theta. \tag{C.20}$$

Then (C.19) can be rewritten as

$$\frac{du}{dt} = -\beta u + \omega v + \alpha_s, \tag{C.21}$$

where

$$\alpha_s := \int_{-\pi}^{\pi} A(\theta) \sin\theta \, d\theta.$$

The result of applying Leibniz's rule again to the newly defined v is a third ordinary differential equation to complement (C.18) and (C.21), namely,

$$\frac{dv}{dt} = -\beta v - \omega u + \alpha_c, \tag{C.22}$$

where α_c is the parameter

$$\alpha_c := \int_{-\pi}^{\pi} A(\theta) \cos\theta \, d\theta.$$

Together (C.18), (C.21), and (C.22) constitute a system of three ordinary differential equations with the three dependent variables ω, u, v:

$$\begin{aligned}
\frac{d\omega}{dt} &= \left(\beta - \frac{\alpha}{m} - k\right)\omega + \frac{g}{mR}u \\
\frac{du}{dt} &= -\beta u + \omega v + \alpha_s \\
\frac{dv}{dt} &= -\beta v - \omega u + \alpha_c,
\end{aligned} \tag{C.23}$$

where m is the mass of the wheel given by formula (C.17). Let us refer to these equations as the ***general water wheel equations***. It was our goal to obtain a system of ordinary differential equations containing the wheel's angular velocity as one of the dependent variables—with (C.23) we have succeeded.

Besides its leaking compartments, the symmetrical shower of water sets the Lorenz water wheel apart from an ordinary water wheel propelled by falling water. Water falls down only one side of an ordinary water wheel to make it rotate continually in the same direction. The Lorenz wheel on the other hand can reverse direction because the water shower is symmetrically distributed on both of its sides, that is, with respect to the x-axis. This symmetry along with the symmetrical construction of the wheel implies that the accretive rate function A is an even function on the interval $[-\pi, \pi]$; in other words,

$$A(-\theta) = A(\theta) \quad \text{for all } -\pi \le \theta \le \pi.$$

In his 1979 paper [55], Lorenz assumed that a point $P(R, \theta)$ gains water at a rate proportional to its height, say above the base of the water wheel. If the distance from the base to the center of the wheel O is d, then this height is $d + R\cos\theta$. So

Lorenz's assumption is

$$A(\theta) = K\,(d + R\cos\theta)$$

for some constant K. Note that this particular function is even. This eliminates the parameter α_s from (C.23) since

$$\alpha_s = \int_{-\pi}^{\pi} A(\theta)\sin\theta\; d\theta = \int_{-\pi}^{\pi} K(d + R\cos\theta)\sin\theta\; d\theta = 0.$$

That $\alpha_s = 0$ follows immediately from the integrand $A(\theta)\sin\theta$ being an odd function.[14]

Instead of Lorenz's assumption, let us merely assume that $A(\theta)$ is periodic with period 2π and that it is an even function on $[-\pi, \pi]$. This will allow us to represent it by the infinite trigonometric series

$$A(\theta) = \frac{1}{2}a_0 + \sum_{n=1}^{\infty} a_n \cos n\theta. \tag{C.24}$$

This series is known as a ***Fourier cosine series*** of the function A.

Before continuing, here is a brief synopsis of Fourier series. For a function f defined on the interval $[-\pi, \pi]$, the trigonometric series

$$\frac{a_0}{2} + \sum_{n=1}^{\infty} (a_n \cos n\theta + b_n \sin n\theta)$$

is called the ***Fourier series*** of f if its coefficients are given by the formulas:

$$a_n = \frac{1}{\pi}\int_{-\pi}^{\pi} f(\theta)\cos n\theta\; d\theta \quad \text{for } n = 0, 1, 2, \ldots$$

and

$$b_n = \frac{1}{\pi}\int_{-\pi}^{\pi} f(\theta)\sin n\theta\; d\theta \quad \text{for } n = 1, 2, 3, \ldots.$$

It can be proven that the Fourier series converges to f at every point θ if f is continuously differentiable and periodic with period 2π. If f is also an even function, then the ***Fourier coefficients*** have the values

$$a_n = \frac{2}{\pi}\int_{0}^{\pi} f(\theta)\cos n\theta\; d\theta \quad \text{for } n = 0, 1, 2, \ldots$$

[14] Recall $\int_{-c}^{c} f(s)\,ds = 0$ if f is an odd function; i.e., a function with the property $f(-s) = -f(s)$.

and

$$b_n = 0 \quad \text{for } n = 1, 2, 3, \ldots .$$

In this case, the Fourier series simplifies to

$$\frac{1}{2}a_0 + \sum_{n=1}^{\infty} a_n \cos n\theta$$

and is called the ***Fourier cosine series of*** f—hence, (C.24).[15]

Integrating to determine α_s, we have

$$\alpha_s = \int_{-\pi}^{\pi} A(\theta)\sin\theta \, d\theta = \int_{-\pi}^{\pi} \left(\frac{1}{2}a_0 + \sum_{n=1}^{\infty} a_n \cos n\theta \right) \sin\theta \, d\theta$$
$$= \frac{1}{2}a_0 \int_{-\pi}^{\pi} \sin\theta \, d\theta + \sum_{n=1}^{\infty} a_n \int_{-\pi}^{\pi} \cos n\theta \sin\theta \, d\theta.$$

Since $\cos n\theta \sin\theta$ is an odd function on $[-\pi, \pi]$, each of the integrals is equal to zero. Therefore, $\alpha_s = 0$ when water falls symmetrically onto the wheel.

There is a further simplification. From (C.17) we see that the wheel's mass consists of the transient term $\frac{\alpha}{\beta}e^{-\beta t}$ and the steady-state term α/β; so it decays exponentially to α/β. In other words, after the transient term has decayed, the torques are acting on a wheel whose mass has reached the constant value α/β, even though water mass is continually being lost, gained, and redistributed around the wheel. Thus, the torque equation in (C.23) simplifies even further to

$$\frac{d\omega}{dt} = -k\omega + \frac{\beta g}{\alpha R}u.$$

To sum up, after the mass of a water wheel in a symmetrical shower has reached a constant value, its motion is governed by the system

$$\begin{aligned} \frac{d\omega}{dt} &= -k\omega + \frac{\beta g}{\alpha R}u \\ \frac{du}{dt} &= -\beta u + \omega v \\ \frac{dv}{dt} &= -\beta v - \omega u + \alpha_c, \end{aligned} \tag{C.25}$$

[15] For more information about Fourier series, see a book on the subject, such as Churchill [20], or certain calculus books, such as Smith-Minton [65, Ch. 8].

Let us call these equations the ***constant mass water wheel equations***.

Next we will show that a linear change of variables will transform the constant mass water wheel equations (C.25) into the Lorenz system (C.1), where $b = 1$. Define the new variables x, y, z, and T by

$$x = -\frac{1}{\beta}\omega, \quad y = -\frac{p}{k}u, \quad z = -\frac{p}{k}v + r, \quad T = \beta t, \tag{C.26}$$

where p and r are new parameters defined by

$$p = \frac{g}{\alpha R} \quad \text{and} \quad r = \frac{p\alpha_c}{\beta k}. \tag{C.27}$$

Using the Chain Rule, the first equation of (C.25) transforms to

$$\begin{aligned}
\frac{dx}{dT} &= \frac{dx}{dt} \cdot \frac{dt}{dT} = \frac{1}{\beta}\frac{dx}{dt} = \frac{1}{\beta}\left(-\frac{1}{\beta}\frac{d\omega}{dt}\right) \\
&= -\frac{1}{\beta^2}\left(-k\omega + \frac{\beta g}{\alpha R}u\right) = -\frac{k}{\beta}\left(-\frac{\omega}{\beta}\right) + \frac{1}{\beta}(-pu) \\
&= -\frac{k}{\beta}x + \frac{1}{\beta}(ky) = \frac{k}{\beta}(y - x).
\end{aligned}$$

Letting $\sigma := k/\beta$, this simplifies to

$$\frac{dx}{dT} = \sigma(y - x).$$

The second equation of (C.25) transforms to

$$\begin{aligned}
\frac{dy}{dT} &= \frac{1}{\beta}\frac{dy}{dt} = \frac{1}{\beta}\left(-\frac{p}{k}\frac{du}{dt}\right) = -\frac{p}{\beta k}(-\beta u + \omega v) \\
&= \frac{p}{k}u + \left(-\frac{\omega}{\beta}\right)\left(\frac{p}{k}v\right) = -y + x(-z + r).
\end{aligned}$$

Thus,

$$\frac{dy}{dT} = rx - y - xz.$$

It is left as an exercise to show that third equation of (C.25) transforms to

$$\frac{dz}{dT} = xy - z.$$

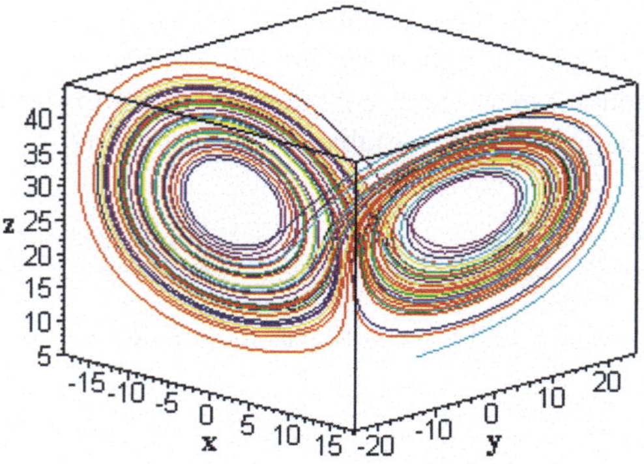

Fig. C.3 A trajectory with the initial point $(x_0, y_0, z_0) = (5, 5, 5)$

In summary, the linear change of variables (C.26) transforms system (C.25) to the system

$$\begin{aligned}\frac{dx}{dT} &= \sigma(y - x)\\ \frac{dy}{dT} &= rx - y - xz\\ \frac{dz}{dT} &= xy - z.\end{aligned} \tag{C.28}$$

Note that if the name of the independent variable is changed to t, then (C.28) is identical to the Lorenz system (C.1) except for the absence of the parameter b in the third equation—so the constant mass water wheel equations are not quite as general as the Lorenz equations.

In his studies of (C.1), Lorenz assigned to the parameters the values $\sigma = 10$, $b = 8/3$, and $r = 28$. The trajectory corresponding to the initial condition $x(0) = 5$, $y(0) = 5$, $z(0) = 5$ is shown in Fig. C.3. If we change the initial value of the x-coordinate to $x(0) = 100$ but keep the initial values of the other two coordinates the same as before, we obtain the trajectory shown in Fig. C.4. Note how the trajectory eventually ends up looking like a pair of butterfly wings. In fact, regardless of the initial conditions, trajectories always are attracted to and wander aimlessly on an extremely complicated set of points resembling butterfly wings. For this reason, it is often called "the butterfly."[16] This is the Lorenz strange attractor that we mentioned earlier.

[16] Lorenz's recollections of the coinage of the terms "the butterfly" and "The Butterfly Effect" are found in his book *The Essence of Chaos* [54, pp. 14–15].

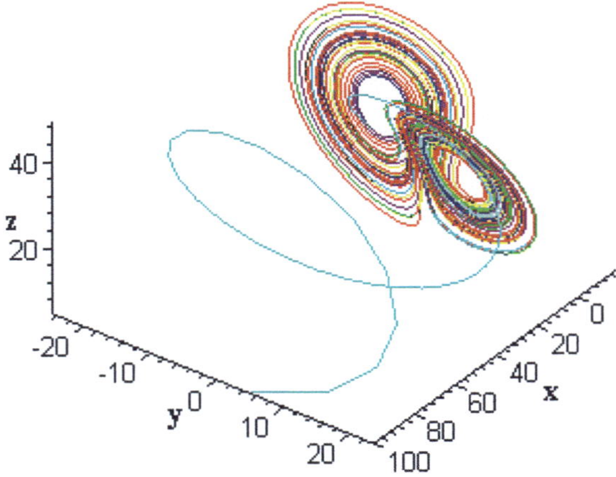

Fig. C.4 A trajectory with the initial point $(x_0, y_0, z_0) = (100, 5, 5)$

Problems

... what is precious is not the reward but the work. And I wish you to understand that. If you work and study in order to get a reward, the work will seem hard to you; but when you work ... if you love the work, you will find your reward in that.

Alexi Alexandrovich's advice to his son Seryozha
Anna Karenina by Leo Tolstoy [73, Ch. XXVII]

The Lorenz Equations

1. Use Leibniz's rule to derive Eq. (C.22).
2. Using the linear transformation (C.26), verify that the third equation of the constant mass water wheel system (C.25) transforms to the third equation of (C.28).
3. Suppose the plane of rotation of the water wheel makes an angle δ with the horizontal so that the wheel rotates in a tilted plane rather than in a vertical plane. For a frame of reference, tilt the one used in the vertical case so that it too makes an angle δ with the horizontal. Assume that the wheel is constructed so that the rates of accretion and loss of mass at a point $P(r, \theta)$ are as in the vertical plane case. Show that the form of the equation of continuity (C.15) does not change. Is the torque equation (C.13) affected? If so, how does that change the constant mass water wheel equations?

Computer Algebra System Problems

4. Use a computer algebra system to view trajectories of the Lorenz system when the values of the parameters are $\sigma = 10$, $b = 8/3$, and $r = 28$. Try other parameter values.
5. Experiment with different parameter values to view trajectories of the general water wheel equations.

Answers to Selected Problems

Chapter 1

1. independent variable: t; dependent variable: N; parameters: r, k; order: 1st.
3. independent variable: t; dependent variable: x; parameters: a, b, γ; order: 4th.
5. independent variable: x; dependent variable: y; parameters: E, I, p, k, c; order: 4th.
7. independent variable: x; dependent variable: ψ; parameters: m, k, $\hbar$, E; order: 2nd.
9. $\partial f/\partial x = 30x - 12x^3y^3$;
$\partial f/\partial y = -9x^4y^2 + \dfrac{10}{3}y^4$
11. $\partial f/\partial x = 1 + 5ye^{2x}\cos(xy)$
$+ 10e^{2x}\sin(xy)$;
$\partial f/\partial y = 5xe^{2x}\cos(xy)$
14. If $C = 0$, $y = x$ is a solution on any interval. If $C \neq 0$, then $y = x/(1 + Cx)$ is a solution on any interval that excludes $x = -1/C$.
17. $y = e^{2x}$ is not a solution.
19. $y = \dfrac{1}{3}\sin 3x + C$
21. $y = \dfrac{1}{\sqrt{5}}\tan^{-1}\left(\dfrac{x}{\sqrt{5}}\right) + C$
23. $y = -\dfrac{1}{5}\ln|\cos 5x| + C$
25. $y = \dfrac{x}{2}e^{2x} - \dfrac{1}{4}e^{2x} + C$
27. $y = -\dfrac{x^2}{2}\cos 2x + \dfrac{x}{2}\sin 2x$
$+ \dfrac{1}{4}\cos 2x + C$
29. $y = \dfrac{3}{4}\ln\left|\dfrac{x}{x+4}\right| + C$
31. $y = 2\ln|x^2 - 5x + 6| + C$
33. $y = 3\tan^{-1}(x + 2) + C$
35. $y = \dfrac{1}{4}\sec^2(2x) + C$
37. $y = 5\sin^{-1}\left(\dfrac{x}{2}\right) + C$
39. $y = \dfrac{1}{6}\sec^{-1}\left(\dfrac{x}{3}\right) + C$
41. $y = \sin(\sqrt{x}) - \sqrt{x}\cos(\sqrt{x}) + C$
44. $y = C - x$
45. $u = 2x^3y + 2xy^{-1} + g(y)$
47. $u = 5y + \frac{1}{2}x\tan^{-1}\left(\frac{y}{2}\right) + h(x)$
49. $u = x^2\cos(\sqrt{y}) + g(y)$
51. $dN/dt = -25$, where $N(t)$ is the number of gallons in the city pool at time t.
53. Let $P(t)$ be the amount of pollutants in the water at time t minutes. Then, $dP/dt = -0.05P$.
57. $dp/dt = bp$, where p is the population of the squirrels at time t and $b > 0$ is the constant of proportionality.
59. $dr/dt = -k$, where r is the radius of the snowball at time t and $k > 0$ is the constant of proportionality.
65. (a) $v = C\sqrt{y}$, where v is the speed of the water through the hole, y is the

© The Author(s), under exclusive license to Springer Nature Switzerland AG 2026

L. C. Becker, *Ordinary Differential Equations: Concepts, Methods, and Models*,
https://doi.org/10.1007/978-3-032-15150-6

water's depth, and C is a positive constant of proportionality.

(b) $dy/dt = -k\sqrt{y}$, where y is the depth of the water in the tank at time t and $k > 0$ is the constant of proportionality.

(c) $dy/dt = -Ky$, where $K > 0$ is the constant of proportionality.

68. From Newton's second law of motion,

$$m\dot{v} = -32m - kv,$$

where m is the mass of the body, v is its velocity at time t, and $k > 0$ is the constant of proportionality.

70. $dz/dt = -\lambda z$, where $z = [O_3]$ denotes the concentration of ozone in the atmosphere at time t and $\lambda > 0$ is the constant of proportionality.

72. $dv/dt = -kv^2/m - g$, where $k > 0$ is the constant of proportionality.

83. (a) $-2/3$ ft/sec

(b) 2075 ft

84. (a) $dV/dt = k_1 p$, where V denotes the volume of fluid that has flowed past a given cross section of the pipe at time t and $k_1 > 0$ is the constant of proportionality.

(e) $\dfrac{dV}{dt} = \lambda \dfrac{pr^4}{l\eta}$, where $\lambda > 0$ denotes the constant of proportionality.

(g) About 34%.

87. $2 - \pi^2/6$

Chapter 2

1. Not separable
3. Not separable
5. Solution
7. Not a solution
11. $y = \pm\sqrt{x^3 + C}$
14. $y = \dfrac{x^2}{1 + Kx^2}$
16. $y + \ln(xy)^2 - 2x = C$
17. $y = \dfrac{1}{C - 2x^{3/2}}$
18. $y = \dfrac{1}{2}\ln|x^2 + 2x| + C$
20. $x = \left(K - \ln|\cos t|\right)^2 - 5$
22. $s = Ce^{2t - \sin t}$
24. $t^2 + \ln(\csc y + \cot y)^2 = C$
26. $y = Ce^{\sin^{-1} x} - 1$
28. $y = \dfrac{1}{2}\ln\left[\ln(x^2 + 4) + C\right]$
31. $x = \tan(t - \frac{1}{2}t^2 + C)$
32. $p = \ln t^2 - te^{-t} - e^{-t} + C$
36. $x = Ce^{-6\sqrt{4 - t^2} - 3\sin^{-1}(t/2)}$
37. $4\sin y - 4y\cos y = 3\sin^{-1}\left(\dfrac{x}{3}\right) + C$
43. $y = 2\tan(2x^2 + C)$
44. $x = \dfrac{\sin t}{1 + C\sin t}$
46. $y = \arcsin\left(Ce^{\arcsin x}\right)$
47. $x = \left(\dfrac{3}{4}t^2 + 24\right)^{2/3}$
48. $y = \pi - \cos^{-1}(x^2)$
49. $y = 5 - 3e^{x^2}$
50. $x = \ln(3e^{2t} - 2)$
51. $y = -\sqrt{15 - \sec x}$
53. $y = \dfrac{1}{\cos x - x^2 + \pi^2}$
54. $y^2 = 34\sqrt{7}\tan^{-1}\left(\dfrac{x-2}{\sqrt{7}}\right) + \dfrac{14(3x - 13)}{x^2 - 4x + 11} + 14$
56. $N(60) = 1250e^{-3.6} = 34.2$ lbs
57. 0.637 hours
60. 4:46 A.M.
62. $A = 100{,}000(1.05)^t$ dollars
64. 29.2 minutes
65. $\alpha = 0.748$, $b = 1.01$, 153.4 million francs
67. $y = \dfrac{x^2}{x + C}$
69. $\sin(y/x) = \ln|x| + C$
71. (c) The solution of $\tan t = t$ for $t \in (0, 2\pi)$, which to 3 decimal places is $t = 4.493$.

(d) $2\pi^2 a$

(f) ka^2, where $k = 117.5958\ldots$

Chapter 3

1. (a) Slope of the direction arrow at (0, 1):

$$\frac{dy}{dx} = \left[(1-x)y - x\right]_{(x,y)=(0,1)}$$
$$= (1-0)(1) - 0 = 1$$

(b)

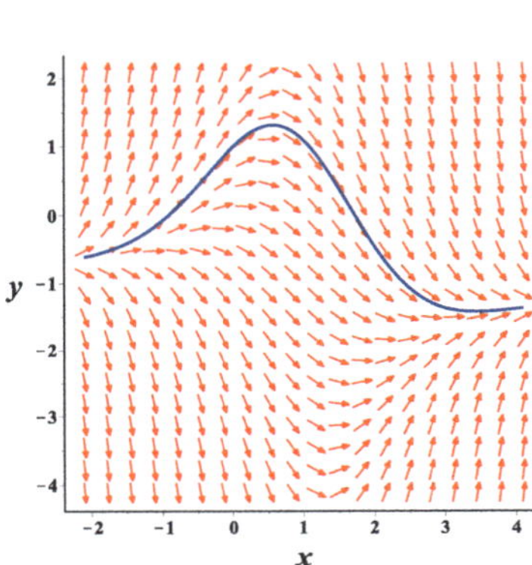

2. (a) Slope of the direction arrow at (−2, 4):

$$\frac{dy}{dx} = \left[\frac{x^2}{2} - \frac{y^2}{5}\right]_{(x,y)=(-2,4)}$$
$$= \frac{(-2)^2}{2} - \frac{4^2}{5} = -1.2$$

(b)

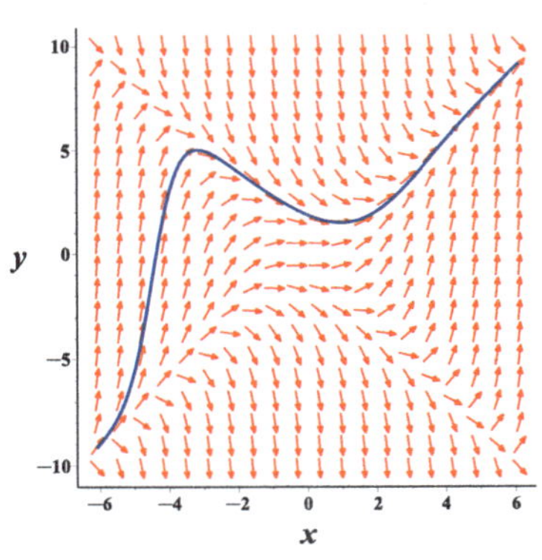

5. (a) 60
(b) $y(x) \equiv 0$ and $y(x) \equiv -4$
(c) positive
(d) negative
(e)

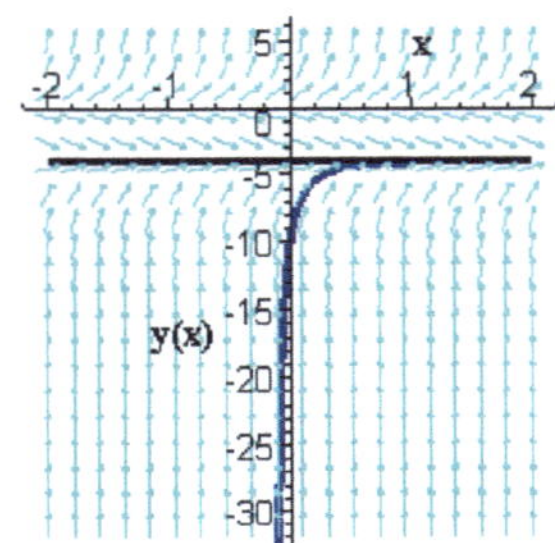

8. (a) $y(x) \equiv -1$ and $y(x) \equiv 2$
(b)

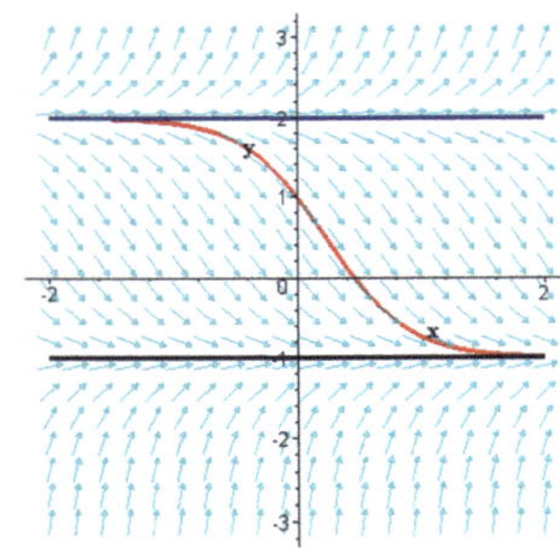

9. (a) $y'' = (4-y)(1-x^2)$
(b)

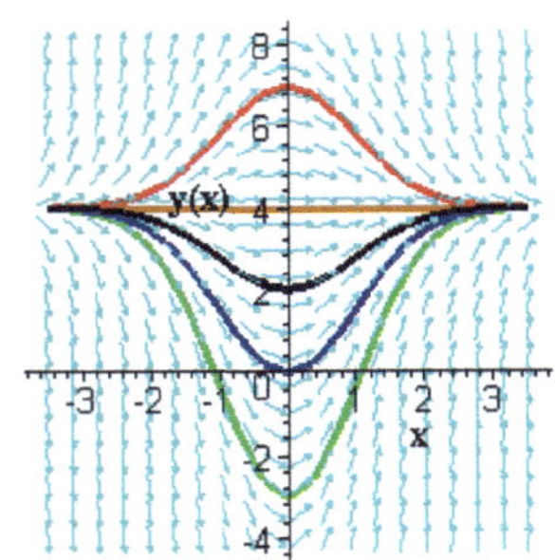

17. $k = 1.529$ lb-hr/mile; 130.8 mph

20. Using (3.3.7), $k = 0.5 \ln 2$ kg/s.
Distance = $50g/(\ln 2)^2$ meters or about 1020 m using $g = 9.8$ m/s^2.

Chapter 4

1. The recursive formulas are

$$x_{n+1} = x_n + 0.5$$
$$y_{n+1} = y_n + 0.5(x_n - y_n^2).$$

Since $x_0 = 0$ and $y_0 = 1$,

$$x_1 = 0 + 0.5 = 0.5$$
$$y_1 = y_0 + 0.5(x_0 - y_0^2)$$
$$= 1 + 0.5(0 - 1^2) = 0.5$$

and

$$x_2 = x_1 + 0.5 = 0.5 + 0.5 = 1$$
$$y_2 = y_1 + 0.5(x_1 - y_1^2)$$
$$= 0.5 + 0.5\big(0.5 - (0.5)^2\big) = 0.625.$$

Thus, with $h = 0.5$, $y(1) \approx 0.625$.

3. (a) $x_{n+1} = x_n + 0.04$
$y_{n+1} = y_n + 0.02 y_n \sqrt{x_n}$
(b) First set $n = 0$:

$$x_1 = x_0 + 0.04 = 0.04$$
$$y_1 = y_0 + 0.02 y_0 \sqrt{x_0}$$
$$= 1 + 0.02(1)\sqrt{0} = 1.$$

Next, let $n = 1$:

$$x_2 = x_1 + 0.04 = 0.08$$
$$y_2 = y_1 + 0.02 y_1 \sqrt{x_1}$$
$$= 1 + 0.02(1)\sqrt{0.04} = 1.004.$$

With $h = 0.04$, $y(0.08) \approx 1.004$.

7. (a) The recursive formulas are

$$x_{n+1} = x_n + 0.5$$
$$y_{n+1} = y_n - 0.5 x_n y_n^{-1}.$$

(b) Set $n = 0$:

$$x_1 = x_0 + 0.5 = 3.5$$
$$y_1 = y_0 - 0.5\left(\frac{x_0}{y_0}\right)$$
$$= 4 - \left(\frac{1}{2}\right)\left(\frac{3}{4}\right) = 3.625.$$

Thus, with $h = 0.5$, $y(3.5) \approx 3.625$.

(c) Separating variables:

$$\int y\, dy = -\int x\, dx + K.$$

Thus,

$$x^2 + y^2 = C.$$

Since $y = 4$ when $x = 3$, $C = 25$. Therefore,

$$y = \sqrt{25 - x^2}.$$

(d) To the nearest thousandth, the value of the solution at $x = 3.5$ is

$$y(3.5) = \sqrt{25 - (3.5)^2} = 3.571.$$

The Euler approximation exceeds the exact value by $3.625 - 3.571 = 0.054$.

10. (c) 2, 2.25, 2.44141, 2.56578, 2.63793, 2.71696, 2.71828
(d) $\lim_{m\to\infty}\left(1 + \frac{1}{m}\right)^m = e$

12. (a) The recursive formulas are

$$x_{n+1} = x_n + 0.25$$
$$y_{n+1}^E = y_n + 0.25(1 - 2x_n y_n)$$
$$y_{n+1} = y_n + 0.125\big(2 - 2x_n y_n - 2x_{n+1} y_{n+1}^E\big).$$

(b) $y(0.25) \approx \frac{15}{64} = 0.234375$
$y(0.50) \approx \frac{1691}{4096} \approx 0.412842$
(c) $y(0.50) \approx 0.424023$

16. (a) $y(0.2) \approx 0.19475069$
(b) $y(1) \approx 0.53807645$

17. (a) $y(-0.8) \approx 1.05811271$
(b) $y(1.5) \approx 2.54679973$

Chapter 5

1. linear
3. both
5. neither
7. both
10. neither
11. $x = \frac{1}{2}e^t + Ce^{-t}$
13. $x = Ce^{2t} - \dfrac{3}{2}t - \dfrac{3}{4}$
15. $y = -2x^3 + Cx^2$
16. $x = 5e^{-t} + Ce^{-2t}$
17. $y = \dfrac{x}{2} + Cx^{-3} \quad (x \neq 0)$
19. $y = x^2 + Cx$
24. linear: $x = \dfrac{8}{3}t^2 + \dfrac{C}{t}$
27. linear: $x = t + Ct^{-2}$
28. separable: $y = \sin^{-1}(x^5 + K)$
29. neither of these
31. separable: $x = C(t + 3)$
33. $y = x \sin x + \pi^{-1}x$
35. $y = \left(3 + 2\sqrt{\pi}\,\mathrm{erf}(t)\right) e^{t^2}$
39. In terms of Dawson's integral $\mathcal{D}(t)$, the solution is

$$v = 32\sqrt{2}\left[e^{32-\frac{1}{2}(t+8)^2}\mathcal{D}(4\sqrt{2}) - \mathcal{D}\left(\tfrac{1}{\sqrt{2}}(t+8)\right)\right].$$

45. (c) Interval of convergence: $(-\infty, \infty)$.
47. $x = \dfrac{\mathrm{Si}(t) - \mathrm{Si}(2) + 2}{t}$
56. $x = \dfrac{9}{9Ce^{6t} + 5e^{-3t}}$.
57. $x = \sqrt[3]{\dfrac{t}{2 + C\sqrt{t}}}$
61. $x = \dfrac{3\cos^2 t}{C + \cos^3 t} - \sec t$
63. $x = \dfrac{2}{Ct^3 - t} + \dfrac{1}{t}$
64. (a) $v^2 = \dfrac{2g}{3}\left(y - \dfrac{{y_0}^3}{y^2}\right)$
(b) $a = \dfrac{g}{3}\left[1 + 2\left(\dfrac{y_0}{y}\right)^3\right]$

Chapter 6

1. $-500 \ln \dfrac{27}{32} \approx 84.9$ minutes
4. 1.05 pounds of salt after 3 minutes
7. 9.21 minutes
8. (b) $h(t) = 3.0059e^{-0.0809t} - 0.0059$
12. $A(t) = A_0 2^{-t/T}$
14. $dx/dt = m - \lambda x$, where $x(t)$ is the amount of radioactive isotope at time t. Its solution is

$$x(t) = \frac{m}{\lambda} + \left(M - \frac{m}{\lambda}\right)e^{-\lambda t}.$$

As $t \to \infty$, $x(t) \to \dfrac{mT}{\ln 2}$.
16. $\ln 2/.0085 \approx 81.5$ years
18. $dp/dt = rp + m$. Its solution is

$$p(t) = \left(p_0 + \frac{m}{r}\right)e^{rt} - \frac{m}{r},$$

where p_0 is the population at $t = 0$.
20. $300{,}000(1.02)^{10} \approx 365{,}698$ people
23. Population: $p(t) = 2000e^{0.20t} + 3000$. So, in 5 years, there will be about 8, 436 prairie dogs.
25. (b) $V(t) = V_0 e^{\frac{a}{b}(1-e^{-bt})}$
(c) $V_0 e^{a/b}$
28. After 3 years, the balance is \$231.53. The APR is 4.879% compounded continuously.
30. Euler's method applied to $dy/dt = f(t, y)$ with $f(t, y) = ry$ (see (4.3.4) and (6.3.16)) yields

$$y_{k+1} = y_k + hf(t_k, y_k) = (1 + hr)y_k.$$

With $y_k = B_k$ and $h = 1/m$, this becomes

$$B_{k+1} = \left(1 + \frac{r}{m}\right) B_k.$$

This implies (6.3.29), where $B_0 = P$.

32. (b) Daily: 9.99863%. For $-1 < x \le 1$,

$$\ln(1 + x) = x - \frac{x^2}{2} + \frac{x^3}{3} - \dots.$$

Thus, for r/m small,

$$m \ln\left(1 + r/m\right) \approx r - r^2/2m.$$

Chapter 7

1. $f(t) = u(t-2) - 6u(t-4)$
3. $h(t) = (t-2)\,H(t-2) - 3(t-4)\,H(t-4) + 2(t-5)\,H(t-5)$
5. $\varphi(t) = e^{2t}[1 - u(t-5)]$
7. $x(t) = \sin t - \sin t \cdot u(t - \pi/2) + (t-2)\,u(t-2) + (3-t)\,u(t-3)$
9. $s(t) = -1 + 2\sum_{k=1}^{\infty}(-1)^{k-1}u(t-k)$
11. $h(t) = 1 + \sum_{j=1}^{\infty} u(t-j)$
12. For $t \ge 0$,

$$x(t) = 1 + \begin{Bmatrix} 0, & t < 1 \\ 1 - t, & t > 1 \end{Bmatrix} = 1 + (1-t)u(t-1).$$

Its formal derivative is

$$x'(t) = (1-t)\delta(t-1) - u(t-1).$$

Since $\delta(t-1) = 0$ for $t \neq 1$, the interpretation is: $x'(t) = 0$ if $t < 1$ whereas $x'(t) = -1$ if $t > 1$. These values are the slopes of the line segment and the line that constitute the graph of x. If $t = 1$, $\delta(t-1)$ is undefined. Consequently, the derivative of x is undefined at $t = 1$ (as it should be since the graph of x has a corner point there).

17. $x(t) = 5e^{-t} + e^{-(t-2)}u(t-2)$. That is,

$$x(t) = \begin{cases} 5e^{-t}, & t < 2 \\ 5e^{-t} + e^{-(t-2)}, & t > 2. \end{cases}$$

19. (a) $x'(t) = 50 - (2t-8) \cdot u(t-4) - (t^2 - 8t + 17) \cdot \delta(t-4)$
    (b) $x(t) = 1000 + 50t - (t^2 - 8t + 17) \cdot u(t-4)$
    (c) $0 \le t \le 71.7$ minutes
21. (a) $R \approx 343$ mustangs per year.
    (b) $R \approx 136$ mustangs per year.
25. 3.33%

Chapter 8

1. $F(x, y) = e^{2x}y;\ \ y = Ce^{-2x}$.
2. $F(x, y) = x^2 \sin 2y$; $y = 0.5 \sin^{-1}\left(\dfrac{C}{x^2}\right)$
3. $F(x, y) = x^2y + 3x - y$; $y = \dfrac{(C - 3x)}{(x^2 - 1)}$.
4. $\dfrac{dF}{dx} = (y - 2x) + (x + e^y)\dfrac{dy}{dx}$
   General solution: $xy - x^2 + e^y = C$.
5. A general solution is $G = C$. That is, $e^y \cos x + x^2 + 3y^{1/3} = C$.
6. $g = C$, i.e., $y + x \ln y = C$.
9. Second-order partial derivatives:

$$f_{xx} = 36x^2y^2 + y^2\cos(xy)$$

$$f_{xy} = 24x^3y + xy\cos(xy) + \sin(xy) + 5\sec^2 y = f_{yx}$$

$$f_{yy} = 6x^4 + x^2\cos(xy) + 10x\sec^2 y \tan y$$

13. $xy - y - x^3 = C \Rightarrow y = \dfrac{x^3 + C}{x - 1}$
15. $x \tan y + \ln|y| - x^2 = C$

17. $\frac{1}{2}e^{2y} - xy - x^3 = C$
18. $\sin(2y) - x^2 - 2xy^2 = K$
19. $y = 2x - 3 \pm \sqrt{(3-2x)^2 - 2\tan x + C}$
20. $5x^3y - y^2 = C$
22. separable: $y = \sin^{-1}\left(\frac{x^5}{5} + C\right)$
24. linear: $y = x^2 + Cx$
26. exact: $x^3 + 2x^2y^2 = C$
30. linear: $y = \dfrac{-3\ln|\cos x| + C}{x+1}$
34. linear (also exact): $y = \dfrac{2x + C}{5x^2 - \cos x}$
35. separable: $y = \ln\left(Ce^{\tan^{-1}x} - 1\right)$
36. none of these types
37. exact: $x^2 + x\sin y + x\cos y = C$
38. exact (also linear): $y = \dfrac{C + e^x}{\sin x}$
40. The integral curves are the circles $x^2 + y^2 = C$ $(C > 0)$ and $(0, 0)$. The solution curve is the graph of $y = -\sqrt{13 - x^2}$ for $|x| < \sqrt{13}$.
42. On an open rectangular region that excludes $y = 0$, the integral curves are defined by the equation $y^2 + \ln y^2 - 4\sin x = C$.
44. $2x^3y + x^2(3y^2 - 1) = C$
46. $2x^2 + xy + y\ln y^2 = Cy$ for $y \neq 0$ (also, $y(x) \equiv 0$)
48. $x^2y^3\sin y = C$

Chapter 9

1. $y_3(x) = 4 + 4x^2 + 2x^4 + \frac{2}{3}x^6$.
Exact solution: $y = 4e^{x^2}$.
The 6th degree Maclaurin polynomial and $y_3(x)$ are equal.
3. $y_3(x) = \frac{1}{2}x^2 + \frac{1}{20}x^5 + \frac{1}{160}x^8 + \frac{1}{4400}x^{11}$.
Approximation: $y_3(0.75) = 0.29375$.
7. The theorem guarantees a unique solution exists on $[-1/2, 1/2]$. The third Picard approximation is

$$y_3(x) = 1 - x + x^2 - x^3 + \frac{2}{3}x^4 - \frac{1}{3}x^5 + \frac{1}{9}x^6 - \frac{1}{63}x^7.$$

The approximations suggest $y = 1/(1+x)$ is the solution. In fact, it is the solution on $(-1, \infty)$, which can be obtained by separating variables and integrating.

10. Picard's recursive formula is

$$w_n(x) = \int_0^x \left[1 + \int_0^u w_{n-1}(v)\,dv\right] du,$$

where $w_0(x) \equiv 0$. The nth Picard approximation is

$$y_n(x) = 1 + \frac{x^2}{2!} + \frac{x^4}{4!} + \cdots + \frac{x^{2n}}{(2n)!}.$$

The solution is

$$y(x) = \sum_{n=0}^{\infty} \frac{x^{2n}}{(2n)!} = \cosh x$$

for $x \in \mathbb{R}$.

Chapter 10

3. $x(t) = c_1 + c_2e^{9t}$
4. $x(t) = c_1e^{2t} + c_2e^{-5t}$
6. $y(x) = c_1e^{-18x/5} + c_2e^{2x/3}$
7. $y = c_1e^{-3t} + c_2e^{t/2}$
8. $y(t) = c_1 + c_2e^{7t/2}$
10. $y(t) = c_1e^{-3t/5} + c_2e^{9t/4}$
12. $x(t) = c_1\cos(t\sqrt{5}) + c_2\sin(t\sqrt{5})$
13. $s(t) = c_1e^{5t/3} + c_2e^{-7t/4}$
16. $x(t) = c_1e^{-2t} + c_2te^{-2t}$
17. $x(t) = c_1e^{t} + c_2e^{-3t/2}$
21. $x = c_1e^{-5t/3} + c_2te^{-5t/3}$
22. $y = c_1e^{-x}\sin(x/3) + c_2e^{-x}\cos(x/3)$
23. $y(x) = c_1e^{5x/2} + c_2xe^{5x/2}$
24. $p(s) = c_1 + c_2e^{5s}$

28. $x(t) = -te^t$

30. $y(t) = e^{-7t} + 4e^{2t}$

35. $x(t) = c_1 \cos(\ln t) + c_2 \sin(\ln t)$

37. $x(t) = c_1 t^{3/5} + c_2 t$

39. (a) $x(t) = c_1 t + c_2 t \ln |t|$
    (b) $x(t) = 3t - 2t \ln t$

40. $x(t) = \frac{2}{t} \cos(\ln t) + \frac{7}{t} \sin(\ln t)$

42. $x(t) = c_1 e^{-2t} + c_2 t e^{-2t} + c_3 e^{3t}$

43. $x(t) = c_1 e^{2t} + c_2 t e^{2t} + c_3 \sin t + c_4 \cos t$

44. $x(t) = c_1 \cos 2t + c_2 \sin 2t + c_3 e^{-t}$

48. $x(t) = 3e^{-2t} - 40e^t \cos(t/2) + 92e^t \sin(t/2)$

51. $x(t) = c_1 + c_2 e^{5t/3} + t^2$

53. $x(t) = c_1 e^{2t} + c_2 t e^{2t} + e^t$

55. $x(t) = c_1 + c_2 e^{-3t} - \frac{12}{13} \cos 2t + \frac{18}{13} \sin 2t$

57. $x(t) = c_1 e^t + c_2 e^{2t} + 5te^{2t} - \frac{3}{10} \cos t - \frac{1}{10} \sin t$

59. $y(x) = c_1 e^{2x} + c_2 e^{-8/5x} - \frac{5}{24} x^2 + \frac{31}{96} x - \frac{131}{768}$

61. $x(t) = c_1 e^{-t/2} + c_2 t e^{-t/2} - \frac{4}{3} e^t + te^t$

63. $y = c_1 e^{-2t} \cos(t\sqrt{3}) + c_2 e^{-2t} \sin(t\sqrt{3}) - t + \frac{18}{7}$

65. $x(t) = c_1 e^{2t} + c_2 t e^{2t} + \frac{1}{2} t^2 e^{2t}$

69. $x(t) = c_1 \cos(\ln t) + c_2 \sin(\ln t) + \frac{1}{2} t$

73. $x(t) = 6t^3 - 23t^3 \ln t - 4$

77. $x(t) = c_1 t^3 + c_2 t - \frac{1}{2} t \ln t$

78. $x(t) = c_1 t^3 + c_2 t^3 \ln t + 2t - \frac{8}{9}$

79. $x(t) = c_1 \cos t + c_2 \sin t + \ln(\tan t) \sin t$

80. $x(t) = \frac{1}{2} - \frac{3}{26} \cos 2t + \frac{1}{13} \sin 2t + e^{-t/2} \left[c_1 \cos\left(\frac{t\sqrt{3}}{2}\right) + c_2 \sin\left(\frac{t\sqrt{3}}{2}\right) \right]$

Chapter 11

1. $\mathcal{L}\{\cos bt\} = \frac{s}{s^2 + b^2}$

2. $\mathcal{L}\{f(t)\} = \frac{1}{s-2}\left(1 - e^{10-5s}\right)$

3. $\mathcal{L}\{g(t)\} = \frac{e^{10-5s}}{s-2} \quad (s > 2)$

5. $\mathcal{L}\{h(t)\} = \frac{1}{s} - \frac{1}{s^2} + e^{-7s}\left(\frac{1}{s^2} + \frac{16}{s}\right)$

7. $\mathcal{L}\{\sin 2t + e^{6t}\} = \frac{2}{s^2 + 2^2} + \frac{1}{s-6}$

9. $\mathcal{L}\{5u(t-2)\} = 5\frac{e^{-2s}}{s}$

12. $\mathcal{L}\left\{\frac{3}{4}\delta(t-5) - 2\cos\frac{4t}{7}\right\} = \frac{3}{4}e^{-5s} - 2 \cdot \frac{s}{s^2 + \left(\frac{4}{7}\right)^2}$

15. $\mathcal{L}\left\{e^{-2t} \sin \frac{t}{3}\right\} = \frac{1/3}{(s+2)^2 + (1/3)^2}$

17. $\mathcal{L}\{t \cos 2t\} = \frac{s^2 - 4}{(s^2 + 4)^2}$

21. $\mathcal{L}\{te^{2t} \sin 3t\} = \frac{6(s-2)}{[(s-2)^2 + 9]^2}$

25. $\mathcal{L}\{t^{10} + 3\delta(t-2)\} = \frac{10!}{s^{11}} + 3e^{-2s}$

27. $\mathcal{L}\{h(t)\} = \frac{3}{s} + \frac{1}{s-5} - \frac{3e^{-s}}{s}$

29. $\mathcal{L}\left\{\cos(3t)u\left(t - \frac{\pi}{6}\right)\right\} = -\frac{3e^{-\pi s/6}}{s^2 + 9}$

41. $x(t) = -\frac{1}{3}e^{2t} + \frac{4}{3}e^{5t}$

44. $x(t) = \frac{3}{13} \cos 2t + \frac{2}{13} \sin 2t + \frac{10}{13} e^{-3t}$

48. $x(t) = 2e^{3t} + 6\,e^{3(t-5)}u(t-5)$

50. $\dfrac{3\sqrt{2}}{2}\sin\left(t\sqrt{2}\right)$

52. $2e^{2t}\cos\left(t\sqrt{14}\right)$

53. $e^{-2t}\cos(4t) - \dfrac{1}{2}e^{-2t}\sin(4t)$

54. $5e^{2t} + \sin(3t)$

56. $3e^{-t}\cos(t/2) - 6e^{-t}\sin(t/2)$

58. $6e^{-4t} - 5\cos\left(t\sqrt{7}\right)$

59. $\dfrac{26}{15}e^{-3t} + \dfrac{11}{10}e^{2t} - \dfrac{5}{6}$

61. $\dfrac{1}{4}\sin(2t) + \dfrac{1}{2}t\cos(2t)$

63. $x(t) = 3e^{3t} - e^{2t}\cos t - 12e^{2t}\sin t$

65. $x(t) = e^{-7t} + 4e^{2t}$

67. $x(t) = 6u(t-2)e^{-2t+4}\sin(t-2) + e^{-2t}\sin t$

68. $x(t) = \dfrac{1}{4} - \dfrac{5}{4}\cos(2t) - \dfrac{1}{2}u(t-5) + \dfrac{1}{2}\cos(2t-10)u(t-5)$

70. $\mathcal{L}\{r(t)\} = \dfrac{2(1-e^{-s})}{s(1+e^{-s})} = \dfrac{2}{s}\tanh\dfrac{s}{2}$

72. $\mathcal{L}\{g(t)\} = \dfrac{s^2e^s - s^2 - 4\pi^2}{s(e^s+1)(s^2+4\pi^2)}$

74. $\mathcal{L}^{-1}\{F(s)\} = f(t)$, where $f(t)$ is periodic with period 2 and

$$f(t) = \begin{cases} t, & 0 < t < 1 \\ 2-t, & 1 < t < 2, \end{cases}$$

which can also be expressed as

$$f(t) = tu(t) + 2\sum_{n=1}^{\infty}(-1)^n \times (t-n)u(t-n).$$

For the graph of $f(t)$, see Fig. 11.7.

76. $\mathcal{L}^{-1}\{F(s)\} = f(t)$, where $f(t+1) = f(t)$ and $f(t) = t$ for $0 < t < 1$. Equivalently,

$$f(t) = \sum_{n=0}^{\infty}(t-n)\big[u(t-n) - u(t-(n+1))\big].$$

The graph of $f(t)$ is given in Fig. 11.2.

79. $x(t) = 1 - 3\sin t + 2\sum_{n=1}^{\infty}(-1)^n\big[1 - \cos(t-n)\big]u(t-n)$

80. $x(t) = \frac{1}{2}g(t) + \sum_{n=1}^{\infty}(-1)^n g(t-n)u(t-n)$, where $g(t) = -\frac{1}{2} + t + \frac{1}{2}e^{-2t}$

81. $x(t) = h(t) - \pi\sum_{n=1}^{\infty} g(t-n\pi)u(t-n\pi)$, where $g(t) = 1 - \frac{1}{15}e^{-t/4}\Big(15\cos\left(\frac{t\sqrt{15}}{4}\right) + \sqrt{15}\sin\left(\frac{t\sqrt{15}}{4}\right)\Big)$ and $h(t) = t - \frac{1}{2} + e^{-t/4}\Big(\frac{1}{2}\cos\left(\frac{t\sqrt{15}}{4}\right) - \frac{7\sqrt{15}}{30}\sin\left(\frac{t\sqrt{15}}{4}\right)\Big)$.

83. $x(t) = C(\sin t - t\cos t) + 2\cos t + 2t\sin t$

85. $x(t) = C(t^3 - 6t^2) + \frac{1}{2}t$

87. $x(t) = t^3 + 6t$

89. $x(t) = \sin(2t) + 2\cos(2t) + \frac{3}{2}e^{2t} + \frac{1}{2}e^{-2t}$

91. $x(t) = e^{-t} + e^{t/2}\Big[\sqrt{3}\sin\left(\frac{t\sqrt{3}}{2}\right) - \cos\left(\frac{t\sqrt{3}}{2}\right)\Big]$

92. $x(t) = \frac{1}{2}\Big[t\sin t + \sum_{n=1}^{\infty}(t-n)\sin(t-n)u(t-n)\Big]$

94. $x(t) = t + \sum_{n=1}^{\infty}(t-n)u(t-n)$

97. $\dfrac{2(1-\cos t)}{t}$

99. $\cos t + \frac{1}{2}t^2 - 1$

104. Graph of the solution $x(t)$ of the initial value problem in Problem 81:

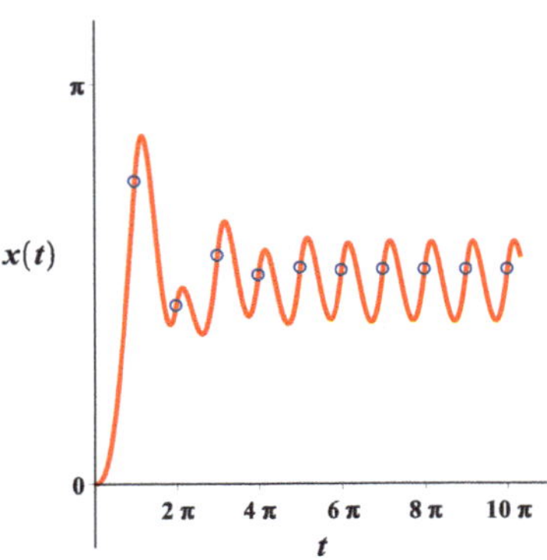

Chapter 12

1. Let $x(t)$ and $y(t)$ be the amounts of sugar (in lbs.) in tanks L and R, resp. The initial conditions are

$$x(0) = 5, \ y(0) = 25.$$

The linear system modeling the amount of sugar in the tanks is

$$\frac{dx}{dt} = -0.14x + 0.04y$$

$$\frac{dy}{dt} = 0.14x - 0.07y.$$

4. $AB = O$ and $BA = \begin{bmatrix} 5 & 5 \\ -5 & -5 \end{bmatrix}$.

5. (a) -42, $\begin{bmatrix} 1/21 & -4/21 \\ -2/21 & -5/42 \end{bmatrix}$
   (c) 0, The matrix is not invertible.

9. The eigenvalues are $\lambda_1 = -2$, $\lambda_2 = 1$. The eigenspace $E(-2)$ are all vectors of the form $c_1\langle 1, -1\rangle$ with $c_1 \in \mathbb{R}$. $E(1)$ consists of all vectors $c_2\langle 5, -2\rangle$ with $c_2 \in \mathbb{R}$.

11. The characteristic equation is

$$\lambda^2 - 3\lambda + 2 = 0$$

since tr $M = 3$ and det $M = 2$.

17. *Hint.* See Problem 16.

20. *Hint.* See Theorem 12.5.20.

23. $W(0) = 8e^{-2t}$

32. $x(t) = c_1e^{5t} + 3c_2e^{-3t}$, $y(t) = -8c_2e^{-3t}$

34. $x(t) = c_1\sqrt{2}e^{(3+2\sqrt{2})t} + c_2\sqrt{2}e^{(3-2\sqrt{2})t}$,
    $y(t) = -c_1e^{(3+2\sqrt{2})t} + c_2e^{(3-2\sqrt{2})t}$

36. $x(t) = 5c_1 \cos t + 5c_2 \sin t$,
    $y(t) = c_1(2\cos t + \sin t) + c_2(2\sin t - \cos t)$

38. $x(t) = c_1e^{5t}(\cos 2t + 2\sin 2t)$
    $+ c_2e^{5t}(\sin 2t - 2\cos 2t)$,
    $y(t) = c_1e^{5t}\cos 2t + c_2e^{5t}\sin 2t$

39. $x(t) = c_1e^{2t} + c_2te^{2t}$,
    $y(t) = -c_1e^{2t} - c_2(t+1)e^{2t}$

41. $x(t) = c_1(\cos 4t - 2\sin 4t)$
    $+ c_2(2\cos 4t + \sin 4t)$,
    $y(t) = 2c_1\cos 4t + 2c_2 \sin 4t$

44. $x(t) = c_1e^{2t} + c_2e^{-5t}$,
    $y(t) = 2c_1e^{2t} - 5c_2e^{-5t}$

46. $x(100) = 1.46$ lbs., $y(100) = 3.85$ lbs.

49. $x(1) = 38°$F, $y(1) = 29°$F

Chapter 13

3. $Z(t, t_0) = e^{\lambda(t-t_0)}\begin{bmatrix} 1 & t-t_0 \\ 0 & 1 \end{bmatrix}$.

10. $J = \begin{bmatrix} -2 & 0 \\ 0 & 1 \end{bmatrix}$, $e^{Jt} = \begin{bmatrix} e^{-2t} & 0 \\ 0 & e^t \end{bmatrix}$.
    Alternatively,
    $J = \begin{bmatrix} 1 & 0 \\ 0 & -2 \end{bmatrix}$, $e^{Jt} = \begin{bmatrix} e^t & 0 \\ 0 & e^{-2t} \end{bmatrix}$.

12. $J = \begin{bmatrix} 0 & 1 \\ -1 & 0 \end{bmatrix}$, $e^{Jt} = \begin{bmatrix} \cos t & \sin t \\ -\sin t & \cos t \end{bmatrix}$.
    Alternatively,
    $J = \begin{bmatrix} 0 & -1 \\ 1 & 0 \end{bmatrix}$, $e^{Jt} = \begin{bmatrix} \cos t & -\sin t \\ \sin t & \cos t \end{bmatrix}$.

14. $J = \begin{bmatrix} 4 & 1 \\ 0 & 4 \end{bmatrix}$, $e^{Jt} = \begin{bmatrix} e^{4t} & te^{4t} \\ 0 & e^{4t} \end{bmatrix}$.

16. $e^{Jt} = \begin{bmatrix} e^{-t} & 0 \\ 0 & e^{5t} \end{bmatrix}$.

$$e^{At} = \frac{1}{2}\begin{bmatrix} e^{5t} + e^{-t} & e^{5t} - e^{-t} \\ e^{5t} - e^{-t} & e^{5t} + e^{-t} \end{bmatrix}.$$

18. $e^{Jt} = \begin{bmatrix} e^{4t} & te^{4t} \\ 0 & e^{4t} \end{bmatrix}$.

$$e^{At} = \begin{bmatrix} e^{4t} & 0 \\ -5te^{4t} & e^{4t} \end{bmatrix}.$$

20. $e^{At} = e^{Jt} = \begin{bmatrix} e^{2t}\cos t & e^{2t}\sin t \\ -e^{2t}\sin t & e^{2t}\cos t \end{bmatrix}$.

21. $e^{Jt} = \begin{bmatrix} e^{-3t} & 0 \\ 0 & e^{2t} \end{bmatrix}$.

$$e^{At} = \frac{1}{5}\begin{bmatrix} -e^{-3t} + 6e^{2t} & -2e^{-3t} + 2e^{2t} \\ 3e^{-3t} - 3e^{2t} & 6e^{-3t} - e^{2t} \end{bmatrix}.$$

25. (a) $e^{At} = \begin{bmatrix} a_{11}(t) & a_{12}(t) \\ a_{21}(t) & a_{22}(t) \end{bmatrix}$, where
    $a_{11}(t) = 1 - t + \frac{5}{2}t^2 - \frac{1}{6}t^3 + \cdots$

$a_{12}(t) = \frac{1}{2}t + \frac{1}{4}t^2 + \frac{7}{12}t^3 + \cdots$
$a_{21}(t) = 8t + 4t^2 + \frac{28}{3}t^3 + \cdots$
$a_{22}(t) = 1 + 2t + 4t^2 + \frac{10}{3}t^3 + \cdots$

(b) $\begin{bmatrix} \frac{4}{5}e^{-2t} + \frac{1}{5}e^{3t} & -\frac{1}{10}e^{-2t} + \frac{1}{10}e^{3t} \\ -\frac{8}{5}e^{-2t} + \frac{8}{5}e^{3t} & \frac{1}{5}e^{-2t} + \frac{4}{5}e^{3t} \end{bmatrix}$.

(c) $\begin{bmatrix} \frac{4(-2)^n}{5} + \frac{3^n}{5} & -\frac{(-2)^n}{10} + \frac{3^n}{10} \\ -\frac{8(-2)^n}{5} + \frac{8\cdot 3^n}{5} & \frac{(-2)^n}{5} + \frac{4\cdot 3^n}{5} \end{bmatrix}$.

27. $\sin 4\theta = 4\cos^3\theta\sin\theta - 4\sin^3\theta\cos\theta$

29. Every point in the x_1x_2-plane is an equilibrium point.

35. Every point on the y_2-axis is an equilibrium point. Phase portrait:

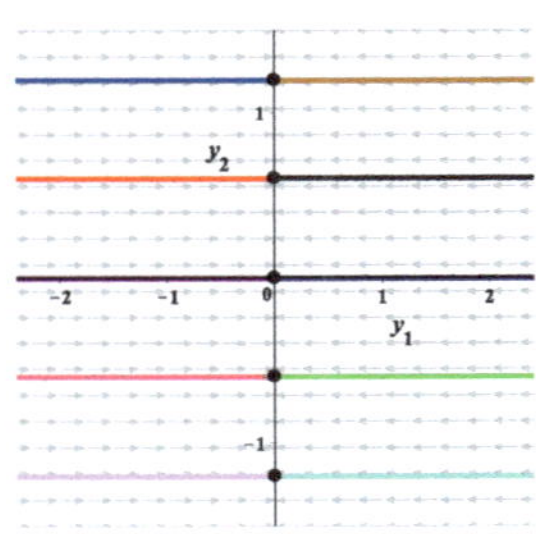

38. (b) $y_1(t) = (a + bt)e^t$, $y_2(t) = be^t$. Phase portrait:

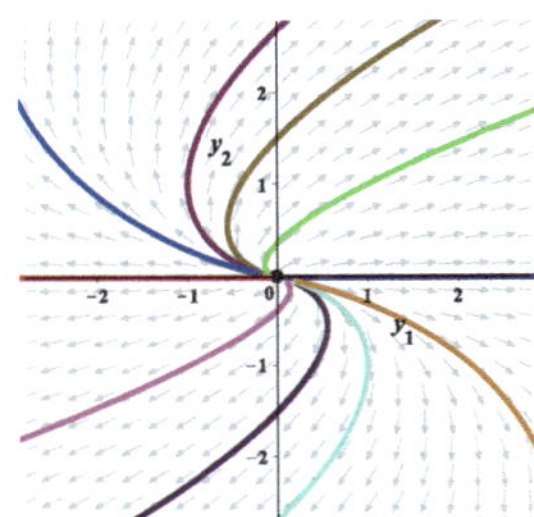

40. (a) $J = \begin{bmatrix} 3 & 0 \\ 0 & 4 \end{bmatrix}$.
$e^{Jt} = \begin{bmatrix} e^{3t} & 0 \\ 0 & e^{4t} \end{bmatrix}$.

(b) $P = \begin{bmatrix} 1 & 5 \\ 1 & 6 \end{bmatrix}$.

(c) $e^{At} = Pe^{Jt}P^{-1} = \begin{bmatrix} 6e^{3t} - 5e^{4t} & -5e^{3t} + 5e^{4t} \\ 6e^{3t} - 6e^{4t} & -5e^{3t} + 6e^{4t} \end{bmatrix}$.

(d) $\mathbf{x}(t) = \langle -20e^{3t} + 20e^{4t}, -20e^{3t} + 24e^{4t} \rangle$.

42. (a) $J = \begin{bmatrix} 0 & 4 \\ -4 & 0 \end{bmatrix}$.
$e^{Jt} = \begin{bmatrix} \cos 4t & \sin 4t \\ -\sin 4t & \cos 4t \end{bmatrix}$.

(b) $P = \begin{bmatrix} 1 & 2 \\ 2 & 0 \end{bmatrix}$.

(c) $e^{At} = Pe^{Jt}P^{-1} = \begin{bmatrix} \cos 4t + \frac{1}{2}\sin 4t & -\frac{5}{4}\sin 4t \\ \sin 4t & \cos 4t + \frac{1}{2}\sin 4t \end{bmatrix}$.

(d) $\mathbf{x}(t) = \langle -5\sin 4t, 4\cos 4t - 2\sin 4t \rangle$.

44. (a) $J = \begin{bmatrix} 0 & 1 \\ 0 & 0 \end{bmatrix}$.
$e^{Jt} = \begin{bmatrix} 1 & t \\ 0 & 1 \end{bmatrix}$.

(b) $P = \begin{bmatrix} 1 & 0 \\ -2 & 1 \end{bmatrix}$.

(c) $e^{At} = Pe^{Jt}P^{-1} = \begin{bmatrix} 1 + 2t & t \\ -4t & 1 - 2t \end{bmatrix}$.

(d) $\mathbf{x}(t) = \langle 4t, -8t + 4 \rangle$.

46. $x_1(t) = -2\cos t + \sin t$,
$x_2(t) = \cos t$.
Trajectory through $(-2, 1)$:
$x_1^2 + 4x_1x_2 + 5x_2^2 = 1$.
Graph:

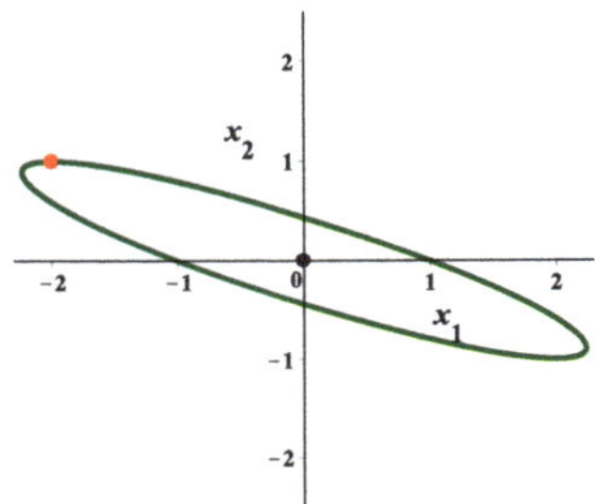

48. (a) $M(x_1, x_2) = a_{21}x_1 + a_{22}x_2$
$N(x_1, x_2) = -a_{11}x_1 - a_{12}x_2$

(b) $a_{21}x_1^2 - 2a_{11}x_1x_2 - a_{12}x_2^2 = K$.

51. $x_1(t) = 2\cos t + 4\sin t - 3e^{2t}$,
$x_2(t) = 3e^{2t} - 3te^t - e^{-t}$.

53. $x_1(t) = 3t^2 + 2t - t\ln t - 3$,
$x_2(t) = 9t^2 - 3t\ln t + \ln t - 8$.

56. $\lim_{t\to\infty} x_1(t) = 356/5$,
$\lim_{t\to\infty} x_2(t) = 452/5$.

References

1. M. Abramowitz and I. A. Stegun, *Handbook of Mathematical Functions: with Formulas, Graphs, and Mathematical Tables* (Dover Books on Mathematics-Revised Edition), Courier, 2012.
2. D. K. Arrowsmith and C. M. Place, *Dynamical Systems: Differential equations, maps and chaotic behavior*, Chapman & Hall, 1992.
3. P. W. Atkins, *Atoms, Electrons, and Change*, Scientific American Library, New York, 1991.
4. Maria Vittoria Barbarossa, Jan Fuhrmann, et al., A first study on the impact of current and future control measures on the spread of COVID-19 in Germany, *medRxiv* preprint doi: https://doi.org/10.1101/2020.04.08.20056630, posted April 11, 2020.
5. Robert G. Bartle, *The Elements of Real Analysis*, 2nd ed., John Wiley & Sons, New York, 1976.
6. Robert G. Bartle and Donald R. Sherbert, *Introduction to Real Analysis*, John Wiley & Sons, New York, 1982.
7. Rubin Battino and Scott E. Wood, *Thermodynamics, An Introduction*, Academic Press, New York, 1968.
8. Leigh C. Becker, Function bounds for solutions of Volterra equations and exponential asymptotic stability, *Nonlinear Analysis: Theory, Methods and Applications* **67**, No. 2 (2007), 382–397.
9. Leigh C. Becker, Principal matrix solutions and variation of parameters for a Volterra integro-differential equation and its adjoint, *Electron. J. Qual. Theory Differ. Equ.*, No. 14 (2006), 1–22.
10. Leigh C. Becker, Constant Delay Differential Equations and the Method of Steps, *Maplesoft Application Center* (2006), Maple worksheet web address: www.maplesoft.com/Applications/Detail.aspx?id=33093.
11. Leigh C. Becker, Setting the Stage with APR Problems, *Mathematics and Computer Science*, Vol. 18 (1984), 189–195.
12. Robert A. Becker, *Introduction to Theoretical Mechanics*, McGraw-Hill, New York, 1954.
13. E. T. Bell, *Men of Mathematics: The Lives and Achievements of the Great Mathematicians from Zeno to Poincaré*, Touchstone, 1986.
14. Richard Bellman and Kenneth L. Cooke, *Modern Elementary Differential Equations*, 2nd ed., Dover Publications, Mineola, New York, 1995.
15. M. Braun, *Differential Equations and Their Applications*, 3rd Edition, Applied Mathematical Sciences, Vol. 15, Springer-Verlag New York, 1983.
16. James W. Brown and Ruel V. Churchill, *Complex Variables and Applications*, 9th ed., McGraw Hill, Boston, 2014.

© The Author(s), under exclusive license to Springer Nature Switzerland AG 2026

L. C. Becker, *Ordinary Differential Equations: Concepts, Methods, and Models*,
https://doi.org/10.1007/978-3-032-15150-6

17. Richard L. Burden, J. Douglas Faires, and Annette M. Burden, *Numerical Analysis*, 10th ed., Cengage Learning, 2015.
18. T. A. Burton, *Stability and Periodic Solutions of Ordinary and Functional Differential Equations*, Dover Publications, Mineola, New York, 2005.
19. T. A. Burton, Fixed Points, Volterra Equations, and Becker's Resolvent, *Acta Math. Hungar.*, **108 (3)** (2005), 261–281.
20. Ruel V. Churchill, *Fourier Series and Boundary Value Problems*, 2nd ed., McGraw-Hill, 1963.
21. Ruel V. Churchill, *Operational Mathematics*, 2nd ed., McGraw-Hill, 1958.
22. Mariana Cook, *Mathematicians: An Outer View of the Inner World*, Princeton University Press, 2009.
23. C. Corduneanu, *Principles of Differential and Integral Equations*, 2nd ed., Chelsea Publishing Co., New York, 1977.
24. Paul Davies, *The Mind of God: The Scientific Basis for a Rational World*, A Touchstone Book, Simon & Schuster, 1992.
25. J. K. Denny and C. A. Yackel, Temperature Models for Ware Hall, *The College Mathematics Journal*, **35 (3)** (2004), 162–170.
26. P. A. M. Dirac, *The Principles of Quantum Mechanics*, 4th ed., Oxford University Press, 1958.
27. Gustav Doetsch, *Theorie und Anwendung der Laplace-Transformation*, Springer, Berlin, 1937.
28. Arthur Conan Doyle, *The Annotated Sherlock Holmes*, Vol. I, Clarkson N. Potter, Inc., 1967.
29. R. D. Driver, Torricelli's Law–An Ideal Example of an Elementary ODE, *Amer. Math. Monthly* **105** (1998), 453–455.
30. R. D. Driver, *Ordinary and Delay Differential Equations*, Applied Mathematical Sciences 20, Springer-Verlag, New York, 1977.
31. Brian Dunning, The Sargasso Sea and the Pacific Garbage Patch, *Skeptoid*, Podcast No. 132 (16 December 2008).
32. Frank L. Dyer and Thomas C. Martin, *Edison: His Life and Inventions*, Harper & Brothers, New York, 1910.
33. Albert Einstein, *Ideas and Opinions*, Crown Publishers, New York, 1954.
34. Albert Einstein, *Mein Weltbild*, Querido Verlag, Amsterdam, 1934.
35. C. S. Forester, *Admiral Hornblower in the West Indies*, Little Brown and Company, New York, 1957 (reissued in paperback by Back Bay Books, 2000).
36. Avner Friedman, *Advanced Calculus*, Dover Publications, Mineola, N.Y., 2007. (unabridged republication of the 1971 edition published by Holt, Rinehart and Winston, N.Y.)
37. W. Fulks, *Advanced Calculus*, Second Edition, John Wiley & Sons, New York, 1969.
38. Edward D. Gaughan, *Introduction to Analysis*, 5th ed., Brookes/Cole, Pacific Grove, 1998.
39. James Gleick, *Chaos: Making a New Science*, Penguin Random House, New York, 2008.
40. T. H. Gronwall, Note on the derivatives with respect to a parameter of the solutions of a system of differential equations, *Annals of Mathematics* **20** (1919), 292–296.
41. Jack K. Hale, *Ordinary Differential Equations*, Wiley-Interscience, New York, 1969.
42. Jack K. Hale and Hüseyin Koçak, *Dynamics and Bifurcations*, Texts in Applied Mathematical Sciences, Vol. 3, Springer-Verlag, New York, 1991.
43. David Halliday and Robert Resnick, *Fundamentals of Physics*, 3rd ed. extended, John Wiley & Sons, New York, 1988.
44. Philip Hartman, *Ordinary Differential Equations*, 2nd ed., SIAM, Philadelphia, 2002.
45. Paul Hoffman, *The Man Who Loved Only Numbers*, Hyperion, New York, 1998.
46. P. D. James, *Devices and Desires*, Vintage Books, New York, 2004.
47. James Jeans, *The Mysterious Universe*, Macmillian Co., New York, 1930.
48. Herbert E. Kasube, A Technique for Integration by Parts, *Amer. Math. Monthly* **90** (1983), 210–211.
49. David M. Knight, *No Power but Love*, His Way Communications, 1999. (See www.immersedinchrist.org).
50. T. W. Körner, *Fourier Analysis*, Cambridge University Press, 1988.
51. Wilhelm Kutta, Beitrag zur näherungsweisen Integration totaler Differentialgleichungen, *Zeitschrift für Mathematik und Physik* **46**, (1901), 435–453.

52. Robert B. Leighton, *Principles of Modern Physics*, McGraw-Hill, 1959.
53. Lyle N. Long and Howard Weiss, The Velocity Dependence of Aerodynamic Drag: A Primer for Mathematicians, *Amer. Math. Monthly* **106** (1999), 127–135.
54. Edward N. Lorenz, *The Essence of Chaos*, University of Washington Press, 1995.
55. Edward N. Lorenz, On the Prevalence of Aperiodicity in Simple Systems, *Global Analysis* (Edited by Grmela, M. and Marsden, J. E.), pp. 53–75, Lecture Notes in Mathematics, **755**, Springer-Verlag, New York, 1979.
56. Edward N. Lorenz, Deterministic Nonperiodic Flow, *Journal of the Atmospheric Sciences*, **20** (1963), 130–141.
57. Jerry B. Marion, *Classical Dynamics of Particles and Systems*, Academic Press, 1965.
58. Jerrold E. Marsden, *Basic Complex Analysis*, W. H. Freeman, San Francisco, 1973.
59. John L. McKenzie, *The Two-Edged Sword, An Interpretation of the Old Testament*, Image Books, 1966.
60. J.J. O'Connor and E.F. Robertson, "Pierre François Verhulst", *MacTutor History of Mathematics Archive*, University of St Andrews, Scotland, Jan. 2014, https://mathshistory.st-andrews.ac.uk/Biographies/Verhulst.
61. William Fogg Osgood, *Mechanics*, Dover, New York, 1965.
62. Kenneth A. Ross, *Elementary Analysis: The Theory of Calculus*, 2nd ed., Springer, New York, 2013.
63. Theodore Shifrin and Malcolm R. Adams, *Linear Algebra: A Geometric Approach*, 2nd ed., W. H. Freeman and Company, New York, 2011.
64. J. Maynard Smith, *Mathematical Ideas in Biology*, Cambridge University, 1968.
65. Robert Smith and Roland Minton, *Calculus: Early Transcendental Functions*, 3rd ed., McGraw-Hill, 2007.
66. Dava Sobel, *Galileo's Daughter: A Historical Memoir of Science, Faith, and Love*, Penquin Books, 2000.
67. Colin Sparrow, *The Lorenz Equations: Bifurcations, Chaos, and Strange Attractors*, Applied Mathematical Sciences 41, Springer-Verlag, New York, 1982.
68. James Stewart, *Single Variable Calculus*, 8th ed., Cengage Learning, 2016.
69. J. Stoer and R. Bulirsch, *Introduction to Numerical Analysis*, 3rd ed., Texts in Applied Mathematics 12, Springer, New York, 2002.
70. Steven H. Strogatz, *Nonlinear Dynamics and Chaos, With Applications to Physics, Biology, Chemistry, and Engineering*, 3rd ed., CRC Press, 2024.
71. George B. Thomas, Jr. (revised by M. D. Weir, Joel Haas, F. R. Giordano), *Thomas' Calculus*, 11th ed., Pearson, 2005.
72. George B. Thomas, Jr. and Ross L. Finney, *Calculus and Analytic Geometry*, 8th ed., Addison-Wesley, 1992.
73. Leo Tolstoy (translated by Richard Pevear and Larissa Volokhonsky), *Anna Karenina*, Penguin Classics, 2000.
74. Arnold Toynbee, *Experiences*, Oxford University Press, 1969.
75. P. F. Verhulst, Notice sur la loi que la population suit dans son accroissement, *Correspondance Mathématique et Physique* **10** (1838), 113–121.
76. Steven Weinberg, *Dreams of a Final Theory*, Vintage Books, 1993.
77. Frank M. White, *Heat Transfer*, Addison-Wesley, 1984.
78. T. H. White, *The Once and Future King*, Ace Books, New York, 1987.
79. Wikipedia, https://en.wikipedia.org/wiki/Logarithmicspiral.
80. Wikipedia, https://en.wikipedia.org/wiki/EgonSchweidler.
81. A. H. Zemanian, *Distribution Theory and Transform Analysis, An Introduction to Generalized Functions, with Applications*, Dover, New York, 1965.

Index

© The Author(s), under exclusive license to Springer Nature Switzerland AG 2026

L. C. Becker, *Ordinary Differential Equations: Concepts, Methods, and Models*,
https://doi.org/10.1007/978-3-032-15150-6

T

U

The manufacturer's authorised representative in the EU is Springer Nature Customer Service Centre GmbH, Europaplatz 3, 69115 Heidelberg, Germany. If you have any concerns regarding our products, please contact ProductSafety@springernature.com

Printed and bound by CPI Group (UK) Ltd, Croydon, CR0 4YY
07/07/2026
02160918-0005